Advances in Magnetic Resonance
Technology and Applications

Volume 11

Quantitative Perfusion MRI

Techniques, Applications and
Practical Considerations

Advances in Magnetic Resonance Technology and Applications Series

Series Editors

In-Young Choi, PhD
Department of Neurology, Department of Radiology, Department of Molecular & Integrative Physiology, Hoglund Biomedical Imaging Center, University of Kansas Medical Center, Kansas City, KS, United States

Peter Jezzard, PhD
Wellcome Centre for Integrative Neuroimaging, Nuffield Department of Clinical Neurosciences University of Oxford, Oxford, United Kingdom

Brian Hargreaves, PhD
Department of Radiology, Department of Electrical Engineering, Department of Bioengineering Stanford University, Stanford, CA, United States

Greg Zaharchuk, MD, PhD
Department of Radiology, Stanford University, Stanford, CA, United States

Titles published:

Visit the Series webpage at https://www.elsevier.com/books/book-series/advances-in-magnetic-resonance-technology-and-applications

Advances in Magnetic Resonance Technology and Applications

Volume 11

Quantitative Perfusion MRI

Techniques, Applications and Practical Considerations

Edited by

Hai-Ling Margaret Cheng

*Institute of Biomedical Engineering; The Edward S. Rogers Sr.
Department of Electrical & Computer Engineering; Ted Rogers
Centre for Heart Research, Translational Biology & Engineering
Program, University of Toronto, Toronto, ON, Canada*

Gustav J. Strijkers

*Department of Biomedical Engineering and Physics,
Amsterdam University Medical Centers,
University of Amsterdam, Amsterdam, The Netherlands*

ELSEVIER

ACADEMIC PRESS

An imprint of Elsevier

Academic Press is an imprint of Elsevier
50 Hampshire Street, 5th Floor, Cambridge, MA 02139, United States
525 B Street, Suite 1650, San Diego, CA 92101, United States
The Boulevard, Langford Lane, Kidlington, Oxford OX5 1GB, United Kingdom
125 London Wall, London, EC2Y 5AS, United Kingdom

Notices
Knowledge and best practice in this field are constantly changing. As new research and experience broaden
our understanding, changes in research methods, professional practices, or medical treatment may
become necessary.

Practitioners and researchers must always rely on their own experience and knowledge in evaluating and
using any information, methods, compounds, or experiments described herein. In using such information
or methods they should be mindful of their own safety and the safety of others, including parties for whom
they have a professional responsibility.

To the fullest extent of the law, neither the Publisher nor the authors, contributors, or editors, assume any
liability for any injury and/or damage to persons or property as a matter of products liability, negligence or
otherwise, or from any use or operation of any methods, products, instructions, or ideas contained in the
material herein.

ISBN: 978-0-323-95209-5
ISSN: 2666-9099

For information on all Academic Press publications visit
our website at https://www.elsevier.com/books-and-journals

Publisher: Mara E. Conner
Acquisitions Editor: Tim Pitts
Developmental Editor: Emily Thomson
Production Project Manager: Surya Narayanan Jayachandran
Cover Designer: Vicky Pearson Esser

Typeset by STRAIVE, India

Working together
to grow libraries in
developing countries

www.elsevier.com • www.bookaid.org

Contents

List of contributors

Fatemeh Adelnia
Department of Radiology and Radiological Sciences, Vanderbilt University Medical Center, Vanderbilt University Institute of Imaging Science, Nashville, TN, United States

Liza Afzali-Hashemi
Department of Radiology and Nuclear Medicine, Amsterdam University Medical Center, Location University of Amsterdam, Amsterdam, The Netherlands

Verónica Aramendía-Vidaurreta
Department of Radiology, Clínica Universidad de Navarra, Pamplona, Spain

Koen P.A. Baas
Department of Radiology and Nuclear Medicine, Amsterdam University Medical Center, Location University of Amsterdam, Amsterdam, The Netherlands

J.D. Biglands
Leeds Institute of Cardiovascular and Metabolic Medicine, University of Leeds; Medical Physics and Engineering, Leeds Teaching Hospitals NHS Trust, Leeds, United Kingdom

D.A. Broadbent
Leeds Institute of Cardiovascular and Metabolic Medicine, University of Leeds; Medical Physics and Engineering, Leeds Teaching Hospitals NHS Trust, Leeds, United Kingdom

Claudia Calcagno
Integrated Research Facility at Fort Detrick, Division of Clinical Research, National Institute of Allergy and Infectious Diseases, National Institutes of Health, Frederick, MD, United States

Donnie Cameron
C.J. Gorter MRI Center, Department of Radiology, Leiden University Medical Center, Leiden, The Netherlands

Hai-Ling Margaret Cheng
Institute of Biomedical Engineering; The Edward S. Rogers Sr. Department of Electrical & Computer Engineering; Ted Rogers Centre for Heart Research, Translational Biology & Engineering Program, University of Toronto, Toronto, ON, Canada

Amedeo Chiribiri
School of Biomedical Engineering and Imaging Sciences, King's College London, London, United Kingdom

Poonam Choudhary
Barrow Neurological Institute, Phoenix, AZ, United States

Walter Dastrù
Department of Molecular Biotechnologies and Health Sciences, University of Turin, Torino, Italy

Ben Dickie
Division of Informatics, Imaging, and Data Science, School of Health Sciences, Faculty of Biology, Medicine and Health, School of Health Sciences, The University of Manchester; Geoffrey Jefferson Brain Research Centre, Manchester Academic Health Science Centre, Northern Care Alliance NHS Group, University of Manchester, Manchester, United Kingdom

Mathijs Dijsselhof
Department of Radiology and Nuclear Medicine, Amsterdam UMC location Vrije Universiteit Amsterdam; Amsterdam Neuroscience, Brain Imaging, Amsterdam, The Netherlands

Rebeca Echeverria-Chasco
Department of Radiology, Clínica Universidad de Navarra; IdiSNA, Instituto de Investigación Sanitaria de Navarra, Pamplona, Spain

Li Feng
Center for Advanced Imaging Innovation and Research (CAI2R), New York University Grossman School of Medicine, New York, NY, United States

Maria A. Fernández-Seara
Department of Radiology, Clínica Universidad de Navarra; IdiSNA, Instituto de Investigación Sanitaria de Navarra, Pamplona, Spain

Philippe Garteiser
University Paris Cité; Laboratory of Imaging Biomarkers, INSERM U1149, Centre for Research on Inflammation, Paris, France

Lucy Elizabeth Kershaw
Edinburgh Imaging; BHF Centre for Cardiovascular Science, University of Edinburgh, Edinburgh, United Kingdom

Linda Knutsson
Department of Medical Radiation Physics, Lund University, Lund, Sweden; Russell H. Morgan Department of Radiology and Radiological Science, Johns Hopkins University School of Medicine; F.M. Kirby Research Center for Functional Brain Imaging, Kennedy Krieger Institute, Baltimore, MD, United States

Frank Kober
Aix Marseille Univ, CNRS, CRMBM, Marseille, France

Ji Hyun Lee
Radiology and Imaging Sciences, Clinical Center, National Institutes of Health, Bethesda, MD, United States

Tim Leiner
Department of Radiology, Mayo Clinic, Minnesota, United States; Department of Radiology, University Medical Center Utrecht, Utrecht, The Netherlands

Emelie Lind
Department of Medical Radiation Physics, Lund University; Department of Medical Imaging and Physiology, Skåne University Hospital, Lund, Sweden

Dario Livio Longo
Institute of Biostructures and Bioimaging (IBB), National Research Council of Italy (CNR), Torino, Italy

Henk J.M.M. Mutsaerts
Department of Radiology and Nuclear Medicine, Amsterdam UMC location Vrije Universiteit Amsterdam; Amsterdam Neuroscience, Brain Imaging, Amsterdam, The Netherlands

Aart J. Nederveen
Department of Radiology and Nuclear Medicine, Amsterdam University Medical Center, Location University of Amsterdam, Amsterdam, The Netherlands

Beatriz Esteves Padrela
Department of Radiology and Nuclear Medicine, Amsterdam UMC location Vrije Universiteit Amsterdam; Amsterdam Neuroscience, Brain Imaging, Amsterdam, The Netherlands

Jan Petr
Department of Radiology and Nuclear Medicine, Amsterdam UMC location Vrije Universiteit Amsterdam; Amsterdam Neuroscience, Brain Imaging, Amsterdam, The Netherlands; Helmholtz-Zentrum Dresden-Rossendorf, Institute of Radiopharmaceutical Cancer Research, Dresden, Germany

S. Plein
Leeds Institute of Cardiovascular and Metabolic Medicine, University of Leeds, Leeds, United Kingdom

Christopher Chad Quarles
The University of Texas MD Anderson Cancer Center, Houston, TX, United States

David A. Reiter
Department of Radiology and Imaging Sciences, Biomedical Engineering, and Orthopedics, Emory University School of Medicine, Atlanta, GA, United States

Maxime Ronot
Department of Radiology, APHP, University Hospitals Paris Nord Val de Seine, Beaujon, Clichy, Hauts-de-Seine; University Paris Cité; Laboratory of Imaging Biomarkers, INSERM U1149, Centre for Research on Inflammation, Paris, France

Michael Salerno
Cardiovascular Medicine, Stanford University, Palo Alto, CA, United States

Giles Santyr
Peter Gilgan Centre for Research and Learning, The Hospital for Sick Children, Department of Medical Biophysics, University of Toronto, Toronto, ON, Canada

Cian M. Scannell
Department of Biomedical Engineering, Eindhoven University of Technology, Eindhoven, The Netherlands; School of Biomedical Engineering and Imaging Sciences, King's College London, London, United Kingdom

Matthias C. Schabel
Advanced Imaging Research Center, Oregon Health & Science University, Portland, OR; Utah Center for Advanced Imaging Research, Department of Radiology, University of Utah Health Sciences Center, Salt Lake City, UT, United States

Eric M. Schrauben
Department of Radiology and Nuclear Medicine, Amsterdam University Medical Center, Location AMC, Amsterdam, The Netherlands

N. Sharrack
Leeds Institute of Cardiovascular and Metabolic Medicine, University of Leeds, Leeds, United Kingdom

Gustav J. Strijkers
Department of Biomedical Engineering and Physics, Amsterdam University Medical Centers, University of Amsterdam, Amsterdam, The Netherlands

Bernard E. Van Beers
Department of Radiology, APHP, University Hospitals Paris Nord Val de Seine, Beaujon, Clichy, Hauts-de-Seine; University Paris Cité; Laboratory of Imaging Biomarkers, INSERM U1149, Centre for Research on Inflammation, Paris, France

Petra J. van Houdt
Department of Radiation Oncology, The Netherlands Cancer Institute, Amsterdam, The Netherlands

Pim van Ooij
Department of Radiology and Nuclear Medicine, Amsterdam University Medical Center, Location AMC; Amsterdam Cardiovascular Sciences; Amsterdam Movement Sciences, Amsterdam University Medical Center, Amsterdam; Department of Paediatric Cardiology, University Medical Center Utrecht, Utrecht, The Netherlands

Nan Wang
Department of Radiology, Stanford University, Stanford, CA, United States

Ronnie Wirestam
Department of Medical Radiation Physics, Lund University, Lund, Sweden

John C. Wood
Division of Hematology, Children's Hospital Los Angeles, Los Angeles, CA, United States

Ruixi Zhou
Artificial Intelligence, Beijing University of Posts and Telecommunications, Beijing, China

Preface

The microvascular network is a key regulator of blood pressure and ensures adequate nutritive flow to every cell in the body. Comprising three types of small vessels—arterioles that control local perfusion by adjusting their diameters; capillaries that allow exchange of gases, nutrients, and wastes; and venules that return blood to veins—this network must be maintained both structurally and functionally for the health of cells and organs. Remarkably, microvascular aberrations manifest in many diseases and injuries, often at early stage, and can be seen in conditions ranging from perfusion deficits due to compromised arteriolar dilation to malformed, leaky capillaries in cancer. Assessing these aberrations, which fall below the resolution limit of clinical imaging modalities, is a greater challenge than angiographic imaging of large blood vessels. Yet, over the past 30 years, physicists and clinicians in the field of magnetic resonance imaging (MRI) have advanced perfusion MRI capabilities to not only infer but also quantify microvascular indices such as volumetric flow rate at the tissue level (i.e., perfusion), blood volume, and capillary leakiness. Compared to its predecessors, namely, perfusion imaging methods based on nuclear medicine or computed tomography, MRI is the only "one-stop shop" modality that offers perfusion imaging in the context of high-resolution, high-contrast structural, functional, physiological, and metabolic imaging. Today, many flavors of perfusion MRI techniques exist, and every part of the body can be imaged.

This book aims to provide a well-rounded coverage of the theory and established techniques for perfusion MRI as well as the latest advances over the past decade, such as approaches to imaging microvessel dynamics or artificial intelligence (AI)-assisted workflows, that are transforming the face of perfusion MRI. Intended for both the MRI physicist and clinician, this book is divided into three sections that collectively serve as a reference textbook, a troubleshooting guide, and a catalyst for further innovation. The first section covers the basic principles underlying perfusion imaging and its four major classes: dynamic contrast enhanced (DCE) MRI, dynamic susceptibility contrast (DSC) MRI, arterial spin labeling (ASL) MRI, and vasoreactivity MRI. The second section provides a unique, detailed examination of the elements of a perfusion MRI protocol that are critical to quantitation accuracy; this section also describes the newest advances in acceleration methods and the integration of AI. The third and final section provides a balanced survey of applications throughout the body, giving equal treatment to all the major organs, including the brain, heart, abdominal organs, and skeletal muscle.

It is our hope that the next generation of scientists and clinicians will continue an endeavor begun more than 30 years ago and ultimately realize robust, quantitative perfusion MRI for every organ in every MRI center. Beyond the effort of the developers and practitioners, this endeavor also calls for the cooperation of MRI vendors and pharmaceutical manufacturers of MRI contrast agents. There is a real need for vendor-agnostic access to the newest acquisition technologies and consistent, open-source analysis platforms. There is also a genuine demand to a greater repertoire of

MRI contrast agents that go beyond nonspecific extravascular agents that are exclusively gadolinium based. Perhaps, in another 30 years, all the materials presented in this book will be discussed as "conventional" and globally available diagnostic methods, and the next frontier to surmount is imaging vascular biology for screening and surveillance.

We conclude by thanking the many chapter authors for their contributions and steadfast efforts, and the editorial staff for bringing this book to fruition. We sincerely hope you enjoy the content herein, whether you are a novice to the field or a seasoned researcher.

Hai-Ling Margaret Cheng,
Gustav J. Strijkers

Basic principles

Basic principles for imaging blood flow

Eric M. Schrauben[a] and Pim van Ooij[a,b,c,d]

[a]*Department of Radiology and Nuclear Medicine, Amsterdam University Medical Center, Location AMC, Amsterdam, The Netherlands*
[b]*Amsterdam Cardiovascular Sciences, Amsterdam University Medical Center, Amsterdam, The Netherlands*
[c]*Amsterdam Movement Sciences, Amsterdam University Medical Center, Amsterdam, The Netherlands*
[b]*Department of Paediatric Cardiology, University Medical Center Utrecht, Utrecht, The Netherlands*

1.1 Introduction

Oxygen and nutrients needed at the cellular level are delivered through a massive circulatory network within the human body, one that if laid out at full length would stretch about 100,000 km. The relative health of tissues throughout the body is largely determined by the movement of blood through this circulatory network, as well as its ability to perfuse and deliver oxygen and nutrients and to remove waste products. In humans, perfusion happens everywhere—it is estimated that capillary surface area in the human body is approximately $120\,m^2$ (Simionescu, 1980). Methods to assess this movement of blood are, therefore, of great interest in clinical and research settings. Over the last 100 years, imaging techniques have emerged to perform this assessment, both for qualitative visualization and quantitative measurements (of *e.g.*, blood volumes, flow rates, and perfusion). As we shall see in this chapter and throughout the book, these techniques have enhanced our understanding of normal physiological processes throughout the body. However, with perfusion there is always more to delve into; modern imaging techniques at the forefront of research permit probing of new perfusion aspects in normal tissues and in many diseases.

There is an important distinction to note in regard to terminology: in large (macrovascular) veins and arteries, blood flow is measured to assess bulk delivery of blood and is expressed as a volumetric rate. Perfusion, however, refers to the delivery of blood at the level of (microvascular) capillaries and is measured as a volumetric rate per unit mass or volume of tissue.

Advances in Magnetic Resonance Technology and Applications, Volume 11, ISSN 2666-9099
https://doi.org/10.1016/B978-0-323-95209-5.00005-2

In the first part of this chapter, we will discuss the mathematical concepts that generally describe flow and perfusion. In the second part, we will briefly provide historical context for blood flow and perfusion imaging, the basics of magnetic resonance imaging (MRI), and several MRI techniques for visualizing and quantifying flow and perfusion.

1.2 Theory
1.2.1 Conservation of mass

Let us consider a theoretical microvascular tissue bed with macrovasculature delivering blood flow (Q_{in}) and draining vessels removing blood from the tissue bed (Q_{out}). Note that flow, which in this context is synonymous with flow rate as opposed to flow velocity, is also frequently indicated by F, but to avoid confusion with *force*, Q is used throughout this book. One of the main and most basic theoretical principles when measuring flow is the conservation of mass (Fig. 1.1), which states that all flow (unit volume per unit time) entering a junction or tissue must equal all flow exiting:

$$Q_{in} = Q_{out} \tag{1.1}$$

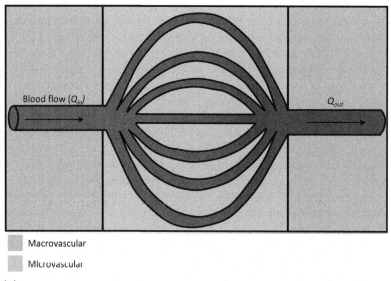

Macrovascular

Microvascular

FIG. 1.1

Blood flows through the macrovascular vessels (Q_{in}) to the microvascular capillary bed to deliver nutrients and oxygen to the tissue and drain waste products through outflow (Q_{out}). The main assumption for macrovascular flow is conservation of mass, that is: $Q_{in} = Q_{out}$.

The conservation of mass also holds for the flow within microvasculature. In contrast to the macrovasculature, microvasculature flow rates and paths are assumed stationary (time invariant). If all possible flow routes are assumed to be connected with the larger delivering and draining vessels, the formal definition of perfusion can be stated as follows:

Perfusion: *the amount of fluid received by a tissue, such as capillary blood flow, in a steady-state manner.*

As we will see later, perfusion can be measured and described in many ways.

1.2.2 The Fick principle

Originally developed and described by Adolf Fick in the 19th century, the Fick principle states that an organ's specific blood flow can be calculated using the conservation of mass, in which one assumes cardiac flow or output (CO, in liters of blood per min [l_{blood}/min]) and oxygen consumption are constant (Fick, 1870). Fick's original "organ" was the entire human body, requiring measurement of three key values—oxygen consumption by the lungs (VO_2, in liters of oxygen per minute, [l_{O_2}/min]), the arterial oxygen concentration (C_a, a percentage, or more precisely [mL_{O_2}/mL_{blood}]), and the venous oxygen concentration (C_v):

$$CO \cdot C_a = CO \cdot C_v + VO_2 \tag{1.2}$$

rearranging Eq. (1.2) allows for the calculation of CO or VO_2:

$$CO = {VO_2}/{(C_a - C_v)} \tag{1.3}$$

$$VO_2 = CO \cdot (C_a - C_v) \tag{1.4}$$

The techniques needed for Fick principle are invasive, requiring sampling of arterial and venous blood and a method for direct capture of VO_2. Thus the Fick principle has declined in clinical significance; however, it presents a simple and elegant method for deriving blood flow based on an indicator's concentration, making it the basis of many modern blood flow imaging techniques. While the indicator in Fick principle is oxygen, we shall see that many other types of indicators are used.

1.2.3 The central volume principle

To better understand the principles behind flow and perfusion, scientists have extensively used the principles of tracer kinetics, or indicator dilution, to measure blood flow. As stated by Stewart (1897), when a known amount of substance is injected (in 1 at $t=0$) into blood at location A (M_A, traditionally indicated by q but we prefer M as to prevent confusion with flow Q), and the diluted concentration is sampled at location B (C_B, in [$ml_{indicator}$/ml_{blood}]) at a known time difference dt, then flow Q is simply:

$$Q = M_A / \int_0^\infty C_B(t)dt \qquad (1.5)$$

where the integral indicates that all tracer will eventually have left the system, resulting in flow Q being constant. This is Fick principle reformulated, describing the difference between quantity at locations A and B. The rate at which the tracer leaves the system is calculated as $QC_B(t)$; it is now convenient to introduce a function that represents the fraction per unit time of tracer (in 1/s) leaving the system to the total injected tracer amount (Meier and Zierler, 1954):

$$h(t) = Q \, C_B(t)/M_A \qquad (1.6)$$

The rate at which the particles enter and leave the system is $Qh(t)dt$. The volume of particles leaving the system can now be expressed as the rate multiplied by the time required for the particles to leave the system, *i.e.*, $dV = tQh(t)dt$ and thus, since Q is stationary:

$$V = Q \int_0^\infty th(t)dt \qquad (1.7)$$

The integral $\int_0^\infty th(t)dt$ can now be thought of as the mean of the time when all particles have arrived at point B, which is termed the mean circulation time or mean transit time τ. This leads to the succinct expression for the central volume principle:

$$V = Q\tau \qquad (1.8)$$

With the assumption that the distribution of traversal times for the tracer particles is the same as for the fluid particles, blood flow can be calculated by measuring the volume of blood at a certain time after injection. See Fig. 1.2 for an illustration of the central volume principle for a single vessel, but as illustrated in Fig. 1.1 the same holds for multiple trajectories. The central volume principle holds for nondiffusible and freely diffusible tracers. However, when a diffusible tracer is injected into the blood vessel, the tracer will diffuse through the vessel wall into the surrounding tissue with a certain diffusion predisposition called extraction fraction λ, which can be calculated by the concentration of the tracer in the tissue C_t divided by the concentration of the tracer in the blood C_b. This is an important consideration for perfusion assessment, since incorrect estimates of extraction fractions (or leaky vessels of study subjects) may introduce errors in the measurements.

1.2.4 Key perfusion parameters

1.2.4.1 Perfusion, perfusion rate, and blood flow

Historically, techniques using radioactive tracers to measure *perfusion* (P) expressed it as *blood flow* (Q) in milliliters of blood per minute [ml/min] normalized to 100 g of tissue mass. This creates the nominal unit for P of [ml/(min · 100 g)]. For the purposes

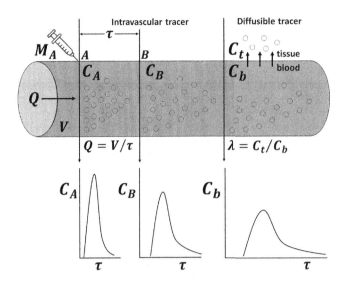

FIG. 1.2

For an intravascular tracer the central volume principle states that the flow Q equals the volume of the blood vessel V divided by mean transit time τ. For a diffusible tracer the extraction fraction λ is defined by the concentration of tracer in the tissue C_t divided by the concentration of tracer in the blood C_b.

of many imaging modalities, in which tissue mass is often not measured, *perfusion rate* (f) is frequently used as a replacement:

$$f = \rho P \tag{1.9}$$

where a given tissue density, ρ, is assumed, most frequently to be 1 g/ml. This produces the rather unintuitive unit for f of $[\frac{1}{\min}]$, though logically this can be thought of as the "rate at which blood volume perfuses into tissue volume." In early scientific imaging literature, all three values $(Q, P, \text{and} f)$ were used interchangeably, and careful attention to reported values was needed. Modern studies, most prevalent in the brain, require prudent definition of measurements in use: (cerebral) blood volume, $(C)BV$ and (cerebral) blood flow, $(C)BF$.

1.2.4.2 Blood volume

Blood volume represents the volume of blood in a given tissue and is, therefore, expressed as relative blood volume per volume of tissue. Contrary to P, blood volume captures the (approximately) *instantaneous* total blood in the region of interest. The volume of interest can simply be represented by the given voxel size of the imaging technique, which makes CBV an especially attractive parameter for investigating local blood supply.

1.2.4.3 Mean transit time

It follows from perfusion/blood flow and blood volume that one can estimate the time required for blood to pass through tissue. Averaged over a volume, this value becomes mean transit time (MTT) and depends on the tracer being studied (intravascular versus freely diffusible). MTT can range from between a few seconds to a few minutes.

1.3 MRI versus other modalities for imaging perfusion

Given the potential importance of blood flow and perfusion, it is no surprise that diagnostic vascular imaging technologies emerged early and have been extensively investigated and refined over the past 100 years. The historical evolution of diagnostic imaging runs in parallel with the desire by scientists and clinicians to better image blood flow and perfusion.

Flow and perfusion inspection consist of two main components: the indicator or tracer, and the technology or detector to resolve its passage. In nonimaging environments, as discussed before, the detector is simply the invasive sampling of arterial and venous blood. The first visualization of blood flow was performed in 1927 by Moniz using the combination of intravascular radio-opaque iodinated contrast and X-ray projection imaging of cerebral vessels (Moniz, 1927; Artico *et al.*, 2017). Successors to this technique employ digital rapid imaging and subtraction of successive X-rays to visualize flowing blood into, for example, blocked coronary arteries and cerebral aneurysms (Crummy *et al.*, 1980). In the late 1940s, Kety and Schmidt introduced a nonimaging quantitative method for measuring brain blood flow based on the Fick principle, which used the inhalation of nitrous oxide and the *in vitro* analysis of arterial and venous samples (Kety and Schmidt, 1948).

Modern imaging techniques for flow and perfusion can generally be split into two categories: those that use ionizing radiation, and those that do not. Radioisotope-based nuclear medicine imaging with single-photon emission computed tomography (SPECT) (Brooks and Di Chiro, 1976) and positron emission tomography (PET) (Lammertsma, 2017) allows for labeling and tracking of molecular uptake *in vivo*, but generally suffer from relatively low spatial and temporal resolution. Computed tomography (CT) (Miles and Griffiths, 2003) combined with radio-opaque contrast agents is more commonly used; however, repeated scans deposit potentially harmful X-rays into the patient. Nonionizing modalities include ultrasound (US) and MRI. Blood flow imaging with US consists of Doppler US, which is commonly used for macrovascular assessment, while perfusion assessment from microbubble contrast enhancement is in its relative infancy (Cosgrove and Lassau, 2010).

MRI provides a comprehensive and noninvasive approach to monitor changes in both vascular structure and function due to its excellent soft tissue contrast and varied contrast mechanisms. While this book focuses on perfusion as assessed by MRI, Fig. 1.3 summarizes MRI-based techniques for assessing both macrovascular blood flow and microvascular perfusion; MRI has the capability to evaluate blood vessels and blood flow in a number of unique and independent ways.

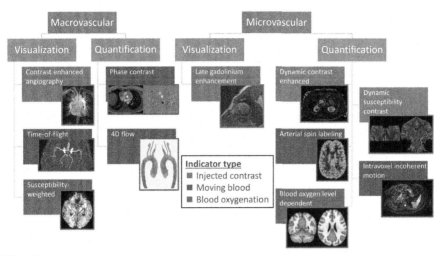

FIG. 1.3

Overview of MRI techniques for visualizing and quantifying macro- and microvascular blood flow. Color-coding denotes indicator for each technique.

Images courtesy of K. Baas, O. Gurney-Champion, A. Kaiser, R.N. Planken, and A. Schrantee—Amsterdam UMC.

1.3.1 Basics of MRI

To understand the principles behind flow and perfusion with MRI, we first describe the very general basics of the imaging technique. MRI is based on the fact that in the presence of a strong magnetic field such as the one in an MRI scanner, the magnetic moment (spin) of hydrogen atoms (protons) in the human body aligns with, acquire magnetization from, and precess around the direction of the magnetic field. A radiofrequency pulse with a frequency equivalent to a certain *flip angle* (*e.g.*, 90 degrees) is used to tip the longitudinal (aligned) magnetization to a transverse (perpendicular to the direction of the magnetic field) magnetization. This precessing transverse magnetization will induce an induction current in a receiver coil which constitutes the measured MRI signal. The contrast in MRI is derived from differential relaxation times in different tissues; each tissue has its own unique T1 (relaxation back to longitudinal equilibrium) and T2 (decay of transverse magnetization) constants. Application of additional magnetic field gradients in *x*-, *y*-, and *z*-directions allows further flexibility in MRI sequence design and multidimensional encoding of quantitative parameters such as diffusion, blood flow velocity, and perfusion.

For micro- and macrovascular assessment using MRI, techniques can be striated based on the type of indicator they use (Chapter 6) and whether they are qualitative or quantitative in nature. See Fig. 1.3. Following is a brief description of the main techniques used in research and in clinical practice.

1.3.2 Indicator: Injected contrast

1.3.2.1 Contrast-enhanced MR angiography

In contrast-enhanced MR angiography (CE-MRA), qualitative angiographic images are produced through the intravenous injection of contrast agent. Vessel signal is high (bright) while background tissues appear relatively dark. This is due to the contrast agent, which shortens the T1 of blood; thus, CE-MRA sequences are heavily T1 weighted. Typical CE-MRA agents are based on chelated gadolinium, the paramagnetic nature of which decreases both T1 and T2. Timing of the acquisition determines whether the CE-MRA will be weighted more heavily toward arterial vasculature, venous vasculature, or a combination (Hartung *et al.*, 2011; Riederer *et al.*, 2018). In clinical practice, CE-MRA is preferred due to its capabilities for high spatial resolution and high signal to noise; however, patients with impaired renal function need to be properly screened to avoid systemic gadolinium-induced fibrosis (Thomsen, 2006; Thomsen *et al.*, 2006).

1.3.2.2 Late-gadolinium enhancement

Because gadolinium is an extracellular agent, it remains in the tissue blood pool. In normal tissues, washout of gadolinium occurs within 1–3 min. Late-gadolinium enhancement makes use of the fact that myocardial infarcts, scars, and inflammation exhibit slower gadolinium washout (on the order of 5–20 min). This late contrast enhancement relative to healthy myocardium allows for the regional discrimination of a number of cardiac disorders, including infarction, myocarditis, and fibrosis (Doltra *et al.*, 2013; Jimenez Juan *et al.*, 2015).

1.3.2.3 Dynamic contrast-enhanced-MRI

Dynamic contrast-enhanced (DCE)-MRI tracks T1 changes induced by contrast in a *dynamic* manner. This technique is performed through serial imaging over the course of several minutes as gadolinium contrast in blood plasma perfuses into organs throughout the body. As shown in Chapter 2, careful measurement paired with modeling of the transfer constant (K^{trans}) between blood flow and permeability allows for the determination of a variety of parameters for the organ of interest. These include tissue volume fractions of extracellular space (v_e) or plasma (v_p) and the rate between the two (k_{ep}) (Tofts, 2010).

1.3.2.4 Dynamic susceptibility contrast-MRI

Rather than tracking T1 changes, dynamic susceptibility contrast (DSC)-MRI (Chapter 3) *dynamically* images T2 and T2* changes induced by contrast injection. Deep lying vasculature will experience local paramagnetic-induced distortions and signal loss caused by gadolinium, which allows determination of blood flows/volumes and transit times. In contrast to DCE, DSC requires imaging of only the first pass of perfusion, so-called bolus tracking, to measure these parameters (Calamante, 2010).

1.3.3 **Indicator: Moving blood**

1.3.3.1 Time-of-flight angiography

Time-of-flight (TOF) is an alternative to CE-MRA that produces angiographic images without the need for contrast injection. Bright vessel signal is generated through flow-related enhancement, in which "fresh," unsaturated blood arrives into a saturated imaging slab. "Saturated" here refers to the equilibrium within an imaging volume caused by repeated and rapid radiofrequency pulses. Optimal contrast can be achieved if thin imaging slabs are oriented perpendicular to the vessel of interest. Thus TOF is primarily used in intracranial and peripheral vasculature, while excessively tortuous vessels or those with slow flow are generally less well visualized (Wehrli *et al.*, 1986; Wheaton and Miyazaki, 2012).

1.3.3.2 2D phase contrast and 4D flow

Phase contrast (PC) MRI allows for the quantification of blood velocity and flow. In this technique, the change of the phase in the image is proportional to the velocities of all tissues within the image. The expected phase change can be captured by a user-defined acquisition parameter, known as velocity encoding, which allows for the direct quantification of faster (relative to static tissue) flowing blood (Moran, 1982; Nayler *et al.*, 1986).

PC MRI is typically performed to encode velocity in a single imaging plane and in a single velocity direction (2D PC), reconstructed over cardiac time. The 4D flow MRI extends this to include flow sensitivity along all spatial dimensions over an imaging volume. This approach can be used for both qualitative and quantitative assessment of whole vasculature regions of interest, such as the thoracic aorta, whole heart, or circle of Willis (Markl *et al.*, 2012).

1.3.3.3 Arterial spin labeling

Arterial spin labeling (ASL) (Chapter 4) is a contrast-free quantitative technique used to give absolute values of blood perfusion within tissues. Blood upstream from the volume of interest is magnetically labeled (or tagged). Imaging is performed both before and after labeling (plus a predefined delay), and these are subtracted to give signal that represents a perfusion map of the imaged volume. As ASL is a low signal technique ($<1\%$ following subtraction), careful protocol development and testing is required (see Chapter 7) (Borogovac and Asllani, 2012).

1.3.3.4 Intravoxel incoherent motion

Intravoxel incoherent motion (IVIM) imaging is a technique that aims to quantify perfusion through the signal loss induced by capillary microcirculation (Le Bihan *et al.*, 1986). As opposed to typical diffusion imaging, which targets the movement of water molecules, IVIM uses weaker MRI gradients and more b-values. IVIM is capable of measuring three main parameters: perfusion fraction, apparent diffusion coefficient, and the pseudo-diffusion coefficient. IVIM precision and accuracy remain challenging, and most current research reflects approaches to improving IVIM quantification before application in clinical practice (Iima and Le Bihan, 2016). For this reason, IVIM is not further addressed in this book.

1.3.4 Indicator: Blood oxygenation

1.3.4.1 Blood oxygen level dependent

Blood oxygen level dependent (BOLD) MRI measures the relative difference between oxygenated and deoxygenated blood in a given tissue (Thulborn *et al.*, 1982). The two appear differently on BOLD, as deoxygenated blood shortens T2 and T2* resulting in lower signal. These differences make BOLD especially useful in studies of cerebral vasoreactivity (Sleight *et al.*, 2021) (Chapter 5). Similar to ASL, MRI acquisition settings play an outsized role in the amount of contrast and the accuracy of the technique (Chapter 7).

1.3.4.2 Susceptibility-weighted imaging

Contrast in susceptibility-weighted imaging (SWI) comes from the magnetic susceptibility-induced signal losses caused by deoxygenated blood. Therefore SWI is primarily used for depicting venous anatomy. Signal losses are further enhanced through processing of both magnitude and phase data. The high resolution, 3D nature of the technique allows for exquisite demonstration of small venous structures through the use of a thick slab minimum intensity projection (Haacke *et al.*, 2004, 2009).

1.4 Clinical relevance

1.4.1 Macrovascular

Cardiovascular disease of the large vessels includes everything from congenital heart disease to carotid atherosclerosis to intracranial aneurysms and peripheral artery disease. Anatomical assessment of diseased vessels is generally performed with MR angiography, with or without contrast administration. In these settings, MR angiography plays a crucial role in clinical management. For example, it offers the possibility of assessing diameters of the aorta in bicuspid aortic valve disease and Marfan syndrome, where surgery is warranted when diameters surpass 5.5 and 4.5 cm, respectively (Erbel *et al.*, 2014; Nishimura *et al.*, 2014). Also, the degree of stenosis in carotid atherosclerosis can be assessed, facilitating the clinical decision to perform endarterectomy. Furthermore, the size of an intracranial aneurysm assessed on MR angiography provides a good indication of when to intervene by coiling or clipping treatment. SWI has the potential to image hemorrhagic lesions, vascular malformations, venous thrombosis, neoplasms, and calcium/iron deposition in the pediatric brain (Tong *et al.*, 2008). The 2D flow MRI can be used clinically to quantify stroke volume, ejection fraction, cardiac output, or unwanted regurgitant flow over the cardiac valves. The volumetric nature 4D flow MRI can provide more detail than 2D flow, especially in the heart and aorta, and allows for qualitative and quantitative assessment of abnormal flow in congenital and acquired heart disease with the added benefit of derivation of energetics and wall shear stress.

1.4.2 **Microvascular**

Techniques to assess microvascular perfusion are particularly useful in visualizing tumors, as tumor angiogenesis drastically alters perfusion compared to surrounding tissue. A classic example is the use of DCE-MRI to predict glioma grade and survival (Law *et al.*, 2006). Microvascular behavior is altered in other diseases as well. For example, ASL has been mainly applied to study impaired cerebral perfusion in brain diseases such as Parkinson's (Shang *et al.*, 2021), Alzheimer's (Hertel *et al.*, 2021), Moyamoya (Agarwal *et al.*, 2021), and other neurological disorders (Bambach *et al.*, 2022). BOLD fMRI provides an indication of hemodynamic reactivity to various stimuli of blood vessels in the brain and is mainly used to study psychiatric disorders (Fornito and Bullmore, 2010). As will be seen in the following chapters, perfusion imaging has a broad range of applications inside and outside of the brain (Gordon *et al.*, 2014), even in the assessment of inflammation seen in abdominal aortic aneurysms (Nguyen *et al.*, 2014).

These macro- and microvascular imaging applications to a plethora of diseases emphasize the versatility of MRI and underline the importance of the continuation of MRI development for improved assessment of flow and perfusion in health and disease.

1.5 **Summary**

In this chapter we described the basic theory of flow imaging, including the conservation of mass, the Fick principle, and the central volume principle. These are leveraged in a number of techniques with MRI to fulfill a crucial clinical role in the assessment of diseases involving the vasculature. In this book, the most important MRI technologies for the assessment of perfusion, its applicability in various anatomical regions, and the technical considerations for use of perfusion MRI are discussed in detail.

References

Agarwal, V., Singh, P., Ahuja, C.K., Gupta, S.K., Aggarwal, A., Narayanan, R., 2021. Non-invasive assessment of cerebral microvascular changes for predicting postoperative cerebral hyperperfusion after surgical revascularisation for moyamoya disease: an arterial spin labelling MRI study. Neuroradiology 63 (4), 563–572. https://doi.org/10.1007/s00234-020-02583-w.

Artico, M., Spoletini, M., Fumagalli, L., *et al.*, 2017. Egas Moniz: 90 years (1927–2017) from cerebral angiography. Front. Neuroanat. 11, 81. https://doi.org/10.3389/fnana.2017.00081.

Bambach, S., Smith, M., Morris, P.P., Campeau, N.G., Ho, M.-L., 2022. Arterial spin labeling applications in pediatric and adult neurologic disorders. J. Magn. Reson. Imaging 55 (3), 698–719. https://doi.org/10.1002/jmri.27438.

Borogovac, A., Asllani, I., 2012. Arterial spin labeling (ASL) fMRI: advantages, theoretical constrains, and experimental challenges in neurosciences. Int. J. Biomed. Imaging 2012, 818456. https://doi.org/10.1155/2012/818456.

Brooks, R.A., Di Chiro, G., 1976. Principles of computer assisted tomography (CAT) in radiographic and radioisotopic imaging. Phys. Med. Biol. 21 (5), 689–732. https://doi.org/10.1088/0031-9155/21/5/001.

Calamante, F., 2010. Perfusion MRI using dynamic-susceptibility contrast MRI: quantification issues in patient studies. Top. Magn. Reson. Imaging 21 (2), 75–85. https://doi.org/10.1097/RMR.0b013e31821e53f5.

Cosgrove, D., Lassau, N., 2010. Imaging of perfusion using ultrasound. Eur. J. Nucl. Med. Mol. Imaging 37 (Suppl. 1), S65–S85. https://doi.org/10.1007/s00259-010-1537-7.

Crummy, A., Strother, C., Sackett, J., et al., 1980. Computerized fluoroscopy: digital subtraction for intravenous angiocardiography and arteriography. Am. J. Roentgenol. 135 (6), 1131–1140. https://doi.org/10.2214/ajr.135.6.1131.

Doltra, A., Amundsen, B.H., Gebker, R., Fleck, E., Kelle, S., 2013. Emerging concepts for myocardial late gadolinium enhancement MRI. Curr. Cardiol. Rev. 9 (3), 185–190. https://doi.org/10.2174/1573403x113099990030.

Erbel, R., Aboyans, V., Boileau, C., et al., 2014. 2014 ESC guidelines on the diagnosis and treatment of aortic diseases: document covering acute and chronic aortic diseases of the thoracic and abdominal aorta of the adult. The Task Force for the Diagnosis and Treatment of Aortic Diseases of the European. Eur. Heart J. 35 (41), 2873–2926. https://doi.org/10.1093/eurheartj/ehu281.

Fick, A., 1870. Uber die Messung des Blutquantums in der Herzventrikeln. *Sitz der Phys. Ges Wurzbg.* 2, 16–28.

Fornito, A., Bullmore, E.T., 2010. What can spontaneous fluctuations of the blood oxygenation-level-dependent signal tell us about psychiatric disorders? Curr. Opin. Psychiatry 23 (3), 239–249. https://doi.org/10.1097/YCO.0b013e328337d78d.

Gordon, Y., Partovi, S., Müller-Eschner, M., et al., 2014. Dynamic contrast-enhanced magnetic resonance imaging: fundamentals and application to the evaluation of the peripheral perfusion. *Cardiovasc. Diagn. Ther.* 4 (2), 147–164. https://doi.org/10.3978/j.issn.2223-3652.2014.03.01.

Haacke, E.M., Xu, Y., Cheng, Y.-C.N., Reichenbach, J.R., 2004. Susceptibility weighted imaging (SWI). Magn. Reson. Med. 52 (3), 612–618. https://doi.org/10.1002/mrm.20198.

Haacke, E.M., Mittal, S., Wu, Z., Neelavalli, J., Cheng, Y.-C.N., 2009. Susceptibility-weighted imaging: technical aspects and clinical applications, part 1. AJNR Am. J. Neuroradiol. 30 (1), 19–30. https://doi.org/10.3174/ajnr.A1400.

Hartung, M.P., Grist, T.M., François, C.J., 2011. Magnetic resonance angiography: current status and future directions. J. Cardiovasc. Magn. Reson. 13, 19. https://doi.org/10.1186/1532-429X-13-19.

Hertel, A., Wenz, H., Al-Zghloul, M., et al., 2021. Crossed cerebellar diaschisis in Alzheimer's disease detected by arterial spin-labelling perfusion MRI. In Vivo 35 (2), 1177–1183. https://doi.org/10.21873/invivo.12366.

Iima, M., Le Bihan, D., 2016. Clinical intravoxel incoherent motion and diffusion MR imaging: past, present, and future. Radiology 278 (1), 13–32. https://doi.org/10.1148/radiol.2015150244.

Jimenez Juan, L., Crean, A.M., Wintersperger, B.J., 2015. Late gadolinium enhancement imaging in assessment of myocardial viability: techniques and clinical applications. Radiol. Clin. N. Am. 53 (2), 397–411. https://doi.org/10.1016/j.rcl.2014.11.004.

Kety, S.S., Schmidt, C.F., 1948. The nitrous oxide method for the quantitative determination of cerebral blood flow in man: theory, procedure and normal values 1. J. Clin. Invest. 27 (4), 476–483. https://doi.org/10.1172/JCI101994.

Lammertsma, A.A., 2017. Forward to the past: the case for quantitative PET imaging. J. Nucl. Med. 58 (7), 1019–1024. https://doi.org/10.2967/jnumed.116.188029.

Law, M., Oh, S., Babb, J.S., *et al.*, 2006. Low-grade gliomas: dynamic susceptibility-weighted contrast-enhanced perfusion MR imaging—prediction of patient clinical response. Radiology 238 (2), 658–667. https://doi.org/10.1148/radiol.2382042180.

Le Bihan, D., Breton, E., Lallemand, D., Grenier, P., Cabanis, E., Laval-Jeantet, M., 1986. MR imaging of intravoxel incoherent motions: application to diffusion and perfusion in neurologic disorders. Radiology 161 (2), 401–407. https://doi.org/10.1148/radiology.161.2.3763909.

Markl, M., Frydrychowicz, A., Kozerke, S., Hope, M., Wieben, O., 2012. 4D flow MRI. J. Magn. Reson. Imaging 36 (5), 1015–1036. https://doi.org/10.1002/jmri.23632.

Meier, P., Zierler, K.L., 1954. On the theory of the indicator-dilution method for measurement of blood flow and volume. J. Appl. Physiol. 6 (12), 731–744. https://doi.org/10.1152/jappl.1954.6.12.731.

Miles, K.A., Griffiths, M.R., 2003. Perfusion CT: a worthwhile enhancement? Br. J. Radiol. 76 (904), 220–231. https://doi.org/10.1259/bjr/13564625.

Moniz, E., 1927. L'encéphalographie artérielle, son importance dans la localisation des tumeurs cérébrales. Rev. Neurol. 2, 72.

Moran, P.R., 1982. A flow velocity zeugmatographic interlace for NMR imaging in humans. Magn. Reson. Imaging 1 (4), 197–203. http://www.sciencedirect.com/science/article/B6T9D-4C06DSJ-V/2/273a4ef3706c21a0d2f2f109629dafa0.

Nayler, G.L., Firmin, D.N., Longmore, D.B., 1986. Blood flow imaging by cine magnetic resonance. J. Comput. Assist. Tomogr. 10 (5), 715–722. http://www.ncbi.nlm.nih.gov/entrez/query.fcgi?cmd=Retrieve&db=PubMed&dopt=Citation&list_uids=3528245.

Nguyen, V.L., Backes, W.H., Kooi, M.E., *et al.*, 2014. Quantification of abdominal aortic aneurysm wall enhancement with dynamic contrast-enhanced MRI: feasibility, reproducibility, and initial experience. J. Magn. Reson. Imaging 39 (6), 1449–1456. https://doi.org/10.1002/jmri.24302.

Nishimura, R.A., Otto, C.M., Bonow, R.O., *et al.*, 2014. 2014 AHA/ACC guideline for the management of patients with valvular heart disease. J. Thorac. Cardiovasc. Surg. 148 (1), e1–e132. https://doi.org/10.1016/j.jtcvs.2014.05.014.

Riederer, S.J., Stinson, E.G., Weavers, P.T., 2018. Technical aspects of contrast-enhanced MR angiography: current status and new applications. Magn. Reson. Med. Sci. 17 (1), 3–12. https://doi.org/10.2463/mrms.rev.2017-0053.

Shang, S., Wu, J., Zhang, H., *et al.*, 2021. Motor asymmetry related cerebral perfusion patterns in Parkinson's disease: an arterial spin labeling study. Hum. Brain Mapp. 42 (2), 298–309. https://doi.org/10.1002/hbm.25223.

Simionescu, M., 1980. Ultrastructural organization of the alveolar-capillary unit. CIBA Found. Symp. 78, 11–36. https://doi.org/10.1002/9780470720615.ch2.

Sleight, E., Stringer, M.S., Marshall, I., Wardlaw, J.M., Thrippleton, M.J., 2021. Cerebrovascular reactivity measurement using magnetic resonance imaging: a systematic review. Front. Physiol. 12, 643468. https://doi.org/10.3389/fphys.2021.643468.

Stewart, G.N., 1897. Researches on the circulation time and on the influences which affect it. J. Physiol. 22 (3), 159–183. https://doi.org/10.1113/jphysiol.1897.sp000684.

Thomsen, H.S., 2006. Nephrogenic systemic fibrosis: a serious late adverse reaction to gadodiamide. Eur. Radiol. 16 (12), 2619–2621. https://doi.org/10.1007/s00330-006-0495-8.

Thomsen, H.S., Morcos, S.K., Dawson, P., 2006. Is there a causal relation between the administration of gadolinium based contrast media and the development of nephrogenic systemic fibrosis (NSF)? Clin. Radiol. 61 (11), 905–906. https://doi.org/10.1016/j.crad.2006.09.003.

Thulborn, K.R., Waterton, J.C., Matthews, P.M., Radda, G.K., 1982. Oxygenation dependence of the transverse relaxation time of water protons in whole blood at high field. Biochim. Biophys. Acta 714 (2), 265–270. https://doi.org/10.1016/0304-4165(82)90333-6.

Tofts, P.S., 2010. T1-weighted DCE imaging concepts: modelling, acquisition and analysis. Signal 500 (450), 400.

Tong, K.A., Ashwal, S., Obenaus, A., Nickerson, J.P., Kido, D., Haacke, E.M., 2008. Susceptibility-weighted MR imaging: a review of clinical applications in children. AJNR Am. J. Neuroradiol. 29 (1), 9–17. https://doi.org/10.3174/ajnr.A0786.

Wehrli, F.W., Shimakawa, A., Gullberg, G.T., MacFall, J.R., 1986. Time-of-flight MR flow imaging: selective saturation recovery with gradient refocusing. Radiology 160 (3), 781–785. https://doi.org/10.1148/radiology.160.3.3526407.

Wheaton, A.J., Miyazaki, M., 2012. Non-contrast enhanced MR angiography: physical principles. J. Magn. Reson. Imaging 36 (2), 286–304. https://doi.org/10.1002/jmri.23641.

Dynamic contrast-enhanced MRI

2

Ben Dickie[a,b] and Petra J. van Houdt[c]

[a]*Division of Informatics, Imaging, and Data Science, School of Health Sciences, Faculty of Biology, Medicine and Health, School of Health Sciences, The University of Manchester, Manchester, United Kingdom*
[b]*Geoffrey Jefferson Brain Research Centre, Manchester Academic Health Science Centre, Northern Care Alliance NHS Group, University of Manchester, Manchester, United Kingdom*
[c]*Department of Radiation Oncology, The Netherlands Cancer Institute, Amsterdam, The Netherlands*

2.1 Introduction

The goal of dynamic contrast-enhanced MRI (DCE-MRI) is to rapidly track the uptake and clearance of T_1-reducing paramagnetic contrast agents in a tissue of interest. The injected contrast agent creates a local reduction in the T_1 relaxation time of nearby water protons, which appears as signal enhancement on T_1-weighted MRI. By acquiring multiple imaging volumes over time, the uptake and clearance of contrast agent produces a signal enhancement time curve, the shape of which depends on tissue perfusion, the permeability of microvessels to the contrast agent, the volume of any leakage/distribution spaces, and the clearance rates of contrast agent from tissue and blood.

The enhancement curves can be analyzed to provide imaging biomarkers (IBs) of tissue perfusion and vascular permeability. The simplest approach quantifies signal time curves using model-free metrics such as time to peak, area under the curve, and peak enhancement (Mayr *et al.*, 1998; Yuh *et al.*, 2009). Metrics derived directly from MRI signal intensities are sensitive to acquisition parameters such as field strength, flip angle, and repetition time. With the aim of reducing this dependence, it is common to convert signal to contrast agent concentration prior to calculating enhancement curve metrics. However, converting signal to contrast agent concentration can introduce additional challenges as it requires knowledge of native R_1, the signal generating model, knowledge of the flip angle distribution, and contrast agent relaxivity, which are not trivial to estimate accurately and can lead to poorer reproducibility (Galbraith *et al.*, 2002). While it may be possible to reduce the impact of these acquisition-related factors, semiquantitative or heuristic metrics are inherently sensitive to confounding patient-specific factors, such as cardiac output, renal

Advances in Magnetic Resonance Technology and Applications, Volume 11, ISSN 2666-9099
https://doi.org/10.1016/B978-0-323-95209-5.00016-7

function, and hematocrit value (Parker *et al.*, 2006) and they lack physiological specificity, making it difficult to interpret how changes in enhancement curve metrics relate to changes in tissue physiology.

Tracer kinetic analysis, also called pharmacokinetic modeling or quantitative DCE-MRI, helps to address variability caused by patient-specific factors and aims to derive more physiologically meaningful metrics compared to semiquantitative analysis (Sourbron and Buckley, 2011). Bulk delivery of contrast agent to tissue must be measured in the form of an arterial input function (AIF) (Calamante, 2013). The kinetic model then describes how the tissue filters the arterial input function to produce the tissue concentration-time curve. By making reasonable assumptions regarding the kinetics of the contrast agent in blood and tissue spaces, parameters related to perfusion and permeability of blood vessels supplying the tissue can be estimated (Tofts *et al.*, 1999). The exact vascular properties that can be measured depend on many factors, including the imaging protocol, the kinetic model, and the tissue itself.

Over the past 30 years, parameters derived from DCE-MRI have found potential as IBs in a wide range of research applications. Oncology is the most well established. Here, DCE-MRI is used to assess perfusion and vascular leakage in solid tumors. In this context, DCE-MRI metrics have shown utility as surrogate endpoints in trials of antivascular and antiangiogenic therapies (Leach *et al.*, 2005; O'Connor *et al.*, 2007; Padhani, 2003; Salem and O'Connor, 2016; Sung *et al.*, 2016) and hold value as predictive biomarkers of early tumor response to chemotherapy and/or radiotherapy (Bernstein *et al.*, 2014; Cao, 2011; Dijkhoff *et al.*, 2017; Gaddikeri *et al.*, 2016; Mazaheri *et al.*, 2017). In the brain, DCE-MRI has been used to assess damage to the blood-brain barrier (BBB), for example, in multiple sclerosis (Kermode *et al.*, 1990), stroke (Villringer *et al.*, 2017), or neurodegenerative diseases such as cerebral small vessel disease (Raja *et al.*, 2018; Thrippleton *et al.*, 2019) and Alzheimer's disease (Dickie *et al.*, 2019, 2021; Farrall and Wardlaw, 2009; Montagne *et al.*, 2015). DCE-MRI has evolved to a robust tool to assess circulatory dysfunction in major and minor coronary artery disease (Coelho-Filho *et al.*, 2013; Saeed *et al.*, 2015; Shehata *et al.*, 2014) or muscle function in pathologies like peripheral arterial disease (Caroca *et al.*, 2021; Howe *et al.*, 2020; Sujlana *et al.*, 2018). Visual assessment of lung DCE-MRI is increasingly being used in several lung diseases, such as COPD or pulmonary hypertension (Gefter *et al.*, 2021). DCE-MRI parameters could have the potential to determine the activity of Crohn's disease and response to subsequent treatment (Bane *et al.*, 2021; Grassi *et al.*, 2022). Another growing application is renal MRI. Renal DCE-MRI can be used to assess kidney function such as perfusion and filtration and offers the advantage of no radiation exposure compared to renal scintigraphy approaches. It has shown utility in detecting changes occurring after kidney transplant and for assessing the severity of renal artery stenosis (Bokacheva *et al.*, 2008; Zhang and Lee, 2020; Zhou *et al.*, 2018).

Many of these applications are ongoing areas of research, and as such almost all are yet to be used routinely in clinical practice for patient benefit. Some areas such as oncology are more established and should perhaps be further along the translational pipeline than they currently are. Efforts to translate imaging biomarkers derived from DCE-MRI are ongoing through a number of international standardization initiatives; however, there are many challenges and barriers that must be surmounted for this to become a reality.

This chapter is intended to provide an introduction to quantitative DCE-MRI, including the basic concepts of acquisition and analysis as well as a discussion of the challenges in clinical implementation. General guidelines regarding acquisition protocols and choices for analysis are given, but these inevitably depend on the application and the goal of the study for which the data are acquired.

2.2 Image acquisition

2.2.1 Imaging protocol

The DCE-MRI examination involves acquiring T_1-weighted volumes before, during, and after intravenous injection of a T_1-reducing paramagnetic contrast agent. For a stable injection of the contrast agent, a power injector is commonly used. The DCE-MRI acquisition is usually performed at the end of the MRI exam as the contrast agent can influence the signal intensities of other scan sequences. Fig. 2.1 shows an example of a DCE-MRI acquisition in prostate cancer. For oncology applications, in which washout of contrast agent is fast, images need only be acquired up to 5–10 min postinjection (Verma *et al.*, 2012). Longer acquisitions of up to 20 min are needed for assessment of contrast agent leakage across the blood-brain barrier in normal appearing brain tissue, since contrast agent leakage is very slow and takes time to reach detectable levels in the brain (Barnes *et al.*, 2016; Heye *et al.*, 2016). In addition, T_1 mapping data is typically acquired before the DCE acquisition, such that the native T_1 of the tissue can be determined before contrast administration. For higher field strengths (3T and higher), a B_1 map is also included.

2.2.2 T_1 mapping acquisition

For quantitative analysis of the DCE data, a T_1 mapping sequence is needed to convert the signal intensities to contrast agent concentration values. This is usually done using inversion recovery (IR) or the variable flip angle (VFA) approach (Bane *et al.*, 2018). The VFA approach is done with a spoiled gradient-echo sequence (SPGR) similar to the DCE acquisition itself. The dependence of T_1 weighting on flip angle makes SPGR highly sensitive to B_1 inhomogeneities (a certain percentage error in the actual flip angle translates to twice the percentage error in the calculated T_1) (Cheng and Wright, 2006; Tofts and Parker, 2013). For higher field strengths (≥ 3 T), the acquisition of a B_1 map is recommended to allow accurate modeling of delivered flip angles (Yarnykh, 2010). Fast inversion recovery or saturation recovery-based sequences that encode T_1 weighting through longer longitudinal

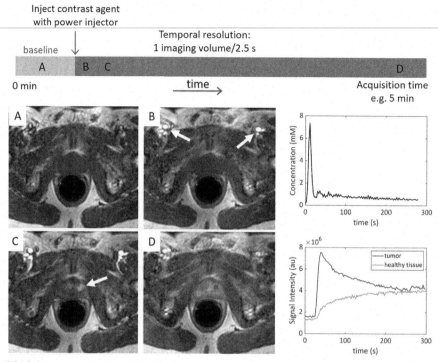

FIG. 2.1

Acquisition of DCE images consists of a series of T_1-weighted images before, during, and after injection of gadolinium-based contrast agent. In this prostate cancer example, the temporal resolution was 2.5 s meaning that every 2.5 s an imaging volume is taken. Almost immediately after contrast agent injection an increase in signal intensity is visible in the main arteries (arrows in B). The measurements of contrast agent concentration in these arteries provide the arterial input function (AIF; top right). During the uptake phase, the contrast first arrives in the tumor area of the prostate (arrow in C) before the rest of the prostate (D). The figure on the bottom right shows the differences in the signal intensity time curves for a tumor voxel (*purple* square; *dark gray* in print version) and a healthy voxel (*orange* square; *black* in print version).

recovery periods can help to reduce the influence of B_1 inhomogeneities on T_1 weighting, but come with the cost of longer acquisition times compared to VFA (Taylor *et al.*, 2016).

2.2.3 DCE acquisition

DCE-MRI acquisition is typically done with a fast 3D T_1-weighted spoiled gradient-echo (SPGR) sequence or magnetization-prepared (IR or SR) SPGR. SPGR is a steady-state gradient-echo sequence that drives longitudinal magnetization to a steady state using a short repetition time (TR). To minimize T_2^* weighting, a short echo time (TE) is used and transverse magnetization is spoiled after each TR using gradient and radiofrequency spoiling (Ernst and Anderson, 1966; Markl and Leupold, 2012). The 3D or slab-selective acquisitions should be used opposed to slice-selective acquisition (2D) to minimize effects of arterial inflow of unsaturated spins (Roberts *et al.*, 2011) and, importantly, to minimize slice profile effects within individual slices. SPGR sequences are available on all commercial systems, known as SPGR for General Electric scanners; FLASH for Siemens and Bruker, T_1-FFE for Philips.

The amount of T_1 weighting for SPGR sequences is controlled by the TR and flip angle. Usually, the TR is kept as short as possible, and the flip angle adjusted to provide the maximum signal (called the Ernst angle). At short TR, using a small flip angle (5–10 degrees) gives more signal sensitivity at lower contrast agent concentrations ($\sim$0.5 mM; lower R_1 values) than larger flip angles (20–30 degrees) (Tofts and Parker, 2013), but less sensitivity when R_1 becomes large (Fig. 2.2C–F). Larger flip angles provide less sensitivity than low flip angles when R_1 is low, but higher sensitivity when R_1 is large (as when measuring the AIF). Intermediate flip angles provide acceptable sensitivity at both low and high R_1 (Fig. 2.2B and D) (Tofts and Parker, 2013) and provide a good compromise between ensuring both tissue and arterial curves can be measured as accurately as possible. Increasing TR and maintaining a low flip angle can improve sensitivity at short R_1, but leads to dismal sensitivity at higher R_1, and will likely lead to saturation of the AIF signal during the first pass. If longer TRs are needed (*e.g.*, to facilitate acquisition of multiple gradient echoes), a larger flip angle will provide similar sensitivity to using short TR and low flip angle.

Optimization of the acquisition parameters often involves a trade-off between the temporal resolution, spatial resolution, spatial coverage requirements, and signal-to-noise ratio (SNR) (De Naeyer *et al.*, 2011; Henderson *et al.*, 1998; Yankeelov and Gore, 2009). For example, large volume coverage reduces the temporal and/or spatial resolution that can be achieved. In the case of prostate imaging, coverage of 60–75 mm in feet–head direction will be sufficient, whereas in case of rectal cancer or cervical cancer typically coverage of more than 100 mm is needed. This difference will result in a lower temporal resolution or thicker slices (or a combination of both) for larger volumes.

Insufficient temporal resolution will impact the accuracy of the quantitative DCE parameters (Crombé *et al.*, 2019; Heisen *et al.*, 2010; Henderson *et al.*, 1998; Kershaw and Cheng, 2010). Low temporal resolution will effectively smooth the measured AIF and tissue curves, making it impossible to resolve changes in curve shape occurring faster than the sampling rate (Henderson *et al.*, 1998; Tofts and Parker, 2013). This will predominantly affect the accuracy with which the AIF peak can be measured, and to a lesser degree, the ability to accurately measure the shape of the tissue enhancement curves. Together, undersampled AIF and tissue curves lead to errors in kinetic parameter fits (Sourbron and Buckley, 2011); these errors typically decrease when the temporal resolution increases (Heisen *et al.*, 2010; Kershaw and Cheng, 2010)—see Fig. 2.3 and the quantitative DCE analysis section (2.3) for further details.

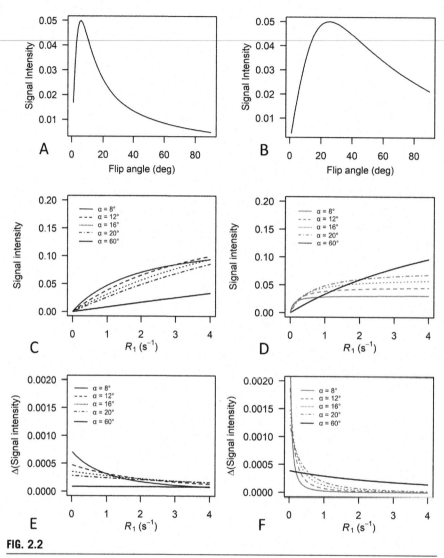

FIG. 2.2

Simulated SPGR signal intensity behavior as a function of flip angle, TR, and R_1. (A and B) SPGR signal as a function of flip angle assuming negligible T_2^* effects (short TE) for TR = 5 ms (left) and TR = 100 ms (right). For shorter TR, the peak of the SPGR curve is narrower, which may increase sensitivity to flip angle errors. (C and D) Signal sensitivity to R_1 as a function of flip angle at TR = 5 ms (left) and TR = 100 ms (right), (E and F) Signal change per unit change in R_1, plotted as a function of R_1 for TR = 5 ms (left) and TR = 100 ms (right). These plots show the gradient of lines in (C) and (D), plotted as a function of R_1. In all plots, the signals have been adjusted such that they represent signal per unit time. For both short and long TR, sensitivity of SPGR signal to changes in R_1 is greatest at low R_1 and lowest for high R_1. In practice, this means sensitivity is highest for detecting

In addition to factors that impact parameter accuracy, SNR of the signal time curves is an important determinant of parameter precision and reliability (Kershaw and Cheng, 2010). Noisy time curves will result in higher uncertainty of the fitted parameters. So even when a high spatial resolution might be achievable with acceptable temporal resolution, this might not be desirable if this results in a low SNR. This may impact the model choice, since more complex models that require higher temporal resolution are typically more sensitive to noise than less complex models (Kershaw and Cheng, 2010).

There are several acquisition options to increase temporal and/or spatial resolution without penalizing SNR. Acceleration of the acquisition by undersampling k-space is one option. In most imaging sequences it is standard to use parallel imaging, like SENSE (Philips) or GRAPPA (Siemens), to accelerate the acquisition. Keyhole sampling or view-sharing has also been used, which involves acquiring a full k-space at the start of the acquisition, but then only updating the center of k-space in subsequent volumes, dramatically reducing the sampling requirements (Bishop *et al.*, 1997; Le *et al.*, 2016; Othman *et al.*, 2015; Sun *et al.*, 2020; Xie *et al.*, 2022). However, careful attention needs to be paid to the settings of the keyhole technique such as the high temporal resolution associated with a lower spatial resolution (Benz *et al.*, 2018; Panek *et al.*, 2016). Recent developments in acceleration techniques such as compressed sensing have been applied to DCE-MRI (Benz *et al.*, 2018; Smith *et al.*, 2011). Compressed sensing exploits the inherent sparsity of medical images to reconstruct high-fidelity data from a much-reduced dataset. More recently deep learning approaches have been applied to accelerate DCE-MRI (Benou *et al.*, 2017; Fang *et al.*, 2021; Ulas *et al.*, 2018). A more pragmatic approach to deal with the balance between spatial and temporal resolution is a two-phase acquisition where spatial resolution is traded for higher temporal resolution during the rapid wash-in phase and temporal resolution traded for higher spatial resolution for the slower washout phase (Evelhoch, 2005; Jelescu *et al.*, 2011; van de Haar *et al.*, 2016).

There are also a number of postprocessing techniques that can improve SNR, allowing more relaxed restrictions on temporal or spatial resolution. Direct

---

changes in R_1 when contrast agent concentration is low (*i.e.*, R_1s are close to their native values) and lowest when it is high (*i.e.*, during bolus passage in the artery). At low flip angles and long TR, the sensitivity to changes in R_1 is highly variable (high at low R_1, low and high R_1). At larger flip angles (*e.g.*, 60 degrees; *blue* line; *black* in print version), sensitivity to R_1 changes is similar to using short TR and low flip angles. The highest sensitivity to R_1 changes is found when longer TR and low flip angles, but this only occurs over a very narrow range of very short R_1 values. For most DCE-MRI applications, it is therefore best to use short TR with relatively low flip angle (between 8 and 20) or long TR with large flip angle (between 50 and 70). Use of long TRs and small flip angles should be avoided; however, this arrangement may be useful for detection of contrast agents in tissues with very short R_1 such as for cerebrospinal fluid in the brain.

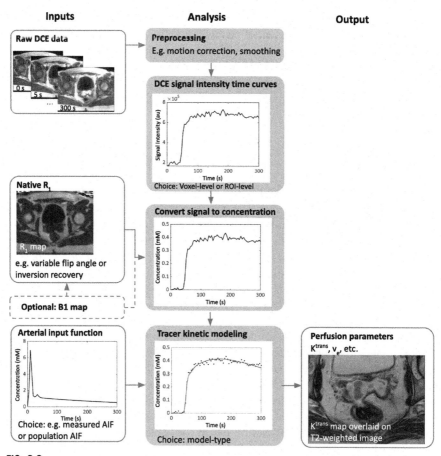

FIG. 2.3

High-level overview of quantitative analysis of DCE-MRI data, describing the most often used analysis steps. The input requires the 4D DCE data, input data for native R_1, for example, variable flip angle scans (VFA) or inversion recovery series (IR). The first step is to perform data preprocessing, such as applying a smoothing filter or motion correction. Next, the signal intensity values need to be converted to concentration values using the native R_1 value. At higher fields, a B_1 correction is recommended to get the correct flip angles. Another input for the pharmacokinetic models is an arterial input function (AIF). This can be an AIF determined from the dataset but may also be a population AIF. If the AIF is derived from the dataset itself it is typically extracted after the conversion from signal to concentration. Deviations from this pipeline are possible.

estimation of parameter maps (Dikaios *et al.*, 2014; Guo *et al.*, 2017) from k-space removes the step to convert k-space to imaging signals, providing statistical benefits and producing higher precision parameter maps than standard approaches (Dikaios *et al.*, 2014; Guo *et al.*, 2017). Denoising signals with simple median or smoothing filters can boost SNR, but the user should bear in mind that small structures may be smoothed out using such an approach. In case of insufficient SNR at voxel level, time curves of voxels within a region of interest should be averaged and regional analysis performed, which could lead to partial volume effects.

More recently, non-Cartesian readouts have been used such as radial ultrashort echo time (UTE) or zero echo time (ZTE) acquisitions and spiral readouts. These approaches sample the center of k-space more frequently than Cartesian readouts, improving the SNR, reducing sensitivity to motion-induced k-space errors, and providing self-navigation for robust motion correction for abdominal and chest imaging (Feng *et al.*, 2016). However, the outer regions of k-space are relatively undersampled, leading to lower spatial resolution (Lin *et al.*, 2008; Vautier *et al.*, 2010). UTE/ZTE sequences are more compatible with undersampling and incoherent k-space patterns required for compressed sensing (Ye, 2019), providing the potential for substantially higher temporal resolution than Cartesian readouts.

2.2.4 Injection protocol

For human applications, the contrast agent is typically injected into the antecubital vein with a power injector at a rate of between 2 and 4 ml/s followed by a saline flush of 20–30 ml (QIBA, 2011). A dose of 0.1 mmol/kg is standard and deemed safe (Kanal, 2016). For rodents, contrast agents are typically injected into the tail vein at a rate of between 0.5 and 1 ml/min.

2.3 Quantitative DCE analysis

Quantitative DCE-MRI analysis involves several analysis steps that culminate in fitting a tracer kinetic model to measure tissue concentration-time curves. To fit a tracer kinetic model, the contrast agent concentration in both the tissue and a feeding artery, also known as the arterial input function (AIF), must be known. Details of each analysis step are given later. Fig. 2.3 shows a basic analysis pipeline which is often used; however, variations on this pipeline are possible.

2.3.1 Preprocessing

Before anything is done, some preprocessing steps might be needed—for example, temporally or spatially smoothing the data to boost the SNR or motion correction to deal with movement during the DCE acquisition.

2.3.2 Converting signal to contrast agent concentration

Next, the raw T_1-weighted signal intensities in the tissue of interest should be converted to estimates of contrast agent concentration. This involves calculating ΔR_1 time courses from the raw signal using knowledge of the signal generation model (*e.g.*, SPGR signal model) and an estimate of native R_1 (Schabel and Parker, 2008).

Next, ΔR_1 is converted to an estimate of contrast agent concentration using the known linear relationship between ΔR_1 and contrast agent concentration (Schabel and Parker, 2008). The proportionality factor, r_1, is known as the contrast agent relaxivity and can be measured experimentally in phantoms or a literature value can be used (Rohrer *et al.*, 2005; Shen *et al.*, 2015).Typical values for r_1 range from approximately 3–6 $(smM)^{-1}$ depending on contrast agent and field strength (r_1 decreases with field strength).

2.3.3 Measuring the AIF

A key requirement of quantitative kinetic modeling is accurate measurement of contrast agent concentration within the main feeding artery. The shape of this concentration-time curve is called the AIF. Measuring an accurate AIF is challenging for several reasons and constitutes one of the main sources of error within quantitative DCE-MRI parameters (Calamante, 2013). The following conditions need to be satisfied:

1. The AIF needs to be sampled from a large vessel in the field of view, free from partial volume effects (Hansen *et al.*, 2009).
2. The large vessel needs to be close to the tissue of interest to minimize dispersion effects (Calamante *et al.*, 2000).
3. The arterial blood should have received a sufficient number of RF pulses before it is sampled (Roberts *et al.*, 2011). This can be mitigated somewhat by ensuring the slice encoding direction is perpendicular to the arterial flow.
4. The AIF should be sampled with a high temporal resolution to accurately capture its shape (Henderson *et al.*, 1998; Kershaw and Cheng, 2010).

A major source of variability among studies arises from the lack of standardization in how the AIF is extracted. Some use manual delineation of an arterial ROI whereas others use automatic or semiautomatic delineation (Huang *et al.*, 2016). Guidelines as to how AIF should be extracted (*i.e.*, number of voxels, location, minimum size of artery, etc.) and guidelines on what concentrations and shape to expect in different applications or populations are lacking.

When the AIF cannot reliably be extracted from the data, it is possible to use literature-based (or population) AIFs or reference-region modeling (which does not require an AIF) (Planey *et al.*, 2009). Literature-based AIFs are averaged concentration-time curves from a population. Several literature AIFs exist derived from human populations, for example, Parker AIF (Parker *et al.*, 2006), Georgiou AIF (Georgiou *et al.*, 2019), or preclinical data (McGrath *et al.*, 2009). When using literature AIFs, care must be taken to ensure they are compatible with the tissue of interest and acquisition protocol used. For example, extrapolating the Parker AIF to long acquisition times using the given functional form substantially underestimates arterial contrast agent concentrations after 6 min. Furthermore, when a literature AIF is used, individual (day-to-day) patient characteristics, such as cardiac output, are ignored. This impacts the accuracy of the DCE parameters on an individual level but may provide more repeatable estimates of kinetic parameters (Parker *et al.*, 2006).

AIF selection depends a great deal on the tissue of interest. In abdominal imaging, it is practical to measure the AIF in the inferior part of the descending aorta.

In contrast, finding a good AIF is more difficult in breast or brain imaging. In brain DCE-MRI, it is difficult to obtain a carotid AIF free from inflow and partial volume effects (Keil *et al.*, 2017). Inflow effects appear as a shortening of T_1 and reduce the dynamic range of detectable concentrations. This may lead to misinterpretation of K^{trans} values in patients with differing carotid blood velocities, for example, between control and Alzheimer's disease groups (Roher *et al.*, 2011). Because of the small diameter of the carotid, partial volume effects are unavoidable, but they can swamp the AIF signal with signal from tissue background, meaning that changes in blood concentration produce smaller or larger changes in the arterial ROI signal than expected (Sourbron *et al.*, 2009). For these reasons, it is common to measure blood contrast agent concentration in the superior sagittal sinus, the main draining vein of the cerebrum. This approach is also not ideal as some dispersion and extraction will have occurred as the bolus travels through the tissue; however, since the majority of the brain will have an intact BBB, even in the case of neoplasms, dispersion and leakage effects should be small.

2.3.4 **Tracer kinetic modeling**

Once the AIF is known, it can be convolved with the tracer kinetic model impulse response function (*IRF*(*t*)) to create a prediction of the measured tissue time course (*C*(*t*)).

$$C(t) = IRF(t) \otimes C_p(t)$$

The tissue impulse response (*IRF*(*t*)) function represents the response of the tissue to an impulse of contrast agent over time. $C_p(t)$ is the AIF. The model is fit by optimizing over different kinetic model parameter values until a predicted curve is found that best matches the measured data.

2.3.4.1 *Choice of model*

Most of the time, a single model is chosen and fit to the data. The model choice should be informed by a priori assumptions (effectively a well-informed guess) or knowledge about the physiology of the tissue of interest. In some cases, when the tissue physiology is not known, or varies across the tissue, multiple models are fit and statistical metrics such as the F-tests or Akaike information criterion used to select the most appropriate or best-fitting model at each pixel or ROI.

The temporal resolution of the acquired data also poses constraints on which models can be applied. Lower temporal resolution data makes it difficult to select an appropriate model, because as temporal resolution decreases, different models tend to fit the data equally well. In this case, the simplest model that provides a good fit to the data, and which provides a reasonable explanation of expected kinetics, should be chosen (Duan *et al.*, 2017). However, it must be noted that even under such circumstances, the model parameters may be inaccurate (Sourbron and Buckley, 2011).

To aid the following discussion, we have simulated a range of concentration-time curves using different kinetic models (Fig. 2.4.). The curves have been created by

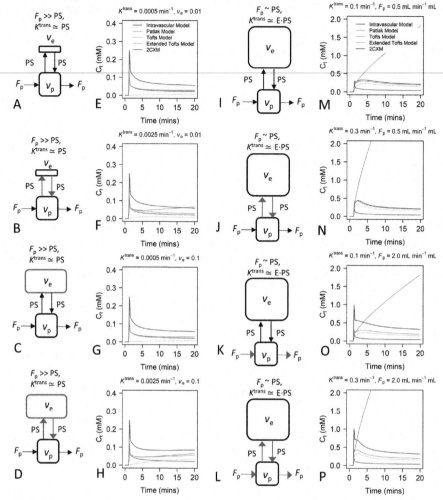

FIG. 2.4

Model diagrams and corresponding simulated tissue concentration-time curves for a range of
kinetic models and tissue types (*i.e.*, normal appearing brain tissue and a pathological
lesion (*e.g.*, tumor/stroke)). Model diagrams (A–D and I–L) show vascular and tissue
compartments and arrows represent exchange of contrast agent. The relative size of
compartments or *arrows* are used to indicate the relative size of volume fractions or contrast
agent permeability. *Red* (*black* in print version) compartments or *arrows* indicate changes
compared to the top row. For each tissue type, concentration-time curves were simulated
using intravascular (*red; black* in print version), Patlak (*cyan; light gray* in print version),
Tofts (*green; dark gray* in print version), Extended Tofts (*yellow; gray* in print version), and
two compartment exchange models (*purple, black* in print version) using the same set of
ground truth values (E–H and M–P). For normal appearing brain tissue, we simulated tissues
with different leakage rates and interstitial volume fractions. Plasma flow F_p was set to
0.5 ml/min/ml, K^{trans} set to 0.0005 or 0.0025 min^{-1}, plasma volume fraction v_p set to
0.03 ml/ml, and interstitial volume fraction v_e set to 0.01 or 0.1 ml/ml. For the pathological
lesion, tissues were simulated with different perfusion and leakage rates. We created curves
with F_p set to 0.5 or 2.0 ml/min/ml, K^{trans} set to 0.1 or 0.3 min^{-1}, v_p set to 0.06 ml/ml, and
interstitial volume fraction v_e set to 0.1 ml/ml.

assuming values for F_p, K^{trans}, v_p, and v_e representative of two key tissue types (see figure caption for full details): (i) normal appearing brain tissue and (ii) a pathological (leaky) lesion. These tissues were chosen because normal appearing brain tissue typically has slow vascular leakage of contrast agent, whereas pathological lesions such as acute stroke or brain tumor have higher leakage rates. In the following, we assume that the two-compartment exchange model represents the true contrast agent kinetics.

The intravascular model is the simplest possible model. It does not allow for leakage of contrast agent and is useful in applications that use intravascular tracers, or in dynamic susceptibility MRI, where leakage of the contrast agent is assumed to be negligible. However, in most applications of DCE-MRI, a model that describes vascular leakage is necessary. The Tofts model allows for uptake and clearance of contrast agent, but assumes the tissue is weakly vascularized (*i.e.*, zero v_p). It was used initially in the brain to study leakage of contrast agent into multiple sclerosis lesions (Tofts and Kermode, 1991), where leakage is relatively high. In tissues with high leakage rate (high K^{trans}), the vascular component of the signal is not such an important determinant of the early curve shape (Fig. 2.4M–P), and so the Tofts model provides a good description of the contrast agent kinetics (compare Tofts model (*green*; *dark gray* in print version) to 2CXM (*purple*; *black* in print version) in Fig. 2.4M). However, it was quickly recognized that the Tofts model fails to accurately describe contrast agent kinetics in normal appearing brain tissue, where leakage of contrast agent into tissue is much slower (Fig. 2.4E; compare Tofts curve (*green*; *dark gray* in print version) to 2CXM (*purple*; *black* in print version)). With the aim of extending the applicability of the Tofts model to tissues with lower levels of leakage, an *ad hoc* vascular term was added, called the extended Tofts model (Tofts, 1997). While not apparent at the time, the additional term changed how K^{trans} must be interpreted for this model. Instead of being a parameter that reflects both perfusion and vascular permeability, Sourbron elegantly showed the extended Tofts model K^{trans} reflects only the *PS* product (Kershaw and Cheng, 2010; Sourbron and Buckley, 2011) and is only valid in highly perfused tissues.

The Patlak model is commonly used instead of the extended Tofts model when contrast agent leakage is so low that it does not return to the blood pool during the measurement period ($v_e \rightarrow 1, C_p \gg C_e$). Under these circumstances the Patlak model and extended Tofts model curves are almost identical (Fig. 2.4E and G), and because the Patlak model has one fewer parameter, it is the preferred choice, particularly when modeling subtle vascular leakage in the brain (Heye *et al.*, 2014). However, the validity of the Patlak model is highly dependent on the balance between the leakage rate and interstitial volume. As can be seen from Fig. 2.4F, the Patlak model becomes an inaccurate description of the 2CXM curve shape when small interstitial volumes are coupled with larger leakage rates.

There are a number of other tissues that use models with active or filtration terms. In the liver, specialized contrast agents (gadoxetate) are used to image hepatocytes, requiring an active uptake model (Berks *et al.*, 2021; Karageorgis *et al.*, 2018). In kidney DCE-MRI, models must be adapted to account for filtration and reabsorption of water in the glomerular and tubular compartments (Sourbron, 2010).

2.3.5 Challenges for clinical adoption of quantitative DCE-MRI

Despite the potential clinical applications of quantitative DCE-MRI parameters, none are routinely used to aid decision making in healthcare systems. Contrast-enhanced MRI is used for high-risk breast cancer screening, but analysis is limited to assessing the shape of enhancement curves, and does not utilize kinetic model derived biomarkers (*Protocols for the Surveillance of Women at Higher Risk of Developing Breast Cancer*, n.d.). Quantitative DCE-MRI biomarkers for many cancer types have passed the first translational gap (*i.e.*, are ready for use in clinical trials as readouts of drug efficacy) as defined by the imaging biomarker roadmap for oncology drug studies (O'Connor *et al.*, 2016), but currently do not have sufficient body of evidence to cross the second translational gap (*i.e.*, for routine use in clinical practice). Applications of quantitative DCE-MRI in noncancer applications are at an even earlier stage.

There are several factors that hinder adoption of quantitative DCE-MRI in the clinic, which are discussed as follows.

2.3.5.1 Biomarker validation

Imaging biomarkers must be validated for their intended use before they can be applied in medical product development or clinical settings. This covers a range of technical and biological requirements reviewed in detail elsewhere (FDA-NIH Biomarker Working Group, 2016; O'Connor *et al.*, 2016). A key consideration when developing imaging biomarkers is reproducibility. It must be possible for two different people to obtain comparable results from the same subject in two separate locations. The reproducibility of quantitative DCE-MRI parameters is currently lower than other imaging biomarkers. For example, the coefficient of variation of K^{trans} is approximately 10%–30%, depending on application (Cutajar *et al.*, 2014; Ng *et al.*, 2012; Roberts *et al.*, 2006; Weber *et al.*, 2022; Wong *et al.*, 2017; Zhang *et al.*, 2014), whereas for others such as scintigraphy-based measurements of cardiac ejection fraction, reproducibility is <5% (Castell-Conesa *et al.*, 2004).

Standardization of imaging protocols and analysis pipelines could help improve the reproducibility. Minor differences in the choice of sequence parameters such as flip angle and TR, or manufacturer differences in B_1 field inhomogeneity, can affect kinetic parameter estimates (Schabel and Parker, 2008; Tofts *et al.*, 1999). Consensus standards are now beginning to emerge through initiatives such as Quantitative Imaging Biomarkers Alliance (QIBA), the International Society for Magnetic Resonance in Medicine (ISMRM), and Cancer Research UK (CRUK). For example, QIBA have produced a DCE profile which provides guidelines for acquisition of DCE-MRI data in a multicenter setting, as well as ideal, target, and accepted settings for DCE-MRI protocols in brain, prostate, and breast applications (Shukla-Dave *et al.*, 2019). The QIBA DCE profile is an ongoing project with future updates to be expected, for example, regarding the use of parallel imaging. There are also disease-specific guidelines, such as COST Action PARENCHIMA (CA16103) for renal MRI (Mendichovszky *et al.*, 2020) and the HARNESS project for harmonization of imaging methods in dementia (Smith *et al.*, 2019).

Quantitative analysis of DCE-MRI data is complex as it involves multiple steps with many choices. Even when the same analysis strategy is followed (*e.g.*, same model choice), the exact computational routines and algorithms to fit these models can vary substantially depending on the implementation, computer operating system, programming language, operator preference, etc. (Beuzit *et al.*, 2016; Heye *et al.*, 2013; Huang *et al.*, 2014; Joint Head and Neck Radiotherapy-MRI Development Cooperative, 2017; Kudo *et al.*, 2013). To address some of these issues, the ISMRM Open Science Initiative for Perfusion Imaging (OSIPI, osipi.ismrm.org), an initiative of the ISMRM Perfusion Study Group, is currently working toward developing a standardized software library for quantitative DCE-MRI.

Many of these consensus guidelines recommend undertaking quality assurance with phantoms before a new trial starts or when the protocol changes. QIBA, for example, proposes to perform regular phantom measurements to assess image SNR, uniformity, and geometric distortion. As direct validation of pharmacokinetic parameter values is difficult, individual aspects of the DCE-MRI protocol should be assessed (QIBA, 2011; Shukla-Dave *et al.*, 2019). One aspect is the accuracy and precision of baseline R_1 measurement, which can be assessed with a suitable phantom (QIBA, 2011). Another aspect is the signal stability, which can be assessed by measuring the signal intensity for at least 6 min to determine magnet stability (QIBA, 2011). Phantom measurements should be repeated regularly, for example, annually, but especially when system upgrades or modifications to the sequences occur (QIBA, 2011). Examples of such static measurements can be found in Bane *et al.* (2018), Foltz *et al.* (2019), Shukla-Dave *et al.* (2019), and van Houdt *et al.* (2020). Dynamic phantom experiments can be used to test and optimize DCE sequences. For example, an in-house built flow phantom was used to assess in-flow and radiofrequency effects on the estimation of the AIF at different flow rates (Foltz *et al.*, 2019). De Deene *et al.* have developed a realistic phantom for the head-and-neck region with which quantitative DCE parameter maps, like K^{trans}, can be derived (De Deene *et al.*, 2022).

Finally, the NIH and FDA Biomarkers, Endpoints, and other Tools (BEST) Resource states a biomarker must also be shown to reflect a defined characteristic that is measured as an indicator of normal biological processes, pathogenic processes, or responses to an exposure or intervention, including therapeutic interventions (FDA-NIH Biomarker Working Group, 2016). Quantitative DCE-MRI parameters have demonstrated potential for diagnosis, prognostication, and treatment monitoring in specific settings, showing coupling to a target disease (*e.g.*, vascular leaks in dementia) or therapy response (*e.g.*, radiotherapy or antiangiogenic therapies in cancer) (Prescott, 2013). However, this evidence is currently accumulated from a small number of poorly powered studies, which often use different imaging protocols and analysis methods. Definitive larger scale studies using standardized imaging protocols are now needed to validate the diagnostic, prognostic, or treatment monitoring value of these biomarkers in specific settings.

2.3.5.2 Lack of standardized terminology

Terminology of quantitative DCE-MRI biomarkers is not standardized, contributing to confusion and misinterpretation of derived parameters by the research and clinical community (Obuchowski *et al.*, 2015). There have been efforts to standardize terms, but this has not been updated since 1999 (Tofts *et al.*, 1999). OSIPI has identified this as a key issue limiting translation of quantitative DCE-MRI and has embarked on defining a lexicon of terms and processes pertinent to DCE-MRI along with reporting guidelines detailing how to use the lexicon when writing manuscripts or protocols. The lexicon provides definitions, units, and a code-based identifier such that quantities and processes can be uniquely identified. The definitions for each quantity and process are reviewed by a panel of DCE-MRI experts (and are continually reviewed by members of OSIPI Task Force 4.2). As such, they represent consensus definitions and should be used when reporting DCE-MRI analyses.

2.3.5.3 Lack of resources

Quantitative DCE-MRI acquisition is nontrivial and relies on local expertise to guide radiography staff. The analysis of data is also complex, requiring use of existing software tools or input from imaging scientists. Both acquisition and analysis require multiple decisions that impact the final result (*e.g.*, which flip angle? which tracer kinetic model to use? or where to sample the AIF?). The work of OSIPI aims to remove some of these barriers by providing well-documented software and educational resources to increase accessibility of DCE-MRI, particularly aimed at those without prior knowledge of the field.

2.4 Summary

Quantitative dynamic contrast-enhanced MRI allows the assessment of tissue perfusion and permeability. The acquisition is typically done by a fast 3D spoiled gradient-echo sequence where images are collected before, during, and after the injection of a T_1-reducing paramagnetic contrast agent over a period of several minutes. The optimization of the scan protocol is often a trade-off between spatial resolution, temporal resolution, spatial coverage, and signal-to-noise ratio. The quantitative analysis involves multiple steps, of which the main ones are the conversion from signal intensity to concentration and tracer kinetic model fitting. In this chapter a basic analysis pipeline was described, but many variations exist. The choices made for the acquisition protocol impact the accuracy of the quantitative DCE parameters. Numerous studies have shown promising applications in many areas of the body. However, current clinical adoption is hindered by lack of reproducibility, nonstandardized terminology, and requirements for imaging scientists to support acquisition and analyses. Currently, several initiatives such as QIBA and OSIPI are aiming to improve the reproducibility of quantitative DCE-MRI, develop recommendations and guidelines to improve data quality, and develop tools to facilitate clinical adoption of this important technology within healthcare systems.

References

Bane, O., Hectors, S.J., Wagner, M., Arlinghaus, L.L., Aryal, M.P., Cao, Y., Chenevert, T.L., Fennessy, F., Huang, W., Hylton, N.M., Kalpathy-Cramer, J., Keenan, K.E., Malyarenko, D.I., Mulkern, R.V., Newitt, D.C., Russek, S.E., Stupic, K.F., Tudorica, A., Wilmes, L.J., *et al.*, 2018. Accuracy, repeatability, and interplatform reproducibility of T quantification methods used for DCE-MRI: results from a multicenter phantom study. Magn. Reson. Med. 79 (5), 2564–2575.

Bane, O., Gee, M.S., Menys, A., Dillman, J.R., Taouli, B., 2021. Emerging imaging biomarkers in Crohn disease. Top. Magn. Reson. Imaging 30 (1), 31–41.

Barnes, S.R., Ng, T.S.C., Montagne, A., Law, M., Zlokovic, B.V., Jacobs, R.E., 2016. Optimal acquisition and modeling parameters for accurate assessment of low Ktrans blood-brain barrier permeability using dynamic contrast-enhanced MRI. Magn. Reson. Med. 75 (5), 1967–1977.

Benou, A., Veksler, R., Friedman, A., Riklin Raviv, T., 2017. Ensemble of expert deep neural networks for spatio-temporal denoising of contrast-enhanced MRI sequences. Med. Image Anal. 42, 145–159.

Benz, M.R., Bongartz, G., Froehlich, J.M., Winkel, D., Boll, D.T., Heye, T., 2018. Acceleration techniques and their impact on arterial input function sampling: non-accelerated versus view-sharing and compressed sensing sequences. Eur. J. Radiol. 104, 8–13.

Berks, M., Little, R.A., Watson, Y., Cheung, S., Datta, A., O'Connor, J.P.B., Scaramuzza, D., Parker, G.J.M., 2021. A model selection framework to quantify microvascular liver function in gadoxetate-enhanced MRI: application to healthy liver, diseased tissue, and hepatocellular carcinoma. Magn. Reson. Med. 86 (4), 1829–1844.

Bernstein, J.M., Homer, J.J., West, C.M., 2014. Dynamic contrast-enhanced magnetic resonance imaging biomarkers in head and neck cancer: potential to guide treatment? A systematic review. Oral Oncol. 50 (10), 963–970.

Beuzit, L., Eliat, P.-A., Brun, V., Ferré, J.-C., Gandon, Y., Bannier, E., Saint-Jalmes, H., 2016. Dynamic contrast-enhanced MRI: study of inter-software accuracy and reproducibility using simulated and clinical data. J. Magn. Reson. Imaging 43 (6), 1288–1300.

Bishop, J.E., Santyr, G.E., Kelcz, F., Plewes, D.B., 1997. Limitations of the keyhole technique for quantitative dynamic contrast-enhanced breast MRI. J. Magn. Reson. Imaging 7 (4), 716–723.

Bokacheva, L., Rusinek, H., Zhang, J.L., Lee, V.S., 2008. Assessment of renal function with dynamic contrast-enhanced MR imaging. Magn. Reson. Imaging 16 (4), 597–611 (viii).

Calamante, F., 2013. Arterial input function in perfusion MRI: a comprehensive review. Prog. Nucl. Magn. Reson. Spectrosc. 74, 1–32.

Calamante, F., Gadian, D.G., Connelly, A., 2000. Delay and dispersion effects in dynamic susceptibility contrast MRI: simulations using singular value decomposition. Magn. Reson. Med. 44 (3), 466–473.

Cao, Y., 2011. The promise of dynamic contrast-enhanced imaging in radiation therapy. Semin. Radiat. Oncol. 21 (2), 147–156.

Caroca, S., Villagran, D., Chabert, S., 2021. Four functional magnetic resonance imaging techniques for skeletal muscle exploration, a systematic review. Eur. J. Radiol. 144, 109995.

Castell-Conesa, J., Aguadé-Bruix, S., García-Burillo, A., González, J.M., Canela, T., Oller, G., Díez, M.J., Roca, I., Candell-Riera, J., 2004. Reproducibility of measurements of left ventricular function with gated myocardial perfusion SPECT and comparison with blood pool radionuclide ventriculography. Rev. Esp. Cardiol. 57 (10), 931–938.

Cheng, H.-L.M., Wright, G.A., 2006. Rapid high-resolution T(1) mapping by variable flip angles: accurate and precise measurements in the presence of radiofrequency field inhomogeneity. Magn. Reson. Med. 55 (3), 566–574.

Coelho-Filho, O.R., Rickers, C., Kwong, R.Y., Jerosch-Herold, M., 2013. MR myocardial perfusion imaging. Radiology 266 (3), 701–715.

Crombé, A., Saut, O., Guigui, J., Italiano, A., Buy, X., Kind, M., 2019. Influence of temporal parameters of DCE-MRI on the quantification of heterogeneity in tumor vascularization. J. Magn. Reson. Imaging 50 (6), 1773–1788.

Cutajar, M., Thomas, D.L., Hales, P.W., Banks, T., Clark, C.A., Gordon, I., 2014. Comparison of ASL and DCE MRI for the non-invasive measurement of renal blood flow: quantification and reproducibility. Eur. Radiol. 24 (6), 1300–1308.

De Deene, Y., Wheatley, M., Greig, T., Hayes, D., Ryder, W., Loh, H., 2022. A multi-modality medical imaging head and neck phantom: Part 1. Design and fabrication. Phys. Med. 96, 166–178.

De Naeyer, D., Verhulst, J., Ceelen, W., Segers, P., De Deene, Y., Verdonck, P., 2011. Flip angle optimization for dynamic contrast-enhanced MRI-studies with spoiled gradient echo pulse sequences. Phys. Med. Biol. 56 (16), 5373–5395.

Dickie, B.R., Vandesquille, M., Ulloa, J., Boutin, H., Parkes, L.M., Parker, G.J.M., 2019. Water-exchange MRI detects subtle blood-brain barrier breakdown in Alzheimer's disease rats. Neuroimage 184, 349–358.

Dickie, B.R., Boutin, H., Parker, G.J.M., Parkes, L.M., 2021. Alzheimer's disease pathology is associated with earlier alterations to blood-brain barrier water permeability compared with healthy ageing in TgF344-AD rats. NMR Biomed. 34 (7), e4510.

Dijkhoff, R.A.P., Beets-Tan, R.G.H., Lambregts, D.M.J., Beets, G.L., Maas, M., 2017. Value of DCE-MRI for staging and response evaluation in rectal cancer: a systematic review. Eur. J. Radiol. 95, 155–168.

Dikaios, N., Arridge, S., Hamy, V., Punwani, S., Atkinson, D., 2014. Direct parametric reconstruction from undersampled (k, t)-space data in dynamic contrast enhanced MRI. Med. Image Anal. 18 (7), 989–1001.

Duan, C., Kallehauge, J.F., Bretthorst, G.L., Tanderup, K., Ackerman, J.J.H., Garbow, J.R., 2017. Are complex DCE-MRI models supported by clinical data? Magn. Reson. Med. 77 (3), 1329–1339.

Ernst, R.R., Anderson, W.A., 1966. Application of Fourier transform spectroscopy to magnetic resonance. Rev. Sci. Instrum. 37 (1), 93–102.

Evelhoch, J.L., 2005. Consensus recommendations for acquisition of dynamic contrasted-enhanced MRI data in oncology. In: Dynamic Contrast-Enhanced Magnetic Resonance Imaging in Oncology. Springer-Verlag, pp. 109–113.

Fang, K., Wang, Z., Li, Z., Wang, B., Han, G., Cheng, Z., Chen, Z., Lan, C., Zhang, Y., Zhao, P., Jin, X., Liu, Y., Bai, R., 2021. Convolutional neural network for accelerating the computation of the extended Tofts model in dynamic contrast-enhanced magnetic resonance imaging. J. Magn. Reson. Imaging 53 (6), 1898–1910.

Farrall, A.J., Wardlaw, J.M., 2009. Blood-brain barrier: ageing and microvascular disease—systematic review and meta-analysis. Neurobiol. Aging 30 (3), 337–352.

FDA-NIH Biomarker Working Group, 2016. BEST (Biomarkers, EndpointS, and other Tools) Resource. https://www.ncbi.nlm.nih.gov/books/NBK326791/.

Feng, L., Axel, L., Chandarana, H., Block, K.T., Sodickson, D.K., Otazo, R., 2016. XD-GRASP: golden-angle radial MRI with reconstruction of extra motion-state dimensions using compressed sensing. Magn. Reson. Med. 75 (2), 775–788.

Foltz, W., Driscoll, B., Laurence Lee, S., Nayak, K., Nallapareddy, N., Fatemi, A., Ménard, C., Coolens, C., Chung, C., 2019. Phantom validation of DCE-MRI magnitude and phase-based vascular input function measurements. Tomography 5 (1), 77–89.

Gaddikeri, S., Gaddikeri, R.S., Tailor, T., Anzai, Y., 2016. Dynamic contrast-enhanced MR imaging in head and neck cancer: techniques and clinical applications. AJNR Am. J. Neuroradiol. 37 (4), 588–595.

Galbraith, S.M., Lodge, M.A., Taylor, N.J., Rustin, G.J.S., Bentzen, S., Stirling, J.J., Padhani, A.R., 2002. Reproducibility of dynamic contrast-enhanced MRI in human muscle and tumours: comparison of quantitative and semi-quantitative analysis. NMR Biomed. 15 (2), 132–142.

Gefter, W.B., Lee, K.S., Schiebler, M.L., Parraga, G., Seo, J.B., Ohno, Y., Hatabu, H., 2021. Pulmonary functional imaging: part 2-state-of-the-art clinical applications and opportunities for improved patient care. Radiology 299 (3), 524–538.

Georgiou, L., Wilson, D.J., Sharma, N., Perren, T.J., Buckley, D.L., 2019. A functional form for a representative individual arterial input function measured from a population using high temporal resolution DCE MRI. Magn. Reson. Med. 81 (3), 1955–1963.

Grassi, G., Laino, M.E., Fantini, M.C., Argiolas, G.M., Cherchi, M.V., Nicola, R., Gerosa, C., Cerrone, G., Mannelli, L., Balestrieri, A., Suri, J.S., Carriero, A., Saba, L., 2022. Advanced imaging and Crohn's disease: an overview of clinical application and the added value of artificial intelligence. Eur. J. Radiol. 157, 110551.

Guo, Y., Lingala, S.G., Zhu, Y., Lebel, R.M., Nayak, K.S., 2017. Direct estimation of tracer-kinetic parameter maps from highly undersampled brain dynamic contrast enhanced MRI. Magn. Reson. Med. 78 (4), 1566–1578.

Hansen, A.E., Pedersen, H., Rostrup, E., Larsson, H.B.W., 2009. Partial volume effect (PVE) on the arterial input function (AIF) in T1-weighted perfusion imaging and limitations of the multiplicative rescaling approach. Magn. Reson. Med. 62 (4), 1055–1059.

Heisen, M., Fan, X., Buurman, J., van Riel, N.A.W., Karczmar, G.S., ter Haar Romeny, B.M., 2010. The influence of temporal resolution in determining pharmacokinetic parameters from DCE-MRI data. Magn. Reson. Med. 63 (3), 811–816.

Henderson, E., Rutt, B.K., Lee, T.Y., 1998. Temporal sampling requirements for the tracer kinetics modeling of breast disease. Magn. Reson. Imaging 16 (9), 1057–1073.

Heye, T., Davenport, M.S., Horvath, J.J., Feuerlein, S., Breault, S.R., Bashir, M.R., Merkle, E.-M., Boll, D.T., 2013. Reproducibility of dynamic contrast-enhanced MR imaging. Part I. Perfusion characteristics in the female pelvis by using multiple computer-aided diagnosis perfusion analysis solutions. Radiology 266 (3), 801–811.

Heye, A.K., Culling, R.D., Valdés Hernández, M.D.C., Thrippleton, M.J., Wardlaw, J.M., 2014. Assessment of blood-brain barrier disruption using dynamic contrast-enhanced MRI. A systematic review. NeuroImage Clin. 6, 262–274.

Heye, A.K., Thrippleton, M.J., Armitage, P.A., Valdés Hernández, M.D.C., Makin, S.D., Glatz, A., Sakka, E., Wardlaw, J.M., 2016. Tracer kinetic modelling for DCE-MRI quantification of subtle blood-brain barrier permeability. Neuroimage 125, 446–455.

Howe, B.M., Broski, S.M., Littrell, L.A., Pepin, K.M., Wenger, D.E., 2020. Quantitative musculoskeletal tumor imaging. Semin. Musculoskelet. Radiol. 24 (4), 428–440.

Huang, W., Li, X., Chen, Y., Li, X., Chang, M.-C., Oborski, M.J., Malyarenko, D.I., Muzi, M., Jajamovich, G.H., Fedorov, A., Tudorica, A., Gupta, S.N., Laymon, C.M., Marro, K.I., Dyvorne, H.A., Miller, J.V., Barbodiak, D.P., Chenevert, T.L., Yankeelov, T.E., et al., 2014. Variations of dynamic contrast-enhanced magnetic resonance imaging in evaluation of breast cancer therapy response: a multicenter data analysis challenge. Transl. Oncol. 7 (1), 153–166.

Huang, W., Chen, Y., Fedorov, A., Li, X., Jajamovich, G.H., Malyarenko, D.I., Aryal, M.P., LaViolette, P.S., Oborski, M.J., O'Sullivan, F., Abramson, R.G., Jafari-Khouzani, K., Afzal, A., Tudorica, A., Moloney, B., Gupta, S.N., Besa, C., Kalpathy-Cramer, J., Mountz, J.M., et al., 2016. The impact of arterial input function determination variations on prostate dynamic contrast-enhanced magnetic resonance imaging pharmacokinetic modeling: a multicenter data analysis challenge. Tomography 2 (1), 56–66.

Jelescu, I.O., Leppert, I.R., Narayanan, S., Araújo, D., Arnold, D.L., Pike, G.B., 2011. Dual-temporal resolution dynamic contrast-enhanced MRI protocol for blood-brain barrier permeability measurement in enhancing multiple sclerosis lesions. J. Magn. Reson. Imaging 33 (6), 1291–1300.

Joint Head and Neck Radiotherapy-MRI Development Cooperative, 2017. A multi-institutional comparison of dynamic contrast-enhanced magnetic resonance imaging parameter calculations. Sci. Rep. 7 (1), 11185.

Kanal, E., 2016. Gadolinium based contrast agents (GBCA): safety overview after 3 decades of clinical experience. Magn. Reson. Imaging 34 (10), 1341–1345.

Karageorgis, A., Lenhard, S.C., Yerby, B., Forsgren, M.F., Liachenko, S., Johansson, E., Pilling, M.A., Peterson, R.A., Yang, X., Williams, D.P., Ungersma, S.E., Morgan, R.E., Brouwer, K.L.R., Jucker, B.M., Hockings, P.D., 2018. A multi-center preclinical study of gadoxetate DCE-MRI in rats as a biomarker of drug induced inhibition of liver transporter function. PLoS One 13 (5), e0197213.

Keil, V.C., Mädler, B., Gieseke, J., Fimmers, R., Hattingen, E., Schild, H.H., Hadizadeh, D.R., 2017. Effects of arterial input function selection on kinetic parameters in brain dynamic contrast-enhanced MRI. Magn. Reson. Imaging 40, 83–90.

Kermode, A.G., Tofts, P.S., Thompson, A.J., MacManus, D.G., Rudge, P., Kendall, B.E., Kingsley, D.P., Moseley, I.F., du Boulay, E.P., McDonald, W.I., 1990. Heterogeneity of blood-brain barrier changes in multiple sclerosis: an MRI study with gadolinium-DTPA enhancement. Neurology 40 (2), 229–235.

Kershaw, L.E., Cheng, H.-L.M., 2010. Temporal resolution and SNR requirements for accurate DCE-MRI data analysis using the AATH model. Magn. Reson. Med. 64 (6), 1772–1780.

Kudo, K., Christensen, S., Sasaki, M., Østergaard, L., Shirato, H., Ogasawara, K., Wintermark, M., Warach, S., Stroke Imaging Repository (STIR) Investigators, 2013. Accuracy and reliability assessment of CT and MR perfusion analysis software using a digital phantom. Radiology 267 (1), 201–211.

Le, Y., Kipfer, H.D., Nickel, D.M., Kroeker, R., Dale, B.M., Holz, S.P., Weiland, E., Lin, C., 2016. Initial experience of applying TWIST-Dixon with flexible view sharing in breast DCE-MRI. Clin. Breast Cancer 16 (3), 202–206.

Leach, M.O., Brindle, K.M., Evelhoch, J.L., Griffiths, J.R., Horsman, M.R., Jackson, A., Jayson, G.C., Judson, I.R., Knopp, M.V., Maxwell, R.J., McIntyre, D., Padhani, A.R., Price, P., Rathbone, R., Rustin, G.J., Tofts, P.S., Tozer, G.M., Vennart, W., Waterton, J.C., Williams, S.R., Workman, P., Pharmacodynamic/Pharmacokinetic Technologies Advisory Committee, Drug Development Office, Cancer Research UK, 2005. The assessment of antiangiogenic and antivascular therapies in early-stage clinical trials using magnetic resonance imaging: issues and recommendations. Br. J. Cancer 92 (9), 1599–1610.

Lin, W., Guo, J., Rosen, M.A., Song, H.K., 2008. Respiratory motion-compensated radial dynamic contrast-enhanced (DCE)-MRI of chest and abdominal lesions. Magn. Reson. Med. 60 (5), 1135–1146.

Markl, M., Leupold, J., 2012. Gradient echo imaging. J. Magn. Reson. Imaging 35 (6), 1274–1289.

Mayr, N.A., Yuh, W.T., Zheng, J., Ehrhardt, J.C., Magnotta, V.A., Sorosky, J.I., Pelsang, R.E., Oberley, L.W., Hussey, D.H., 1998. Prediction of tumor control in patients with cervical cancer: analysis of combined volume and dynamic enhancement pattern by MR imaging. AJR Am. J. Roentgenol. 170 (1), 177–182.

Mazaheri, Y., Akin, O., Hricak, H., 2017. Dynamic contrast-enhanced magnetic resonance imaging of prostate cancer: a review of current methods and applications. World J. Radiol. 9 (12), 416–425.

McGrath, D.M., Bradley, D.P., Tessier, J.L., Lacey, T., Taylor, C.J., Parker, G.J.M., 2009. Comparison of model-based arterial input functions for dynamic contrast-enhanced MRI in tumor bearing rats. Magn. Reson. Med. 61 (5), 1173–1184.

Mendichovszky, I., Pullens, P., Dekkers, I., Nery, F., Bane, O., Pohlmann, A., de Boer, A., Ljimani, A., Odudu, A., Buchanan, C., Sharma, K., Laustsen, C., Harteveld, A., Golay, X., Pedrosa, I., Alsop, D., Fain, S., Caroli, A., Prasad, P., et al., 2020. Technical recommendations for clinical translation of renal MRI: a consensus project of the cooperation in science and technology action PARENCHIMA. MAGMA 33 (1), 131–140.

Montagne, A., Barnes, S.R., Sweeney, M.D., Halliday, M.R., Sagare, A.P., Zhao, Z., Toga, A.W., Jacobs, R.E., Liu, C.Y., Amezcua, L., Harrington, M.G., Chui, H.C., Law, M., Zlokovic, B.V., 2015. Blood-brain barrier breakdown in the aging human hippocampus. Neuron 85 (2), 296–302.

Ng, C.S., Waterton, J.C., Kundra, V., Brammer, D., Ravoori, M., Han, L., Wei, W., Klumpp, S., Johnson, V.E., Jackson, E.F., 2012. Reproducibility and comparison of DCE-MRI and DCE-CT perfusion parameters in a rat tumor model. Technol. Cancer Res. Treat. 11 (3), 279–288.

O'Connor, J.P.B., Jackson, A., Parker, G.J.M., Jayson, G.C., 2007. DCE-MRI biomarkers in the clinical evaluation of antiangiogenic and vascular disrupting agents. Br. J. Cancer 96 (2), 189–195.

O'Connor, J.P.B., Aboagye, E.O., Adams, J.E., Aerts, H.J.W.L., Barrington, S.F., Beer, A.J., Boellaard, R., Bohndiek, S.E., Brady, M., Brown, G., Buckley, D.L., Chenevert, T.L., Clarke, L.P., Collette, S., Cook, G.J., deSouza, N.M., Dickson, J.C., Dive, C., Evelhoch, J.L., et al., 2016. Imaging biomarker roadmap for cancer studies. Nat. Rev. Clin. Oncol. 14 (3), 169–186.

Obuchowski, N.A., Barnhart, H.X., Buckler, A.J., Pennello, G., Wang, X.-F., Kalpathy-Cramer, J., Kim, H.J.G., Reeves, A.P., Case Example Working Group, 2015. Statistical issues in the comparison of quantitative imaging biomarker algorithms using pulmonary nodule volume as an example. Stat. Methods Med. Res. 24 (1), 107–140.

Othman, A.E., Martirosian, P., Schraml, C., Taron, J., Weiss, J., Bier, G., Schwentner, C., Nickel, D., Bamberg, F., Kramer, U., Nikolaou, K., Notohamiprodjo, M., 2015. Feasibility of CAIPIRINHA-Dixon-TWIST-VIBE for dynamic contrast-enhanced MRI of the prostate. Eur. J. Radiol. 84 (11), 2110–2116.

Padhani, A.R., 2003. MRI for assessing antivascular cancer treatments. Br. J. Radiol. 76 Spec No 1, S60–S80.

Panek, R., Schmidt, M.A., Borri, M., Koh, D.-M., Riddell, A., Welsh, L., Dunlop, A., Powell, C., Bhide, S.A., Nutting, C.M., Harrington, K.J., Newbold, K.L., Leach, M.O., 2016. Time-resolved angiography with stochastic trajectories for dynamic contrast-enhanced MRI in head and neck cancer: are pharmacokinetic parameters affected? Med. Phys. 43 (11), 6024.

Parker, G.J.M., Roberts, C., Macdonald, A., Buonaccorsi, G.A., Cheung, S., Buckley, D.L., Jackson, A., Watson, Y., Davies, K., Jayson, G.C., 2006. Experimentally-derived functional form for a population-averaged high-temporal-resolution arterial input function for dynamic contrast-enhanced MRI. Magn. Reson. Med. 56 (5), 993–1000.

Planey, C.R., Welch, E.B., Xu, L., Chakravarthy, A.B., Gatenby, J.C., Freehardt, D., Mayer, I., Meszeoly, I., Kelley, M., Means-Powell, J., Gore, J.C., Yankeelov, T.E., 2009. Temporal sampling requirements for reference region modeling of DCE-MRI data in human breast cancer. J. Magn. Reson. Imaging 30 (1), 121–134.

Prescott, J.W., 2013. Quantitative imaging biomarkers: the application of advanced image processing and analysis to clinical and preclinical decision making. J. Digit. Imaging 26 (1), 97–108.

Protocols for the Surveillance of Women at Higher Risk of Developing Breast Cancer. (n.d.). GOV.UK. Retrieved February 28, 2023, from https://www.gov.uk/government/publications/breast-screening-higher-risk-women-surveillance-protocols/protocols-for-surveillance-of-women-at-higher-risk-of-developing-breast-cancer.

QIBA, 2011 December 13. Profile: DCE-MRI Quantification. https://qibawiki.rsna.org/images/7/7b/DCEMRIProfile_v1_6-20111213.pdf.

Raja, R., Rosenberg, G.A., Caprihan, A., 2018. MRI measurements of blood-brain barrier function in dementia: a review of recent studies. Neuropharmacology 134 (Pt B), 259–271.

Roberts, C., Issa, B., Stone, A., Jackson, A., Waterton, J.C., Parker, G.J.M., 2006. Comparative study into the robustness of compartmental modeling and model-free analysis in DCE-MRI studies. J. Magn. Reson. Imaging 23 (4), 554–563.

Roberts, C., Little, R., Watson, Y., Zhao, S., Buckley, D.L., Parker, G.J.M., 2011. The effect of blood inflow and B(1)-field inhomogeneity on measurement of the arterial input function in axial 3D spoiled gradient echo dynamic contrast-enhanced MRI. Magn. Reson. Med. 65 (1), 108–119.

Roher, A.E., Garami, Z., Tyas, S.L., Maarouf, C.L., Kokjohn, T.A., Belohlavek, M., Vedders, L.J., Connor, D., Sabbagh, M.N., Beach, T.G., Emmerling, M.R., 2011. Transcranial doppler ultrasound blood flow velocity and pulsatility index as systemic indicators for Alzheimer's disease. Alzheimers Dement. 7 (4), 445–455.

Rohrer, M., Bauer, H., Mintorovitch, J., Requardt, M., Weinmann, H.-J., 2005. Comparison of magnetic properties of MRI contrast media solutions at different magnetic field strengths. Invest. Radiol. 40 (11), 715–724.

Saeed, M., Van, T.A., Krug, R., Hetts, S.W., Wilson, M.W., 2015. Cardiac MR imaging: current status and future direction. Cardiovasc Diagn Ther 5 (4), 290–310.

Salem, A., O'Connor, J.P.B., 2016. Assessment of tumor angiogenesis: dynamic contrast-enhanced MR imaging and beyond. Magn. Reson. Imaging Clin. N. Am. 24 (1), 45–56.

Schabel, M.C., Parker, D.L., 2008. Uncertainty and bias in contrast concentration measurements using spoiled gradient echo pulse sequences. Phys. Med. Biol. 53 (9), 2345–2373.

Shehata, M.L., Basha, T.A., Hayeri, M.R., Hartung, D., Teytelboym, O.M., Vogel-Claussen, J., 2014. MR myocardial perfusion imaging: insights on techniques, analysis, interpretation, and findings. Radiographics 34 (6), 1636–1657.

Shen, Y., Goerner, F.L., Snyder, C., Morelli, J.N., Hao, D., Hu, D., Li, X., Runge, V.M., 2015. T1 relaxivities of gadolinium-based magnetic resonance contrast agents in human whole blood at 1.5, 3, and 7 T. Invest. Radiol. 50 (5), 330–338.

Shukla-Dave, A., Obuchowski, N.A., Chenevert, T.L., Jambawalikar, S., Schwartz, L.H., Malyarenko, D., Huang, W., Noworolski, S.M., Young, R.J., Shiroishi, M.S., Kim, H., Coolens, C., Laue, H., Chung, C., Rosen, M., Boss, M., Jackson, E.F., 2019. Quantitative imaging biomarkers alliance (QIBA) recommendations for improved precision of DWI and DCE-MRI derived biomarkers in multicenter oncology trials. J. Magn. Reson. Imaging 49 (7), e101–e121.

Smith, D.S., Welch, E.B., Li, X., Arlinghaus, L.R., Loveless, M.E., Koyama, T., Gore, J.C., Yankeelov, T.E., 2011. Quantitative effects of using compressed sensing in dynamic contrast enhanced MRI. Phys. Med. Biol. 56 (15), 4933–4946.

Smith, E.E., Biessels, G.J., De Guio, F., de Leeuw, F.E., Duchesne, S., Düring, M., Frayne, R., Ikram, M.A., Jouvent, E., MacIntosh, B.J., Thrippleton, M.J., Vernooij, M.W., Adams, H., Backes, W.H., Ballerini, L., Black, S.E., Chen, C., Corriveau, R., DeCarli, C., et al., 2019. Harmonizing brain magnetic resonance imaging methods for vascular contributions to neurodegeneration. Alzheimers Dement. 11, 191–204.

Sourbron, S., 2010. Compartmental modelling for magnetic resonance renography. Z. Med. Phys. 20 (2), 101–114.

Sourbron, S.P., Buckley, D.L., 2011. On the scope and interpretation of the Tofts models for DCE-MRI. Magn. Reson. Med. 66 (3), 735–745.

Sourbron, S., Ingrisch, M., Siefert, A., Reiser, M., Herrmann, K., 2009. Quantification of cerebral blood flow, cerebral blood volume, and blood-brain-barrier leakage with DCE-MRI. Magn. Reson. Med. 62 (1), 205–217.

Sujlana, P., Skrok, J., Fayad, L.M., 2018. Review of dynamic contrast-enhanced MRI: technical aspects and applications in the musculoskeletal system. J. Magn. Reson. Imaging 47 (4), 875–890.

Sun, K., Zhu, H., Chai, W., Zhan, Y., Nickel, D., Grimm, R., Fu, C., Yan, F., 2020. Whole-lesion histogram and texture analyses of breast lesions on inline quantitative DCE mapping with CAIPIRINHA-Dixon-TWIST-VIBE. Eur. Radiol. 30 (1), 57–65.

Sung, Y.S., Park, B., Choi, Y., Lim, H.-S., Woo, D.-C., Kim, K.W., Kim, J.K., 2016. Dynamic contrast-enhanced MRI for oncology drug development. J. Magn. Reson. Imaging 44 (2), 251–264.

Taylor, A.J., Salerno, M., Dharmakumar, R., Jerosch-Herold, M., 2016. T1 mapping: basic techniques and clinical applications. J. Am. Coll. Cardiol. Img. 9 (1), 67–81.

Thrippleton, M.J., Backes, W.H., Sourbron, S., Ingrisch, M., van Osch, M.J.P., Dichgans, M., Fazekas, F., Ropele, S., Frayne, R., van Oostenbrugge, R.J., Smith, E.E., Wardlaw, J.M., 2019. Quantifying blood-brain barrier leakage in small vessel disease: review and consensus recommendations. Alzheimers Dement. 15 (6), 840–858.

Tofts, P.S., 1997. Modeling tracer kinetics in dynamic Gd-DTPA MR imaging. J. Magn. Reson. Imaging 7 (1), 91–101.

Tofts, P.S., Kermode, A.G., 1991. Measurement of the blood-brain barrier permeability and leakage space using dynamic MR imaging. 1. Fundamental concepts. Magn. Reson. Med. 17 (2), 357–367.

Tofts, P.S., Parker, G.J.M., 2013. DCE-MRI: acquisition and analysis techniques. In: Barker, P., Golay, X., Zaharchuk, G. (Eds.), Clinical Perfusion MRI. Cambridge University Press, pp. 58–74.

Tofts, P.S., Brix, G., Buckley, D.L., Evelhoch, J.L., Henderson, E., Knopp, M.V., Larsson, H.B., Lee, T.Y., Mayr, N.A., Parker, G.J., Port, R.E., Taylor, J., Weisskoff, R.M., 1999. Estimating kinetic parameters from dynamic contrast-enhanced T(1)-weighted MRI of a diffusable tracer: standardized quantities and symbols. J. Magn. Reson. Imaging 10 (3), 223–232.

Ulas, C., Das, D., Thrippleton, M.J., Valdés Hernández, M.D.C., Armitage, P.A., Makin, S.D., Wardlaw, J.M., Menze, B.H., 2018. Convolutional neural networks for direct inference of pharmacokinetic parameters: application to stroke dynamic contrast-enhanced MRI. Front. Neurol. 9, 1147.

van de Haar, H.J., Burgmans, S., Jansen, J.F.A., van Osch, M.J.P., van Buchem, M.A., Muller, M., Hofman, P.A.M., Verhey, F.R.J., Backes, W.H., 2016. Blood-brain barrier leakage in patients with early Alzheimer disease. Radiology 281 (2), 527–535.

van Houdt, P.J., Kallehauge, J.F., Tanderup, K., Nout, R., Zaletelj, M., Tadic, T., van Kesteren, Z.J., van den Berg, C.A.T., Georg, D., Côté, J.-C., Levesque, I.R., Swamidas, J., Malinen, E., Telliskivi, S., Brynolfsson, P., Mahmood, F., van der Heide, U.A., EMBRACE Collaborative Group, 2020. Phantom-based quality assurance for multicenter quantitative MRI in locally advanced cervical cancer. Radiother. Oncol. 153, 114–121.

Vautier, J., Heilmann, M., Walczak, C., Mispelter, J., Volk, A., 2010. 2D and 3D radial multi-gradient-echo DCE MRI in murine tumor models with dynamic R2*-corrected R1 mapping. Magn. Reson. Med. 64 (1), 313–318.

Verma, S., Turkbey, B., Muradyan, N., Rajesh, A., Cornud, F., Haider, M.A., Choyke, P.L., Harisinghani, M., 2012. Overview of dynamic contrast-enhanced MRI in prostate cancer diagnosis and management. AJR Am. J. Roentgenol. 198 (6), 1277–1288.

Villringer, K., Sanz Cuesta, B.E., Ostwaldt, A.-C., Grittner, U., Brunecker, P., Khalil, A.A., Schindler, K., Eisenblätter, O., Audebert, H., Fiebach, J.B., 2017. DCE-MRI blood-brain barrier assessment in acute ischemic stroke. Neurology 88 (5), 433–440.

Weber, J.-P.D., Spiro, J.E., Scheffler, M., Wolf, J., Nogova, L., Tittgemeyer, M., Maintz, D., Laue, H., Persigehl, T., 2022. Reproducibility of dynamic contrast enhanced MRI derived transfer coefficient Ktrans in lung cancer. PLoS One 17 (3), e0265056.

Wong, S.M., Jansen, J.F.A., Zhang, C.E., Staals, J., Hofman, P.A.M., van Oostenbrugge, R.J., Jeukens, C.R.L.P.N., Backes, W.H., 2017. Measuring subtle leakage of the blood-brain barrier in cerebrovascular disease with DCE-MRI: test-retest reproducibility and its influencing factors. J. Magn. Reson. Imaging 46 (1), 159–166.

Xie, T., Jiang, T., Zhao, Q., Fu, C., Nickel, M.D., Peng, W., Gu, Y., 2022. Model-free and model-based parameters derived from CAIPIRINHA-Dixon-TWIST-VIBE DCE-MRI: associations with prognostic factors and molecular subtypes of invasive ductal breast cancer. J. Magn. Reson. Imaging.

Yankeelov, T.E., Gore, J.C., 2009. Dynamic contrast enhanced magnetic resonance imaging in oncology: theory, data acquisition, analysis, and examples. Curr. Med. Imaging Rev. 3 (2), 91–107.

Yarnykh, V.L., 2010. Optimal radiofrequency and gradient spoiling for improved accuracy of T1 and B1 measurements using fast steady-state techniques. Magn. Reson. Med. 63 (6), 1610–1626.

Ye, J.C., 2019. Compressed sensing MRI: a review from signal processing perspective. BMC Biomed. Eng. 1, 8.

Yuh, W.T.C., Mayr, N.A., Jarjoura, D., Wu, D., Grecula, J.C., Lo, S.S., Edwards, S.M., Magnotta, V.A., Sammet, S., Zhang, H., Montebello, J.F., Fowler, J., Knopp, M.V., Wang, J.Z., 2009. Predicting control of primary tumor and survival by DCE MRI during early therapy in cervical cancer. Invest. Radiol. 44 (6), 343–350.

Zhang, J.L., Lee, V.S., 2020. Renal perfusion imaging by MRI. J. Magn. Reson. Imaging 52 (2), 369–379.

Zhang, X., Pagel, M.D., Baker, A.F., Gillies, R.J., 2014. Reproducibility of magnetic resonance perfusion imaging. PLoS One 9 (2), e89797.

Zhou, J.-Y., Wang, Y.-C., Zeng, C.-H., Ju, S.-H., 2018. Renal functional MRI and its application. J. Magn. Reson. Imaging 48 (4), 863–881.

Dynamic susceptibility contrast MRI

Christopher Chad Quarles[a] and Poonam Choudhary[b]

[a]*The University of Texas MD Anderson Cancer Center, Houston, TX, United States*
[b]*Barrow Neurological Institute, Phoenix, AZ, United States*

3.1 Introduction

Dynamic susceptibility contrast (DSC) magnetic resonance imaging (MRI) is one of the most frequently used methods to interrogate hemodynamic and vascular properties of brain tissue, including neuropathologies such as tumors, stroke, and aging-related diseases. DSC-MRI involves the intravenous administration of contrast agent (CA) and dynamically tracking its passage through the cerebral circulation. The passage of CA shortens the intrinsic relaxation times ($T1$, $T2$, and $T2^*$) of tissue water, leading to detectable changes in the MRI signal. Biophysically, DSC-MRI acquisition protocols are optimized to detect the serial changes in CA-induced $T2$ (or $T2^*$) changes (Willats and Calamante, 2013), while dynamic contrast-enhanced (DCE) MRI protocols involve tracking $T1$ changes in tissue. Pharmacokinetic and biophysical models are applied as part of postprocessing methods to estimate structural and functional features of the tissue vasculature. Numerous studies have established the utility of DSC-MRI-based biomarkers for diagnosis, prognosis, and therapeutic response assessment (Bester *et al.*, 2015; Binnewijzend *et al.*, 2016; Nielsen *et al.*, 2017; Ryu *et al.*, 2017). The maturation of DSC-MRI as a perfusion imaging technique and its increased clinical use have led to multiple national and international initiatives aimed at increasing repeatability and reproducibility through the standardization of acquisition and analysis protocols, and establishing benchmarks for clinical software (Welker *et al.*, 2015; Schmainda *et al.*, 2019; Boxerman *et al.*, 2020; Henriksen *et al.*, 2022).

The clinical applicability, interpretation, and fidelity of DSC-MRI primarily depend on the following factors: (1) its biophysical basis, (2) data acquisition, and (3) postprocessing and kinetic analysis. Over two decades of extensive research has provided a systematic characterization of the fundamental contrast mechanisms underlying DSC-MRI. The development and optimization of pulse sequences to enhance contrast-to-noise ratio (CNR) and temporal and spatial resolution, and the refinement of postprocessing pipelines that maximize the accuracy and clinical utility of the extracted hemodynamic parameters such as cerebral blood flow (CBF),

Advances in Magnetic Resonance Technology and Applications, Volume 11, ISSN 2666-9099
https://doi.org/10.1016/B978-0-323-95209-5.00001-5

cerebral blood volume (CBV), and mean transit time (MTT). The advancement and integration of these three elements, along with improved MR hardware, have enabled the simultaneous interrogation of a range of structural and physiological biomarkers of interest. This chapter discusses these three essential elements of DSC-MRI methodology, focusing on recent advances in the field. This chapter also provides insight into the selection of appropriate biophysical methods and the postprocessing steps to ensure robust parameter derivation. Representative applications of advanced DSC-MRI to characterize pathologic conditions such as tumors, stroke, aging, and multiple sclerosis are highlighted.

3.2 Theory

Unlike other imaging modalities (*e.g.*, CT, PET), DSC-MRI and DCE-MRI are unique because they do not directly detect the injected *CA*. Instead, the passage of the CA through tissue dynamically shortens the $T1$, $T2$, and $T2^*$ relaxation times of water, leading to a corresponding change in the local MRI signal. As illustrated in Fig. 3.1A, CA injection alters the relaxation rates in two ways: (1) the direct microscopic (on the order of a molecule size) interaction of CA with water can shorten $T1$ and $T2$ and (2) through-space mesoscopic (on the order of the cell or vessel size) effects induced by CA compartmentalization reduce $T2$ and $T2^*$. The extent to which tissue $T1$, $T2$, and $T2^*$ (Fig. 3.1) are shortened reflects a combination of the CA concentration, magnetic properties of CA, distribution of CA within tissue compartments, the local microstructure (*e.g.*, vascular and cellular architecture), water diffusion, and water exchange across cellular membranes.

3.2.1 Paramagnetic contrast agents, dipole–dipole interactions, and magnetic susceptibility effects

Gd-based chelates, *e.g.*, Gd-diethylenetriaminepentaacetate, Gd-DPTA, are the most prominent clinically used MRI contrast agents. Such agents comprise a metal ion with seven unpaired electrons and are paramagnetic. The theoretical and molecular basis of dipole–dipole interactions and lanthanide-based CAs is described extensively in the literature (Aime *et al.*, 1998; Caravan *et al.*, 1999; Belorizky *et al.*, 2008). While dipole–dipole interactions are the dominant source of CA-induced $T1$ shortening effects, DSC related $T2$, $T2^*$ shortening effects arise from susceptibility differences between tissue's intra- and extravascular space that induce mesoscopic magnetic field gradients (Villringer *et al.*, 1988; Boxerman *et al.*, 1995; Fröhlich *et al.*, 2005; Kiselev and Novikov, 2018). These field inhomogeneities accelerate the loss of phase coherence spanning over the distance of a few nanometers and reduce MRI signal on $T2$, $T2^*$ weighted (w) images due to increased variability in local resonance frequencies experienced by protons and their diffusional motion. DSC-MRI pulse sequence parameters are configured to optimize sensitivity to these susceptibility effects while also permitting high temporal resolution imaging

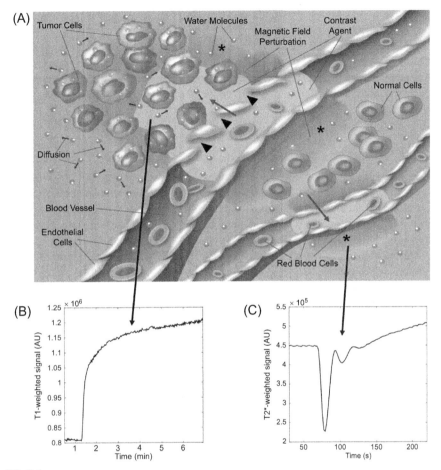

FIG. 3.1

Illustration of CA distribution within tissue, its interaction with water protons (A), and the induced $T1$-weighted (B) or $T2^*$-weighted (C) signal changes. With an intact blood–brain barrier, as illustrated in the lower blood vessel, the CA only has direct interactions with intravascular water protons (red arrow) so that the associated change in the effective tissue $T1$ is small, due to the relative size of the vascular compartment in comparison to the whole tissue. However, when the blood–brain barrier is disrupted (top blood vessel, black triangles) the CA is free to distribute within the extravascular space (red arrow) and its interaction with water protons substantially decreases tissue $T1$ and increases the $T1$-weighted signal (B). As highlighted by the asterisks, the compartmentalization of CA in blood or in the extravascular extracellular space gives rise to magnetic field gradients surrounding these compartments. The diffusion of water through these fields (small black arrows) decreases $T2^*$ and the $T2^*$-weighted signal (C).

Adapted, with permission, from Quarles, C.C., Bell, L.C., Stokes, A.M., 2019. Imaging vascular and hemodynamic features of the brain using dynamic susceptibility contrast and dynamic contrast enhanced MRI. Neuroimage 187, 32–55. doi:10.1016/j.neuroimage.2018.04.069.

($\sim$1–1.5 s) to record the dynamics of the CA's first passage through the tissue when concentration is highest.

Fundamental to DSC-MRI is relating the dynamic $T2$ and/or $T2^*$ changes, and associated signal changes, to local contrast agent concentration. The relaxivity of a CA is a measure of its ability to modulate proton relaxation rates (e.g., $R_2 = 1/T2$) as a function of concentration (Toth et al., 2002), expressed in units of $mM^{-1}s^{-1}$ and defined by Eq. (3.1),

$$R_i(t) = r_i \, C_t(t) + R_{io} \tag{3.1}$$

where $i = 1$, 2, or 2^* denoting three different relaxation rates (R_1, R_2, and R_2^*); R_{io} and $R_i(t)$ are the pre- and postcontrast relaxation rates, respectively; $C_t(t)$ is the dynamic CA concentration in the tissue; and r_i is the relaxivity of the CA. Each relaxation rate (R_1, R_2, and R_2^*) mechanism has its corresponding contrast agent relaxivities r_1, r_2, and r_2^* and their values depend on the magnetic field strength. For comparison, while $T1$ relaxivity of clinical Gd-based CAs is around $4\,mM^{-1}\,s^{-1}$, when compartmentalized within capillaries, the $T2^*$ relaxivity of clinical Gd-based CAs is approximately $80\,mM^{-1}\,s^{-1}$ (Kjølby et al., 2006).

3.2.2 Contrast dose, injection, and risks

The injected dose of CA is selected to balance patient safety and a suitable contrast-to-noise ratio (CNR) needed for robust measures of the derived hemodynamic parameters. Risks of Gd-based CAs include nephrogenic systemic fibrosis (NSF) in patients with renal insufficiency (Leiner and Kucharczyk, 2009) and concerns about Gd accumulation over time in normal brain regions as a function of dose. Gradient-echo (GRE) based DSC-MRI is typically acquired using a 0.1 mmol/kg body weight, while spin echo (SE) may require higher doses, up to 0.3 mmol/kg, to induce similar MR signal drops (Hu et al., 2010). The injection may be performed in the right arm to prevent CA flux into venous structures. The dose is administered with a MR compatible power injector at a rate greater than 4 ml/s, followed by a saline flush (range, 20–30 ml) at the same rate (Essig et al., 2013). Slower injections may lead to underestimation of perfusion parameters (van Osch et al., 2003). However, the rate of contrast injection rate will also depend on the underlying condition of the patient. For example, a modest rate of 3 ml/s is usually recommended to avoid any risk to the nephrogenic system in patients with low estimated glomerular filtration rate (eGFR) (less than $30\,ml/min/1.73\,m^2$) (Leiner and Kucharczyk, 2009).

In patients with chronic kidney disease and impaired renal function, iron oxide nanoparticle-based CAs are a potential alternative to Gd-based CAs (Toth et al., 2017). Iron oxide nanoparticle-based CAs are based on ultrasmall superparamagnetic iron oxide particles with relaxivities 10–20 times higher than Gd-based CAs (Knobloch et al., 2018). Examples of such agents that have been clinically evaluated as MRI CAs include Ferumoxides (Feridex IV, Berlex Laboratories), Ferucarbotran (Resovist, Bayer Healthcare), Ferumoxtran-10 (AMI-227 or

Code-7227, Combidex, AMAG Pharma; Sinerem, Guerbet), and NC100150 (Clariscan, Nycomed) (Daldrup-Link, 2017). In North America, ferumoxytol (Feraheme, AMAG Pharmaceuticals) is approved for use as an iron supplement in patients with iron-deficiency anemia and is increasingly used, off-label, as an MRI contrast agent. In 2015 the FDA issued a black box warning due to fatal and serious hypersensitivity reactions, leading to recommendations including the administration of ferumoxytol as an IV infusion over 15 min (previously rapid bolus injection was allowed), patient monitoring for 30 min following infusion, and dilution prior to infusion. The safety of alternative bolus protocols for dynamic imaging in clinically warranted cases has been studied in 671 MR studies across 331 patients (Varallyay et al., 2017). Importantly, ferumoxytol-related adverse events occurred in 10% of infusions, which is consistent with that reported in prior randomized trials for iron replacement and equivalent to ionic iodinated contrast media. More recently, a multicenter observational registry safety experience in 3215 patients who received 4240 injections showed no systematic changes in vital signs after ferumoxytol administration and was reported to be well tolerated with no serious adverse events (Nguyen et al., 2019). As such studies and trials continue, iron oxide nanoparticles should mature into an alternative CA for perfusion imaging in kidney-sensitive patients.

3.3 Acquisition of DSC-MRI data

Given the maturation of DSC-MRI and its clinical use, disease-specific consensus recommendations have been established for acute stroke (Wintermark et al., 2008) and brain cancer (e.g., Boxerman et al., 2020; QIBA_DSC-MRI_Stage1_Profile. pdf, n.d.). The standardized protocols were informed by evidence-based best practices centered on clinical utility and performance, parameter accuracy, reproducibility, multisite feasibility, and technical and practical feasibility. These recommendations are summarized later.

3.3.1 The MR pulse sequence: Choice of acquisition readout

Given the transient nature of CA passing through the brain's vasculature, fast imaging readout methods are needed to accurately sample contrast-induced MR signal changes (Stehling et al., 1991). Insufficient temporal resolution can lead to underestimation of hemodynamic parameters (Knutsson et al., 2004). Consequently, a temporal resolution equal to or less than 1.5 s is required for robust DSC-MRI in acute stroke and brain cancer (Wintermark et al., 2008; Boxerman et al., 2020; QIBA_DSC-MRI_Stage1_Profile.pdf, n.d.). Due to the rapid temporal requirements, the most common acquisition strategy for DSC-MRI is single-shot, gradient-echo, echo planar imaging (EPI). EPI also enables whole-brain coverage of typically 15–25 slices at reasonable signal-to-noise ratios (SNRs) (van Gelderen et al., 2000).

The main drawbacks to EPI-based acquisitions are signal dropouts and geometric distortions caused by susceptibility artifacts around, for example, sinuses, air–tissue interfaces, and resection cavities (such as those following tumor removal). Although not widely available, alternative advanced sequences may provide significant advantages over single-shot EPI. Using segmented or interleaved EPI readout by splitting the k-space traversal can manage off-resonance effects during the long readouts required to fully sample k-space (Liu *et al.*, 1993). The three-dimensional "Principle of Echo Shifting with a Train of Observations" (PRESTO) sequence reduces distortions and can acquire images at very high temporal resolution (van Gelderen *et al.*, 2000; Pedersen *et al.*, 2004). Another reliable multiecho interleaved EPI method with parallel imaging is PERfusion with Multiple Echoes and Temporal Enhancement (PERMEATE). This approach reduces image distortion, blurring, and saturation artifacts while maintaining motion sensitivity and SNR (Jochimsen *et al.*, 2007; Newbould *et al.*, 2007). Additionally, more advanced readout options have also been proposed, including non-Cartesian readouts such as spiral or radial acquisitions (Jonathan *et al.*, 2014; Paulson *et al.*, 2016).

While single-echo gradient-echo EPI is recommended and widely used for the acquisition of conventional DSC-MRI parameters, including CBF, CBV, MTT, the use of multiple gradient echoes enables multicontrast imaging. This feature stems from the finding that multiecho DSC-MRI enables separation and quantification of contrast agent-induced $T1$ and $T2^*$ effects (Vonken *et al.*, 2000). The dynamic $T1$ data can be leveraged to derive DCE-MRI kinetic parameters, enabling assessment of blood–brain barrier permeability (Quarles *et al.*, 2012). Further, in the context of brain tumors, multiecho acquisitions enable removal of the confounding $T1$ effects associated with contrast agent extravasation, enabling accurate brain tumor CBV mapping (Stokes *et al.*, 2021a,b).

3.3.2 The image contrast: Gradient and spin echo

DSC-MRI data may be acquired using either spin- (SE) and gradient-echo (GRE) sequences or combination of both (Fig. 3.2) (Donahue *et al.*, 2000; Schmiedeskamp *et al.*, 2012b; Stokes and Quarles, 2016; Stokes *et al.*, 2021a,b). Biophysically, GRE acquisitions have higher sensitivity toward vessels of all sizes, while SE is maximally sensitive to capillary-sized vessels (Boxerman *et al.*, 1995). Therefore GRE has higher contrast-to-noise ratio (CNR) per dose of CA as compared to SE sequences and, thus, is clinically preferred over SE. In general, GRE acquisitions have shorter TEs, enabling higher temporal resolution. The high vascular sensitivity of GRE may increase its sensitivity to susceptibility and blooming artifacts, resulting in overestimation of perfusion parameters (Varallyay *et al.*, 2018; Maral *et al.*, 2020). In contrast, SE sequences are considered more consistent with microvascular perfusion measurements in reference to positron emission tomography (Ostergaard *et al.*, 1998).

The combined application of GRE and SE may increase the clinical value of DSC acquisitions owing to their complementary sensitivity to different vessel populations.

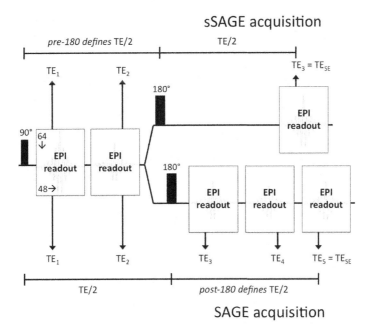

FIG. 3.2

Pulse sequence diagram for the combined spin-echo and gradient-echo approach. SAGE (bottom) involves the acquisition of two gradient, two asymmetric, and one spin echo, while sSAGE (top) acquires two gradient and one spin echo.

Adapted, with permission, from Stokes, A., Skinner, J., Yankeelov, T., Quarles, C., 2016. Assessment of a simplified spin and gradient echo (sSAGE) approach for human brain tumor perfusion imaging. Magn. Reson. Imaging 34. doi:10.1016/j.mri.2016.07.004.

A combined single echo, spin- and gradient-echo acquisition was shown to enhance differentiation of brain tumor grades (Donahue *et al.*, 2000). More recently, Schmiedeskamp *et al.* (2012b) presented a multi-echo, spin- and gradient-echo-planar imaging (SAGE EPI) MRI sequence (Fig. 3.2) which enables simultaneous measurements of total vascular and capillary CBF, CBV, and MTT. This combined GRE, SE acquisition strategy retained the benefits of multiecho sequences (both SE and GRE). The SAGE sequence includes five readout trains (2 GREs followed by 2 asymmetric SEs, and a single SE) with a spin-echo TE of less than 100 ms. Also, Stokes *et al.* proposed a simplified SAGE approach (sSAGE) comprised of two gradient echoes followed by a spin echo as shown in Fig. 3.2 (Stokes and Quarles, 2016). The advantage of implementing sSAGE over SAGE is that it similarly provides $T1$-insensitive GE and SE hemodynamic parameters while also reducing the computation time by 450 times, as it relies on linear fitting, as opposed to nonlinear, for derivation of ΔR_2^* and ΔR_2 time curves.

Another advantage of simultaneous GRE and SE readouts is that average vessel size can be estimated from ΔR_2^* and ΔR_2 time series according to theoretical predictions provided either by Monte Carlo simulations (Boxerman *et al.*, 1995; Tropres *et al.*, 2001) or by analytical modeling (Yablonskiy and Haacke, 1994; Kiselev and Posse, 1998, 1999). The vessel size measures on MRI accurately reflect histologic measures of vessel caliber (Chakhoyan *et al.*, 2018). Recently, Emblem *et al.* proposed a novel approach to characterize blood vessel architecture using ΔR_2^* and ΔR_2 time series, an approach termed vessel architectural imaging (VAI) (Emblem *et al.*, 2013). VAI was demonstrated to identify response to antiangiogenic therapy in brain cancer patients (Emblem *et al.*, 2013) and vessel type determination to improve early tumor progression and pseudo-progression (PP) (Kim *et al.*, 2021). VAI relies on characterizing plots of $\Delta R_2^*(t)$ and $\Delta R_2(t)$, forming a hysteresis loop (Fig. 3.3), where its slope and direction provide estimates of vessel size and type, respectively (Kim *et al.*, 2021).

3.3.3 Pulse sequence: Parameter selection and acceleration methods

The choice of echo time (TE) in single-echo GRE DSC-MRI influences the degree of $T2^*$ weighting and SNR. With longer TEs, significant signal saturation near large vessels is induced due to higher $T2^*$ weighting (Johnson *et al.*, 2004). On the other hand, shorter TEs result in higher image SNR and increased sensitivity to $T1$ effects (Thilmann *et al.*, 2004). In acute stroke, the recommended range of TEs is 25–30 ms (at 3 T) and 35–45 ms (at 1.5 T) (Wintermark *et al.*, 2008). In brain tumors, the recommended range of TEs is 25–35 ms (at 3 T) and 40–50 ms (at 1.5 T) (Boxerman *et al.*, 2020). While recommendations for SE echo times have not been formally established, the minimal echo time given by the other desired sequence parameters is often selected.

The temporal sampling rate, the equivalent of repetition time (TR) for single-shot EPI acquisitions, is selected to effectively sample the first pass of CA through tissue. Regardless of the disease involved, the TR should be less than 1.5 s. Higher flip angles increase the image SNR but induce more $T1$-weighting, while lower flip angles lead to lower SNR and reduced CBF accuracy. Recommendations for acute stroke specify flip angles of 60–70° (at 1.5 T) and 60° (at 3 T) (Wintermark *et al.*, 2008). In the context of brain tumors, the selection of flip angle is more biophysically relevant due to the potential for pronounced CA-induced $T1$ leakage effects and associated reduced CBV accuracy (Boxerman *et al.*, 2020). For nearly twenty years, preloads of contrast agent were used to reduce the $T1$ effects (in conjunction with moderate to high flip angles), but this required double the standard dose of CA. Recent computational studies (Leu *et al.*, 2016; Semmineh *et al.*, 2017) verified that high CBV accuracy was achievable with a single CA dose when a low flip angle acquisition was paired with pharmacokinetic model-based leakage correction. These computational results were validated in patients in a subsequent multisite study (Schmainda *et al.*, 2019). Example data from this study is highlighted in Fig. 3.4. Given the results of these computational and patient studies, the consensus recommendations for single-dose DSC-MRI in brain tumors include the use

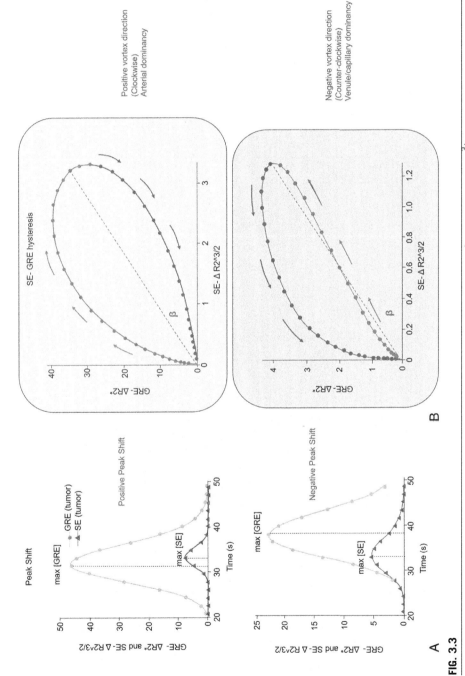

FIG. 3.3

Illustration of parameters derived from vessel architectural imaging. (A) GRE-based ΔR_2^* and SE-based $\Delta R_2^{3/2}$ time series are fit to γ-variate function and the peak shift is computed. A positive peak shift indicates a GRE peak preceding the SE peak (top), while a negative peak shift indicates the SE peak preceding the GRE peak (bottom). (B) The fitted curves are also plotted with $\Delta R_2^{3/2}$ (SE based) along the x-axis and ΔR_2^* (GRE based) along the y-axis, forming a hysteresis loop. A clockwise rotation (top) reflects that the changes in GRE data precede those in SE data, reflecting arterial dominance in the underlying vasculature function. A counterclockwise rotation (bottom) shows that changes in SE data precede GRE data, reflecting venous dominance.

Adapted, with permission, from Kim, M., Park, J.E., Emblem, K., Bjørnerud, A., Kim, H.S., 2021. Vessel type determined by vessel architectural imaging improves differentiation between early tumor progression and pseudoprogression in glioblastoma. AJNR Am. J. Neuroradiol. 42, 663–670. doi:10.3174/ajnr.A6984.

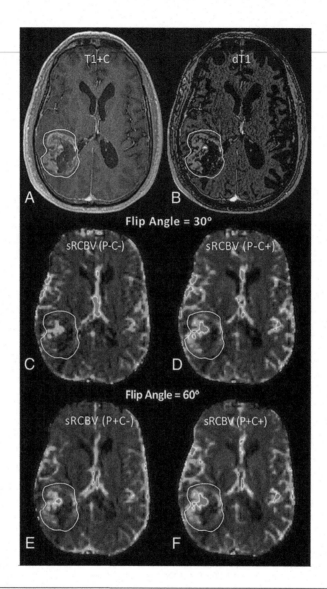

FIG. 3.4

Anatomic images and standardized, relative CBV (sRCBV) maps in a glioblastoma patient. (A) Postcontrast *T*1-weighted (*T*1 + C); (B) difference between post- and pre-*T*1-weighted images (d*T*1); (C and D) Corresponding sRCBV maps obtained from the first contrast agent injection (C) without preload (P−) and without leakage correction (C−), and (D) without preload (P−) but after application of leakage correction using the Boxerman–Schmainda–Weisskoff (BSW) algorithm (C+). (E and F) The sRCBVs obtained during the second contrast dose and thus after the preload are shown: (E) without (P+C−) and (F) with (P+C+) leakage correction.

Adapted, with permission, from Schmainda, K., Prah, M., Hu, L., Quarles, C., Semmineh, N., Rand, S., Connelly, J., Anderies, B., Zhou, Y., Liu, Y., Logan, B., Stokes, A., Baird, G., Boxerman, J., 2019. Moving toward a consensus DSC-MRI protocol: validation of a low-flip angle single-dose option as a reference standard for brain tumors. AJNR Am. J. Neuroradiol. 40, 626–633. doi:10.3174/ajnr.A6015.

of a 30–35° degree flip angle at both 1.5 T and 3 T (Boxerman *et al.*, 2020). While the selection of the flip angle in these efforts was based on optimizing CBV accuracy, it should be noted that the DSC-MRI-derived metric, percent signal recovery (PSR), which may aid in preoperative diagnosis of cerebral tumors, is more effective with higher flip angles (*e.g.*, 75–90°) as this enhances the signal's sensitivity to contrast agent $T1$ leakage effects (Mangla *et al.*, 2011; Chakravorty *et al.*, 2015).

The importance of whole-brain (for acute stroke) or tumor coverage and spatial resolution must be addressed in identifying and characterizing abnormal tissue. The recommendations for spatial resolution are generally 1–3 mm in-plane and 3–5 mm through-plane, with adequate slice coverage, which can be achieved with a 24-cm field of view and 128×128 matrix. The trade-off between spatial and temporal resolution may be adjusted based on the clinical application. Per the consensus recommendations, the duration of image acquisition (and number of baseline images before injection) for acute stroke and brain tumors is 90–120 s (10–12 images) and at least 120 s (30–50 images), respectively (Wintermark *et al.*, 2008; Boxerman *et al.*, 2020). Although longer scan durations can influence tracer recirculation, this does not generally impact the resulting perfusion metrics. On the other hand, smaller acquisition durations can lead to errors of up to 50% in CBF and CBV errors due to the truncation of the recirculation of contrast bolus (Kosior and Frayne, 2010).

The high spatiotemporal resolution for DSC-MRI requires image acceleration. Parallel imaging strategies such as SENSitivity Encoding (SENSE) (Pruessmann *et al.*, 1999; Zhang *et al.*, 2019a), GeneRalized Autocalibrating Partial Parallel Acquisition (GRAPPA) (Tsao *et al.*, 2003; Huang *et al.*, 2005; Jung *et al.*, 2008; Bilgic *et al.*, 2018), and wave-controlled Aliasing In Parallel Imaging (CAIPI) (Bilgic *et al.*, 2015; Polak *et al.*, 2018) can be implemented to reduce acquisition time and provide higher resolution images. Among other undersampling methods, partial Fourier acquisitions (undersampled high-frequency k-space is reconstructed from remaining k-space), view-sharing methods (undersampled data reconstructed from other dynamic time frames), and sliding window view-sharing method (k-space is split into consecutive windows combined for image reconstruction) have been used to improve DSC-MRI temporal resolution (d'Arcy *et al.*, 2002; Jonathan *et al.*, 2013). Keyhole view-sharing techniques involve acquiring center low-frequency k-space at each dynamic, while remaining high-frequency information is obtained from a reference image and is shared among the undersampled dynamic images (Oesterle *et al.*, 2000). Advances in simultaneous multi-slice (SMS) acquisitions (also termed multiband imaging), which utilizes multiband RF pulses to simultaneously acquire multiple slices and then separated using parallel imaging, can also be leveraged for improved spatiotemporal resolution for DSC-MRI. The SMS method was shown to improve slice coverage with improved spatial or temporal resolution combined GRE and SE-based DSC-MRI sequences with minimum SNR penalty (Eichner *et al.*, 2014; Chakhoyan *et al.*, 2018).

3.3.4 **Data preprocessing for acquisition: Motion, noise, artifacts**

In the clinical setting, DSC-MRI data may get corrupted by factors such as patient motion, low SNR, and failed contrast injection. Therefore careful DSC-MRI data quality control is vital for accurate subsequent analysis. A general recommendation for mitigating motion artifacts is to register the images across time (Kosior *et al.*, 2007b). Perfusion parameter, particularly in hypoperfused low-contrast areas, is directly affected by the amount of noise in the DSC-MRI. Spatial smoothing in each time frame independent of other time frames using a Gaussian filter (Willats *et al.*, 2006) is among few preprocessing methods that have been proposed to improve SNR. Filtering in the temporal dimension can minimize possible distortion of the signal–time curves using wavelet denoising (Wirestam and Ståhlberg, 2005), partial differential equations (Lysaker *et al.*, 2003), and four-dimensional nonlinear filtering (Kosior *et al.*, 2007a). Low SNR can be managed by analyzing multiple voxels simultaneously using Bayesian random effects models (King *et al.*, 2011), which are known to be more efficient due to their information-borrowing behavior.

3.4 **Perfusion parameters**

As noted before, the conventional hemodynamic parameters calculated from DSC-MRI are CBV, CBF, and MTT. The following section describes the theory behind the derivation of hemodynamic parameters obtained using perfusion imaging and are based on the assumption that CA is nondiffusible from the intravascular space.

3.4.1 **Conversion of MR signal to concentration-time curves**

To convert the MR signal into CA concentration passing through a capillary, it is necessary to obtain an accurate estimate of the baseline MR signal intensity, $S(0)$, before the arrival of *CA*. To maximize SNR, the precontrast baseline signal is calculated as the average signal over 30–60 time points after establishing a steady state. For a single-shot GRE-EPI sequence, the MR signal as a function of time, $S(t)$, can be approximated using:

$$S(t) = M_0 \frac{\sin(\alpha)[1 - \exp(-TR \times R_1(t))]}{1 - \cos(\alpha)\exp(-TR \times R_1(t))} \exp\left(-TE \times R_2^*(t)\right) \tag{3.2}$$

where M_0 is the equilibrium longitudinal magnetization; α is the flip angle; and R_i ($i = 1,2$) is the relevant relaxation rate ($R_i = 1/T_i$).

The change in *CA* concentration is assumed to be linearly proportional to the change in transverse relaxation rate, $\Delta R_2^*(t)$, within the tissue, with a proportionality constant equal to the contrast agent relaxivity, r_2^* (Kiselev, 2001). Unlike the contrast agent *T*1 relaxivity (r_1), the $T2^*$ relaxivity (r_2^*) is dependent on the local vascular geometry; thus, r_2^* is spatially variable across the brain and is likely altered in cases of pathology. In conventional DSC-MRI, since r_2^* is not known a priori, it is often set to unity.

3.4.1.1 Single-echo DSC-MRI acquisition

With single-echo GRE-EPI, $\Delta R_2^*(t)$ can be computed as follows:

$$\Delta R_2^*(t) = R_2^*(t) - R_2^*(0) = \frac{1}{TE} \ln \left[\frac{S(t)}{S(0)} + \frac{E_1(0)}{E_1(t)} \right] \approx -\frac{1}{TE} \ln \left[\frac{S(t)}{S(0)} \right] \quad (3.3)$$

where $S(0)$ is the baseline signal before CA arrival. Typically, $T1$ effects (E_1) are assumed to be negligible. In cases of disrupted *BBB*, specialized acquisition methods such as multiecho acquisitions, preload dosing, and advanced postprocessing methods can be employed.

3.4.1.2 Dual-echo DSC-MRI acquisition

With a dual-echo DSC acquisition, contrast agent-induced $T1$ effects can be removed (Vonken *et al.*, 1999, 2000) through the computation of $\Delta R_2^*(t)$:

$$\Delta R_2^*(t) = R_2^*(t) - R_2^*(0) = \frac{1}{TE_1 - TE_2} \ln \left[\frac{S_{TE_1}(t)}{S_{TE_2}(t)} + \frac{S_{TE_1}(0)}{S_{TE_2}(0)} \right] \quad (3.4)$$

3.4.1.3 SAGE and sSAGE DSC-MRI acquisition

The original SAGE analysis (Schmiedeskamp *et al.*, 2012a; Skinner *et al.*, 2014) is based on nonlinear fitting of all five echoes to derive $R_2(t)$ and $R_2^*(t)$. The simplified SAGE (sSAGE) analysis utilizes the acquisition of three echoes and an analytical solution for $\Delta R_2(t)$ and $\Delta R_2^*(t)$ (Stokes *et al.*, 2016; Stokes and Quarles, 2016). The ΔR_2^* (t) for sSAGE can be obtained using Eq. (3.5). A similar expression can be written for $\Delta R_2(t)$.

$$\Delta R_2^*(t) = \frac{1}{TE_{SE} - TE_2} \left(\ln \left(\frac{S_{TE_{SE}}, pre}{S_{TE_{SE}}(t)} \right) - \ln \left(\frac{S_{TE=0}, pre}{S_{TE=0}(t)} \right) \right) \quad (3.5)$$

3.4.2 DSC-MRI kinetic models

The hemodynamic parameters of DSC-MRI are derived from indicator dilution theory (Axel, 1980; Meier and Zierler, 1954). It is based on a one-compartment model assuming that CA concentration is spatially uniform within the volume of distribution and compartmentalized within blood vessels. The following functions help in deriving hemodynamic parameters:

(1) *The transport function h(t)*—The function h(t) is the probability density function of transit time t (following an instantaneous input of tracer molecules at t=0). This function reflects the distribution of transit times through the voxel and is dependent on vascular flow and structure.

(2) *The residue function R(t)*—The residue function represents the fraction of CA present in the volume of interest at t after an ideal instantaneous unit bolus injection at $t = 0$

$$R(t) = \left[1 - \int_0^t h(\tau)d\tau \right] \Rightarrow h(t) = -\frac{dR(t)}{dt} \qquad (3.6)$$

where the integral term represents the fraction of the CA that has left the volume of interest (VOI), and $R(t=0)=1$, that is, all of the CA is present at time $t = 0$.

(3) *The arterial input function (AIF)*—The arterial input function, $C_{\mathrm{AIF}}(t)$, characterizes the system's input into the volume of interest. Physiologically, each voxel in the brain will have a slightly different input function due to variations in vascular supply, particularly in cases of vascular pathology. In practice, a single AIF (or the average across multiple voxels) is typically measured in a major cerebral artery or its branches to minimize partial volume effects (Zaharchuk et al., 2009). Most DSC-MRI studies use a single, *global* AIF approach in contrast to *local* or *regional* AIF which involves multiple 'local' inputs with each tissue voxel based on the assumption that the tissue is fed by spatially proximal arterial branches. The latter approach reduces the effects of contrast agent delay and dispersion. Due to the potential influence of vascular pathology on the measured AIF, such as in cerebrovascular disease patients, measuring the AIF near small proximal arteries is preferred, as there may be a greater likelihood of bolus delay and dispersion between major feeding arteries and the tissue of interest. In other pathologies, selecting an AIF in a larger artery is reasonable. Selecting the AIF voxels can be performed manually but is time intensive and subjective. More commonly, fully automatic and semiautomatic methods employing algorithms based on different identification criterion are utilized in AIF determination (Lorenz et al., 2006; Kim et al., 2010).

(4) *The tissue contrast agent concentration-time curve*—In a DSC-MRI study, the signal in each voxel reflects the tissue contrast agent concentration, $C_t(t)$ which is related to $C_{\mathrm{AIF}}(t)$ and $R(t)$ written in terms of a convolution equation (Østergaard et al., 1996):

$$C_t(t) = \frac{\rho}{k_H} \cdot \mathrm{CBF} \cdot \int_0^t C_{\mathrm{AIF}}(\tau) R(t - \tau) dt \qquad (3.7)$$

where ρ is the density of brain tissue (typically, 1 g/ml tissue) and k_H is the hematocrit factor (typically, 0.67) which accounts for the difference in hematocrit between the AIF and the capillary system (Rempp et al., 1994) and $k_H = \frac{1-H_{art}}{1-H_{cap}}$ is the difference in hematocrit (H) between capillaries and large arteries, as only the plasma volume is accessible to the *CA*. In practice, $C_t(t)$ and $C_{AIF}(\tau)$ are derived from DSC-MRI data, and a deconvolution is performed to compute $CBF \bullet R(t)$. The CBF (ml of blood/100 g) is determined from the initial height of the residue function, since $R(0) = 1$. For an intact BBB, maps of CBV (ml of blood/100 g) can be derived using:

$$\text{CBV} = \frac{k_H}{\rho} \cdot \frac{\int_{-\infty}^{+\infty} C_{\text{tissue}}(t)dt}{\int_{-\infty}^{+\infty} C_{\text{AIF}}(t)dt} \tag{3.8}$$

Note that this assumes that the contrast agent $T2^*$ relaxivity (r_2^*) is equivalent between the AIF and the tissue of interest (Kiselev, 2001). Thus, in practice, relative brain tumor CBV (rCBV) is often computed using the integral of the voxel-wise ΔR_2^* time series: $rCBV = \int_0^t \Delta R_2^* dt$. Classically, MTT (s), which is the average time for CA to pass through the tissue after an ideal bolus injection, can be calculated from CBV and CBF using the central volume theorem (Stewart, 1893; Meier and Zierler, 1954):

$$\text{MTT} = \frac{\text{CBV}}{\text{CBF}} \tag{3.9}$$

3.5 DSC-MRI data postprocessing possibilities

Similar to acquisition protocols, the postprocessing methods used to analyze DSC-MRI data can significantly influence the fidelity of the derived parameters. We now summarize current best practices and emerging methods.

3.5.1 Multivendor postprocessing software applications

In order to facilitate the use of DSC-MRI in clinical practice, several software packages are available for automatically analyzing perfusion imaging data (Mokli *et al.*, 2019). *RapidAI* (www.rapidai.com) is a cloud-based tool for analysis of perfusion data acquired in acute stroke setting. It delivers quantified, color-coded perfusion maps showing severity of delays in contrast arrival time, volume of tissue delays, and hypoperfusion intensity ratio. For brain tumor DSC-MRI analysis, *IB Neuro* (Imaging Biometrics, Elm Grove, Wisconsin), *NordicICE* (NordicNeuroLab, Bergen, Norway), and *Olea Sphere* (Olea Medical, Cambridge, MA) are options that include leakage correction for more robust CBV mapping. Note that *IB Neuro* also includes analysis options for multiecho-based, simultaneous DSC/DCE-MRI. The MRI vendors also offer platforms for perfusion analysis, including *dStream* (Philips), *FuncTool Performance* (General Electric), *syngo*.via (Siemens), and *Olea Sphere* (Canon). The reproducibility of in-house and clinical software for perfusion analysis has been previously evaluated (Hu *et al.*, 2015; Bell *et al.*, 2019; Bell *et al.*, 2020a,b).

3.5.2 Magnetic field inhomogeneity correction

The presence of susceptibility artifacts in DSC-MRI-based images leads to inaccurate spatial correlation with corresponding undistorted anatomical scans, potentially limiting the usefulness of DSC-MRI-based information for stereotactic biopsy and surgery (Knopp *et al.*, 1999). The method of reverse phase encoding (PE) (Chang and Fitzpatrick, 1992) comprises traversing *k*-space in the opposite direction along the PE axis, which reverses distortions and can improve anatomic fidelity. This correction of B_0 distortions using two SE-EPI prescans with opposite PE polarity improves accuracy relative to the uncorrected data (Vardal *et al.*, 2014).

3.5.3 Deconvolution

The quantification of CBF $R(t)$ in Eq. (3.7) requires a deconvolution, which is an ill-posed inverse problem that is highly sensitive to noise that can reduce the accuracy of the derived CBF values. Both model-dependent (*e.g.*, smooth monotonically decreasing estimate of $R(t)$) and model-independent (no analytical constraints on data or assumptions about shape of $R(t)$) approaches to solving this equation have been proposed (Gobbel and Fike, 1994; Østergaard *et al.*, 1996; Wu *et al.*, 2003; Mouridsen *et al.*, 2006; Mehndiratta *et al.*, 2013). Model-independent methods are preferred over model-dependent methods as the residue function likely exhibits significant heterogeneity between healthy and diseased tissue. Among model-independent methods, the singular value decomposition (SVD) and regularization proposed by Østergaard *et al.* estimated higher SNR at high flow rates in both gray and white matter (Østergaard *et al.*, 1996). A temporal delay between the AIF and local voxel-wise contrast agent concentration-time profile, for example in acute stroke, can also introduce inaccuracies in the computed CBF. To compensate for this, Wu *et al.* proposed the use of a delay-insensitive block circulant matrix for deconvolution and also incorporating an oscillation index to limit nonphysiological oscillations of the derived residue function (Wu *et al.*, 2003). The implementation of Tikhonov regularization (Calamante *et al.*, 2003) with L-curve criterion improves the characterization of $R(t)$ (Sourbron *et al.*, 2007). The vascular model proposed by Mouridsen *et al.* (2006) used Bayesian framework to model the tracer transit times with gamma probability distribution functions. The nonparametric control point interpolation (CPI) deconvolution method implemented in Bayesian framework improved the estimation of CBF by estimating smoother residue function compared to circular SVD (Mehndiratta *et al.*, 2013). Recently, a nonparametric deconvolution approach based on Bézier curves showed robustness toward different underlying residue function shapes in comparison to SVD and Bayesian methods (Chakwizira *et al.*, 2022). The perfusion estimates of MTT, CBF involving corrections for delay, dispersion, and (delay + dispersion) using Bézier deconvolution (BzD) were also more accurate (Chakwizira *et al.*, 2022).

3.5.4 Absolute or relative quantification

The topic of absolute or relative quantification parameters has been a matter of debate and research in perfusion imaging. Many clinical applications rely on relative values of DSC-MRI perfusion measures (*e.g.*, rCBV, rCBF), particularly if radiographic visualization of regional perfusion status is sufficient. For semiquantitative analysis, relative values are often normalized to normal-appearing white or gray matter tissue, as this facilitates evaluation of the degree of perfusion abnormalities within and across subjects. Note that there is no consensus on the optimal approach for normalization (*e.g.*, manual versus automated, region of interest location and size), but a recent study has systematically investigated these factors (Cho *et al.*, 2022). More recently, image standardization (Madabhushi and Udupa, 2006) has been applied to the computed CBV maps and has shown improved predictive performance and inter- and intrapatient repeatability (Bedekar *et al.*, 2010; Prah *et al.*, 2015). The standardization of rCBV also eliminates the need for user-based input for normalization, an important factor toward workflow optimization and consensus methodology (Hoxworth *et al.*, 2020a,b).

Absolute quantification of hemodynamic parameters also depends on many considerations. Susceptibility artifacts can reduce regional CNR and accuracy of parameter derivation. Although values for the proportionality constants between ΔR_2^* and CA concentration may be assumed, their values are likely to vary between (and within) individuals and, particularly, in pathological tissue (Pathak *et al.*, 2003). While a linear relationship reasonably describes the association in tissue voxels, Calamante *et al.* (2009) showed that the use of a quadratic relationship when computing the AIF improved CBF accuracy. Post hoc calibration may also be applied to the CBF values and matched with techniques, such as positron emission tomography, arterial spin labeling (ASL), and/or normalizing to whole-brain blood flow measured with phase-contrast magnetic resonance angiography (Zaro-Weber *et al.*, 2010; Bonekamp *et al.*, 2011).

3.5.5 Leakage correction

A particular challenge in using DSC-MRI in brain tumors for the determination of rCBV is that the presence of a leaky BBB may confound measurements (Paulson and Schmainda, 2008). In regions with a disrupted BBB, the CA leaks into the extravascular extracellular space and reduces the reliability of the computed rCBV values and their clinical utility (Kluge *et al.*, 2016; Semmineh *et al.*, 2017). While many biophysical and kinetic models have been proposed to address the CA leakage issue in DSC-MRI, the accuracy and clinical utility of Boxerman–Schmainda–Weisskoff (BSW) (Boxerman *et al.*, 2006) (Fig. 3.4) approach has been validated using computational modeling and image-guided histopathology (Hoxworth *et al.*, 2020a,b) and it is included in the recent consensus recommendations (Boxerman *et al.*, 2020).

3.5.6 Standardization efforts

Several initiatives have focused on optimizing and standardizing DSC-MRI protocols and postprocessing steps (Boxerman *et al.*, 2020; QIBA_DSC-MRI_Stage1_Profile.pdf, n.d.). The availability of public imaging databases such as The Cancer Imaging Archives [TCIA] that include pathology and other clinical endpoints can be used to investigate various postprocessing steps and therefore can be assessed for their impact on the diagnostic and predictive potential of imaging parameters. The development of patient data-driven digital reference objects (DRO) (Semmineh et al., 2017) and simulation studies (Leu et al., 2016) in order to identify optimal DSC-MRI acquisition and analysis protocols for clinical trials are promising initiatives that motivated the clinical trials (Schmainda et al., 2019) and led to the most recent consensus recommendations (Boxerman et al., 2020). These standardization efforts for both acquisition and postprocessing may improve the fidelity of DSC biomarkers and their reproducibility in clinical trials and patient care settings.

3.6 Advanced biomarkers in DSC-MRI

While CBV, CBF, and MTT are the primary hemodynamic biomarkers for most DSC-MRI clinical applications, more advanced biomarkers (e.g., intravoxel capillary transit time heterogeneity (CTH), oxygen extraction fraction (OEF), VSI, and VAI) can be derived from sophisticated kinetic models and analysis techniques (Figs. 3.3 and 3.5).

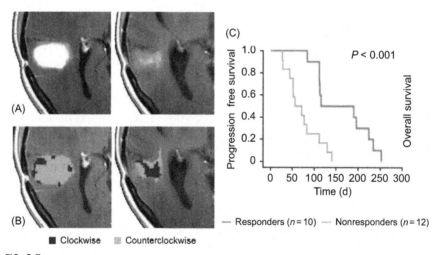

FIG. 3.5

Application of vessel architectural imaging VAI analysis to evaluate antiangiogenic therapy in recurrent glioblastoma patients. (A) Postcontrast $T1$-weighted images and (B) maps of vortex direction (left) before and (right) 28 days after therapy onset. Note that treatment shifted the predominantly counterclockwise direction to clockwise, indicating a positive response to therapy. (C) Progression-free survival curves in responding (median = 153 days) and nonresponding patients (median = 64 days) that were identified based on vortex direction (Emblem et al., 2013).

3.6.1 Capillary transit time heterogeneity and oxygen extraction function

Capillary transit time heterogeneity is changed during functional activation and in some disease states, therefore affecting the extraction efficacy of oxygen from blood. Several biophysical models of oxygen delivery developed by Jespersen and Østergaard (2012), Angleys *et al.* (2015) and Rasmussen *et al.* (2015) suggested that the redistribution of blood flow indexed by the extent of CTH alters the effective permeability of the capillary surface area and, hence, the extraction efficacy of freely diffusible molecules (Angleys *et al.*, 2016). The CTH can be estimated by modifying the $h(\tau)$ (see Eqs. 3.6 and 3.10) by parametric gamma variate modeling (α and β, gamma variate parameters) of the transit time distribution for better estimation of the underlying microvasculature (Østergaard *et al.*, 1996; Mouridsen *et al.*, 2014):

$$h(\tau; \alpha, \beta) = \frac{1}{\beta^\alpha \Gamma(\alpha)} \tau^{\alpha-1} e^{-\tau/\beta} \tag{3.10}$$

Following the new $h(\tau; \alpha, \beta)$, new $MTT - \alpha\beta$ and the standard deviation (SD) of the distribution is referred to as the capillary transit time heterogeneity ($CTH = \beta\sqrt{\alpha}$). Further addition of a delay parameter, δ, in the model can account for the delay between the AIF and tissue time curves.

Capillary flow heterogeneity is known to reduce the extraction of diffusible solutes from the capillary bed (Jespersen and Østergaard, 2012). It is also well known that flow pattern homogenization can increase oxygen extraction efficacy (Kuschinsky and Paulson, 1992; Østergaard *et al.*, 2000). According to the method developed by Jespersen *et al.* (Jespersen and Østergaard, 2012), oxygen extraction fraction (OEF) can be calculated by integrating oxygen extraction over the distribution of transit times, Eq. (3.11), where $Q(\tau)$, a function of transit time τ, is the oxygen extraction from a single capillary. Note that $Q(\tau)$ based on the assumption that the solubilities of oxygen in plasma and tissue are similar.

$$OEF^{max} = \int_0^\infty h(\tau)Q(\tau)d\tau \tag{3.11}$$

3.6.2 Vessel size and architecture imaging

As mentioned before (see Section 3.4.1.3), the simultaneous spin- and gradient-echo readouts allow the measurement of ΔR_2^* and ΔR_2 time series, enabling the estimation of vascular features extending beyond perfusion. Dennie *et al.* (1998) proposed the ratio of $\Delta R_2^* / \Delta R_2$ as a relative measure of the mean vessel size in a voxel. Later, Jensen *et al.* demonstrated that the index, Q, is a sensitive marker of microvascular density (Jensen and Chandra, 2000):

$$Q = \frac{\Delta R_2}{\Delta R_2^{*2/3}} \tag{3.12}$$

Tropes *et al.* proposed that quantitative measures of vessel size index (VSI) can be computed using (Tropres *et al.*, 2001):

$$VSI = 0.425 \left(\frac{ADC}{\gamma \Delta \chi B_0} \right)^{1/2} \left(\frac{\Delta R_2^*}{\Delta R_2} \right)^{3/2} \tag{3.13}$$

Here, ADC is the apparent diffusion coefficient, typically averaged over three directions with a diffusion-weighted imaging pulse sequence, γ is the gyromagnetic ratio of the protons, $\Delta \chi$ is the increased susceptibility difference between blood and the surrounding tissue due to the presence of contrast agent, and B_0 is the magnetic field strength. In practice, $\Delta \chi$ is measured *in vitro* (Boxerman *et al.*, 2016) or estimated *in vivo* by assuming a blood volume fraction in normal tissue (Lemasson *et al.*, 2013). It has been reported that vessel size measures accurately reflected histologic measures and, therefore, may be able to provide insight into the mechanism of PP and to differentiate tumoral angiogenesis seen in early tumor patients and treatment-induced change seen in PP (Fig. 3.5) (Chakhoyan *et al.*, 2018; Kim *et al.*, 2021).

Most recently, VAI has emerged as a new DSC-MR-based analysis approach for *in vivo* assessment of vascular architecture and oxygen saturation status (Emblem *et al.*, 2013; Zhang *et al.*, 2019b; Kim *et al.*, 2021). VAI relies on the relative shift in the shape and peak position of the relaxation rate curves from the GRE and SE echo signals. VAI has found utility in providing information about glioma grading, response to antiangiogenic therapy, and roadmap for *in vivo* and per-voxel determination of vascular status (Carrete *et al.*, 2022). To generate VAI maps, the pixel-wise GRE and SE dynamic signals are converted to R_2^* and R_2 changes, respectively, and fit to a Gaussian. The fitted curves are then plotted with $\Delta R_2^{3/2}$ along the *x*-axis and ΔR_2^* along the *y*-axis, forming hysteresis loops (Fig. 3.3). The slope and direction of hysteresis can provide estimates of vessel size and type, respectively. The positive (clockwise) and negative (anticlockwise) directions represent arterial and venous/capillary dominancy, respectively.

3.7 Applications

For more than two decades, researchers and clinicians have advocated DSC-MRI to noninvasively investigate the vascular structure, function, and integrity of neurologic disorders, stroke, and brain tumors. Conventional DSC-MRI is increasingly employed in clinical trials to assess early and more specific responses to therapy. This section discusses examples of advanced applications using recent pulse sequences and expanded biophysical and kinetic models that allow the derivation of novel biomarkers of cerebral structure and function.

3.7.1 DSC-MRI in brain tumors

DSC-MRI-derived measures of CBV are increasingly used in clinical practice, owing to its ability to reliably detect and differentiate the abnormal vascular characteristics in brain tumors (Liu *et al.*, 2017; Iv *et al.*, 2019; van Dijken *et al.*, 2019;

Connelly *et al.*, 2021). Multiple image-guided tissue histopathology studies, across sites and scanners, have demonstrated that tumor CBV accurately differentiates tumor recurrence and posttreatment radiation effects (Lund *et al.*, 2005; Hu *et al.*, 2012; Hoxworth *et al.*, 2020a,b). These studies formed the basis for a new DSC-MRI derived map, termed fractional tumor burden (FTB), which is defined as the volume fraction of tumor voxels above a specified rCBV threshold (M. Iv *et al.*, 2019; Connelly *et al.*, 2021). The FTB maps enable efficient visualization of regional tumor content and informing clinical decisions. Emblem *et al.* published *VAI* data in recurrent glioblastoma patients receiving antiangiogenic therapy with Cediranib (Emblem *et al.*, 2013). The vortex direction of the VAI parameters was reported to differentiate between responding (clockwise vortex direction increase with therapy) and nonresponding (clockwise vortex direction decrease with therapy) patients and predict progression-free survival as illustrated in Fig. 3.5. Unregulated tumor angiogenesis gives rise to subvoxel capillary transit time and flow heterogeneity due to increased vessel tortuosity, edema, and thrombosis. The ability of CTH, quantified as the standard deviation of the transit time distribution, to predict tumor grade was also reported recently (Tietze *et al.*, 2015; Bell *et al.*, 2020a,b). The combination of CTH and CBV data provided the best delineation of tumor grade (AUC = 0.88), as compared to CBV alone (AUC = 0.78).

3.7.2 DSC-MRI in acute stroke

MR perfusion imaging offers the prospective for measuring brain perfusion in acute stroke patients at a time when treatment decisions based on these measurements may influence the outcomes dramatically. Advances in both acute stroke therapy and perfusion imaging techniques have resulted in continuing reevaluation of the perfusion imaging aspects in patient management. The primary goal of DSC-MRI, or perfusion CT, in acute stroke is to identify patients who are candidates for intravenous tissue plasminogen activator (tPA) or endovascular therapy to restore blood flow. A recent meta-analysis published by Ryu *et al.* (2017), comprised of 994 patients with perfusion scans and 1819 without, reported that perfusion imaging improves the identification of patients who will benefit from combined intravenous tPA and endovascular thrombectomy. As a result of the approach involving perfusion imaging, patients had nearly twice the probability of achieving independent functional status at three months compared to those without perfusion scans (Ryu *et al.*, 2017). As mentioned in Section 3.6, OEF and CTH are perfusion metrics employed for exploring flow heterogeneity among patients with anterior circulation strokes. In the stroke patient study, parametric MTT and CTH maps clearly distinguished the hypoperfused tissue from surrounding, unaffected tissue and abnormal capillary heterogeneity provided prognostic information as well (Mouridsen *et al.*, 2014).

3.7.3 DSC-MRI in aging

Diseases of aging are often associated with impaired cerebral microvasculature and increased BBB permeability. In Alzheimer's disease (AD), mild cognitive impairment (MCI), and other forms of dementia, vascular changes are reported as contributing factors in the pathological cascade leading to cognitive decline. Cerebral hypoperfusion occurs early in the MCI trajectory, gradually leading to AD, preceding both structural brain changes and clinical symptoms (Binnewijzend et al., 2016; Chao et al., 2010). The advanced DSC-MRI methods not only provide information about hypoperfusion but also extend our understanding to age-related cerebrovascular changes by providing more sensitive biomarkers of pathology-specific vascular changes. The progression of AD is associated with alterations in microvessel density and capillary morphology, both assessed advanced DSC-MRI biomarkers (Section 3.6) (Østergaard et al., 2013). Using sequential GRE and SE DSC-MRI, declines in cognitive function and regional brain atrophy were associated with hypoperfusion measured in terms of reduced CBF, CBV, and elevated MTT and CTH (Nielsen et al., 2017). Overall, metrics that incorporate the range of vascular parameters afforded by DSC-MRI may provide additional diagnostic and prognostic information on cerebral microcirculatory insufficiencies in AD.

3.7.4 DSC-MRI in multiple sclerosis

Multiple sclerosis is an autoimmune disease characterized by neuroinflammation, eventually leading to neurodegeneration accompanied by axonal and neuronal loss. The MS lesions are known to predominantly develop around small central veins, preceding perivenular inflammatory infiltration (Lapointe et al., 2018). Several studies have shown normal appearing white matter (NAWM) hyperperfusion (Bester et al., 2015; Inglese et al., 2008; Sowa et al., 2015) and hypoperfusion (Bester et al., 2015) across brains of MS patients, reflecting the spatiotemporal dynamics of MS disease progression. The perfusion in gray matter lesions is reported to be altered compared to normal appearing gray matter (NAGM) (Inglese et al., 2008; Peruzzo et al., 2013). Recent studies have also suggested that while hypoperfusion reflects persistent low-grade inflammation, metabolic or vascular dysfunction, neuronal loss, or even primary ischemia, increased perfusion may indicate a high-inflammatory phase or increased metabolic activity. The perfusion changes in brain regions are found to be directly correlated with clinical disability (Adhya et al., 2006; Bester et al., 2015) and neuropsychological dysfunction (Francis et al., 2013; Inglese et al., 2008). A recently reported study done in 48 patients with relapsing remitting MS (RRMS) showed that SAGE-DSC-derived global and capillary-sized DSC-MRI parameters revealed decreased CBV and CBF and increased MTT in lesion ROIs accompanied by altered WM microstructural integrity (Sisco et al., 2021). Overall, these studies suggest that DSC-MRI biomarkers may play a critical role in revealing the underlying pathophysiology and etiology of MS.

3.8 Summary

The noninvasive assessment of morphological and functional features of the brain vasculature using DSC-MRI is already having a considerable impact on the care of patients with brain tumors and acute stroke. With recent consensus efforts for acquisition and analysis protocols, the role of DSC-MRI will continue to expand as multisite studies leveraging these protocols establish use-case guidelines for specific clinical challenges. While consensus recommendations represent current best practices, they are not intended to limit technological developments, and the field will certainly see continued innovation in pulse sequence design, analysis strategies, and the range of biomarkers that can be derived.

References

Adhya, S., Johnson, G., Herbert, J., Jaggi, H., Babb, J.S., Grossman, R.I., Inglese, M., 2006. Pattern of hemodynamic impairment in multiple sclerosis: dynamic susceptibility contrast perfusion MR imaging at 3.0 T. Neuroimage 33, 1029–1035. https://doi.org/10.1016/j.neuroimage.2006.08.008.

Aime, S., Botta, M., Fasano, M., Terreno, E., 1998. Lanthanide(III) chelates for NMR biomedical applications. Chem. Soc. Rev. 27, 19–29. https://doi.org/10.1039/A827019Z.

Angleys, H., Østergaard, L., Jespersen, S.N., 2015. The effects of capillary transit time heterogeneity (CTH) on brain oxygenation. J. Cereb. Blood Flow Metab. 35, 806–817.

Angleys, H., Jespersen, S.N., Østergaard, L., 2016. The effects of capillary transit time heterogeneity (CTH) on the cerebral uptake of glucose and glucose analogs: application to FDG and comparison to oxygen uptake. Front. Comput. Neurosci. 10, 103. https://doi.org/10.3389/fncom.2016.00103.

Axel, L., 1980. Cerebral blood flow determination by rapid-sequence computed tomography: theoretical analysis. Radiology 137, 679–686.

Bedekar, D., Jensen, T., Schmainda, K.M., 2010. Standardization of relative cerebral blood volume (rCBV) image maps for ease of both inter-and intrapatient comparisons. Magn. Reson. Med. 64, 907–913.

Bell, L.C., Semmineh, N., An, H., Eldeniz, C., Wahl, R., Schmainda, K.M., Prah, M.A., Erickson, B.J., Korfiatis, P., Wu, C., Sorace, A.G., Yankeelov, T.E., Rutledge, N., Chenevert, T.L., Malyarenko, D., Liu, Y., Brenner, A., Hu, L.S., Zhou, Y., Boxerman, J.L., Yen, Y.-F., Kalpathy-Cramer, J., Beers, A.L., Muzi, M., Madhuranthakam, A.J., Pinho, M., Johnson, B., Quarles, C.C., 2019. Evaluating multisite rCBV consistency from DSC-MRI imaging protocols and postprocessing software across the NCI quantitative imaging network sites using a digital reference object (DRO). Tomography 5, 110–117. https://doi.org/10.18383/j.tom.2018.00041.

Bell, L.C., Semmineh, N., An, H., Eldeniz, C., Wahl, R., Schmainda, K.M., Prah, M.A., Erickson, B.J., Korfiatis, P., Wu, C., Sorace, A.G., Yankeelov, T.E., Rutledge, N., Chenevert, T.L., Malyarenko, D., Liu, Y., Brenner, A., Hu, L.S., Zhou, Y., Boxerman, J.L., Yen, Y.-F., Kalpathy-Cramer, J., Beers, A.L., Muzi, M., Madhuranthakam, A.J., Pinho, M., Johnson, B., Quarles, C.C., 2020a. Evaluating the use of rCBV as a tumor grade and treatment response classifier across NCI quantitative imaging network sites: Part II of the DSC-MRI digital reference object (DRO) challenge. Tomography 6, 203–208. https://doi.org/10.18383/j.tom.2020.00012.

Bell, L.C., Stokes, A.M., Quarles, C.C., 2020b. Analysis of postprocessing steps for residue function dependent dynamic susceptibility contrast (DSC)-MRI biomarkers and their clinical impact on glioma grading for both 1.5 and 3T. J. Magn. Reson. Imaging 51, 547–553. https://doi.org/10.1002/jmri.26837.

Belorizky, E., Fries, P.H., Helm, L., Kowalewski, J., Kruk, D., Sharp, R.R., Westlund, P.-O., 2008. Comparison of different methods for calculating the paramagnetic relaxation enhancement of nuclear spins as a function of the magnetic field. J. Chem. Phys. 128, 052315. https://doi.org/10.1063/1.2833957.

Bester, M., Forkert, N.D., Stellmann, J.P., Aly, L., Drabik, A., Young, K.L., Heesen, C., Fiehler, J., Siemonsen, S., 2015. Increased perfusion in normal appearing white matter in high inflammatory multiple sclerosis patients. PloS One 10, e0119356.

Bilgic, B., Gagoski, B.A., Cauley, S.F., Fan, A.P., Polimeni, J.R., Grant, P.E., Wald, L.L., Setsompop, K., 2015. Wave-CAIPI for highly accelerated 3D imaging. Magn. Reson. Med. 73, 2152–2162. https://doi.org/10.1002/mrm.25347.

Bilgic, B., Kim, T.H., Liao, C., Manhard, M.K., Wald, L.L., Haldar, J.P., Setsompop, K., 2018. Improving parallel imaging by jointly reconstructing multi-contrast data. Magn. Reson. Med. 80, 619–632.

Binnewijzend, M.A.A., Benedictus, M.R., Kuijer, J.P.A., van der Flier, W.M., Teunissen, C.E., Prins, N.D., Wattjes, M.P., van Berckel, B.N.M., Scheltens, P., Barkhof, F., 2016. Cerebral perfusion in the predementia stages of Alzheimer's disease. Eur. Radiol. 26, 506–514. https://doi.org/10.1007/s00330-015-3834-9.

Bonekamp, D., Degaonkar, M., Barker, P.B., 2011. Quantitative cerebral blood flow in dynamic susceptibility contrast MRI using total cerebral flow from phase contrast magnetic resonance angiography. Magn. Reson. Med. 66, 57–66.

Boxerman, J.L., Hamberg, L.M., Rosen, B.R., Weisskoff, R.M., 1995. MR contrast due to intravascular magnetic susceptibility perturbations. Magn. Reson. Med. 34, 555–566.

Boxerman, J.L., Schmainda, K.M., Weisskoff, R.M., 2006. Relative cerebral blood volume maps corrected for contrast agent extravasation significantly correlate with glioma tumor grade, whereas uncorrected maps do not. Am. J. Neuroradiol. 27, 859.

Boxerman, J.L., Shiroishi, M.S., Ellingson, B.M., Pope, W.B., 2016. Dynamic susceptibility contrast MR imaging in glioma: review of current clinical practice. Magn. Reson. Imaging Clin. N. Am. 24 (4), 649–670. https://doi.org/10.1016/j.mric.2016.06.005. Epub 2016.

Boxerman, J.L., Quarles, C.C., Hu, L.S., Erickson, B.J., Gerstner, E.R., Smits, M., Kaufmann, T.J., Barboriak, D.P., Huang, R.H., Wick, W., Weller, M., Galanis, E., Kalpathy-Cramer, J., Shankar, L., Jacobs, P., Chung, C., van den Bent, M.J., Chang, S., Al Yung, W.K., Cloughesy, T.F., Wen, P.Y., Gilbert, M.R., Rosen, B.R., Ellingson, B.M., Schmainda, K.M., 2020. Consensus recommendations for a dynamic susceptibility contrast MRI protocol for use in high-grade gliomas. Neuro Oncol. 22, 1262–1275. https://doi.org/10.1093/neuonc/noaa141.

Calamante, F., Gadian, D.G., Connelly, A., 2003. Quantification of bolus-tracking MRI: improved characterization of the tissue residue function using Tikhonov regularization. Magn. Reson. Med. 50, 1237–1247. https://doi.org/10.1002/mrm.10643.

Calamante, F., Connelly, A., van Osch, M.J., 2009. Nonlinear ΔR effects in perfusion quantification using bolus-tracking MRI. Magn. Reson. Med. 61, 486–492.

Caravan, P., Ellison, J.J., McMurry, T.J., Lauffer, R.B., 1999. Gadolinium(III) chelates as MRI contrast agents: structure, dynamics, and applications. Chem. Rev. 99, 2293–2352. https://doi.org/10.1021/cr980440x.

Carrete, L.R., Young, J.S., Cha, S., 2022. Advanced imaging techniques for newly diagnosed and recurrent gliomas. Front. Neurosci. 16, 787755. https://doi.org/10.3389/fnins.2022.787755.

Chakhoyan, A., Leu, K., Pope, W.B., Cloughesy, T.F., Ellingson, B.M., 2018. Improved spatiotemporal resolution of dynamic susceptibility contrast perfusion MRI in brain tumors using simultaneous multi-slice echo-planar imaging. AJNR Am. J. Neuroradiol. 39, 43–45. https://doi.org/10.3174/ajnr.A5433.

Chakravorty, A., Steel, T., Chaganti, J., 2015. Accuracy of percentage of signal intensity recovery and relative cerebral blood volume derived from dynamic susceptibility-weighted, contrast-enhanced MRI in the preoperative diagnosis of cerebral tumours. Neuroradiol. J. 28, 574–583. https://doi.org/10.1177/1971400915611916.

Chakwizira, A., Ahlgren, A., Knutsson, L., Wirestam, R., 2022. Non-parametric deconvolution using Bézier curves for quantification of cerebral perfusion in dynamic susceptibility contrast MRI. MAGMA 35, 791–804. https://doi.org/10.1007/s10334-021-00995-0.

Chang, H., Fitzpatrick, J.M., 1992. A technique for accurate magnetic resonance imaging in the presence of field inhomogeneities. IEEE Trans. Med. Imaging 11, 319–329.

Chao, L.L., Buckley, S.T., Kornak, J., Schuff, N., Madison, C., Yaffe, K., Miller, B.L., Kramer, J.H., Weiner, M.W., 2010. ASL perfusion MRI predicts cognitive decline and conversion from MCI to dementia. Alzheimer Dis. Assoc. Disord. 24, 19–27. https://doi.org/10.1097/WAD.0b013e3181b4f736.

Cho, N.S., Hagiwara, A., Sanvito, F., Ellingson, B.M., 2022. A multi-reader comparison of normal-appearing white matter normalization techniques for perfusion and diffusion MRI in brain tumors. Neuroradiology. https://doi.org/10.1007/s00234-022-03072-y.

Connelly, J.M., Prah, M.A., Santos-Pinheiro, F., Mueller, W., Cochran, E., Schmainda, K.M., 2021. Magnetic resonance imaging mapping of brain tumor burden: clinical implications for neurosurgical management: case report. Neurosurg Open 2, okab029. https://doi.org/10.1093/neuopn/okab029.

d'Arcy, J., Collins, D., Rowland, I., Padhani, A., Leach, M., 2002. Applications of sliding window reconstruction with cartesian sampling for dynamic contrast enhanced MRI. NMR Biomed. 15, 174–183.

Daldrup-Link, H.E., 2017. Ten things you might not know about Iron oxide nanoparticles. Radiology 284, 616–629. https://doi.org/10.1148/radiol.2017162759.

Dennie, J., Mandeville, J.B., Boxerman, J.L., Packard, S.D., Rosen, B.R., Weisskoff, R.M., 1998. NMR imaging of changes in vascular morphology due to tumor angiogenesis. Magn. Reson. Med. 40, 793–799.

Donahue, K.M., Krouwer, H.G., Rand, S.D., Pathak, A.P., Marszalkowski, C.S., Censky, S.C., Prost, R.W., 2000. Utility of simultaneously acquired gradient-echo and spin-echo cerebral blood volume and morphology maps in brain tumor patients. Magn. Reson. Med. 43, 845–853. https://doi.org/10.1002/1522-2594(200006)43:6<845::aid-mrm10>3.0.co;2-j.

Eichner, C., Jafari-Khouzani, K., Cauley, S., Bhat, H., Polaskova, P., Andronesi, O.C., Rapalino, O., Turner, R., Wald, L.L., Stufflebeam, S., 2014. Slice accelerated gradient-echo spin-echo dynamic susceptibility contrast imaging with blipped CAIPI for increased slice coverage. Magn. Reson. Med. 72, 770–778.

Emblem, K.E., Mouridsen, K., Bjornerud, A., Farrar, C.T., Jennings, D., Borra, R.J.H., Wen, P.Y., Ivy, P., Batchelor, T.T., Rosen, B.R., Jain, R.K., Sorensen, A.G., 2013. Vessel architectural imaging identifies cancer patient responders to anti-angiogenic therapy. Nat. Med. 19, 1178–1183. https://doi.org/10.1038/nm.3289.

Essig, M., Shiroishi, M.S., Nguyen, T.B., Saake, M., Provenzale, J.M., Enterline, D., Anzalone, N., Dörfler, A., Rovira, A., Wintermark, M., Law, M., 2013. Perfusion MRI: the five most frequently asked technical questions. AJR Am. J. Roentgenol. 200, 24–34. https://doi.org/10.2214/AJR.12.9543.

Francis, P.L., Jakubovic, R., O'Connor, P., Zhang, L., Eilaghi, A., Lee, L., Carroll, T.J., Mouannes-Srour, J., Feinstein, A., Aviv, R.I., 2013. Robust perfusion deficits in cognitively impaired patients with secondary-progressive multiple sclerosis. AJNR Am. J. Neuroradiol. 34, 62–67. https://doi.org/10.3174/ajnr.A3148.

Frøhlich, A.F., Østergaard, L., Kiselev, V.G., 2005. Theory of susceptibility-induced transverse relaxation in the capillary network in the diffusion narrowing regime. Magn. Reson. Med. 53, 564–573. https://doi.org/10.1002/mrm.20394.

Gobbel, G.T., Fike, J.R., 1994. A deconvolution method for evaluating indicator-dilution curves. Phys. Med. Biol. 39, 1833–1854. https://doi.org/10.1088/0031-9155/39/11/004.

Henriksen, O.M., del Mar Álvarez-Torres, M., Figueiredo, P., Hangel, G., Keil, V.C., Nechifor, R.E., Riemer, F., Schmainda, K.M., Warnert, E.A.H., Wiegers, E.C., Booth, T.C., 2022. High-grade glioma treatment response monitoring biomarkers: a position statement on the evidence supporting the use of advanced MRI techniques in the clinic, and the latest bench-to-bedside developments. Part 1: perfusion and diffusion techniques. Front. Oncol. 12.

Hoxworth, J.M., Eschbacher, J.M., Gonzales, A.C., Singleton, K.W., Leon, G.D., Smith, K.A., Stokes, A.M., Zhou, Y., Mazza, G.L., Porter, A.B., Mrugala, M.M., Zimmerman, R.S., Bendok, B.R., Patra, D.P., Krishna, C., Boxerman, J.L., Baxter, L.C., Swanson, K.R., Quarles, C.C., Schmainda, K.M., Hu, L.S., 2020a. Performance of standardized relative CBV for quantifying regional histologic tumor burden in recurrent high-grade glioma: comparison against normalized relative CBV using image-localized stereotactic biopsies. Am. J. Neuroradiol. https://doi.org/10.3174/ajnr.A6486.

Hoxworth, J.M., Eschbacher, J.M., Gonzales, A.C., Singleton, K.W., Leon, G.D., Smith, K.A., Stokes, A.M., Zhou, Y., Mazza, G.L., Porter, A.B., Mrugala, M.M., Zimmerman, R.S., Bendok, B.R., Patra, D.P., Krishna, C., Boxerman, J.L., Baxter, L.C., Swanson, K.R., Quarles, C.C., Schmainda, K.M., Hu, L.S., 2020b. Performance of standardized relative CBV for quantifying regional histologic tumor burden in recurrent high-grade glioma: comparison against normalized relative CBV using image-localized stereotactic biopsies. AJNR Am. J. Neuroradiol. 41, 408–415. https://doi.org/10.3174/ajnr.A6486.

Hu, L.S., Baxter, L.C., Pinnaduwage, D.S., Paine, T.L., Karis, J.P., Feuerstein, B.G., Schmainda, K.M., Dueck, A.C., Debbins, J., Smith, K.A., Nakaji, P., Eschbacher, J.M., Coons, S.W., Heiserman, J.E., 2010. Optimized preload leakage-correction methods to improve the diagnostic accuracy of dynamic susceptibility-weighted contrast-enhanced perfusion MR imaging in posttreatment gliomas. Am. J. Neuroradiol. 31 (1), 40–48. https://doi.org/10.3174/ajnr.A1787. Epub 2009 Sep 12. PMID: 19749223. PMCID: PMC4323177.

Hu, L.S., Eschbacher, J.M., Heiserman, J.E., Dueck, A.C., Shapiro, W.R., Liu, S., Karis, J.P., Smith, K.A., Coons, S.W., Nakaji, P., Spetzler, R.F., Feuerstein, B.G., Debbins, J., Baxter, L.C., 2012. Reevaluating the imaging definition of tumor progression: perfusion MRI quantifies recurrent glioblastoma tumor fraction, pseudoprogression, and radiation necrosis to predict survival. Neuro Oncol. 14, 919–930. https://doi.org/10.1093/neuonc/nos112.

Hu, L.S., Kelm, Z., Korfiatis, P., Dueck, A.C., Elrod, C., Ellingson, B.M., Kaufmann, T.J., Eschbacher, J.M., Karis, J.P., Smith, K., Nakaji, P., Brinkman, D., Pafundi, D., Baxter, L.C., Erickson, B.J., 2015. Impact of software modeling on the accuracy of perfusion MRI in glioma. AJNR Am. J. Neuroradiol. 36, 2242–2249. https://doi.org/10.3174/ajnr.A4451.

Huang, F., Akao, J., Vijayakumar, S., Duensing, G.R., Limkeman, M., 2005. k-t GRAPPA. a k-space implementation for dynamic MRI with high reduction factor. Magn. Reson. Med. 54, 1172–1184.

Inglese, M., Adhya, S., Johnson, G., Babb, J.S., Miles, L., Jaggi, H., Herbert, J., Grossman, R.I., 2008. Perfusion magnetic resonance imaging correlates of neuropsychological impairment in multiple sclerosis. J. Cereb. Blood Flow Metab. 28, 164–171. https://doi.org/10.1038/sj.jcbfm.9600504.

Iv, M., Liu, X., Lavezo, J., Gentles, A.J., Ghanem, R., Lummus, S., Born, D.E., Soltys, S.G., Nagpal, S., Thomas, R., Recht, L., Fischbein, N., 2019. Perfusion MRI-based fractional tumor burden differentiates between tumor and treatment effect in recurrent glioblastomas and informs clinical decision-making. AJNR Am. J. Neuroradiol. 40, 1649–1657. https://doi.org/10.3174/ajnr.A6211.

Jensen, J., Chandra, R., 2000. MR imaging of microvasculature. Magn. Reson. Med. 44, 224–230.

Jespersen, S.N., Østergaard, L., 2012. The roles of cerebral blood flow, capillary transit time heterogeneity, and oxygen tension in brain oxygenation and metabolism. J. Cereb. Blood Flow Metab. 32, 264–277.

Jochimsen, T.H., Newbould, R.D., Skare, S.T., Clayton, D.B., Albers, G.W., Moseley, M.E., Bammer, R., 2007. Identifying systematic errors in quantitative dynamic-susceptibility contrast perfusion imaging by high-resolution multi-echo parallel EPI. NMR Biomed. 20, 429–438. https://doi.org/10.1002/nbm.1107.

Johnson, G., Wetzel, S.G., Cha, S., Babb, J., Tofts, P.S., 2004. Measuring blood volume and vascular transfer constant from dynamic, T(2)*-weighted contrast-enhanced MRI. Magn. Reson. Med. 51 (5), 961–968. https://doi.org/10.1002/mrm.20049. PMID: 15122678.

Jonathan, S.V., Vakil, P., Jeong, Y., Ansari, S., Hurley, M., Bendok, B., Carroll, T.J., 2013. A radial 3D GRE-EPI pulse sequence with kz blip encoding for whole-brain isotropic 3D perfusion using DSC-MRI bolus tracking with sliding window reconstruction (3D RAZIR). In: Presented at the Proceedings of the 21st Annual Meeting of ISMRM, p. 582.

Jonathan, S.V., Vakil, P., Jeong, Y.I., Menon, R.G., Ansari, S.A., Carroll, T.J., 2014. RAZER: a pulse sequence for whole-brain bolus tracking at high frame rates. Magn. Reson. Med. 71, 2127–2138. https://doi.org/10.1002/mrm.24882.

Jung, B., Ullmann, P., Honal, M., Bauer, S., Hennig, J., Markl, M., 2008. Parallel MRI with extended and averaged GRAPPA kernels (PEAK-GRAPPA): optimized spatiotemporal dynamic imaging. J. Magn. Reson. Imaging 28, 1226–1232.

Kim, J., Leira, E.C., Callison, R.C., Ludwig, B., Moritani, T., Magnotta, V.A., Madsen, M.T., 2010. Toward fully automated processing of dynamic susceptibility contrast perfusion MRI for acute ischemic cerebral stroke. Comput. Methods Programs Biomed. 98, 204–213. https://doi.org/10.1016/j.cmpb.2009.12.005.

Kim, M., Park, J.E., Emblem, K., Bjørnerud, A., Kim, H.S., 2021. Vessel type determined by vessel architectural imaging improves differentiation between early tumor progression and pseudoprogression in glioblastoma. AJNR Am. J. Neuroradiol. 42, 663–670. https://doi.org/10.3174/ajnr.A6984.

King, M.D., Calamente, F., Clark, C.A., Gadian, D.G., 2011. Markov chain Monte Carlo random effects modeling in magnetic resonance image processing using the BRugs interface to WinBUGS. J. Stat. Softw. 44, 1–23.

Kiselev, V.G., 2001. On the theoretical basis of perfusion measurements by dynamic susceptibility contrast MRI. Magn. Reson. Med. 46, 1113–1122. https://doi.org/10.1002/mrm.1307.

Kiselev, V.G., Novikov, D.S., 2018. Transverse NMR relaxation in biological tissues. NeuroImage 182, 149–168. https://doi.org/10.1016/j.neuroimage.2018.06.002.

Kiselev, V.G., Posse, S., 1998. Analytical theory of susceptibility induced NMR signal dephasing in a cerebrovascular network. Phys. Rev. Lett. 81, 5696.

Kiselev, V., Posse, S., 1999. Analytical model of susceptibility-induced MR signal dephasing: effect of diffusion in a microvascular network. Magn. Reson. Med. 41, 499–509.

Kjølby, B.F., Østergaard, L., Kiselev, V.G., 2006. Theoretical model of intravascular paramagnetic tracers effect on tissue relaxation. Magn. Reson. Med. 56, 187–197. https://doi.org/10.1002/mrm.20920.

Kluge, A., Lukas, M., Toth, V., Pyka, T., Zimmer, C., Preibisch, C., 2016. Analysis of three leakage-correction methods for DSC-based measurement of relative cerebral blood volume with respect to heterogeneity in human gliomas. Magn. Reson. Imaging 34, 410–421.

Knobloch, G., Colgan, T., Wiens, C.N., Wang, X., Schubert, T., Hernando, D., Sharma, S.D., Reeder, S.B., 2018. Relaxivity of Ferumoxytol at 1.5 T and 3.0 T. Invest. Radiol. 53, 257–263. https://doi.org/10.1097/RLI.0000000000000434.

Knopp, E.A., Cha, S., Johnson, G., Mazumdar, A., Golfinos, J.G., Zagzag, D., Miller, D.C., Kelly, P.J., Kricheff, I.I., 1999. Glial neoplasms: dynamic contrast-enhanced T2*-weighted MR imaging. Radiology 211, 791–798. https://doi.org/10.1148/radiology.211.3.r99jn46791.

Knutsson, L., Ståhlberg, F., Wirestam, R., 2004. Aspects on the accuracy of cerebral perfusion parameters obtained by dynamic susceptibility contrast MRI: a simulation study. Magn. Reson. Imaging 22, 789–798.

Kosior, J.C., Frayne, R., 2010. Perfusion parameters derived from bolus-tracking perfusion imaging are immune to tracer recirculation. J. Magn. Reson. Imaging 31, 753–756. https://doi.org/10.1002/jmri.22052.

Kosior, J.C., Kosior, R.K., Frayne, R., 2007a. Robust dynamic susceptibility contrast MR perfusion using 4D nonlinear noise filters. J. Magn. Reson. Imaging 26, 1514–1522.

Kosior, R.K., Kosior, J.C., Frayne, R., 2007b. Improved dynamic susceptibility contrast (DSC)-MR perfusion estimates by motion correction. J. Magn. Reson. Imaging 26, 1167–1172. https://doi.org/10.1002/jmri.21128.

Kuschinsky, W., Paulson, O., 1992. Capillary circulation in the brain. Cerebrovasc. Brain Metab. Rev. 4, 261–286.

Lapointe, E., Li, D.K.B., Traboulsee, A.L., Rauscher, A., 2018. What have we learned from perfusion MRI in multiple sclerosis? AJNR Am. J. Neuroradiol. 39, 994–1000. https://doi.org/10.3174/ajnr.A5504.

Leiner, T., Kucharczyk, W., 2009. NSF prevention in clinical practice: summary of recommendations and guidelines in the United States, Canada, and Europe. J. Magn. Reson. Imaging 30, 1357–1363. https://doi.org/10.1002/jmri.22021.

Lemasson, B., Valable, S., Farion, R., Krainik, A., Rémy, C., Barbier, E.L., 2013. In vivo imaging of vessel diameter, size, and density: a comparative study between MRI and histology. Magn. Reson. Med. 69, 18–26. https://doi.org/10.1002/mrm.24218.

Leu, K., Boxerman, J.L., Cloughesy, T.F., Lai, A., Nghiemphu, P.L., Liau, L.M., Pope, W.B., Ellingson, B.M., 2016. Improved leakage correction for single-echo dynamic susceptibility contrast perfusion MRI estimates of relative cerebral blood volume in high-grade gliomas by accounting for bidirectional contrast agent exchange. Am. J. Neuroradiol. 37, 1440–1446.

Liu, G., Sobering, G., Duyn, J., Moonen, C.T., 1993. A functional MRI technique combining principles of echo-shifting with a train of observations (PRESTO). Magn. Reson. Med. 30, 764–768.

Liu, T.T., Achrol, A.S., Mitchell, L.A., Rodriguez, S.A., Feroze, A., Iv, M., Kim, C., Chaudhary, N., Gevaert, O., Stuart, J.M., Harsh, G.R., Chang, S.D., Rubin, D.L., 2017. Magnetic resonance perfusion image features uncover an angiogenic subgroup of glioblastoma patients with poor survival and better response to antiangiogenic treatment. Neuro Oncol. 19, 997–1007. https://doi.org/10.1093/neuonc/now270.

Lorenz, C., Benner, T., Chen, P.J., Lopez, C.J., Ay, H., Zhu, M.W., Menezes, N.M., Aronen, H., Karonen, J., Liu, Y., Nuutinen, J., Sorensen, A.G., 2006. Automated perfusion-weighted MRI using localized arterial input functions. J. Magn. Reson. Imaging 24, 1133–1139. https://doi.org/10.1002/jmri.20717.

Lund, R., Rand, S., Krouwer, H., Schultz, C., Schmainda, K., 2005. Using rCBV to distinguish radiation necrosis from tumor recurrence in malignant gliomas. Int. J. Radiat. Oncol. Biol. Phys. 63, S65–S66. https://doi.org/10.1016/j.ijrobp.2005.07.114.

Lysaker, M., Lundervold, A., Tai, X.-C., 2003. Noise removal using fourth-order partial differential equation with applications to medical magnetic resonance images in space and time. IEEE Trans. Image Process. 12, 1579–1590.

Madabhushi, A., Udupa, J.K., 2006. New methods of MR image intensity standardization via generalized scale. Med. Phys. 33, 3426–3434. https://doi.org/10.1118/1.2335487.

Mangla, R., Kolar, B., Zhu, T., Zhong, J., Almast, J., Ekholm, S., 2011. Percentage signal recovery derived from MR dynamic susceptibility contrast imaging is useful to differentiate common enhancing malignant lesions of the brain. Am. J. Neuroradiol. 32, 1004. https://doi.org/10.3174/ajnr.A2441.

Maral, H., Ertekin, E., Tunçyürek, Ö., Özsunar, Y., 2020. Effects of susceptibility artifacts on perfusion MRI in patients with primary brain tumor: a comparison of arterial spin-labeling versus DSC. AJNR Am. J. Neuroradiol. 41, 255–261. https://doi.org/10.3174/ajnr.A6384.

Mehndiratta, A., MacIntosh, B.J., Crane, D.E., Payne, S.J., Chappell, M.A., 2013. A control point interpolation method for the non-parametric quantification of cerebral haemodynamics from dynamic susceptibility contrast MRI. Neuroimage 64, 560–570. https://doi.org/10.1016/j.neuroimage.2012.08.083.

Meier, P., Zierler, K.L., 1954. On the theory of the indicator-dilution method for measurement of blood flow and volume. J. Appl. Physiol. 6, 731–744.

Mokli, Y., Pfaff, J., Dos Santos, D.P., Herweh, C., Nagel, S., 2019. Computer-aided imaging analysis in acute ischemic stroke–background and clinical applications. Neurol. Res. Pract. 1, 1–13.

Mouridsen, K., Friston, K., Hjort, N., Gyldensted, L., Østergaard, L., Kiebel, S., 2006. Bayesian estimation of cerebral perfusion using a physiological model of microvasculature. Neuroimage 33, 570–579. https://doi.org/10.1016/j.neuroimage.2006.06.015.

Mouridsen, K., Hansen, M.B., Østergaard, L., Jespersen, S.N., 2014. Reliable estimation of capillary transit time distributions using DSC-MRI. J. Cereb. Blood Flow Metab. 34, 1511–1521. https://doi.org/10.1038/jcbfm.2014.111.

Newbould, R.D., Skare, S.T., Jochimsen, T.H., Alley, M.T., Moseley, M.E., Albers, G.W., Bammer, R., 2007. Perfusion mapping with multiecho multishot parallel imaging EPI. Magn. Reson. Med. 58, 70–81. https://doi.org/10.1002/mrm.21255.

Nguyen, K.-L., Yoshida, T., Kathuria-Prakash, N., Zaki, I.H., Varallyay, C.G., Semple, S.I., Saouaf, R., Rigsby, C.K., Stoumpos, S., Whitehead, K.K., Griffin, L.M., Saloner, D., Hope, M.D., Prince, M.R., Fogel, M.A., Schiebler, M.L., Roditi, G.H., Radjenovic, A., Newby, D.E., Neuwelt, E.A., Bashir, M.R., Hu, P., Finn, J.P., 2019. Multicenter safety and practice for off-label diagnostic use of ferumoxytol in MRI. Radiology 293, 554–564. https://doi.org/10.1148/radiol.2019190477.

Nielsen, R.B., Egefjord, L., Angleys, H., Mouridsen, K., Gejl, M., Møller, A., Brock, B., Brændgaard, H., Gottrup, H., Rungby, J., Eskildsen, S.F., Østergaard, L., 2017. Capillary dysfunction is associated with symptom severity and neurodegeneration in Alzheimer's disease. Alzheimers Dement. 13, 1143–1153. https://doi.org/10.1016/j.jalz.2017.02.007.

Oesterle, C., Strohschein, R., Köhler, M., Schnell, M., Hennig, J., 2000. Benefits and pitfalls of keyhole imaging, especially in first-pass perfusion studies. J. Magn. Reson. Imaging 11, 312–323.

Østergaard, L., Sorensen, A.G., Kwong, K.K., Weisskoff, R.M., Gyldensted, C., Rosen, B.R., 1996. High resolution measurement of cerebral blood flow using intravascular tracer bolus passages. Part II: experimental comparison and preliminary results. Magn. Reson. Med. 36, 726–736.

Ostergaard, L., Smith, D.F., Vestergaard-Poulsen, P., Hansen, S.B., Gee, A.D., Gjedde, A., Gyldensted, C., 1998. Absolute cerebral blood flow and blood volume measured by magnetic resonance imaging bolus tracking: comparison with positron emission tomography values. J. Cereb. Blood Flow Metab. 18, 425–432. https://doi.org/10.1097/00004647-199804000-00011.

Østergaard, L., Sorensen, A.G., Chesler, D.A., Weisskoff, R.M., Koroshetz, W.J., Wu, O., Gyldensted, C., Rosen, B.R., 2000. Combined diffusion-weighted and perfusion-weighted flow heterogeneity magnetic resonance imaging in acute stroke. Stroke 31, 1097–1103.

Østergaard, L., Aamand, R., Gutiérrez-Jiménez, E., Ho, Y.-C.L., Blicher, J.U., Madsen, S.M., Nagenthiraja, K., Dalby, R.B., Drasbek, K.R., Møller, A., Brændgaard, H., Mouridsen, K., Jespersen, S.N., Jensen, M.S., West, M.J., 2013. The capillary dysfunction hypothesis of Alzheimer's disease. Neurobiol. Aging 34, 1018–1031. https://doi.org/10.1016/j.neurobiolaging.2012.09.011.

Pathak, A.P., Rand, S.D., Schmainda, K.M., 2003. The effect of brain tumor angiogenesis on the in vivo relationship between the gradient-echo relaxation rate change ($\Delta R2^*$) and contrast agent (MION) dose. J. Magn. Reson. Imaging 18, 397–403. https://doi.org/10.1002/jmri.10371.

Paulson, E.S., Schmainda, K.M., 2008. Comparison of dynamic susceptibility-weighted contrast-enhanced MR methods: recommendations for measuring relative cerebral blood volume in brain tumors. Radiology 249, 601–613. https://doi.org/10.1148/radiol.2492071659.

Paulson, E.S., Prah, D.E., Schmainda, K.M., 2016. Spiral perfusion imaging with consecutive echoes (SPICE™) for the simultaneous mapping of DSC- and DCE-MRI parameters in brain tumor patients: theory and initial feasibility. Tomography 2, 295–307. https://doi.org/10.18383/j.tom.2016.00217.

Pedersen, M., Klarhöfer, M., Christensen, S., Ouallet, J.-C., Østergaard, L., Dousset, V., Moonen, C., 2004. Quantitative cerebral perfusion using the PRESTO acquisition scheme. J. Magn. Reson. Imaging 20, 930–940. https://doi.org/10.1002/jmri.20206.

Peruzzo, D., Castellaro, M., Calabrese, M., Veronese, E., Rinaldi, F., Bernardi, V., Favaretto, A., Gallo, P., Bertoldo, A., 2013. Heterogeneity of cortical lesions in multiple sclerosis: an MRI perfusion study. J. Cereb. Blood Flow Metab. 33, 457–463. https://doi.org/10.1038/jcbfm.2012.192.

Polak, D., Setsompop, K., Cauley, S.F., Gagoski, B.A., Bhat, H., Maier, F., Bachert, P., Wald, L.L., Bilgic, B., 2018. Wave-CAIPI for highly accelerated MP-RAGE imaging. Magn. Reson. Med. 79, 401–406. https://doi.org/10.1002/mrm.26649.

Prah, M., Stufflebeam, S., Paulson, E., Kalpathy-Cramer, J., Gerstner, E., Batchelor, T., Barboriak, D., Rosen, B., Schmainda, K., 2015. Repeatability of standardized and normalized relative CBV in patients with newly diagnosed glioblastoma. Am. J. Neuroradiol. 36, 1654–1661.

Pruessmann, K.P., Weiger, M., Scheidegger, M.B., Boesiger, P., 1999. SENSE: sensitivity encoding for fast MRI. Magn. Reson. Med. 42, 952–962.

QIBA_DSC-MRI_Stage1_Profile.pdf, n.d. QIBA_DSC-MRI_Stage1_Profile.pdf.

Quarles, C.C., Gore, J.C., Xu, L., Yankeelov, T.E., 2012. Comparison of dual-echo DSC-MRI- and DCE-MRI-derived contrast agent kinetic parameters. Magn. Reson. Imaging 30, 944–953. https://doi.org/10.1016/j.mri.2012.03.008.

Rasmussen, P.M., Jespersen, S.N., Østergaard, L., 2015. The effects of transit time heterogeneity on brain oxygenation during rest and functional activation. J. Cereb. Blood Flow Metab. 35, 432–442.

Rempp, K.A., Brix, G., Wenz, F., Becker, C.R., Gückel, F., Lorenz, W.J., 1994. Quantification of regional cerebral blood flow and volume with dynamic susceptibility contrast-enhanced MR imaging. Radiology 193, 637–641.

Ryu, W.H.A., Avery, M.B., Dharampal, N., Allen, I.E., Hetts, S.W., 2017. Utility of perfusion imaging in acute stroke treatment: a systematic review and meta-analysis. J. Neurointerv. Surg. 9, 1012–1016. https://doi.org/10.1136/neurintsurg-2016-012751.

Schmainda, K., Prah, M., Hu, L., Quarles, C., Semmineh, N., Rand, S., Connelly, J., Anderies, B., Zhou, Y., Liu, Y., Logan, B., Stokes, A., Baird, G., Boxerman, J., 2019. Moving toward a consensus DSC-MRI protocol: validation of a low-flip angle single-dose option as a reference standard for brain tumors. AJNR Am. J. Neuroradiol. 40, 626–633. https://doi.org/10.3174/ajnr.A6015.

Schmiedeskamp, H., Straka, M., Bammer, R., 2012a. Compensation of slice profile mismatch in combined spin- and gradient-echo echo-planar imaging pulse sequences. Magn. Reson. Med. 67, 378–388. https://doi.org/10.1002/mrm.23012.

Schmiedeskamp, H., Straka, M., Newbould, R.D., Zaharchuk, G., Andre, J.B., Olivot, J.-M., Moseley, M.E., Albers, G.W., Bammer, R., 2012b. Combined spin- and gradient-echo perfusion-weighted imaging. Magn. Reson. Med. 68, 30–40. https://doi.org/10.1002/mrm.23195.

Semmineh, N.B., Stokes, A.M., Bell, L.C., Boxerman, J.L., Quarles, C.C., 2017. A population-based digital reference object (DRO) for optimizing dynamic susceptibility contrast (DSC)-MRI methods for clinical trials. Tomography 3, 41–49.

Sisco, N.J., Borazanci, A., Dortch, R., Stokes, A.M., 2021. Investigating the relationship between multi-scale perfusion and white matter microstructural integrity in patients with relapsing-remitting MS. Mult. Scler. J. Exp. Transl. Clin. 7, 20552173211037000. https://doi.org/10.1177/20552173211037002.

Skinner, J.T., Robison, R.K., Elder, C.P., Newton, A.T., Damon, B.M., Quarles, C.C., 2014. Evaluation of a multiple spin- and gradient-echo (SAGE) EPI acquisition with SENSE acceleration: applications for perfusion imaging in and outside the brain. Magn. Reson. Imaging 32, 1171–1180. https://doi.org/10.1016/j.mri.2014.08.032.

Sourbron, S., Dujardin, M., Makkat, S., Luypaert, R., 2007. Pixel-by-pixel deconvolution of bolus-tracking data: optimization and implementation. Phys. Med. Biol. 52, 429–447. https://doi.org/10.1088/0031-9155/52/2/009.

Sowa, P., Bjørnerud, A., Nygaard, G.O., Damangir, S., Spulber, G., Celius, E.G., Due-Tønnessen, P., Harbo, H.F., Beyer, M.K., 2015. Reduced perfusion in white matter lesions in multiple sclerosis. Eur. J. Radiol. 84, 2605–2612. https://doi.org/10.1016/j.ejrad.2015.09.007.

Stehling, M., Turner, R., Mansfield, P., 1991. Echo-planar imaging—magnetic resonance imaging in a fraction of a second. Science 254, 43–50. https://doi.org/10.1126/science.1925560.

Stewart, G., 1893. Researches on the circulation time in organs and on the influences which affect it: Parts I.—III. J. Physiol. 15, 1.

Stokes, A.M., Quarles, C.C., 2016. A simplified spin and gradient echo approach for brain tumor perfusion imaging. Magn. Reson. Med. 75, 356–362. https://doi.org/10.1002/mrm.25591.

Stokes, A., Skinner, J., Yankeelov, T., Quarles, C., 2016. Assessment of a simplified spin and gradient echo (sSAGE) approach for human brain tumor perfusion imaging. Magn. Reson. Imaging 34. https://doi.org/10.1016/j.mri.2016.07.004.

Stokes, A.M., Bergamino, M., Alhilali, L., Hu, L.S., Karis, J.P., Baxter, L.C., Bell, L.C., Quarles, C.C., 2021a. Evaluation of single bolus, dual-echo dynamic susceptibility contrast MRI protocols in brain tumor patients. J. Cereb. Blood Flow Metab. 41, 3378–3390. https://doi.org/10.1177/0271678X211039597.

Stokes, A.M., Ragunathan, S., Robison, R.K., Fuentes, A., Bell, L.C., Karis, J.P., Pipe, J.G., Quarles, C.C., 2021b. Development of a spiral spin- and gradient-echo (spiral-SAGE) approach for improved multi-parametric dynamic contrast neuroimaging. Magn. Reson. Med. 86, 3082–3095. https://doi.org/10.1002/mrm.28933.

Thilmann, O., Larsson, E.-M., Björkman-Burtscher, I., Ståhlberg, F., Wirestam, R., 2004. Effects of echo time variation on perfusion assessment using dynamic susceptibility contrast MR imaging at 3 tesla. Magn. Reson. Imaging 22, 929–935.

Tietze, A., Mouridsen, K., Lassen-Ramshad, Y., Østergaard, L., 2015. Perfusion MRI derived indices of microvascular shunting and flow control correlate with tumor grade and outcome in patients with cerebral glioma. PloS One 10, e0123044. https://doi.org/10.1371/journal.pone.0123044.

Toth, E., Helm, L., Merbach, A., 2002. Relaxivity of MRI contrast agents. In: Contrast Agents I, pp. 61–101, https://doi.org/10.1007/3-540-45733-X_3.

Toth, G.B., Varallyay, C.G., Horvath, A., Bashir, M.R., Choyke, P.L., Daldrup-Link, H.E., Dosa, E., Finn, J.P., Gahramanov, S., Harisinghani, M., Macdougall, I., Neuwelt, A., Vasanawala, S.S., Ambady, P., Barajas, R., Cetas, J.S., Ciporen, J., DeLoughery, T.J., Doolittle, N.D., Fu, R., Grinstead, J., Guimaraes, A.R., Hamilton, B.E., Li, X., McConnell, H.L., Muldoon, L.L., Nesbit, G., Netto, J.P., Petterson, D., Rooney, W.D., Schwartz, D., Szidonya, L., Neuwelt, E.A., 2017. Current and potential imaging applications of ferumoxytol for magnetic resonance imaging. Kidney Int. 92, 47–66. https://doi.org/10.1016/j.kint.2016.12.037.

Tropres, I., Grimault, S., Vaeth, A., Grillon, E., Julien, C., Payen, J., Lamalle, L., Décorps, M., 2001. Vessel size imaging. Magn. Reson. Med. 45, 397–408.

Tsao, J., Boesiger, P., Pruessmann, K.P., 2003. k-t BLAST and k-t SENSE: dynamic MRI with high frame rate exploiting spatiotemporal correlations. Magn. Reson. Med. 50, 1031–1042.

van Dijken, B.R.J., van Laar, P.J., Smits, M., Dankbaar, J.W., Enting, R.H., van der Hoorn, A., 2019. Perfusion MRI in treatment evaluation of glioblastomas: clinical relevance of current and future techniques. J. Magn. Reson. Imaging 49, 11–22. https://doi.org/10.1002/jmri.26306.

van Gelderen, P., Grandin, C., Petrella, J.R., Moonen, C.T., 2000. Rapid three-dimensional MR imaging method for tracking a bolus of contrast agent through the brain. Radiology 216, 603–608. https://doi.org/10.1148/radiology.216.2.r00au27603.

van Osch, M.J.P., Vonken, E.-J.P.A., Wu, O., Viergever, M.A., van der Grond, J., Bakker, C.J.G., 2003. Model of the human vasculature for studying the influence of contrast injection speed on cerebral perfusion MRI. Magn. Reson. Med. 50, 614–622. https://doi.org/10.1002/mrm.10567.

Varallyay, C.G., Toth, G.B., Fu, R., Netto, J.P., Firkins, J., Ambady, P., Neuwelt, E.A., 2017. What does the boxed warning tell us? safe practice of using ferumoxytol as an MRI contrast agent. Am. J. Neuroradiol. 38, 1297. https://doi.org/10.3174/ajnr.A5188.

Varallyay, C.G., Nesbit, E., Horvath, A., Varallyay, P., Fu, R., Gahramanov, S., Muldoon, L.L., Li, X., Rooney, W.D., Neuwelt, E.A., 2018. Cerebral blood volume mapping with ferumoxytol in dynamic susceptibility contrast perfusion MRI: comparison to standard of care. J. Magn. Reson. Imaging 48, 441–448. https://doi.org/10.1002/jmri.25943.

Vardal, J., Salo, R.A., Larsson, C., Dale, A.M., Holland, D., Groote, I.R., Bjørnerud, A., 2014. Correction of B0-distortions in echo-planar-imaging–based perfusion-weighted MRI. J. Magn. Reson. Imaging 39, 722–728. https://doi.org/10.1002/jmri.24213.

Villringer, A., Rosen, B.R., Belliveau, J.W., Ackerman, J.L., Lauffer, R.B., Buxton, R.B., Chao, Y.-S., Wedeenand, V.J., Brady, T.J., 1988. Dynamic imaging with lanthanide chelates in normal brain: contrast due to magnetic susceptibility effects. Magn. Reson. Med. 6, 164–174. https://doi.org/10.1002/mrm.1910060205.

Vonken, E.J., van Osch, M.J., Bakker, C.J., Viergever, M.A., 1999. Measurement of cerebral perfusion with dual-echo multi-slice quantitative dynamic susceptibility contrast MRI. J. Magn. Reson. Imaging 10 (2), 109–117. https://doi.org/10.1002/(sici)1522-2586(199908) 10:2<109::aid-jmri1>3.0.co;2-#. PMID: 10441012.

Vonken, E.P.A., van Osch, M.J.P., Bakker, C.J.G., Viergever, M.A., 2000. Simultaneous quantitative cerebral perfusion and Gd-DTPA extravasation measurement with dual-echo dynamic susceptibility contrast MRI. Magn. Reson. Med. 43, 820–827. https://doi.org/ 10.1002/1522-2594(200006)43:6<820::AID-MRM7>3.0.CO;2-F.

Welker, K., Boxerman, J., Kalnin, A., Kaufmann, T., Shiroishi, M., Wintermark, M., 2015. ASFNR recommendations for clinical performance of MR dynamic susceptibility contrast perfusion imaging of the brain. AJNR Am. J. Neuroradiol. 36, E41–E51. https://doi.org/ 10.3174/ajnr.A4341.

Willats, L., Calamante, F., 2013. The 39 steps: evading error and deciphering the secrets for accurate dynamic susceptibility contrast MRI. NMR Biomed. 26, 913–931.

Willats, L., Connelly, A., Calamante, F., 2006. Improved deconvolution of perfusion MRI data in the presence of bolus delay and dispersion. Magn. Reson. Med. 56, 146–156. https://doi.org/10.1002/mrm.20940.

Wintermark, M., Albers, G.W., Alexandrov, A.V., Alger, J.R., Bammer, R., Baron, J.-C., Davis, S., Demaerschalk, B.M., Derdeyn, C.P., Donnan, G.A., 2008. Acute stroke imaging research roadmap. Stroke 39, 1621–1628.

Wirestam, R., Ståhlberg, F., 2005. Wavelet-based noise reduction for improved deconvolution of time-series data in dynamic susceptibility-contrast MRI. MAGMA 18, 113–118.

Wu, O., Østergaard, L., Weisskoff, R.M., Benner, T., Rosen, B.R., Sorensen, A.G., 2003. Tracer arrival timing-insensitive technique for estimating flow in MR perfusion-weighted imaging using singular value decomposition with a block-circulant deconvolution matrix. Magn. Reson. Med. 50, 164–174. https://doi.org/10.1002/mrm.10522.

Yablonskiy, D.A., Haacke, E.M., 1994. Theory of NMR signal behavior in magnetically inhomogeneous tissues: the static dephasing regime. Magn. Reson. Med. 32, 749–763.

Zaharchuk, G., Bammer, R., Straka, M., Newbould, R.D., Rosenberg, J., Olivot, J.-M., Mlynash, M., Lansberg, M.G., Schwartz, N.E., Marks, M.M., Albers, G.W., Moseley, M.E., 2009. Improving dynamic susceptibility contrast MRI measurement of quantitative cerebral blood flow using corrections for partial volume and nonlinear contrast relaxivity: a xenon computed tomographic comparative study. J. Magn. Reson. Imaging 30, 743–752. https:// doi.org/10.1002/jmri.21908.

Zaro-Weber, O., Moeller-Hartmann, W., Heiss, W.-D., Sobesky, J., 2010. A simple positron emission tomography-based calibration for perfusion-weighted magnetic resonance maps to optimize penumbral flow detection in acute stroke. Stroke 41, 1939–1945.

Zhang, J., Chu, Y., Ding, W., Kang, L., Xia, L., Jaiswal, S., Wang, Z., Chen, Z., 2019a. HF-SENSE: an improved partially parallel imaging using a high-pass filter. BMC Med. Imaging 19, 27. https://doi.org/10.1186/s12880-019-0327-3.

Zhang, K., Yun, S.D., Triphan, S.M.F., Sturm, V.J., Buschle, L.R., Hahn, A., Heiland, S., Bendszus, M., Schlemmer, H.-P., Shah, N.J., Ziener, C.H., Kurz, F.T., 2019b. Vessel architecture imaging using multiband gradient-echo/spin-echo EPI. PloS One 14, e0220939. https://doi.org/10.1371/journal.pone.0220939.

Further reading

Aronen, H.J., Gazit, I.E., Louis, D.N., Buchbinder, B.R., Pardo, F.S., Weisskoff, R.M., Harsh, G.R., Cosgrove, G., Halpern, E.F., Hochberg, F.H., 1994. Cerebral blood volume maps of gliomas: comparison with tumor grade and histologic findings. Radiology 191, 41–51.

Baydas, S., Karakas, B., 2019. Defining a curve as a Bezier curve. J. Taibah Univ. Sci. 13, 522–528. https://doi.org/10.1080/16583655.2019.1601913.

Calamante, F., Thomas, D.L., Pell, G.S., Wiersma, J., Turner, R., 1999. Measuring cerebral blood flow using magnetic resonance imaging techniques. J. Cereb. Blood Flow Metab. 19, 701–735. https://doi.org/10.1097/00004647-199907000-00001.

Caramia, F., Huang, Z., Hamberg, L., Weisskoff, R., Zaharchuk, G., Moskowitz, M., Cavagna, F., Rosen, B., 1998. Mismatch between cerebral blood volume and flow index during transient focal ischemia studied with MRI and GD-BOPTA. Magn. Reson. Imaging 16, 97–103.

Davis, T.L., Kwong, K.K., Weisskoff, R.M., Rosen, B.R., 1998. Calibrated functional MRI: mapping the dynamics of oxidative metabolism. Proc. Natl. Acad. Sci. 95, 1834–1839.

Edelman, R.R., Mattle, H.P., Atkinson, D.J., Hill, T., Finn, J., Mayman, C., Ronthal, M., Hoogewoud, H., Kleefield, J., 1990. Cerebral blood flow: assessment with dynamic contrast-enhanced T2*-weighted MR imaging at 1.5 T. Radiology 176, 211–220.

Fitter, H.N., Pandey, A.B., Patel, D.D., Mistry, J.M., 2014. A review on approaches for handling Bezier curves in CAD for manufacturing. Procedia Eng. 97, 1155–1166. https://doi.org/10.1016/j.proeng.2014.12.394.

Ostergaard, L., Weisskoff, R.M., Chesler, D.A., Gyldensted, C., Rosen, B.R., 1996. High resolution measurement of cerebral blood flow using intravascular tracer bolus passages. Part I: mathematical approach and statistical analysis. Magn. Reson. Med. 36, 715–725. https://doi.org/10.1002/mrm.1910360510.

Quarles, C.C., Bell, L.C., Stokes, A.M., 2019. Imaging vascular and hemodynamic features of the brain using dynamic susceptibility contrast and dynamic contrast enhanced MRI. Neuroimage 187, 32–55. https://doi.org/10.1016/j.neuroimage.2018.04.069.

Rosen, B.R., Belliveau, J.W., Buchbinder, B.R., McKinstry, R.C., Porkka, L.M., Kennedy, D.N., Neuder, M.S., Fisel, C.R., Aronen, H.J., Kwong, K.K., 1991. Contrast agents and cerebral hemodynamics. Magn. Reson. Med. 19, 285–292.

Siegal, T., Rubinstein, R., Tzuk-Shina, T., Gomori, J.M., 1997. Utility of relative cerebral blood volume mapping derived from perfusion magnetic resonance imaging in the routine follow up of brain tumors. J. Neurosurg. 86, 22–27.

Sorensen, A., Wray, S., Weisskoff, R., Boxerman, J., Davis, T., Caramia, F., Kwong, K., Stern, C., Baker, J., Breiter, H., 1995. Functional MR of brain activity and perfusion in patients with chronic cortical stroke. Am. J. Neuroradiol. 16, 1753–1762.

Tong, D., Yenari, M., Albers, G., O'brien, M., Marks, M., Moseley, M., 1998. Correlation of perfusion-and diffusion-weighted MRI with NIHSS score in acute (< 6.5 hour) ischemic stroke. Neurology 50, 864–869.

Tsuchiya, K., Inaoka, S., Mizutani, Y., Hachiya, J., 1998. Echo-planar perfusion MR of moyamoya disease. Am. J. Neuroradiol. 19, 211–216.

Tzika, A., Massoth, R., Ball Jr., W., Majumdar, S., Dunn, R., Kirks, D., 1993. Cerebral perfusion in children: detection with dynamic contrast-enhanced T2*-weighted MR images. Radiology 187, 449–458.

Tzika, A., Robertson, R.L., Barnes, P.D., Vajapeyam, S., Burrows, P.E., Treves, S., Scott, R.M., 1997. Childhood moyamoya disease: hemodynamic MRI. Pediatr. Radiol. 27, 727–735.

Warach, S., Li, W., Ronthal, M., Edelman, R.R., 1992. Acute cerebral ischemia: evaluation with dynamic contrast-enhanced MR imaging and MR angiography. Radiology 182, 41–47.

Warach, S., Dashe, J.F., Edelman, R.R., 1996. Clinical outcome in ischemic stroke predicted by early diffusion-weighted and perfusion magnetic resonance imaging: a preliminary analysis. J. Cereb. Blood Flow Metab. 16, 53–59.

Wenz, F., Rempp, K., Hess, T., Debus, J., Brix, G., Engenhart, R., Knopp, M., Van Kaick, G., Wannenmacher, M., 1996. Effect of radiation on blood volume in low-grade astrocytomas and normal brain tissue: quantification with dynamic susceptibility contrast MR imaging. AJR Am. J. Roentgenol. 166, 187–193.

Zierler, K.L., 1962. Theoretical basis of indicator-dilution methods for measuring flow and volume. Circ. Res. 10, 393–407.

Zierler, K.L., 1965. Equations for measuring blood flow by external monitoring of radioisotopes. Circ. Res. 16, 309–321.

Arterial spin labeling MRI

Mathijs Dijsselhof[a,b], Beatriz Esteves Padrela[a,b], Jan Petr[a,b,c], and Henk J.M.M. Mutsaerts[a,b]

[a]*Department of Radiology and Nuclear Medicine, Amsterdam UMC location Vrije Universiteit Amsterdam, Amsterdam, The Netherlands*
[b]*Amsterdam Neuroscience, Brain Imaging, Amsterdam, The Netherlands*
[c]*Helmholtz-Zentrum Dresden-Rossendorf, Institute of Radiopharmaceutical Cancer Research, Dresden, Germany*

Abbreviations

AD	Alzheimer's disease
ASL	arterial spin labeling
ATA	arterial transit artifact
ATT	arterial transit time
BBB	blood-brain barrier
CASL	continuous ASL
CBF	cerebral blood flow
DSC	dynamic susceptibility contrast-weighted imaging
DWI	diffusion-weighted imaging
FDA	Food and Drug Administration
GM	gray matter
M0	reference magnetization
MCI	mild cognitive impairment
MNI	Montreal Neurological Institute
OSIPI	open science initiative for perfusion imaging
PASL	pulsed ASL
PCASL	pseudo-continuous ASL
PET	positron emission tomography
PLD	postlabeling delay
SNR	signal-to-noise ratio
SPECT	single-proton emission computed tomography

Advances in Magnetic Resonance Technology and Applications, Volume 11, ISSN 2666-9099
https://doi.org/10.1016/B978-0-323-95209-5.00007-6

SVD	small vessel disease
TE	echo time
TR	repetition time
WM	white matter

4.1 Introduction

ASL was invented for preclinical applications in 1990 (Detre *et al.*, 1992). The first human applications appeared in 1996 when the postlabeling delay (PLD) was introduced to compensate for the larger distance between labeling in the neck and readout in the brain (Alsop and Detre, 1996). The initially low clinical reliability improved with the implementation of background suppression in 1999 (Alsop and Detre, 1999; Ye *et al.*, 2000), the increasing availability of 3T MRI, and the invention of pseudo-continuous ASL in 2008 (Dai *et al.*, 2008). In 2012 the European COST-action BM1103 "ASL In Dementia" was founded and worked toward reducing the ASL differences between MRI scanners to improve between-center reproducibility. Together with other ISMRM ASL investigators, this resulted in the 2014 consensus paper that recommended single-PLD PCASL with a 3D readout, background suppression, no vascular crushing, the acquisition of a separate M0 image, and a simplified single-compartment quantification model for clinical ASL (Alsop *et al.*, 2015).

ASL measures cerebral perfusion, referred to as cerebral blood flow (CBF). Unlike many MRI methods that provide relative values, such as fMRI or DTI, ASL is used for the direct measurement and absolute quantification of brain physiology (Oliver-Taylor *et al.*, 2017). Therefore the within- and between-session reproducibility of ASL had to be established (Chen *et al.*, 2011a; Gevers *et al.*, 2011; Heijtel *et al.*, 2014; Mutsaerts *et al.*, 2014, 2015; Petersen *et al.*, 2010) and compared to gold-standard CBF measurement using $[^{15}O]$-H_2O positron emission tomography (PET) (Heijtel *et al.*, 2014). These studies showed that ASL techniques had matured to the point where the acquisition variability was lower than physiological perfusion fluctuations (Joris *et al.*, 2018; Clement *et al.*, 2017). After proof-of-principle studies in patients with cerebrovascular and neurodegenerative diseases (Detre *et al.*, 1998; Alsop *et al.*, 2000), epilepsy (Liu *et al.*, 2001), brain tumors (Warmuth *et al.*, 2003), and pharmacological applications (Wang *et al.*, 2011; MacIntosh *et al.*, 2008; Handley *et al.*, 2013), ASL became ready for large multicenter observational studies and clinical trials (Marcus *et al.*, 2007; Jack *et al.*, 2010; Lorenzini *et al.*, 2022; Almeida *et al.*, 2018).

At the 2019 University of Michigan ISMRM ASL meeting, the community concluded that: (1) the 2014 consensus recommendations were a success: its recommendations were implemented by most clinical MRI scanners and investigators; (2) while advanced ASL techniques could outperform the 2014 consensus technique in certain clinical cases, they still required validation and could be daunting to implement: so it was decided not to update the consensus recommendation; (3) further clinical translation of ASL required education on its merits and challenges. Recently, four new consensus reviews were written on the application of ASL in different clinical use cases (Lindner *et al.*, 2023), on the use of velocity-encoded ASL (Qin *et al.*, 2022) and multi-PLD ASL (Woods *et al.*, 2023), and on the potential of new advanced ASL techniques (Hernandez-Garcia *et al.*, 2022). Furthermore, the Open Science Initiative for Perfusion Imaging (OSIPI) was initiated by the ISMRM Perfusion Study Group to obtain consensus, harmonization, and resources on perfusion MRI image processing (Bell *et al.*, 2023).

This chapter focuses on the use of ASL in the human brain, for obtaining CBF and arterial transit time (ATT), in clinical practice and research. Following the above-mentioned consensus literature, we introduce clinically applied ASL acquisition techniques, discuss important ASL-specific image processing steps, and clinical applications of ASL. Finally, we address the most promising upcoming ASL techniques.

4.2 Acquisition

As with other perfusion techniques, ASL aims to assess the volume of blood delivered to a unit volume of tissue per given time. Unlike most previous imaging techniques, however, ASL does not rely on the administration of exogenous paramagnetic contrast agents or radiopharmaceuticals. Instead, ASL alters the magnetization of water molecules in blood endogenously and then measures the effect it has on the signal in tissue that is supplied with this labeled blood (Williams *et al.*, 1992).

In a basic ASL sequence, arterial blood water is labeled by an inversion pulse proximal to the imaging region. After a short delay on the order of seconds, called the postlabeling delay, an image is acquired in the brain, which will be affected by the inflow of inverted spins in blood. To gain sufficient signal, the labeling and acquisition part is repeated several times, alternating control and label image acquisition. The control image, acquired without any prior labeling, contains signals from both blood and tissue. The label image acquired with an identical readout but after labeling contains the tissue signal but the signal of the inflowing blood has an inverted magnetization. The control-label image difference thus theoretically completely subtracts the tissue signal and then only contains the signal from the inflowing blood and, thus, allows quantifying the volume of labeled blood delivered to each voxel (Detre *et al.*, 1992).

The recommended field strength for ASL is 3T (Alsop *et al.*, 2015); however, the use of different field strengths is possible as long as care is taken to obtain data with sufficient signal-to-noise ratio (SNR). The main driver of quality in ASL is the T1 of blood. With higher field strength, the T1 time of blood is higher and, consequently, SNR is higher (Lu *et al.*, 2004). Nonetheless, operating at 1.5T with an inherently lower SNR can be compensated for by lowering the spatial resolution, yielding similar quantitative accuracy between 1.5T and 3T (Baas *et al.*, 2021). While 7T should in theory offer a superior image quality, ASL at 7T faces severe challenges with labeling efficiency due to B1 inhomogeneity (Zimmer *et al.*, 2016), especially for pseudo-continuous labeling (Wang *et al.*, 2022).

4.2.1 Labeling strategies

In ASL, labeling is the process where the longitudinal magnetization of blood water is inverted by 180° to trace it through the macro- and microvasculature and, ultimately, quantify the volume that flows through tissue. Labeling strategies can roughly be separated into velocity selective—that is, based on the velocity or acceleration of blood—or spatially selective. The clinically most used labeling strategy is spatially selective, aiming to suppress the arterial blood water signal only in the feeding vessels of the brain, either by instantaneously labeling a thick slab—pulsed ASL (PASL)—or by continuously labeling a thin plane—(pseudo-)continuous ASL ((P)CASL).

In PASL, a 15–20 cm slab is labeled using a single-RF pulse in the neck (Wong, 2005). PASL's advantages include a low specific absorption ratio (SAR), minimal magnetization transfer effects, a high labeling efficiency (98%) insensitive to the vascular anatomy (Wong, 2014), and shorter acquisition times (~2–4 s per volume) due to instantaneous labeling. As PASL only defines the spatial width of the bolus, the effective labeling duration has to be defined by additional saturation pulses (common implementations are QUIPSS II and Q2TIPS) to allow accurate quantification (Luh *et al.*, 1999; Wong *et al.*, 1998a). The preferred saturation time is 800 ms, as the bolus duration will not exceed 1 s for a 20-cm slab and blood velocity in large arteries around 20 cm/s. Disadvantages of PASL are, thus, a relatively short labeling duration (Wong *et al.*, 1998b) and a longer effect of the T1 decay on the blood from the distal side of the inverted slab, which jointly leads to a lower SNR of PASL.

To address the issues of PASL, CASL was introduced (Alsop and Detre, 1998) with a constant RF and gradient, achieving a flow-driven adiabatic inversion (Dixon *et al.*, 1986). CASL has relatively high SAR and magnetization transfer effects (McLaughlin *et al.*, 1997; Hernandez-Garcia *et al.*, 2007), and a practical dependency on separate labeling coils (Zaharchuk *et al.*, 1999; O'Gorman *et al.*, 2006). Therefore the currently preferred method (Alsop *et al.*, 2015) is PCASL, which uses a series of hundreds of 1–2 ms pulses instead (Dai *et al.*, 2008; Wu *et al.*, 2007) to alleviate CASL's magnetization transfer issues (Wong, 2014). While the labeling efficiency of PCASL (85%) is lower than PASL (98%), the sustained labeling for 1.8–2 s (compared to 800-ms PASL) generates higher SNR (Chen *et al.*, 2011a).

4.2.1.1 Labeling efficiency

Labeling efficiency depends on the labeling technique used and on the location of the labeling plane. For most PASL techniques (Wang *et al.*, 2002; Wong *et al.*, 1997; Golay *et al.*, 2005), keeping a 1–2 cm distance between the labeling slab and imaging volume is sufficient. The PCASL plane requires a more careful positioning, perpendicularly to the feeding vessels and without overlapping with the imaging volume (*e.g.*, at the level of the 2nd or 3rd cervical vertebrae (Zhao *et al.*, 2017) or 4 cm below the base of the cerebellum and parallel with the anterior-commissure posterior-commissure (AC-PC) line (Aslan *et al.*, 2010). In case the position or rotation of the labeling plane can be freely chosen, a vessel scout is preferred to both guide the placement and inspect the vessel—placing the labeling plane at a vessel at an angle or at a location of high vessel tortuosity will decrease labeling efficiency.

While literature values for labeling efficiency (Dai *et al.*, 2008) may suffice in many clinical cases (Heijtel *et al.*, 2014), individual correction can be beneficial for specific populations (Václavů *et al.*, 2018). For these, phase-contrast MRI can improve the CBF quantification by calibrating CBF based on total flow through the brain-feeding arteries (Aslan and Lu, 2010; Ambarki *et al.*, 2015) or modeling the labeling efficiency based on the velocity in the labeling plane (Václavů *et al.*, 2018). Compared to ASL, drawbacks of phase-contrast MRI include its lower reproducibility for whole-brain CBF estimates (Dolui *et al.*, 2016) and its lower agreement with PET (Puig *et al.*, 2018).

4.2.1.2 Postlabeling delay

One major confounder of ASL is the ATT, that is, the time it takes labeled blood to transit from the labeling plane/slab to the imaging voxel. ATT varies widely within and between patients, mainly with age, sex, and cardiovascular or cerebrovascular health. Common clinical ASL implementations employ a single-PLD ASL implementation that assumes the entire label has arrived at the tissue after the PLD. While PLD recommendations vary between young and older patients (Alsop *et al.*, 2015), there are many cases where distal regions in old patients show incomplete label arrival. In such cases, CBF can be underestimated in distal and overestimated in proximal regions. Longer PLDs can address this problem at the cost of lower image quality due to label relaxation with blood T1.

ATT's confounding effect can be alleviated by obtaining ASL images at multiple PLDs. A dual PLD approach (*e.g.*, at 1500 and 2500 ms) acquires information on both vascular reconfiguration and CBF. A multi-PLD approach either has different PLDs lengths (Mezue *et al.*, 2014) or intermittent labeling and control phases ordered according to a Hadamard matrix (von Samson-Himmelstjerna *et al.*, 2016; Dai *et al.*, 2013). Typically, 3–10 PLDs are acquired for an accurate ATT measurement. While multi-PLD can be more accurate, it can suffer from lower SNR and motion artifacts and requires a longer measurement time.

4.2.2 Readout

Both 2D and 3D acquisition techniques are available for ASL. Initially, 2D echo-planar imaging (EPI) was preferred for readout, as it allowed a whole-brain image within a reasonable time before the labeled signal decreases due to blood T1 relaxation (Williams *et al.*, 1992). Currently, 3D segmented readouts are preferred due to higher SNR and are widely available as product sequence on all MRI scanners as stack-of-spirals fast spin echo (3D spiral) in GE (Chang *et al.*, 2017), and 3D gradient-spin-echo (3D GRaSE) in Philips and Siemens (Feinberg and Oshio, 1991).

The 2D EPI still has several advantages: its spatial resolution is sufficient in a single shot, whereas for 3D sequences multiple shots are usually obtained to limit spatial blurring. A typical 2D EPI control-label pair takes ~8 s, whereas a 3D GRaSE acquisition (*e.g.*, Philips/Siemens) with 4 segments is 32 s and a typical 3D spiral acquisition with 8 arms is 72 s (*e.g.*, GE). Therefore 2D EPI can still be less sensitive to motion and offer the temporal resolution needed for functional ASL. The 3D acquisitions use a single excitation pulse, thus virtually having a single effective PLD for all slices (Günther *et al.*, 2005). For 2D, the PLD increases slice-wise as each slice takes ~40 ms to be acquired; this needs to be accounted for in the quantification.

Because of its relatively low SNR, ASL is typically acquired at a resolution of approximately $4 \times 4 \times 4 \, \text{mm}^3$ (Edelman *et al.*, 1994; Yang *et al.*, 1998). It should be noted that the nominal spatial resolution can seem higher than the effective spatial resolution, especially for some 3D acquisitions (Petr *et al.*, 2018). A major issue of 3D ASL readouts is that T2 and T2* decay causes blurring for spin echo (*e.g.*, 3D spiral) and gradient echo (*e.g.*, 3D GRaSE), respectively (Vidorreta *et al.*, 2013, 2014). While recent advances can reduce the amount of blurring (Chang *et al.*, 2017; Liang *et al.*, 2014; Zhao *et al.*, 2018; Vidorreta *et al.*, 2017), the effective spatial resolution of 3D sequences is often lower because of their increased sensitivity to head motion and smoothing in the reconstruction (Petr *et al.*, 2018). The 2D EPI, and to some extent 3D GRaSE, suffers from susceptibility-induced geometric distortion in air-tissue transit regions, mainly in the orbitofrontal and inferior-temporal cortices. Use of an extra M0 scan with reversed phase-encoding direction is advised to compensate for this, especially if such regions are of interest, as is the case for frontotemporal dementia (Mutsaerts *et al.*, 2019).

The current recommendation is a moderately segmented 3D readout (Alsop *et al.*, 2015), combined with parallel imaging at 2–3 times acceleration (Hernandez-Garcia *et al.*, 2022). Notably, acquisition resolution should not be confused with reconstructed resolution, which is often on the order of 2 mm. In general, whole-brain coverage with an acquisition time of 4–5 min is deemed sufficient for diagnostic use in gray matter (CBF) CBF or 2 min in acute situations with a reduced resolution to obtain a reasonable SNR (Lindner *et al.*, 2023). ASL in white matter (WM) suffers from much lower SNR due to long ATT and low CBF compared to GM (van Osch *et al.*, 2009; Skurdal *et al.*, 2015). Therefore if the WM is the region of interest, longer labeling duration and PLD—and possibly longer measurement—are required.

4.2.2.1 Background suppression

One major limitation of the use of blood as an endogenous perfusion tracer is that blood signal is only 0%–3% of the signal in an imaging voxel (Alsop *et al.*, 2015), the other 97% being brain tissue. Consequently, head motion can reduce ASL SNR dramatically. To reduce this influence, static tissue signal suppression called background suppression is important, especially for the relatively long control-label duration of 3D segmented sequences. Background suppression is usually implemented using an even number of spatially nonselective 180° pulses affecting both labeled blood and static tissue. The even number ensures that the longitudinal magnetization of the labeled blood is not inverted. Background suppression pulses are timed to effect—together with the T1 signal decay of GM and WM—a near-complete reduction of static tissue signal at the readout. However, this comes at the cost of a reduced effective labeling efficiency to 83% or 75% for 2 or 4 pulses, respectively. A 50–100 ms delay from the perfect suppression timing is recommended for 3D sequences to ensure the static tissue signal crosses zero and does not lead to negative control-label subtraction, also acknowledging the between-patient variability of T1 times. While background suppression is equal across the brain at 3D sequences, for 2D EPI, it is optimized for the first slice, and the efficiency of suppression decreases in the more superior slices (Fig. 4.1). At 3T MRI, the T1 of blood is 1.65 s (Lu *et al.*, 2004; Zhang *et al.*, 2013), whereas the tissue T1 is even shorter—1.2 s for gray matter (GM), 0.9 s for white matter (WM) (Weiskopf *et al.*, 2013). Note that gadolinium-based contrast agents drastically shorten blood T1, and, thus, ASL should always be performed before administering contrast agent to avoid the resulting ASL SNR decrease.

4.2.3 **M0 acquisition**

MRI data by itself has no meaningful physical units; therefore, to derive absolute quantitative perfusion values from ASL, the measurement must be calibrated by obtaining the value of equilibrium magnetization (M0) of arterial blood. Due to the relatively low spatial resolution of ASL, a voxel filled fully with blood is difficult to find. Therefore brain tissue equilibrium magnetization is typically measured instead and converted to arterial blood magnetization using known tissue parameters (Çavuşoğlu *et al.*, 2009). Acquisition of a separate M0 image, either in a standalone sequence or integrated with ASL, is thus highly recommended (Alsop *et al.*, 2015). While a TR around 4–6 s, similar to that of ASL, is recommended, there are also recommendations to use a long TR of around 10 s with nearly full relaxation or a short TR of ~2 s to decrease GM/WM difference or an inversion recovery sequence. However, no consensus or widely accepted comparison of those approaches exists.

4.3 **Analysis**

4.3.1 **Motion correction and outlier removal**

As ASL is a subtraction technique, even subtle head motion can introduce large artifacts in the perfusion-weighted image, especially in locations with large differences in signal intensities, such as between the brain and skull/air or between GM/WM and CSF. Motion correction between all ASL volumes is thus necessary, ideally considering the control-label differences (Wang *et al.*, 2008).

Individual control-label pairs with excessive artifacts beyond repair (such as excessive head motion or labeling failure) can be excluded as outliers, as most ASL sequences consist of time series. A single control-label pair with significant artifact

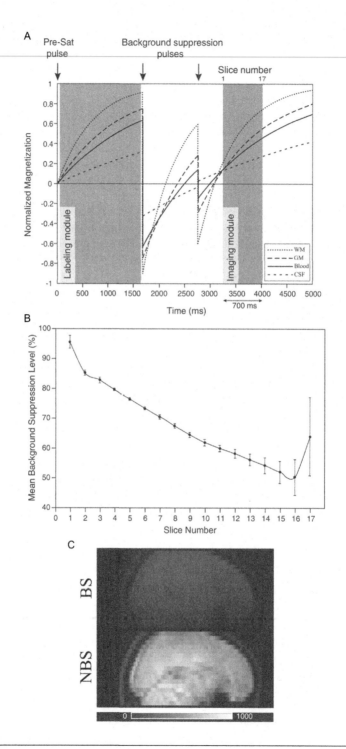

FIG. 4.1

See figure legend on opposite page.

or motion can ruin the entire CBF calculation; therefore, excluding a relatively small number of pairs with outliers can improve the quality of the resulting CBF image. Several ASL outlier detection methods exist (Dolui *et al.*, 2017; Shirzadi *et al.*, 2018; Maumet *et al.*, 2014), although there is no validated consensus on the best approach. Both motion correction and outlier removal might be difficult with 3D segmented readout with limited repetitions or when the scanner does not output individual images.

4.3.2 Segmentation and registration

Typically, mean CBF values are evaluated within the GM of specific anatomical regions. Because brains differ in shape and size, tissue type segmentation and registration to anatomical atlases are needed. Because of the relatively poor resolution of ASL scans, high-resolution anatomical scans are normally used for segmentation and spatial normalization.

Rigid-body registration of ASL and anatomical scans are usually sufficient. In the case of susceptibility-related geometric distortion common for 2D EPI and 3D GRaSE readouts, a second M0 scan with reversed phase-encoding direction is recommended for distortion correction, for example, with TOP-UP (Andersson *et al.*, 2003). Anatomical scans are segmented into GM, WM, and cerebrospinal fluid (CSF) probabilistic maps using intensity- and template-based methods such as FSL, SPM12/CAT12, or Freesurfer. These probabilistic tissue-class maps are then used by ASL for partial volume correction and may be nonlinearly registered to a common-space brain template like the Montreal Neurological Institute (MNI) template (Mazziotta *et al.*, 1995) using, for example, DARTEL or FSL-FIRST (Ashburner, 2007; Klein *et al.*, 2009). During spatial normalization, each part of the brain is remodeled (referred to as "warped") to the size and shape of an average brain. Poor GM-WM contrast or anomalies like lesions or tumors can cause incorrect segmentation. Because of ASL's dependence on structural processing for partial volume correction and registration, incorrect segmentation can lead to wrongly estimated CBF values.

FIG. 4.1—Cont'd

Demonstrates the effect of background suppression. (A) shows its implementation for a pseudo-continuous labeling module with a multislice 2D readout and the resulting evolution of tissue magnetization over time for gray matter (GM), white matter (WM), blood, and cerebrospinal fluid (CSF). The second gray surface represents the readout phase during which all slices are consecutively acquired, during which the "background" tissue signal returns to its original positive magnetization, reducing the background suppression level (B). This can be visually appreciated in (C), in which sagittal projections of raw EPI images are shown with (BS, upper pane) and without background suppression (NBS, lower pane). The bar below the picture indicates the signal intensity.

From Ghariq, E., Chappell, M.A., Schmid, S., Teeuwisse, W.M., van Osch, M.J.P., 2014. Effects of background suppression on the sensitivity of dual-echo arterial spin labeling MRI for BOLD and CBF signal changes. NeuroImage 103, 316–322. https://doi.org/10.1016/j.neuroimage.2014.09.051.

4.3.3 **Quantification**

ASL raw images are "quantified"—that is, converted to physiological values—to obtain a CBF image in ml/100 g/min. ASL quantification starts with a control-label subtraction and continues by applying several factors to account, for example, labeling efficiency and duration, M0 of blood, and labeling relaxation (Alsop *et al.*, 2015). Literature values can be used as these factors or they can be measured individually in ancillary acquisitions. Using a measured factor can increase CBF accuracy at the expense of loss of precision. This trade-off, together with the additional scanning time, has resulted in the consensus to clinically use a simplified quantification model with a fixed value of most of these factors (Alsop *et al.*, 2015).

The simplified single-PLD quantification model (Alsop *et al.*, 2015) assumes that all labeled blood has arrived in the capillaries/tissue before the start of the readout (*i.e.*, ATT shorter than PLD) and that label has only decayed with blood T1: ~1.65 s at 3 T. This is often not the case in older and diseased populations, and longer ATT leads to proximal macrovascular artifacts (*i.e.*, labeled blood is present in proximal arteries rather than in the capillary bed or tissue) and distal incorrect quantification displayed as apparent hypoperfusion. The use of a general kinetic model in multi-PLD acquisitions (Wang *et al.*, 2013; Buxton *et al.*, 1998) can improve this estimation (Wang *et al.*, 2013; Buxton *et al.*, 1998).

While the use of a single M0 value is possible—for example, derived from the ventricular CSF on a control image or previously acquired calibration scan on the same scanner—the acquisition of a separate M0 map is preferred, because it cancels out acquisition specific effects such as receiver coil inhomogeneity, or spatial T2 or T2* variability (Alsop *et al.*, 2015). M0 images are typically smoothed and masked, but the extent of manipulation varies between investigators (Mutsaerts *et al.*, 2020a; Pinto *et al.*, 2020). A control image can be used instead of a separate M0 scan if no background suppression was applied.

ASL's voxel size—typically between 2×2 and $4 \times 4\,\text{mm}^2$ in-plane with 4–8 mm slice thickness—is usually larger than the GM thickness (~2.5 mm), especially in the cortex. Most ASL voxels contain a signal mixture of GM, WM, and CSF. As GM CBF is ~3 times as high as WM CBF (Asllani *et al.*, 2008; Pohmann, 2010), the measured voxel-wise CBF is highly dependent on the structural composition, which is referred to as partial volume (PV) effects. PV effects differ between ASL readouts and can be a major source of bias if not accounted for. This is why PV correction (PVC) is typically based on the GM, WM, and CSF segmentations from high-resolution structural images. Even without PVC, the ROI-average GM CBF is calculated from voxels that have a high GM partial volume (*e.g.*, more than 70%). However, in cases of atrophy, this will leave a few voxels that can be included when calculating GM CBF (Chappell *et al.*, 2021). This decreases the SNR and thus the statistical power and may bias the GM CBF calculation toward regions with relatively thick GM. Therefore PVC is recommended for studies focusing on GM, especially if atrophy differences are expected between participants or cohorts (Hernandez-Garcia *et al.*, 2022).

Several algorithms correct PV effects at the voxel level, using GM and WM PV maps obtained from downsampling the segmented structural images to the ASL space, assuming locally homogeneous GM and WM CBF (Asllani *et al.*, 2008). These PVC ASL maps can be improved by leveraging the ATT differences between GM and WM (Chappell *et al.*, 2011). However, these errors may similarly affect non PV corrected GM CBF evaluation by including more or fewer voxels in the GM mask (Petr *et al.*, 2018). PVC typically assumes each ASL voxel to be composed of GM, WM, or CSF and may lead to incorrect CBF estimates with high blood volume, such as in macrovascular artifacts. An alternative to PVC is to use GM volume as a covariate in the statistical analysis (Chen *et al.*, 2011b).

4.3.4 Physiological variability and confounders

Two nuisance components of the variability of ASL-based CBF measurements can be distinguished: instrumental (measurement noise) and physiological (perfusion fluctuations within and between subjects) (Clement *et al.*, 2017; Parkes *et al.*, 2004). Both components vary significantly between patients' brain regions. In patients with high CBF—such as children, especially with sickle cell disease—SNR optimizations have low priority (Gevers *et al.*, 2012). On the other hand, in patients with low CBF—such as elderly with cerebrovascular or neurodegenerative disease—it is important to focus on sustaining sufficiently high SNR.

ASL CBF is a so-called hot tracer, meaning its baseline value is nonzero and has a relatively large normal variability. In healthy volunteers, the within-measurement variability (considered mostly instrumental) has a within-subject coefficient of variation (wsCV) of $\sim5\%$, and the between-measurement variability (considered mostly physiological) has a wsCV $\sim10\%$ (Chen *et al.*, 2011a; Heijtel *et al.*, 2014; Baas *et al.*, 2021). Major physiological CBF modifiers include age (Asllani *et al.*, 2009; Ances *et al.*, 2009), sex (Parkes *et al.*, 2004), medication (especially vasoactive drugs, antihypertensives, and diuretics) (van Dalen *et al.*, 2021; Hu *et al.*, 2017), and sedation (MacIntosh *et al.*, 2008). Even the time of day (Elvsåshagen *et al.*, 2018) and caffeine use (Haller *et al.*, 2014) or food intake can strongly affect CBF. Although standard operating procedures have been proposed (Clement *et al.*, 2017), there is no consensus if the subject should continue usual behavior—to represent the normal subject's baseline and avoid compensatory CBF mechanisms or to restrict food and caffeine intake. Considering that caffeine not only lowers CBF by $\sim20\%$ but subsequently also ASL SNR, it is typically recommended to restrict caffeine for at least two hours prior to scanning (Lindner *et al.*, 2023).

One major CBF modifier is the hematocrit, which differs between subjects but may also differ within subjects over time (Elvsåshagen *et al.*, 2018). Hematocrit has both an instrumental and a physiological effect. Firstly, blood T1 is inversely correlated with hematocrit, leading to an under- or overestimation of CBF if this is not accounted for (instrumental effect). Secondly, low hematocrit leads to a compensatory CBF increase (physiological effect). Hematocrit can thus be a source of bias in certain populations if not corrected, for example, hematocrit is higher in men than women, increased in type 2 diabetes mellitus, and reduced in anemia (Ibaraki *et al.*, 2010; Václavů *et al.*, 2020; Xu *et al.*, 2018). Blood T1 can be measured venously in the sagittal sinus (Wu *et al.*, 2010; Varela *et al.*, 2011) or arterially in the carotids (Li *et al.*, 2017), but can also be inferred from hematocrit values (Lu *et al.*, 2004; Hales *et al.*, 2016). Hematocrit effects can also be added as a covariate in statistical models to account for its physiological effect on CBF (Smith *et al.*, 2019).

4.3.5 Image processing software

While basic image processing and quantification are available on most MR scanners, it is recommended to use dedicated software for higher quality results, especially for population-based research studies. Several open-source ASL processing pipelines are available, for example, ASLtoolbox, ASLprep, BASIL/OxfordASL, Explor-eASL, and MRIcloud (Wang *et al.*, 2008; Mutsaerts *et al.*, 2020a; Li *et al.*, 2018; Adebimpe *et al.*, 2021; Arteaga *et al.*, 2017). A list of pipelines is provided at ISMRM OSIPI (ASL pipeline inventory, n.d.), including pipeline functionality, features, and user-friendliness. Because ASL implementations differ between MRI vendors and imaging centers, it is important to know the specific acquisition and quantification parameters to allow comparability between centers and studies. Many important parameters are not saved as DICOM metadata. The brain imaging data structure (BIDS) (Gorgolewski *et al.*, 2016) was recently extended with ASL-BIDS (Clement *et al.*, 2022), which supports the most common ASL sequence types. The use of the BIDS format for data sharing and processing is recommended to increase reproducibility.

4.4 Clinical applications

Clinically, ASL is mainly used to detect changes in blood supply or demand. A phenomenon called "neurovascular coupling" states that there is a relation between the demand of blood flow and neuronal activity (Thompson *et al.*, 2003). From this perspective, ASL CBF is a potential proxy biomarker of the activity of brain regions—a diseased part of the brain that has reduced activity will 'demand' lower CBF. On the other hand, cerebrovascular impairment can also influence the supply of blood to the different regions of the brain. With ASL it is possible to investigate these synergetic effects and study the possible mechanisms underlying cognitive decline.

4.4.1 Cerebrovascular disease

ASL can be a biomarker of problems in blood supply to the brain due to malformed, occluded, or damaged vessels. These vessels transfer less labeled blood and at a lower speed than do normal vessels. Steno-occlusive or other forms of cerebrovascular disease typically manifest as regions with delayed blood arrival and reduced perfusion (Detre *et al.*, 1998; Ramachandran *et al.*, 2022; Bambach *et al.*, 2022). When the PLD is not adapted according to the prolonged ATT, the labeled blood partly still resides in the feeding vessels at the time of imaging (Alsop *et al.*, 2015). This manifests as hyperintense serpentinous macrovascular ASL signal in the proximal feeding vessels—referred to as arterial transit artifacts (ATAs)—and apparent distal hypoperfusion. These ATAs can compromise CBF estimates in distal regions, making it infeasible to differentiate between true hypoperfusion and delayed label arrival. On the other hand, delayed ATT and ATAs themselves can be a potent cerebrovascular biomarker (Roach *et al.*, 2016; Di Napoli *et al.*, 2020; Shirzadi *et al.*, 2019). ATAs can be visually rated using a modified ASPECTS score (Roach *et al.*, 2016) or quantified from the distribution of ASL signal across a perfusion territory, using the spatial coefficient of variation (sCoV) (Mutsaerts *et al.*, 2020b).

A typical application of perfusion imaging in acute ischemic stroke is comparison of the infarct core and its penumbra (Fig. 4.2). A lesion visible on diffusion-weighted imaging (DWI) is then the unsalvageable necrotic infarct core, whereas the penumbra represents an area of hypoperfusion that would become necrotic without timely treatment. A prognostic factor of stroke treatment outcome is the presence of collaterals, which can be visible on (early) PLDs of an ASL scan in the form of ATAs.

Another common pathology that may benefit from the use of ASL as an early biomarker is cerebral small vessel disease (SVD), an aging-related condition that affects the cerebral microvessels (Wardlaw *et al.*, 2019). Because white matter lesions, a major SVD biomarker, are believed to be partly caused by hypoperfusion, ASL may contribute to the understanding of SVD pathophysiology (Zhang *et al.*, 2022; van Dalen *et al.*, 2016). No consistent ASL CBF patterns were found yet in SVD (Shi *et al.*, 2016; Stewart *et al.*, 2021), which is hypothesized to be due to the existence of three different SVD subtypes (Fig. 4.3) (Lu *et al.*, 2022).

4.4.2 Brain tumors

ASL can be used for diagnosing, grading, or monitoring primary (*e.g.*, glioma) or secondary (*i.e.*, metastasis) brain tumors (Henriksen *et al.*, 2022). Tumors can exhibit hyperperfusion because of increased metabolism or neovascularization (Falk Delgado *et al.*, 2018; Dangouloff-Ros *et al.*, 2016). Recommended indications of ASL include the differentiation between tumor and nontumoral pathologies (Smits, 2021) or between various types of brain tumors (Sunwoo *et al.*, 2016; Weber *et al.*, 2006; Suh *et al.*, 2018; Abdel Razek *et al.*, 2019), assessing the spatial extent of the tumor, grading and malignant transformation of primary brain tumors (Alsaedi *et al.*, 2019), and the differentiation between tumor progression and treatment-related radiological abnormalities (Choi *et al.*, 2013; Wang *et al.*, 2020; Manning *et al.*, 2020). Another promising application is distinguishing between tumor progression and pseudoprogression (Choi *et al.*, 2013; Wang *et al.*, 2020; Manning *et al.*, 2020; Song *et al.*, 2013), where structural MRI alone may not be sufficient. CBF normalization to healthy tissue is recommended to reduce variability in labeling efficiency between subjects (Alsaedi *et al.*, 2019). Most metastases

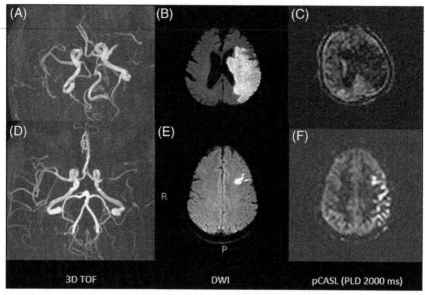

3D TOF	DWI	pCASL (PLD 2000 ms)

FIG. 4.2

Two patients with an AIS in the left MCA territory. The first patient in the upper row (A–C) has poor collaterals. The TOF angiogram (A) shows occlusion of the left M1 segment, which has resulted in a large infarct on DWI (B). On the perfusion-weighted ASL image with a PLD of 2000 ms (C), there is no visible ATA—hyperintensities in the larger vessels—indicating a lack of collateral vessels. The second patient (D–F) has robust collaterals. The TOF (D) shows severe narrowing/near occlusion of the left M1 segment and DWI (E) demonstrates a much smaller acute infarct than in the first patient. On the perfusion-weighted ASL images with a PLD of 2000 ms (F), there is a serpiginous high signal overlying the left hemisphere. These reflect ATA and presumably correspond to labeled spins in leptomeningeal collateral vessels, which have not reached the brain parenchyma at the 2000-ms PLD, yet provide adequate blood supply to prevent a larger infarction (at least at the time of imaging). Images are shown as perfusion values.

From Lindner, T., Bolar, D.S., Achten, E., et al., 2023. Current state and guidance on arterial spin labeling perfusion MRI in clinical neuroimaging. Magn. Reson. Med. 89, 2024–2047. https://doi.org/10. 1002/mrm.29572.

appear isointense to hypoperfused, except for hypervascular metastases, which show intratumoral hyperperfusion (Sunwoo *et al.*, 2016).

While ASL and DSC perfusion imaging can be used for similar purposes (Xiao *et al.*, 2015), no extensive multicenter comparison or metaanalysis exists, and ASL has not been validated to the same extent as DSC (Boxerman *et al.*, 2020). ASL offers distinct advantages for quantitative CBF measurements as blood-brain barrier (BBB) leakage does not need to be modeled as would be the case for perfusion MRI using a gadolinium-based contrast agent (Maral *et al.*, 2020).

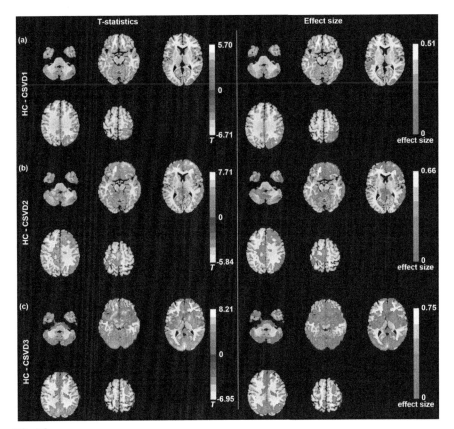

FIG. 4.3

Patterns of CBF identifying the three CSVD subtypes in comparison with healthy controls. Comparison of CBF map between healthy controls and (A) CSVD subtype 1, (B) CSVD subtype 2, (C) CSVD subtype 3, and corresponding absolute effect size map.

From Lu, W., Yu, C., Wang, L., Wang, F., Qiu, J., 2022. Perfusion heterogeneity of cerebral small vessel disease revealed via arterial spin labeling MRI and machine learning. Neuroimage Clin. 36, 103165. https://doi.org/10.1016/j.nicl.2022.103165.

Moreover, DSC can fail in regions of susceptibility, for example, due to blood products after tumor resection and near air-tissue interfaces of the skull base (Fig. 4.4).

4.4.3 **Neurodegenerative disease**

In addition to amyloid-β (Aβ), tau pathology, and neuron loss, AD is associated with early neurovascular dysfunction, which contributes to disease pathogenesis, as indicated by recent epidemiological, clinical, pathological, and experimental studies (Zlokovic, 2011; Toledo *et al.*, 2013; Soto-Rojas *et al.*, 2021; Garnier-Crussard

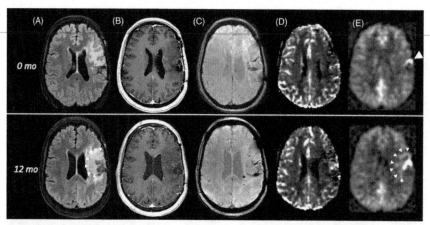

FIG. 4.4

Forty-year-old woman with previously resected and radiated left frontoparietal grade III anaplastic astrocytoma. FLAIR (A), postcontrast (B), SWI (C), DSC perfusion (D), and ASL perfusion (E) are shown at time = 0 month (top row) and 12 months (bottom row). At time = 0, ASL demonstrates a small focus of hyperperfusion along the resection cavity margins (*yellow* arrowhead; *white* in print version) that raises suspicion for recurrent tumor despite lack of mass-like FLAIR abnormality or suspicious enhancement. Importantly, no convincing abnormality is seen on DSC, likely due to susceptibility in the setting of chronic postsurgical blood products (C). At 12 months, there has been marked interval growth, subtly seen on FLAIR and more easily identified by ASL hyperperfusion (*yellow* arrowheads; *white* in print version). DSC again is of poor diagnostic utility and shows only minimal perfusion abnormality.

From Lindner, T., Bolar, D.S., Achten, E., et al., 2023. Current state and guidance on arterial spin labeling perfusion MRI in clinical neuroimaging. Magn. Reson. Med. 89, 2024–2047. https://doi.org/10.1002/mrm.29572.

et al., 2022). Diagnosis and evaluation of neurodegenerative disease is a rapidly expanding research application of ASL (Lu *et al.*, 2022; Brayne and Davis, 2012; Dolui *et al.*, 2020). ASL has the potential to detect perfusion changes in AD and related dementias (Alsop *et al.*, 2000; Lindner *et al.*, 2023; Binnewijzend *et al.*, 2013; Lee *et al.*, 2009; Sandson *et al.*, 1996), before macroscopic volume loss or other structural changes are evident. Since clinical dementia MRI protocols are typically performed without contrast, ASL is the ideal candidate for measuring perfusion in these patients. ASL CBF may help predict and monitor disease progression and the efficacy of clinical trials.

ASL captures AD hypoperfusion patterns similar to those observed using positron emission tomography (PET) and single photon emission computed tomography (SPECT) (Alsop *et al.*, 2000; Johnson *et al.*, 2005; Tosun *et al.*, 2016; Dai *et al.*, 2009) and may predict conversion from mild cognitive impairment (MCI) to AD (Dolui *et al.*, 2020; Chao *et al.*, 2010; Li *et al.*, 2020). Interestingly, patients with

MCI and early stage AD can sometimes present with increased perfusion in the hippocampus and other subcortical regions, potentially attributable to a compensatory mechanism to neuronal injury (Lee *et al.*, 2009). ASL also captures patterns of regional perfusion abnormalities in frontotemporal dementia (FTD) (Mutsaerts *et al.*, 2019), dementia with Lewy bodies (Dolui *et al.*, 2020; Nedelska *et al.*, 2018), and Parkinson's disease (Kamagata *et al.*, 2011) and may help differentiate between AD and FTD (Fällmar *et al.*, 2017; Steketee *et al.*, 2016; Du *et al.*, 2006; Hu *et al.*, 2011). In addition, ASL is potentially a useful method to evaluate disease progression (Li *et al.*, 2020; Mak *et al.*, 2021; Kim *et al.*, 2013; Xekardaki *et al.*, 2015).

4.4.4 Epilepsy

Seizures are a common neurological presentation, ranging from a first-time acute seizure to chronic seizure disorders. Patients who experience seizures are often referred for imaging, with MRI being the preferred method for detecting structural abnormalities in the neocortex, such as tumors, infectious or inflammatory lesions, or vascular malformations. In 20%–40% of cases, traditional MRI does not reveal significant findings in individuals evaluated for epilepsy surgery (Leeman-Markowski, 2016). Functional imaging techniques including ASL may then be used to identify the location of the epileptogenic focus (Gajdoš *et al.*, 2021) and to understand different phases of the disease (*e.g.*, early ictal, ictal, postictal, and interictal) (Pizzini *et al.*, 2013). Seizures or epilepsy often manifest on ASL as (peri-)ictal hyperperfusion or interictal hypoperfusion (Fig. 4.5).

4.4.5 Pediatric diseases

ASL is particularly useful in children because of its noninvasive nature (Bambach *et al.*, 2022). ASL can even be performed in preterm neonates, for whom gadolinium injection is not FDA approved (although off-label use is allowed). Another significant advantage of ASL is that it can be repeated immediately in infants and children who are moving.

In newborns, both premature and full-term, CBF in the brainstem, thalami, basal ganglia, and sensorimotor cortices is much higher than in other parts of the cortex. Consideration needs to be taken when studying this age group since the development of CBF is heterogeneous (Ouyang *et al.*, 2017). CBF is low in newborns and increases to almost double that of an adult brain during the first years of life (Kim *et al.*, 2018). CBF then peaks at 8 years, after which it steadily decreases. To avoid head motion, young children are often sedated or fully anesthetized, which influences CBF (Morris *et al.*, 2018). Major clinical applications of multi-PLD ASL in the pediatric brain, including stroke, vasculopathy, hypoxic-ischemic injury, epilepsy, migraine, tumor, infection, and metabolic disease, were recently reviewed (Golay and Ho, 2022).

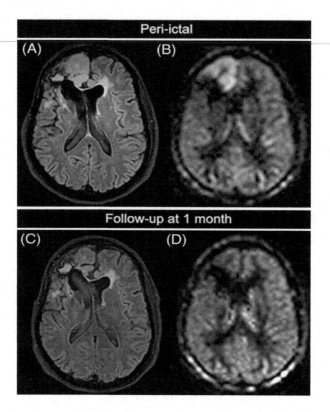

FIG. 4.5

Seizure activity on ASL. A patient with a history of frontal ganglioneuroma resection 20 years before this presentation had an epileptic seizure, presumably due to encephalomalacia/gliosis. On the peri-ictal MRI, the FLAIR image (A) shows a swollen cortical ribbon in the right medial frontal lobe, with corresponding increased perfusion signal on 3D ASL (B). Both the FLAIR abnormality and the hyperperfusion are limited to the cortex. Note the contrast between the perfusion signal in the cortex and the subcortical white matter. Lateral to this finding is the stable-appearing surgical cavity with fluid-fluid level and ex vacuo dilation of the right lateral ventricle. At follow-up, the cortical abnormality has almost completely resolved: the cortex has normal size, and only minimal FLAIR hyperintensity is visible (C). The ASL scan (D) shows normalized perfusion in the affected region, similar to the left frontal lobe.

From Lindner, T., Bolar, D.S., Achten, E., et al., 2023. Current state and guidance on arterial spin labeling perfusion MRI in clinical neuroimaging. Magn. Reson. Med. 89, 2024–2047. https://doi.org/10.1002/mrm.29572.

4.5 Advanced techniques and future directions

Advanced ASL techniques can be subdivided into those that improve the robustness of ASL's CBF measurement and those that obtain additional hemodynamic information.

4.5.1 **More robust ASL techniques**

New MRI acquisitions allow measuring labeling efficiency, which needs to be accounted for when the labeling plane is not perpendicular to the feeding vessels or in patients with abnormal blood flow velocity (Chen *et al.*, 2011a; O'Gorman *et al.*, 2006; Lorenz *et al.*, 2018). Velocity-selective ASL techniques label blood water in the brain, much closer to the location of CBF readout than the standard ASL techniques with a spatially defined labeling plane (*i.e.*, spatial selective). Velocity-selective labeling exploits the fact that blood flow is slower in capillaries than in larger arteries or veins (Schmid *et al.*, 2014, 2015, 2017; Wong *et al.*, 2006). A promising future approach could be to combine velocity-selective and standard ASL, which can image both ATT and CBF even for very long ATTs (Woods *et al.*, 2022; Qiu *et al.*, 2012).

4.5.2 **Additional hemodynamic information**

ASL-based MR angiography (MRA) has advantages compared to other MRA techniques. Vessels can be labeled selectively, and the temporal and spatial resolution can be optimized (Hernandez-Garcia *et al.*, 2022; Suzuki *et al.*, 2020). The sparsity of the acquired signal makes ASL-MRA well suited for accelerated acquisition using undersampling (Schauman *et al.*, 2020). Other interesting developments are those that combine several of the abovementioned techniques. A 3D-MRA can be acquired with PCASL (Koktzoglou *et al.*, 2015), and a 4D MRA with blood-inflow visualization can be acquired with Look-Locker PASL or PCASL (Suzuki *et al.*, 2019, 2020; Jang *et al.*, 2014), possibly together with the CBF image (Suzuki *et al.*, 2018), potentially replacing current angiographic techniques with better temporal and spatial resolution.

Other new ASL sequences allow measuring properties of the labeled blood water as it crosses the blood-brain barrier (BBB). These are able to map the BBB water permeability, either based on diffusion or T2 relaxation time differences between the intra- and extravascular ASL signal (Dickie *et al.*, 2020; Shao *et al.*, 2019; Stanisz *et al.*, 2005). These two properties can be used to separate labeled water compartments using diffusion-weighted (Shao *et al.*, 2023) or multiecho time (Gregori *et al.*, 2013) ASL readouts. Preclinical results suggest that BBB permeability is associated with aging (Ohene *et al.*, 2019, 2020) and SVD (Fujima *et al.*, 2020).

On the processing side, there are newer algorithms that investigate changes in CBF patterns, beyond separately investigating individual regions. Machine learning techniques were shown to detect CBF patterns unique to major depression patients (Ramasubbu *et al.*, 2019) and dementia patients (Yamashita *et al.*, 2010). CBF may also improve the prediction of the brain's biological age (MacDonald *et al.*, 2020; Rokicki *et al.*, 2021; Dijsselhof *et al.*, 2023). Such approaches require large multi-center training datasets to achieve generalizability. Harmonizing ASL acquisitions and image processing, but also improvements in data sharing are essential for these developments (Mutsaerts *et al.*, 2020a). The recent ASL extension of the brain imaging data structure (BIDS) may facilitate data aggregation (Clement *et al.*, 2022).

4.6 Conclusion

Since its invention in 1990, ASL has developed from a preclinical research sequence to a standardized clinical sequence available on most MRI scanners (Alsop *et al.*, 2015). We have discussed the developments that allowed clinical research and the most important clinical applications.

Examples of ASL in vascular disease are assessment of infarctions and the resulting penumbra, decreases in perfusion due to damaged or occluded vessels, or changes in blood-brain barrier integrity. ASL in tumors offers the opportunity to differentiate between benign or malignant neoplasms, the type of tumor, and to determine its grade and progression while removing the need for contrast agents. Neurodegenerative diseases are associated with changes in perfusion and blood-brain barrier integrity, all measurable by ASL. Epilepsia, and associated attacks or seizures, may manifest as changes in perfusion and can be visualized by ASL. We discussed the advanced ASL techniques that are being developed and validated that we believe have the highest ability to help future clinical ASL applications.

We have focused this chapter on applications of ASL in the brain, as these are the most established and clinically used. Further clinical ASL applications are likely to occur in the kidney, which, similar to the brain, has relatively high perfusion and thus ASL SNR. Challenges for ASL in the body are breathing motion and the localization of the labeling plane.

References

Abdel Razek, A.A.K., Talaat, M., El-Serougy, L., Gaballa, G., Abdelsalam, M., 2019. Clinical applications of arterial spin labeling in brain tumors. J. Comput. Assist. Tomogr. 43, 525–532.

Adebimpe, A., *et al.*, 2021. ASLPrep: a generalizable platform for processing of arterial spin labeled MRI and quantification of regional brain perfusion. bioRxiv. https://doi.org/10.1101/2021.05.20.444998. 2021.05.20.444998.

Almeida, J.R.C., *et al.*, 2018. Test-retest reliability of cerebral blood flow in healthy individuals using arterial spin labeling: findings from the EMBARC study. Magn. Reson. Imaging 45, 26–33.

Alsaedi, A., *et al.*, 2019. The value of arterial spin labelling in adults glioma grading: systematic review and meta-analysis. Oncotarget 10, 1589–1601.

Alsop, D.C., Detre, J.A., 1996. Reduced transit-time sensitivity in noninvasive magnetic resonance imaging of human cerebral blood flow. J. Cereb. Blood Flow Metab. 16, 1236–1249.

Alsop, D.C., Detre, J.A., 1998. Multisection cerebral blood flow MR imaging with continuous arterial spin labeling. Radiology 208, 410–416. Preprint at https://doi.org/10.1148/radiology.208.2.9680569.

Alsop, D.C., Detre, J.A., 1999. Background suppressed 3D RARE ASL perfusion imaging. In: International Society for Magnetic Resonance in Medicine, p. 601.

Alsop, D.C., Detre, J.A., Grossman, M., 2000. Assessment of cerebral blood flow in Alzheimer's disease by spin-labeled magnetic resonance imaging. Ann. Neurol. 47, 93–100.

Alsop, D.C., *et al.*, 2015. Recommended implementation of arterial spin-labeled perfusion MRI for clinical applications: a consensus of the ISMRM perfusion study group and the European consortium for ASL in dementia. Magn. Reson. Med. 73, 102–116.

Ambarki, K., *et al.*, 2015. Accuracy of parenchymal cerebral blood flow measurements using pseudocontinuous arterial spin-labeling in healthy volunteers. AJNR Am. J. Neuroradiol. 36, 1816–1821.

Ances, B.M., *et al.*, 2009. Effects of aging on cerebral blood flow, oxygen metabolism, and blood oxygenation level dependent responses to visual stimulation. Hum. Brain Mapp. 30, 1120–1132.

Andersson, J.L.R., Skare, S., Ashburner, J., 2003. How to correct susceptibility distortions in spin-echo echo-planar images: application to diffusion tensor imaging. NeuroImage 20, 870–888.

Arteaga, D.F., *et al.*, 2017. Planning-free cerebral blood flow territory mapping in patients with intracranial arterial stenosis. J. Cereb. Blood Flow Metab. 37, 1944–1958.

Ashburner, J., 2007. A fast diffeomorphic image registration algorithm. NeuroImage 38, 95–113.

ASL pipeline inventory. https://docs.google.com/document/d/e/2PACX-1vQ-1GF2fmz6Q4Iu kuKP_-57H-xi872Xq_uBlX5P0Cwpj4RYd_t73pvZ64UqXegPaVpQJhQQrVRJRPro/pub.

Aslan, S., Lu, H., 2010. On the sensitivity of ASL MRI in detecting regional differences in cerebral blood flow. Magn. Reson. Imaging 28, 928–935.

Aslan, S., *et al.*, 2010. Estimation of labeling efficiency in pseudocontinuous arterial spin labeling. Magn. Reson. Med. 63, 765–771.

Asllani, I., Borogovac, A., Brown, T.R., 2008. Regression algorithm correcting for partial volume effects in arterial spin labeling MRI. Magn. Reson. Med. 60, 1362–1371.

Asllani, I., *et al.*, 2009. Separating function from structure in perfusion imaging of the aging brain. Hum. Brain Mapp. 30, 2927–2935.

Baas, K.P.A., *et al.*, 2021. Effects of acquisition parameter modifications and field strength on the reproducibility of brain perfusion measurements using arterial spin-labeling. AJNR Am. J. Neuroradiol. 42, 109–115.

Bambach, S., Smith, M., Morris, P.P., Campeau, N.G., Ho, M.-L., 2022. Arterial spin labeling applications in pediatric and adult neurologic disorders. J. Magn. Reson. Imaging 55, 698–719.

Bell, L.C., Suzuki, Y., Petr, J., Van Houdt, P.J., Sourbron, S., Mutsaerts, H.J.M.M., 2023. The road to the ISMRM open science initiative for perfusion imaging (OSIPI): a community-led initiative for reproducible perfusion MRI, MRM (in print).

Binnewijzend, M.A.A., *et al.*, 2013. Cerebral blood flow measured with 3D pseudocontinuous arterial spin-labeling MR imaging in Alzheimer disease and mild cognitive impairment: a marker for disease severity. Radiology 267, 221–230.

Boxerman, J.L., *et al.*, 2020. Consensus recommendations for a dynamic susceptibility contrast MRI protocol for use in high-grade gliomas. Neuro-Oncology 22, 1262–1275.

Brayne, C., Davis, D., 2012. Making Alzheimer's and dementia research fit for populations. Lancet vol. 380, 1441–1443. Preprint at https://doi.org/10.1016/s0140-6736(12)61803-0.

Buxton, R.B., *et al.*, 1998. A general kinetic model for quantitative perfusion imaging with arterial spin labeling. Magn. Reson. Med. 40, 383–396.

Çavuşoğlu, M., Pfeuffer, J., Uğurbil, K., Uludağ, K., 2009. Comparison of pulsed arterial spin labeling encoding schemes and absolute perfusion quantification. Magn. Reson. Imaging 27, 1039–1045.

Chang, Y.V., Vidorreta, M., Wang, Z., Detre, J.A., 2017. 3D-accelerated, stack-of-spirals acquisitions and reconstruction of arterial spin labeling MRI. Magn. Reson. Med. 78, 1405–1419.

Chao, L.L., *et al.*, 2010. ASL perfusion MRI predicts cognitive decline and conversion from MCI to dementia. Alzheimer Dis. Assoc. Disord. 24, 19–27.

Chappell, M.A., *et al.*, 2011. Partial volume correction of multiple inversion time arterial spin labeling MRI data. Magn. Reson. Med. 65, 1173–1183.

Chappell, M.A., *et al.*, 2021. Partial volume correction in arterial spin labeling perfusion MRI: a method to disentangle anatomy from physiology or an analysis step too far? NeuroImage 238, 118236.

Chen, Y., Wang, D.J.J., Detre, J.A., 2011a. Test-retest reliability of arterial spin labeling with common labeling strategies. J. Magn. Reson. Imaging 33, 940–949.

Chen, J.J., Rosas, H.D., Salat, D.H., 2011b. Age-associated reductions in cerebral blood flow are independent from regional atrophy. NeuroImage 55, 468–478.

Choi, Y.J., Kim, H.S., Jahng, G.-H., Kim, S.J., Suh, D.C., 2013. Pseudoprogression in patients with glioblastoma: added value of arterial spin labeling to dynamic susceptibility contrast perfusion MR imaging. Acta Radiol. 54, 448–454.

Clement, P., *et al.*, 2017. Variability of physiological brain perfusion in healthy subjects—a systematic review of modifiers. Considerations for multi-center ASL studies. J. Cereb. Blood Flow Metab. 271678X17702156.

Clement, P., *et al.*, 2022. ASL-BIDS, the brain imaging data structure extension for arterial spin labeling. Sci. Data 9, 543.

Dai, W., Garcia, D., de Bazelaire, C., Alsop, D.C., 2008. Continuous flow-driven inversion for arterial spin labeling using pulsed radio frequency and gradient fields. Magn. Reson. Med. 60, 1488–1497.

Dai, W., *et al.*, 2009. Mild cognitive impairment and Alzheimer disease: patterns of altered cerebral blood flow at MR imaging. Radiology 250, 856–866.

Dai, W., Shankaranarayanan, A., Alsop, D.C., 2013. Volumetric measurement of perfusion and arterial transit delay using hadamard encoded continuous arterial spin labeling. Magn. Reson. Med. 69, 1014–1022.

Dangouloff-Ros, V., *et al.*, 2016. Arterial spin labeling to predict brain tumor grading in children: correlations between histopathologic vascular density and perfusion MR imaging. Radiology 281, 553–566.

Detre, J.A., Leigh, J.S., Williams, D.S., Koretsky, A.P., 1992. Perfusion imaging. Magn. Reson. Med. 23, 37–45.

Detre, J.A., *et al.*, 1998. Noninvasive MRI evaluation of cerebral blood flow in cerebrovascular disease. Neurology 50, 633–641.

Di Napoli, A., *et al.*, 2020. Arterial spin labeling MRI in carotid stenosis: arterial transit artifacts may predict symptoms. Radiology 297, 652–660.

Dickie, B.R., Parker, G.J.M., Parkes, L.M., 2020. Measuring water exchange across the blood-brain barrier using MRI. Prog. Nucl. Magn. Reson. Spectrosc. 116, 19–39.

Dijsselhof, M.B.J., *et al.*, 2023. The value of arterial spin labelling perfusion MRI in brain age prediction. Hum. Brain Mapp. https://doi.org/10.1002/hbm.26242.

Dixon, W.T., Du, L.N., Faul, D.D., Gado, M., Rossnick, S., 1986. Projection angiograms of blood labeled by adiabatic fast passage. Magn. Reson. Med. 3, 454–462.

Dolui, S., *et al.*, 2016. Comparison of non-invasive MRI measurements of cerebral blood flow in a large multisite cohort. J. Cereb. Blood Flow Metab. 36, 1244–1256.

Dolui, S., *et al.*, 2017. Structural correlation-based outlier rejection (SCORE) algorithm for arterial spin labeling time series. J. Magn. Reson. Imaging 45, 1786–1797.

Dolui, S., Li, Z., Nasrallah, I.M., Detre, J.A., Wolk, D.A., 2020. Arterial spin labeling versus F-FDG-PET to identify mild cognitive impairment. Neuroimage Clin. 25, 102146.

Du, A.T., et al., 2006. Hypoperfusion in frontotemporal dementia and Alzheimer disease by arterial spin labeling MRI. Neurology 67, 1215–1220.

Edelman, R.R., et al., 1994. Qualitative mapping of cerebral blood flow and functional localization with echo-planar MR imaging and signal targeting with alternating radio frequency. Radiology vol. 192, 513–520. Preprint at https://www.ncbi.nlm.nih.gov/pubmed/8029425.

Elvsåshagen, T., et al., 2018. Cerebral blood flow changes after a day of wake, sleep, and sleep deprivation. NeuroImage. S1053811918321104.

Falk Delgado, A., De Luca, F., van Westen, D., Falk Delgado, A., 2018. Arterial spin labeling MR imaging for differentiation between high- and low-grade glioma-a meta-analysis. Neuro-Oncology 20, 1450–1461.

Fällmar, D., et al., 2017. Arterial spin labeling-based Z-maps have high specificity and positive predictive value for neurodegenerative dementia compared to FDG-PET. Eur. Radiol. 27, 4237–4246.

Feinberg, D.A., Oshio, K., 1991. GRASE (gradient- and spin-echo) MR imaging: a new fast clinical imaging technique. Radiology 181, 597–602.

Fujima, N., et al., 2020. Utility of a diffusion-weighted arterial spin labeling (DW-ASL) technique for evaluating the progression of brain white matter lesions. Magn. Reson. Imaging 69, 81–87.

Gajdoš, M., et al., 2021. Epileptogenic zone detection in MRI negative epilepsy using adaptive thresholding of arterial spin labeling data. Sci. Rep. 11, 10904.

Garnier-Crussard, A., et al., 2022. White matter hyperintensity topography in Alzheimer's disease and links to cognition. Alzheimers Dement. 18, 422–433.

Gevers, S., et al., 2011. Intra-and multicenter reproducibility of pulsed, continuous and pseudo-continuous arterial spin labeling methods for measuring cerebral perfusion. J. Cereb. Blood Flow Metab. https://doi.org/10.1038/jcbfm.2011.10.

Gevers, S., et al., 2012. Arterial spin labeling measurement of cerebral perfusion in children with sickle cell disease. J. Magn. Reson. Imaging 35, 779–787.

Golay, X., Ho, M.-L., 2022. Multidelay ASL of the pediatric brain. Br. J. Radiol. 95, 20220034.

Golay, X., Petersen, E.T., Hui, F., 2005. Pulsed star labeling of arterial regions (PULSAR): a robust regional perfusion technique for high field imaging. Magn. Reson. Med. 53, 15–21.

Gorgolewski, K.J., et al., 2016. The brain imaging data structure, a format for organizing and describing outputs of neuroimaging experiments. Sci. Data 3, 160044.

Gregori, J., Schuff, N., Kern, R., Günther, M., 2013. T2-based arterial spin labeling measurements of blood to tissue water transfer in human brain. J. Magn. Reson. Imaging 37, 332–342.

Günther, M., Oshio, K., Feinberg, D.A., 2005. Single-shot 3D imaging techniques improve arterial spin labeling perfusion measurements. Magn. Reson. Med. 54, 491–498.

Hales, P.W., Kirkham, F.J., Clark, C.A., 2016. A general model to calculate the spin-lattice (T1) relaxation time of blood, accounting for haematocrit, oxygen saturation and magnetic field strength. J. Cereb. Blood Flow Metab. 36, 370–374.

Haller, S., et al., 2014. Acute caffeine administration effect on brain activation patterns in mild cognitive impairment. J. Alzheimers Dis. 41, 101–112.

Handley, R., et al., 2013. Acute effects of single-dose aripiprazole and haloperidol on resting cerebral blood flow (rCBF) in the human brain. Hum. Brain Mapp. 34, 272–282.

Heijtel, D.F.R., *et al.*, 2014. Accuracy and precision of pseudo-continuous arterial spin labeling perfusion during baseline and hypercapnia: a head-to-head comparison with ^{15}O H$_2$O positron emission tomography. NeuroImage 92, 182–192.

Henriksen, O.M., *et al.*, 2022. High-grade glioma treatment response monitoring biomarkers: a position statement on the evidence supporting the use of advanced MRI techniques in the clinic, and the latest bench-to-bedside developments. Part 1: perfusion and diffusion techniques. Front. Oncol. 12, 810263.

Hernandez-Garcia, L., Lewis, D.P., Moffat, B., Branch, C.A., 2007. Magnetization transfer effects on the efficiency of flow-driven adiabatic fast passage inversion of arterial blood. NMR Biomed. 20, 733–742.

Hernandez-Garcia, L., *et al.*, 2022. Recent technical developments in ASL: a review of the state of the art. Magn. Reson. Med. vol. 88, 2021–2042. Preprint at https://doi.org/10.1002/mrm.29381.

Hu, W.T., Wang, Z., Lee, V.M., Hu, W.T., Lee, V.M., 2011. Distinct cerebral perfusion patterns in FTLD and AD. https://doi.org/10.1212/WNL.0b013e3181f11e35.

Hu, H.H., *et al.*, 2017. Assessment of cerebral blood perfusion reserve with acetazolamide using 3D spiral ASL MRI: preliminary experience in pediatric patients. Magn. Reson. Imaging 35, 132–140.

Ibaraki, M., *et al.*, 2010. Interindividual variations of cerebral blood flow, oxygen delivery, and metabolism in relation to hemoglobin concentration measured by positron emission tomography in humans. J. Cereb. Blood Flow Metab. 30, 1296–1305.

Jack, C.R., *et al.*, 2010. Update on the magnetic resonance imaging core of the Alzheimer's disease neuroimaging initiative. Alzheimers Dement. 6, 212–220.

Jang, J., *et al.*, 2014. Non-contrast-enhanced 4D MR angiography with STAR spin labeling and variable flip angle sampling: a feasibility study for the assessment of Dural Arteriovenous Fistula. Neuroradiology 56, 305–314.

Johnson, N.A., *et al.*, 2005. Pattern of cerebral hypoperfusion in Alzheimer disease and mild cognitive impairment measured with arterial spin-labeling MR imaging: initial experience. Radiology 234, 851–859.

Joris, P.J., Mensink, R.P., Adam Tanja, C., Liu, T.T., 2018. Cerebral blood flow measurements in adults: a review on the effects of dietary factors and exercise. Nutrients 10, 1–15.

Kamagata, K., *et al.*, 2011. Posterior hypoperfusion in Parkinson's disease with and without dementia measured with arterial spin labeling MRI. J. Magn. Reson. Imaging 33, 803–807.

Kim, S.M., *et al.*, 2013. Regional cerebral perfusion in patients with Alzheimer's disease and mild cognitive impairment: effect of APOE epsilon4 allele. Neuroradiology 55, 25–34.

Kim, H.G., *et al.*, 2018. Multidelay arterial spin-labeling MRI in neonates and infants: cerebral perfusion changes during brain maturation. AJNR Am. J. Neuroradiol. 39, 1912–1918.

Klein, A., *et al.*, 2009. Evaluation of 14 nonlinear deformation algorithms applied to human brain MRI registration. NeuroImage 46, 786–802.

Koktzoglou, I., *et al.*, 2015. Nonenhanced arterial spin labeled carotid MR angiography using three-dimensional radial balanced steady-state free precession imaging. J. Magn. Reson. Imaging 41, 1150–1156.

Lee, C., *et al.*, 2009. Imaging cerebral blood flow in the cognitively normal aging brain with arterial spin labeling: implications for imaging of neurodegenerative disease. J. Neuroimaging 19, 344–352.

Leeman-Markowski, B., 2016. Review of MRI-negative epilepsy. JAMA Neurol. 73, 1377.

Li, W., *et al.*, 2017. Fast measurement of blood T1 in the human carotid artery at 3T: accuracy, precision, and reproducibility. Magn. Reson. Med. 77, 2296–2302.

Li, Y., *et al.*, 2018. ASL-MRICloud: an online tool for the processing of ASL MRI data. NMR Biomed. e4051.

Li, D., *et al.*, 2020. Quantitative study of the changes in cerebral blood flow and iron deposition during progression of Alzheimer's disease. J. Alzheimers Dis. 78, 439–452.

Liang, X., Connelly, A., Tournier, J.-D., Calamante, F., 2014. A variable flip angle-based method for reducing blurring in 3D GRASE ASL. Phys. Med. Biol. 59, 5559–5573.

Lindner, T., Achten, R., Barkhof, F., Bolar, D., Detre, J., Golay, X., Gunther, M., Haller, S., Ingala, S., Jager, H.R., Jahng, G.H., Juttukonda, M., Kimura, H., Lequin, M., Lou, X., Petr, J., Pinter, N., Smits, M., Sokolska, M., Zaharchuk, G., Mutsaerts, H.J.M.M., 2023. A positioning statement on how to protocol and interpret ASL perfusion images in clinical neuroimaging applications. Magn. Reason. Med. 89, 2024–2027.

Liu, H.-L., *et al.*, 2001. Perfusion-weighted imaging of interictal hypoperfusion in temporal lobe epilepsy using FAIR-HASTE: Comparison with H2150 PET measurements. Magn. Reson. Med. 45, 431–435.

Lorenz, K., Mildner, T., Schlumm, T., Möller, H.E., 2018. Characterization of pseudo-continuous arterial spin labeling: simulations and experimental validation. Magn. Reson. Med. 79, 1638–1649.

Lorenzini, L., *et al.*, 2022. The Open-Access European Prevention of Alzheimer's Dementia (EPAD) MRI dataset and processing workflow. Neuroimage Clin. 35, 103106.

Lu, H., Clingman, C., Golay, X., van Zijl, P.C.M., 2004. Determining the longitudinal relaxation time (T1) of blood at 3.0 Tesla. Magn. Reson. Med. 52, 679–682.

Lu, W., Yu, C., Wang, L., Wang, F., Qiu, J., 2022. Perfusion heterogeneity of cerebral small vessel disease revealed via arterial spin labeling MRI and machine learning. Neuroimage Clin. 36, 103165.

Luh, W.-M., Wong, E.C., Bandettini, P.A., Hyde, J.S., 1999. QUIPSS II with thin-slice TI1 periodic saturation: a method for improving accuracy of quantitative perfusion imaging using pulsed arterial spin labeling. Magn. Reson. Med. vol. 41, 1246–1254. Preprint at doi:10.1002/(sici)1522-2594(199906)41:6<1246::aid-mrm22>3.0.co;2-n.

MacDonald, M.E., *et al.*, 2020. Age-related differences in cerebral blood flow and cortical thickness with an application to age prediction. Neurobiol. Aging 95, 131–142.

MacIntosh, B.J., *et al.*, 2008. Measuring the effects of remifentanil on cerebral blood flow and arterial arrival time using 3D GRASE MRI with pulsed arterial spin labelling. J. Cereb. Blood Flow Metab. 28, 1514–1522.

Mak, E., *et al.*, 2021. Proximity to dementia onset and multi-modal neuroimaging changes: The prevent-dementia study. NeuroImage 229, 117749.

Manning, P., *et al.*, 2020. Differentiation of progressive disease from pseudoprogression using 3D PCASL and DSC perfusion MRI in patients with glioblastoma. J. Neuro-Oncol. 147, 681–690.

Maral, H., Ertekin, E., Tunçyürek, Ö., Özsunar, Y., 2020. Effects of susceptibility artifacts on perfusion MRI in patients with primary brain tumor: a comparison of arterial spin-labeling versus DSC. AJNR Am. J. Neuroradiol. 41, 255–261.

Marcus, D.S., *et al.*, 2007. Open access series of imaging studies (OASIS): cross-sectional MRI data in young, middle aged, nondemented, and demented older adults. J. Cogn. Neurosci. 19, 1498–1507.

Maumet, C., Maurel, P., Ferre, J.C., Barillot, C., 2014. Robust estimation of the cerebral blood flow in arterial spin labelling. Magn. Reson. Imaging 32, 497–504.

Mazziotta, J.C., Toga, A.W., Evans, A., Fox, P., Lancaster, J., 1995. A probabilistic atlas of the human brain: theory and rationale for its development. NeuroImage 2, 89–101.

McLaughlin, A.C., Ye, F.Q., Pekar, J.J., Santha, A.K., Frank, J.A., 1997. Effect of magnetization transfer on the measurement of cerebral blood flow using steady-state arterial spin tagging approaches: a theoretical investigation. Magn. Reson. Med. 37, 501–510.

Mezue, M., et al., 2014. Optimization and reliability of multiple postlabeling delay pseudo-continuous arterial spin labeling during rest and stimulus-induced functional task activation. J. Cereb. Blood Flow Metab. 34, 1919–1927.

Morris, E.A., et al., 2018. Elevated brain oxygen extraction fraction in preterm newborns with anemia measured using noninvasive MRI. J. Perinatol. 38, 1636–1643.

Mutsaerts, H.J.M.M., et al., 2014. Inter-vendor reproducibility of pseudo-continuous arterial spin labeling at 3 Tesla. PLoS One 9, e104108.

Mutsaerts, H.J.M.M., et al., 2015. Multi-vendor reliability of arterial spin labeling perfusion MRI using a near-identical sequence: implications for multi-center studies. NeuroImage 113.

Mutsaerts, H.J.M.M., et al., 2019. Cerebral perfusion changes in presymptomatic genetic frontotemporal dementia: a GENFI study. Brain 142, 1108–1120.

Mutsaerts, H.J.M.M., et al., 2020a. ExploreASL: an image processing pipeline for multi-center ASL perfusion MRI studies. NeuroImage, 117031.

Mutsaerts, H.J.M.M., et al., 2020b. Spatial coefficient of variation of arterial spin labeling MRI as a cerebrovascular correlate of carotid occlusive disease. PLoS One 15, e0229444.

Nedelska, Z., et al., 2018. Regional cortical perfusion on arterial spin labeling MRI in dementia with Lewy bodies: associations with clinical severity, glucose metabolism and tau PET. Neuroimage Clin. 19, 939–947.

O'Gorman, R.L., et al., 2006. In vivo estimation of the flow-driven adiabatic inversion efficiency for continuous arterial spin labeling: a method using phase contrast magnetic resonance angiography. Magn. Reson. Med. 55, 1291–1297.

Ohene, Y., et al., 2019. Non-invasive MRI of brain clearance pathways using multiple echo time arterial spin labelling: an aquaporin-4 study. NeuroImage 188, 515–523.

Ohene, Y., Harrison, I., Evans, P.E., Thomas, D.L., Lythgoe, M.F., Wells, J., 2020. Increased blood-brain interface permeability to water in the ageing brain detected using non-invasive multi-TE ASL MRI. Magn. Reson. Med. 85(1), 326–333.

Oliver-Taylor, A., et al., 2017. A calibrated perfusion phantom for quality assurance of quantitative arterial spin labelling. In: *ISMRM'17: Proceedings of the 25th Scientific Meeting and Exhibition of International Society for Magnetic Resonance in Medicine* vol. 25, 681. ISMRM.

Ouyang, M., et al., 2017. Heterogeneous increases of regional cerebral blood flow during preterm brain development: preliminary assessment with pseudo-continuous arterial spin labeled perfusion MRI. NeuroImage 147, 233–242.

Parkes, L.M., Rashid, W., Chard, D.T., Tofts, P.S., 2004. Normal cerebral perfusion measurements using arterial spin labeling: reproducibility, stability, and age and gender effects. Magn. Reson. Med. 51, 736–743.

Petersen, E.T., et al., 2010. The QUASAR reproducibility study, part II: results from a multi-center arterial spin labeling test-retest study. NeuroImage 49, 104–113.

Petr, J., *et al.*, 2018. Effects of systematic partial volume errors on the estimation of gray matter cerebral blood flow with arterial spin labeling MRI. MAGMA 31, 725–734.

Pinto, J., *et al.*, 2020. Calibration of arterial spin labeling data-potential pitfalls in post-processing. Magn. Reson. Med. 83, 1222–1234.

Pizzini, F.B., *et al.*, 2013. Cerebral perfusion alterations in epileptic patients during peri-ictal and post-ictal phase: PASL vs DSC-MRI. Magn. Reson. Imaging 31, 1001–1005.

Pohmann, R., 2010. Accurate, localized quantification of white matter perfusion with single-voxel ASL. Magn. Reson. Med. 64, 1109–1113.

Puig, O., *et al.*, 2018. Phase contrast mapping MRI measurements of global cerebral blood flow across different perfusion states—a direct comparison with O-H2O positron emission tom77ography using a hybrid PET/MR system. J. Cereb. Blood Flow Metab. 1–11.

Qin, Q., *et al.*, 2022. Velocity-selective arterial spin labeling perfusion MRI: a review of the state of the art and recommendations for clinical implementation. Magn. Reson. Med. 88, 1528–1547.

Qiu, D., *et al.*, 2012. CBF measurements using multidelay pseudocontinuous and velocity-selective arterial spin labeling in patients with long arterial transit delays: comparison with xenon CT CBF. J. Magn. Reson. Imaging 36, 110–119.

Ramachandran, S., *et al.*, 2022. Feasibility of arterial spin labeling in evaluating high- and low-flow peripheral vascular malformations: a case series. BJR Case Rep. 8, 20210083.

Ramasubbu, R., Brown, E.C., Marcil, L.D., Talai, A.S., Forkert, N.D., 2019. Automatic classification of major depression disorder using arterial spin labeling MRI perfusion measurements. Psychiatry Clin. Neurosci. 73, 486–493.

Roach, B.A., *et al.*, 2016. Interrogating the functional correlates of collateralization in patients with intracranial stenosis using multimodal hemodynamic imaging. AJNR Am. J. Neuroradiol. 37, 1132–1138.

Rokicki, J., *et al.*, 2021. Multimodal imaging improves brain age prediction and reveals distinct abnormalities in patients with psychiatric and neurological disorders. Hum. Brain Mapp. 42, 1714–1726.

Sandson, T.A., O'Connor, M., Sperling, R.A., Edelman, R.R., Warach, S., 1996. Noninvasive perfusion MRI in Alzheimer's disease: a preliminary report. Neurology 47, 1339–1342.

Schauman, S.S., Chiew, M., Okell, T.W., 2020. Highly accelerated vessel-selective arterial spin labeling angiography using sparsity and smoothness constraints. Magn. Reson. Med. 83, 892–905.

Schmid, S., Ghariq, E., Teeuwisse, W.M., Webb, A., van Osch, M.J.P., 2014. Acceleration-selective arterial spin labeling. Magn. Reson. Med. 71, 191–199.

Schmid, S., *et al.*, 2015. Comparison of velocity- and acceleration-selective arterial spin labeling with [15O]H2O positron emission tomography. J. Cereb. Blood Flow Metab. 35, 1–8.

Schmid, S., Petersen, E.T., Van Osch, M.J.P., 2017. Insight into the labeling mechanism of acceleration selective arterial spin labeling. MAGMA 30, 165–174.

Shao, X., *et al.*, 2019. Mapping water exchange across the blood-brain barrier using 3D diffusion-prepared arterial spin labeled perfusion MRI. Magn. Reson. Med. 81, 3065–3079.

Shao, X., Zhao, C., Shou, Q., St Lawrence, K.S., Wang, D.J.J., 2023. Quantification of blood-brain barrier water exchange and permeability with multidelay diffusion-weighted pseudo-continuous arterial spin labeling. Magn. Reson. Med. https://doi.org/10.1002/mrm.29581.

Shi, Y., *et al.*, 2016. Cerebral blood flow in small vessel disease: a systematic review and meta-analysis. J. Cereb. Blood Flow Metab. 36, 1653–1667.

Shirzadi, Z., *et al.*, 2018. Enhancement of automated blood flow estimates (ENABLE) from arterial spin-labeled MRI. J. Magn. Reson. Imaging 47, 647–655.

Shirzadi, Z., *et al.*, 2019. Classifying cognitive impairment based on the spatial heterogeneity of cerebral blood flow images. J. Magn. Reson. Imaging. https://doi.org/10.1002/jmri.26650.

Skurdal, M.J., *et al.*, 2015. Voxel-wise perfusion assessment in cerebral white matter with PCASL at 3T; Is it possible and how long does it take? PLoS One vol. 10, e0135596. Preprint at https://www.ncbi.nlm.nih.gov/pubmed/26267661.

Smith, L.A., *et al.*, 2019. Cortical cerebral blood flow in ageing: effects of haematocrit, sex, ethnicity and diabetes. Eur. Radiol. 29, 5549–5558.

Smits, M., 2021. MRI biomarkers in neuro-oncology. Nat. Rev. Neurol. https://doi.org/10.1038/s41582-021-00510-y.

Song, Y.S., *et al.*, 2013. True progression versus pseudoprogression in the treatment of glioblastomas: a comparison study of normalized cerebral blood volume and apparent diffusion coefficient by histogram analysis. Korean J. Radiol. 14, 662–672.

Soto-Rojas, L.O., *et al.*, 2021. The neurovascular unit dysfunction in Alzheimer's disease. Int. J. Mol. Sci. 22.

Stanisz, G.J., *et al.*, 2005. T1, T2 relaxation and magnetization transfer in tissue at 3T. Magn. Reson. Med. 54, 507–512.

Steketee, R.M.E., *et al.*, 2016. Early-stage differentiation between presenile Alzheimer's disease and frontotemporal dementia using arterial spin labeling MRI. Eur. Radiol. 26, 244–253.

Stewart, C.R., Stringer, M.S., Shi, Y., Thrippleton, M.J., Wardlaw, J.M., 2021. Associations between white matter hyperintensity burden, cerebral blood flow and transit time in small vessel disease: an updated meta-analysis. Front. Neurol. 12, 647848.

Suh, C.H., Kim, H.S., Jung, S.C., Choi, C.G., Kim, S.J., 2018. Perfusion MRI as a diagnostic biomarker for differentiating glioma from brain metastasis: a systematic review and meta-analysis. Eur. Radiol. 28, 3819–3831.

Sunwoo, L., *et al.*, 2016. Differentiation of glioblastoma from brain metastasis: qualitative and quantitative analysis using arterial spin labeling MR imaging. PLoS One 11, e0166662.

Suzuki, Y., Helle, M., Koken, P., Van Cauteren, M., van Osch, M.J.P., 2018. Simultaneous acquisition of perfusion image and dynamic MR angiography using time-encoded pseudo-continuous ASL. Magn. Reson. Med. 79, 2676–2684.

Suzuki, Y., Okell, T.W., Fujima, N., van Osch, M.J.P., 2019. Acceleration of vessel-selective dynamic MR angiography by pseudocontinuous arterial spin labeling in combination with acquisition of control and labeled images in the same shot (ACTRESS). Magn. Reson. Med. 81, 2995–3006.

Suzuki, Y., Fujima, N., van Osch, M.J.P., 2020. Intracranial 3D and 4D MR angiography using arterial spin labeling: technical considerations. Magn. Reson. Med. Sci. 19, 294–309.

Thompson, J.K., Peterson, M.R., Freeman, R.D., 2003. Single-neuron activity and tissue oxygenation in the cerebral cortex. Science 299, 1070–1072.

Toledo, J.B., *et al.*, 2013. Clinical and multimodal biomarker correlates of ADNI neuropathological findings. Acta Neuropathol. Commun. 1, 65.

Tosun, D., *et al.*, 2016. Diagnostic utility of ASL-MRI and FDG-PET in the behavioral variant of FTD and AD. Ann. Clin. Transl. Neurol. 3, 740–751.

Václavů, L., *et al.*, 2018. Hemodynamic provocation with acetazolamide shows impaired cerebrovascular reserve in adults with sickle cell disease. Haematologica. https://doi.org/10.3324/haematol.2018.206094.

Václavů, L., *et al.*, 2020. Cerebral oxygen metabolism in adults with sickle cell disease. Am. J. Hematol. 95, 401–412.

van Dalen, J.W., *et al.*, 2016. White matter hyperintensity volume and cerebral perfusion in older individuals with hypertension using arterial spin-labeling. AJNR Am. J. Neuroradiol. 37, 1824–1830.

van Dalen, J.W., *et al.*, 2021. Longitudinal relation between blood pressure, antihypertensive use and cerebral blood flow, using arterial spin labelling MRI. J. Cereb. Blood Flow Metab. 41, 1756–1766.

van Osch, M.J.P., *et al.*, 2009. Can arterial spin labeling detect white matter perfusion signal? Magn. Reson. Med. 62, 165–173.

Varela, M., *et al.*, 2011. A method for rapid *in vivo* measurement of blood T_1. NMR Biomed. vol. 24, 80–88. Preprint at https://doi.org/10.1002/nbm.1559.

Vidorreta, M., *et al.*, 2013. Comparison of 2D and 3D single-shot ASL perfusion fMRI sequences. NeuroImage 66, 662–671.

Vidorreta, M., *et al.*, 2014. Evaluation of segmented 3D acquisition schemes for whole-brain high-resolution arterial spin labeling at 3 T. NMR Biomed. 27, 1387–1396.

Vidorreta, M., *et al.*, 2017. Whole-brain background-suppressed pCASL MRI with 1D-accelerated 3D RARE stack-of-spirals readout. PLoS One 12, e0183762.

von Samson-Himmelstjerna, F., Madai, V.I., Sobesky, J., Guenther, M., 2016. Walsh-ordered hadamard time-encoded pseudocontinuous ASL (WH pCASL). Magn. Reson. Med. 76, 1814–1824.

Wang, J., *et al.*, 2002. Comparison of quantitative perfusion imaging using arterial spin labeling at 1.5 and 4.0 Tesla. Magn. Reson. Med. 48, 242–254.

Wang, Z., *et al.*, 2008. Empirical optimization of ASL data analysis using an ASL data processing toolbox: ASLtbx. Magn. Reson. Imaging 26, 261–269.

Wang, D.J.J., *et al.*, 2011. Potentials and challenges for arterial spin labeling in pharmacological magnetic resonance imaging. J. Pharmacol. Exp. Ther. 337, 359–366.

Wang, D.J.J., *et al.*, 2013. Multi-delay multi-parametric arterial spin-labeled perfusion MRI in acute ischemic stroke—comparison with dynamic susceptibility contrast enhanced perfusion imaging. Neuroimage Clin. 3, 1–7.

Wang, L., *et al.*, 2020. Evaluation of perfusion MRI value for tumor progression assessment after glioma radiotherapy: a systematic review and meta-analysis. Medicine 99, e23766.

Wang, K., *et al.*, 2022. Optimization of pseudo-continuous arterial spin labeling at 7T with parallel transmission B1 shimming. Magn. Reson. Med. 87, 249–262.

Wardlaw, J.M., Smith, C., Dichgans, M., 2019. Small vessel disease: mechanisms and clinical implications. Lancet Neurol. 18, 684–696.

Warmuth, C., Günther, M., Zimmer, C., 2003. Quantification of blood flow in brain tumors: comparison of arterial spin labeling and dynamic susceptibility-weighted contrast-enhanced MR imaging. Radiology 228, 523–532.

Weber, M.A., *et al.*, 2006. Diagnostic performance of spectroscopic and perfusion MRI for distinction of brain tumors. Neurology 66, 1899–1906.

Weiskopf, N., *et al.*, 2013. Quantitative multi-parameter mapping of R1, PD(*), MT, and R2(*) at 3T: a multi-center validation. Front. Neurosci. vol. 7, 95. Preprint at https://www.ncbi.nlm.nih.gov/pubmed/23772204.

Williams, D.S., Detre, J.A., Leigh, J.S., Koretsky, A.P., 1992. Magnetic resonance imaging of perfusion using spin inversion of arterial water. Proc. Natl. Acad. Sci. U. S. A. 89, 212–216.

Wong, E.C., 2005. Quantifying CBF with pulsed ASL: technical and pulse sequence factors. J. Magn. Reson. Imaging 22, 727–731.

Wong, E.C., 2014. An introduction to ASL labeling techniques. J. Magn. Reson. Imaging 40, 1–10.

Wong, E.C., Buxton, R.B., Frank, L.R., 1997. Implementation of quantitative perfusion imaging techniques for functional brain mapping using pulsed arterial spin labeling. NMR Biomed. 10, 237–249.

Wong, E.C., Buxton, R.B., Frank, L.R., 1998a. Quantitative imaging of perfusion using a single subtraction (QUIPSS and QUIPSS II). Magn. Reson. Med. 39, 702–708.

Wong, E.C., Buxton, R.B., Frank, L.R., 1998b. A theoretical and experimental comparison of continuous and pulsed arterial spin labeling techniques for quantitative perfusion imaging. Magn. Reson. Med. 40, 348–355.

Wong, E.C., et al., 2006. Velocity-selective arterial spin labeling. Magn. Reson. Med. 55, 1334–1341.

Woods, J.G., Achten, E., Asllani, I., Bolar, D.S., Weying, D., Detre, J.A., Fan, A., Fernández-Seara, M., Golay, X., Guo, J., Hernandez-Garcia, L., Ho, M.-L., Juttukonda, M.R., Lu, H., MacIntosh, B.J., Madhuranthakam, A., Mutsaerts, H.J.M.M., Okell, T.O., Parkes, L.M., Pinter, N., Pinto, J., Qin, Q., Smits, M., Suzuki, Y., Thomas, D.L., Van Osch, M.J.P., Wang, D.J.J., Warnert, E.A.H., Zaharchuk, G., Zelaya, F., Zhao, M., Chappell, M.A., 2023. Quantitative cerebral perfusion MRI using multi-timepoint arterial spin labeling: recommendations and clinical applications (in submission).

Woods, J.G., Wong, E.C., Boyd, E.C., Bolar, D.S., 2022. VESPA ASL: VElocity and SPAtially selective arterial spin labeling. Magn. Reson. Med. 87, 2667–2684.

Wu, W.-C., Fernández-Seara, M., Detre, J., a., Wehrli, F. W. & Wang, J., 2007. A theoretical and experimental investigation of the tagging efficiency of pseudocontinuous arterial spin labeling. Magn. Reson. Med. 58, 1020–1027.

Wu, W.-C., et al., 2010. In vivo venous blood T1 measurement using inversion recovery true-FISP in children and adults. Magn. Reson. Med. 64, 1140–1147.

Xekardaki, A., et al., 2015. Arterial spin labeling may contribute to the prediction of cognitive deterioration in healthy elderly individuals. Radiology 274, 490–499.

Xiao, H.-F., et al., 2015. Astrocytic tumour grading: a comparative study of three-dimensional pseudocontinuous arterial spin labelling, dynamic susceptibility contrast-enhanced perfusion-weighted imaging, and diffusion-weighted imaging. Eur. Radiol. 25, 3423–3430.

Xu, F., et al., 2018. Accounting for the role of hematocrit in between-subject variations of MRI-derived baseline cerebral hemodynamic parameters and functional BOLD responses. Hum. Brain Mapp. 39, 344–353.

Yamashita, Y., et al., 2010. Computer-aided classification of patients with dementia of Alzheimer's type based on cerebral blood flow determined with arterial spin labeling technique. SPIE Proc. https://doi.org/10.1117/12.845530.

Yang, Y., et al., 1998. Multislice imaging of quantitative cerebral perfusion with pulsed arterial spin labeling. Magn. Reson. Med. vol. 39, 825–832. Preprint at https://www.ncbi.nlm.nih.gov/pubmed/9581614.

Ye, F.Q., Frank, J.A., Weinberger, D.R., McLaughlin, A.C., 2000. Noise reduction in 3D perfusion imaging by attenuating the static signal in arterial spin tagging (ASSIST). Magn. Reson. Med. 44, 92–100.

Zaharchuk, G., et al., 1999. Multislice perfusion and perfusion territory imaging in humans with separate label and image coils. Magn. Reson. Med. 41, 1093–1098.

Zhang, X., *et al.*, 2013. In vivo blood T(1) measurements at 1.5 T, 3 T, and 7 T. Magn. Reson. Med. 70, 1082–1086.

Zhang, R., *et al.*, 2022. Decreased cerebral blood flow and delayed arterial transit are independently associated with white matter hyperintensity. Front. Aging Neurosci. 14, 762745.

Zhao, L., Vidorreta, M., Soman, S., Detre, J.A., Alsop, D.C., 2017. Improving the robustness of pseudo-continuous arterial spin labeling to off-resonance and pulsatile flow velocity. Magn. Reson. Med. vol. 78, 1342–1351. Preprint at https://doi.org/10.1002/mrm.26513.

Zhao, L., Chang, C.-D., Alsop, D.C., 2018. Controlling T2 blurring in 3D RARE arterial spin labeling acquisition through optimal combination of variable flip angles and k-space filtering. Magn. Reson. Med. 80, 1391–1401.

Zimmer, F., *et al.*, 2016. Pulsed arterial spin labelling at ultra-high field with a B1+-optimised adiabatic labelling pulse. Magn. Reson. Mater. Phys. Biol. Med. 29, 463–473.

Zlokovic, B.V., 2011. Neurovascular pathways to neurodegeneration in Alzheimer's disease and other disorders. Nat. Rev. Neurosci. 12, 723–738.

Recommended further reading

Alsop, D.C., *et al.*, 2015. Recommended implementation of arterial spin-labeled perfusion MRI for clinical applications: a consensus of the ISMRM perfusion study group and the European consortium for ASL in dementia. Magn. Reson. Med. 73, 102–116.

Clement, P., *et al.*, 2022. A beginner's guide to arterial spin labeling (ASL) image processing. Front. Radiol. 2.

Hernandez-Garcia, *et al.*, 2022. Recent technical developments in ASL: a review of the state of the art. Magn. Reson. Med. 88 (5), 2021–2042. https://doi.org/10.1002/mrm.29381.

Lindner, *et al.*, 2023. Current state and guidance on arterial spin labeling perfusion MRI in clinical neuroimaging. Magn. Reson. Med. https://doi.org/10.1002/mrm.29572.

Barker, P.B., Golay, X., Zaharchuk, G. (Eds.), 2013. Clinical Clinical Perfusion MRI: Techniques and Applications. Cambridge University Press.

Qin, *et al.*, 2022. Velocity-selective arterial spin labeling perfusion MRI: a review of the state of the art and recommendations for clinical implementation. Magn. Reson. Med. 88 (5), 1528–1547. https://doi.org/10.1002/mrm.29371.

Woods, *et al.*, 2023. Recommendations for quantitative perfusion MRI with multi-timepoint arterial spin labeling, in preparation for submission to Magn. Reson. Med.

Vasoreactivity MRI

Hai-Ling Margaret Cheng

Institute of Biomedical Engineering, University of Toronto, Toronto, ON, Canada
The Edward S. Rogers Sr. Department of Electrical & Computer Engineering, University of Toronto, Toronto, ON, Canada
Ted Rogers Centre for Heart Research, Translational Biology & Engineering Program, University of Toronto, Toronto, ON, Canada

Abbreviations

ASL	arterial spin labeling
BOLD	blood-oxygen-level-dependent
cAMP	cyclic adenosine monophosphate
CO_2	carbon dioxide
CVR	cerebrovascular reactivity
EDHFs	endothelium-derived hyperpolarizing factors
FMD	flow-mediated dilation
Hb	deoxyhemoglobin
HbO_2	oxyhemoglobin
MRI	magnetic resonance imaging
NO	nitric oxide
O_2	oxygen
$PaCO_2$	arterial partial pressure of CO_2
PET	positron emission tomography
$P_{ET}CO_2$	end-tidal partial pressure of CO_2
SNR	signal-to-noise ratio
VSMCs	vascular smooth muscle cells

5.1 Introduction

The preceding chapters surveyed the central MRI techniques for measuring static perfusion, namely, blood flow in the absence of a vasoactive perturbation. These perfusion MRI technologies have found broad clinical and research utility, precisely because adequate nutritive flow is vital to tissue health and perfusion levels substantially lower or higher than nominal values expected in healthy tissues often signify

Advances in Magnetic Resonance Technology and Applications, Volume 11, ISSN 2666-9099
https://doi.org/10.1016/B978-0-323-95209-5.00014-3

disease or injury. Indeed, the term "perfusion imaging" largely refers to the static scenario. However, blood vessels are dynamic entities and are constantly responding to changing local metabolic demands. When a person exercises on a treadmill, for instance, blood flow to the skeletal muscle increases substantially and is diverted away from the digestive organs. This ability to adapt the vascular response appropriately indicates intact vasoreactivity, an ability predicated on normal vessel architecture, function, and milieu.

Compromised vasoreactivity manifests in many diseases and conditions and often precedes overt static perfusion deficits. For example, in mice fed a high-fat diet, the coronary microvasculature presented the earliest abnormality in the form of a diminished response to a vasoconstrictor, which preceded even alterations in cardiac function (Kwiatkowski *et al.*, 2021). Conversely, restoration of normal vasoreactivity can be a better predictor of prognosis than full recovery of baseline perfusion level. A case in point is a ^{133}Xe SPECT imaging study of patients treated for internal carotid artery occlusion (Kuroda *et al.*, 1993). All patients who showed improvement and did not suffer postsurgery ischemia after a superficial temporal artery–middle cerebral artery double anastomosis shared a curious commonality on a follow-up vasodilator test: cerebral vasoreactivity was invariably restored or maintained, but depressed cerebral blood flow was not necessarily rescued.

In this chapter, we will review MRI methods suited to imaging vasoreactivity. In contrast to the dynamic methods described in previous chapters for measuring static perfusion, vasoreactivity MRI remains underutilized. This relative paucity is a direct consequence of the complexity of the technique, from both an imaging perspective—as dynamic imaging must now capture vessel dynamics—and a pharmacological perspective—the stimulus applied must elicit the specific functional deficit suspected in the tissue of interest. We will briefly review the physiology of vasoreactivity and the stimuli safe to use on humans. To encourage future exploitation of this powerful tool, we will also discuss the earliest and still predominant applications in the brain, as well as emerging roles and new frontiers yet to be explored. An atlas of this chapter is shown in Fig. 5.1.

5.2 Theory

5.2.1 The physiology of vasomotor control

The vascular system is an organized network of blood vessels comprised of arteries feeding the microvascular bed before connecting to veins. The microvascular bed consists of three main vessel types: arterioles that control the bulk of our systemic hemodynamic resistance, capillaries where nutrient and gas exchange occur, and venules that serve as a capacitive buffer. At a cellular level, the endothelium and vascular smooth muscle cells (VSMCs) play a dominant role in the normal functioning of blood vessels. Enveloping all vessels except capillaries and venules, VSMCs control vascular tone by relaxing or contracting. They are especially abundant on

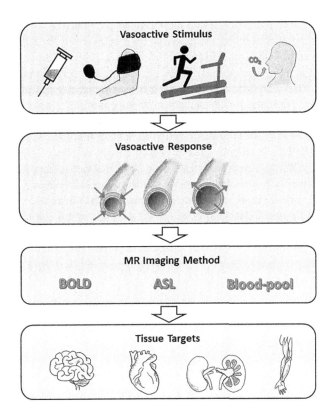

FIG. 5.1

An atlas for imaging vasoreactivity. A vasoactive stimulus is applied to induce a vasodilatory or vasoconstrictive response. Capturing vasoactive dynamics can be performed using blood oxygen level-dependent (BOLD) MRI, arterial spin label (ASL) perfusion, or blood-pool steady-state T1 imaging.

arterioles and respond to a multitude of humoral factors released by the endothelium. In healthy vessels, VSMCs are generally in a quiescent, nonproliferative state. However, proliferation, hypertrophy, or even damage of VSMCs is seen in certain conditions, such as hypertension and atherosclerosis (Bacakova, 2018). Disease can also disrupt the normal function of the endothelium and its production of vasoactive factors such as nitric oxide (Rush *et al.*, 2005). Such changes can and do adversely affect vasoreactivity, the ability of blood vessels to adjust their tone appropriately to local demands. In the following, we review the key mediators of vasoreactivity.

Vascular tone is regulated by both VSMCs and the endothelium via neuronal and humoral control. Endogenous nerve cells on blood vessels can directly stimulate VSMCs to constrict or, less frequently, to relax, depending on the type of adrenoreceptors stimulated on VSMCs (Rath *et al.*, 2012). Independent of neuronal control,

VSMCs also respond to slower-acting humoral signals, many of which are derived from the inner endothelial lining. Nitric oxide (NO) released by the endothelium is a well-known vasodilatory agent discovered in the late 1980s (Palmer *et al.*, 1987). The role of NO as a key mediator of VSMC relaxation is so central that its abnormal production or bioavailability is a key culprit of microvascular dysfunction arising from a diseased endothelium and seen across many conditions, including hypertension and diabetes. In addition to NO, other endothelium-derived vasodilators exist within the broader class of endothelium-derived hyperpolarizing factors (EDHFs). Interestingly, EDHFs mediate greater vasodilation as vessel size decreases (Luksha *et al.*, 2009). Furthermore, they have been shown to dilate resistance-size arteries even when NO is abolished and to exhibit a sex dependency, with EDHF-mediated dilation being more important in females and NO-mediated dilation being more important in males (Scotland *et al.*, 2005). Not all endothelial-derived factors are vasodilating, however. The peptide endothelin 1 is an example of a potent vasoconstrictor produced by vascular endothelial cells. An exhaustive review of established and emerging players in vasomodulation can be found in Brozovich *et al.* (2016).

5.2.2 **Vasoactive stimuli**

5.2.2.1 *Pharmacological agents*

Some of the pharmacological agents used today have been known since the early 1900s (Ewins, 1914; Davis, 1931), and insight into their vasoactive properties predates that for nonpharmacological stimuli (Koppanyi, 1948). Acetylcholine is a neurotransmitter first recognized in 1914 for its blood pressure-lowering ability at low doses (Dale, 1914). It exerts vasodilation by binding directly to muscarinic receptors on the endothelium to subsequently release the vasodilator NO (Kellogg *et al.*, 2005). Salbutamol is another drug that stimulates NO release, but its use has been much less prevalent compared to that of acetylcholine and is indicated primarily for respiratory ailments, such as asthma and chronic obstructive pulmonary disease.

A different class of vasoactive compounds exerts their action independent of the endothelium. Adenosine, frequently used as a coronary vasodilator in clinical stress tests, blocks irregular electrical signals in the heart and is the most common endothelium-independent vasodilator. Adenosine brings about vasodilation via a cyclic adenosine monophosphate (cAMP)-dependent mechanism in VSMCs (Gagnon *et al.*, 1980). Other endothelium-independent drugs are also available, such as NO donors— sodium nitroprusside and nitroglycerin. A detailed exploration of pharmacological agents and their mechanisms of action is found in Troy and Cheng (2021).

Pharmacological intervention is well established, but they also carry limitations and risks. Perhaps the greatest concern is that drug action cannot be reversed at will should an adverse event arise, such as hypotension, bradycardia, or occlusive spasm (Parker *et al.*, 2007; Tio *et al.*, 2002). For this reason, "reversible" alternatives have emerged for the express purpose of stimulating a transient and relatively safe vasoactive response.

5.2.2.2 Flow-mediated dilation

Flow-mediated dilation (FMD) involves measuring vascular response after an interval of time over which blood flow is limited to a part of the vascular tree. Typically, in humans, the brachial artery (arm) is occluded, although the popliteal artery (leg) has also been used. In animals such as rodents, the femoral artery provides the easiest access. Upon removal of the occlusion, conduit blood flow in healthy tissue transiently rises in response to conduit artery dilation and increased microvascular perfusion (also known as reactive hyperemia). It is well established that the flow velocity surge following the removal of occlusion triggers a shear stress-mediated NO release that effects conduit vessel dilation (Green *et al.*, 2014). Indeed, FMD has been used widely as a surrogate marker of vascular health and an independent prognostic indicator of cardiovascular or neurovascular diseases (Thijssen *et al.*, 2011).

It is crucial to recognize, however, that different levels of the vascular tree, and vasculature in different organs, have specific responses—in other words, the response in one conduit vessel cannot directly represent that of the microvasculature in all tissues. A 2011 population-based study of 5000 men and women compared three noninvasive tests of vascular function and found that reduced brachial artery FMD correlated strongly with cardiovascular risks in women, whereas hyperemic response variables in the fingers were weakly correlated (Schnabel *et al.*, 2011). To date, the mechanisms underlying reactive hyperemia remain elusive amid tenuous relationships involving different pathways. The critical takeaway message is that despite the utility of FMD, the concomitant hyperemic responses have distinct, tissue-specific pathophysiological underpinning that requires detection methods sensitive not to large peripheral vessels but to the microvasculature of interest.

5.2.2.3 Exercise

Akin to FMD, exercise is a noninvasive stimulus that has been utilized for vasoreactivity assessment. During exercise, perfusion increases in key organs with heightened oxygen demands: the heart and skeletal muscle receive the bulk of increased cardiac output (Joyner and Casey, 2015). Healthy microvasculature in these organs can adapt local perfusion levels to increased metabolic demands. Failure to increase perfusion adequately is a sign of microvascular dysfunction and compromised reactivity. A study on 27 patients without significant epicardial artery stenosis discovered that patients who had exercise-induced myocardial ischemia on SPECT imaging were also the ones who exhibited a blunted coronary perfusion response to acetylcholine (Zeiher *et al.*, 1995). These results uncovered impaired endothelium-dependent coronary vasodilation as the underlying pathology. They also showcased exercise as a viable surrogate stimulus in place of acetylcholine.

5.2.2.4 Respiratory gas challenge

Inhaling altered levels of carbon dioxide (CO_2) to evoke vasoreactivity is the most recent of all stimuli in our repertoire. It offers the important advantage of patient safety over pharmacological agents, as a gas challenge can be immediately terminated and normal room air restored should an adverse event surface. At elevated levels, CO_2 is a potent vasodilator. In the brain, cerebral perfusion increases 2.5% per mmHg increase in arterial partial pressure of CO_2 ($PaCO_2$) over the range 20–70 mmHg (Willie et al., 2014). In the heart, myocardial perfusion doubles over the end-tidal partial pressure of CO_2 ($P_{ET}CO_2$) range of 35–60 mmHg (Pelletier-Galarneau et al., 2018). Note that $P_{ET}CO_2$ approaches $PaCO_2$ with complete rebreathing (Duffin and McAvoy, 1988). There is limited data in other human vascular beds, but in vivo rodent studies reveal similar effects in the kidneys, liver, and skeletal muscle, with vasoconstriction setting in at higher CO_2 levels (Ganesh et al., 2016). Side effects are few; only at $PaCO_2$ levels acutely above 80 mmHg does confusion or unconsciousness become a concern (Sobczyk et al., 2021).

Implementation of a gas challenge experiment must be conducted judiciously. Simply having the subject inspire a fixed CO_2 level is less effective, particularly in sedated subjects, because a compensatory increase in respiratory rate will counteract the impact of elevated inspired CO_2. Breath-hold is another technique that has become increasingly utilized (Alwatban et al., 2018), but this approach is difficult for patients with underlying cardiorespiratory conditions. The most precise and widely amenable manipulation is to control the $PaCO_2$ level to the desired target level via a sequential gas delivery circuit that involves rebreathing exhaled air (Ito et al., 2008). With this method, the $P_{ET}CO_2$ level, which now tracks the $PaCO_2$ level very closely, can be measured noninvasively and used to alter the inspired gas levels. It is worth mentioning that lower oxygen (O_2) levels, known as hypoxia, also elicit vasodilation in cerebral, myocardial, renal, and skeletal microvasculature. In the brain, cerebral perfusion increases 35% when inspired O_2 is reduced from 21% to 10% (Kety and Schmidt, 1948), while the other three vascular beds see a reduction in resistance below 40 mmHg O_2 (Daugherty et al., 1967).

Gas challenge provides an analogue to pharmacological intervention due to their systemic effect, which contrasts starkly to FMD and even exercise, where only certain tissue beds are stimulated. Consequently, gas challenge is an attractive approach for directly probing vasoreactivity throughout the body.

A comparison of the four types of vasoactive stimuli and their pros and cons is provided in Table 5.1.

5.2.3 MRI of vasoreactivity

A spectrum of imaging modalities has been applied alongside different vasoactive stimuli for assessing vasoreactivity. Positron emission tomography (PET), considered the gold standard for perfusion imaging, is a highly sensitive modality for detecting perfusion changes. Laser Doppler flowmetry is an optical method that detects a Doppler shift from moving red blood cells in shallow tissue. There are other options available, including contrast-enhanced ultrasound and alternative optical methods. However, each modality given before has a significant limitation (e.g., PET suffers from low spatial resolution, and optical methods cannot penetrate deep tissue) that tempers their value for imaging vasoreactivity with high spatial definition and without depth limitations. MRI has the capability to address all these concerns.

Table 5.1 Vasoactive stimuli: Comparison of methods.

Vasoactive stimulus	Pros	Cons
Pharmacological agents	Well established Permits targeted mechanism of action Can test both vasodilation and constriction	Not reversible Potential for adverse reaction Potential interaction with other drugs Contraindicated for some patients
Flow-mediated dilation	Simple implementation Reversible and safe Amenable to all patients	Variability in implementation (*e.g.*, cuff placement, occlusion duration) Information on peripheral vasculature only; cannot extrapolate to other tissues No targeting of specific mechanism of action
Exercise	Simple implementation Reversible and safe	Not recommended for some patients No targeting of specific mechanism of action Exercise during MRI scan requires specialized equipment
Respiratory gas challenge	Permits some targeting of mechanism of action Reversible and safe Amenable to all patients Can test both vasodilation and constriction	Implementation requires specialized equipment for gas delivery

Three approaches have been described in the literature for measuring vasoreactivity on MRI. Blood oxygen level-dependent (BOLD) MRI is by far the most common, followed by arterial spin labeling (ASL) (Sleight *et al.*, 2021). ASL is sensitive to bulk perfusion changes in response to vasodilation or constriction, whereas BOLD is sensitive to the ensuing change in blood oxygen saturation. However, neither index directly measures vasoreactivity due to competing effects from a variety of sources, including fluctuations in blood velocity, heart rate, and tissue oxygen consumption. On the other hand, blood-pool imaging provides an index that is directly related to the microvascular blood volume and is, therefore, the most specific approach proposed to date for assessing vasoreactivity (Ganesh *et al.*, 2017). Table 5.2 summarizes the pros and cons of all the three methods.

5.2.3.1 BOLD-MRI

The BOLD signal is obtained using a T2*-weighted gradient echo sequence and is sensitive to the ratio of oxyhemoglobin (HbO_2) to total hemoglobin (*i.e.*, HbO_2 + deoxyhemoglobin (Hb)), otherwise known as blood oxygen saturation. When microvessels dilate and perfusion increases, the venous $HbO_2/(HbO_2 + Hb)$ ratio increases whereas the arterial $HbO_2/(HbO_2 + Hb)$ ratio remains largely

Table 5.2 MRI of vasoreactivity: Comparison of methods.

MRI platform	Pros	Cons
BOLD MRI	No exogenous contrast agent Imaging can be repeated	Not a direct measurement of vascular volume—confounders from blood oxygenation saturation and tissue oxygen metabolism Low spatial resolution Low signal to noise
ASL MRI	No exogenous contrast agent Imaging can be repeated	Not a direct measurement of vascular volume—confounders from flow velocity and errors in transit times Low spatial resolution Low signal to noise
Blood-pool imaging	Most direct measurement of vascular volume High spatial resolution High signal to noise Imaging can be repeated during stable profile of blood-pool signal Quantitative analysis possible for estimating degree of vasomodulation	Requires injection of blood-pool agent

unaltered. As a result of a lower concentration of the paramagnetic species Hb, susceptibility effect decreases and T2*-weighted signal increases (Ogawa et al., 1990).

One of the earliest demonstrations of the BOLD effect for evaluating vasoreactivity was in evaluating ovarian carcinoma xenografts in response to a hypercapnic stimulus—lower vasoreactivity was seen in immature and poorly formed tumor vessels compared to normal vasculature (Gilad et al., 2005). In this and many later vasoreactivity studies (Cohen-Adad et al., 2010; Venkatraghavan et al., 2018), a simple evaluation of signal evolution on a T2*-weighted scan provided the rapid acquisition required for capturing dynamics. The BOLD neurovascular response in an individual patient can even be compared against the population average to determine if the response falls within or outside the "normal" range (Fisher and Mikulis, 2021). Quantitative T2* mapping of vasoreactivity has also been successfully demonstrated (Mürtz et al., 2010; Ganesh et al., 2016), but the extra complexity and much longer acquisitions involved in obtaining high-quality T2* maps have precluded their widespread adoption, especially considering the sensitivity advantage of a quantitative approach has yet to be established. However, in one study where abdominal tissue response to hypercapnia was evaluated on T2* mapping (Ganesh et al., 2016) and later compared against blood volume changes (Ganesh et al., 2017), the authors discovered that the magnitude of perfusion changes in the liver and kidneys measured from T2* differences was not consistent with the degree of vasoconstriction. Competing T2* effects from a higher heart rate, unknown oxygen consumption rates, and lower arterial oxygen saturation rendered it technically impossible to ascribe the T2* relaxation time strictly to a vasoactive response.

5.2.3.2 Arterial spin labeling

ASL offers the advantage of a contrast-free approach to assessing perfusion dynamics. It provides also a more interpretable assessment of vasoreactivity compared to BOLD, because it is sensitive to bulk perfusion only and not to blood oxygen saturation. However, ASL uses larger voxels to counter an inherently lower signal-to-noise ratio (SNR) and is technically more challenging, requiring tagging arterial blood upstream of the tissue region of interest. Furthermore, to achieve accurate quantitation of perfusion values, arterial transit times must be known accurately and so must the blood T1, especially if hyperoxia is administered (Zaharchuk et al., 2008). Recent strategies to improve spatial resolution and cope with long transit delays are available, and the reader is referred to Chapter 4 for specifics related to sequence implementation and analysis. Nonetheless, there are specific clinical scenarios where ASL remains unsuitable. One example is the presence of significant stenosis in the feeding artery; if the carotid arteries were stenosed, full replacement of saturated blood by tagged blood in the brain may be impossible. Another example is patients with heterogeneous transit delays due to abnormal collateral pathways.

Similar to BOLD, applications of ASL to assess vasoreactivity have focused primarily on the brain. Stroke is a classic example where vasoreactivity assessment has consistently shown value. In a 2009 study of patients with ischemic stroke (Zhao et al., 2009), it was discovered that increased perfusion response to hypercapnia in stroke patients was blunted compared to that in healthy controls, but vasoconstriction was preserved. In the years since, many more neurovascular studies involving ASL have confirmed the value of the technique in assessing a range of neurological disorders. Perhaps not unexpectedly, in one of the few head-to-head comparisons of ASL versus BOLD MRI in assessing vasoreactivity, ASL maps revealed larger affected regions of paradoxical cerebrovascular reactivity relative to BOLD maps (Mandell et al., 2008) (Fig. 5.2). The question becomes which is the more sensitive method for delineating compromised vasoreactivity? ASL has also been combined with BOLD to arrive at a better understanding of the underlying physiological changes (Gauthier et al., 2011). Yet, neither method provides a direct readout on vasoconstriction and vasodilation.

5.2.3.3 Blood-pool steady-state T1 imaging

The concept of using dynamic blood-pool imaging to assess vasoreactivity (Ganesh et al., 2017) was conceived to address the limitations of BOLD and ASL, with the goal of providing the most direct measurement possible of a changing microvascular volume from vasoconstriction or vasodilation. Such an approach would obviate the need to correct for differences in blood oxygen saturation, blood velocity, oxygen consumption, and hematocrit. What is measured is simply the degree of

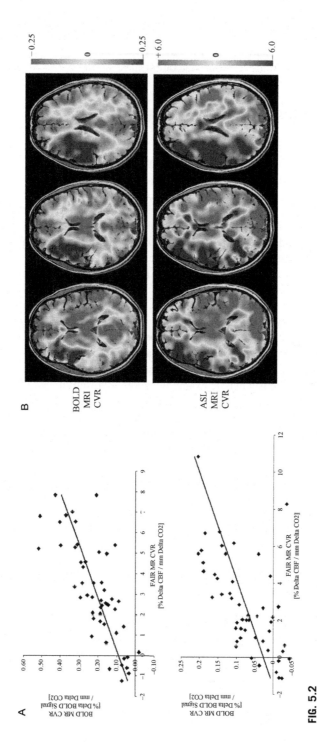

FIG. 5.2

Comparison of BOLD and ASL. Correlation of (A) gray matter and (E) white matter cerebrovascular reactivity (CVR) measured by BCLD versus ASL MRI for 50 hemispheres in 25 patients.

Credit: Reprinted with permission from Wolters Kluwer Health Inc. (Mandell, D.M., et al. (2008). Mapping cerebrovascular reactivity using blood oxygen level-dependent MRI in patients with arterial steno-occlusive disease: comparison with arterial spin labeling MRI. Stroke 39(7) 2021–2028. https://doi.org/10.1161/STROKEAHA.107.5G6709. Figs. 1 and 3).

vasomodulation. Measuring blood volume is not straightforward in MRI, and the impact of water exchange means there is always a bias in the measurements we make of vascular volume. However, if a T1 blood-pool MR contrast agent can be employed, it will serve three important functions: (1) it sensitizes T1 relaxation times to the intravascular compartment, (2) it saturates potential T1 effects from variations in dissolved oxygen content in blood plasma, and (3) it provides a much longer window of stable blood signal (relative to extracellular contrast agents) over which gas challenge stimuli can be manipulated. Qualitative T1-weighted gradient echo imaging or quantitative 3D T1 mapping can be used to dynamically track tissue response with a temporal resolution of 1–2 min. This temporal resolution is, in fact, sufficient, given the slow dynamics of vasomodulation. If quantitative T1 mapping is performed, the T1 values mapped pre- and postgas challenge can be used to estimate how much the microvascular volume has altered.

Using a clinically approved gadolinium-based blood-pool agent, gadofosveset, Ganesh *et al.* gave the first demonstration of a blood-pool imaging platform to visualize the dynamics of vasomodulation in abdominal organs, including the liver, kidneys, and skeletal muscle (Ganesh *et al.*, 2017) (Fig. 5.3). Unlike BOLD or ASL techniques, blood-pool imaging afforded superior SNR and clear evidence of vaso-dilation or vasoconstriction on T1-weighted images alone and dramatically extended the range of T1 changes obtained in the absence of a blood-pool contrast agent (Ganesh *et al.*, 2016). Quantitative T1 analysis provided a deeper insight into the degree of vasomodulation, but quantitation was not necessary. In a follow-up study of an ischemia–reperfusion injury in the rat leg, the blood-pool imaging method revealed compromised vasodilatory capacity in the injured leg, whereas standard perfusion imaging did not note any deficits (Ganesh *et al.*, 2019). It is also important to highlight that this method is amenable to both high- and low-flow organs, and special analysis can be applied to low-perfusion tissues such as skeletal muscle (Zakher *et al.*, 2020).

It is noteworthy that contrast administration has been the gold standard for cardiac stress perfusion tests on MRI. However, the traditional protocol applies two injections of an extravascular gadolinium agent: one injection for calculating rest perfusion and the second injection for calculating stress perfusion (Ge *et al.*, 2020). There are a few shortcomings with the gold-standard cardiac stress MRI test. First, there is no room for error when it comes to contrast injection, and since there is only one window of opportunity to capture the first passage, the acquisition sequence and the contrast injection must be timed perfectly. Second, since the interval between the two injections is short relative to the elimination half-life of gadolinium chelate, an elevated baseline signal for the second injection will inevi-tably introduce errors in quantitative perfusion estimation. Even if T1 mapping is performed before each injection to convert signal intensity to gadolinium concen-tration, the pharmacokinetics is altered for the second injection due to residual contrast from the first. A final consideration is that the total contrast dose is double that of typical contrast-enhanced MRI exams.

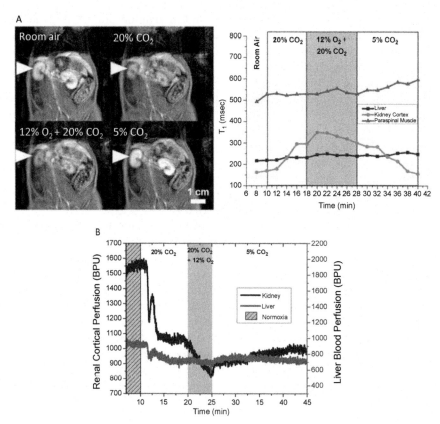

FIG. 5.3

Blood-pool steady-state T1 imaging. Abdominal tissue response to hypercapnic and hypoxic gas challenges in a healthy rat. Use of a blood-pool T1 agent, gadofosveset, enables stark visualization of vasoconstriction followed by vasodilation (A) on T1-weighted images (arrowhead points to renal cortex) and T1 relaxation time measurements. (B) Optical perfusion validation in the renal cortex and liver confirms MRI results.

Credit: Creative Commons license from Nature Publishing Group (Ganesh, T., et al. (2017). A non-invasive magnetic resonance imaging approach for assessment of real-time microcirculation dynamics. Sci. Rep. 7(1) 7468. https://doi.org/10.1038/s41598-017-06983-6. Figs. 2 and 8).

More recently, and likely to eliminate quantification errors arising from double contrast injection protocols, new studies have adopted the single-injection, blood-pool T1 imaging approach to assess microvascular reactivity. A back-to-back study on swine hearts applied the concept of a blood-pool agent for extending the range of T1 reactivity. Instead of Ablavar®, however, whose production was discontinued in 2017 due to low sales, these papers used Feraheme® (ferumoxytol), an iron-based compound currently still approved in the U.S. as an iron supplement and used off-label as a blood-pool MRI contrast agent. These papers demonstrated that with the use of ferumoxytol, the blunted reactivity to adenosine in the hypoperfused region could be better distinguished from reactivity in the remote region and normal myocardium (Nguyen *et al.*, 2019; Colbert *et al.*, 2021).

5.3 **Applications**

5.3.1 **Cerebral steno-occlusive disease**

Cerebral steno-occlusive diseases are by far the most common application of vasoreactivity MRI. A 2021 systematic review of cerebral reactivity studies showed that carotid stenosis/occlusive diseases made up 31% of the 235 brain clinical studies reviewed (Sleight *et al.*, 2021). This prevalence points to both the relevance of vasoreactivity measurements in this setting—where lower reactivity and longer delays are found in affected brain regions (Fig. 5.4)—and the reproducibility of the imaging

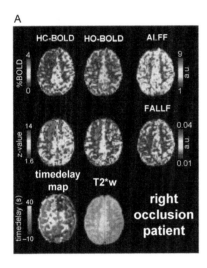
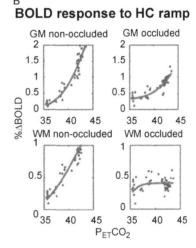

FIG. 5.4

Stenosis in a right occlusion patient. (A) Hypercapnic (HC)-BOLD, hyperoxic (HO)-BOLD, amplitude of low-frequency fluctuations (ALFF), fractional ALFF (fALFF), and time delay map; a single slice is shown. Reduced HC-BOLD and prolonged time delays are seen in the affected hemisphere, whereas for HO-BOLD, ALFF, and ALFF, no clear interhemispheric differences are observed. (B) The effect of right occlusion is also apparent in the BOLD response to progressive hypercapnia with pronounced reductions in the response amplitude in both gray matter (GM) and white matter (WM) regions ipsilateral to occlusion. Prolonged time delays in GM manifest as delayed onset in BOLD response (B—top right) as a stronger HC stimulus is required to increase vessel diameter in presumably predilated regions.

Credit: Reprinted with permission from Elsevier (De Vis, J.B., et al. (2018). Effect sizes of BOLD CVR, resting-state signal fluctuations and time delay measures for the assessment of hemodynamic impairment in carotid occlusion patients. NeuroImage 179, 530–539. https://doi.org/10.1016/ j.neuroimage.2018.06.017. Fig. 2).

strategies employed (Hartkamp *et al.*, 2017; Duffin *et al.*, 2018; De Vis *et al.*, 2018). Machine learning is even being investigated for classifying cerebrovascular impairment on the basis of altered vasoreactivity (Waddle *et al.*, 2020).

5.3.2 Dementia and cognitive impairment

Cognitive impairment is the second most prevalent application of vasoreactivity studies in the brain (Sleight *et al.*, 2021). Lower vasoreactivity in the impaired brain has been consistently observed (Cantin *et al.*, 2011; Yezhuvath *et al.*, 2012) (Fig. 5.5). However, it is also important to calibrate against the lower vasoreactivity that comes with old age even in healthy subjects (Liu *et al.*, 2013).

5.3.3 Small vessel disease

Small vessel disease accounts for 3% of all brain clinical studies reviewed (Sleight *et al.*, 2021). Similar to the conditions described before, the vasoreactive response is lower in patients, who have a higher white matter hyperintensity burden (Liem *et al.*, 2009). Incidentally, lower cerebral vasoreactivity is also associated with increased number of risk factors, such as hypertension, diabetes, and elevated cholesterol levels, all of which adversely affect blood vessels at all levels of the vascular tree (Tchistiakova *et al.*, 2015).

5.3.4 Diabetes

Despite the vast majority of brain vasoreactivity studies on impairment or disease originating in the brain, there have also been studies examining the impact of systemic disease. Diabetes is an example of a systemic condition, one that alters microvascular structure and function throughout the body. In a 2-year prospective study of patients with type 2 diabetes, cerebral vasoreactivity was found to decrease compared to baseline, and the decrease was significantly associated with a decline in executive function and quality of life (Chung *et al.*, 2015) (Fig. 5.6). Remarkably, despite the known pervasive damage wreaked on the systemic vasculature (*e.g.*, kidneys, heart, retina), no literature report has investigated the potential value of vasoreactivity MRI in assessing the extracranial tissues of diabetic subjects.

5.3.5 Cardiovascular disease

Imaging vasoreactivity in the myocardium is one of the newest and most exciting frontiers. Cardiac MRI is associated with its unique set of challenges, not the least of which is attaining volumetric coverage while maintaining good spatial and temporal resolution in the presence of both cardiac and respiratory motion. Despite these challenges, an increasing number of papers are reporting vasoreactivity measurements in the myocardium. In one study of ischemic heart disease in canines, interrogation with hypercapnia and BOLD allowed assessment of myocardial oxygenation (Yang *et al.*, 2019) (Fig. 5.7). In another study of patients with hypertension, hypercapnia alongside quantitative T2/T2* mapping revealed diverging

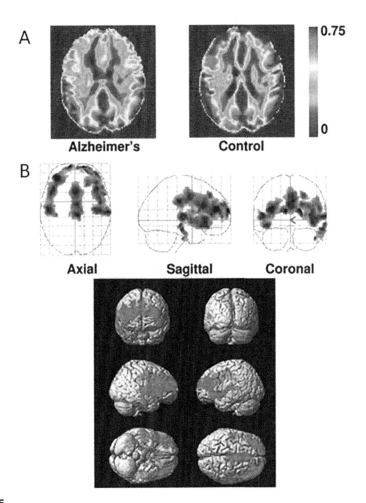

FIG. 5.5

Comparison of relative cerebrovascular reactivity (CVR) maps in Alzheimer's disease (AD) and controls. (A) Averaged CVR maps in the patient ($N = 12$) and control ($N = 13$) groups. Warmer color indicates a higher CVR value. Color bar shows the range of 0–0.75 times the cerebellar CVR. (B) Voxel-based maps of age- and vascular risk factor-corrected CVR differences between AD and controls. The top panel shows the glass brain overlay and the bottom panel shows the rendering on the Montreal Neurological Institute brain template. Colored voxels indicate brain regions with low CVR. These include prefrontal cortex, anterior cingulate cortex, and insular cortex.

Credit: Reprinted with permission from Elsevier (Yezhuvath, U.S., et al. (2012). Forebrain-dominant deficit in cerebrovascular reactivity in Alzheimer's disease. Neurobiol. Aging 33(1) 75–82. https://doi.org/10.1016/j.neurobiolaging.2010.02.005. Fig. 1).

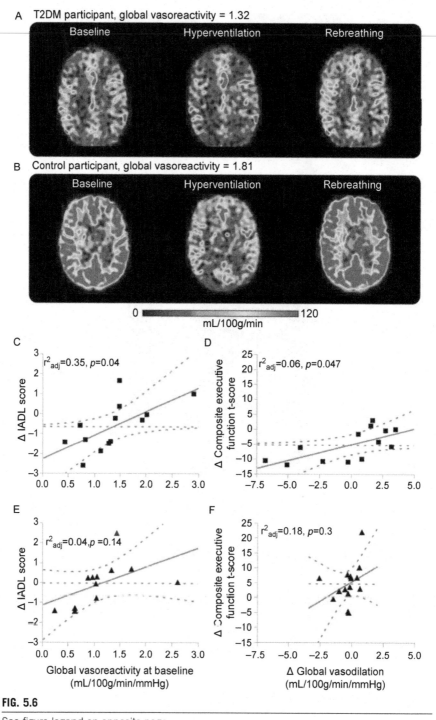

FIG. 5.6

See figure legend on opposite page.

responses for healthy versus hypertensive subjects (van den Boomen *et al.*, 2020). Finally, in a multicenter cardiac MRI study of heart failure patients, microvascular dysfunction assessed on reduced coronary microvascular reactivity to acetylcholine was found in 66% of participants (Rush *et al.*, 2021).

5.4 **Future directions**

The value of vasoreactivity MRI rests on its ability to detect functional deficit in blood vessels long before morphological abnormalities set it. This ability provides a deeper insight into the pathophysiology of disease, as well as a potential biomarker for early detection. As we have seen in this chapter, a normal, functional blood supply depends as much on the different components of a blood vessel (*e.g.*, endothelium, perivascular smooth muscle cells) and their ability to respond to neuronal and humoral vasoactive factors as on the structural integrity of the vessels themselves. In many conditions, vessel structure does not alter until much later in disease progression. For instance, in heart failure, by the time microvascular rarefaction appears in the heart, disease has already advanced to a late stage, accompanied often by myocardial hypertrophy, stiffness, and fibrosis. The true value for vasoreactivity MRI, therefore, lies in its potential for early diagnosis across a multitude of vessel-related conditions throughout the body—the list is vast and includes diabetes, hypertension, ischemia–reperfusion injury, heart disease, vascular disease, arthritis, and chronic inflammation.

Vasoreactivity MRI had its genesis in neurological imaging, which remains the dominant area of application to date. There have been very few attempts to transfer the technique to other organs to benefit nonneurological conditions. A few possible reasons exist for this inertia. One possibility is that vascular response is specific to the stimulus selected and to the target organ. Every organ behaves differently, and because the brain response, as well as effective stimuli, has been so well characterized over the decades to a variety of vasoactive stimuli, the barrier to entry is substantially

FIG. 5.6—Cont'd

Associations between cerebral vasoregulation and decline in cognition. (A and B) Perfusion maps for 2 representative participants using a 3 T, 3-dimensional CASL MRI. (A) Participant with type 2 diabetes mellitus (T2DM) who has lower global vasoreactivity. (B) Participant without T2DM who has higher global vasoreactivity. (C) Lower baseline global vasoreactivity is associated with greater decline in Instrumental Activities of Daily Living (IADL) scores, and (D) lower baseline global vasodilation is associated with greater decline in composite executive function T scores after the 2-year follow-up in the T2DM group only. No similar effect was observed for the IADL scores (E) and executive function (F) in the control group. Best fit = red solid line; confidence interval = red dashed lines; mean = blue dashed line.
Credit: Reprinted with permission from Wolters Kluwer Health Inc. (Chung, C.-C., et al. (2015). Inflammation-associated declines in cerebral vasoreactivity and cognition in type 2 diabetes. Neurology 85(5) 450–458. https://doi.org/10.1212/WNL.0000000000001820. Fig. 1).

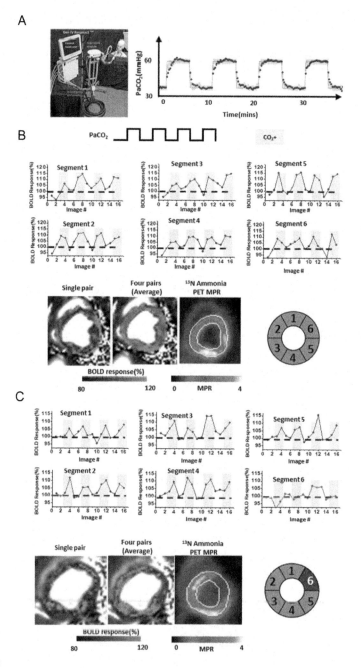

FIG. 5.7

See figure legend on opposite page.

lower. Another possible reason is that the available MRI techniques for assessing vasoreactivity, mainly BOLD and ASL, are much simpler to implement in the brain than in organs that suffer from motion, low baseline perfusion, or modest vasoactive response. Furthermore, the physiological vasoactive response must be characterized thoroughly in extracranial organs before one can determine if the response attainable meets the sensitivity threshold for detection. Opportunity is ripe for explorations beyond the brain in terms of new imaging approaches and reproducible protocols for vasoactive stimuli.

Aside from making inroads in other organs outside the brain, several other challenges need to be addressed. One challenge is to ask oneself whether the existing MRI techniques for measuring vasoreactivity are sufficiently specific to detecting compromised vasoreactivity and not a different, unrelated pathology. This question may be relevant in specific pathologies under certain stimuli. For example, tumor blood vessels generally are not vasoresponsive but vessels in surrounding normal tissue are, in which case tumor blood flow may adjust to compensate and give an incorrect readout on BOLD or ASL. A method to image the blood volume is, thus, the most specific approach, but the requirement for a blood-pool contrast agent also brings challenges, as gadofosveset was pulled from the market in 2017 due to poor sales (Morgan and Murphy, 2020) after having occupied a pivotal role in MR angiography for many years. Currently, the only blood-pool agent still clinically available is ferumoxytol, but it is an iron oxide nanoparticle that is clinically approved only for treating anemia; its use as a contrast agent has been performed off-label. Whether or not ferumoxytol will remain accessible is an open question. It can also be argued that the advent of faster sequences had diminished the necessity of agents that remained in the vasculature for an extended interval. However, acceleration is not the answer for all applications, especially ones where the timescale of the physiological response being measured is relatively long. Specific imaging of vasoreactivity is one example that can benefit immeasurably from the reinstatement of blood-pool agents.

FIG. 5.7—Cont'd

Myocardial BOLD response. (A) Prospective control of $PaCO_2$. The image shows the system (computer-controlled gas control, input for source gases, and disposable breathing circuit) for prospectively modulating $PaCO_2$. The graph shows a $PaCO_2$ trace achieved during scanning. Light blue trace represents the targeted $P_{ET}CO_2$, and dark blue points denote actual $P_{ET}CO_2$ values. (B) Representative results in a healthy dog during repeated intermittent hypercapnia (four stimulations). Corresponding location of the AHA segments is shown in a bull's-eye plot. (C) Representative results in a dog with left anterior descending coronary artery (LAD) stenosis during repeated intermittent hypercapnia (four stimulations). The LAD territory highlighted with a blue shade indicates the presence of coronary stenosis.

Credit: Reprinted with permission from The American Association for the Advancement of Science (Yang, H-J., et al. (2019). Accurate needle-free assessment of myocardial oxygenation for ischemic heart disease in canines using magnetic resonance imaging. Sci. Transl. Med. 11(494). https://doi.org/10.1126/scitranslmed.aat4407. Fig. 1).

5.5 **Summary**

The microvasculature is a vital systemic highway that keeps our bodies healthy and thriving, delicately balancing the time-varying metabolic demands in different organs. A complex interplay of neuronal and local humoral factors signals to blood vessels when and where perfusion needs to be adjusted and by what amount. When any of the players in this complex system becomes dysfunctional, it is usually an early symptom in disease progression across a range of conditions. For this reason, detecting this dysfunction, or compromised vasoreactivity, can be a tool for early detection and diagnosis. Fortunately, there are several vasoactive stimuli at our disposal for probing whether microvessels can adjust perfusion appropriately; these include pharmacological agents, flow-mediated dilation, exercise, and respiratory challenges. We also have MRI approaches to assess perfusion response to these stimuli. The most popular approaches, BOLD and ASL, provide adequate detection sensitivity, despite lacking specificity related to vasomodulation. Numerous applications of these methods in brain pathologies— *e.g.*, ischemia, dementia, small vessel disease—have been demonstrated. Looking to the future, methodological innovations should develop "blood-pool-based imaging" that is truly specific to vasomodulation, whereas applications need to break through new frontiers in cardiovascular, oncological, metabolic, and inflammatory conditions. Fig. 5.8 presents a vision of the role of vasoreactivity MRI in the future MRI clinic.

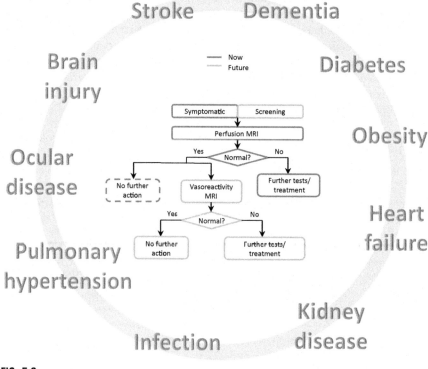

FIG. 5.8

Potential role of vasoreactivity MRI in future clinical workflow.

References

Alwatban, M., et al., 2018. The breath-hold acceleration index: a new method to evaluate cerebrovascular reactivity using transcranial Doppler. J. Neuroimaging 28 (4), 429–435. https://doi.org/10.1111/jon.12508.

Bacakova, L., 2018. In: Travnickova, M. (Ed.), The Role of Vascular Smooth Muscle Cells in the Physiology and Pathophysiology of Blood Vessels. IntechOpen, Rijeka, https://doi.org/10.5772/intechopen.77115. Ch. 12.

Brozovich, F.V., et al., 2016. Mechanisms of vascular smooth muscle contraction and the basis for pharmacologic treatment of smooth muscle disorders. Pharmacol. Rev. 68 (2), 476–532. https://doi.org/10.1124/pr.115.010652.

Cantin, S., et al., 2011. Impaired cerebral vasoreactivity to CO2 in Alzheimer's disease using BOLD fMRI. NeuroImage 58 (2), 579–587. https://doi.org/10.1016/j.neuroimage.2011.06.070.

Chung, C.-C., et al., 2015. Inflammation-associated declines in cerebral vasoreactivity and cognition in type 2 diabetes. Neurology 85 (5), 450–458. https://doi.org/10.1212/WNL.0000000000001820.

Cohen-Adad, J., et al., 2010. BOLD signal responses to controlled hypercapnia in human spinal cord. NeuroImage 50 (3), 1074–1084. https://doi.org/10.1016/j.neuroimage.2009.12.122.

Colbert, C.M., et al., 2021. Ferumoxytol-enhanced magnetic resonance T1 reactivity for depiction of myocardial hypoperfusion. NMR Biomed. 34 (7), e4518. https://doi.org/10.1002/nbm.4518.

Dale, H.H., 1914. The action of certain esters and ethers of choline, and their relation to muscarine. J. Pharmacol. Exp. Ther. 6 (2), 147–190.

Daugherty, R.M.J., et al., 1967. Local effects of O2 and CO2 on limb, renal, and coronary vascular resistances. Am. J. Phys. 213 (5), 1102–1110. https://doi.org/10.1152/ajplegacy.1967.213.5.1102.

Davis, E., 1931. Relations between the actions of adrenaline, acetylcholine, and ions, on the perfused heart. J. Physiol. 71 (4), 431–441. https://doi.org/10.1113/jphysiol.1931.sp002748.

De Vis, J.B., et al., 2018. Effect sizes of BOLD CVR, resting-state signal fluctuations and time delay measures for the assessment of hemodynamic impairment in carotid occlusion patients. NeuroImage 179, 530–539. https://doi.org/10.1016/j.neuroimage.2018.06.017.

Duffin, J., McAvoy, G.V., 1988. The peripheral-chemoreceptor threshold to carbon dioxide in man. J. Physiol. 406, 15–26. https://doi.org/10.1113/jphysiol.1988.sp017365.

Duffin, J., et al., 2018. Cerebrovascular resistance: the basis of cerebrovascular reactivity. Front. Neurosci. 12, 409. https://doi.org/10.3389/fnins.2018.00409.

Ewins, A.J., 1914. Acetylcholine, a new active principle of ergot. Biochem. J. 8 (1), 44–49. https://doi.org/10.1042/bj0080044.

Fisher, J.A., Mikulis, D.J., 2021. Cerebrovascular reactivity: purpose, optimizing methods, and limitations to interpretation—a personal 20-year odyssey of (re)searching. Front. Physiol. 12, 629651. https://doi.org/10.3389/fphys.2021.629651.

Gagnon, G., Regoli, D., Rioux, F., 1980. Studies on the mechanism of action of various vasodilators. Br. J. Pharmacol. 70 (2), 219–227. https://doi.org/10.1111/j.1476-5381.1980.tb07927.x.

Ganesh, T., et al., 2016. T2* and T1 assessment of abdominal tissue response to graded hypoxia and hypercapnia using a controlled gas mixing circuit for small animals. J. Magn. Reson. Imaging: JMRI 44 (2), 305–316. https://doi.org/10.1002/jmri.25169.

Ganesh, T., *et al.*, 2017. A non-invasive magnetic resonance imaging approach for assessment of real-time microcirculation dynamics. Sci. Rep. 7 (1), 7468. https://doi.org/10.1038/s41598-017-06983-6.

Ganesh, T., *et al.*, 2019. Assessment of microvascular dysfunction in acute limb ischemia-reperfusion injury. J. Magn. Reson. Imaging: JMRI 49 (4), 1174–1185. https://doi.org/10.1002/jmri.26308.

Gauthier, C.J., *et al.*, 2011. Elimination of visually evoked BOLD responses during carbogen inhalation: implications for calibrated MRI. NeuroImage 54 (2), 1001–1011. https://doi.org/10.1016/j.neuroimage.2010.09.059.

Ge, Y., *et al.*, 2020. Stress cardiac MRI in stable coronary artery disease. Curr. Opin. Cardiol. 35 (5), 566–573. https://doi.org/10.1097/HCO.0000000000000776.

Gilad, A.A., *et al.*, 2005. Functional and molecular mapping of uncoupling between vascular permeability and loss of vascular maturation in ovarian carcinoma xenografts: the role of stroma cells in tumor angiogenesis. Int. J. Cancer 117 (2), 202–211. https://doi.org/10.1002/ijc.21179.

Green, D.J., *et al.*, 2014. Is flow-mediated dilation nitric oxide mediated? A meta-analysis. Hypertension (Dallas, Tex.: 1979) 63 (2), 376–382. https://doi.org/10.1161/HYPERTENSIONAHA.113.02044.

Hartkamp, N.S., *et al.*, 2017. Cerebrovascular reactivity in the caudate nucleus, lentiform nucleus and thalamus in patients with carotid artery disease. J. Neuroradiol. = Journal de neuroradiologie 44 (2), 143–150. https://doi.org/10.1016/j.neurad.2016.07.003.

Ito, S., *et al.*, 2008. Non-invasive prospective targeting of arterial P(CO2) in subjects at rest. J. Physiol. 586 (15), 3675–3682. https://doi.org/10.1113/jphysiol.2008.154716.

Joyner, M.J., Casey, D.P., 2015. Regulation of increased blood flow (hyperemia) to muscles during exercise: A hierarchy of competing physiological needs. Physiol. Rev. 95 (2), 549–601. https://doi.org/10.1152/physrev.00035.2013.

Kellogg, D.L.J., *et al.*, 2005. Acetylcholine-induced vasodilation is mediated by nitric oxide and prostaglandins in human skin. J. Appl. Physiol. (Bethesda, Md.: 1985) 98 (2), 629–632. https://doi.org/10.1152/japplphysiol.00728.2004.

Kety, S.S., Schmidt, C.F., 1948. The effects of altered arterial tensions of carbon dioxide and oxygen on cerebral blood flow and cerebral oxygen consumption of normal young men. J. Clin. Invest. 27 (4), 484–492. https://doi.org/10.1172/JCI101995.

Koppanyi, T., 1948. Acetylcholine as a pharmacological agent. Bull. Johns Hopkins Hosp. 83 (6), 532–567.

Kuroda, S., *et al.*, 1993. Acetazolamide test in detecting reduced cerebral perfusion reserve and predicting long-term prognosis in patients with internal carotid artery occlusion. Neurosurgery 32 (6), 912–919. https://doi.org/10.1227/00006123-199306000-00005.

Kwiatkowski, G., *et al.*, 2021. MRI-based in vivo detection of coronary microvascular dysfunction before alterations in cardiac function induced by short-term high-fat diet in mice. Sci. Rep. 11 (1), 18915. https://doi.org/10.1038/s41598-021-98401-1.

Liem, M.K., *et al.*, 2009. Cerebrovascular reactivity is a main determinant of white matter hyperintensity progression in CADASIL. AJNR Am. J. Neuroradiol. 30 (6), 1244–1247. https://doi.org/10.3174/ajnr.A1533.

Liu, P., *et al.*, 2013. Age-related differences in memory-encoding fMRI responses after accounting for decline in vascular reactivity. NeuroImage 78, 415–425. https://doi.org/10.1016/j.neuroimage.2013.04.053.

Luksha, L., Agewall, S., Kublickiene, K., 2009. Endothelium-derived hyperpolarizing factor in vascular physiology and cardiovascular disease. Atherosclerosis 202 (2), 330–344. https://doi.org/10.1016/j.atherosclerosis.2008.06.008.

Mandell, D.M., *et al.*, 2008. Mapping cerebrovascular reactivity using blood oxygen level-dependent MRI in patients with arterial steno-occlusive disease: Comparison with arterial spin labeling MRI. Stroke 39 (7), 2021–2028. https://doi.org/10.1161/STROKEAHA.107.506709.

Morgan, M., Murphy, A., 2020. Gadofosveset Trisodium. Radiopaedia.org, https://doi.org/10.53347/rID-31767.

Mürtz, P., *et al.*, 2010. Changes in the MR relaxation rate R(2)* induced by respiratory challenges at 3.0 T: a comparison of two quantification methods. NMR Biomed. 23 (9), 1053–1060. https://doi.org/10.1002/nbm.1532.

Nguyen, K.-L., *et al.,* 2019. Ferumoxytol-enhanced CMR for vasodilator stress testing: a feasibility study. J. Am. Coll. Cardiol. Img. 1582–1584. https://doi.org/10.1016/j.jcmg.2019.01.024.

Ogawa, S., *et al.*, 1990. Oxygenation-sensitive contrast in magnetic resonance image of rodent brain at high magnetic fields. Magn. Reson. Med. 14 (1), 68–78. https://doi.org/10.1002/mrm.1910140108.

Palmer, R.M., Ferrige, A.G., Moncada, S., 1987. Nitric oxide release accounts for the biological activity of endothelium-derived relaxing factor. Nature 327 (6122), 524–526. https://doi.org/10.1038/327524a0.

Parker, J.D., *et al.*, 2007. Safety of intravenous nitroglycerin after administration of sildenafil citrate to men with coronary artery disease: A double-blind, placebo-controlled, randomized, crossover trial. Crit. Care Med. 35 (8), 1863–1868. https://doi.org/10.1097/01.CCM.0000269371.70738.30.

Pelletier-Galarneau, M., *et al.*, 2018. Effects of hypercapnia on myocardial blood flow in healthy human subjects. J. Nucl. Med. 59 (1), 100–106. https://doi.org/10.2967/jnumed.117.194308.

Rath, G., Balligand, J.-L., Chantal, D., 2012. Vasodilatory mechanisms of beta receptor blockade. Curr. Hypertens. Rep. 14 (4), 310–317. https://doi.org/10.1007/s11906-012-0278-3.

Rush, J.W.E., Denniss, S.G., Graham, D.A., 2005. Vascular nitric oxide and oxidative stress: Determinants of endothelial adaptations to cardiovascular disease and to physical activity. Can. J. Appl. Physiol. = Revue canadienne de physiologie appliquee 30 (4), 442–474. https://doi.org/10.1139/h05-133.

Rush, C.J., *et al.*, 2021. Prevalence of coronary artery disease and coronary microvascular dysfunction in patients with heart failure with preserved ejection fraction. JAMA Cardiol. 6 (10), 1130–1143. https://doi.org/10.1001/jamacardio.2021.1825.

Schnabel, R.B., *et al.*, 2011. Noninvasive vascular function measurement in the community: cross-sectional relations and comparison of methods. Circ. Cardiovasc. Imaging 4 (4), 371–380. https://doi.org/10.1161/CIRCIMAGING.110.961557.

Scotland, R.S., *et al.*, 2005. Investigation of vascular responses in endothelial nitric oxide synthase/cyclooxygenase-1 double-knockout mice: key role for endothelium-derived hyperpolarizing factor in the regulation of blood pressure in vivo. Circulation 111 (6), 796–803. https://doi.org/10.1161/01.CIR.0000155238.70797.4E.

Sleight, E., *et al.*, 2021. Cerebrovascular reactivity measurement using magnetic resonance imaging: a systematic review. Front. Physiol., 643468. https://doi.org/10.3389/fphys.2021.643468.

Sobczyk, O., *et al.*, 2021. Measuring cerebrovascular reactivity: sixteen avoidable pitfalls. Front. Physiol. 12, 665049. https://doi.org/10.3389/fphys.2021.665049.

Tchistiakova, E., *et al.*, 2015. Vascular risk factor burden correlates with cerebrovascular reactivity but not resting state coactivation in the default mode network. J. Magn. Reson. Imaging: JMRI 42 (5), 1369–1376. https://doi.org/10.1002/jmri.24917.

Thijssen, D.H.J., *et al.*, 2011. Assessment of flow-mediated dilation in humans: a methodological and physiological guideline. Am. J. Physiol. Heart Circ. Physiol. 300 (1), H2–12. https://doi.org/10.1152/ajpheart.00471.2010.

Tio, R.A., *et al*., 2002. Safety evaluation of routine intracoronary acetylcholine infusion in patients undergoing a first diagnostic coronary angiogram. J. Investig. Med. 50 (2), 133–139. https://doi.org/10.2310/6650.2002.31305.

Troy, A.M., Cheng, H.-L.M., 2021. Human microvascular reactivity: a review of vasomodulating stimuli and non-invasive imaging assessment. Physiol. Meas. 42 (9). https://doi.org/10.1088/1361-6579/ac18fd.

van den Boomen, M., *et al*., 2020. Blood oxygen level-dependent MRI of the myocardium with multiecho gradient-echo spin-echo imaging. Radiology 294 (3), 538–545. https://doi.org/10.1148/radiol.2020191845.

Venkatraghavan, L., *et al*., 2018. Measurement of cerebrovascular reactivity as blood oxygen level-dependent magnetic resonance imaging signal response to a hypercapnic stimulus in mechanically ventilated patients. J. Stroke Cerebrovasc. Dis. 27 (2), 301–308. https://doi.org/10.1016/j.jstrokecerebrovasdis.2017.08.035.

Waddle, S.L., *et al*., 2020. Classifying intracranial stenosis disease severity from functional MRI data using machine learning. J. Cereb. Blood Flow Metab. 40 (4), 705–719. https://doi.org/10.1177/0271678X19848098.

Willie, C.K., *et al*., 2014. Integrative regulation of human brain blood flow. J. Physiol. 592 (5), 841–859. https://doi.org/10.1113/jphysiol.2013.268953.

Yang, H.-J., *et al.*, 2019. Accurate needle-free assessment of myocardial oxygenation for ischemic heart disease in canines using magnetic resonance imaging. Sci. Transl. Med. 11 (494). https://doi.org/10.1126/scitranslmed.aat4407.

Yezhuvath, U.S., *et al.*, 2012. Forebrain-dominant deficit in cerebrovascular reactivity in Alzheimer's disease. Neurobiol. Aging 33 (1), 75–82. https://doi.org/10.1016/j.neurobiolaging.2010.02.005.

Zaharchuk, G., Martin, A.J., Dillon, W.P., 2008. Noninvasive imaging of quantitative cerebral blood flow changes during 100% oxygen inhalation using arterial spin-labeling MR imaging. AJNR Am. J. Neuroradiol. 29 (4), 663–667. https://doi.org/10.3174/ajnr.A0896.

Zakher, E., Ganesh, T., Cheng, H.-L.M., 2020. A novel MRI analysis for assessment of microvascular vasomodulation in low-perfusion skeletal muscle. Sci. Rep. 10 (1), 4705. https://doi.org/10.1038/s41598-020-61682-z.

Zeiher, A.M., *et al.*, 1995. Impaired endothelium-dependent vasodilation of coronary resistance vessels is associated with exercise-induced myocardial ischemia. Circulation 91 (9), 2345–2352. https://doi.org/10.1161/01.cir.91.9.2345.

Zhao, P., *et al.*, 2009. Vasoreactivity and peri-infarct hyperintensities in stroke. Neurology 72 (7), 643–649. https://doi.org/10.1212/01.wnl.0000342473.65373.80.

Further reading

Chen, J., Fierstra, J. (Eds.), 2022. Cerebrovascular Reactivity: Methodological Advances and Clinical Applications. Humana Press, p. 230.

Technical considerations

MR contrast agents for perfusion imaging

Claudia Calcagno[a], Ji Hyun Lee[b], and Gustav J. Strijkers[c]

[a]*Integrated Research Facility at Fort Detrick, Division of Clinical Research, National Institute of Allergy and Infectious Diseases, National Institutes of Health, Frederick, MD, United States*
[b]*Radiology and Imaging Sciences, Clinical Center, National Institutes of Health, Bethesda, MD, United States*
[c]*Department of Biomedical Engineering and Physics, Amsterdam University Medical Centers, University of Amsterdam, Amsterdam, The Netherlands*

6.1 Introduction

Perfusion-weighted MRI extracts quantitative measures of organ perfusion and permeability by analyzing transient changes in tissue MR signal intensity caused by the injection of a contrast agent (Aime and Caravan, 2009; Wahsner *et al.*, 2019).

Most commonly, perfusion MR contrast agents are chelates of the lanthanide metal gadolinium (Gd^{3+}). In both humans and animals, perfusion MR contrast agents are typically injected intravenously (i.v.), and, after injection, they rapidly circulate in the venous and arterial system before distributing to smaller vessels in target organs (Aime and Caravan, 2009; Wahsner *et al.*, 2019). Perfusion MR contrast agents can be categorized as either "intravascular" or "extracellular" based on their distribution in the body. "Intravascular" contrast agents remain in the bloodstream for longer and allow specific quantification of blood flow and microvascular volume (Aime and Caravan, 2009; Wahsner *et al.*, 2019). "Extracellular" contrast agents extravasate into the extravascular extracellular space (EES) and also provide information on tissue permeability and EES volume (Aime and Caravan, 2009; Wahsner *et al.*, 2019). These parameters are typically derived from quick changes in T1 and T2/T2* relaxation times (and therefore MR signal) caused by the contrast agent first pass in tissues, as well as their transient persistence in the EES, by applying specific kinetic modeling techniques (Tofts *et al.*, 1999; Eyal and Degani, 2009). Alternatively, an estimate of tissue perfusion and permeability can be obtained using nonmodel-based parameters, such as area under the curve (AUC) and time-to-peak of tissue contrast agent uptake (Eyal and Degani, 2009).

Since the first introduction of Gd^{3+}-based contrast agents on the market, a plethora of novel compounds with different physicochemical properties, organ or tissue distribution and tropism, and safety profiles have been developed for MR

Advances in Magnetic Resonance Technology and Applications, Volume 11, ISSN 2666-9099
https://doi.org/10.1016/B978-0-323-95209-5.00018-0

perfusion imaging (Aime and Caravan, 2009; Wahsner *et al.*, 2019). In this chapter, we first provide a classification of MR perfusion contrast agents based on chemical composition, magnetic properties, and biodistribution. We also describe organ-specific applications, focusing on the rationale for use of agents with different properties in different organs and tissues. Finally, we address safety considerations that are relevant to the use of these agents.

6.2 Classification of MR contrast agents for perfusion imaging

MR perfusion contrast agents can be classified according to different characteristics, such as their physicochemical and magnetic properties, biodistribution, and application to specific body parts. While MR contrast agents in general may also be classified by administration route, agents used for perfusion imaging are usually all administered intravenously (i.v.). In the following subsections, we provide a classification of the available MR perfusion contrast agents. A more in-depth description of contrast agent applications in different organs will be provided in a following section.

6.2.1 Classification based on chemical composition

MR perfusion contrast agents can be roughly subdivided into paramagnetic and superparamagnetic compounds based on their chemical composition (Aime and Caravan, 2009; Wahsner *et al.*, 2019).

Paramagnetic agents owe their properties to the presence of unpaired electrons in their nucleus. The atom most commonly used to develop paramagnetic MR contrast agents is the lanthanide metal gadolinium (Gd^{3+}), which is considered the most stable ion with seven unpaired electrons. Other paramagnetic elements that can be used to develop MR contrast agents are dysprosium (Dy^{3+}, with 4 unpaired electrons) and the transition metal manganese (Mn^{2+}, with 5 unpaired electrons) (Aime and Caravan, 2009; Wahsner *et al.*, 2019). Since metal ions are potentially toxic due to their accumulation in the liver, spleen, and bone, their use as MR contrast agents typically requires complexation with a chelating agent. Three main types of Gd^{3+}-based chelates are currently available for use in clinical practice: (i) ionic, hydrophilic complexes, such as gadolinium (III) diethylenetriamine pentaacetate (Gd-DTPA, gadopentetate dimeglumine), Gd(III) 1,4,7,10-tetrazacyclododecane *NN'N''N'''*-tetra-acetate (Gd-DOTA, gadoterate), and Gd(III) polyaspartate; (ii) non-ionic, hydrophilic complexes, such as Gd3-diethylenetriamine pentaacetate-bis(methylamide) (Gd-DTPA-BMA, gadodiamide) and a macrocyclic chelate analog of Gd-DOTA (Gd-HP-DO3A, gadoteridol); and (iii) ionic, lipophilic complexes, including Gd benzyl-oxy-methyl derivative of diethylenetriamine pentaacetate dimethylglucamine salt (Gd-BOPTA, gadobenate dimeglumine) and Gd ethoxy-benzyl diethylenetriamine pentaacetate (Gd-EOB-DTPA, gadoxetate) (Xiao *et al.*, 2016). Table 6.1 contains examples of Gd^{3+}-based contrast agents used for MR

Table 6.1 Example of Gd^{3+}-based contrast agents that have been used for MR perfusion in the clinics (currently commercially available, suspended, or not marketed anymore).

Short chemical name	Generic name	Brand name
Gd-DTPA[a]	Gadopentetate dimeglumine	Magnevist
Gd-DOTA	Gadoterate meglumine	Dotarem, Clariscan
Gd-BOPTA[b]	Gadobenate dimeglumine	MultiHance
Gd-DTPA-BMA[a]	Gadodiamide	Omniscan
Gd-DTPA-BMEA[a]	Gadoversetamide	Optimark
Gd-HP-DO3A	Gadoteridol	ProHance
Gd-DO3A-butrol[a]	Gadobutrol	Gadovist, Gadavist
Gd-EOB-DTPA[b]	Gadoxetic acid disodium	Primovist, Eovist
MS-325[c]	Gadofosveset trisodium	Ablavar (formerly: Vasovist, Angiomark)

[a]*Agents suspended by the European Medicines Agency in 2017.*
[b]*Agent available for limited, liver-specific indications in the EU.*
[c]*Approved for magnetic resonance angiography, but no longer commercially available.*

perfusion, while Fig. 6.1 shows their chemical structure (Aime and Caravan, 2009; Wahsner *et al.*, 2019).

Due to increasing concerns related to the long-term safety profile of certain Gd^{3+}-based contrast agents, other metals such as Mn^{2+} (Pan *et al.*, 2011) and Dy^{3+} are also being investigated as alternative MR contrast agents, also with some applications in perfusion imaging.

Unlike Gd^{3+}, Mn^{2+} is a metal that naturally occurs in our bodies, where it often acts as a cofactor for enzymes and receptors. As a contrast agent, Mn^{2+} acts similarly to Gd^{3+}, by shortening tissue T1 relaxation time. Mn^{2+}-based agents (Daksh *et al.*, 2022) have been investigated to quantify myocardial perfusion (Schaefer *et al.*, 1989; Eriksson *et al.*, 2006; Pomeroy *et al.*, 1989; Hu *et al.*, 2004; Yang *et al.*, 2009; Storey *et al.*, 2003; Natanzon *et al.*, 2005; Saeed *et al.*, 1989; Singh *et al.*, 2023) and, more recently, brain perfusion (Grillon *et al.*, 2008) and pancreatic β-cell function (Dhyani *et al.*, 2013). A Mn^{2+}-based chelate has recently been investigated in comparison with Gd-DTPA for use in contrast-enhanced MR angiography (Gale *et al.*, 2018). Mn^{2+}-based agents have also been used for organ-specific applications, especially in the liver (Toft *et al.*, 1997).

While less common, Dy^{3+}-based agents have also found applications for MR perfusion imaging. For example, dysprosium diethylenetriaminepentaacetic acid bis(methylamide) (Dy-DTPA-BMA) has been investigated as an alternative to (Beache *et al.*, 1998; Nilsson *et al.*, 1995a, 1996a; Arteaga *et al.*, 1999; Wendland *et al.*, 1993) or in combination with (Nilsson *et al.*, 1995b, 1996b; Wikstrom *et al.*, 1993) Gd^{3+}-based agents to quantify myocardial perfusion. The same Dy^{3+} chelate has also been investigated to quantify brain perfusion (Haraldseth *et al.*, 1996; Kucharczyk *et al.*, 1991).

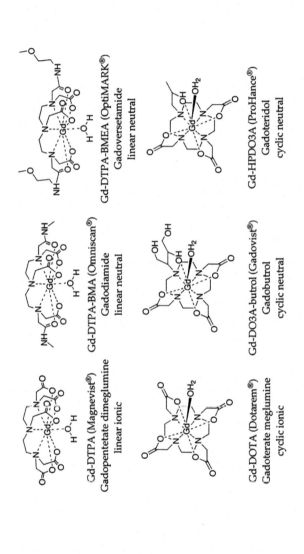

FIG. 6.1

Chemical structure of Gd^{3+}-based contrast agents most commonly used for MR perfusion.

Reprinted with permission from Aime, S., Caravan, P., 2009. Biodistribution of gadolinium-based contrast agents, including gadolinium depcsition. J. Magn. Reson. Imaging 30, 1259–1267.

In contrast to the chelated structures described before, superparamagnetic agents consist of suspended colloids of iron oxide nanoparticles. These include superparamagnetic iron oxide (SPIO) and ultrasmall superparamagnetic iron oxide (USPIO) particles (Aime and Caravan, 2009; Wahsner *et al*., 2019). SPIO (Saito *et al*., 2009) (>50 nm) agents are composed of maghemite and magnetite (γFe_2O_3, Fe_3O_4) and coating materials. Hydrodynamic size, crystal structures, and coating materials can affect the biodistribution and application of SPIOs. Dextran is the most used coating material for SPIOs due to its higher biocompatibility among other polysaccharides. USPIO (<50 nm) agents have a long blood half-life due to improved coating and are normally ingested by macrophages. SPIOs and USPIOs have been investigated and used to study organ perfusion for more than 20 years, with main applications in the brain (Haraldseth *et al*., 1996; Gharagouzloo *et al*., 2017; Loubeyre *et al*., 1999; Reimer *et al*., 1995; Reith *et al*., 1995; Varallyay *et al*., 2009, 2013; Zimmer *et al*., 1995; Bentzen *et al*., 2005; Dosa *et al*., 2011; Gahramanov *et al*., 2013; Netto *et al*., 2016; Neuwelt *et al*., 2007; Pannetier *et al*., 2012), heart (Beache *et al*., 1998; Bjerner *et al*., 2001, 2004a,b; Bjornerud *et al*., 2003; Bjornerud and Johansson, 2004; Canet *et al*., 1993, 1995; Colbert *et al*., 2021a,b; Nguyen *et al*., 2019), liver (Saito *et al*., 2009; Caramella *et al*., 1996; Hahn *et al*., 1990; Kurata *et al*., 2022; Sahani *et al*., 2001; Wersebe *et al*., 2006), kidneys (Sahani *et al*., 2001; Morell *et al*., 2008; Niendorf *et al*., 2020; Trillaud *et al*., 1993, 1995; Bachmann *et al*., 2002; Yang *et al*., 2001; Bjornerud *et al*., 2001, 2002; Schoenberg *et al*., 2003; Aumann *et al*., 2003; Wang *et al*., 2014; Daly *et al*., 1989; Cantow *et al*., 2016), placenta (Deloison *et al*., 2012; Ludwig *et al*., 2019), and cancer (Varallyay *et al*., 2009; Bentzen *et al*., 2005; Dosa *et al*., 2011; Gahramanov *et al*., 2011, 2013; Netto *et al*., 2016; Neuwelt *et al*., 2007; Pannetier *et al*., 2012; Hahn *et al*., 1990; Bjornerud *et al*., 2001; Ichikawa *et al*., 1999; Kato *et al*., 2004; Kostourou *et al*., 2003; Melemenidis *et al*., 2015; Nasseri *et al*., 2014; Pathak, 2009; Persigehl *et al*., 2010; Pike *et al*., 2009; Robinson *et al*., 2003, 2007, 2017; Saito *et al*., 2020; Thompson *et al*., 2012), and both in experimental animal models and patients. Because of significant long-term concerns over the use of Gd chelates, recent years have seen increasing interest on investigating these agents to quantify organ perfusion, with specific emphasis on ferumoxytol, a compound used to treat certain kind of anemias, which is also used off-label as an MR contrast agent (Bashir *et al*., 2015; Vasanawala *et al*., 2016). Table 6.2 contains examples of iron oxide contrast agents used for MR perfusion.

6.2.2 Classification based on magnetic properties

As mentioned in the introduction, MR contrast agents act by shortening the T1 (longitudinal) and T2 (transverse) relaxation times of neighboring water protons (Aime and Caravan, 2009; Wahsner *et al*., 2019). The contrast agent concentration and the associated changes in tissue relaxation are linked by a linear relationship, whose slope is called relaxivity (namely r_1 or r_2 for either T1 or T2 relaxation)

Table 6.2 Examples of iron oxide MR contrast agents that have been used for MR perfusion.

Short chemical name	Generic name	Brand name
AMI-25	Ferumoxides (SPIO)	Endorem, Feridex
AMI-277	Ferumoxtran-10 (USPIO)	Sinerem, Combidex
SHU 555A	Ferucarbotran (SPIO)	Resovist, Cliavist
Ferumoxytol	Ferumoxytol	Feraheme

(Aime and Caravan, 2009; Wahsner *et al.*, 2019). Shortening of T1 relaxation time is usually captured using T1-weighted MR sequences and results in brighter tissue signal in areas where the contrast agent accumulates. Therefore MR contrast agents that mainly affect the T1 relaxation time are usually referred to as "positive" contrast agents. Shortening of T2 (or T2*) relaxation times is instead visualized using T2 (or T2*) weighted MR sequences, and, with notable exceptions, typically results in negative contrast in areas of contrast agent accumulation (Aime and Caravan, 2009; Wahsner *et al.*, 2019). Therefore T2 (or T2*) shortening agents are typically referred to as "negative" agents. More specifically, T1-weighted MR imaging of tissue perfusion is typically referred to as dynamic contrast-enhanced (DCE) MRI, while T2/T2*-weighted perfusion imaging as dynamic susceptibility contrast (DSC) MRI.

Paramagnetic and superparamagnetic contrast agents do affect both T1 and T2 relaxation times, although to a widely different extent based on their chemical structure, composition, and tissue concentration (Aime and Caravan, 2009; Wahsner *et al.*, 2019).

Among paramagnetic agents, Gd- and Mn-based compounds shorten the T1 relaxation time at lower concentrations, while they affect T2 at higher concentrations (or, in other words, their r_1 relaxivity is usually higher than their r_2 relaxivity). Therefore they are usually referred to as "positive" agents. In contrast, Dy-based compounds have very low r_1 relaxivity and typically exhibit predominant T2 shortening with minimal effects on T1 relaxation time across a wide range of concentrations, with their tissue accumulation being usually visualized as signal loss.

Superparamagnetic agents are typically characterized by much higher r_2 relaxivity compared to paramagnetic agents and, therefore, shorten T2/T2* relaxation times at much lower concentrations. However, at lower concentrations, some of these agents, such as ferumoxytol, exhibit significant T1 shortening properties, which can be exploited for specific applications (such as angiography studies) (Bashir *et al.*, 2015).

6.2.3 Classification based on biodistribution

MR contrast agents for perfusion imaging can be generally classified into intravascular (or blood-pool), extracellular, or organ-specific agents (Aime and Caravan, 2009; Wahsner *et al.*, 2019). Extravascular agents have typically lower molecular weight and are commonly used for quantifying tissue perfusion and permeability. After injection into the blood circulation, they rapidly extravasate into tissues (during the first pass), and they distribute into the extracellular space. Because of this behavior, these agents allow quantification of tissue perfusion and permeability. On the contrary, blood-pool agents remain in the vasculature much longer than extravascular agents, making them suitable for angiography studies or for quantifying organ blood flow and perfusion (but not permeability). Their longer residence in the blood circulation is achieved either by binding to plasma proteins (typically albumin, either before injection or upon the agent entering the blood circulation) or by the agents themselves having higher molecular weight and size (by incorporating polymers or liposomes, *e.g.*), which slow down or prevent tissue extravasation. Some MR contrast agents used for perfusion imaging may also provide an assessment of specific organ functions. For example, the liver MR contrast agent gadolinium ethoxybenzyl dimeglumine or gadoxetate dimeglumine (Gd-EOB-DTPA) can provide information on liver vascularity and also on hepatobiliary function (since about 50% is taken up by hepatocytes) (Aime and Caravan, 2009; Wahsner *et al.*, 2019). Fig. 6.2 shows the possible biodistribution patterns of contrast agents for MR perfusion, including excretion routes.

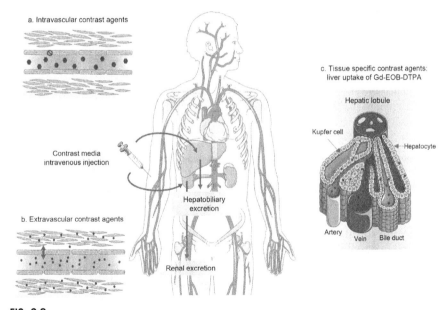

FIG. 6.2

Schematics of contrast agent biodistribution. (A) intravascular agents. (B) extravascular agents. (C) example of the behavior of a tissue-specific (liver) agent. *Blue* hexagon (*black* in print version), contrast agent.

Parts of the figure were drawn by using pictures from Servier Medical Art. Available from: https://smart.servier. com/. Servier Medical Art by Servier is licensed under a Creative Commons Attribution 3.0 Unported License (https://creativecommons.org/licenses/by/3.0/).

6.3 Clinical applications

Depending on their physicochemical and magnetic properties and biodistribution, different contrast agents are better suited to investigate organ pathophysiology in different vascular districts or conditions. While the quantification of perfusion and permeability in specific organs and tissues is covered in great details in other chapters, here we provide a brief perspective on the rationale for use of specific agents to quantify these parameters in different organs and condition.

MR perfusion imaging is commonly used in oncology to quantify vascularity in different types of cancerous lesions, from brain gliomas (Shukla *et al.*, 2017) to breast (Iacconi, 2010) and prostate (Somford *et al.*, 2008) cancer, to liver (Thng *et al.*, 2014) or head and neck cancer (Kabadi *et al.*, 2018). During tumor growth, an extended network of leaky, friable microvessels develops in response to hypoxia. The extent and leakiness of tumor vascularization is a pathophysiological feature related to tumor grade and stage, and its changes upon therapeutic intervention can be monitored as an indicator of treatment response. In this context, gadolinium-based extravascular contrast agents are usually used in combination with either T2*- or T1-weighted MR sequences to perform either DSC- or DCE-MRI of cancerous lesions and quantify tumor perfusion and/or permeability-related parameters, such as blood volume, blood flow, tissue permeability, and EES volume. MR perfusion imaging can also complement 18F fluorodeoxyglucose (18F-FDG) positron emission tomography (PET) in evaluating organ inflammation and immune system activation in infectious diseases (Jayakumar *et al.*, 2013). In addition, MR perfusion imaging is employed to assess changes in tissue vascularity in a variety of organ-specific diseases. For example, in the brain, gadolinium-based extravascular agents have been used in combination with DCE-MRI to quantify blood-brain-barrier disruption in multiple sclerosis lesions (Oghabian *et al.*, 2022), in the physiologically aging human brain (Montagne *et al.*, 2022), and in several forms of dementia (Raja *et al.*, 2018; Chagnot *et al.*, 2021), psychiatric diseases (Goldstein and MacIntosh, 2022), and traumatic brain injury (Brooks *et al.*, 2019). Extravascular contrast agents are also used to evaluate hypoperfusion and ischemia both in the brain (stroke) and in the heart (myocardial infarction) (Russo *et al.*, 2020). Cardiac perfusion measurements are typically performed using T1-weighted dynamic MR acquisitions in combination with the injection of gadolinium-based extracellular agents. These agents are also used for late gadolinium enhancement (LGE) imaging to characterize viable and nonviable myocardium in ischemic or nonischemic cardiomyopathies (Coelho-Filho *et al.*, 2013). Although intravascular contrast agents have been tested for myocardial perfusion MR studies, they do not show a significant difference compared to extracellular agents (Kraitchman *et al.*, 2002; Nacif *et al.*, 2018; Ritter *et al.*, 2011). In body imaging, and excluding oncological applications, MR perfusion imaging is often used to characterize, for example, liver and kidney disease. From the perspective of this imaging technique, the liver is interesting to mention since it can be investigated by using conventional gadolinium-based extravascular agents but also hepatocyte-specific tracers, such as Gd-EOB-DTPA, to gain information about vascularity and hepatobiliary function in a variety of liver diseases and pathophysiological conditions, such as steatosis, fibrosis, and cirrhosis (Poetter-Lang *et al.*, 2020). Lung perfusion with DCE-MRI using gadolinium-based agents is less commonly employed but has been used for quantification in children and adolescent after congenital diaphragmatic hernia repair (Weis *et al.*, 2016a,b). While gadolinium-based contrast agents are most commonly used in MR perfusion imaging, specific applications for Dy^{3+}- and Mn^{2+}-based chelates, as well as iron oxides, have been already discussed in the previous sections.

6.4 Safety issues

Perfusion-weighted MRI acquisitions are generally considered relatively safe and well tolerated. However, some safety aspects related to these procedures still need to be considered. Both DCE- and DSC-MRI require the injection of contrast agents intravenously, at fairly rapid injection rates (ACR manual on contrast media, 2023), to deliver the contrast medium as a tight bolus within 5–10 s (Sourbron, 2010). Although no significant adverse effects have been reported in relation to the fast injection rate, potential side effects at the injection site, such as minor pain or extravasation, have been reported (ACR manual on contrast media, 2023).

The acute adverse event rate following Gd^{3+} chelates injection is low (0.07%–2.4%). Mild reactions, such as nausea, hives, and taste disturbance, are typically managed by careful observation, while more severe allergic reactions and, very rarely, anaphylactic shock, are managed with pharmacological intervention (administration of epinephrine) and cardiopulmonary support (ACR manual on contrast media, 2023). In the past, the administration of Gd^{3+} chelates was considered safe in the long run, to the point that, in many cases, it was common practice to administer double or triple the recommended clinical dose of Gd^{3+} chelates (even in repeated longitudinal examinations) to achieve better signal enhancement and lesion delineation (Tombach *et al.*, 2003). However, starting in 2006, contrast agents, and especially certain Gd^{3+} chelates, have been associated with the risk of nephrogenic systemic fibrosis (NSF) in patients with renal dysfunction or end-stage renal disease (ACR manual on contrast media, 2023). Although NSF is an extremely rare condition in subjects with normal or moderately impaired renal function, individuals with impaired renal function are at higher risk for this condition, especially when being injected with linear Gd^{3+} chelates as opposed to their macrocyclic counterparts (Marckmann *et al.*, 2006; Broome *et al.*, 2007). Therefore guidelines from the American College of Radiology (ACR) recommend extreme caution when administering any Gd^{3+} chelate in patients with impaired kidney function and to use macrocyclic agents (group II agents from Table 6.3) in this group of at-risk individuals when possible. These guidelines have greatly reduced the incidence of new NSF cases (ACR manual on contrast media, 2023). In addition, in 2013, the incidental finding of Gd^{3+} chelates deposition in the brain of subjects with normal renal function was reported after multiple contrast agent injections (Kanda *et al.*, 2014). The long-term accumulation of Gd^{3+} chelates in bone tissue has also been documented (Murata *et al.*, 2016). The mechanism

Table 6.3 American College of Radiology manual classification of gadolinium-based contrast agents relative to NSF (ACR manual on contrast media, 2023).

Group 1 Agents associated with the greatest number of NSF cases	Group II Agents associated with few, if any, unconfounded cases of NSF	Group III Agents for which data remain limited regarding NSF risk, but for which few, if any unconfounded cases of NSF have been reported
Gadodiamide (Omniscan—GE Healthcare) Gadopentetate dimeglumine (Magnevist—Bayer HealthCare Pharmaceuticals) Gadoversetamide (OptiMARK—Guerbet)	Gadobenate dimeglumine (MultiHance—Bracco Diagnostics) Gadobutrol (Gadavist—Bayer HealthCare Pharmaceuticals; Gadovist in many countries) Gadoteric acid (Dotarem—Guerbet, Clariscan—GE Healthcare) Gadoteridol (ProHance—Bracco Diagnostics)	Gadoxetate disodium (Eovist—Bayer HealthCare Pharmaceuticals; Primovist in many countries)

underlying Gd^{3+} chelates tissue accumulation and its long-term effects are not clear at this stage. Because of these potential long-term safety concerns with Gd^{3+}-based contrast agents, and as mentioned in previous sections, other paramagnetic (Dy^{3+}- and Mn^{2+}-based chelates (Daksh *et al.*, 2022)) or superparamagnetic (iron oxides, such as ferumoxytol) compounds are being currently explored as perfusion MR contrast agent. To date, postmarketing safety data are only available for therapeutic use of ferumoxytol. While the risk of acute adverse events after ferumoxytol injection is higher than for Gd^{3+}-based agents (Vasanawala *et al.*, 2016), the long-term risk of nephrotoxicity and NSF risks associated with Gd^{3+}-based agents in patients with severe renal disease is even higher and can be fatal (Vasanawala *et al.*, 2016). In addition, noncontrast-enhanced MRI techniques to quantify tissue perfusion (Falk Delgado *et al.*, 2019), such as arterial spin labeling (ASL) (Overton *et al.*, 2020) and intravoxel incoherent motion (IVIM) (Federau, 2021), also continue to be developed.

Lastly, other techniques based on nuclei other than hydrogen are being developed to characterize tissue perfusion. For example, hyperpolarized 129 Xenon (^{129}Xe) MR imaging has recently been proposed to investigate regional brain and lung perfusion (Shepelytskyi *et al.*, 2020; Driehuys *et al.*, 2009).

6.5 Learning and knowledge outcomes

Perfusion MRI is a powerful technique that can give deep insights into organ function and pathophysiology. Because of its relevance, the scientific community continues to invest significant efforts in either validating or investigating the utility of existing agents (experimental or clinically approved) to quantify organ perfusion by MRI, as well as in developing new contrast agents with more favorable properties and better safety profiles. This chapter provides a basic understanding of the physicochemical and relaxation properties of contrast agents used for MR perfusion imaging, as well as several relevant examples for their usage in different conditions, and information on their safety profile. With this knowledge, healthcare professionals can make informed decisions about the use of contrast agents for perfusion MRI, select the most appropriate agent for a particular clinical scenario, and manage potential adverse effects.

Disclaimer

The content of this publication does not necessarily reflect the views or policies of the US Department of Health and Human Services (DHHS) or of the institutions and companies affiliated with the authors. This project has been funded in whole or in part with Federal funds from the National Institute of Allergy and Infectious Diseases, National Institutes of Health, Department of Health and Human Services, under Contract No. HHSN272201800013C. C.C. performed this work as employee of Tunnell Government Services, Inc., and as subcontractor Laulima Government Solutions, LLC.

References

ACR manual on contrast media, 2023. Available from: https://www.acr.org/-/media/ACR/Files/Clinical-Resources/Contrast_Media.pdf.

Aime, S., Caravan, P., 2009. Biodistribution of gadolinium-based contrast agents, including gadolinium deposition. J. Magn. Reson. Imaging 30 (6), 1259–1267.

Arteaga, C., et al., 1999. Myocardial "low reflow" assessed by Dy-DTPA-BMA-enhanced first-pass MR imaging in a dog model. J. Magn. Reson. Imaging 9 (5), 679–684.

Aumann, S., et al., 2003. Quantification of renal perfusion using an intravascular contrast agent (part 1): results in a canine model. Magn. Reson. Med. 49 (2), 276–287.

Bachmann, R., et al., 2002. Evaluation of a new ultrasmall superparamagnetic iron oxide contrast agent Clariscan, (NC100150) for MRI of renal perfusion: experimental study in an animal model. J. Magn. Reson. Imaging 16 (2), 190–195.

Bashir, M.R., et al., 2015. Emerging applications for ferumoxytol as a contrast agent in MRI. J. Magn. Reson. Imaging 41 (4), 884–898.

Beache, G.M., et al., 1998. Imaging perfusion deficits in ischemic heart disease with susceptibility-enhanced T2-weighted MRI: preliminary human studies. Magn. Reson. Imaging 16 (1), 19–27.

Bentzen, L., et al., 2005. Intravascular contrast agent-enhanced MRI measuring contrast clearance and tumor blood volume and the effects of vascular modifiers in an experimental tumor. Int. J. Radiat. Oncol. Biol. Phys. 61 (4), 1208–1215.

Bjerner, T., et al., 2001. First-pass myocardial perfusion MR imaging with outer-volume suppression and the intravascular contrast agent NC100150 injection: preliminary results in eight patients. Radiology 221 (3), 822–826.

Bjerner, T., et al., 2004a. In and ex vivo MR evaluation of acute myocardial ischemia in pigs by determining R1 in steady state after the administration of the intravascular contrast agent NC100150 injection. Investig. Radiol. 39 (8), 479–486.

Bjerner, T., et al., 2004b. High in-plane resolution T2-weighted magnetic resonance imaging of acute myocardial ischemia in pigs using the intravascular contrast agent NC100150 injection. Investig. Radiol. 39 (8), 470–478.

Bjornerud, A., Johansson, L., 2004. The utility of superparamagnetic contrast agents in MRI: theoretical consideration and applications in the cardiovascular system. NMR Biomed. 17 (7), 465–477.

Bjornerud, A., Johansson, L.O., Ahlstrom, H.K., 2001. Pre-clinical results with Clariscan (NC100150 Injection); experience from different disease models. MAGMA 12 (2-3), 99–103.

Bjornerud, A., Johansson, L.O., Ahlstrom, H.K., 2002. Renal T(*)(2) perfusion using an iron oxide nanoparticle contrast agent—influence of T(1) relaxation on the first-pass response. Magn. Reson. Med. 47 (2), 298–304.

Bjornerud, A., et al., 2003. Assessment of myocardial blood volume and water exchange: theoretical considerations and in vivo results. Magn. Reson. Med. 49 (5), 828–837.

Brooks, B.L., et al., 2019. Cerebral blood flow in children and adolescents several years after concussion. Brain Inj. 33 (2), 233–241.

Broome, D.R., et al., 2007. Gadodiamide-associated nephrogenic systemic fibrosis: why radiologists should be concerned. AJR Am. J. Roentgenol. 188 (2), 586–592.

Canet, E., et al., 1993. Superparamagnetic iron oxide particles and positive enhancement for myocardial perfusion studies assessed by subsecond T1-weighted MRI. Magn. Reson. Imaging 11 (8), 1139–1145.

Canet, E., et al., 1995. Noninvasive assessment of no-reflow phenomenon in a canine model of reperfused infarction by contrast-enhanced magnetic resonance imaging. Am. Heart J. 130 (5), 949–956.

Cantow, K., et al., 2016. Acute effects of ferumoxytol on regulation of renal hemodynamics and oxygenation. Sci. Rep. 6, 29965.

Caramella, D., et al., 1996. Liver and spleen enhancement after intravenous injection of carboxydextran magnetite: effect of dose, delay of imaging, and field strength in an ex vivo model. MAGMA 4 (3-4), 225–230.

Chagnot, A., Barnes, S.R., Montagne, A., 2021. Magnetic resonance imaging of blood-brain barrier permeability in dementia. Neuroscience 474, 14–29.

Coelho-Filho, O.R., et al., 2013. MR myocardial perfusion imaging. Radiology 266 (3), 701–715.

Colbert, C.M., et al., 2021a. Ferumoxytol-enhanced magnetic resonance T1 reactivity for depiction of myocardial hypoperfusion. NMR Biomed. 34 (7), e4518.

Colbert, C.M., et al., 2021b. Estimation of fractional myocardial blood volume and water exchange using ferumoxytol-enhanced magnetic resonance imaging. J. Magn. Reson. Imaging 53 (6), 1699–1709.

Daksh, S., et al., 2022. Current advancement in the development of manganese complexes as magnetic resonance imaging probes. J. Inorg. Biochem. 237, 112018.

Daly, P.F., et al., 1989. Rapid MR imaging of renal perfusion: a comparative study of GdDTPA, albumin-(GdDTPA), and magnetite. Am. J. Physiol. Imaging 4 (4), 165–174.

Deloison, B., et al., 2012. SPIO-enhanced magnetic resonance imaging study of placental perfusion in a rat model of intrauterine growth restriction. BJOG 119 (5), 626–633.

Dhyani, A.H., et al., 2013. Empirical mathematical model for dynamic manganese-enhanced MRI of the murine pancreas for assessment of beta-cell function. Magn. Reson. Imaging 31 (4), 508–514.

Dosa, E., *et al.*, 2011. MRI using ferumoxytol improves the visualization of central nervous system vascular malformations. Stroke 42 (6), 1581–1588.

Driehuys, B., *et al.*, 2009. Pulmonary perfusion and xenon gas exchange in rats: MR imaging with intravenous injection of hyperpolarized 129Xe. Radiology 252 (2), 386–393.

Eriksson, R., *et al.*, 2006. Contrast enhancement of manganese-hydroxypropyl-tetraacetic acid, an MR contrast agent with potential for detecting differences in myocardial blood flow. J. Magn. Reson. Imaging 24 (4), 858–863.

Eyal, E., Degani, H., 2009. Model-based and model-free parametric analysis of breast dynamic-contrast-enhanced MRI. NMR Biomed. 22 (1), 40–53.

Falk Delgado, A., *et al.*, 2019. Diagnostic value of alternative techniques to gadolinium-based contrast agents in MR neuroimaging-a comprehensive overview. Insights Imaging 10 (1), 84.

Federau, C., 2021. Measuring perfusion: intravoxel incoherent motion MR imaging. Magn. Reson. Imaging Clin. N. Am. 29 (2), 233–242.

Gahramanov, S., *et al.*, 2011. Improved perfusion MR imaging assessment of intracerebral tumor blood volume and antiangiogenic therapy efficacy in a rat model with ferumoxytol. Radiology 261 (3), 796–804.

Gahramanov, S., *et al.*, 2013. Pseudoprogression of glioblastoma after chemo- and radiation therapy: diagnosis by using dynamic susceptibility-weighted contrast-enhanced perfusion MR imaging with ferumoxytol versus gadoteridol and correlation with survival. Radiology 266 (3), 842–852.

Gale, E.M., *et al.*, 2018. A manganese-based alternative to gadolinium: contrast-enhanced MR angiography, excretion, pharmacokinetics, and metabolism. Radiology 286 (3), 865–872.

Gharagouzloo, C.A., *et al.*, 2017. Quantitative vascular neuroimaging of the rat brain using superparamagnetic nanoparticles: new insights on vascular organization and brain function. NeuroImage 163, 24–33.

Goldstein, B.I., MacIntosh, B.J., 2022. The unrealized promise of cerebrovascular magnetic resonance imaging in psychiatric research across the lifespan. Eur. Neuropsychopharmacol. 55, 11–13.

Grillon, E., *et al.*, 2008. Blood-brain barrier permeability to manganese and to Gd-DOTA in a rat model of transient cerebral ischaemia. NMR Biomed. 21 (5), 427–436.

Hahn, P.F., *et al.*, 1990. Clinical application of superparamagnetic iron oxide to MR imaging of tissue perfusion in vascular liver tumors. Radiology 174 (2), 361–366.

Haraldseth, O., *et al.*, 1996. Comparison of dysprosium DTPA BMA and superparamagnetic iron oxide particles as susceptibility contrast agents for perfusion imaging of regional cerebral ischemia in the rat. J. Magn. Reson. Imaging 6 (5), 714–717.

Hu, T.C., *et al.*, 2004. Simultaneous assessment of left-ventricular infarction size, function and tissue viability in a murine model of myocardial infarction by cardiac manganese-enhanced magnetic resonance imaging (MEMRI). NMR Biomed. 17 (8), 620–626.

Iacconi, C., 2010. Diffusion and perfusion of the breast. Eur. J. Radiol. 76 (3), 386–390.

Ichikawa, T., *et al.*, 1999. Perfusion MR imaging with a superparamagnetic iron oxide using T2-weighted and susceptibility-sensitive echoplanar sequences: evaluation of tumor vascularity in hepatocellular carcinoma. AJR Am. J. Roentgenol. 173 (1), 207–213.

Jayakumar, P.N., Chandrashekar, H.S., Ellika, S., 2013. Imaging of parasitic infections of the central nervous system. Handb. Clin. Neurol. 114, 37–64.

Kabadi, S.J., *et al.*, 2018. Dynamic contrast-enhanced MR imaging in head and neck cancer. Magn. Reson. Imaging Clin. N. Am. 26 (1), 135–149.

Kanda, T., et al., 2014. High signal intensity in the dentate nucleus and globus pallidus on unenhanced T1-weighted MR images: relationship with increasing cumulative dose of a gadolinium-based contrast material. Radiology 270 (3), 834–841.

Kato, H., et al., 2004. Ferumoxide-enhanced MR imaging of hepatocellular carcinoma: correlation with histologic tumor grade and tumor vascularity. J. Magn. Reson. Imaging 19 (1), 76–81.

Kostourou, V., et al., 2003. Effects of overexpression of dimethylarginine dimethylaminohydrolase on tumor angiogenesis assessed by susceptibility magnetic resonance imaging. Cancer Res. 63 (16), 4960–4966.

Kraitchman, D.L., et al., 2002. MRI detection of myocardial perfusion defects due to coronary artery stenosis with MS-325. J. Magn. Reson. Imaging 15 (2), 149–158.

Kucharczyk, J., et al., 1991. Magnetic resonance imaging of brain perfusion using the nonionic contrast agents Dy-DTPA-BMA and Gd-DTPA-BMA. Investig. Radiol. 26 (Suppl. 1), S250–S252; discussion S253-4.

Kurata, C., et al., 2022. The feasibility of superparamagnetic iron oxide-enhanced magnetic resonance imaging for assessing liver lesions in patients with contraindications for iodine CT contrast media or gadolinium-based MR contrast media: a retrospective case-control study. Quant. Imaging Med. Surg. 12 (9), 4612–4621.

Loubeyre, P., et al., 1999. Comparison of iron oxide particles (AMI 227) with a gadolinium complex (Gd-DOTA) in dynamic susceptibility contrast MR imagings (FLASH and EPI) for both phantom and rat brain at 1.5 Tesla. J. Magn. Reson. Imaging 9 (3), 447–453.

Ludwig, K.D., et al., 2019. Perfusion of the placenta assessed using arterial spin labeling and ferumoxytol dynamic contrast enhanced magnetic resonance imaging in the rhesus macaque. Magn. Reson. Med. 81 (3), 1964–1978.

Marckmann, P., et al., 2006. Nephrogenic systemic fibrosis: suspected causative role of gadodiamide used for contrast-enhanced magnetic resonance imaging. J. Am. Soc. Nephrol. 17 (9), 2359–2362.

Melemenidis, S., et al., 2015. Molecular magnetic resonance imaging of angiogenesis in vivo using polyvalent cyclic RGD-iron oxide microparticle conjugates. Theranostics 5 (5), 515–529.

Montagne, A., et al., 2022. Imaging subtle leaks in the blood-brain barrier in the aging human brain: potential pitfalls, challenges, and possible solutions. Geroscience 44 (3), 1339–1351.

Morell, A., et al., 2008. Quantitative renal cortical perfusion in human subjects with magnetic resonance imaging using iron-oxide nanoparticles: influence of T1 shortening. Acta Radiol. 49 (8), 955–962.

Murata, N., et al., 2016. Macrocyclic and other non-group 1 gadolinium contrast agents deposit low levels of gadolinium in brain and bone tissue: preliminary results from 9 patients with normal renal function. Investig. Radiol. 51 (7), 447–453.

Nacif, M.S., et al., 2018. Myocardial T1 mapping and determination of partition coefficients at 3 tesla: comparison between gadobenate dimeglumine and gadofosveset trisodium. Radiol. Bras. 51 (1), 13–19.

Nasseri, M., et al., 2014. Evaluation of pseudoprogression in patients with glioblastoma multiforme using dynamic magnetic resonance imaging with ferumoxytol calls RANO criteria into question. Neuro-Oncology 16 (8), 1146–1154.

Natanzon, A., et al., 2005. Determining canine myocardial area at risk with manganese-enhanced MR imaging. Radiology 236 (3), 859–866.

Netto, J.P., *et al.*, 2016. Misleading early blood volume changes obtained using ferumoxytol-based magnetic resonance imaging perfusion in high grade glial neoplasms treated with bevacizumab. Fluids Barriers CNS 13 (1), 23.

Neuwelt, E.A., *et al.*, 2007. The potential of ferumoxytol nanoparticle magnetic resonance imaging, perfusion, and angiography in central nervous system malignancy: a pilot study. Neurosurgery 60 (4), 601–611; discussion 611-2.

Nguyen, K.L., *et al.*, 2019. Ferumoxytol-enhanced CMR for vasodilator stress testing: a feasibility study. JACC Cardiovasc. Imaging 12 (8 Pt 1), 1582–1584.

Niendorf, T., *et al.*, 2020. Probing renal blood volume with magnetic resonance imaging. Acta Physiol (Oxford) 228 (4), e13435.

Nilsson, S., *et al.*, 1995a. Dy-DTPA-BMA as an indicator of tissue viability in MR imaging. An experimental study in the pig. Acta Radiol. 36 (4), 338–345.

Nilsson, S., *et al.*, 1995b. MR imaging of double-contrast enhanced porcine myocardial infarction. Correlation with microdialysis. Acta Radiol. 36 (4), 346–352.

Nilsson, S., *et al.*, 1996a. Myocardial cell death in reperfused and nonreperfused myocardial infarctions. MR imaging with dysprosioum-DTPA-BMA in the pig. Acta Radiol. 37 (1), 18–26.

Nilsson, S., *et al.*, 1996b. Double-contrast MR imaging of reperfused porcine myocardial infarction. An experimental study using Gd-DTAA and Dy-DTPA-BMA. Acta Radiol. 37 (1), 27–35.

Oghabian, M.A., Fatemidokht, A., Haririchian, M.H., 2022. Quantification of blood-brain-barrier permeability dysregulation and inflammatory activity in MS lesions by dynamic-contrast enhanced MR imaging. Basic Clin. Neurosci. 13 (1), 117–128.

Overton, D.J., *et al.*, 2020. Identifying psychosis spectrum youth using support vector machines and cerebral blood perfusion as measured by arterial spin labeled fMRI. Neuroimage Clin. 27, 102304.

Pan, D., *et al.*, 2011. Manganese-based MRI contrast agents: past, present and future. Tetrahedron 67 (44), 8431–8444.

Pannetier, N., *et al.*, 2012. Vessel size index measurements in a rat model of glioma: comparison of the dynamic (Gd) and steady-state (iron-oxide) susceptibility contrast MRI approaches. NMR Biomed. 25 (2), 218–226.

Pathak, A.P., 2009. Magnetic resonance susceptibility based perfusion imaging of tumors using iron oxide nanoparticles. Wiley Interdiscip. Rev. Nanomed. Nanobiotechnol. 1 (1), 84–97.

Persigehl, T., *et al.*, 2010. Tumor blood volume determination by using susceptibility-corrected DeltaR2* multiecho MR. Radiology 255 (3), 781–789.

Pike, M.M., *et al.*, 2009. High-resolution longitudinal assessment of flow and permeability in mouse glioma vasculature: Sequential small molecule and SPIO dynamic contrast agent MRI. Magn. Reson. Med. 61 (3), 615–625.

Poetter-Lang, S., *et al.*, 2020. Quantification of liver function using gadoxetic acid-enhanced MRI. Abdom. Radiol. (NY) 45 (11), 3532–3544.

Pomeroy, O.H., *et al.*, 1989. Magnetic resonance imaging of acute myocardial ischemia using a manganese chelate, Mn-DPDP. Invest. Radiol. 24 (7), 531–536.

Raja, R., Rosenberg, G.A., Caprihan, A., 2018. MRI measurements of blood-brain barrier function in dementia: a review of recent studies. Neuropharmacology 134 (Pt B), 259–271.

Reimer, P., *et al.*, 1995. Application of a superparamagnetic iron oxide (resovist) for MR imaging of human cerebral blood volume. Magn. Reson. Med. 34 (5), 694–697.

Reith, W., *et al.*, 1995. Early MR detection of experimentally induced cerebral ischemia using magnetic susceptibility contrast agents: comparison between gadopentetate dimeglumine and iron oxide particles. AJNR Am. J. Neuroradiol. 16 (1), 53–60.

Ritter, C.O., *et al.*, 2011. Comparison of intravascular and extracellular contrast media for absolute quantification of myocardial rest-perfusion using high-resolution MRI. J. Magn. Reson. Imaging 33 (5), 1047–1051.

Robinson, S.P., *et al.*, 2003. Tumor vascular architecture and function evaluated by non-invasive susceptibility MRI methods and immunohistochemistry. J. Magn. Reson. Imaging 17 (4), 445–454.

Robinson, S.P., *et al.*, 2007. Susceptibility contrast magnetic resonance imaging determination of fractional tumor blood volume: a noninvasive imaging biomarker of response to the vascular disrupting agent ZD6126. Int. J. Radiat. Oncol. Biol. Phys. 69 (3), 872–879.

Robinson, S.P., *et al.*, 2017. Monitoring the vascular response and resistance to sunitinib in renal cell carcinoma in vivo with susceptibility contrast MRI. Cancer Res. 77 (15), 4127–4134.

Russo, V., Lovato, L., Ligabue, G., 2020. Cardiac MRI: technical basis. Radiol. Med. 125 (11), 1040–1055.

Saeed, M., *et al.*, 1989. Occlusive and reperfused myocardial infarcts: differentiation with Mn-DPDP—enhanced MR imaging. Radiology 172 (1), 59–64.

Sahani, D., *et al.*, 2001. Dynamic T1-weighted ferumoxides enhanced MRI for imaging liver hemangiomas: preliminary observations. Abdom. Imaging 26 (2), 166–170.

Saito, K., *et al.*, 2009. Perfusion study of liver lesions with superparamagnetic iron oxide: distinguishing hepatocellular carcinoma from focal nodular hyperplasia. Clin. Imaging 33 (6), 447–453.

Saito, K., *et al.*, 2020. Validation study of perfusion parameter in hypervascular hepatocellular carcinoma and focal nodular hyperplasia using dynamic susceptibility magnetic resonance imaging with super-paramagnetic iron oxide: comparison with single level dynamic CT arteriography. Quant. Imaging Med. Surg. 10 (6), 1298–1306.

Schaefer, S., *et al.*, 1989. In vivo nuclear magnetic resonance imaging of myocardial perfusion using the paramagnetic contrast agent manganese gluconate. J. Am. Coll. Cardiol. 14 (2), 472–480.

Schoenberg, S.O., *et al.*, 2003. Quantification of renal perfusion abnormalities using an intravascular contrast agent (part 2): results in animals and humans with renal artery stenosis. Magn. Reson. Med. 49 (2), 288–298.

Shepelytskyi, Y., *et al.*, 2020. Hyperpolarized (129)Xe time-of-flight MR imaging of perfusion and brain function. Diagnostics (Basel) 10 (9).

Shukla, G., *et al.*, 2017. Advanced magnetic resonance imaging in glioblastoma: a review. Chin. Clin. Oncol. 6 (4), 40.

Singh, T., *et al.*, 2023. Manganese-enhanced magnetic resonance imaging of the heart. J. Magn. Reson. Imaging 57 (4), 1011–1028.

Somford, D.M., *et al.*, 2008. Diffusion and perfusion MR imaging of the prostate. Magn. Reson. Imaging Clin. N. Am. 16 (4), 685–695. ix.

Sourbron, S., 2010. Technical aspects of MR perfusion. Eur. J. Radiol. 76 (3), 304–313.

Storey, P., *et al.*, 2003. Preliminary evaluation of EVP 1001-1: a new cardiac-specific magnetic resonance contrast agent with kinetics suitable for steady-state imaging of the ischemic heart. Investig. Radiol. 38 (10), 642–652.

Thng, C.H., *et al.*, 2014. Perfusion imaging in liver MRI. Magn. Reson. Imaging Clin. N. Am. 22 (3), 417–432.

Thompson, E.M., et al., 2012. Dual contrast perfusion MRI in a single imaging session for assessment of pediatric brain tumors. J. Neuro-Oncol. 109 (1), 105–114.

Toft, K.G., et al., 1997. Metabolism and pharmacokinetics of MnDPDP in man. Acta Radiol. 38 (4 Pt 2), 677–689.

Tofts, P.S., et al., 1999. Estimating kinetic parameters from dynamic contrast-enhanced T (1)-weighted MRI of a diffusable tracer: standardized quantities and symbols. J. Magn. Reson. Imaging 10 (3), 223–232.

Tombach, B., et al., 2003. Do highly concentrated gadolinium chelates improve MR brain perfusion imaging? Intraindividually controlled randomized crossover concentration comparison study of 0.5 versus 1.0 mol/L gadobutrol. Radiology 226 (3), 880–888.

Trillaud, H., et al., 1993. First-pass evaluation of renal perfusion with TurboFLASH MR imaging and superparamagnetic iron oxide particles. J. Magn. Reson. Imaging 3 (1), 83–91.

Trillaud, H., et al., 1995. Evaluation of experimentally induced renal hypoperfusion using iron oxide particles and fast magnetic resonance imaging. Acad. Radiol. 2 (4), 293–299.

Varallyay, C.G., et al., 2009. Dynamic MRI using iron oxide nanoparticles to assess early vascular effects of antiangiogenic versus corticosteroid treatment in a glioma model. J. Cereb. Blood Flow Metab. 29 (4), 853–860.

Varallyay, C.G., et al., 2013. High-resolution steady-state cerebral blood volume maps in patients with central nervous system neoplasms using ferumoxytol, a superparamagnetic iron oxide nanoparticle. J. Cereb. Blood Flow Metab. 33 (5), 780–786.

Vasanawala, S.S., et al., 2016. Safety and technique of ferumoxytol administration for MRI. Magn. Reson. Med. 75 (5), 2107–2111.

Wahsner, J., et al., 2019. Chemistry of MRI contrast agents: current challenges and new frontiers. Chem. Rev. 119 (2), 957–1057.

Wang, F., et al., 2014. Repeatability and sensitivity of high resolution blood volume mapping in mouse kidney disease. J. Magn. Reson. Imaging 39 (4), 866–871.

Weis, M., et al., 2016a. Region of interest-based versus whole-lung segmentation-based approach for MR lung perfusion quantification in 2-year-old children after congenital diaphragmatic hernia repair. Eur. Radiol. 26 (12), 4231–4238.

Weis, M., et al., 2016b. Lung perfusion MRI after congenital diaphragmatic hernia repair in 2-year-old children with and without extracorporeal membrane oxygenation therapy. AJR Am. J. Roentgenol. 206 (6), 1315–1320.

Wendland, M.F., et al., 1993. First pass of an MR susceptibility contrast agent through normal and ischemic heart: gradient-recalled echo-planar imaging. J. Magn. Reson. Imaging 3 (5), 755–760.

Wersebe, A., et al., 2006. Comparison of gadolinium-BOPTA and ferucarbotran-enhanced three-dimensional T1-weighted dynamic liver magnetic resonance imaging in the same patient. Investig. Radiol. 41 (3), 264–271.

Wikstrom, M., et al., 1993. Double-contrast enhanced MR imaging of myocardial infarction in the pig. Acta Radiol. 34 (1), 64–71.

Xiao, Y.D., et al., 2016. MRI contrast agents: classification and application (Review). Int. J. Mol. Med. 38 (5), 1319–1326.

Yang, D., et al., 2001. USPIO-enhanced dynamic MRI: evaluation of normal and transplanted rat kidneys. Magn. Reson. Med. 46 (6), 1152–1163.

Yang, Y., et al., 2009. Manganese-enhanced MRI of acute cardiac ischemia and chronic infarction in pig hearts: kinetic analysis of enhancement development. NMR Biomed. 22 (2), 165–173.

Zimmer, C., et al., 1995. Cerebral iron oxide distribution: in vivo mapping with MR imaging. Radiology 196 (2), 521–527.

Protocol requirements for quantitation accuracy

Lucy Elizabeth Kershaw[a,b]

[a]*Edinburgh Imaging, University of Edinburgh, Edinburgh, United Kingdom*
[b]*BHF Centre for Cardiovascular Science, University of Edinburgh, Edinburgh, United Kingdom*

7.1 Introduction

In this chapter, we discuss the protocol requirements for accurate quantification of three different perfusion techniques described in earlier chapters: DCE-MRI, DSC-MRI, and ASL. For DCE- and DSC-MRI, we assume the use of a gadolinium-based contrast agent.

Before designing a perfusion MRI acquisition protocol, one should carefully consider what measures are desired and need to be derived from the data. In general, measurement parameters can be divided into those with direct physical meaning (quantitative analysis of blood flow or vessel permeability) and those that characterize the tissue but which do not have a straightforward physical interpretation (semiquantitative analysis of relative blood flow or area under the contrast agent uptake curve). Semiquantitative parameters often impose less stringent constraints on the acquisition and analysis protocols and tend not to be comparable across studies, but they can, nevertheless, provide useful information (Gigli *et al.*, 2019; Bernal *et al.*, 2020; Ziayee *et al.*, 2021). Quantitative parameters are usually derived from tracer kinetics modeling, and accuracy is heavily dependent on the quality of the imaging data.

The acquisition protocol is designed to track the tracer, whether it is an exogenous contrast agent or endogenously labeled blood, as it enters the tissue of interest. This requires rapid, repeated imaging over an extended time interval. For accurate quantitation, the key requirements are as follows:

i. Accurately measuring contrast agent concentration from the MR signal.
ii. Imaging rapidly enough to capture the passage of tracer, especially high, rapidly changing concentrations of tracer in the blood.
iii. Imaging with a spatial resolution suitable for the expected size of features in the tissue of interest.
iv. Achieving a signal-to-noise ratio that minimizes uncertainty in the resultant perfusion parameters.
v. Compensating for motion in the data.

Advances in Magnetic Resonance Technology and Applications, Volume 11, ISSN 2666-9099
https://doi.org/10.1016/B978-0-323-95209-5.00006-4

There is no single acquisition strategy that perfectly addresses these issues for all perfusion applications. Rather, the acquisition should be tailored to the particular tissue of interest and the parameters to be measured. We begin by discussing the ideal dynamic acquisition and then make recommendations for the best sequence implementation. Finally, we add some practical tips to maximize measurement accuracy.

7.2 Dynamic acquisition: The ideal sequence

The ideal acquisition protocol for accurate quantitative measurement of tissue characteristics would have the properties shown in Fig. 7.1.

7.2.1 Measurement of tracer concentration

The pixel values for the images acquired would ideally be a direct measurement of the concentration of the tracer used in the experiment. This is not possible in the techniques discussed here because we measure the effect of the tracer on the relaxation time of water protons in the vicinity of the tracer instead of the tracer itself, as in PET. For DSC-MRI and DCE-MRI, we use the relationship between T2 and T1, respectively, and contrast agent concentration:

$$[CA] = \frac{1}{r_k}\left\{\frac{1}{T_{k0}} - \frac{1}{T_k}\right\} \tag{7.1}$$

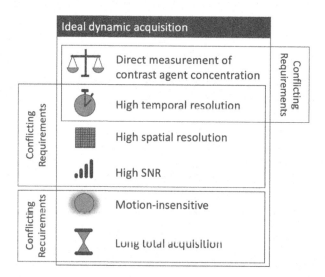

FIG. 7.1

The ideal dynamic sequence, namely one with high spatial resolution, temporal resolution and SNR, a direct measurement of contrast agent concentration, motion insensitivity, and a long total acquisition duration.

where [CA] is the concentration of contrast agent in mM, subscript k represents T1 or T2 in s, r is the relaxivity for the contrast agent of interest in mM^{-1} s^{-1}, and 0 denotes the native T_k before the addition of contrast agent. This can be written succinctly as follows:

$$[CA] = \frac{\Delta R_k}{r_k} \qquad (7.2)$$

where R_k is $1/T_k$ in s^{-1}.

This presents a further problem with the choice of sequence for the dynamic acquisition. Not only can we not measure [CA] directly, we also contend with measuring T1 and T2 quickly enough to satisfy temporal resolution requirements for the experiment (see the following section).

7.2.2 High temporal resolution and complete washout

The temporal resolution of the dynamic sequence should be high enough to allow sampling of the dynamic process of tracer delivery, including both uptake and washout in the tissue of interest, without ambiguity. The appropriate temporal resolution will vary depending on the particular tissue to be imaged and the intended analysis method, but the main consideration for temporal sampling in DCE-MRI and DSC-MRI is measurement of the arterial input function (AIF). The AIF is measured from a feeding blood vessel close to the tissue of interest and represents the concentration of tracer delivered to the tissue as a function of time. After bolus injection of a contrast agent, this curve tends to rise rapidly with a sharp peak followed by an exponential decay, possibly with an additional smaller peak due to recirculation. Undersampling of this curve means that the AIF is poorly captured, which leads to poor parameter estimation because the tissue uptake curve generated from an undersampled AIF for data fitting is corrupted (Kershaw and Cheng, 2010; Sourbron, 2010; Sourbron and Buckley, 2011). Fig. 7.2 shows the model curves generated from an AIF simulated with different temporal resolutions and the effect that this has on the simulated uptake curve. Poor temporal sampling of the AIF leads to calculation of a curve that no longer resembles the curve calculated using a well-sampled AIF using the same parameters. This means that nonlinear curve fitting using a poorly sampled AIF is very unlikely to reflect the true underlying tissue parameters. Even if the AIF is not required for analysis, rapid acquisition is necessary to circumvent motion-related errors. Tissues that undergo significant motion, such as the liver (due to breathing), require rapid acquisition to freeze motion during the acquisition of an individual frame. In addition to the temporal resolution, the total sampling time for the dynamic series (i.e., the number of frames) also requires careful consideration. A complete description of the passage of contrast agent would require imaging to continue until washout is complete. It is clearly impractical to do this with an exogenous tracer because it would take several hours; later in this chapter, we will examine practical solutions to selecting the imaging interval.

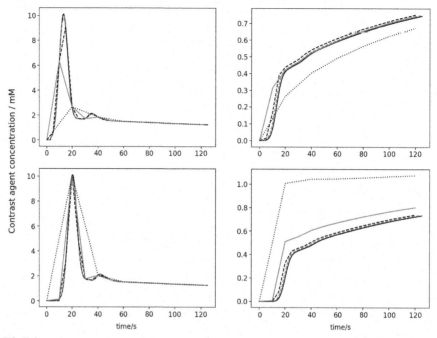

FIG. 7.2

Schematic to show effect of AIF undersampling. *Top row*: *Left*: AIF simulated using Parker AIF function (Parker *et al.*, 2006) from the OSIPI code collection (OSIPI DCE-DSC-MRI Code Collection, 2022) using temporal resolutions of 0.1 s (*black solid line*), 2 s (*gray dashed line*), 5 s (*black dashed line*), 10 s (*gray solid line*), and 20 s (*black dotted line*) and delay of 3 s before contrast agent arrives in the vessel. *Right*: Two-compartment exchange model curve simulated from these AIFs with parameters $E = 0.5$, $F_p = 0.005$ mL/mL tissue/s, $v_p = 0.5$, $v_e = 0.5$. *Bottom row*: as top row but with AIF delay of 10 s. A poorly sampled AIF (particularly *gray solid line* and *black dotted line*) leads to corruption of the tissue curve shape when the AIF is used in the modeling. This means that data fitting is unlikely to reflect the true tissue parameters.

7.2.3 High spatial resolution

The spatial resolution needed in the dynamic acquisition is dependent on the level of detail required for the tissue of interest and whether the AIF is being measured. In breast dynamic imaging, for example, the detection of small lesions is of critical importance; therefore, spatial resolution must be prioritized (Kuhl *et al.*, 2005; Sardanelli *et al.*, 2010). The requirement for the AIF to be measured in a voxel containing only blood can lead to stringent constraints on spatial resolution if the feeding vessel is small. For example, the carotid artery used in brain perfusion

MRI is around 5 mm in diameter (Krejza *et al.*, 2006) (also note that vessel diameters are smaller in women than in men and vary with age), which may represent only two pixels (when measuring perpendicular to the vessel) or a single slice (when measuring parallel to the vessel) uncorrupted by partial volume effects. However, ultimately, the spatial resolution cannot be infinitely high, as it must be balanced against acceptable levels of signal-to-noise (SNR) and temporal resolution. Therefore circumstances may necessitate that the spatial resolution be lower than desired, with the AIF inevitably contaminated by partial volume effects, resulting in an underestimated peak height and propagation of errors into model parameters.

7.2.4 High signal-to-noise ratio

The previously mentioned constraints of high spatial and temporal resolutions can result in images with very poor SNR, which is clearly suboptimal for accurate quantitative assessment and results in measurements that have large uncertainties associated with them. This is a particular problem for model fitting, which tends to be done through nonlinear least squares optimization. With noisy measured data, large uncertainties in the optimum fitted parameters will emerge, because a wide range of parameter sets can result in similar looking curves that have the same goodness of fit. Similarly, the number of model parameters that can be fitted with confidence depends on the information contained within the data. Adding extra model parameters will always improve the goodness-of-fit measure for a fitted curve, but this might be due to a better fit to the noise. There are statistical measures that allow assessment of the appropriate number of free parameters to fit for a given noisy curve, for example the Akaike information criterion (Akaike, 1974) and the F-test (used in Donaldson *et al.*, 2010).

7.2.5 Motion compensation

The ideal sequence would be insensitive to motion, which can be an additional source of corruption in the dynamic data and lead to misalignment of tissues between frames. In the body, breathing and cardiac motion are the most common issues, but bowel motion and bladder filling, for example, can also be problematic. In the brain, motion is less likely to cause problems, but smaller voxels tend to be used here which means that even minor head movements can result in alignment errors. For AIF measurement, inflow effects on the signal can cause quantification problems, because spins that have not yet reached the dynamic equilibrium associated with the signal in a gradient echo sequence flow into the imaging volume (Roberts *et al.*, 2011). Eq. (7.3) does not apply to these spins, and concentration of contrast agent cannot be determined accurately.

7.3 Dynamic acquisition: Practical solutions

No protocol can satisfy all of the requirements presented before simultaneously. In this section, we discuss the compromises that must be made to allow practical implementation of the dynamic acquisition and the influence of these compromises on measurement accuracy.

7.3.1 DCE-MRI

In DCE-MRI, rather than measuring tracer concentration or T1 directly at every measurement timepoint, tracer concentration is inferred from a baseline (i.e., precontrast) measurement of T1 (see Section 7.4.2) followed by a T1-weighted dynamic acquisition, usually a 3D spoiled gradient echo sequence. A 3D nonselective excitation is chosen over a 2D multi-slice acquisition to minimize errors in T1 estimation from slice profile effects. The signal intensity (SI) equation for a spoiled gradient echo sequence is given as follows:

$$SI = S_0 \frac{1 - \exp(-TR/T1)}{1 - \cos(\alpha) \exp(-TR/T1)} \qquad (7.3)$$

where S_0 is a constant incorporating scanner gain, equilibrium magnetization, and a factor of $\sin(\alpha)$; α is the excitation flip angle; and TR is the time between successive excitation pulses. This form of the equation assumes $TE \ll T2$. Signal intensity in the image is, therefore, related to T1. The calculation of contrast agent concentration can be done using the following steps:

1) Measure baseline T1, and hence baseline R1 = 1/T1.
2) Invert Eq. (7.3) to calculate the constant S_0 using the baseline R1 and baseline signal intensity before contrast agent arrives in the tissue.
3) Rearrange Eq. (7.3) to calculate R1 at each timepoint using the measured signal intensity and the calculated S_0.
4) Calculate the change in R1 from the baseline value.
5) Use Eq. (7.2) to calculate concentration of contrast agent.

This calculation relies on an accurate and precise measurement of the baseline signal intensity before any contrast agent arrives in the tissue. It is important, therefore, to acquire several dynamic frames before injection to obtain a less noisy average baseline. Fig. 7.3 shows a curve converted from SI to concentration using a varying number of baseline points. If too few points are used, the concentration vs time curve can become badly corrupted. Making sure that the baseline SI is well estimated is important both for modeling and for simpler empirical measures such as the area-under-the-concentration curve.

The balance between SNR and acceptable spatial and temporal resolutions will be heavily dependent on the area of the body being imaged. This will, in turn, influence the choice of α, TR, and TE, but there are additional considerations for these parameter choices. The image must be T1 weighted, requiring a short TE and TR

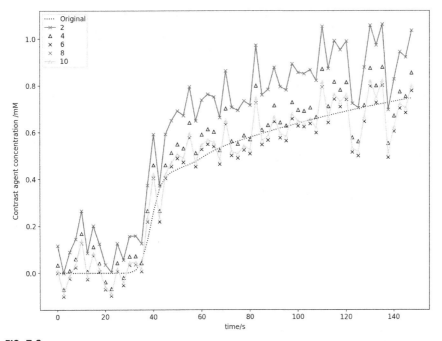

FIG. 7.3

Influence of baseline. Two-compartment exchange model curve simulated from the Parker AIF with parameters $E = 0.5$, $F_p = 0.005$ mL/mL tissue/s, $v_p = 0.5$, $v_e = 0.5$ and temporal resolution of 2.5 s, converted to signal intensity using TR $= 2$ ms, $\alpha = 15°$, baseline T1 $= 1.4$ s, and $S_0 = 10,000$. Normally distributed random noise was added to the signal intensity curve with a standard deviation of 10% of the baseline SI. The curve was then converted back to concentration using 2–10 baseline points as shown and the noiseless curve plotted in black. If an inadequate number of baseline points are used the curve can become badly corrupted (*dark gray line with crosses*, compare with *light gray triangles* from 10 baseline points).

with good gradient and RF spoiling. The optimal flip angle will be guided by the Ernst angle for the native tissue but will be further dictated by the expected contrast change; a larger angle is necessary to accommodate a larger range of contrast concentrations and to avoid saturation. The maximum practical flip angle is likely to be determined by limits on specific absorption ratio (SAR) to avoid overheating the tissue.

In general, accurate sampling of the AIF requires a temporal resolution of less than ~2.5 s (Henderson *et al.*, 1998; Ingrisch *et al.*, 2010; Sourbron, 2010), but this will depend to an extent on the injection protocol (see Section 7.4.4). There have been attempts to use two separate acquisitions to image the AIF (high temporal resolution) and the tissue (lower temporal resolution), but this has not seen widespread adoption. This is, likely due to the additional complexities of accurate contrast agent dilution and increased overall imaging time (Kershaw and Cheng, 2011; Jajamovich *et al.*, 2014; Scannell *et al.*, 2020). Measurement of the AIF using phase images has also been proposed but requires some complex postprocessing (Bleeker *et al.*, 2010; Cron *et al.*, 2011; Foltz *et al.*, 2019).

All dynamic sequences will need to make use of well-established techniques for k-space undersampling, including parallel imaging, to optimize temporal versus spatial resolution. Sharing of k-space data across temporal phases has been used for some time in MR angiography (Van Vaals *et al.*, 1993) and is available as standard sequences (e.g., TWIST (Laub and Randall, 2006), TRICKS (Korosec *et al.*, 1996), TRAK (Willinek *et al.*, 2008)) for use in dynamic MRI. More recent advances have provided a large array of possible acceleration schemes for DCE-MRI which are discussed in more details in Chapter 11. Examples include non-Cartesian sampling (Lu *et al.*, 2019), multiband acquisitions (where more than one slice is collected at the same time (Eickel *et al.*, 2018)), compressed sensing (Kim *et al.*, 2021), or a combination of these options (Othman *et al.*, 2015). More recently, deep learning (Shaul *et al.*, 2020) has seen rapid development for reconstruction of undersampled MRI datasets (see Chapter 12).

Total acquisition duration can be chosen based on the anticipated analysis to be applied. If the volume of the extravascular extracellular space is to be measured, a long total acquisition duration is important, because the total acquisition duration must be sufficient to observe the backflux from this space, which may take some time for slow leakage (Sourbron, 2010; Ingrisch and Sourbron, 2013; Thrippleton *et al.*, 2019). This timescale will vary depending on the tissue being imaged but can be estimated from the literature or from pilot data by imaging for long enough that the uptake curve begins to decrease. Acquisition durations are typically around 5 min, (Verma *et al.*, 2016; Keil *et al.*, 2021; Reavey *et al.*, 2021) depending on the tissue type (potentially shorter for first pass myocardial perfusion (Kramer *et al.*, 2020)). When measuring semiquantitative parameters such as the initial slope of the curve, peak height, or area under the curve at 60 s, then the acquisition duration need only capture these curve features with some space for individual variation.

7.3.2 DSC-MRI

DSC MRI is used almost exclusively as a brain imaging technique and has similar considerations to DCE-MRI in terms of SNR vs spatial and temporal resolution. The AIF must be measured to quantify cerebral blood flow (CBF) and cerebral blood volume (CBV). These can also be presented as relative measures (relative to the contralateral half of the brain).

The major difference between DSC-MRI and DCE-MRI is that DSC-MRI utilizes the changes in T2 or T2* rather than in T1. To satisfy temporal resolution requirements, echo planar imaging (EPI) is used, which can be based on spin-echo (T2) or gradient echo (T2*) acquisitions. Gradient echo offers superior SNR and is the most commonly used approach, because the change in T2* is larger than the change in T2 (Boxerman *et al.*, 1995), and the linear relationship between change in R2* and contrast agent concentration is valid over a larger range of vessel sizes (Weisskoff *et al.*, 1994). Total acquisition duration can be as short as 1 min, because analysis focuses on the first pass of the bolus.

Unlike DCE-MRI, DSC-MRI assumes that the tracer remains intravascular; when that assumption fails, attempts are made to correct for leakage. In healthy brain vasculature, contrast agents do not cross the intact blood-brain barrier (BBB) during the course of the DSC experiment. However, if the BBB is disrupted by a tumor or stroke, for example, this leakage leads to contrast agent concentration vs time curves that do not return to zero after the first pass (Calamante, 2010). If T1 effects dominate, the curve dips below zero at the end of the first pass. This can be minimized by using a low flip angle in the acquisition (Paulson and Schmainda, 2008), administering a small 'preload' dose of contrast agent (Hu *et al.*, 2010), or by utilizing a dual-echo sequence that allows calculation of T2* directly (Vonken *et al.*, 1999). For additional correction of leakage effects from both T1 and T2* (which lead to a concentration above zero after the first pass), a postprocessing method can be applied as proposed, for example, in Boxerman *et al.* (2006). No standard method for DSC-MRI has been adopted, though there are some consensus recommendations for high-grade gliomas (Boxerman *et al.*, 2020).

7.3.3 **ASL**

Arterial spin labeling uses flowing blood as an endogenous contrast agent. Blood is labeled by an RF excitation pulse before it flows into the tissue of interest (Golay *et al.*, 2004). Although ASL was developed as a brain imaging technique, it has been used in other body areas, for example in the myocardium and the kidneys (Kober *et al.*, 2016; Nery *et al.*, 2020). ASL protocol design is slightly different to DSC- and DCE-MRI in that the accuracy of an ASL perfusion measurement is heavily dependent on the particular labeling scheme used. In general, blood spins are inverted or saturated in an artery feeding the tissue of interest; then, after a delay time, the tissue of interest is imaged. An image is also taken without the labeling, and the difference between these two acquisitions allows quantification of perfusion.

ASL is an inherently low SNR technique, as flowing blood represents only a small proportion of the signal acquired from the tissue of interest (Golay *et al.*, 2004). This drawback is compounded in tissue with low perfusion, such as muscle. To compensate for low SNR, a common approach is to use larger voxels (which may lead to undesired partial volume effects) and greater averaging (which increases acquisition time and, consequently, motion sensitivity). Partial volume effects pose a challenge for accurate perfusion measurements, particularly in the brain where tissues with very different perfusion (gray and white matter) exist in close proximity (Chappell *et al.*, 2021), but this can be corrected to some extent. Background suppression is often applied (Maleki *et al.*, 2012; Lindner *et al.*, 2022) to reduce errors in subtraction of labeled and unlabeled images due to motion.

In a conventional acquisition, the delay between labeling and image acquisition tends to be fixed. This assumes that all labeled blood reaches the tissue before the imaging begins, but blood velocity is an unknown and can change with pathology (Tsujikawa *et al.*, 2016). Leaving a long delay to ensure that this assumption is met reduces quantification errors (Alsop and Detre, 1996), but it compounds problems with low SNR because the label decays away with T1. This can be mitigated to an extent by modeling the transit time between the labeling and imaging volumes but requires careful consideration of the pathology involved (Fan *et al.*, 2017). More advanced labeling strategies are discussed in Chapter 4.

7.4 Additional considerations for accurate quantification

7.4.1 Structural imaging

It can be difficult to draw regions of interest directly onto the dynamic images, because contrast is poor at the beginning of the series and resolution tends to be worse than would be acceptable for clinical interpretation. To aid in region drawing and interpretation of the results (for example to overlay parameter maps onto the relevant anatomy), it is common to acquire structural images of the dynamic volume. Often these are acquired as part of the clinical examination, but it can be useful to acquire a matched anatomical image just before the dynamic series to make sure that the anatomy is aligned for region drawing. The optimal contrast weighting depends on the tissue to be imaged; for example, in the prostate, a T2-weighted image tends to demonstrate the anatomy most clearly. In the brain, an MPRAGE (T1-weighted) sequence may be more appropriate.

7.4.2 T1 measurement

A detailed discussion of T1 measurement methods is beyond the scope of this chapter, as there are many different sequences and methods available. Modern scanners often have dedicated sequences and postprocessing that allow T1 maps to be generated automatically. Briefly, the variable flip angle method (Fram *et al.*, 1987) has proved a popular choice, because it is much faster than inversion recovery or saturation recovery methods. However, this method can be heavily influenced by B1 inhomogeneity, which causes errors in the applied flip angle. It is important to acquire a B1 map to correct for these errors (Cheng and Wright, 2006). In the heart, the MOLLI family of sequences (Messroghli *et al.*, 2004; Piechnik *et al.*, 2010) are used extensively, because they are designed to be used with cardiac gating. For brain applications, there are a number of options (e.g., Deoni, 2007; Marques *et al.*, 2010), optimized for high-resolution T1 mapping in a clinically acceptable time. In the body, the variable flip angle method has also been used, as well as saturation recovery and inversion recovery-based sequences (Brix *et al.*, 1990; Schmitt *et al.*, 2004), but these can take some time to acquire. An assessment of the accuracy of any mapping method should be carried out using a phantom with known T1 values (Lerski *et al.*, 1993; Boss *et al.*, 2018).

7.4.3 Water exchange

Water exchange is often overlooked in designing perfusion imaging protocols. Movement of water protons between different tissue compartments has an effect on the measured relaxation properties of the tissue. This is considered in detail in

Chapter 17, but in general problems with quantification can be avoided for DCE-MRI by choosing a spoiled gradient echo sequence with a short TR and a large flip angle (Donahue *et al.*, 1996). Water exchange effects on T2 relaxation are rarely considered for DSC-MRI, because it is a small source of error (Donahue *et al.*, 1997). For ASL, it has been shown that inclusion of water exchange as a model parameter improves the accuracy of CBF measurements (Parkes and Tofts, 2002).

7.4.4 **Practical considerations**

Some practical aspects of the examination can potentially have a large impact on quantification accuracy (see Box 7.1 for a summary of practical suggestions). The issue of motion is one that is not easy to solve for applications in the abdomen and thorax, but careful patient setup can help. The patient should empty their bladder before the imaging session and eat and drink lightly. Bowel motion can be reduced with the use of Buscopan; when given intravenously, its effects last around 20 min (Gutzeit *et al.*, 2012). The order of sequences in the protocol should be prioritized appropriately.

Contrast agent is present in blood plasma but does not enter red blood cells. Quantitative analysis, therefore, relies on the measurement of hematocrit (the proportion of blood volume occupied by cells). This allows conversion between measures in blood plasma and whole blood, for example to convert between v_p (the plasma volume) and v_b (the blood volume). A normal value for hematocrit can be assumed (males: 40%–54%; females 36%–48% (Billett, 1990)), but it is preferable to measure this for an individual, especially if they have a condition that can skew this value from the expected normal (Shahid, 2016).

Choice of coil should also be carefully considered. If the AIF is to be measured in the aorta, for example, coil coverage above the bifurcation will allow measurement in the largest part of the vessel, minimizing both partial volume effects and in-flow effects (Cheng, 2007). For oncology applications, it may be beneficial to image with the patient in the radiotherapy treatment position; therefore, coils will need to be positioned to avoid distortion of the skin surface.

BOX 7.1 Practical tips for dynamic imaging.

- for body acquisitions consider a drinking protocol and buscopan
- check for artifacts using a short test dynamic run
- ensure adequate number of baseline points
- take care when using kvo option
- if examination is long, consider sequence ordering to collect most important data first
- look at the images and curves before analysis

The injection of contrast agent should be done using a power injector, followed by a saline chaser to flush the injection line. Typical injection rates for DCE-MRI are around 2–4 mL/s (Essig *et al.*, 2013) or higher for DSC-MRI (van Osch *et al.*, 2003). A faster rate will require a dynamic acquisition with a higher temporal resolution to sample the peak properly, but a very slow rate would result in an equilibrium between extravascular and intravascular concentrations that would violate assumptions in some of the tracer kinetics models and compromise model validity and quantification accuracy (Sourbron, 2010). The 'keep vein open' (KVO) option trickles a small amount of saline though the cannula to keep the line patent. This should be used with care; even this small amount of saline can result in substantial bladder filling and, therefore, motion if the dynamic series comes at the end of a long examination.

The appropriate imaging plane requires careful thought. Sampling the AIF with the blood flowing perpendicular to the imaging plane can lead to errors due to inflow effect, so it can be advantageous to image parallel to the vessel that the AIF is to be taken from (Cheng, 2007). In practice, this is probably only suitable for imaging the aorta because the typical slice thicknesses used for DCE-MRI are likely to be too large to capture a voxel that is only in blood if the vessel is small. To use the aorta with a sagittal or coronal acquisition will need careful planning of the imaging volume to ensure that there is a slice running directly through the middle of the vessel, which can be difficult if the aorta is not completely straight. The introduction of parallel imaging and interpolation will lead to slice thicknesses that are larger than the reported thickness, so some partial volume effect is likely to be present. Inevitably, the choice of imaging plane will be a balance between the requirement to sample the AIF and the most appropriate orientation to see the anatomy in the tissue of interest.

The phase-encoding direction should be chosen so that there is no ghosting artifact running through the tissue of interest from vessels or breathing. However, the individual dynamic images are not intended to be diagnostic, so it can be acceptable to have fold-over artifacts if this does not overlap with the tissue of interest. It can be useful to acquire a small number of test dynamic frames before the main dynamic sequence to check for artifacts and image wrap.

7.5 Future directions

MRI sequences are constantly improving in terms of spatial resolution, temporal resolution, and SNR. Future technological developments such as deep learning reconstruction (Chapter 12) will allow improvements in all these key areas (Zeng *et al.*, 2021). Motion correction has not typically been applied in dynamic imaging because the changes in contrast over the dynamic series make this particularly challenging. Future focus should be on solving this issue, either in postprocessing (images with different contrasts can be coregistered (Islam *et al.*, 2021)) or during acquisition (Johansson *et al.*, 2018). Sharing of data and analysis methods (OSIPI, 2022) will allow benchmarking of data quality and comparison between analysis methods in a transparent way.

7.6 Summary

When designing an acquisition protocol for accurate quantification, it is vital to decide which parameters are to be measured from the data. No acquisition protocol can simultaneously have high temporal resolution, high spatial resolution, and high SNR, so the acquisition must be tailored for a particular application and the desired output parameters. For example, measurement of an AIF puts stringent constraints on temporal resolution, but these might be relaxed for semiquantitative measures in DCE-MRI where the AIF is not needed. In ASL, the inherently low SNR of the technique often requires use of a lower spatial resolution than other methods. Tailoring the acquisition to each application allows the best possible measurements to be made, and by minimizing sources of error such as motion and artifacts, perfusion imaging can be used to make robust, useful physiological measurements.

References

Akaike, H., 1974. A new look at the statistical model identification. IEEE Trans. Autom. Control 19 (6), 716–723. https://doi.org/10.1109/TAC.1974.1100705.

Alsop, D.C., Detre, J.A., 1996. Reduced transit-time sensitivity in noninvasive magnetic resonance imaging of human cerebral blood flow. J. Cereb. Blood Flow Metab. 16 (6), 1236–1249. https://doi.org/10.1097/00004647-199611000-00019.

Bernal, J., et al., 2020. Examining the relationship between semiquantitative methods analysing concentration-time and enhancement-time curves from dynamic-contrast enhanced magnetic resonance imaging and cerebrovascular dysfunction in small vessel disease. J. Imaging 6 (6). https://doi.org/10.3390/jimaging6060043.

Billett, H., 1990. Hemoglobin and hematocrit. In: Walker, H.K., Hall, W.D., H. J. (Eds.), Clinical Methods: The History, Physical, and Laboratory Examinations, 3rd ed., Butterworths, Boston.

Bleeker, E.J.W., et al., 2010. Phase-based arterial input function measurements for dynamic susceptibility contrast MRI. Magn. Reson. Med. 64 (2), 358–368. https://doi.org/10.1002/mrm.22420.

Boss, M.A., et al., 2018. Magnetic Resonance Imaging Biomarker Calibration Service: Proton Spin Relaxation Times. Special Publication (NIST SP) - 250-97, Gaithersburg, MD, https://doi.org/10.6028/NIST.SP.250-97.

Boxerman, J.L., et al., 1995. Mr contrast due to intravascular magnetic susceptibility perturbations. Magn. Reson. Med. 34 (4), 555–566. https://doi.org/10.1002/mrm.1910340412.

Boxerman, J.L., Schmainda, K.M., Weisskoff, R.M., 2006. Relative cerebral blood volume maps corrected for contrast agent extravasation significantly correlate with glioma tumor grade, whereas uncorrected maps do not. Am. J. Neuroradiol. 27 (4), 859–867. Available at: http://www.ncbi.nlm.nih.gov/pubmed/16611779.

Boxerman, J.L., et al., 2020. Consensus recommendations for a dynamic susceptibility contrast MRI protocol for use in high-grade gliomas. Neuro Oncol. 22 (9), 1262–1275. https://doi.org/10.1093/neuonc/noaa141.

Brix, G., et al., 1990. Fast and precise T1 imaging using a TOMROP sequence. Magn. Reson. Imaging 8 (4), 351–356.

Calamante, F., 2010. Perfusion MRI using dynamic-susceptibility contrast MRI: quantification issues in patient studies. Top. Magn. Reson. Imaging 21 (2), 75–85. https://doi.org/10.1097/RMR.0b013e31821e53f5.

Chappell, M.A., et al., 2021. Partial volume correction in arterial spin labeling perfusion MRI: a method to disentangle anatomy from physiology or an analysis step too far? Neuroimage 238, 118236. https://doi.org/10.1016/j.neuroimage.2021.118236.

Cheng, H.-L.M.L., 2007. T1 measurement of flowing blood and arterial input function determination for quantitative 3D T1-weighted DCE-MRI. J. Magn. Reson. Imaging 25 (5), 1073–1078. https://doi.org/10.1002/jmri.20898.

Cheng, H.L., Wright, G.A., 2006. Rapid high-resolution T(1) mapping by variable flip angles: accurate and precise measurements in the presence of radiofrequency field inhomogeneity. Magn. Reson. Med. 55 (3), 566–574.

Cron, G.O., et al., 2011. Arterial input functions determined from MR signal magnitude and phase for quantitative dynamic contrast-enhanced MRI in the human pelvis. Magn. Reson. Med. 66 (2), 498–504. https://doi.org/10.1002/mrm.22856.

Deoni, S.C.L., 2007. High-resolution T1 mapping of the brain at 3T with driven equilibrium single pulse observation of T1 with high-speed incorporation of RF field inhomogeneities (DESPOT1-HIFI). J. Magn. Reson. Imaging 26 (4), 1106–1111. https://doi.org/10.1002/jmri.21130.

Donahue, K.M., et al., 1996. Improving MR quantification of regional blood volume with intravascular T1 contrast agents: accuracy, precision, and water exchange. Magn. Reson. Med. 36 (6), 858–867.

Donahue, K.M., Weisskoff, R.M., Burstein, D., 1997. Water diffusion and exchange as they influence contrast enhancement. J. Magn. Reson. Imaging 7 (1), 102–110. https://doi.org/10.1002/jmri.1880070114.

Donaldson, S.B., et al., 2010. A comparison of tracer kinetic models for T1-weighted dynamic contrast-enhanced MRI: application in carcinoma of the cervix. Magn. Reson. Med. 63 (3), 691–700. https://doi.org/10.1002/mrm.22217.

Eickel, K., et al., 2018. Simultaneous multislice acquisition with multi-contrast segmented EPI for separation of signal contributions in dynamic contrast-enhanced imaging. PloS One 13 (8), 1–22. https://doi.org/10.1371/journal.pone.0202673.

Essig, M., et al., 2013. Perfusion MRI: the five most frequently asked technical questions. Am. J. Roentgenol. 200 (1), 24–34. https://doi.org/10.2214/AJR.12.9543.

Fan, A.P., et al., 2017. Long-delay arterial spin labeling provides more accurate cerebral blood flow measurements in Moyamoya patients: a simultaneous positron emission tomography/MRI study. Stroke 48 (9), 2441–2449. https://doi.org/10.1161/STROKEAHA.117.017773.

Foltz, W., et al., 2019. Phantom validation of dce-mri magnitude and phase-based vascular input function measurements. Tomography 5 (1), 77–89. https://doi.org/10.18383/j.tom.2019.00001.

Fram, E.K., et al., 1987. Rapid calculation of T1 using variable flip angle gradient refocused imaging. Magn. Reson. Imaging 5 (3), 201–208.

Gigli, S., et al., 2019. Morphological and semiquantitative kinetic analysis on dynamic contrast enhanced MRI in triple negative breast cancer patients. Acad. Radiol. 26 (5), 620–625. https://doi.org/10.1016/j.acra.2018.06.014.

Golay, X., Hendrikse, J., Lim, T.C.C., 2004. Perfusion imaging using arterial spin labeling. Top. Magn. Reson. Imaging 15 (1), 10–27. https://doi.org/10.1097/00002142-200402000-00003.

Gutzeit, A., et al., 2012. Evaluation of the anti-peristaltic effect of glucagon and hyoscine on the small bowel: comparison of intravenous and intramuscular drug administration. Eur. Radiol. 22 (6), 1186–1194. https://doi.org/10.1007/s00330-011-2366-1.

Henderson, E., Rutt, B.K., Lee, T.Y., 1998. Temporal sampling requirements for the tracer kinetics modeling of breast disease. Magn. Reson. Imaging 16 (9), 1057–1073.

Hu, L.S., et al., 2010. Optimized preload leakage-correction methods to improve the diagnostic accuracy of dynamic susceptibility-weighted contrast-enhanced perfusion MR imaging in posttreatment gliomas. Am. J. Neuroradiol. 31 (1), 40–48. https://doi.org/10.3174/ajnr.A1787.

Ingrisch, M., Sourbron, S., 2013. Tracer-kinetic modeling of dynamic contrast-enhanced MRI and CT: a primer. J. Pharmacokinet. Pharmacodyn. 40 (3), 281–300. https://doi.org/10.1007/s10928-013-9315-3.

Ingrisch, M., et al., 2010. Quantitative pulmonary perfusion magnetic resonance imaging: influence of temporal resolution and signal-to-noise ratio. Invest. Radiol. 45 (1), 7–14. https://doi.org/10.1097/RLI.0b013e3181bc2d0c.

Islam, K.T., Wijewickrema, S., O'Leary, S., 2021. A deep learning based framework for the registration of three dimensional multi-modal medical images of the head. Sci. Rep. 11 (1), 1–13. https://doi.org/10.1038/s41598-021-81044-7.

Jajamovich, G.H., et al., 2014. DCE-MRI of the liver: reconstruction of the arterial input function using a low dose pre-bolus contrast injection. PLoS One 9 (12), 1–15. https://doi.org/10.1371/journal.pone.0115667.

Johansson, A., Balter, J., Cao, Y., 2018. Rigid-body motion correction of the liver in image reconstruction for golden-angle stack-of-stars DCE MRI. Magn. Reson. Med. 79 (3), 1345–1353. https://doi.org/10.1002/mrm.26782.

Keil, V.C., et al., 2021. DCE-MRI in glioma, infiltration zone and healthy brain to assess angiogenesis: a biopsy study. Clin. Neuroradiol. 31 (4), 1049–1058. https://doi.org/10.1007/s00062-021-01015-3.

Kershaw, L.E., Cheng, H.-L.M., 2010. Temporal resolution and SNR requirements for accurate DCE-MRI data analysis using the AATH model. Magn. Reson. Med. 64 (6), 1772–1780. https://doi.org/10.1002/mrm.22573.

Kershaw, L.E., Cheng, H.-L.L.M., 2011. A general dual-bolus approach for quantitative DCE-MRI. Magn. Reson. Imaging 29 (2), 160–166. https://doi.org/10.1016/j.mri.2010.08.009.

Kim, J.J., et al., 2021. Ultrafast dynamic contrast-enhanced MRI using compressed sensing: associations of early kinetic parameters with prognostic factors of breast cancer. Am. J. Roentgenol. 217 (1), 56–63. https://doi.org/10.2214/AJR.20.23457.

Kober, F., et al., 2016. Myocardial arterial spin labeling. J. Cardiovasc. Magn. Reson. 18 (1), 22. https://doi.org/10.1186/s12968-016-0235-4.

Korosec, F.R., et al., 1996. Time-resolved contrast-enhanced 3D MR angiography. Magn. Reson. Med. 36 (3), 345–351. Available at: http://www.ncbi.nlm.nih.gov/pubmed/12354982.

Kramer, C.M., et al., 2020. Standardized cardiovascular magnetic resonance imaging (CMR) protocols: 2020 update. J. Cardiovasc. Magn. Reson. 22 (1), 1–18. https://doi.org/10.1186/s12968-020-00607-1.

Krejza, J., et al., 2006. Carotid artery diameter in men and women and the relation to body and neck size. Stroke 37 (4), 1103–1105. https://doi.org/10.1161/01.STR.0000206440.48756.f7.

Kuhl, C.K., Schild, H.H., Morakkabati, N., 2005. Dynamic bilateral contrast-enhanced MR imaging of the breast: trade-off between spatial and temporal resolution. Radiology 236 (3), 789–800. https://doi.org/10.1148/radiol.2363040811.

Laub, G., Randall, K., 2006. Syngo TWIST for Dynamic Time-Resolved MR Angiography. MAGNETOM Flash, pp. 92–95. Available at: http://scholar.google.com/scholar?hl=en&btnG=Search&q=intitle:syngo+TWIST+for+Dynamic+Time-Resolved+MR+Angiography#0.

Lerski, R.A., De Certaines, J., de Certaines, J.D., 1993. Performance assessment and quality control in MRI by eurospin test objects and protocols R.A. Magn. Reson. Imaging 11 (6), 817–833. Available at: http://www.sciencedirect.com/science/article/pii/0730725X9390199N. (Accessed: 20 March 2014).

Lindner, T., *et al.*, 2022. Individualized arterial spin labeling background suppression by rapid T1 mapping during acquisition. Eur. Radiol., 4521–4526. https://doi.org/10.1007/s00330-022-08550-8.

Lu, Y., *et al.*, 2019. The value of GRASP on DCE-MRI for assessing response to neoadjuvant chemotherapy in patients with esophageal cancer. BMC Cancer 19 (1), 1–9. https://doi.org/10.1186/s12885-019-6247-3.

Maleki, N., Dai, W., Alsop, D.C., 2012. Optimization of background suppression for arterial spin labeling perfusion imaging. MAGMA 25 (2), 127–133. https://doi.org/10.1007/s10334-011-0286-3.

Marques, J.P., *et al.*, 2010. MP2RAGE, a self bias-field corrected sequence for improved segmentation and T1-mapping at high field. Neuroimage 49 (2), 1271–1281. https://doi.org/10.1016/j.neuroimage.2009.10.002.

Messroghli, D.R., *et al.*, 2004. Modified look-locker inversion recovery (MOLLI) for high-resolution T1 mapping of the heart. Magn. Reson. Med. 52 (1), 141–146. https://doi.org/10.1002/mrm.20110.

Nery, F., *et al.*, 2020. Consensus-based technical recommendations for clinical translation of renal ASL MRI. Biol. Med. 33 (1), 141–161. https://doi.org/10.1007/s10334-019-00800-z.

OSIPI. 2022. Available at: https://osipi.ismrm.org (accessed 01.12.22).

OSIPI DCE-DSC-MRI Code Collection. 2022. Available at: https://github.com/OSIPI/DCE-DSC-MRI_CodeCollection (accessed 01.12.22).

Othman, A.E., *et al.*, 2015. Feasibility of CAIPIRINHA-Dixon-TWIST-VIBE for dynamic contrast-enhanced MRI of the prostate. Eur. J. Radiol. 84 (11), 2110–2116. https://doi.org/10.1016/j.ejrad.2015.08.013.

Parker, G.J.M., *et al.*, 2006. Experimentally-derived functional form for a population-averaged high-temporal-resolution arterial input function for dynamic contrast-enhanced MRI. Magn. Reson. Med. 56 (5), 993–1000. https://doi.org/10.1002/mrm.21066.

Parkes, L.M., Tofts, P.S., 2002. Improved accuracy of human cerebral blood perfusion measurements using arterial spin labeling: accounting for capillary water permeability. Magn. Reson. Med. 48 (1), 27–41. https://doi.org/10.1002/mrm.10180.

Paulson, E.S., Schmainda, K.M., 2008. Comparison of dynamic susceptibility-weighted contrast-enhanced MR methods: recommendations for measuring relative cerebral blood volume in brain tumors. Radiology 249 (2), 601–613. https://doi.org/10.1148/radiol.2492071659.

Piechnik, S.K., *et al.*, 2010. Shortened modified look-locker inversion recovery (ShMOLLI) for clinical myocardial T1-mapping at 1.5 and 3 T within a 9 heartbeat breathhold. J. Cardiovasc. Magn. Reson. 12 (1). https://doi.org/10.1186/1532-429X-12-69.

Reavey, J.J., *et al.*, 2021. Markers of human endometrial hypoxia can be detected in vivo and ex vivo during physiological menstruation. Hum. Reprod. 36 (4), 941–950. https://doi.org/10.1093/humrep/deaa379.

Roberts, C., *et al.*, 2011. The effect of blood inflow and B(1)-field inhomogeneity on measurement of the arterial input function in axial 3D spoiled gradient echo dynamic contrast-enhanced MRI. Magn. Reson. Med. 65 (1), 108–119. https://doi.org/10.1002/mrm.22593.

Sardanelli, F., *et al.*, 2010. Magnetic resonance imaging of the breast: recommendations from the EUSOMA working group. Eur. J. Cancer 46 (8), 1296–1316. https://doi.org/10.1016/j.ejca.2010.02.015.

Scannell, C.M., *et al.*, 2020. Feasibility of free-breathing quantitative myocardial perfusion using multi-echo Dixon magnetic resonance imaging. Sci. Rep. 10 (1), 1–11. https://doi.org/10.1038/s41598-020-69747-9.

Schmitt, P., *et al.*, 2004. Inversion recovery TrueFISP: quantification of T1, T 2, and spin density. Magn. Reson. Med. 51 (4), 661–667. https://doi.org/10.1002/mrm.20058.

Shahid, S., 2016. Review of hematological indices of cancer patients receiving combined chemotherapy & radiotherapy or receiving radiotherapy alone. Crit. Rev. Oncol. Hematol., 145–155. https://doi.org/10.1016/j.critrevonc.2016.06.001.

Shaul, R., *et al.*, 2020. Subsampled brain MRI reconstruction by generative adversarial neural networks. Med. Image Anal. 65, 101747. https://doi.org/10.1016/j.media.2020.101747.

Sourbron, S., 2010. Technical aspects of MR perfusion. Eur. J. Radiol. 76 (3), 304–313. https://doi.org/10.1016/j.ejrad.2010.02.017.

Sourbron, S.P., Buckley, D.L., 2011. On the scope and interpretation of the Tofts models for DCE-MRI. Magn. Reson. Med. 66 (3), 735–745. https://doi.org/10.1002/mrm.22861.

Thrippleton, M.J., *et al.*, 2019. Quantifying blood-brain barrier leakage in small vessel disease: review and consensus recommendations. Alzheimer's Dementia 15 (6), 840–858. https://doi.org/10.1016/j.jalz.2019.01.013.

Tsujikawa, T., *et al.*, 2016. Arterial transit time mapping obtained by pulsed continuous 3D ASL imaging with multiple post-label delay acquisitions: comparative study with PET-CBF in patients with chronic occlusive cerebrovascular disease. PLoS One 11 (6). https://doi.org/10.1371/journal.pone.0156005.

van Osch, M.J.P., *et al.*, 2003. Model of the human vasculature for studying the influence of contrast injection speed on cerebral perfusion MRI. Magn. Reson. Med. 50 (3), 614–622. https://doi.org/10.1002/mrm.10567.

Van Vaals, J.J., *et al.*, 1993. "Keyhole" method for accelerating imaging of contrast agent uptake. J. Magn. Reson. Imaging 3 (4), 671–675. https://doi.org/10.1002/jmri.1880030419.

Verma, S., Turkbey, B., Muradyan, N., *et al.*, 2016. Overview of dynamic contrast-enhanced MRI in prostate cancer diagnosis and management. Am. J. Roentgenol. 118 (24), 6072–6078. https://doi.org/10.2214/AJR.12.8510.Overview.

Vonken, E.J.P.A., *et al.*, 1999. Measurement of cerebral perfusion with dual-echo multi-slice quantitative dynamic susceptibility contrast MRI. J. Magn. Reson. Imaging 10 (2), 109–117. https://doi.org/10.1002/(sici)1522-2586(199908)10:2<109::aid-jmri1->3.0.co;2-%23.

Weisskoff, R.M., *et al.*, 1994. Microscopic susceptibility variation and transverse relaxation: theory and experiment. Magn. Reson. Med. 31 (6), 601–610.

Willinek, W.A., *et al.*, 2008. 4D time-resolved MR angiography with keyhole (4D-TRAK): more than 60 times accelerated MRA using a combination of CENTRA, keyhole, and SENSE at 3.0T. J. Magn. Reson. Imaging 27 (6), 1455–1460. https://doi.org/10.1002/jmri.21354.

Zeng, G., *et al.*, 2021. A review on deep learning MRI reconstruction without fully sampled k-space. BMC Med. Imaging 21 (1), 1–11. https://doi.org/10.1186/s12880-021-00727-9.

Ziayee, F., *et al.*, 2021. Impact of qualitative, semi-quantitative, and quantitative analyses of dynamic contrast-enhanced magnet resonance imaging on prostate cancer detection. PLoS One 16, 1–12. https://doi.org/10.1371/journal.pone.0249532.

Arterial input function:
A friend or a foe?

Linda Knutsson[a,b,c], Ronnie Wirestam[a], and Emelie Lind[a,d]

[a]*Department of Medical Radiation Physics, Lund University, Lund, Sweden*
[b]*Russell H. Morgan Department of Radiology and Radiological Science, Johns Hopkins University School of Medicine, Baltimore, MD, United States*
[c]*F.M. Kirby Research Center for Functional Brain Imaging, Kennedy Krieger Institute, Baltimore, MD, United States*
[d]*Department of Medical Imaging and Physiology, Skåne University Hospital, Lund, Sweden*

8.1 Introduction

The relevance and importance of monitoring arterial concentrations of tracer in connection with measurements of tissue blood flow are obvious from the most fundamental tracer-kinetic relationships, such as the Fick principle (Fick, 1870). Furthermore, the theoretical requirement of dynamically monitoring the arterial tracer concentration time course, *i.e.*, the arterial input function (AIF), is evident from several pharmacokinetic models related to quantification of tissue blood flow (*e.g.*, the Kety equation), as well as from the basic tracer kinetics convolution relationship:

$$c(t) = Q[c_a(t) \otimes R(t)] = Q \int_0^t c_a(\tau)R(t-\tau)d\tau, \qquad (8.1)$$

where Q is the tissue blood flow rate, t is the time, $c(t)$ is the tissue tracer concentration, $R(t)$ is the tissue residue function, and $c_a(t)$ is the arterial tracer concentration, *i.e.*, the AIF.

Monitoring of arterial tracer concentration is not straightforward, and in early human cerebral blood flow (CBF) imaging measurements, based on external detection of radioactive tracers, AIF registration was circumvented by employing rapid intraarterial injection of tracer, in order for the true AIF to approximate a Dirac delta function (Hoedt-Rasmussen *et al.*, 1966). A less invasive approach to avoid the difficulties related to the acquisition of subject-specific AIFs has been to use semipopulation- or standard population-based AIFs (Ashton *et al.*, 2008; Parker *et al.*, 2006).

Advances in Magnetic Resonance Technology and Applications, Volume 11, ISSN 2666-9099
https://doi.org/10.1016/B978-0-323-95209-5.00011-8

Registration of arterial tracer concentration data can, in principle, be accomplished by continuous or discrete arterial blood sampling or by dynamic imaging of an arterial signal time series. Dynamic measurement of the arterial tracer concentration, by obtaining blood samples in animals, is documented in scientific publications from the early 1960s (Harvey and Brothers, 1962). In the field of medical imaging, dynamic registration of AIFs, after inhalation or intravenous injection of tracer, started to appear in CBF experiments carried out with nuclear medicine modalities around 1980, for example, in ^{133}Xe single photon emission tomography (SPECT) (Miura et al., 1981) and $H_2^{15}O$ positron emission tomography (PET) (Raichle et al., 1983).

In MRI, the two major perfusion-related applications involving AIFs, introduced around 1990, are (i) T2*-weighted dynamic susceptibility contrast (DSC) MRI for cerebral perfusion imaging relying on the theory of intravascular tracers, and (ii) T1-weighted dynamic contrast-enhanced (DCE) MRI, based on the theory of diffusible tracers. Most DSC- and DCE-MRI implementations are based on a rapid intravenous injection of a compact gadolinium contrast agent volume, referred to as a bolus injection (although infusion-based pharmacokinetic models for DCE-MRI also exist). In both these techniques, one needs to consider that the MRI signal is generally not linearly dependent on tracer concentration, and the required conversion from MRI signal to tracer concentration is not always trivial.

DSC-MRI is used for cerebral perfusion imaging, and CBF estimation by DSC-MRI is based on deconvolution of the first-pass tissue concentration curve with the AIF, according to the convolution relationship in Eq. (8.1). Hence, DSC-MRI relies heavily on an accurate AIF, sampled at high temporal resolution. It is probably fair to state that the vast majority of all DSC-MRI experiments include registration of a subject-specific AIF. Unfortunately, AIF quantification in DSC-MRI is notorious for numerous methodological complications, as further discussed later, related to, for example, blood relaxivity issues, partial volume effects (PVEs), and signal clipping and signal relocation at peak concentration. In DCE-MRI, distribution of tracer from the vascular to the extravascular extracellular space (EES) occurs, implying that a slower process than in DSC-MRI is to be monitored. Hence, the acquisition of tissue tracer information in DCE-MRI is often characterized by a longer measurement duration and lower temporal resolution. The low temporal resolution, in combination with other impediments such as low signal-to-noise ratio (SNR) and limited volume coverage, constitutes an obvious obstacle to accurate AIF registration, and the use of population-based AIFs has historically been more common in DCE-MRI applications (Parker et al., 2006). The third major methodological category of perfusion MRI techniques in human studies is arterial spin labeling (ASL), in which magnetically labeled arterial blood water protons are employed as an endogenous tracer. In ASL, CBF quantification is most commonly accomplished by the general kinetic model (Buxton et al., 1998), in principle relying on the convolution relationship in Eq. (8.1). Normally, the shape of the AIF (often referred to as the delivery function) is analytically determined by the temporal distribution of arterially labeled spins when delivered to the tissue of interest, and the AIF is, thus, not measured.

However, ASL with and without flow suppression gradients, proposed in the 'quantitative STAR labeling of arterial regions (QUASAR)' technique, enables separation of large vessel and tissue signals (Petersen *et al.*, 2006). Hence, local AIFs can be measured and perfusion analysis by model-free deconvolution can be employed.

In this chapter, aspects of AIF data acquisition and concentration quantification are reviewed, as well as routines for appropriate AIF selection. Relevant difficulties and pitfalls are identified, and potential solutions and practical actions to optimize AIF registration are provided.

8.2 Pitfalls and possibilities

8.2.1 Acquisition of the AIF

8.2.1.1 Dynamic susceptibility contrast MRI

In DSC-MRI, the aim is to quantify the arterial concentration versus time curve, *i.e.*, the AIF, during the first passage of a contrast agent (CA) bolus, by measuring the increase in transverse relaxation rate ($\Delta R2^*$ or $\Delta R2$) in a blood vessel that supplies the tissue of interest. Due to issues in converting $\Delta R2$ or $\Delta R2^*$ to concentration (as further discussed later), the phase of the MRI signal is sometimes used to obtain the AIF, as concentration is proportional to the phase shift induced by the CA. However, acquisition of phase-based AIFs is still not standard procedure, mainly because additional, nontrivial, postprocessing steps are needed, for example, to account for motion-induced phase shifts and B0 drift during scanning as well as for the shape and orientation of the artery.

Correct measurement of the AIF in DSC-MRI is troublesome due to several factors: rapid passage of the CA bolus in comparison with the temporal resolution, high maximal concentration levels, and small vessel sizes in comparison with typical spatial resolutions. Different scanner settings for acquisition of DSC-MRI data will affect the appearance of the measured AIF and, thus, the resulting quantitative parameters related to perfusion and microvasculature. For correct perfusion estimation, both the area under the curve (*i.e.*, the time integral) and the shape of the measured AIF need to represent the true input function at the inlet of the local capillary network. In general, an applied AIF that is narrower than the true AIF will cause underestimated CBF (and overestimated MTT) while an AIF showing a broader appearance than the true AIF will cause overestimated CBF (and underestimated MTT), see Fig. 8.1.

Because of the difficulties in performing a correct measurement of the AIF, and to avoid the deconvolution step, parameters that describe the measured signal versus time or concentration versus time curve in tissue are sometimes used instead (often referred to as summary parameters). In such cases, parameters such as the time to peak (TTP), maximum peak concentration (MPC), bolus arrival time (BAT), full width at half maximum (FWHM), peak area, or the first moment of the peak (FM) are extracted. These parameters can, however, be difficult to interpret as they

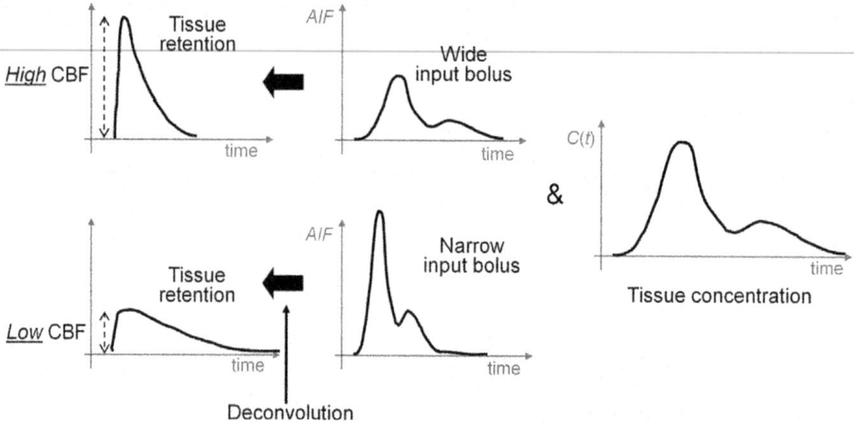

FIG. 8.1

An illustration of how the shape of the AIF will affect the CBF estimates. For a given concentration versus time curve in tissue (shown to the right), different shapes of the AIF (shown in the middle) will result in different CBF estimates (shown to the left). In general, a wide AIF will result in high CBF values and a narrow AIF result in low CBF values, assuming the same tissue curve.

Figure parts reproduced with permission from Calamante, F., 2010. Perfusion MRI using dynamic-susceptibility contrast MRI: quantification issues in patient studies. Top. Magn. Reson. Imaging, 21, 75–85.

depend both on the relevant physiological parameters (such as CBF, CBV, and MTT) and on how the measurement was performed (*i.e.*, CA injection rate and volume) as well as on the function of larger vessels in the vascular system (*e.g.*, cardiac output). Another approach that has been suggested to circumvent AIF measurement is to use a CA concentration curve from healthy white matter as input function (Kosior *et al.*, 2009), resulting in relative perfusion measures.

Still, despite the obstacles in DSC-MRI AIF measurements, the subject-specific AIF is normally measured, as accurately as possible, to obtain quantitative parameters that are minimally influenced by measurement setup and by the condition of the subject's main circulatory system.

DSC-MRI is normally performed at 1.5 or 3 T. Higher magnetic field strength means higher SNR at baseline (*i.e.*, before the CA arrives at the brain), but it also implies more pronounced susceptibility artifacts. For a given dosage of CA and similar TE, the maximal signal drop will also be larger at higher field strengths, which is an advantage for the concentration-to-noise ratio of the tissue concentration curves, but a disadvantage in AIF measurements for which the signal tends to drop to the noise level at peak concentrations (as will be further discussed later).

A standard DSC-MRI experiment is typically performed using a single-echo gradient echo (GRE) with single-shot echo planar imaging (EPI) readout, to obtain a combination of sufficiently high temporal resolution and full brain coverage.

However, other pulse sequences have also been used for DSC-MRI, including double-echo GRE, spin-echo (SE) EPI, and a multiecho with both GRE and SE (referred to as SAGE EPI) (Schmiedeskamp *et al.*, 2012). Regarding the choice of sequence for AIF measurements, neither the GRE nor the SE is ideal. The SE shows a pronounced peak in CA relaxivity at capillary vessel diameters (Boxerman *et al.*, 1995), implying inferior measurement sensitivity at larger diameters (*i.e.*, for arterial vessels) although the sensitivity to the CA in large vessels has showed some variability between different studies (Boxerman *et al.*, 1995; Wilson *et al.*, 2017; Van Dorth *et al.*, 2022). On the other hand, the generally higher relaxivity of GRE implies a risk of signal extinction at arterial peak concentrations. To obtain sufficient signal drop in tissue, a GRE echo time of approximately 40–45 ms at 1.5 T and 25–35 ms at 3 T (Welker *et al.*, 2015) is normally used (and an even longer TE for SE), which, when using a standard dose (0.1 mmol/kg body weight) of CA, results in such a large signal drop in blood that the signal reaches the noise floor at high concentrations of CA (Calamante, 2013). A power injector is employed, as the CA needs to be injected at a sufficiently high rate (at least 3 ml/s) to avoid underestimation of CBF, and a narrow bolus will further increase the peak concentration. An injection rate of approximately 3–5 ml/s is normally used (Knutsson *et al.*, 2004), because an even higher injection rate does not improve the results.

Hence, from the perspective of AIF quantification using GRE, either a lower dose of CA or a shorter TE would have been preferable. Different approaches have been proposed to solve this issue, including the use of a smaller bolus of CA, injected before the main DSC-MRI experiment (a so-called prebolus), providing the opportunity to correct the saturated AIF measured in the main DSC-MRI experiment (Knutsson *et al.*, 2014). The use of a single slice over the carotids in the neck (Kellner *et al.*, 2013), imaged with a much shorter echo time (and higher temporal resolution), has also been proposed. Similarly, a double echo is sometimes employed (at the expense of brain coverage), where the shorter TE is used to measure the AIF while the longer TE is used to obtain the tissue concentration curve. Both prebolus injection and double-echo acquisition can also reduce the so-called T1 effect, resulting from the fact that the employed sequence is not purely T2/T2*-weighted, implying that the CA causes less signal reduction than expected (or even an increased signal), because the CA, in addition to the desired shortening of T2/T2*, will also shorten T1. Neither prebolus injections nor double echoes are used in a standard DSC-MRI protocol.

Rapid CA injection implies that high temporal resolution is needed to capture the shape of the AIF correctly. Hence, a temporal resolution corresponding to no more than 1.5 s between images in the time series is recommended (Knutsson *et al.*, 2004; Welker *et al.*, 2015), which implies some compromises with image quality, *i.e.*, the spatial resolution is limited and the use of EPI readout may cause distortions. It should be noted that too high a temporal resolution is not recommended as the T1 effect will increase. In GRE, to reduce the T1 effect, a low flip angle can be used, albeit at the cost of reduced SNR, and a flip angle of 60–70 degrees is generally recommended (Welker *et al.*, 2015).

To achieve sufficient SNR and coverage, the spatial resolution in a DSC-MRI experiment is typically around $2 \cdot 2 \cdot 5 \, mm^3$. In the context of MRI, this is a rather poor spatial resolution implying that the voxel or voxels chosen to measure the AIF will almost always exhibit PVEs.

8.2.1.2 Dynamic contrast-enhanced MRI

While DSC-MRI relies on changes in transverse relaxation rate ($\Delta R2^*$ or $\Delta R2$) during the first passage of the CA bolus to determine the AIF, DCE-MRI relies on the change in longitudinal relaxation rate R1. Acquisition of the AIF is not trivial and several factors, described later, need to be considered.

SNR is a decisive factor in DCE-MRI, as inadequate SNR leads to errors in the calculated pharmacokinetic parameters. DCE-MRI is also inherently associated with modest SNR, because a short acquisition time is required to capture each time point in the dynamic process. Hence, higher field strengths should be beneficial, but they are also associated with potential disadvantages, including increased static and RF field inhomogeneities. In addition, a higher field strength leads to faster T2* relaxation, and this can violate the assumption that the recorded signal change is purely T1 weighted (see Fig. 8.2). This assumption of negligible T2* effects is even more

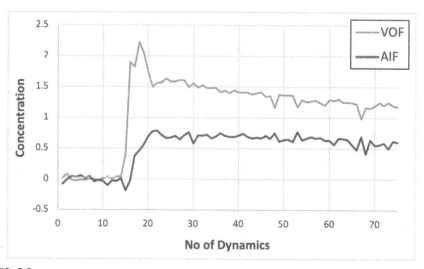

FIG. 8.2

Calculated AIF and VOF using the LC method (see Section 8.2.3). The AIF and VOF are obtained from the internal carotid artery and the superior sagittal sinus, respectively, at ultrahigh field strength (7 T) using an SPGR sequence with an echo time of 1.3 ms, temporal resolution of 3.2 s, and spatial resolution of $2 \cdot 2 \cdot 3 \, mm^3$. Note the substantial T2* effect, especially at the first passage of the bolus in the artery, even at an echo time of 1.3 ms. In addition, notice that there is a higher degree of PVEs in the AIF than in the VOF, manifested as a smaller area of the arterial concentration–time curve.

problematic for high CA concentrations. Since the AIF is presumed to be sampled from a large vessel, where a high CA concentration exists during the first-pass bolus passage, any significant T2* decay will result in an incorrect shape of the AIF. This, in turn, will lead to errors in the estimation of the pharmacokinetic parameters. Several suggestions have been made on how to minimize this problem. The simplest solution is to use short echo times (<2.5 ms), but due to hardware limitations, it is not technically feasible to obtain an extremely short TE. In addition, at ultrahigh field strengths (≥7 T) there is a substantial T2* effect even at short TEs (1–4 ms), and this has led to the development of additional methods to reduce T2* effects. One method is to model the true peak of the AIF by using appropriate postprocessing (Wang and Cao, 2012). Another is to use a double-echo sequence enabling the estimated $\Delta R2^*$ to be included in the derivation of the AIF (Kleppesto et al., 2014). This method is easy to implement but requires an extra analysis in the postprocessing step. An alternative approach is to acquire phase data (Garpebring et al., 2011), but DCE-MRI sequences tend to be rather insensitive to CA-induced phase shifts, due to the very short TE, and a separate prebolus injection will, therefore, be necessary. Such an approach will, however, provide an advantage in that the selection of the AIF is not restricted to the limited spatial coverage of the tissue of interest, neither to the original spatial and time resolutions. Disadvantages of AIFs obtained with phase data are that they are sensitive to B0 shifts and motion. Often these issues can be reduced by subtracting the background phase from vessel phase if the phase shifts in the vessel and the background tissue correlate temporally.

The temporal resolution is also vital for the sampled AIF shape from the DCE-MRI experiment. If the sampling rate is too low, the shape of the AIF may be inaccurate (see Fig. 8.3). It has been shown that using a time interval, *i.e.*, the time

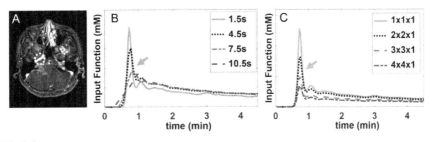

FIG. 8.3

The dependence of AIF measurements, obtained from the internal carotid artery (A), on temporal resolution (B), and spatial resolution (C). In this study, high-definition reference data (temporal resolution: 1.5 s, spatial resolution: 1.0 mm³) were used to emulate different temporal resolutions while keeping the spatial resolution at 1.0 mm³. The data were also used to emulate different in-plane spatial resolutions while maintaining the temporal resolution at 1.5 s.

Figure parts reproduced with permission from Park, J.S., Lim, E., Choi, S.H., et al., 2020. Model-based high-definition dynamic contrast enhanced MRI for concurrent estimation of perfusion and microvascular permeability. Med. Image Anal. 59, 101566.

between two dynamic acquisitions, exceeding nine seconds leads to a poorly sampled AIF and hence large errors in the pharmacokinetic parameters (Roberts *et al.*, 2006). The QIBA DCE-MRI recommendations from 2017 (QIBA, 2017) state that the ideal temporal resolution for DCE-MRI in the brain is less than five seconds and that it should not exceed ten seconds. However, even a temporal resolution of five seconds might not be enough for retrieving the correct shape of the AIF and it has been suggested that 2 s is warranted for having an adequate sampling of the AIF (Sourbron, 2010). For other organs, the optimal temporal resolution may differ, but a high temporal resolution is generally warranted to capture the true shape of the AIF. However, a high temporal resolution often comes at the expense of the spatial resolution. A low spatial resolution will result in AIFs that include signal components from tissue surrounding the vessel (PVEs), normally leading to an underestimated area of the AIF (see Fig. 8.3). Partial volume correction can be performed by sampling the concentration–time curve in a large vein, where the voxel is completely embedded in the vessel, and using this to correct for the PVE in the AIF (Sourbron *et al.*, 2009). An additional problem with low spatial resolution is that it may necessitate the use of a distal AIF location, since the AIFs that are close to the tissue of interest are not visible, and this may result in additional problems in the quantification due to delay and dispersion effects that are not reflected in the distal AIF.

DCE-MRI usually employs 2D or 3D sequences with a spoiled gradient-echo readout (SPGR). For DCE-MRI, the choice of sequence parameters such as TE, TR, and flip angle will determine the acquired signal level (see Table 8.1 for sequence parameters requirements). Additionally, the temporal resolution will depend on the TR. Thus, to obtain an acceptable T1-weighted signal from a large vessel, the TR should be as short as possible (ideally <3 ms). Furthermore, the signal change is approximately proportional to the CA concentration when the flip angle is higher. As a result, the flip angle should be selected as high as the SAR limit allows to prevent signal saturation at high concentrations. At higher field strengths (>3 T), there is an increased degree of nonuniformity in the RF field which may cause flip angle variations. As a result of these flip angle-related errors, the concentration–time curves, including the AIF, will show an incorrect shape. By incorporating a separate B1+ mapping, the concentration–time curves and subsequently the AIF shape can be improved. The B1+ mapping provides scaling factors to correct the flip angles, thereby improving the quantitative DCE-MRI analysis (Sengupta *et al.*, 2017).

Inflow effects are a common problem compromising AIF measurements. Flowing blood mimics the effects of higher relaxation rates, because fresh inflowing spins are completely relaxed. The inflow effect is minimized by using nonselective inversion or saturation pulses in 2D acquisitions. For a 3D acquisition, the imaging slab is usually placed to ensure that the inflowing spins in the arteries have traveled a sufficient distance in the slab when they have reached the location for the AIF measurement (Roberts *et al.*, 2011).

Table 8.1 Requirements for image acquisition settings with respect to AIF measurement.

Pulse sequence	SPGR or similar
Spatial resolution (in-plane)	1–2 mm
Spatial resolution (through-plane)	ideal ≤5 mm, acceptable ≤8 mm
Temporal resolution	ideal ≤2 s, acceptable ≤5 s
Echo time	As short as possible, often ranging from 1 to 3 ms
Repetition time	3–8 ms, considering temporal resolution and coverage
Flip angle	Ranging from 25 to 35 degree (1.5 T)/10 to 15 degree (3 T)
B1 mapping	Yes, at 3 T and above

8.2.1.3 Arterial spin labeling

Unlike DSC- and DCE-MRI, which require the measurement of an AIF for pharmacokinetic quantification, standard ASL methods typically rely on assumptions that determine the AIF analytically. However, this approach makes the quantification prone to errors, especially in patients with a wide transit time distribution. As an attempt to avoid this problem, the QUASAR sequence was introduced. In this approach, after labeling, images are acquired at different inversion times, thereby allowing the entire signal difference curve over time to be acquired (Petersen et al., 2006). The application of flow suppression gradients allows for estimation of AIFs by subtracting the flow-suppressed ASL data from the corresponding non-flow-suppressed data. CBF can then be quantified by deconvolving the signal difference time curve, obtained from the flow-suppressed ASL data, with the nearest AIF, using the same mathematical procedure as in DSC-MRI.

As a consequence of using endogenous blood water as a tracer, comprising only a small fraction of the brain tissue water signal, ASL shows an inherently low SNR. In QUASAR, labeling is performed using a short RF inversion pulse that is applied to a thick slab through the brain-feeding arteries (so-called pulsed ASL, PASL). Hence, the entire bolus of labeled blood water is created instantaneously, with the amount of labeled blood determined by the coverage of the inversion slab. It should be noted that QUASAR shows a lower SNR than, for example, pseudo-continuous ASL (PCASL), which is the labeling approach currently recommended for clinical applications (Alsop et al., 2015). PCASL uses a long train of short RF pulses and magnetic field gradients for continuous labeling (∼2 s). Similar to all ASL methods, QUASAR would seem to benefit from ultrahigh field (≥7 T) scanners due to increased SNR and prolonged T1. However, RF and static field inhomogeneities are more significant at 7 T than at lower field strengths, possibly leading to artifacts and spatial variations in the labeling, and ASL is, therefore, usually performed at 3 T.

Since QUASAR is a bolus tracking technique, it is important to ensure adequate temporal resolution to capture the complete dynamic process. However, if the sampling time interval is too small, the labeled blood water will not be refreshed between two consecutive excitations, which in turn gives rise to the AIF area being underestimated. A similar underestimation occurs if the excitation flip angle is too large; it has been observed that the AIF area was 20% lower when using a 25 degree flip angle compared to a 10 degree flip angle (Petersen et al., 2006). It is, therefore, important to choose the temporal resolution, the slice thickness, and the flip angle carefully to avoid saturation of the inflowing labeled blood water.

Since the AIF is obtained by subtracting the flow-suppressed ASL data from non-flow-suppressed ASL data, the strength of the flow suppression gradient will influence the result. The blood velocity in the capillary bed is approximately 0.2–5 mm/s. Therefore if a velocity encoding (V_{enc}) of less than 1 cm/s is used, the capillary blood water will be highly suppressed and the resulting AIFs will include contributions also from the capillary pool. On the other hand, if the V_{enc} is higher than 5 cm/s there will be less suppression in the larger arteries/arterioles which may lead to a reduction of measured AIF values. To avoid this issue, Petersen et al. (2006) and Petersen et al. (2010) set the V_{enc} to 3 or 4 cm/s.

8.2.2 Selection of the AIF

8.2.2.1 Dynamic susceptibility contrast MRI

Ideally, a local AIF should be measured as close as possible to the tissue element of interest, i.e., the AIF should be measured in the small artery that supplies the local tissue element with blood. The reason for this is that the CA bolus will be increasingly dispersed during transport in the large blood vessels, and the actual AIF will differ between different positions in the brain. However, as the spatial resolution is limited, measurement of AIFs in small vessels will suffer from substantial PVEs. The consequences of PVEs are, normally, that both the AIF shape and the area under the curve become incorrectly measured, and this will substantially affect the quantification of perfusion parameters in absolute terms. In the presence of PVEs, blood and tissue signal components are combined into a vector sum of the respective signal components, and the vector summation can lead to a complex result that is difficult to predict, and PVEs can cause either higher or lower concentration estimates than the true pure blood concentration (see Fig. 8.4). However, because the concentration in tissue is much lower than in the arterial blood, PVEs tend generally to cause an underestimation of the AIF as shown in Fig. 8.5, which, in turn, will lead to an overestimation of both CBF and CBV. Importantly, the contrast agent residing in the blood will also cause magnetic field distortions outside the vessel, leading to spin-phase dispersion and signal loss in the compartment surrounding the vessel. It has actually been shown that, in terms of the shape of the AIF, it is more advantageous to measure the AIF completely outside of the vessel, provided that specific conditions with regard to measurement position are fulfilled (Bleeker et al., 2009). Such an AIF will provide reliable relative perfusion values, but incorrect absolute quantitative values because the area under the AIF curve will be underestimated.

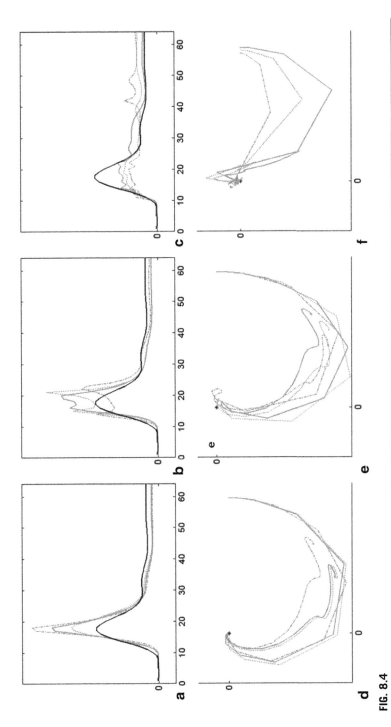

FIG. 8.4

Illustration of how partial volume conditions can have different effects on the estimated concentration. The voxel arterial concentration [a.u.] versus time [s] is shown in (A–C) and the corresponding complex MR signal (imaginary MR signal part versus real MR signal part) is shown in (E–F). PVEs can cause either overestimation of the concentration (A), truncated cr double peaks (B), or underestimated concentration levels (C). The black line indicates the true value, and all curves are rescaled to show the same area under the curve. In (D–F), the complex MR signal for different contrast agent concentrations is shown, each line corresponding to one AIF illustrating how the signal evolves when the contrast agent passes the artery.

Figure reproduced, with permission from Bleeker, E.J., Van Osch, M.J., Connelly, A., et al., 2011. New criterion to aid manual and automatic selection of the arterial input function in dynamic susceptibility contrast MRI. Magn. Reson. Med. 65, 448–56.

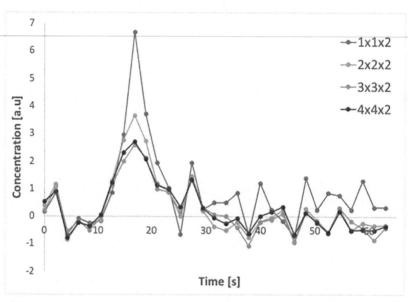

FIG. 8.5

The effects on the AIF caused by altering the spatial resolution. High spatial resolution data (spatial resolution: $1 \cdot 1 \cdot 2\,mm^3$) were used to emulate different in-plane spatial resolutions while maintaining the temporal resolution at 2.1 s. DSC-MRI data acquisition using high spatial resolution is prone to low SNR which explains the noise in the concentration–time curves.

Courtesy of Dr. Jun Hua, Johns Hopkins University, Baltimore, US.

The most common approach in DSC-MRI is to use a global AIF, registered in a fairly large vessel, instead of local AIFs. The global AIF is represented by the concentration time series of one arterial voxel, or by the mean concentration values from a few arterial voxels, and this single AIF is used in the calculation of perfusion estimates for each tissue voxel in the entire brain. This methodology is likely to decrease the degree of PVEs compared with the use of local AIFs from smaller arteries. It is, however, important to point out that in a straightforward DSC-MRI experiment, with a typical spatial resolution, single-shot EPI readout, normal TE, and a standard dose of CA, a substantial amount of PVE is likely to be present even for global AIFs. In fact, whether the AIF is selected manually or with an automatic algorithm, voxels with 100% blood will (or at least should) normally be discarded because the shape of such an AIF will be judged to be unreasonable. One reason is that, using a standard dose of CA and normal TE (*i.e.*, 35–45 ms at 1.5 T and 25–30 ms at 3 T for GRE), the CA will cause complete signal loss in blood at the highest concentrations during the first passage (Calamante, 2013). As the signal reaches the noise floor, it is not possible for the signal to decrease further, and the estimated concentration reaches a maximal level. Hence, the resulting

AIF (based on calculated $\Delta R2^*$ or $\Delta R2$) will be truncated, *i.e.*, the estimates at maximum concentration levels are underestimated and the peak is flattened. Another possible scenario at high concentration levels is that the noise contribution happens to result in a signal that is very close to zero, which will be interpreted as an extremely high concentration value, often at a single time point, resulting in an AIF with a very narrow peak. Another reason that may render high-concentration voxels useless for AIF registration is related to the low bandwidth of single-shot EPI readout in the phase-encoding direction. At high concentrations of CA, the high magnetic susceptibility of the arterial voxel may, depending on vessel orientation, cause a magnetic field shift so large that it causes a signal pixel shift in the phase-encoding direction, *i.e.*, the arterial signal may be displaced a distance corresponding to several pixels (Rausch *et al.*, 2000). Because the AIF is registered by tracking the temporal signal development in one specific pixel in the image matrix, low concentration values will be estimated at true peak concentrations, and such an AIF will also be excluded when appropriate voxels for AIF estimation are selected.

Furthermore, the use of global AIFs is, as discussed before, not entirely unproblematic as a global AIF does not accurately represent the true local arterial input function. One issue, referred to as arterial delay, is that the global AIF is measured at time points that occur either before or after the arrival of the true AIF at the tissue of interest. Some deconvolution algorithms are sensitive to time shifts and may return incorrect perfusion values. However, nowadays, most deconvolution algorithms either include a time-shift correction or are inherently insensitive to such arterial delays, as, for example, the block-circulant singular value decomposition (SVD) and Fourier transform-based algorithms. Another more fundamental issue, which cannot easily be dealt with by the deconvolution algorithm, is the arterial dispersion of the bolus that takes place between the site of the global AIF measurement (*i.e.*, in a larger vessel, upstream in the vascular tree) and the position of the true AIF (a smaller vessel at the inlet of the tissue of interest). Hence, when using a global AIF, the measured AIF is, in most normal cases, narrower than the true AIF, and CBF will, accordingly, be underestimated and MTT overestimated. In some deconvolution algorithms, modeling of the dispersion is accomplished by including a so-called vascular transfer function, but such attempts to compensate for arterial dispersion often encounter difficulties in separating the effects of delay and dispersion.

8.2.2.2 Dynamic contrast-enhanced MRI

The selection of an AIF in DCE-MRI is not trivial, and the choice will affect quantification of the pharmacokinetic parameters. There are some general criteria that should be fulfilled to allow a suitable AIF to be selected for the DCE-MRI experiment: the image volume must include an appropriate artery, the spatial resolution must be sufficient to minimize PVEs, and the temporal resolution must be adequate for the shape of the AIF to be correctly sampled. Apart from the injection scheme, the shape of the AIF depends on the subject's cardiac output and on vascular autoregulation. Therefore choosing an AIF close to the tissue of interest should, in theory, be the gold standard approach. In pathologies causing arterial bolus dispersion, it may be beneficial to use a local AIF from that region instead of one global AIF. However, when using local AIFs for multiple regions, one needs to take into consideration that they can have various amounts of PVEs, thereby introducing errors in the quantification.

Manual AIF selection introduces operator dependence, which may increase the variability between raters or between follow-up studies or between sites when assessing pharmacokinetic parameters. One way of minimizing operator variability is to use semiautomatic or automatic AIF identification, and this will also reduce the time for postprocessing of the DCE-MRI data. A comparison of manual and automatic AIF selections is shown in Fig. 8.6. In automated approaches, an AIF is generally

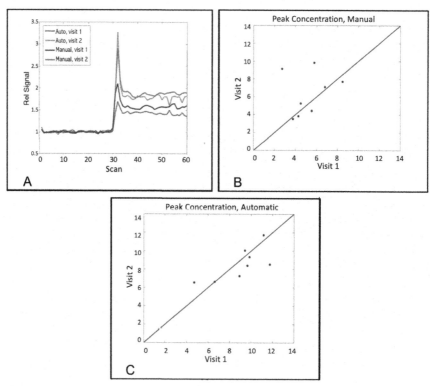

FIG. 8.6

Repeatability measurement of manual and automatic AIF selection, using K-means cluster analysis, from DCE-MRI data in eight patients with glioblastoma. The AIFs from the patients were averaged (A) and the repeatability of peak concentration, both using manual (B) and automatic (C) AIF selection, was examined. It was observed that automated selection led to better peak-signal repeatability compared to manual selection.

Figure parts reproduced with permission from Mouridsen, K., Jennings, D., Gelasca, E., et al., 2010. Towards Robust and Automated Identification of Vascular Input Function in DCE-MRI. International Society of Magnetic Resonance in Medicine, Stockholm, p. 18.

selected based on various characteristics of the concentration–time curves, such as peak height, peak width, arrival time, and initial slope (Ashton *et al.*, 2005; Rijpkema *et al.*, 2001). Data-driven methods such as principal component analysis (Sanz-Requena *et al.*, 2015) or deep learning (Nalepa *et al.*, 2020) are also used. The QIBA (2017) recommendation is to use a semiautomatic method instead of a manual or fully automatic method (QIBA, 2017). In addition, it is stated that the input function should be determined from a slice located at least 3 cm away from the first slice to prevent inflow effects (Roberts *et al.*, 2011).

A simplified approach for AIF selection is to use a population-based AIF. This has been advocated for instances when it is impossible to obtain an AIF from the acquired data, for example, because of the location of the acquired data volume or because of insufficient temporal resolution, T2* effects, motion artifacts, or flow effects (Parker *et al.*, 2006). Some studies have demonstrated that the use of a population-based AIF can improve the repeatability (Parker *et al.*, 2006; Koopman *et al.*, 2020), while other studies have reported the opposite (Rijpkema *et al.*, 2001; Ashton *et al.*, 2008). Fig. 8.7 shows the difference in K^{trans} values when using patient-specific AIFs selected by three domain experts and a population-based AIF. The QIBA (2017) recommendations state that a population-based AIF should only be used when a patient-specific AIF is not available (QIBA, 2017).

A more recent approach is to select a venous concentration curve from a major vein, such as the superior sagittal sinus or the jugular vein, as a substitute for an input function from arterial vessels. Using the venous concentration curve as a surrogate vascular input function (VIF) minimizes the PVE problem. However, a drawback is that this approach usually requires this surrogate VIF to be measured from an additional DCE-MRI experiment using a smaller dose of the CA (a so-called prebolus experiment) before the regular DCE-MRI experiment. In addition, this method

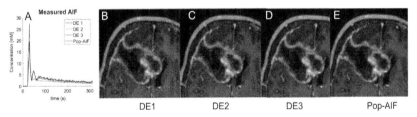

FIG. 8.7

AIFs registered by three domain experts (DE) compared with a time-aligned population AIF (pop-AIF) (A). Corresponding tumor K^{trans} maps overlaid on structural postcontrast T1-weighted images (B–E). The K^{trans} color scale ranges from 0.0 to 0.4 min^{-1}. Median K^{trans} estimates of the whole tumor for DE 1, DE 2, DE 3, and pop-AIF were 0.049, 0.053, 0.044, and 0.101 min^{-1}, respectively.

Figure parts reproduced with permission from Kleppesto, M., Bjornerud, A., Groote, I. R., et al., 2022. Operator dependency of arterial input function in dynamic contrast-enhanced MRI. MAGMA 35, 105–112.

assumes that the venous and arterial contrast-agent concentrations are equal and that the effects of the delay and/or dispersion of the bolus occurring between the arterial and venous site are negligible in the context of the pharmacokinetic modeling. Relying on these assumptions can introduce errors, although the delay issue can be reduced by recording the bolus arrival time of the concentration–time curve in each tissue voxel and then time shifting the venous concentration curve accordingly before the pharmacokinetic analysis (Lewis *et al.*, 2022).

Instead of using a large vessel as input function, the reference region model (RRM) uses the concentration–time curve from a tissue instead (Yankeelov *et al.*, 2005). It is then assumed that the reference tissues of all subjects have the same extravascular extracellular volume (EEV) fraction. However, since the EEV can differ between subjects there can be significant variability of the quantified pharmacokinetic parameters, and it has been shown that the reproducibility is low (Yankeelov *et al.*, 2007).

8.2.2.3 Arterial spin labeling

In the original QUASAR paper (Petersen *et al.*, 2006), AIFs were identified by selecting arterial signal curves with an arterial blood volume (aBV) above a certain threshold (1.2%). The aBV was given by the total area of the arterial signal divided by the bolus area:

$$\text{aBV} = \frac{\int_{-\infty}^{\infty} (\Delta M_{ns}(t) - \Delta M_s(t)) e^{\frac{t}{T1a}} dt}{2 \cdot M_{a,0} \cdot \tau_b \cdot \alpha} \tag{8.2}$$

where T1a is the longitudinal relaxation time of arterial blood, τ_b is the duration of the labeling, and α is the inversion efficiency. ΔM_{ns} and ΔM_s correspond to the non-suppressed and suppressed image data, respectively, and $M_{a,0}$ corresponds to the equilibrium magnetization in arterial blood. As can be seen from Eq. (8.2), $M_{a,0}$, τ_b, and T1a also need to be estimated, which makes the aBV calculation prone to errors. For instance, there is a risk that important AIF locations will be missed when using an aBV threshold. When the arterial signal curves are saturated, such as in the most superior slices or in the posterior areas of the slice, the aBV thresholding method will fail to detect AIFs that correspond to aBV values below the fixed threshold. In principle, saturation effects can be minimized by increasing ΔTI, but this could result in errors in the perfusion estimates due to incorrect shape of the AIF. An alternative approach, taking the shape and timing of the arterial signal time curves into consideration, is to use data-driven methods for curve classification, for example, factor analysis of dynamic studies (FADS) (Knutsson *et al.*, 2008). When comparing the FADS technique with the aBV threshold method, it was observed that some curves classified as AIFs by the aBV threshold method, but not by the FADS approach, showed lower amplitude and higher degree of dispersion, possibly explained by tissue contamination (see Fig. 8.8).

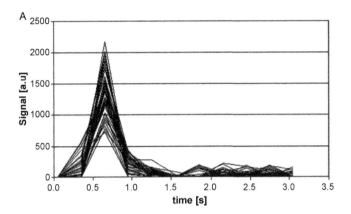

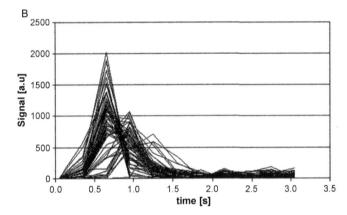

FIG. 8.8

The AIFs identified at different locations within one slice by FADS (A) and aBV thresholding (B). Note that the use of aBV thresholding approach returned AIFs that had lower amplitude and higher degree of dispersion.

Figure reproduced, with permission from Knutsson, L., Markenroth Bloch, K., Holtas, S., et al., 2008. Model-free arterial spin labelling for cerebral blood flow quantification: introduction of regional arterial input functions identified by factor analysis. Magn. Reson. Imaging 26, 554–559.

8.2.3 Calculation of AIF concentration

8.2.3.1 Dynamic susceptibility contrast MRI

To quantify the AIF, the signal drop induced by the CA needs to be converted to concentration. Assuming GRE acquisition, $\Delta R2^*$ is normally calculated as $\Delta R2^* = -\ln[S(t)/S_0]/TE$, where $S(t)$ is the signal at time point t and S_0 is the baseline signal. The concentration is assumed to be proportional to $\Delta R2^*$, *i.e.*, the relationship $\Delta R2^* = r2^* \cdot [CA]$ is assumed, where $r2^*$ is the transverse relaxivity of the CA and

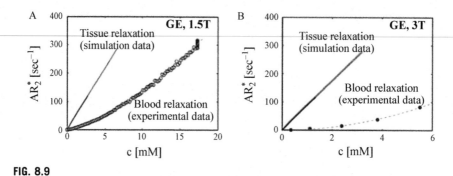

FIG. 8.9

$\Delta R2^*$ as a function of concentration in tissue and blood for a GRE sequence at 1.5 T (A) and 3 T (B), showing that the relaxivity in tissue is higher than the relaxivity in blood.

Figure is reproduced with permission from Kjølby, B.F., Østergaard, L., Kiselev, V.G., 2006. Theoretical model of intravascular paramagnetic tracers effect on tissue relaxation. Magn. Reson. Med. 56, 187–97.

[CA] is the concentration of CA. However, *in vivo*, the r2* relaxivity is normally not known and in standard implementations, the relaxivities in tissue and blood are assumed to be equal. The relaxivity is then cancelled out in the equations describing CBF and CBV. There are, however, several complications related to this assumption. First, the relaxivities in blood and tissue are actually not the same (see Fig. 8.9). In fact, the relaxivity in tissue is higher than in whole blood, *i.e.*, the signal drop for a certain concentration of CA is larger in tissue than in blood. Thus the AIF will be underestimated compared to the tissue concentration curve and the relaxivity difference will cause overestimated CBF and CBV if not accounted for. This relaxivity difference between blood and tissue increases with field strength and CBF and CBV estimates from data acquired at 3 T are more overestimated than at 1.5 T. If the relaxivity is actually a constant, *i.e.*, a linear relationship between $\Delta R2$ or $\Delta R2^*$ and the concentration exists, this relaxivity difference would still result in reliable relative values. The situation is, however, further complicated by the fact that the relationship between $\Delta R2$ or $\Delta R2^*$ and the CA concentration has been shown to be nonlinear (Van Dorth et al., 2022; Kjølby et al., 2006) and often described by a quadratic relationship in pure whole blood as illustrated by Fig. 8.10. Hence, if measuring dynamically in blood over a range of concentrations, and assuming a linear relationship, the shape of the calculated AIF will be incorrect. However, as already discussed, the AIF is, in practice, not likely to be measured in a voxel with 100% blood, and for an AIF voxel in which a large fraction of the signal actually derives from tissue, a linear relationship might be a good approximation (Knutsson et al., 2013). The situation is even more complex as the relaxivity of the CA in blood also depends on the hematocrit (Hct) of the blood, where higher Hct levels result in higher relaxivity (Fig. 8.10) (Van Dorth et al., 2022). However, as the Hct in a standard DSC-MRI experiment is generally unknown, this effect is normally neglected.

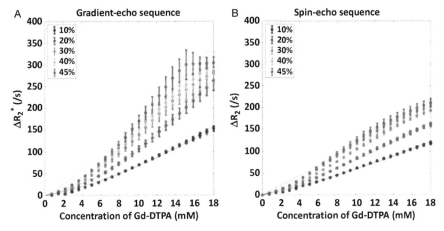

FIG. 8.10

Simulated ΔR2* (A) and ΔR2 (B) in blood as a function of concentration, assuming a GRE with echo time 20 ms (A) and a SE with echo time 40 ms (B) at 1.5 T. Data are shown for different Hct levels (10%–45%).

Figure is reproduced with permission from Van Dorth, D., Venugopal, K., Poot, D.H.J., et al., 2022. Dependency of R2 and R2* relaxation on Gd-DTPA concentration in arterial blood: influence of hematocrit and magnetic field strength. NMR Biomed. 35, e4653.

As mentioned before, the phase of the MRI signal has also been employed to estimate the concentration of the AIF. Apart from altered relaxation parameters, the CA also causes a local magnetic field shift due to the high bulk susceptibility of the CA. Hence, the CA will affect the phase of the MRI signal in proportion to the voxel concentration. Phase-based AIFs are, indeed, promising, as the relationship between the phase shift and concentration is linear (and also the same in tissue and blood). It has also been shown that the phase shift caused by CA is independent of the Hct level (Akbudak *et al.*, 1998). However, there are other practical problems with the use of phase data that need to be solved in order for phase AIF registration to become a standard method. For example, during scanning, a drift of the static magnetic field is commonly seen, which needs to be corrected for. To obtain quantitative values of concentration, it must be reasonable to apply the infinite-cylinder approximation to employed artery, and the angle relative to the main magnetic field in the MRI scanner must be known. Additionally, the problem with PVEs is relevant also for phase data, and the effects of partial volumes on phase data can be even more intricate as the true phase depends on the geometry of the vessel. A more direct approach is to estimate the source of the phase shift, *i.e.*, the magnetic susceptibility of the CA in the vessel, by performing a deconvolution of the measured phase shift with the unit magnetic dipole kernel in a process called quantitative susceptibility mapping (QSM) (Ruetten *et al.*, 2019). The advantage of using QSM is that there are no requirements on the geometry and orientation of the artery, but issues such as magnetic field drift during the scanning, identification of a reliable reference to be able to compare estimates from different time points in the dynamic study, and partial volume effects still need to be resolved.

8.2.3.2 Dynamic contrast-enhanced MRI

To be able to apply a model and retrieve relevant pharmacokinetic parameters, the signal–time curve in each voxel, including the AIF, needs to be converted to a concentration–time curve. There are two common ways to achieve this: linear conversion (LC) and nonlinear conversion (NLC). True to its name, LC is performed by assuming a linear relationship between signal and concentration:

$$C(t) = \frac{1}{r1 T10} \left(\frac{S(t)}{S_0} - 1 \right), \tag{8.3}$$

where S_0 is the mean precontrast signal intensity, $r1$ is the longitudinal relaxivity, and T10 is the tissue T1 before CA injection. For NLC, the signal equation corrects for the pulse sequence specifics by including the sequence parameters TR and flip angle α, in addition to T10, leading to:

$$C(t) = \frac{1}{r1} \left(\frac{1}{T1(t)} - \frac{1}{T10} \right), \tag{8.4}$$

where T1(t) can be directly derived from the signal dependence on TR and flip angle. For instance, using an SPGR sequence, we have:

$$S(t) \propto M_0 \frac{1 - e^{-\frac{TR}{T1(t)}}}{1 - \cos(\alpha) \cdot e^{-\frac{TR}{T1(t)}}} \cdot \sin(\alpha), \tag{8.5}$$

where M_0 is the fully relaxed longitudinal magnetization for a 90 degree pulse when TR >> T10. Assuming that TE is short enough for the $T2^*$ effect to be negligible, the following approximation is valid:

$$T1(t) \approx \frac{-TR}{\ln \left(A - \frac{S(t)}{S_0} \right) - \ln \left(A - \frac{S(t)}{S_0} \right) \cos(\alpha)}, \tag{8.6}$$

where

$$A = \left(1 - \cos(\alpha) e^{-TR} \right) / \left(1 - e^{-TR} \right) \tag{8.7}$$

It has been observed that, at high CA concentrations, the LC will underestimate the concentration value. Since the concentration of the CA is high in the artery, especially during the first passage, the use of LC may distort the shape of the AIF, leading to erroneous quantitative pharmacokinetic values (see Fig. 8.11).

From the equations given before, it can also be deduced that any variation in the precontrast tissue T1 relaxation time (T10) will affect the calculation of the concentration–time curve and, thus, the pharmacokinetic parameter quantification.

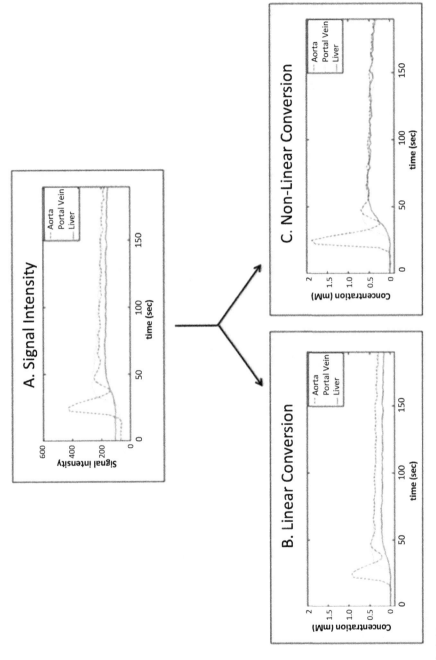

FIG. 8.11

Signal intensity (SI) versus time curves for abdominal aorta (in *red*; gray in print version), portal vein (in *blue*; dark gray in print version), and liver parenchyma (in *green*; light gray in print version) are shown in (A). Concentration vs. time curves for the aorta, portal vein, and liver parenchyma using LC and NLC are shown in (B and C), respectively. There are clear differences in the curves obtained using the different conversion methods. The most prominent observation is that a saturation is seen for the aortic curve using the LC method.

Figure adapted from Aronhime, S., Calcagno, C., Jajamovich, G.H., et al., 2014. DCE-MRI of the liver: effect of linear and nonlinear conversions on hepatic perfusion quantification and reproducibility. J. Magn. Reson. Imaging 40, 90–98.

The volume transfer constant K^{trans} and the interstitial volume V_e are especially sensitive to the accuracy of T10, and reliability of the T10 estimation is crucial when DCE-MRI is used to predict treatment outcome and to monitor effects of therapy. To assess T10, one can either (i) measure it before the CA injection, usually with a variable flip angle gradient-echo sequence or (ii) use a fixed T10 reference value, generally literature based. The first method allows for a voxel-based T10 measurement but is heavily dependent on flip angle accuracy. The second method applies the same T10 to all voxels. It has been shown that, when a precontrast T1 measurement is not available, the uncertainty in the kinetic parameter estimation can be reduced by using a low flip angle and the LC method with a population-based AIF (Wake *et al.*, 2018). In addition, QIBA has recommended that if accurate T10 values cannot be reproduced, it is better to use literature values.

It has been shown that water exchange between compartments (the intravascular space, extracellular extravascular space EES, and intracellular space) may not be fast, *i.e.*, a spatial variation of the CA-affected water exists which creates a nonlinear relationship between the measured T1 and CA concentration. However, for the intravascular compartment, the water exchange can probably be assumed to be fast between the red blood cells and the plasma, since water resides in red blood cells during a very short time (10–14 ms) compared to the duration of the delays for T1 estimation (Herbst and Goldstein, 1989). Notice that if there is tissue mixing in the voxel, creating PVEs, the water exchange may also affect the longitudinal relaxation in the voxel representing the AIF and thereby create inaccuracies in the pharmacokinetic values.

The Hct also contributes to uncertainty when calculating the AIF. In most cases, a literature value is used, which may introduce errors because the Hct varies in subjects over time (morning versus evening, summer versus winter). Hct also differs between men and women; in healthy females the Hct range is lower (0.36–0.48) than in healthy males (0.40–0.53). In addition, some diseases such as anemia and sickle cell disease affect the Hct level.

8.2.3.3 Arterial spin labeling

Buxton *et al.* described ASL from a general tracer kinetic modeling perspective (*cf.* Eq. 8.1), referred to as the general kinetic model for ASL (Buxton *et al.*, 1998). Using this approach, the difference in magnetization ΔM between the control and labeled images can be written as follows:

$$\Delta M = 2 \cdot M_{a,0} \cdot f \int_0^t c(\tau) r(t - \tau) m(t - \tau) d\tau, \tag{8.8}$$

where $M_{a,0}$ is the equilibrium magnetization in blood, f is the blood flow, $c(\tau)$ is the fractional arterial input function, and $r(t - \tau)$ is the residue function describing the fraction of labeled spins arriving at a voxel at time τ that still remains within the voxel at time t. The magnetization relaxation term, $m(t - \tau)$, quantifies the longitudinal magnetization fraction of the labeled spins that arrives at a voxel at time τ and remains at time t.

The AIF, as measured in a voxel representing arterial blood, is obtained by:

$$\text{AIF}(t) = 2 \cdot M_{a,0} \cdot c(t),\qquad(8.9)$$

where

$$c(t) = \frac{(\Delta M_{\text{ns}}(t) - \Delta M_{\text{s}}(t))e^{\frac{t}{T1a}}}{\int_{-\infty}^{\infty}(\Delta M_{\text{ns}}(t) - \Delta M_{\text{s}}(t))e^{\frac{t}{T1a}}dt}\, e^{\frac{-(1-(\tau_m-\tau_a))}{T1a}} \cdot \tau_b \qquad(8.10)$$

Eq. (8.9) can then be rewritten as follows:

$$\text{AIF}(t) = \frac{(\Delta M_{\text{ns}}(t) - \Delta M_{\text{s}}(t))e^{\frac{t}{T1a}}}{\text{aBV}}\, e^{\frac{-(1-(\tau_m-\tau_a))}{T1a}} \qquad(8.11)$$

The calculation of the AIF relies on several parameters that need to be estimated (see also above when discussing aBV), and this makes it highly prone to errors which will subsequently affect the perfusion quantification.

8.3 Summary

The degree of friendship that needs to be established with the AIF depends, in most cases, on the level of ambition that is sought with regard to accurate absolute quantification of an established perfusion- or permeability-related quantity. In ASL, a comparison of model-based and model-free analysis methods for QUASAR perfusion quantification showed that both approaches enabled comparable CBF estimates and uncertainties (Chappell *et al.*, 2013). In DSC-MRI (Withey *et al.*, 2016) and, in particular, DCE-MRI (*e.g.*, Ashton *et al.*, 2008; Parker *et al.*, 2006), several studies have addressed the issues of subject-specific versus population-based AIFs, and it has been questioned whether to deconvolve or not to deconvolve in DSC-MRI (Meijs *et al.*, 2016). In some applications, population-based AIFs may have advantages, for example, to increase the precision/reproducibility in DCE-MRI pharmacokinetic parameters (Parker *et al.*, 2006), and strategies for modification of a population-based AIF to incorporate individual variations have also been suggested. Sometimes, the diagnosis of a disease or the characterization of a lesion can be accomplished without full quantification (Meijs *et al.*, 2016), and the use of perfusion indices or so-called summary parameters should not be disregarded. Attempts to find solutions to manage without the AIF are relevant and can definitely be pragmatically reasonable or even necessary. However, in order to obtain metrics that are comparable among individuals and over time, as well as between different sites and installations and between different medical imaging modalities, well-established physiological parameters, quantified in absolute terms, are indeed warranted. To achieve that, from a theoretical point of view, we do need our old friend—the AIF. Admittedly, a phrase that might come to mind, after reading this chapter, is "with friends like the AIF, who needs enemies?" No doubt, the AIF is a bit of an elusive, evasive, or even reluctant friend at times. Nevertheless, pursuing a close and long-lasting friendship with the AIF is likely to be the most rewarding path to follow in perfusion MRI, at least in DSC-MRI and DCE-MRI applications.

References

Akbudak, E., Hsu, R., Li, Y., et al., 1998. ΔR^* and $\Delta \varphi$ Contrast Agent Perfusion Effects in Blood: Quantitation and Linearity Assessment. International Society of Magnetic Resonance in Medicine, Sydney, p. 1197.

Alsop, D.C., Detre, J.A., Golay, X., et al., 2015. Recommended implementation of arterial spin-labeled perfusion MRI for clinical applications: a consensus of the ISMRM perfusion study group and the European consortium for ASL in dementia. Magn. Reson. Med. 73, 102–116.

Ashton, E., Mcshane, T., Evelhoch, J., 2005. Inter-operator variability in perfusion assessment of tumors in MRI using automated AIF detection. Med. Image Comput. Comput. Assist. Interv. 8, 451–458.

Ashton, E., Raunig, D., Ng, C., et al., 2008. Scan-rescan variability in perfusion assessment of tumors in MRI using both model and data-derived arterial input functions. J. Magn. Reson. Imaging 28, 791–796.

Bleeker, E.J., Van Buchem, M.A., Van Osch, M.J., 2009. Optimal location for arterial input function measurements near the middle cerebral artery in first-pass perfusion MRI. J. Cereb. Blood Flow Metab. 29, 840–852.

Boxerman, J.L., Bandettini, P.A., Kwong, K.K., et al., 1995. The intravascular contribution to fMRI signal change: Monte Carlo modeling and diffusion-weighted studies in vivo. Magn. Reson. Med. 34, 4–10.

Buxton, R.B., Frank, L.R., Wong, E.C., et al., 1998. A general kinetic model for quantitative perfusion imaging with arterial spin labeling. Magn. Reson. Med. 40, 383–396.

Calamante, F., 2013. Arterial input function in perfusion MRI: a comprehensive review. Prog. Nucl. Magn. Reson. Spectrosc. 74, 1–32.

Chappell, M.A., Woolrich, M.W., Petersen, E.T., et al., 2013. Comparing model-based and model-free analysis methods for QUASAR arterial spin labeling perfusion quantification. Magn. Reson. Med. 69, 1466–1475.

DCE-MRI Technical Committee, 2017. DCE MRI Quantification. Quantitative Imaging Biomarkers Alliance. Available https://qibawiki.rsna.org/images/1/1f/QIBA_DCE-MRI_Profile-Stage_1-Public_Comment.pdf.

Fick, A., 1870. Über die messung des blutquantums in der herzventrikel. Seitung der Physikalisches und Medicinisches Gesellschaft zu Würzburg.

Garpebring, A., Wirestam, R., Yu, J., et al., 2011. Phase-based arterial input functions in humans applied to dynamic contrast-enhanced MRI: potential usefulness and limitations. MAGMA 24, 233–245.

Harvey, R.B., Brothers, A.J., 1962. Renal extraction of para-aminohippurate and creatinine measured by continuous in vivo sampling of arterial and renal-vein blood. Ann. N. Y. Acad. Sci. 102, 46–54.

Herbst, M.D., Goldstein, J.H., 1989. A review of water diffusion measurement by NMR in human red blood cells. Am. J. Phys. 256, C1097–C1104.

Hoedt-Rasmussen, K., Sveinsdottir, E., Lassen, N.A., 1966. Regional cerebral blood flow in man determined by intra-arterial injection of radioactive inert gas. Circ. Res. 18, 237–247.

Kellner, E., Mader, I., Mix, M., *et al.*, 2013. Arterial input function measurements for bolus tracking perfusion imaging in the brain. Magn. Reson. Med. 69, 771–780.

Kjølby, B.F., Østergaard, L., Kiselev, V.G., 2006. Theoretical model of intravascular paramagnetic tracers effect on tissue relaxation. Magn. Reson. Med. 56, 187–197.

Kleppesto, M., Larsson, C., Groote, I., *et al.*, 2014. T2*-correction in dynamic contrast-enhanced MRI from double-echo acquisitions. J. Magn. Reson. Imaging 39, 1314–1319.

Knutsson, L., Stahlberg, F., Wirestam, R., 2004. Aspects on the accuracy of cerebral perfusion parameters obtained by dynamic susceptibility contrast MRI: a simulation study. Magn. Reson. Imaging 22, 789–798.

Knutsson, L., Markenroth Bloch, K., Holtas, S., *et al.*, 2008. Model-free arterial spin labelling for cerebral blood flow quantification: introduction of regional arterial input functions identified by factor analysis. Magn. Reson. Imaging 26, 554–559.

Knutsson, L., Stahlberg, F., Wirestam, R., *et al.*, 2013. Effects of blood $\Delta R2*$ non-linearity on absolute perfusion quantification using DSC-MRI: comparison with Xe-133 SPECT. Magn. Reson. Imaging 31, 651–655.

Knutsson, L., Lindgren, E., Ahlgren, A., *et al.*, 2014. Dynamic susceptibility contrast MRI with a prebolus contrast agent administration design for improved absolute quantification of perfusion. Magn. Reson. Med. 72, 996–1006.

Koopman, T., Martens, R.M., Lavini, C., *et al.*, 2020. Repeatability of arterial input functions and kinetic parameters in muscle obtained by dynamic contrast enhanced MR imaging of the head and neck. Magn. Reson. Imaging 68, 1–8.

Kosior, J.C., Smith, M.R., Kosior, R.K., *et al.*, 2009. Cerebral blood flow estimation in vivo using local tissue reference functions. J. Magn. Reson. Imaging 29, 183–188.

Lewis, D., Zhu, X., Coope, D.J., *et al.*, 2022. Surrogate vascular input function measurements from the superior sagittal sinus are repeatable and provide tissue-validated kinetic parameters in brain DCE-MRI. Sci. Rep. 12, 8737.

Meijs, M., Christensen, S., Lansberg, M.G., *et al.*, 2016. Analysis of perfusion MRI in stroke: to deconvolve, or not to deconvolve. Magn. Reson. Med. 76, 1282–1290.

Miura, Y., Kanno, I., Miura, S., *et al.*, 1981. Measurement of regional cerebral blood flow by 133Xe inhalation method -experimental system and its evaluation of data analysis by simulation study (author's transl). Radioisotopes 30, 92–98.

Nalepa, J., Ribalta Lorenzo, P., Marcinkiewicz, M., *et al.*, 2020. Fully-automated deep learning-powered system for DCE-MRI analysis of brain tumors. Artif. Intell. Med. 102, 101769.

Parker, G.J., Roberts, C., Macdonald, A., *et al.*, 2006. Experimentally-derived functional form for a population-averaged high-temporal-resolution arterial input function for dynamic contrast-enhanced MRI. Magn. Reson. Med. 56, 993–1000.

Petersen, E.T., Lim, T., Golay, X., 2006. Model-free arterial spin labeling quantification approach for perfusion MRI. Magn. Reson. Med. 55, 219–232.

Petersen, E.T., Mouridsen, K., Golay, X., *et al.*, 2010. The QUASAR reproducibility study, Part II: results from a multi-center arterial spin labeling test-retest study. NeuroImage 49, 104–113.

Raichle, M.E., Martin, W.R., Herscovitch, P., *et al.*, 1983. Brain blood flow measured with intravenous H2(15)O. II. Implementation and validation. J. Nucl. Med. 24, 790–798.

Rausch, M., Scheffler, K., Rudin, M., *et al.*, 2000. Analysis of input functions from different arterial branches with gamma variate functions and cluster analysis for quantitative blood volume measurements. Magn. Reson. Imaging 18, 1235–1243.

Rijpkema, M., Kaanders, J.H., Joosten, F.B., *et al.*, 2001. Method for quantitative mapping of dynamic MRI contrast agent uptake in human tumors. J. Magn. Reson. Imaging 14, 457–463.

Roberts, C., Buckley, D.L., Parker, G.J., 2006. Comparison of errors associated with single- and multi-bolus injection protocols in low-temporal-resolution dynamic contrast-enhanced tracer kinetic analysis. Magn. Reson. Med. 56, 611–619.

Roberts, C., Little, R., Watson, Y., *et al.*, 2011. The effect of blood inflow and B(1)-field inhomogeneity on measurement of the arterial input function in axial 3D spoiled gradient echo dynamic contrast-enhanced MRI. Magn. Reson. Med. 65, 108–119.

Ruetten, P.P.R., Gillard, J.H., Graves, M.J., 2019. Introduction to quantitative susceptibility mapping and susceptibility weighted imaging. Br. J. Radiol. 92 (1101), 20181016.

Sanz-Requena, R., Prats-Montalban, J.M., Marti-Bonmati, L., *et al.*, 2015. Automatic individual arterial input functions calculated from PCA outperform manual and population-averaged approaches for the pharmacokinetic modeling of DCE-MR images. J. Magn. Reson. Imaging 42, 477–487.

Schmiedeskamp, H., Straka, M., Newbould, R.D., *et al.*, 2012. Combined spin- and gradient-echo perfusion-weighted imaging. Magn. Reson. Med. 68, 30–40.

Sengupta, A., Gupta, R.K., Singh, A., 2017. Evaluation of B1 inhomogeneity effect on DCE-MRI data analysis of brain tumor patients at 3T. J. Transl. Med. 15, 242.

Sourbron, S., 2010. Technical aspects of MR perfusion. Eur. J. Radiol. 76, 304–313.

Sourbron, S., Ingrisch, M., Siefert, A., *et al.*, 2009. Quantification of cerebral blood flow, cerebral blood volume, and blood-brain-barrier leakage with DCE-MRI. Magn. Reson. Med. 62, 205–217.

Van Dorth, D., Venugopal, K., Poot, D.H.J., *et al.*, 2022. Dependency of R2 and R2 * relaxation on Gd-DTPA concentration in arterial blood: influence of hematocrit and magnetic field strength. NMR Biomed. 35, e4653.

Wake, N., Chandarana, H., Rusinek, H., *et al.*, 2018. Accuracy and precision of quantitative DCE-MRI parameters: how should one estimate contrast concentration? Magn. Reson. Imaging 52, 16–23.

Wang, H., Cao, Y., 2012. Correction of arterial input function in dynamic contrast-enhanced MRI of the liver. J. Magn. Reson. Imaging 36, 411–421.

Welker, K., Boxerman, J., Kalnin, A., *et al.*, 2015. ASFNR recommendations for clinical performance of MR dynamic susceptibility contrast perfusion imaging of the brain. AJNR Am. J. Neuroradiol. 36, E41–E51.

Wilson, G.J., Springer Jr., C.S., Bastawrous, S., *et al.*, 2017. Human whole blood (1) H2 O transverse relaxation with gadolinium-based contrast reagents: magnetic susceptibility and transmembrane water exchange. Magn. Reson. Med. 77, 2015–2027.

Withey, S.B., Novak, J., Macpherson, L., *et al.*, 2016. Arterial input function and gray matter cerebral blood volume measurements in children. J. Magn. Reson. Imaging 43, 981–989.

Yankeelov, T.E., Luci, J.J., Lepage, M., *et al.*, 2005. Quantitative pharmacokinetic analysis of DCE-MRI data without an arterial input function: a reference region model. Magn. Reson. Imaging 23, 519–529.

Yankeelov, T.E., Cron, G.O., Addison, C.L., *et al.*, 2007. Comparison of a reference region model with direct measurement of an AIF in the analysis of DCE-MRI data. Magn. Reson. Med. 57, 353–361.

Motion compensation strategies

Michael Salerno[a] and Ruixi Zhou[b]

[a]Cardiovascular Medicine, Stanford University, Palo Alto, CA, United States
[b]Artificial Intelligence, Beijing University of Posts and Telecommunications, Beijing, China

9.1 Introduction

Perfusion is the passage of arterial blood through the microvasculature in biological tissue. It contains one of the most important physiological parameters that can be measured noninvasively with MRI. Straightforward qualitative visual assessment can detect reduction and altered timing of the contrast enhancement, which directly reflect regional perfusion abnormalities. As such visual assessment is subjective, recent advances in quantitative perfusion imaging have gained significant interest to provide objective, accurate, and reproducible analysis of the functional status of various tissues. However, both qualitative and quantitative perfusion imaging techniques suffer from motion, which has been problematic since their introduction in the clinical examination procedure. Although the sensitivity to certain intravoxel incoherent motion of blood can be availed to provide powerful image contrast (Federau, 2017), bulk motion contributes to significant problems in the current clinical perfusion imaging setting, which is the source of the majority of artifacts. Perfusion measurements are important in a number of organs, including the brain, heart, liver, and kidney, and routinely utilized for an expanding number of indications. Cerebral perfusion imaging can reveal ischemic conditions like the ischemic penumbra (Karonen et al., 1999) and moyamoya disease (Noguchi et al., 2011), neoplasms such as brain tumor (Zonari et al., 2007), along with neurodegenerative diseases. Cardiac perfusion imaging can demonstrate known or suspected coronary artery disease and microvascular disease (Manka et al., 2010; Zorach et al., 2016). Abdominal perfusion imaging can help diagnose diseases including but not limited to liver cirrhosis, pancreatitis, and renal microvascular disease (Pandharipande et al., 2005). Different organs exhibit various types of motion, which lead to distinct motion artifacts in the reconstructed image. In some cases, such as arterial spin labeling (ASL) imaging,

Advances in Magnetic Resonance Technology and Applications, Volume 11, ISSN 2666-9099
https://doi.org/10.1016/B978-0-323-95209-5.00017-9

where the final results are generated by subtraction, even subtle motion can degrade image quality to a large extent. On the other hand, as contrast-enhanced perfusion imaging is a dynamic image technique, the passage of the contrast bolus can only be assessed once, and thus significant motion during perfusion assessment could significantly impact clinical utility particularly for images reconstructed with advanced techniques, which rely on the temporal correlation in the data. Accordingly, it is particularly important and challenging to obtain accurate measurements during each perfusion image acquisition. Since most motion cannot be avoided, motion compensation strategies become a necessary key to guarantee good perfusion image quality.

9.2 Acquisition strategies

The most common strategies to characterize perfusion with MRI can be categorized into two major approaches. One is application of an exogenous contrast agent, which is usually a gadolinium (Gd)-based chelate. Within this category, dynamic susceptibility contrast (DSC) imaging focuses on the susceptibility effects of contrast agents on signal intensity, typically using a T2- or T2*-weighted imaging sequence, while dynamic contrast-enhanced (DCE) imaging is based on the T1 relaxivity effects of the contrast typically acquired using a T1-weighted acquisition strategy (Essig et al., 2013). Both techniques are performed before, during, and after administration of contrast agent, monitoring the signal changes induced by the contrast agent tracer as a function of time. Currently, the DSC technique is mostly used in the brain, whereas DCE is often performed outside the brain, in organs such as the heart, liver, kidneys, and skeletal muscle. Since both imaging techniques track the dynamic signal intensity with time, they intrinsically suffer from motion-related artifacts.

By comparison, the other main approach, arterial spin labeling (ASL), uses water as endogenous contrast agent. In this technique, magnetically labeled arterial blood water is used as a flow tracer. It is carried out by an acquisition of labeling arterial protons and a control acquisition. Typically, radiofrequency pulses are used to magnetically label the arterial protons upstream from the volume of interest. Then, after the labeled protons migrate via arterial vessels during the inversion time, which corresponds to the time required for the labeled protons to perfuse the tissue of interest, the labeled images will be acquired. On the contrary, the control acquisition is performed without labeling which leaves the volume of interest unaffected. The final output image is obtained by subtraction of the label and control acquisitions, where the static tissue signal can be suppressed and perfusion-weighted image will be created (Detre et al., 1992). However, considering the low intrinsic SNR of this technique, this label and control acquisition pair is usually performed multiple times to attain sufficient SNR. With such a long and pair-type acquisition, any motion occurring during the acquisition will greatly influence the final image quality.

9.3 **Types of motion and related artifacts**

Among the multiple factors that can affect the accuracy and precision of perfusion imaging, motion-related artifacts play an important role in determining image quality. These types of artifacts not only affect the perfusion image quality when performing visual diagnosis, but also are problematic when quantitative analysis is needed.

Typically, motion-induced image quality degradation may arise from several effects, including: blurring of sharp contrast or edges, ghosting arising from moving objects, signal loss and appearance of undesired signals due to spin dephasing, and nonideal magnetization evolution (Zaitsev *et al.*, 2015). Depending on the sampling strategy, motion-induced artifacts can have different spatial distributions. Motion in segmented Cartesian acquisition typically results in ghosting in the phase-encoding direction, whereas motion in single-shot imaging can result in ringing artifacts such as the dark rim artifact. With respiratory motion, radial trajectories tend to show blurring and streaking artifacts, while spiral trajectories typically display blurring and image swirling artifacts, as shown in Fig. 9.1.

Depending on the organ that is being imaged, motion-induced artifacts can stem from one or a combination of factors, including both voluntary and involuntary motion. For brain perfusion imaging, voluntary head motion and sudden involuntary movements, including yawning and sneezing, can degrade image quality. Particularly given that patients undergoing brain imaging are more likely to move because of neurological diseases (Kieran and Brunbergz, 1997). Those movements are usually unpredictable and nonperiodic. Particularly for ASL imaging, where images are obtained by subtraction, even subtle head movements can significantly impact the image quality. On the other hand, periodic involuntary respiratory motion plays an important role in both brain and body imaging. As respiratory motion is induced by both diaphragm and chest wall movements, lung and abdominal organs, such as liver and kidney, greatly suffer from displacements and/or distortion due to breathing. Moreover, their image qualities may also be affected by some involuntary nonperiodic motion, such as the up-down/in-out motion of the lung, and motion of the gastrointestinal tract. For cardiac perfusion imaging, several image artifacts can arise from both cardiac motion and respiratory motion. Cardiac motion usually only affects the region of the heart, whereas respiratory motion tends to affect the whole image to a larger extent. Generally, based on the scanning sequence and motion type, motion-related artifacts manifest as blurring, ghosting, and signal variation in the reconstructed images. For 2D imaging, through-plane motion can result in a change of the anatomy that is imaged. The most notable and challenging artifact for cardiac perfusion imaging is the dark rim artifact. It typically appears as a dark ring-like artifact at the border of myocardium and blood pool along the phase-encoding direction, which can mimic the perfusion defects at the subendocardial regions. This can easily lead to false positive diagnosis unless visualized by trained professionals. In addition to cardiac motion, other effects such as partial volume effects, and magnetic susceptibility or a combination of these factors, may contribute to the artifacts.

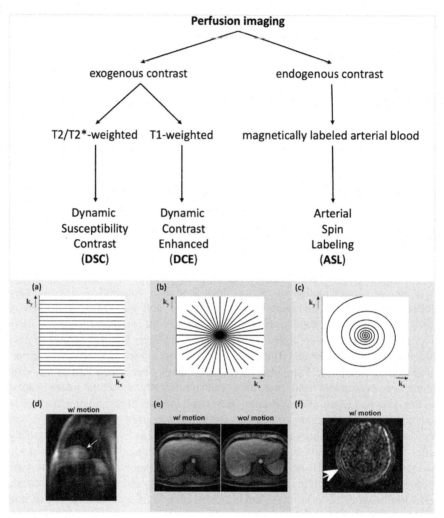

FIG. 9.1

Acquisition trajectory and motion artifacts in perfusion images. (A) Cartesian sampling trajectory, (B) radial sampling trajectory, (C) spiral sampling trajectory, (D) motion artifact with blurring demonstrated in cardiac first-pass perfusion image with Cartesian sampling, (E) motion artifact with signal reduction demonstrated in liver DCE with radial sampling, (F) motion artifact with ghosting shown in brain ASL image with spiral sampling.

9.4 Potential strategies to mitigate motion-related artifacts

9.4.1 Head motion

During brain perfusion imaging, the object to be scanned is constrained within the head coil, which does not allow for large movements. The amount of head motion during brain imaging highly depends on patient cooperation and, thus, varies individually. Noncompliant subjects, such as children, are likely to move more. Normally, this type of motion includes both in-plane rotation and translation, as well as through-plane motion. Researchers have been investigating methods to suppress head motion for a long time. For example, Pipe developed the PROPELLER (Periodically Rotated Overlapping ParallEL Lines with Enhanced Reconstruction) technique in the late 1990s (Pipe, 1999). The idea was to acquire data in concentric rectangular strips rotated about the k-space origin; thus, a central circle around the k-space center is resampled within each block and this redundant central data can be used for motion tracking. The disadvantages of this technique are a reduced sampling efficiency, as part of the sampling points are used as navigators, and only motion between blocks can be corrected. Alternatively, researchers have used single-shot non-Cartesian sampling strategy with improved efficiency to be more robust to motion (Zhao et al., 2015). Instead of suppressing head motion during acquisition, considering that head movements mainly result in rigid motion, studies have investigated using rigid registration during reconstruction. An example is shown in Fig. 9.2 (Kosior et al., 2007), which demonstrates how motion correction effectively restored the white matter, gray matter, and ischemic region definition in cerebral blood flow maps by reducing partial volume effect.

9.4.2 Respiratory motion for body imaging

As shown in Fig. 9.3, there are several strategies to minimize breathing motion. The most commonly used method is to acquire data with breath-holds. Although this method is quite intuitive and straightforward, it has distinct limitations. Breath-holding relies on good patient cooperation and requires more operator involvement. Even for cooperative patients, the breath-holding patterns can be difficult to reproduce; thus, misalignment may occur arising from the variance among multiple breath-holds, exhausting the subjects and leading to image degradation. In addition, a long breath-hold may cause unperceived diaphragmatic drift or result in a change in the heart rate (Sharrack et al., 2022). For acquisitions such as first-pass perfusion, where the acquisition may be upwards of 40 seconds, patients may not be able to hold their breath for the full duration of the scan. This is practically challenging and may generate motion-related artifacts if the breath-holds are poor. If multiple breath-holds are used, there may be discrepancies between breath-holds.

Another idea is to synchronize the perfusion imaging acquisition to the respiratory state by utilizing respiratory or spirometric triggering and gating strategies. This can be achieved using either external hardware or the MR signal itself. The MR manufactures often provide respiratory bellows or pneumotachographs for this purpose (Arnold et al., 2007). Alternatively, navigator echoes can be applied to measure breathing motion directly from MR data, where a dedicated pencil beam readout can be used to measure the MR signal at the diaphragm region in the navigator sequence. Furthermore, self-navigators have also been proposed to be integrated into the imaging sequence. Those signals can be used either as navigators to select the signal at certain motion state or as triggers to only acquire data at particular motion state. In general, the major problem of gating and triggering strategy is

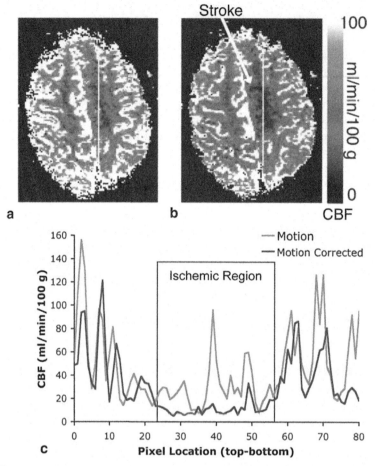

FIG. 9.2

Motion correction in patient cerebral blood flow (CBF) map. (A) Motion-corrupted data,
(B) motion-corrected data. The motion-corrected CBF map has a larger, more severe
ischemic region with more clearly defined tissue boundaries compared to the motion-
corrupted CBF map, as reflected by the (C) line profiles that show lower CBF values in
the ischemic region in the motion-corrected data.

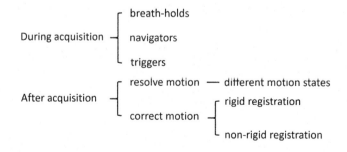

FIG. 9.3

Strategies to deal with respiratory motion.

the loss of acquisition efficiency, where a certain number of acquired data cannot be utilized or part of imaging time will be wasted.

Moreover, perfusion acquisition can be carefully designed to alleviate the effects caused by motion. A study has shown that increasing the spatial resolution can help to partially mitigate dark rim artifacts (Di Bella *et al.*, 2005); however, increasing spatial resolution may worsen the temporal footprint leading to worsening of motion-induced dark rim artifact.

Instead of intervening during the acquisition, alternative strategies to avoid motion-related artifacts are to step in at the time of image reconstruction and post-processing, by either correcting or resolving the motion. The first potential approach is to perform motion compensation on the data, which generates the image in a single motion state using all the acquired data; while resolving the motion is to accept the motion and separate the acquired data into different motion states, thus generate multiple images, respectively. This is a challenging task for perfusion imaging, considering that the dramatic contrast changes add more complexity to the signal evolution in addition to motion. As shown in Fig. 9.4, to correct the respiratory motion, people have investigated the correction of bulk motion using rigid models (Zhou *et al.*, 2018) during image reconstruction or applying more complex nonrigid deformation (Benovoy *et al.*, 2017; Scannell *et al.*, 2018; Schmidt *et al.*, 2014; Wollny *et al.*, 2012) during and after image reconstruction before quantifying perfusion. These motion registration methods have shown good ability when correcting respiratory motion, although most of them require certain computation power and processing time, especially for nonrigid approaches. On the other hand, instead of correcting motion, people have also investigated resolving the motion during perfusion imaging. Studies (Feng *et al.*, 2016) have shown good potential to separate and reconstruct the k-space data based on different respiratory motion states when applied to 3D liver dynamic contrast-enhanced (DCE) imaging, where the motion information is derived from radial self-navigator signals. This approach is difficult for perfusion imaging, where the aperiodic nature of the contrast enhancement prevents binning data into multiple respiratory states.

9.4.3 Motion around the heart

As mentioned in the previous section, respiratory motion is a big challenge in cardiac perfusion imaging. Particularly, for 2D imaging, the effect of through-plane motion and blood flow on cardiac perfusion images is nearly impossible to completely correct using traditional registration techniques, due to the complex motion of the heart and blood flow in the LV cavity. Therefore, when the subject takes deep breaths during acquisition, it poses a challenge for accurate perfusion quantification using 2D imaging techniques. Approaches such as slice tracking (Basha *et al.*, 2014), where a leading navigator is added immediately before each 2D slice acquisition to track the respiratory motion and update the slice location in real time, have shown promise for correcting through-plane motion in 2D first-pass cardiac perfusion imaging. By comparison, 3D imaging techniques may reduce the effect of

(a)

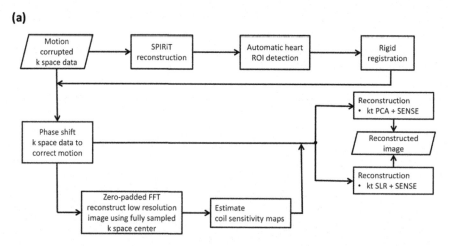

Simple motion correction strategy reduces respiratory-induced motion artifacts for k-t accelerated and compressed-sensing cardiovascular magnetic resonance perfusion imaging, Zhou et al, JCMR, 2018

(b)

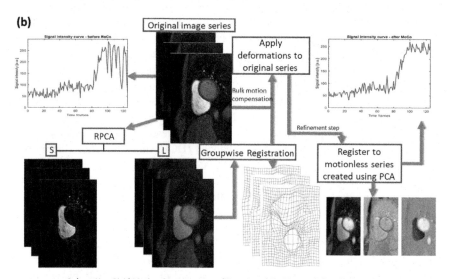

Robust Non-Rigid Motion Compensation of Free-Breathing Myocardial Perfusion MRI Data, Scannell et al, IEEE TMI, 2019

FIG. 9.4

Motion correction strategies in perfusion imaging. (A) Rigid motion correction strategy, (B) nonrigid motion correction strategy.

through-plane motion on perfusion imaging, but they bring their own set of challenges, including an increased temporal footprint.

Alleviating motion-induced artifact is particularly challenging for cardiac applications, where cardiac motion adds more complexity for this dynamic contrast-induced technique in addition to respiratory motion. In most cases, ECG signals are used to gate the cardiac motion during perfusion. However, this method has some limitations considering that the ECG signals might be distorted by the gradients used for perfusion imaging, and the quality of the ECG triggering varies from subject to subject and is highly dependent on the lead placement and heart orientation. Therefore researchers have tried to extract cardiac motion signal from the k-space data itself instead of relying on ECG signals. For example, one approach is to resolve the data along multiple mathematical dimensions. By binning the data along a cardiac motion dimension, a saturation recovery time dimension, and conventional DCE time dimension, and applying a low-rank tensor reconstruction, no ECG myocardial perfusion imaging can be performed (Christodoulou et al., 2018). Another novel approach is to use a nonimaging-based navigator such as the Pilot tone. The Pilot tone (Speier et al., 2015) is a separately generated radio-frequency signal that can be used to monitor physiological motion. Its frequency is outside of the frequency range of the image and, thus, can be separated from the MRI signals. As it is embedded in the body array coil, its contactless nature can save time and circumvent the challenges of ECG placement and respiratory bellows setup. The Pilot tone technique has been successfully applied to cine (Ludwig et al., 2021) and flow (Falcão et al., 2022) imaging, and it has great future potential for perfusion imaging.

9.5 New approaches and future directions

Taking the motion that might happen within each frame into account, improving temporal resolution to certain degree can explicitly circumvent the burden of motion compensation. Recently, Feng has proposed a 4D GRASP MRI technique at subsecond temporal resolution, which can acquire 3D image in less than one second (Feng, 2022). Thus intraframe respiratory blurring can be greatly reduced for body perfusion imaging. Beyond that, with rapid development of artificial intelligence (AI) in medical imaging, techniques using AI have been increasingly applied in the field of image reconstruction and postprocessing, including MRI image motion correction. These approaches have either tried to use convolutional neural networks to derive the spatial transform between fixed and moving images, and then apply the displacement vector fields on unseen image pairs with registration (Lv et al., 2018), or use adversarial autoencoder network to learn the identity map for breathing motion-corrupted images with preserved structure to remove the motion artifacts (Ghodrati et al., 2021; Qi et al., 2021). However, in the field of perfusion imaging, this application of deep learning techniques on motion correction is still emerging. Researchers have proposed to automatically segment the left ventricle during myocardial first-pass perfusion using a convolutional neural network (CNN) (Meyer et al., 2022). The contours can be further analyzed to extract respiratory motion information to retrospectively suppress the motion during postprocessing.

With the urgent need of fast and simple MRI, multiparametric imaging including perfusion has become one of the most popular topics in the field. This type of strategy brings both challenges and opportunity for motion compensation. Comparing to single contrast image, multiparametric imaging technique involves at least two different signal evolutions besides motion during the acquisition, which adds more complexity to the signal. On the other hand, when more than one type of sequence is involved, as motion happens consistently during the whole imaging process, motion-related information can be extracted and utilized across the multiparametric dimensions.

One important consideration is the need to be able to see images immediately on the console during perfusion imaging, particularly when imaging is performed during stress. Considering that most motion correction strategies require certain amount of computation time and power, researchers have been investigating using cloud-based technology to achieve rapid inline feedback, such as Gadgetron (Hansen and Sørensen, 2013) or FIRE (Chow *et al.*, 2021), where acquired raw data were immediately transferred to the cloud server, then reconstruction and post-processing were performed in the cloud and transferred back the images to the local computer in real time. This strategy can also contribute to popularize deep learning-based motion correction techniques, for which neural network model structures and trained parameters can be stored and employed in the cloud. Furthermore, AI techniques also have the processing speed advantage that traditional methods cannot match. Therefore there is great potential in the use of deep learning techniques to correct motion in perfusion images in the future.

9.6 Summary

This chapter gives an overview of motion compensation strategies in the perfusion imaging. Motion plays an important role during the acquisition, which can significantly affect the reconstructed image quality. As different organs are affected by different types of motion, the effects of motion on perfusion images will depend on both the technique and the organ that is being imaged. The main types of motion include bulk patient motion, and motion due to respiratory and cardiac physiology. The appearance of the motion artifacts will also vary with sampling pattern and can result in ghosting, blurring, and unwanted signal variations. Therefore the motion compensation strategies need to be carefully tailored for the particular organ to be imaged and the acquisition strategy. There is no perfect way to eliminate the effects of motion, but typical strategies either intervene during image acquisition, such as the use of slice tracking, diaphragmatic navigation, and self-gating, or during image reconstruction and postprocessing steps. Newer multiparametric techniques attempt to resolve cardiac and respiratory motion and show promise for perfusion imaging. The high computational demand for motion-corrected techniques will likely necessitate greater use of cloud computing and deep learning approaches, so that motion-compensated perfusion imaging can be immediately available on the console.

References

Arnold, J.F.T., Mörchel, P., Glaser, E., Pracht, E.D., Jakob, P.M., 2007. Lung MRI using an MR-compatible active breathing control (MR-ABC). Magn. Reson. Med. 58, 1092–1098. https://doi.org/10.1002/mrm.21424.

Basha, T.A., Roujol, S., Kissinger, K.V., Goddu, B., Berg, S., Manning, W.J., Nezafat, R., 2014. Free-breathing cardiac MR stress perfusion with real-time slice tracking. Magn. Reson. Med. 72, 689–698. https://doi.org/10.1002/mrm.24977.

Benovoy, M., Jacobs, M., Cheriet, F., Dahdah, N., Arai, A.E., Hsu, L.Y., 2017. Robust universal nonrigid motion correction framework for first-pass cardiac MR perfusion imaging. J. Magn. Reson. Imaging 46, 1060–1072. https://doi.org/10.1002/jmri.25659.

Chow, K., Kellman, P., Xue, H., 2021. Prototyping image reconstruction and analysis with FIRE. In: SCMR 24th Annual Scientific Sessions.

Christodoulou, A.G., Shaw, J.L., Nguyen, C., Yang, Q., Xie, Y., Wang, N., Li, D., 2018. Magnetic resonance multitasking for motion-resolved quantitative cardiovascular imaging. Nat. Biomed. Eng. 2, 215–226. https://doi.org/10.1038/s41551-018-0217-y.

Detre, J.A., Leigh, J.S., Williams, D.S., Koretsky, A.P., 1992. Perfusion imaging. Magn. Reson. Med. 23, 37–45. https://doi.org/10.1002/mrm.1910230106.

Di Bella, E.V.R., Parker, D.L., Sinusas, A.J., 2005. On the dark rim artifact in dynamic contrast-enhanced MRI myocardial perfusion studies. Magn. Reson. Med. 54, 1295–1299. https://doi.org/10.1002/mrm.20666.

Essig, M., Shiroishi, M.S., Nguyen, T.B., Saake, M., Provenzale, J.M., Enterline, D., Anzalone, N., Dörfler, A., Rovira, À., Wintermark, M., Law, M., 2013. Perfusion MRI: The five most frequently asked technical questions. Am. J. Roentgenol. 200, 24–34. https://doi.org/10.2214/AJR.12.9543.

Falcão, M.B.L., Di Sopra, L., Ma, L., Bacher, M., Yerly, J., Speier, P., Rutz, T., Prša, M., Markl, M., Stuber, M., Roy, C.W., 2022. Pilot tone navigation for respiratory and cardiac motion-resolved free-running 5D flow MRI. Magn. Reson. Med. 87, 718–732. https://doi.org/10.1002/mrm.29023.

Federau, C., 2017. Intravoxel incoherent motion MRI as a means to measure in vivo perfusion: a review of the evidence. NMR Biomed. 30, 1–15. https://doi.org/10.1002/nbm.3780.

Feng, L., 2022. 4D GRASP MRI at sub-second temporal resolution. NMR Biomed. https://doi.org/10.1002/nbm.4844. 0–2.

Feng, L., Axel, L., Chandarana, H., Block, K.T., Sodickson, D.K., Otazo, R., 2016. XD-GRASP: Golden-angle radial MRI with reconstruction of extra motion-state dimensions using compressed sensing. Magn. Reson. Med. 75, 775–788. https://doi.org/10.1002/mrm.25665.

Ghodrati, V., Bydder, M., Ali, F., Gao, C., Prosper, A., Nguyen, K.L., Hu, P., 2021. Retrospective respiratory motion correction in cardiac cine MRI reconstruction using adversarial autoencoder and unsupervised learning. NMR Biomed. 34, 1–14. https://doi.org/10.1002/nbm.4433.

Hansen, M.S., Sørensen, T.S., 2013. Gadgetron: An open source framework for medical image reconstruction. Magn. Reson. Med. 69, 1768–1776. https://doi.org/10.1002/mrm.24389.

Karonen, J.O., Vanninen, R.L., Liu, Y., Østergaard, L., Kuikka, J.T., Nuutinen, J., Vanninen, E.J., Partanen, P.L.K., Vainio, P.A., Korhonen, K., Perki, J., Roivainen, R., Sivenius, J., Aronen, H.J., 1999. Combined diffusion and perfusion MRI with correlation to single-photon emission CT in acute ischemic stroke ischemic penumbra predicts infarct growth. Stroke 30, 1583–1590.

Kieran, K.J., Brunbergz, J.A., 1997. Adult claustrophobia, anxiety and sedation in MRI. Magn. Reson. Imaging 15, 51–54. https://doi.org/10.1016/S0730-725X(96)00351-7.

Kosior, R.K., Kosior, J.C., Frayne, R., 2007. Improved dynamic susceptibility contrast (DSC)-MR perfusion estimates by motion correction. J. Magn. Reson. Imaging 26, 1167–1172. https://doi.org/10.1002/jmri.21128.

Ludwig, J., Speier, P., Seifert, F., Schaeffter, T., Kolbitsch, C., 2021. Pilot tone-based motion correction for prospective respiratory compensated cardiac cine MRI. Magn. Reson. Med. 85, 2403–2416. https://doi.org/10.1002/mrm.28580.

Lv, J., Yang, M., Zhang, J., Wang, X., 2018. Respiratory motion correction for free-breathing 3D abdominal MRI using CNN-based image registration: a feasibility study. Br. J. Radiol. 91, 1–9. https://doi.org/10.1259/bjr.20170788.

Manka, R., Vitanis, V., Boesiger, P., Flammer, A.J., Plein, S., Kozerke, S., 2010. Clinical feasibility of accelerated, high spatial resolution myocardial perfusion imaging. JACC Cardiovasc. Imaging 3, 710–717. https://doi.org/10.1016/j.jcmg.2010.03.009.

Meyer, C.H., Feng, X., Salerno, M., 2022. System and method for fully automatic lv segmentation of myocardial first-pass perfusion images.

Noguchi, T., Kawashima, M., Irie, H., Ootsuka, T., Nishihara, M., Matsushima, T., Kudo, S., 2011. Arterial spin-labeling MR imaging in moyamoya disease compared with SPECT imaging. Eur. J. Radiol. 80, e557–e562. https://doi.org/10.1016/j.ejrad.2011.01.016.

Pandharipande, P.V., Krinsky, G.A., Rusinek, H., Lee, V.S., 2005. Perfusion imaging of the liver: current challenges and future goals. Radiology 234, 661–673. https://doi.org/10.1148/radiol.2343031362.

Pipe, J.G., 1999. Motion correction with PROPELLER MRI: Application to head motion and free-breathing cardiac imaging. Magn. Reson. Med. 42, 963–969. https://doi.org/10.1002/(SICI)1522-2594(199911)42:5<963::AID-MRM17>3.0.CO;2-L.

Qi, H., Fuin, N., Cruz, G., Pan, J., Kuestner, T., Bustin, A., Botnar, R.M., Prieto, C., 2021. Non-rigid respiratory motion estimation of whole-heart coronary MR images using unsupervised deep learning. IEEE Trans. Med. Imaging 40, 444–454. https://doi.org/10.1109/TMI.2020.3029205.

Scannell, C., Villa, A., Lee, J., Breeuwer, M., Chiribiri, A., 2018. Robust non-rigid motion compensation of free-breathing myocardial perfusion MRI data. IEEE Trans. Med. 1–47.

Schmidt, J.F.M., Wissmann, L., Manka, R., Kozerke, S., 2014. Iterative k-t principal component analysis with nonrigid motion correction for dynamic three-dimensional cardiac perfusion imaging. Magn. Reson. Med. 72, 68–79. https://doi.org/10.1002/mrm.24894.

Sharrack, N., Chiribiri, A., Schwitter, J., Plein, S., 2022. How to do quantitative myocardial perfusion cardiovascular magnetic resonance. Eur. Heart J. Cardiovasc. Imaging 23, 315–318. https://doi.org/10.1093/ehjci/jeab193.

Speier, P., Fenchel, M., Rehner, R., 2015. PT-Nav: A novel respiratory navigation method for continuous acquisition based on modulation of a pilot tone on the MR-receiver. In: Proceedings of the 32nd Annual Scientific Meeting of ESMRMB, 129, pp. 97–98.

Wollny, G., Kellman, P., Santos, A., Ledesma, M.-J., 2012. Nonrigid motion compensation of free breathing acquired myocardial perfusion data. Med. Image Anal. 16, 84–88. https://doi.org/10.1007/978-3-642-19335-4.

Zaitsev, M., Maclaren, J., Herbst, M., 2015. Motion artefacts in MRI: a complex problem with many partial solutions. J. Magn. Reson. Imaging 42, 887–901. https://doi.org/10.1002/jmri.24850.Motion.

Zhao, L., Fielden, S.W., Feng, X., Wintermark, M., Mugler, J.P., Meyer, C.H., 2015. Rapid 3D dynamic arterial spin labeling with a sparse model-based image reconstruction. Neuroimage 121, 205–216. https://doi.org/10.1016/j.neuroimage.2015.07.018.

Zhou, R., Huang, W., Yang, Y., Chen, X., Weller, D.S., Kramer, C.M., Kozerke, S., Salerno, M., 2018. Simple motion correction strategy reduces respiratory-induced motion artifacts for k-t accelerated and compressed-sensing cardiovascular magnetic resonance perfusion imaging. J. Cardiovasc. Magn. Reson. 20, 1–13. https://doi.org/10.1186/s12968-018-0427-1.

Zonari, P., Baraldi, P., Crisi, G., 2007. Multimodal MRI in the characterization of glial neoplasms: the combined role of single-voxel MR spectroscopy, diffusion imaging and echo-planar perfusion imaging. Neuroradiology 49, 795–803. https://doi.org/10.1007/s00234-007-0253-x.

Zorach, B., Shaw, P.W., Bourque, J., Kuruvilla Jr., S., Balfour, P.C., Yang, Y., Mathew, R., Pan, J., Gonzalez, J.A., Taylor, A.M., Meyer, C.H., Epstein, F.H., Kramer, C.M., Salerno, M., 2016. Quantitative CMR perfusion imaging identifies reduced flow reserve in microvascular coronary artery disease. J. Cardiovasc. Magn. Reson. 18, 1–8. https://doi.org/10.1186/1532-429x-18-s1-p79.

Practical considerations for water exchange modeling in DCE-MRI

10

Matthias C. Schabel

Advanced Imaging Research Center, Oregon Health & Science University, Portland, OR, United States
Utah Center for Advanced Imaging Research, Department of Radiology, University of Utah Health Sciences Center, Salt Lake City, UT, United States

10.1 Introduction

Dynamic contrast-enhanced magnetic resonance imaging (DCE-MRI) is widely used for *in vivo* characterization of the physiology of normal and pathologic tissues in which the temporal dynamics of the uptake and washout of an injected contrast agent are studied. The characteristic length scale at which DCE-MRI experiments measure the response to an injected bolus of a contrast agent is almost invariably much larger than the characteristic length scale of the tissue microstructure. As a result, the measured signal is a composite of contributions from many smaller constituents with dimensions on the order of the intercapillary spacing. In order to render this extremely complex coupled physiological/biochemical system down to a manageable quantitative model, individual imaging voxels are typically viewed as comprised of a multitude of parallel, uncoupled, and well-mixed tissue units (Bassingthwaighte *et al.*, 1989). A prototypical model for these microscopic tissue units is the one-dimensional (1D) Krogh cylinder, which considers a single blood-filled capillary coupled to the tissue via a semipermeable vessel wall, and provides a mechanism for relating capillary blood flow, permeability-surface area product, and first-pass extraction via the Renkin-Crone equation (Renkin, 1959; Crone, 1963). These constituent subunits are then typically aggregated and treated as a single homogeneous compartment for the purposes of modeling.

Compartment modeling is a well-developed discipline, and there is an extensive literature surrounding the application of various compartment models to a wide range of dynamic imaging modalities (Jacquez, 1985; Muzic and Cornelius, 2001; Turco *et al.*, 2016). It is critical to recognize that the organs and tissues of the body can manifest dramatic differences in capillary and tissue microstructural architecture and physiology, so a model that is appropriate for intact gray or white matter in the brain may not be appropriate for brain tumors, and a model that accurately depicts pancreatic tissue may not be applicable to, for example, the liver or kidney.

Advances in Magnetic Resonance Technology and Applications, Volume 11, ISSN 2666-9099
https://doi.org/10.1016/B978-0-323-95209-5.00009-X

Numerous variant models specialized for different tissues and/or physiological regimes have been proposed and applied to the analysis of DCE-MRI data (Tofts, 1997; St. Lawrence and Lee, 1998a, b; Tofts et al., 1999; Koh et al., 2003; Beers et al., 2003; Brix et al., 2004; Sourbron and Buckley, 2011, 2013; Pedersen et al., 2021), many derived from similar models in nuclear imaging (Kuikka et al., 1991; Lorthois et al., 2014; Gjedde and Wong, 2022). In the following, we avoid the question of model selection and appropriateness, and presume that a model describing contrast agent biodistribution that is well-suited to the intended tissue of interest has been selected based on objective criteria.

Unlike nuclear imaging, where signals originating from radioactive decay can be treated if they were uniformly distributed throughout the imaging voxel, contrast-enhanced MRI depends both on the details of compartmental distribution of contrast agent molecules and on the distribution and movement of water molecules between compartments (Conlon and Outhred, 1972; Donahue et al., 1994). The underlying signal arises from the magnetic polarization of ^{1}H in water, lipids, and, to a lesser degree, proteins, and other biomolecules, and is determined by the static magnetic field strength and the chosen image acquisition parameters. This signal is, in turn, indirectly and nonlinearly affected by the type, concentration, and compartmental distribution of contrast agent(s), as well as the degree to which protons are able to move between tissue domains, limited by the presence of structural barriers such as microvascular endothelium and cell membranes. The latter effects arise because protons experiencing different local chemical environments are theoretically distinguishable based on their differential relaxation rates, both longitudinal and transverse (Labadie et al., 1994). Mobile protons found within the tissue are almost exclusively bound to water, with the large size of other molecules leading to vastly lower rates of passive and/or active transport, so the effects of restricted movement primarily manifest as finite water exchange lifetime. As a result, the already complex challenge of quantifying and modeling the distribution of exogenous tracer molecules is further complicated by the need to model, in parallel, the compartmentation of water within the imaging voxel. Detection of these effects is ultimately contingent on the presence of excursions in the relaxation rates between tissue compartments that are large relative to the intracompartmental water residence times, sensitizing the signal to alterations in the local chemical environment arising from the presence of relaxation-enhancing contrast agent molecules (Bains et al., 2010).

A simple way to visualize the effect of finite water exchange rates is to consider an imaging volume comprised of two completely isolated compartments, one containing pure water and the second containing water with a significant concentration of gadolinium-based contrast agent (GBCA). Because the compartments are isolated from one another, the signal clearly is the sum of two contributions with different relaxation rates, one having the native relaxation rate of pure water and the second having a much larger relaxation rate as a result of the presence of the GBCA. In this case, an accurate measurement of the relaxation curve of the entire volume will show biexponential behavior with a signal contribution from each proportional to the relative volume of that compartment. Conversely, if the compartment containing GBCA is separated from the pure water compartment by a membrane that is impermeable to the contrast agent molecules but highly permeable to water, and the characteristic diffusion length of water molecules on the time scale of the measurement is comparable to or longer than the distance between the compartments, then the entire voxel will be equally visible to all of the water molecules within it and they will experience a change in their relaxation rate corresponding to the average GBCA concentration within the entire voxel.

A somewhat more realistic depiction of a simplified four-compartment representation of tissue, comprised of erythrocytes, blood plasma, extracellular extravascular space (EES) consisting primarily of interstitial fluid, and cell bodies comprising the intracellular space (ICS), is shown in Fig. 10.1. The primary barriers to transport of contrast agent molecules are the cell membranes of the erythrocytes and other tissues, which are generally completely impermeable to GBCA, and the microvascular endothelium, which varies significantly in permeability from essentially zero in intact blood-brain barrier (BBB) to quite high in normal tissues having fenestrated

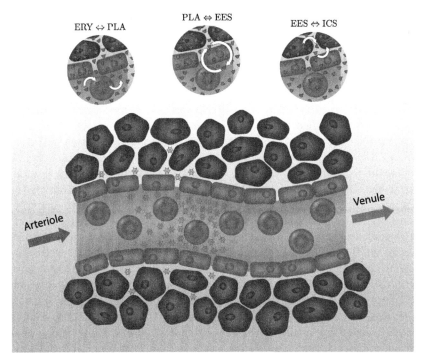

FIG. 10.1

A schematic illustration of a four-compartment catenary model of tissue including erythrocytes (ERY), blood plasma (PLA), extracellular extravascular space (EES), and intracellular space (ICS). The exchange of contrast agent molecules occurs transiently between the plasma and the EES in tissues where microvascular endothelium is permeable to them, while water exchange occurs irrespective of the presence of a contrast agent but is only detectable when the water exchange rates are slow enough and the local relaxation rates differ significantly between adjacent tissue compartments.

Figure credit: Maia Schabel.

capillaries and sinusoids (*e.g.*, kidney, pancreas, intestines, endocrine glands, spleen, liver). Elevated capillary permeability also is often observed in inflammation, malignant tumors, and other disease processes, making measurement of this quantity of interest in the diagnosis and monitoring of various cancers. Unlike GBCA, water is quite readily transported between the various intra- and extracellular spaces in the body by a range of passive and active mechanisms and, being much smaller and more mobile, tissue permeabilities to water are much higher than those for GBCA (Springer *et al.*, 2022a, b; Shao *et al.*, 2023).

Given the heterogeneous distribution of contrast agent molecules within tissue compartments and the presence of numerous barriers to water transport within and between tissues (Yazdani *et al.*, 2019), the *a priori* assumption that water compartmentation and concomitant restriction of intercompartmental water exchange are potentially significant is at least as plausible as the fast exchange paradigm for DCE-MRI (Donahue *et al.*, 1997). Furthermore, there is ample evidence from *in vitro* studies that water exchange across cell membranes (transcytolemmal exchange) is appreciably hindered (Labadie *et al.*, 1994; Quirk *et al.*, 2003; Zhang *et al.*, 2011; Springer *et al.*, 2022b). However, identification of these effects in measured data requires both adequate sensitivity and sufficient data quality to allow their discrimination from other potential confounding effects, and necessitates the use of more complex modeling methods to quantitatively characterize them (Landis *et al.*, 1999; Li *et al.*, 2005; Bains *et al.*, 2010). In the ideal case, experiments intended to interrogate water exchange would thoroughly characterize the system being studied by the acquisition of a complete set of relaxographic measurements (Labadie *et al.*, 1994). If these measurements have sufficiently high signal-to-noise and sufficiently dense sampling in the relevant relaxation rate domain, techniques like the inverse Laplace transform (ILT) may be used to estimate the distribution of underlying relaxation rates directly from the data (Bi *et al.*, 2022). Such approaches are feasible in nonbiological specimens and, to some degree, in *in vitro* studies where the compartmental distribution of contrast agent is static and extended imaging times are feasible, although the ill-conditioned nature of the ILT is a well-known limiting factor in the quantitative extraction of components of multiexponential decay curves (Craig *et al.*, 1994). However, the lengthy data acquisition times required for full characterization of tissue relaxation, particularly relative to the characteristic time scales of contrast uptake and elimination, make such approaches impractical for *in vivo* studies and necessitate the adoption of less direct methods. The added complexity incurred in modeling water exchange effects also represents a significant impediment to their broader adoption. As a result, despite the existence of a longstanding and active community of researchers who have focused on developing and refining the theory and practice of water exchange modeling, studies of tracer kinetics using DCE-MRI most often neglect the impact of water exchange. In the following, we review the underlying theory, discuss practical considerations accompanying the incorporation of water exchange in DCE-MRI data modeling, and consider some methodological and statistical considerations pertinent to these types of studies.

10.2 **Modeling water exchange in DCE-MRI**

In order to quantitatively estimate parameters characterizing tissue physiology and water exchange dynamics from DCE-MRI measurements, one must first select a forward model at the appropriate level of detail to capture all the relevant effects and idiosyncrasies of the organ system being imaged. In the presence of finite measurement noise, the choice of a sufficiently parsimonious set of free model parameters is crucial to prevent covariances from leading to ill-posed solutions of the inverse problem. In systems such as malignant tumors where relative homogeneity of tissue characteristics cannot be reasonably assumed, hierarchical modeling with a ladder of models of increasing complexity along with a statistical assessment of quality-of-fit relative to model degrees of freedom or more sophisticated model selection methods such as nested or Bayesian modeling may be necessary to avoid overfitting or underfitting (Brix *et al.*, 2009; Donaldson *et al.*, 2010; Ewing and Bagher-Ebadian, 2013; Kallehauge *et al.*, 2014; Duan *et al.*, 2016; Berks *et al.*, 2021).

The following discussion assumes that a suitable pharmacokinetic model with n compartments has been chosen for the tissue of interest, and that the time-dependent solutions for compartmental concentrations of contrast agent ($C_i(t)$) can be computed from the injected contrast bolus ($C_{\mathrm{AIF}}(t)$, generally referred to as the arterial input function or AIF) and the vector of model parameters characterizing the tissue itself, $\mathbf{p}$:

$$
\begin{aligned}
C_1(t) &= C_1(C_{\mathrm{AIF}}(t), \mathbf{p}, t) \\
C_2(t) &= C_2(C_{\mathrm{AIF}}(t), \mathbf{p}, t) \\
&\ \ \vdots \\
C_n(t) &= C_n(C_{\mathrm{AIF}}(t), \mathbf{p}, t).
\end{aligned}
\tag{10.1}
$$

Because the impact of water exchange depends sensitively on the nonlinearity of the relationship between compartment contrast concentrations and signal, in addition to the requirement that the chosen model be appropriate for the tissue and contrast agent, it is critical that the scaling of the AIF be accurate so that the modeled $C_i(t)$ are quantitatively accurate.

Prior to the analysis of the more complex question of finite intercompartmental water exchange rate, we consider the conceptually simpler limiting cases represented by the no exchange limit (NXL) and the fast exchange limit (FXL). In the NXL, the contrast agent and concentration-dependent longitudinal relaxation rate, R_1, of each compartment is computed separately:

$$
\begin{aligned}
R_{1,1}(t) &= R_{10,1} + \Delta R_1(C_1(t), \mathbf{q}) \\
R_{1,2}(t) &= R_{10,2} + \Delta R_1(C_2(t), \mathbf{q}) \\
&\ \ \vdots \\
R_{1,n}(t) &= R_{10,n} + \Delta R_1(C_n(t), \mathbf{q}),
\end{aligned}
\tag{10.2}
$$

where $R_{10,i}$ represents the intrinsic relaxation rate of the tissue in that compartment in the absence of an exogenous contrast agent, and ΔR_1 represents the dependence of the compartmental relaxation rate on contrast agent concentration and other parameters (*e.g.*, relaxivity, blood oxygen saturation) given by $\mathbf{q}$. Since the water in each compartment is entirely restricted to that space, the no exchange limit results in multiexponential signal behavior with each compartment relaxing independently. In the FXL, in contrast, water moves between compartments so quickly that protons occupy a single homogenized compartment with an effective concentration equal to the volume-weighted average of the individual compartment concentrations and a single relaxation rate shared by all compartments in fast exchange with one another:

$$\bar{C}(t) = \sum_{i=1}^{n} v_i C_i(t), \tag{10.3}$$

with the sum of compartment volumes constrained to unity:

$$\sum_i v_i = 1. \tag{10.4}$$

In this limit, the overall relaxation rate is

$$\bar{R}_1(t) = \bar{R}_{10} + \Delta R_1(\bar{C}(t), \mathbf{q}), \tag{10.5}$$

where

$$\bar{R}_{10} = \sum_{i=1}^{n} v_i R_{10,i}. \tag{10.6}$$

In the limit where concentrations and signal are linearly dependent (*i.e.*, where intercompartmental relaxation rate gradients are small), the fast exchange limit always holds. Similar expressions to those given above will also apply to the transverse relaxation rates (R_2/R_2^*) in the different water exchange regimes, although macroscopic dephasing effects can become a significant nonlinear and nonlocal factor in long T_E data acquisitions typical in dynamic susceptibility contrast (DSC)-MRI studies (Donahue *et al.*, 1997). Such considerations lie outside the scope of this review. Computation of the predicted MRI signal in the FXL and NXL limits is relatively straightforward. Assuming pulse sequence parameters $\mathbf{x}$, the NXL signal is simply the volume-weighted sum of the independent contributions from each compartment

$$S^{\text{NXL}} = M_0 \sum_{i=1}^{n} v_i S(\mathbf{x}, R_{1,i}, R_{2,i}, R_{2,i}^*), \tag{10.7}$$

while the corresponding expression in the FXL is

$$S^{\text{FXL}} = M_0 S(\mathbf{x}, \bar{R}_1, \bar{R}_2, \bar{R}_2^*). \tag{10.8}$$

We can account for the possibility of "occult" tissue that neither contains water nor takes up contrast agent (*e.g.*, avascular adipose tissue), but potentially has MRI visible protons contributing to the measured signal, by simply adding one or more compartments, uncoupled to any of the other compartments, with their own volumes and intrinsic relaxation rates.

To model the impact of finite intercompartmental water exchange, we follow the Spencer and Fishbein treatment based on the Bloch-McConnell theory (Spencer and Fishbein, 2000; Li *et al.*, 2005). Assuming that any pair of compartments can potentially exchange water, the exchange matrix can be expressed in terms of the rate constants k_{ij} governing exchange from compartment i to compartment j as

$$\mathbb{K} = \begin{pmatrix} k_{11} & k_{21} & \cdots & k_{n1} \\ k_{12} & k_{22} & \cdots & k_{n2} \\ \vdots & \vdots & \ddots & \vdots \\ k_{1n} & k_{2n} & \cdots & k_{nn} \end{pmatrix}, \tag{10.9}$$

where steady-state equilibrium requires that the volume-weighted exchange rates balance

$$\sum_{j \neq i} k_{ij} v_i = \sum_{j \neq i} k_{ji} v_j \tag{10.10}$$

and conservation requires

$$k_{ii} = -\sum_{j \neq i} k_{ij}. \tag{10.11}$$

The water lifetime within a compartment can be expressed in terms of the exchange rates as

$$\tau_i = \frac{1}{\sum_{j \neq i} k_{ij}}. \tag{10.12}$$

The longitudinal relaxation rate matrix is

$$\mathbb{R}_1 = \text{diag}(R_{1,1}, R_{1,2}, \cdots, R_{1,n}), \tag{10.13}$$

and we can define an effective relaxation rate matrix incorporating the effect of finite water exchange as

$$\mathbb{R}_1^\dagger = \mathbb{R}_1 - \mathbb{K}. \tag{10.14}$$

In this analysis, we have not considered the interplay of water exchange and transverse relaxation. Incorporating the effect of water exchange on both longitudinal and transverse relaxation simultaneously is, in general, significantly more complex. However, because characteristic transverse relaxation times are much shorter than longitudinal relaxation times in cases of interest *in vivo*, it is generally reasonable to neglect the coupling of water exchange with transverse magnetization. Furthermore, we assume that the time rates of change of the compartmental relaxation rates, $dR_{1,i}(t)/dt$, are small enough that they can be viewed as static on time scales relevant for water exchange.

For a spin echo inversion recovery measurement (with inversion time T_I, repetition time T_R, echo time T_E, and flip angle θ), the resulting signal vector, representing the separate signal contribution from each compartment, can be written (Dixon and Ekstrand, 1982; Strijkers *et al.*, 2009):

$$\mathbf{M} = M_0 \left(\mathbb{I} - (1 - \cos\theta)e^{-T_I \mathbb{R}_1^\dagger} - e^{-T_R \mathbb{R}_1^\dagger} \right) \left(\mathbb{R}_1^\dagger \right)^{-1} (\mathbb{R}_1 \mathbb{V}) e^{-T_E \mathbb{R}_2}, \quad (10.15)$$

where $\mathbb{I}$ is the diagonal identity matrix, $\mathbb{V} = \mathrm{diag}(v_1, v_2, \ldots, v_n)$, and $\mathbb{R}_2 = \mathrm{diag}(R_{2,1}, R_{2,2}, \ldots, R_{2,n})$. Similarly, for spoiled gradient recalled echo (SPGR) acquisitions (by far the most common type of pulse sequence used in DCE-MRI), the signal vector is given by (Spencer and Fishbein, 2000; Li *et al.*, 2005)

$$\mathbf{M} = M_0 \sin\theta \left(\mathbb{I} - e^{-T_R \mathbb{R}_1^\dagger} \cos\theta \right)^{-1} \left(\mathbb{I} - e^{-T_R \mathbb{R}_1^\dagger} \right) \left(\mathbb{R}_1^\dagger \right)^{-1} (\mathbb{R}_1 \mathbb{V}) e^{-T_E \mathbb{R}_2^\star}, \quad (10.16)$$

with $\mathbb{R}_2^\star = \mathrm{diag}(R_{2,1}^\star, R_{2,2}^\star, \cdots, R_{2,n}^\star)$. In this step, it is crucial to observe that the matrix exponential function, $e^{\mathbb{X}}$, is only equivalent to the element-wise exponential when $\mathbb{X}$ is diagonal. This occurs in the NXL where $\mathbb{K} \to 0$, so $\mathbb{R}_1^\dagger = \mathbb{R}_1$ and this equation reduces to the simpler expression for S^{NXL} from Eq. (10.7).

10.2.1 Four-compartment catenary model

To help make the preceding, somewhat abstract, discussion more concrete, we consider as an example the specific case of a catenary (linear) four-compartment model with separate compartments for erythrocytes (ERY), blood plasma (PLA), EES, and ICS. As discussed earlier, while the details of the compartment model are critical in realistically depicting specific tissues and/or organ systems, for the purposes of this exposition it is only necessary that the tissue compartment volumes and concentrations are well-described. The compartments, associated volumes, and concentrations are denoted as follows:

Compartment 1 (erythrocytes):	v_{ERY},	$C_{\mathrm{ERY}}(t)$,
Compartment 2 (blood plasma):	v_{PLA},	$C_{\mathrm{PLA}}(t)$,
Compartment 3 (extracellular extravascular space):	v_{EES},	$C_{\mathrm{EES}}(t)$,
Compartment 4 (intracellular space).	v_{ICS},	$C_{\mathrm{ICS}}(t)$.

Typical physiological parameters governing the distribution of contrast agent within compartments might include, in addition to compartmental volumes, variables such as plasma flow (F_p), first-pass extraction (E), delay time (t_d), capillary transit time (t_c), capillary transit time heterogeneity (α^{-1}), etc. (Schabel, 2012; Sourbron and Buckley, 2013).

A conventional GBCA with distribution restricted to the PLA and EES compartments causes increases in relaxation rate that are linearly related to its concentration and the relaxivity, r_1, which is an intrinsic property of the agent itself. In the NXL, there are four distinct longitudinal relaxation rates for the tissue compartments:

$$\begin{aligned}
R_{1,\text{ERY}}(t) &= R_{10,\text{ERY}}, \\
R_{1,\text{PLA}}(t) &= R_{10,\text{PLA}} + r_1 C_{\text{PLA}}(t), \\
R_{1,\text{EES}}(t) &= R_{10,\text{EES}} + r_1 C_{\text{EES}}(t), \\
R_{1,\text{ICS}}(t) &= R_{10,\text{ICS}},
\end{aligned} \tag{10.17}$$

and similarly for transverse relaxation:

$$\begin{aligned}
R_{2,\text{ERY}}^{\star}(t) &= R_{20,\text{ERY}}^{\star}, \\
R_{2,\text{PLA}}^{\star}(t) &= R_{20,\text{PLA}}^{\star} + r_2^{\star} C_{\text{PLA}}(t), \\
R_{2,\text{EES}}^{\star}(t) &= R_{20,\text{EES}}^{\star} + r_2^{\star} C_{\text{EES}}(t), \\
R_{2,\text{ICS}}^{\star}(t) &= R_{20,\text{ICS}}^{\star}.
\end{aligned} \tag{10.18}$$

In the FXL, the overall relaxation rate is determined from the mean contrast concentration:

$$\bar{C}(t) = v_{\text{PLA}} C_{\text{PLA}}(t) + v_{\text{EES}} C_{\text{EES}}(t) \tag{10.19}$$

and the mean intrinsic relaxation rate of the tissue compartments:

$$\bar{R}_{10} = \left(v_{\text{ERY}} R_{10,\text{ERY}} + v_{\text{PLA}} R_{10,\text{PLA}} + v_{\text{EES}} R_{10,\text{EES}} + v_{\text{ICS}} R_{10,\text{ICS}} \right), \tag{10.20}$$

$$\bar{R}_{20}^{\star} = \left(v_{\text{ERY}} R_{20,\text{ERY}}^{\star} + v_{\text{PLA}} R_{20,\text{PLA}}^{\star} + v_{\text{EES}} R_{20,\text{EES}}^{\star} + v_{\text{ICS}} R_{20,\text{ICS}}^{\star} \right), \tag{10.21}$$

with resulting time-dependent relaxation rates

$$\bar{R}_1(t) = \bar{R}_{10} + r_1 \bar{C}(t), \tag{10.22}$$

$$\bar{R}_2^{\star}(t) = \bar{R}_{20}^{\star} + r_2^{\star} \bar{C}(t). \tag{10.23}$$

We also consider a mixed exchange limit (MXL) model in which the ERY and PLA compartments are in the FXL and form a single blood (BLD) compartment, as are the EES and ICS compartments, forming a tissue (TIS) compartment, but there is no appreciable water exchange between BLD and TIS. In this case, we have

$$\begin{aligned}
R_{1,\text{BLD}}(t) &= v_{\text{ERY}} R_{10,\text{ERY}} + v_{\text{PLA}} \left(R_{10,\text{PLA}} + r_1 C_{\text{PLA}}(t) \right), \\
R_{1,\text{TIS}}(t) &= v_{\text{ICS}} R_{10,\text{ICS}} + v_{\text{EES}} \left(R_{10,\text{EES}} + r_1 C_{\text{EES}}(t) \right).
\end{aligned} \tag{10.24}$$

For strictly intravascular contrast agents such as ferumoxytol, $C_{\text{EES}}(t) = 0$ (excepting hepatic, splenic, and similar tissues where the fenestrations are sufficiently large for these nanoparticles to escape the capillary vasculature). More complex relaxation relationships exist for contrast agents like gadobenate dimeglumine (Multihance) or gadofosveset trisodium (Ablavar) that partially and reversibly bind serum albumin in a concentration-dependent way, resulting in a nonlinear relaxation rate equation, where the relaxivity itself may depend on the compartmental contrast agent concentrations (Pintaske *et al*., 2006). In these cases, it is critical to have a thorough understanding of the properties of the contrast agents employed (Wilson *et al.*, 2013; Richardson *et al.*, 2014).

Water exchange in this model is restricted to adjacent compartments, so, once the steady state and conservation constraints are incorporated, the exchange rate matrix becomes tridiagonal:

$$
\mathbb{K} =
\begin{pmatrix}
-\left(\dfrac{v_2}{v_1}\right)k_{21} & k_{21} & 0 & 0 \\[2mm]
\left(\dfrac{v_2}{v_1}\right)k_{21} & -k_{21} - \left(\dfrac{v_3}{v_2}\right)k_{32} & k_{32} & 0 \\[2mm]
0 & \left(\dfrac{v_3}{v_2}\right)k_{32} & -k_{32} - \left(\dfrac{v_4}{v_3}\right)k_{43} & k_{43} \\[2mm]
0 & 0 & \left(\dfrac{v_4}{v_3}\right)k_{43} & -k_{43}
\end{pmatrix},
\quad (10.25)
$$

with compartment volumes

$$
\begin{aligned}
v_1 &\equiv v_{\mathrm{ERY}}, \\
v_2 &\equiv v_{\mathrm{PLA}}, \\
v_3 &\equiv v_{\mathrm{EES}}, \\
v_4 &\equiv v_{\mathrm{ICS}},
\end{aligned}
\quad (10.26)
$$

and the three free exchange rate constants

$$
\begin{aligned}
k_{21} &\equiv k_{\mathrm{PLA}\to\mathrm{ERY}}, \\
k_{32} &\equiv k_{\mathrm{EES}\to\mathrm{PLA}}, \\
k_{43} &\equiv k_{\mathrm{ICS}\to\mathrm{EES}}.
\end{aligned}
\quad (10.27)
$$

10.3 Signal modeling in the presence of water exchange

Any practical mesoscale model of living tissue will be a gross simplification of the complex and dynamic underlying biological system. The ultimate objective of DCE-MRI studies is to recover, with acceptable accuracy and precision, a set of physiological model parameters characterizing the tissue of interest at an appropriate level of fidelity, with the remaining unmodeled effects being considered "nuisance" parameters that contribute to biases and/or uncertainties in model parameter estimates. What constitutes a parameter of interest versus a nuisance parameter is, to some degree, a matter of choice on the part of the experimenter, ideally guided by current art and past experience and knowledge. Water exchange can be considered to lie in either the former or the latter category depending on the study objectives and measurement methods. When the complicating effects of water exchange are not taken into consideration, it is common to convert the measured signal, via various analytical approximations or numerical methods, into an estimated contrast agent concentration and the resulting approximate concentration-time curves are directly fit with the chosen compartment model to extract parameter estimates. If the FXL assumption is valid, this procedure has the advantage of simplicity and ease of implementation, with water exchange effects potentially adding biases and uncertainties to the measured parameters. A more comprehensive and flexible approach directly models the signal, incorporating the tissue model, the signal equation, contrast relaxation, and the effect of water exchange into a combined regression model. Table 10.1 presents a complete list of the various parameters that appear in the full signal equation for the four-compartment model described above, all of which must be specified in order to compute a predicted signal-time curve. Clearly, there is significant added complexity associated with signal modeling in this way, but it is generally possible to measure or estimate many of these parameters and thus constrain

Table 10.1 Parameters characterizing the complete signal model for an SPGR acquisition with a four-compartment gamma capillary transit time pharmacokinetic model coupled to a four-compartment catenary water exchange model.

Symbol	Parameter	
$C_{AIF}(t)$	Arterial input function (plasma concentration, mmol)	
v_{ERY}	Erythrocyte volume (ml ml^{-1})	
v_{PLA}	Plasma volume (ml ml^{-1})	
v_{EES}	Extracellular extravascular volume (ml ml^{-1})	
v_{ICS}	Intracellular volume (ml ml^{-1})	
HCT	Hematocrit	$= v_{ERY}/(v_{ERY} + v_{PLA})$
F_p	Plasma flow (ml ml^{-1} min^{-1})	$= v_{PLA}/t_c$
F_b	Blood flow (ml ml^{-1} min^{-1})	$= (v_{ERY} + v_{PLA})/t_c$
K^{trans}	Volume transfer constant (ml ml^{-1} min^{-1})	$= E F_p$
		$= F_p(1 - \exp(-PS/F_p))$
k_{ep}	Elimination rate constant (min^{-1})	$= K^{trans}/v_{EES}$
f_e	Extracellular extravascular volume fraction	$= v_e/(1 - v_b)$
t_c	Capillary transit time (min)	
α^{-1}	Capillary transit time heterogeneity	
t_d	Delay time (min)	
E	First-pass capillary extraction	$= (1 - \exp(-PS/F_p))$
PS	Capillary permeability-surface area product (ml ml^{-1} min^{-1})	$= -F_p \log(1 - E)$

Continued

Table 10.1 Parameters characterizing the complete signal model for an SPGR acquisition with a four-compartment gamma capillary transit time pharmacokinetic model coupled to a four-compartment catenary water exchange model.—cont'd

Symbol	Parameter	
$R_{10,ERY}$	Erythrocyte longitudinal relaxation rate (ms^{-1})	
$R_{10,PLA}$	Plasma longitudinal relaxation rate (ms^{-1})	
$R_{10,EES}$	EES longitudinal relaxation rate (ms^{-1})	
$R_{10,ICS}$	ICS longitudinal relaxation rate (ms^{-1})	
$R_{20,ERY}^{\star}$	Erythrocyte transverse relaxation rate (ms^{-1})	
$R_{20,PLA}^{\star}$	Plasma transverse relaxation rate (ms^{-1})	
$R_{20,EES}^{\star}$	EES transverse relaxation rate (ms^{-1})	
$R_{20,ICS}^{\star}$	ICS transverse relaxation rate (ms^{-1})	
r_1	Contrast agent longitudinal relaxivity (ms^{-1} $mmol^{-1}$)	
$r_2^{\star}$	Contrast agent transverse relaxivity (ms^{-1} $mmol^{-1}$)	
M_0	Equilibrium magnetization	
B_0	Magnetic field strength (T)	
T_R	Repetition time (ms)	
T_E	Echo time (ms)	
θ	Flip angle (degrees)	
Δt	Temporal sampling interval (s)	
$k_{PLA \rightarrow ERY}$	Plasma to erythrocyte water exchange rate constant (ms^{-1})	
$k_{EES \rightarrow PLA}$	EES to plasma exchange rate constant (ms^{-1})	
$k_{ICS \rightarrow EES}$	ICS to plasma exchange rate constant (ms^{-1})	

the number of free model parameters. In any case, if they are not explicitly modeled then the corresponding uncertainties are simply being "swept under the rug." Another benefit gained from using a direct signal modeling approach is that it facilitates more complete characterization of the impact of various assumptions on parameter estimates. In this section, we systematically and individually consider the parameters that enter into a complete signal model and describe potential methods for independently estimating them when feasible.

10.3.1 Arterial input function

Quantitative measurement of the time-dependent concentration of contrast agent in the blood plasma (the arterial input function [AIF]) is crucial for the computation of model tissue uptake curves. Despite significant efforts aimed at improving the ability to characterize the AIF directly from DCE-MRI data, this objective is confounded by a wide range of experimental and analytical challenges (Parker *et al.*, 2006; Cheng, 2007;

Cron *et al.*, 2011; Gwilliam *et al.*, 2021). In order to maximize the identifiability of the blood pool versus tissue uptake of contrast, bolus injections are typically administered, resulting in a temporally narrow peak (FWHM <30 s) localized within the large arterial vessels, with a correspondingly high peak concentration (perhaps as high as 20 mM or more for rapid injections and high-contrast doses). From a measurement perspective, at least five, and ideally more, measured time points during the first pass are needed to capture the peak of the first-pass bolus and obtain some information about its shape, placing a stringent limit on the required temporal resolution of the acquisition. In addition, the inflow of blood from outside the imaging volume results in spatially varying mixing of blood in different magnetization states (Zhang *et al.*, 2009). Signal saturation at high-contrast concentrations is another issue further complicating AIF quantification, becoming more pronounced at lower flip angles (which are often necessitated by SAR limits, particularly in scanners operating at high field strengths, *e.g.*, ≥3T) (Schabel and Parker, 2008).

A useful approach to ameliorate some of these difficulties is to constrain the first-pass amplitude of the AIF using the Stewart-Hamilton equation

$$\int_{FP} C_{AIF}(t)dt = \frac{D}{CO} \tag{10.28}$$

describing the relationship between the integrated area under the first-pass bolus peak, the administered contrast dose (D in mmol), and the cardiac output (CO in 1 min^{-1}), allowing a calibrated scaling that is more representative of the true AIF in the presence of signal saturation, partial volume effects, etc. (Zhang *et al.*, 2009). Given that numerical integration (perhaps in conjunction with model fitting) of the area under the first pass is relatively straightforward and the administered contrast dose should be accurately known (assuming an uncomplicated bolus injection), the main uncertainty remaining is in the cardiac output. While measurement of cardiac output is unlikely to become a routine part of DCE-MRI experiments (aside from studies of myocardial perfusion in which it is trivial to estimate), it is possible to estimate this parameter based on patient size and weight, sex, age, and heart rate (Collis *et al.*, 2001; Dawson, 2014), making it a useful method of decreasing the variance in estimates of bolus amplitude.

It is not always possible to measure a high-quality AIF, and there has been significant interest in developing various algorithms for blind estimation that use measured data along with regularization methods to directly estimate the AIF from tissue curves (Fluckiger *et al.*, 2009, 2010; Schabel *et al.*, 2010a, b; Kratochvíla *et al.*, 2016; Guo *et al.*, 2017; Mazaheri *et al.*, 2022). Unfortunately, while promising results have been obtained in various studies, these methods are complex, their performance is incompletely understood, and they are typically reliant on details of the forward model chosen to perform the inversion. While measured AIFs are generally obtained from large vessels within the imaging field of view, there will be a finite amount of delay and dispersion of the bolus as it passes from the major arterials through smaller feeding arterioles to ultimately arrive at the tissue being imaged. Depending on the details of the upstream blood supply, the local AIF may take on a significantly different form than it does in the larger vessels, and may even vary from voxel to voxel within the imaged tissues (Fluckiger *et al.*, 2010; Schabel, 2012). As a result of the numerous difficulties surrounding the quantification of the AIF, it remains a significant factor contributing to uncertainty in quantitative DCE-MRI studies (Gwilliam *et al.*, 2021).

10.3.2 Compartment volumes

Both total tissue blood volume, given by $v_b = v_{ERY} + v_{PLA}$, and the relative partitioning of extracellular to intracellular compartment volumes are primary parameters of interest in DCE-MRI measurements, and separate volume estimates for the erythrocyte and plasma compartments are needed both for the modeling of water exchange and for the computation of permeability related parameters (PS, K^{trans}). There is a simple relationship between the plasma volume, erythrocyte volume, and hematocrit:

$$v_{ERY} = v_{PLA}\left(\frac{HCT}{1 - HCT}\right). \tag{10.29}$$

Hematocrit is easily measured by standard laboratory techniques, so can be, in principle, determined independently prior or subsequent to a DCE-MRI study. However, this is not routine and it is quite common to use an assumed population-average value for HCT, typically in the range of 0.42–0.45. In either case, there will be finite uncertainty in this value. Furthermore, large vessel and small vessel hematocrit values can differ significantly, with capillary hematocrit being notably lower (Gaehtgens, 1981). Within the compartment modeling paradigm, one can either assume an average value or compute the average over some assumed distribution of arterioles, capillaries, and venules. It is also possible to incorporate one or more additional nonexchanging compartments to account for the volume(s) of nonaqueous compounds (which is equivalent to specifying aqueous compartment fractions) (Strijkers *et al.*, 2009).

10.3.3 Pharmacokinetic parameters

In most common pharmacokinetic models, free parameters fall into three categories. The first category consists of the various compartment volumes, as discussed in Section 10.3.2. The second category is composed of hemodynamic parameters related to vascular architecture. These include plasma and/or blood flow (F_p and F_b, respectively), capillary transit time (t_c), capillary transit time heterogeneity (α^{-1}), and bolus delay time (t_d). Transfer constant, K^{trans}, is the most commonly reported tissue-specific parameter in clinical and preclinical DCE-MRI studies. As shown in Table 10.1, K^{trans} is a composite of the microvascular permeability-surface area product (PS), or the closely related first-pass extraction (E), and the plasma flow, F_p, with the contribution of each varying depending on the relative magnitude of the ratio of PS to F_p. As a consequence, identical K^{trans} values may be associated with dramatically different tissue physiologies and this parameter should be regarded as primarily phenomenological in quantitative applications. The very high peak concentrations of contrast agent within the blood pool during the first pass result in even relatively small tissue blood volumes having a significant impact on the shape of the contrast uptake curves during the initial arrival phase. Extended transit times are often observed, particularly in pathologic tissues (Kershaw and Buckley, 2006; Donaldson et al., 2010; Schabel, 2012; Jafari et al., 2017), and there is growing evidence of the observability of differences in the distribution of these transit times (Honig et al., 1977; Jespersen and Østergaard, 2011; Schabel, 2012; Angleys et al., 2015; Kunze et al., 2016; Larsson et al., 2016), making accurate modeling of blood within the tissue microvasculature an important consideration in many circumstances. As long as the assumptions of indicator dilution theory hold, these parameters are independent of the details of the injected contrast agent. The third category comprises parameters relating to the biodistribution of the contrast agent itself. In general, clinically used MRI contrast agents are passive tracers unlike, for example, radiolabeled tracers used in PET and SPECT, some of which bind to specific receptors or are metabolically processed. This is not a fundamental limitation of MRI, however, and a number of interesting active MRI agents for molecular imaging have been developed (Louie et al., 2000; Weinmann et al., 2003; James and Gambhir, 2012).

10.3.4 Intrinsic relaxation rates

When considering water exchange, the intrinsic relaxation rate of each compartment must be separately specified. Longitudinal relaxation rates in human blood and plasma have been thoroughly characterized over a wide range of physiological and experimental conditions (Grgac et al., 2012; Li et al., 2015). The theory is well developed and extensive, enabling the prediction of quantitative relaxation rates over a range of blood oxygen levels and magnetic field strengths, so values for $R_{10,\text{ERY}}$ and $R_{10,\text{PLA}}$ can be reasonably assumed to be known. Given the near similarity in composition between blood plasma and interstitial fluid (also cerebrospinal fluid [CSF] in the brain) in most tissues, it is reasonable to assume that $R_{10,\text{EES}} = R_{10,\text{PLA}}$, although in certain circumstances (e.g., bile, chyme, and other fluids associated with the digestive system), the presence of significant concentrations of lipids may lead to much larger relaxation rates. Similarly, the presence of blood products, such as may accumulate due to lysed erythrocytes in necrotic regions of solid tumors or various diseases of iron deposition, may result in nonnegligible alterations of the native relaxation rates of tissues and fluids.

It is generally considered best practice to make a preinjection baseline measurement of the longitudinal relaxation time when performing DCE-MRI studies, often via variable flip angle (VFA) data acquisition for reasons of time efficiency (Schabel and Morrell, 2008). Because the differences in intrinsic tissue relaxation rates are comparatively small, assuming fast relaxation in this case is unlikely to result in significant errors and one can use the measured $\bar{R}_{10}(= 1/\bar{T}_{10})$ and Eq. (10.20) to estimate the unknown relaxation rate for the ICS based on the (unknown) compartment volumes:

$$R_{10,\text{ICS}} = \frac{\bar{R}_{10} - \left(v_{\text{ERY}}R_{10,\text{ERY}} + v_{\text{PLA}}R_{10,\text{PLA}} + v_{\text{EES}}R_{10,\text{EES}}\right)}{v_{\text{ICS}}}. \tag{10.30}$$

It is important to note that this implies that the estimated intracellular tissue relaxation rate will vary with model parameter estimates. While effective transverse relaxation rates are generally less well characterized than their longitudinal counterparts, for the short echo time acquisitions typically used in DCE-MRI studies the sensitivity of signal to these values is also much smaller (Wilson *et al.*, 2016; Li and Zijl, 2020). As a result, the basic process for estimating these intrinsic relaxation rates follows the same approach described here but precision in the values chosen is also of less importance for most cases of interest.

10.3.5 Contrast agent relaxivities

The primary mechanism by which MRI contrast agents lead to differential enhancement of tissues is by increasing the intrinsic $^1\text{H}_2\text{O}$ relaxation rates (both longitudinal and transverse) of the compartments in which they are distributed. A wide range of different agents for clinical use in MRI have been developed, essentially all of which (with the notable exception of the iron nanoparticle-based agents such as ferumoxytol [Feraheme]) are various gadolinium chelates that biodistribute in the blood plasma and interstitial fluid but do not significantly permeate either erythrocyte or cell membranes. Gadolinium chelates fall into two broad categories, those that behave like pure tracer molecules and those that transiently bind to various proteins such as albumin. Agents in the first category, which is the most widely used, are well described by a linear relaxivity relationship for longitudinal relaxation rate over a broad range of contrast concentrations (Wilson *et al.*, 2013):

$$R_1 = R_{10} + r_1 C, \tag{10.31}$$

and similar expressions are generally assumed for the effective transverse relaxation rate (Wilson *et al.*, 2016):

$$R_2 = R_{20} + r_2 C, \tag{10.32}$$

$$R_2^\star = R_{20}^\star + r_2^\star C. \tag{10.33}$$

Agents in the second category, such as gadobenate dimeglumine (Multihance) and gadofosveset trisodium (Ablavar), transiently bind serum albumin, thereby increasing their relaxivity. The relaxation rate changes are, however, dependent both on the concentration of albumin and the concentration of the contrast agent itself, resulting in a nonlinear relationship between relaxation rate and concentration that may distort the shape of the measured uptake curves, particularly in situations where the temporal gradient of concentration is large (*e.g.*, during the first pass of the injected contrast bolus) (Pintaske *et al.*, 2006; Wilson *et al.*, 2013). Studies using these, or other similar, contrast agents should, at a minimum, consider the impact of the effective presence of two pseudoagents with time-varying concentrations (corresponding to the unbound and protein bound phases) on the measured AIF. Transverse relaxation is normally not a major contributor to uncertainty in DCE-MRI studies so modeling this term with high accuracy is of relatively low importance (Schabel and Parker, 2008).

10.3.6 **Measurement parameters**

As for any quantitative magnetic resonance measurement, deviations from steady state, reconstruction errors, compressed sensing (CS) reconstruction artifacts, motion, etc. can all lead to inaccuracies in the measured signal in DCE-MRI studies. While these considerations are nontrivial in some cases, for example, in highly undersampled CS approaches where the shape of curves, especially those with high blood volume, can be significantly distorted, the discussion here assumes that measured signal curves can be well represented by theoretical pulse sequence equations or that nonidealities have been incorporated in the modeling at the signal level. The equilibrium magnetization term, M_0, which represents the overall proton density scaled by receive sensitivity, is not typically of particular interest in DCE-MRI studies. A simple method for eliminating this term is to convert the dynamic signal measurements into relative enhancement by subtracting and rescaling by the average baseline signal prior to contrast injection:

$$\Xi(t) = \frac{S(t) - \bar{S}(t_{baseline})}{\bar{S}(t_{baseline})}. \tag{10.34}$$

The static magnetic field strength, B_0, is generally quite accurately known for given scanner hardware. It has a modest impact on DCE-MRI modeling through the dependence of tissue relaxation rates (Section 10.3.4) and contrast agent relaxivities (Section 10.3.5) on the magnetic field. B_0 also has an indirect effect on measured data through the larger spatial excursions of RF flip angle (θ) from the nominal specified value at high field strengths as a result of dielectric resonance effects. Aside from the general considerations that apply to deviations of actual MRI data acquisition from the idealized case, neither repetition time (T_R) nor echo time (T_E) is generally subject to significant uncertainty and neither is typically expected to contribute meaningfully to measurement errors so can be assumed to be fixed at nominal values.

In contrast, variation in the flip angle (θ), as governed by spatial variation in RF amplitude (B_1), can be quite significant (Sung *et al.*, 2013). In three-dimensional (3D) acquisitions, the finite slab selection pulse duration results in a flip angle falloff at the slab boundaries that can easily affect the peripheral 10%–20% of slices for rapid data acquisition protocols. At high field strengths (3T and above), the shortening of RF wavelength results in dielectric resonance effects that lead to additional flip angle variability throughout the imaging volume. Because the nonlinear dependence of MRI signal on relaxation rate is itself quite strongly dependent on θ, with smaller values leading to signal saturation at lower R_1s, accurate characterization (via separate B_1 measurement), and/or realistic estimation of biases and uncertainties in flip angle is particularly important in modeling situations where the blood pool and/or tissue are suspected of being closer to the NXL than the FXL.

Finally, the temporal sampling interval, Δt, which is dependent on T_R, the resolution in the phase encoding and slice encoding directions, and the details of the k-space subsampling pattern when k-space is not densely sampled, is an important determinant of the ability to adequately resolve rapid changes in image contrast. In general, the most rapidly varying temporal curves are those with significant arterial blood contribution as the first pass of the injected bolus generally forms a high and narrow peak. If the temporal resolution employed during data acquisition is inadequate to resolve the bolus, subsequent modeling will generally be unable to effectively separate the contributions of blood and EES.

10.3.7 Water exchange rate constants

Our model formulation contains three free water exchange rate parameters, governing the transport of water between adjacent compartments. It is highly improbable that all three of these parameters will be measurable in a single experiment. As a result, it is important to carefully consider means of constraining some (or all, in the case of experiments where water exchange rates are viewed as confounding parameters rather than experimental endpoints in their own right) of these values. The least controversial rate constant, and the one for which the most consistent experimental data exists, is that between erythrocytes and blood plasma, with a generally accepted erythrocyte water lifetime in the range of 10–20 ms that is well into the fast exchange limit for any reasonable *in vivo* DCE-MRI acquisition parameters (Conlon and Outhred, 1972; Gianolio *et al.*, 2016). Even here, there are pathologies that may alter this value. In sickle cell disease, for example, a water diffusional permeability decrease of 60% was reported in Fung *et al.* (1989). Nevertheless, the assumption of fast exchange between the erythrocyte and plasma compartments is likely to be a reasonable one in most DCE-MRI studies. Capillary endothelial barriers pose significant potential impediments for the movement of water between the blood plasma and the EES in many organs (Bains *et al.*, 2010), although others (such as the liver and spleen) have a sufficiently porous capillary architecture that the blood plasma and EES compartments are relatively tightly coupled. Particularly in the intact brain, where the tight junctions of the BBB block the transport of most small molecules, evidence suggests that transcapillary water exchange across the BBB is significantly hindered (Anderson *et al.*, 2011; Dickie *et al.*, 2020). Evidence also exists for restricted water exchange in myelin (Dortch *et al.*, 2012; Nilsson *et al.*, 2013; Brusini *et al.*, 2019), and across the myocardial microvascular endothelium (Judd *et al.*, 1999; Larsson *et al.*, 2001; Bane *et al.*, 2014). Of the three water exchange rate parameters, the magnitude and measurability of the intracellular to extravascular (transcytolemmal) term is the most contentious (Landis *et al.*, 1999, 2000; Buckley, 2002, 2018; Quirk *et al.*, 2003; Buckley *et al.*, 2008; Springer *et al.*, 2002, 2022b). In the following section, we use simulated data to clarify the impact of finite water exchange rates on the shape of DCE-MRI enhancement curves, particularly focusing on the transcapillary and transcytolemmal terms.

10.4 Simulation of water exchange effects in DCE-MRI

Here, we present a number of simulations using the four-compartment catenary model with water exchange that provide insight into the impact of varying water exchange rates on DCE-MRI data. We consider two specific tissue types to demonstrate these effects, rather than attempting to comprehensively survey the enormous range of tissues that might be of interest. Both tissues (normal skeletal muscle and anaplastic meningioma, a form of brain tumor with an anomalous microvasculature) are modeled using realistic parameters characterizing the tissue with the gamma capillary transit time (GCTT) model (Schabel, 2012; Hindel *et al.*, 2017). The GCTT model is an analytically soluble formulation of the broader class of distributed parameter models incorporating a distribution of transit times for the passage of blood through the capillary microvasculature (Koh *et al.*, 2003). By smoothly interpolating between the plug flow adiabatic tissue homogeneity (ATH) model with monodisperse transit times described by St. Lawrence and Lee (1998a, b), and the maximally disordered two-compartment exchange (2CX) model, which is characterized by an exponential distribution of transit times (Brix *et al.*, 2004), the GCTT model allows us to consider all the pharmacokinetic models most commonly applied to DCE-MRI analysis as limiting cases of a single more general model. A comprehensive list of all parameters used in the simulations is presented in Table 10.2. In the following, a particular trajectory in the space of water exchange rates was followed, shown schematically in Fig. 10.2. This trajectory begins in the no exchange limit (NXL) with $(k_{PLA \rightarrow ERY}, k_{EES \rightarrow PLA}, k_{ICS \rightarrow EES}) = (10^{-5}, 10^{-5}, 10^{-5})$ ms^{-1}, increases $k_{PLA \rightarrow ERY}$ from 10^{-5} to 10^{5} ms^{-1} (blue circles extending vertically from the NXL), then increases $k_{ICS \rightarrow EES}$ from 10^{-5} to 10^{5} ms^{-1} (green circles extending right to the MXL) to reach the mixed exchange limit (MXL), and finally increases $k_{EES \rightarrow PLA}$ from 10^{-5} to 10^{5} ms^{-1} (red circles extending from the MXL to the FXL) to reach the fast exchange limit (FXL).

Fig. 10.3 plots simulated curves of relative enhancement (Ξ) in skeletal muscle (upper panel) for pulse sequence parameters characteristic of a typical fast and heavily T_1-weighted SPGR acquisition ($T_R = 3$ms, $T_E = 1$ms, $\theta = 25$ degrees) along the trajectory from Fig. 10.2. Beginning from the NXL (blue circles, bottommost) and allowing the water exchange rate between plasma and erythrocytes to increase to the fast exchange limit (blue curves), we see a small but noticeable effect on the first-pass bolus peak and a more modest effect on the curve tails. Gradually

Table 10.2 Parameters used in full signal model simulations including water exchange effects for skeletal muscle and anaplastic meningioma.

	Skeletal muscle	Anaplastic meningioma
HCT	0.45	0.45
Y	0.97	0.97
$[Hb]$ (g dl^{-1})	15	15
f_{metHb}	0.01	0.01
CO (l min^{-1})	6.22	6.22
D_G (mmol)	4.5	4.5
D_F (mmol)	1.6	1.6
v_{ICS}	0.916	0.715
v_{EES}	0.064	0.157
v_{PLA}	0.011	0.070
v_{ERY}	0.009	0.058
F_b (ml ml^{-1} min^{-1})	0.260	0.314
t_c (s)	4.614	24.42
α^{-1}	1.00	0.78
E_G	0.307	0.190
E_F	0.000	0.000
$T_{10,\ ERY}$ (ms^{-1})	1932	1932
$T_{10,\ PLA}$ (ms^{-1})	2243	2243
$T_{10,\ EES}$ (ms^{-1})	2243	2243
$T_{10,\ ICS}$ (ms^{-1})	936	1327
$T^{*}_{20,\ ERY}$ (ms^{-1})	50	50
$T^{*}_{20,\ PLA}$ (ms^{-1})	200	200
$T^{*}_{20,\ EES}$ (ms^{-1})	100	100
$T^{*}_{20,\ ICS}$ (ms^{-1})	50	50
$r_{1,\ G}$ (ms^{-1} mmol^{-1})	6.3×10^{-3}	6.3×10^{-3}
$r^{*}_{2,\ G}$ (ms^{-1} mmol^{-1})	8.7×10^{-3}	8.7×10^{-3}
$r_{1,\ F}$ (ms^{-1} mmol^{-1})	23.6×10^{-3}	23.6×10^{-3}
$r^{*}_{2,\ F}$ (ms^{-1} mmol^{-1})	53.7×10^{-3}	53.7×10^{-3}
M_0	1.0	1.0
B_0 (T)	1.5	1.5
T_R (ms)	3.0	3.0
T_E (ms)	1.0	1.0
θ (degrees)	10, 25	10, 25
Δt (s)	1	1
	NXL/MXL/FXL	NXL/MXL/FXL
$k_{PLA \rightarrow ERY}$ (ms^{-1})	$10^{-5}/10^{+5}/10^{+5}$	$10^{-5}/10^{+5}/10^{+5}$
$k_{EES \rightarrow PLA}$ (ms^{-1})	$10^{-5}/10^{-5}/10^{+5}$	$10^{-5}/10^{-5}/10^{+5}$
$k_{ICS \rightarrow EES}$ (ms^{-1})	$10^{-5}/10^{+5}/10^{+5}$	$10^{-5}/10^{+5}/10^{+5}$

Longitudinal relaxation rates for blood plasma and erythrocytes were computed using the methods described by Li et al. (2015). Where relevant, parameters specific to the gadolinium-based contrast agent are indicated by a subscript of "G" and those specific to the intravascular ferumoxytol contrast agent are indicated by a subscript of "F."

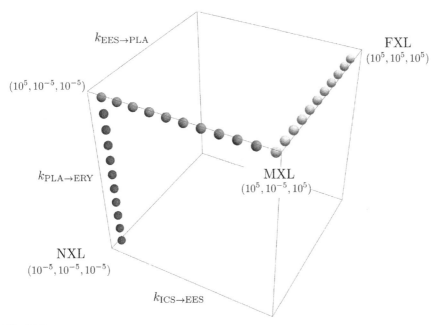

FIG. 10.2

Trajectory of simulated water exchange rates from the no exchange limit (NXL) where water exchange between compartments is negligible, through the mixed exchange limit (MXL) where water movement is only restricted across the microvascular endothelium, to the fast exchange limit (FXL) in which water is able to freely exchange between all compartments.

increasing the exchange rate between the ICS and the EES (thus allowing the trans-cytolemmal water exchange to traverse the range from no exchange to fast exchange, green curves) reveals, as one would expect, minimal impact on the bolus phase, but a larger shift during the tissue uptake phase as we approach the MXL (green circles, second from top). Finally, allowing exchange across the capillary endothelium to progress from no exchange to fast exchange (red curves) results in a significant increase in the conspicuity of the bolus peak as we reach the FXL (red circles, top-most). We also plot a curve in the intermediate exchange regime (IXR), with values of water exchange rates in the transitional range of $(k_{PLA \to ERY}, k_{EES \to PLA}, k_{ICS \to EES})$ $= (10^{-1}, 10^{-3}, 10^{-2})$ ms^{-1} in magenta circles (second from bottom). This type of acquisition is considered to be relatively insensitive to the effects of water exchange, although the difference in curve shape between the NXL and the FXL is still conspicuous, particularly during the first pass of the bolus. The two inset plots show the dependence of Ξ on the three water exchange rate parameters at $t = 1.25$ min (15 s postinjection, left inset) and $t = 6$ min (5 min postinjection, right inset), revealing a sigmoid behavior with the transition between no exchange and fast exchange

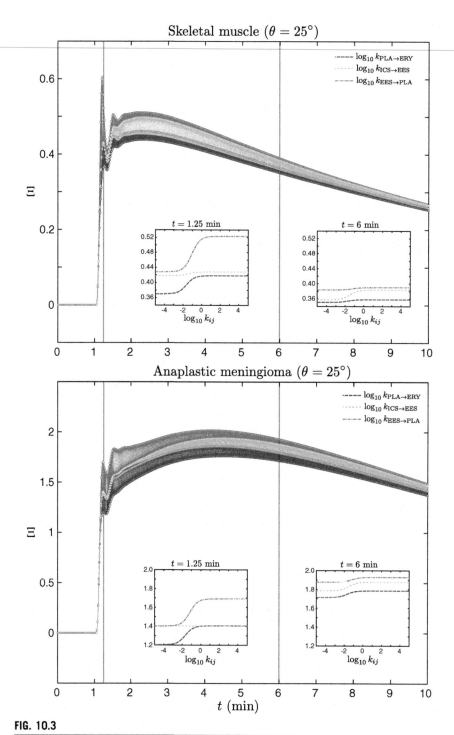

FIG. 10.3

See figure legend on opposite page.

for each compartment occurring for exchange rates between approximately 10^{-3} and 10^1 ms^{-1}. At 15 s postinjection, Ξ_{FXL} is 52% greater than Ξ_{NXL} and 29% greater than Ξ_{MXL}, decreasing to 11% and 1.6%, respectively, at 5 min postinjection. The lower panel plots corresponding simulations for a brain tumor (anaplastic meningioma), revealing similar behavior: an increase in Ξ of 42% between the NXL and FXL limits and 21% between the MXL and MXL limits at 15 s postinjection and 12% increase between NXL and FXL and 2.8% between MXL and FXL at 5 min.

It is well-established that heavily T_1-weighted pulse sequences like the one used for the simulations shown in Fig. 10.3 tend to minimize sensitivity to water exchange by acquiring data in a regime where the relaxation rate dependence on contrast agent concentration is only weakly nonlinear and has a comparatively high saturation concentration (Schabel and Parker, 2008). This is obviously desirable when water exchange is viewed as a potential confounder rather than a primary endpoint. However, specific absorbed radiation (SAR) also increases rapidly at high RF flip angles, particularly when high field scanners (3T and above) are employed, potentially limiting the ability to achieve the desired value. In studies where water exchange is an endpoint of interest, a common method for improving sensitivity is the acquisition of data at a decreased flip angle, amplifying the nonlinear relationship between contrast concentration and relaxation rate. Fig. 10.4 shows simulated data for model parameters identical to those used to generate Fig. 10.3, with the exception that the flip angle has been decreased from 25 degrees to 10 degrees. The resulting curves show modestly decreased overall enhancement corresponding to lower sensitivity to the injected contrast agent but also show a significantly increased water exchange effect. For skeletal muscle, Ξ_{FXL} is 142% greater than Ξ_{NXL} and 82% greater than Ξ_{MXL} at 15 s postinjection and Ξ_{FXL} is 52% greater than Ξ_{NXL} and 6% greater than Ξ_{MXL} at 5 min postinjection. Similarly, for the anaplastic meningioma, Ξ_{FXL} is 180% greater than Ξ_{NXL} and 96% greater than Ξ_{MXL} at 15 s postinjection and Ξ_{FXL} is 56% greater than Ξ_{NXL} and 10% greater than Ξ_{MXL} at 5 min postinjection.

FIG. 10.3—Cont'd

Simulated relative enhancement (Ξ) curves with the injection of a gadolinium contrast agent for skeletal muscle (*upper panel*) and brain tumor (anaplastic meningioma, *lower panel*), using parameters from Table 10.2 at a flip angle of $\theta = 25$ degrees. Curves are plotted from the no exchange limit (NXL, *blue circles*, bottommost) through the mixed exchange limit (MXL, *green circles*, second from top) to the fast exchange limit (FXL, *red circles*, topmost) along the trajectories shown in Fig. 10.2, with an increment of 0.1 in $\log_{10} k_{ij}$. Curves show the effect of changing the plasma-erythrocyte exchange rates (*blue*), the intracellular-extracellular exchange rates (*green*), and the extracellular-plasma exchange rates (*red*). A simulated curve in an intermediate exchange regime (IXR, *magenta* circles, second from bottom), with $k_{PLA \to ERY} = 10^2$ ms^{-1}, $k_{EES \to PLA} = 10^{-3}$ ms^{-1}, and $k_{ICS \to EES} = 10^{-2}$ ms^{-1} is plotted in *magenta*. The two *inset* plots show the dependence of Ξ on $\log_{10} k_{ij}$ for two time points (indicated by the *gray vertical lines*): a time point 15 s postinjection (*left inset*) and a time point 5 min postinjection (*right inset*).

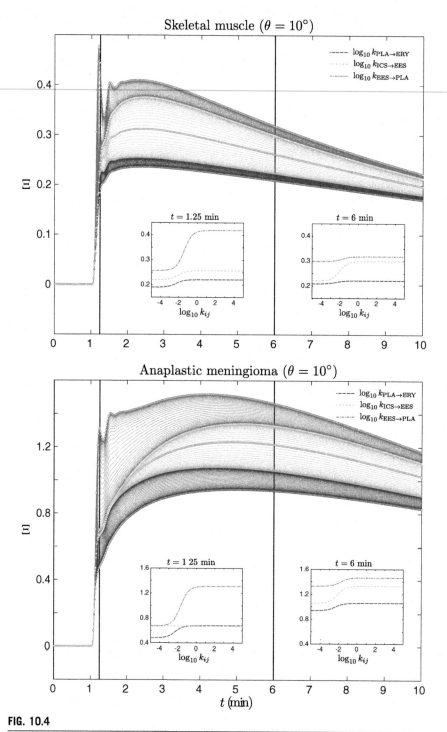

FIG. 10.4

Simulated relative enhancement (Ξ) curves with the injection of a gadolinium contrast agent for skeletal muscle (*upper panel*) and brain tumor (*lower panel*) at a flip angle of $\theta = 10$ degrees. All other parameters were held at the values used to generate the simulated curves in Fig. 10.3.

10.5 **Model parameter estimation**

The ultimate objective of tracer kinetic studies and pharmacokinetic modeling is the extraction of the underlying hemodynamic and microstructural parameters characterizing the tissues of interest with adequate accuracy and precision. Whether one is primarily concerned with quantifying parameters related to water exchange or with assessing realistic uncertainties that might be ascribed to the neglect of these effects, it is crucial to use models and methods that are appropriately tuned to the tissue(s) of interest. Parsimony of model parameters is key to avoid overfitting, particularly if water exchange rates are included. At the same time, the use of a model with fewer free parameters implies, in general, that one has accurate prior information allowing the unmodeled parameters to be fixed at known values. In this section, we investigate the sensitivity of simulated data to various parameters of interest in the presence of measurement noise.

10.5.1 **Bootstrapping initial parameter guesses**

One of the major challenges in multiparametric nonlinear regression modeling is the problem of local minima. Gradient-based optimization algorithms are quite efficient at converging to minima that are near the initial guess, but they offer no guarantees of convergence to a global optimum. In the case of relatively low-dimensional parameter spaces, particularly those where the parameter values are bounded by physical constraints, it may be feasible to perform a grid search to increase the likelihood of converging to the true optimum. However, as the dimension of the parameter space increases with increasing number of free model parameters, such approaches rapidly become unmanageable. There are a number of methods for approaching the global optimization problem in high dimensions, but these tend to be statistical in nature, add significant computational complexity, and are unable to guarantee global convergence. Although it is common to use assumed values for initial parameter guesses, data-driven approaches are generally better performing and significantly increase the likelihood of stable convergence, particularly when using complex models such as those that incorporate detailed tissue microstructure and water exchange effects where there can be significant covariance between model parameters. Using known constraints that are physically reasonable for specific applications (*e.g.*, setting contrast agent first-pass extraction, E, to zero for intact brain tissue or when using an intravascular tracer or setting the erythrocyte-plasma water exchange the fast exchange limit), on the other hand, is an effective means of increasing model parsimony when such constraints can be justified.

Linearized solutions have been developed for the extended Tofts-Kety (ETK) and two-compartment exchange (2CX) pharmacokinetic models (Murase, 2004; Flouri *et al.*, 2015), with the main motivation being improvement in computational efficiency relative to iterative nonlinear regression algorithms. A secondary benefit to these linearization approaches is that they provide a unique solution for the model parameters for a given AIF and tissue concentration curve. This fact can be used as the basis of an effective method for bootstrapping initialization of nonlinear regressions to more complex models, which proceed as follows: starting with the measured time-dependent signal, $S(t)$, compute the corresponding relative signal enhancement, $\Xi(t)$. From the nominal repetition time, echo time, and flip angle, θ (ideally corrected using a measured B_1 map), along with measured or assumed values for R_{10} and $R_{20}^\star$ and known contrast agent relaxivities (r_1 and $r_2^\star$) appropriate for the administered contrast agent and measurement field strength, use methods such as those described in Schabel and Parker (2008) (specifically Section 10.2.1) to compute an estimated average concentration, $\bar{C}(t)$, for the FXL. Fitting this curve via linear regression then provides an initial estimate for parameter values that may be used to initialize more complex models incorporating additional tissue pharmacokinetic parameters, water exchange parameters, or both by, for example, leveraging known relationships between various impulse response models (Schabel, 2012; Sourbron and Buckley, 2013).

Bootstrapping can also be profitably combined with hierarchical modeling, in which various models of differing levels of complexity are applied in the data analysis process and the quality of fit compared via direct chi-squared testing for models with the same number of parameters and, for example, information-theoretic methods such as the AIC and BIC for models of differing numbers of parameters (Brix *et al.*, 2009; Banerji *et al.*, 2011; Luypaert *et al.*, 2012; Gaa *et al.*, 2017). Some care must be taken when performing such comparisons, particularly in the presence of partial decoupling of the rapid vascular signal from the relatively slower contrast uptake and elimination in the tissue spaces. Because typical data acquisitions in DCE-MRI studies can extend 5–10 min postinjection, while the vascular phase, even in tissues with extended transit times, is normally restricted to the first minute or less after bolus arrival, the tail of the measured data is much longer and is dominated by the tissue and blood washout phase. As a result, the tail may be heavily overweighted both in the regression fitting process and in the analysis of model residuals. The model fitting can be stabilized by the introduction of a curve weighting contribution that more heavily emphasizes the vascular phase at first and is then gradually equalized over successive iterations of model regression, where the model parameter guesses are updated at each step. Similarly, in model intercomparisons, particularly when looking at models that primarily differ in their representation of the shape of the vascular component, it may be more relevant to treat the vascular and tissue phases separately by, for example, using a hard or soft threshold centered at the time point where the two terms cross over in concentration.

10.5.2 Model parameter constraints and reparameterization

There are a number of theoretically equivalent ways of expressing pharmacokinetic models in terms of different underlying parameters, as can be seen from the various interrelationships given in the third column of Table 10.2. While any formulation of a given model will be mathematically equivalent to any other, from a numerical standpoint there are some considerations that make optimization algorithms more reliable. The first consideration is to simplify the parameter bounds to ensure that resulting model values lie within a physically achievable realm. For example, it is common to formulate the extended Tofts-Kety model in terms of the transfer constant, K^{trans}, the washout rate constant, k_{ep}, and the blood volume, v_b. In this case, it is not obvious what bounds, beyond nonnegativity, can be applied to the two former parameters even though their ratio, $K^{trans}/k_{ep} = v_e$, can only take on values between 0 and 1. Reparameterizing in terms of the interstitial volume, v_e, thus simplifies the optimization problem by eliminating a complex optimization bound and replacing it with the simpler conservation of volume constraint: $v_b + v_e = 1$. It is possible to take one further step, by defining interstitial volume fraction, f_e, as the fraction of the nonvascular volume that is occupied by the extracellular extravascular compartment so that $v_e = f_e(1 - v_b)$. While reframing the model in this way seems (and, mathematically speaking, is) trivial, by doing so we have gone from a numerical optimization problem with two nonlinearly coupled nonnegativity constraints: $K^{trans} \in [0, \infty)$ and $k_{ep} \in [0, \infty)$, and one simple bound constraint: $v_b \in [0, 1]$, to one with a simple nonnegativity constraint and two simple bound constraints: $K^{trans} \in [0, \infty)$, $v_b \in [0, 1]$, and $f_e \in [0, 1]$. Similarly, while the range of water exchange rate constants that are considered potentially realistic *in vivo* spans nearly four orders of magnitude from 10^{-4} to 1 ms^{-1}, posing a scaling challenge for many nonlinear optimization algorithms, simply logarithmically transforming these rate constants results in much smoother and "better behaved" functions as can be seen in the insets of Figs. 10.3 and 10.4.

10.5.3 Identifiability of effects of finite water exchange

The simulation results plotted in Figs. 10.3 and 10.4 clearly demonstrate the impact of finite water exchange on both the shape and amplitude of relative enhancement curves, with the effects being most pronounced in the (exchange sensitized) low flip angle acquisitions. While a measurable change in signal with changes in water exchange rates is necessary for reliable determination of the underlying parameters, it is not sufficient, particularly in the presence of measurement noise. Accurate estimation of model parameters in situations where covariance is significant places additional identifiability constraints, namely that effects of changing individual model parameters are sufficiently decoupled from those arising from changes in the other parameters that the resulting modeled curves are distinguishable (Jacquez and Perry, 1990; Novikov *et al.*, 2018). In this section, we consider the identifiability problem for water exchange parameters using Monte Carlo simulations, following a general approach that has been used by other authors (Zhang and Kim, 2013; Buckley, 2018). Rather than attempting to comprehensively investigate all of the possible sources of uncertainty and bias that might be found in data measured *in vivo*, we focus on the interplay between parameters characterizing the tissue pharmacokinetics and parameters characterizing water exchange in the simplified situation where the same model used to generate the simulated signal curves is used for nonlinear regressions.

Model data were generated using the reference parameters for the anaplastic meningioma from Table 10.2 for three different water exchange rate regimes: the FXL, the MXL, and the IXR. Parameters from the brain tumor were chosen for this comparison specifically because fitting these data generally requires the use of models with a more sophisticated treatment of the vascular contribution, precluding the use of simplifying assumptions such as neglecting the tissue blood pool entirely or fixing the transit time to an assumed value. Gaussian distributed random noise with a standard deviation of 0.03 (corresponding to an SNR of 33.3) was used to generate 1000 noisy simulated curves of relative signal enhancement (Ξ) for each model at both $\theta = 25$ degrees and $\theta = 10$ degrees. Each curve was then fit using both the FXL and MXL models, with the initial parameter guesses being determined using the bootstrapping method as described earlier. The results of this modeling approach are shown for representative curves in Fig. 10.5, with the left-hand panel showing the conventional acquisition ($\theta = 25$ degrees) and the right-hand panel showing the low flip angle acquisition ($\theta = 10$ degrees). Model data points are shown in black, with the FXL model fit in plotted as a dashed red line, the MXL model fit plotted as a solid green line, and the ground truth generated with the IXR parameters shown by the dot-dashed magenta line. Fit residuals for the FXL (red, middle panel) and MXL (green, bottom panel) are plotted in the two lower panels, with the 3σ noise envelope plotted by the dashed gray upper and lower bounds. Inset tables show the model fit parameters, along with the true values. Here, the identifiability problem is already quite apparent on visual inspection: while estimated model pharmacokinetic parameters for the FXL and MXL models differ modestly for $\theta = 25$ degrees and more significantly for $\theta = 10$ degrees, the regressions result in model enhancement curves in both cases that are essentially indistinguishable from each other or from the IXR reference truth, indicating significant covariance between tissue model parameters and water exchange rate parameters.

Box plots of χ^2 and the individual model parameters (v_e, v_b, E, t_c, α^{-1}, t_d, and $K^{trans} = Ev_b/t_c/(1 - \text{HCT})$) for the two different acquisitions are plotted in Fig. 10.6, with $\theta = 25$ degrees in the left-hand panels and $\theta = 10$ degrees in the right-hand ones. Because both models used in the fitting process had an identical number of degrees of freedom (k), with the assumed value of the $k_{\text{EES}\rightarrow\text{PLA}}$ rate constant being the only difference between the two, direct comparison of the chi-squared sum for model fits is adequate to assess the identifiability of the water exchange term. In our case, we have 601 measured time points and 6 free model parameters, resulting in $k = 595$. For a theoretical chi-squared distribution with uncorrelated normal Gaussian noise, the expected mean and standard deviation are

$$E[\chi^2(k)] = k, \tag{10.35}$$

$$\text{SD}[\chi^2(k)] = \sqrt{2k}. \tag{10.36}$$

In the box plot of χ^2 for $\theta = 25$ degrees plotted in the top panel of the figure, the expectation value is indicated by the heavy gray line, the $1\sigma (= \text{E} \pm \text{SD})$ bounds

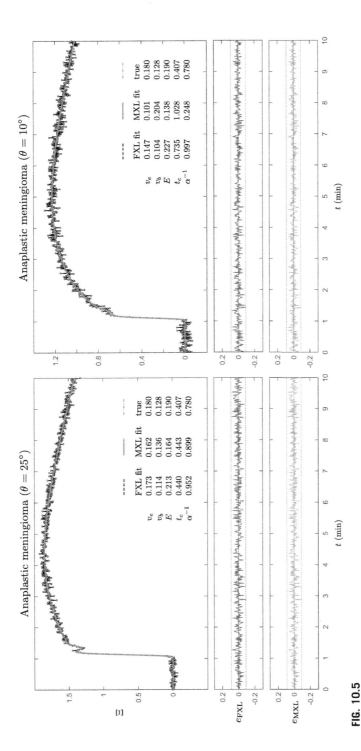

FIG. 10.5

Examples of typical nonlinear model regressions to simulated brain tumor data (*black*) for $\theta = 25$ degrees (*left panel*) and $\theta = 10$ degrees (*right panel*). Data were simulated in the IXR (*dot-dashed magenta line*, see caption of Fig. 10.3) for water exchange rate constants with Gaussian noise having $\sigma_\Xi = 0.03$, corresponding to an SNR of 33.3. FXL model (*dashed red line*) and MXL model (*solid green line*) regressions are shown in the upper panels, with FXL residuals in middle panels and MXL residuals in the bottom panels.

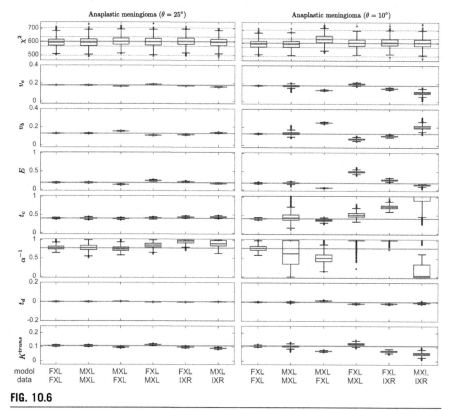

FIG. 10.6

Box plots summarizing the distributions of model regression χ^2 and tissue pharmacokinetic parameters for various model/data combinations, computed for 1000 realizations of noisy simulated data with $\sigma_\Xi = 0.03$ for $\theta = 25$ degrees (left-hand side) and $\theta = 10$ degrees (right-hand side). Median parameter values are indicated by heavy red lines within the blue boxes encompassing the 25th–75th percentile intervals. Black whiskers extend 1.5 times the interquartile range above and below 25th and 75th percentile values, with outlier data points beyond the whiskers indicated by the *blue circles*. In the χ^2 plots shown in the *upper panels*, $E[\chi^2(k)]$ is indicated by *thick gray line*, with the 1σ and 3σ intervals shown by *light gray* and *dashed light gray lines*, respectively. True tissue parameter values in the other panels are indicated by *thick gray lines*.

by the thin solid gray lines, and the $3\sigma(= E \pm 3SD)$ bounds by the thin dashed gray lines. This verifies our intuition that water exchange is essentially impossible to statistically identify in this simulated data. It is clear that the distributions of χ^2 appear to all be perfectly consistent with expectations for high-quality model regression: there are no outliers across the model/data combinations that are either anomalously low, which would signify overfitting resulting in unexpectedly low-fit residuals, or anomalously high, which would signify poor data modeling. In particular, model parameter estimates when the simulated data and regression model are matched

Table 10.3 Model estimated parameter means and uncertainties derived from nonlinear regression to 1000 simulated IXR curves for brain tumor with FXL and MXL models (see Figs. 10.5 and 10.7), with true parameter values given in the rightmost column.

	Gd θ = 25 degrees		Gd θ = 10 degrees		Fe/Gd θ = 25 degrees		
	FXL fit	**MXL fit**	**FXL fit**	**MXL fit**	**FXL fit**	**MXL fit**	**Truth**
v_e	0.173 ($\pm$0.001)	0.162 ($\pm$0.002)	0.147 ($\pm$0.004)	0.101 ($\pm$0.032)	0.169 ($\pm$0.001)	0.162 ($\pm$0.001)	0.180
v_b	0.114 ($\pm$0.002)	0.136 ($\pm$0.003)	0.104 ($\pm$0.006)	0.204 ($\pm$0.041)	0.094 ($\pm$0.000)	0.125 ($\pm$0.000)	0.128
E	0.213 ($\pm$0.007)	0.164 ($\pm$0.006)	0.277 ($\pm$0.020)	0.138 ($\pm$0.027)	0.295 ($\pm$0.005)	0.187 ($\pm$0.004)	0.190
t_c	0.440 ($\pm$0.011)	0.443 ($\pm$0.015)	0.735 ($\pm$0.053)	1.028 ($\pm$0.374)	0.570 ($\pm$0.007)	0.419 ($\pm$0.008)	0.407
α^{-1}	0.952 ($\pm$0.056)	0.899 ($\pm$0.084)	0.997 ($\pm$0.018)	0.248 ($\pm$0.362)	0.452 ($\pm$0.017)	0.730 ($\pm$0.047)	0.780

(*e.g.*, FXL data fit with an FXL model) show minimal bias and small uncertainties in both water exchange domains, with mixed data/model pairs revealing modest biases and only slightly larger uncertainties (specific numerical values for parameter biases and uncertainties for these simulations can be found in Table 10.3). Thus, on the basis of the model regressions to the simulated data, it is not possible to reliably discriminate between the FXL, the MXL, and the IXR. If we now consider the box plots for the θ = 10 degrees acquisition, we see similar behavior in χ^2, with the exception that the MXL model regressions to FXL data show an upward shift in their mean of approximately 1σ. While this would not be regarded as a strong basis for rejecting this model, it does demonstrate a somewhat increased sensitivity of the low flip angle acquisitions to water exchange. Unfortunately, this sensitivity comes at the cost of substantially elevating the biases and uncertainties in parameter estimates. Because the low flip angle acquisitions approach signal saturation at substantially lower contrast concentrations than high flip angle acquisitions, their sensitivity to the microvascular component of the signal is also diminished. This can be seen in the MXL model/MXL data distributions of transit time (t_c) and transit time heterogeneity (α^{-1}). Both of these parameters are particularly sensitive to the details of the passage of the bolus, and they show large increases in their measurement uncertainties relative to the θ = 25 degrees data. So, by attempting to increase our sensitivity to water exchange effects, we incur nontrivial additional bias and uncertainty in our estimates of the tissue pharmacokinetic parameters without, in fact, materially improving our ability to discriminate data in widely differing water exchange regimes.

10.6 Methods of enhancing sensitivity to water exchange

As described in the previous section, the identification of parameters pertaining to water exchange is quite difficult without the use of ancillary constraints such as fixing parameters. Given the high degree of heterogeneity due to physiological variation within normal tissues, the use of *a priori* parameter assumptions may stabilize model regressions and enable the estimation of water exchange parameters with a decreased variance but with potentially large and unknown biases in their values.

10.6.1 Dual flip angle imaging

Recent work by Zhang and Kim (2019) and Kiser *et al.* (2023) leverages the differential sensitivity of measured signal enhancement curves to changes in RF flip angle (θ) to increase sensitivity to transcytolemmal water exchange with a double flip angle (DFA) data acquisition protocol that acquires data for 200 s with $\theta = 8$ degrees, followed by 200 s of $\theta = 25$ degrees acquisition, followed by a final 200 s of imaging with $\theta = 8$ degrees. By acquiring experimental data with a very high signal-to-noise ratio (SNR ≈ 57) and incorporating the variation in flip angle into their regression model, they were able to demonstrate significant improvement in the ability to determine $k_{EES \rightarrow PLA}$ in their simulations and found that rate constants estimated from measured data were much more consistent than those determined using conventional single flip angle measurements.

10.6.2 Multiagent imaging

An alternative potential method for increasing the sensitivity of DCE-MRI measurements to the effects of water exchange is to sequentially inject boluses of two different types of contrast agents: one an intravascular agent such as ferumoxytol that is confined to blood plasma and the other a conventional GBCA with finite extraction from the blood plasma to the EES. By applying equality constraints on model pharmacokinetic parameters across the two boluses, it is possible to extract additional information from DCE-MRI studies, with the intravascular agent serving as a means of disambiguating the vascular contribution to the early uptake phase of the GBCA curve. Multiagent DCE-MRI methods using self-consistent constrained modeling methods have been demonstrated to improve the quantitative determination of both vascular and tissue parameters in both simulation and in an animal tumor model with three sequential injections of agents of differing molecular weight (Jacobs *et al.*, 2016). Fig. 10.7 shows a simulated multiagent data set with constrained FXL and MXL model regressions, generated in a method analogous to that described earlier and plotted in Fig. 10.5. In these simulations, a bolus of ferumoxytol (with the injected volume scaled down by a factor of 3 to compensate for the differential in relaxivity for ferumoxytol vs. GBCA) is injected prior to the GBCA bolus. Here, unlike in the single-agent simulations plotted in Fig. 10.5, we note that the constrained modeling prevents the FXL model (dashed red curve) from achieving a good

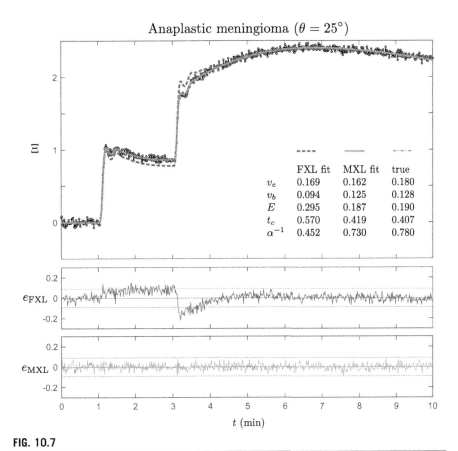

Anaplastic meningioma ($\theta = 25°$)

	FXL fit	MXL fit	true
v_e	0.169	0.162	0.180
v_b	0.094	0.125	0.128
E	0.295	0.187	0.190
t_c	0.570	0.419	0.407
α^{-1}	0.452	0.730	0.780

FIG. 10.7

Nonlinear regression for the constrained multiagent model with $\theta = 25$ degrees, plotted as in Fig. 10.5.

fit of the simulated measurements, while the MXL model (solid green curve) remains basically identical to the IXR reference (dot-dashed magenta curve).

Box plots for the multiagent simulation, analogous to those plotted in Fig. 10.6 for the single-agent simulations, are shown in Fig. 10.8. It is immediately apparent from inspection of the distributions of χ^2 in the top row that there are three combinations of model and data that can be immediately rejected based on their dramatically increased fit residuals: MXL model/FXL data, FXL model/MXL data, and FXL model/IXR data. This leaves us in the situation where we are able to distinguish between the FXL and MXL limits, and also able to distinguish between FXL and IXR, but we are unable to differentiate between the MXL and IXR. In addition, parameter biases and uncertainties for the three situations where model/data agreement is good are noticeably smaller than the corresponding values for single-agent modeling, so the multiagent approach appears to enhance sensitivity to water exchange without compromising the ability to estimate other model parameters.

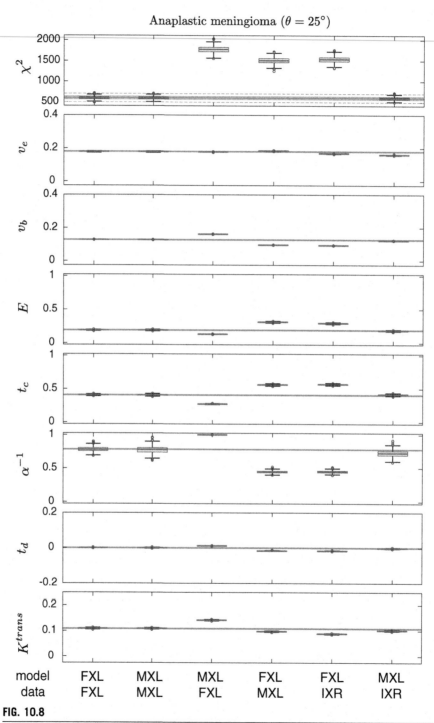

FIG. 10.8

Box plots for the constrained multiagent model regressions at $\theta = 25$ degrees, plotted as in Fig. 10.6.

10.7 Conclusions

Sensitivity to finite water exchange rates is a simultaneously intriguing and vexing property of quantitative dynamic contrast-enhanced MRI. For typical experimental conditions, with rapid T_1-weighted acquisition and injection of a conventional GBCA, even straightforward identification of the presence of restricted water exchange via comparison of pharmacokinetic models in various limiting cases is quite difficult. The silver lining here is that the insensitivity of these acquisitions to water exchange also limits the impact of neglecting it. While acquiring data at a decreased flip angle enables one to increase sensitivity to water exchange, the benefit of this appears to be offset by decreased identifiability of tissue parameters. In situations where it is important to be able to quantify water exchange rates from *in vivo* measurements it is essential to perform simulations that are representative of the type and quality of data that is expected, so that one can assess the likelihood of being able to identify these effects. In addition to acquiring comprehensive data with the highest possible SNR, experiments attempting to measure water exchange rates with DCE-MRI may benefit from more sophisticated approaches such as the dual flip angle or multiagent methods.

References

Anderson, V.C., Lenar, D.P., Quinn, J.F., Rooney, W.D., 2011. The blood-brain barrier and microvascular water exchange in Alzheimer's disease. Cardiovasc. Psychiatry Neurol. 2011, 1–9. https://doi.org/10.1155/2011/615829.

Angleys, H., Østergaard, L., Jespersen, S.N., 2015. The effects of capillary transit time heterogeneity (CTH) on brain oxygenation. J. Cereb. Blood Flow Metab. 35 (5), 806–817. https://doi.org/10.1038/jcbfm.2014.254.

Bains, L.J., McGrath, D.M., Naish, J.H., Cheung, S., Watson, Y., Taylor, M.B., Logue, J.P., Parker, G.J.M., Waterton, J.C., Buckley, D.L., 2010. Tracer kinetic analysis of dynamic contrast-enhanced MRI and CT bladder cancer data: a preliminary comparison to assess the magnitude of water exchange effects. Magn. Reson. Med. 64 (2), 595–603. https://doi.org/10.1002/mrm.22430.

Bane, O., Lee, D.C., Benefield, B.C., Harris, K.R., Chatterjee, N.R., Carr, J.C., Carroll, T.J., 2014. Leakage and water exchange characterization of gadofosveset in the myocardium. Magn. Reson. Imaging 32 (3), 224–235. https://doi.org/10.1016/j.mri.2013.10.014.

Banerji, A., Naish, J.H., Watson, Y., Jayson, G.C., Buonaccorsi, G.A., Parker, G.J.M., 2011. DCE-MRI model selection for investigating disruption of microvascular function in livers with metastatic disease. J. Magn. Reson. Imaging 35 (1), 196–203. https://doi.org/10.1002/jmri.22692.

Bassingthwaighte, J.B., Wang, C.Y., Chan, I.S., 1989. Blood-tissue exchange via transport and transformation by capillary endothelial cells. Circ. Res. 65 (4), 997–1020. https://doi.org/10.1161/01.res.65.4.997.

Beers, B.E.V., Materne, R., Annet, L., Hermoye, L., Sempoux, C., Peeters, F., Smith, A.M., Jamart, J., Horsmans, Y., 2003. Capillarization of the sinusoids in liver fibrosis: noninvasive assessment with contrast-enhanced MRI in the rabbit. Magn. Reson. Med. 49 (4), 692–699. https://doi.org/10.1002/mrm.10420.

Berks, M., Little, R.A., Watson, Y., Cheung, S., Datta, A., O'Connor, J.P.B., Scaramuzza, D., Parker, G.J.M., 2021. A model selection framework to quantify microvascular liver function in gadoxetate-enhanced MRI: application to healthy liver, diseased tissue, and hepatocellular carcinoma. Magn. Reson. Med. 86 (4), 1829–1844. https://doi.org/10.1002/mrm.28798.

Bi, C., Fishbein, K., Bouhrara, M., Spencer, R.G., 2022. Stabilization of parameter estimates from multiexponential decay through extension into higher dimensions. Sci. Rep. 12 (1). https://doi.org/10.1038/s41598-022-08638-7.

Brix, G., Kiessling, F., Lucht, R., Darai, S., Wasser, K., Delorme, S., Griebel, J., 2004. Microcirculation and microvasculature in breast tumors: pharmacokinetic analysis of dynamic MR image series. Magn. Reson. Med. 52 (2), 420–429. https://doi.org/10.1002/mrm.20161.

Brix, G., Zwick, S., Kiessling, F., Griebel, J., 2009. Pharmacokinetic analysis of tissue microcirculation using nested models: multimodel inference and parameter identifiability. Med. Phys. 36 (7), 2923–2933. https://doi.org/10.1118/1.3147145.

Brusini, L., Menegaz, G., Nilsson, M., 2019. Monte Carlo simulations of water exchange through Myelin wraps: implications for diffusion MRI. IEEE Trans. Med. Imaging 38 (6), 1438–1445. https://doi.org/10.1109/tmi.2019.2894398.

Buckley, D.L., 2002. Transcytolemmal water exchange and its affect on the determination of contrast agent concentration in vivo. Magn. Reson. Med. 47 (2), 420–421. https://doi.org/10.1002/mrm.10098.

Buckley, D.L., 2018. Shutter-speed dynamic contrast-enhanced MRI: is it fit for purpose? Magn. Reson. Med. 81 (2), 976–988. https://doi.org/10.1002/mrm.27456.

Buckley, D.L., Kershaw, L.E., Stanisz, G.J., 2008. Cellular-interstitial water exchange and its effect on the determination of contrast agent concentration in vivo: dynamic contrast-enhanced MRI of human internal obturator muscle. Magn. Reson. Med. 60 (5), 1011–1019. https://doi.org/10.1002/mrm.21748.

Cheng, H.-L.M., 2007. T_1 measurement of flowing blood and arterial input function determination for quantitative 3D T_1-weighted DCE-MRI. J. Magn. Reson. Imaging 25 (5), 1073–1078. https://doi.org/10.1002/jmri.20898.

Collis, T., Devereux, R.B., Roman, M.J., de Simone, G., Yeh, J.-L., Howard, B.V., Fabsitz, R.R., Welty, T.K., 2001. Relations of stroke volume and cardiac output to body composition. Circulation 103 (6), 820–825. https://doi.org/10.1161/01.cir.103.6.820.

Conlon, T., Outhred, R., 1972. Water diffusion permeability of erythrocytes using an NMR technique. Biochim. Biophys. Acta Biomembr. 288 (2), 354–361. https://doi.org/10.1016/0005-2736(72)90256-8.

Craig, I.J.D., Thompson, A.M., Thompson, W.J., 1994. Practical numerical algorithms why Laplace transforms are difficult to invert numerically. Comput. Phys. 8 (6), 648. https://doi.org/10.1063/1.4823347.

Cron, G.O., Foottit, C., Yankeelov, T.E., Avruch, L.I., Schweitzer, M.E., Cameron, I., 2011. Arterial input functions determined from MR signal magnitude and phase for quantitative dynamic contrast-enhanced MRI in the human pelvis. Magn. Reson. Med. 66 (2), 498–504. https://doi.org/10.1002/mrm.22856.

Crone, C., 1963. The permeability of capillaries in various organs as determined by use of the 'indicator diffusion' method. Acta Physiol. Scand. 58 (4), 292–305. https://doi.org/10.1111/j.1748-1716.1963.tb02652.x.

Dawson, T., 2014. Allometric relations and scaling laws for the cardiovascular system of mammals. Systems 2 (2), 168–185. https://doi.org/10.3390/systems2020168.

Dickie, B.R., Parker, G.J.M., Parkes, L.M., 2020. Measuring water exchange across the blood-brain barrier using MRI. Prog. Nucl. Magn. Reson. Spectrosc. 116, 19–39. https://doi.org/10.1016/j.pnmrs.2019.09.002.

Dixon, R.L., Ekstrand, K.E., 1982. The physics of proton NMR. Med. Phys. 9 (6), 807–818. https://doi.org/10.1118/1.595189.

Donahue, K.M., Burstein, D., Manning, W.J., Gray, M.L., 1994. Studies of gd-DTPA relaxivity and proton exchange rates in tissue. Magn. Reson. Med. 32 (1), 66–76. https://doi.org/10.1002/mrm.1910320110.

Donahue, K.M., Weisskoff, R.M., Burstein, D., 1997. Water diffusion and exchange as they influence contrast enhancement. J. Magn. Reson. Imaging 7 (1), 102–110. https://doi.org/10.1002/jmri.1880070114.

Donaldson, S.B., West, C.M.L., Davidson, S.E., Carrington, B.M., Hutchison, G., Jones, A.P., Sourbron, S.P., Buckley, D.L., 2010. A comparison of tracer kinetic models for T_1-weighted dynamic contrast-enhanced MRI: application in carcinoma of the cervix. Magn. Reson. Med. 63 (3), 691–700. https://doi.org/10.1002/mrm.22217.

Dortch, R.D., Harkins, K.D., Juttukonda, M.R., Gore, J.C., Does, M.D., 2012. Characterizing inter-compartmental water exchange in myelinated tissue using relaxation exchange spectroscopy. Magn. Reson. Med. 70 (5), 1450–1459. https://doi.org/10.1002/mrm.24571.

Duan, C., Kallehauge, J.F., Bretthorst, G.L., Tanderup, K., Ackerman, J.J.H., Garbow, J.R., 2016. Are complex DCE-MRI models supported by clinical data? Magn. Reson. Med. 77 (3), 1329–1339. https://doi.org/10.1002/mrm.26189.

Ewing, J.R., Bagher-Ebadian, H., 2013. Model selection in measures of vascular parameters using dynamic contrast-enhanced MRI: experimental and clinical applications. NMR Biomed. 26 (8), 1028–1041. https://doi.org/10.1002/nbm.2996.

Flouri, D., Lesnic, D., Sourbron, S.P., 2015. Fitting the two-compartment model in DCE-MRI by linear inversion. Magn. Reson. Med. 76 (3), 998–1006. https://doi.org/10.1002/mrm.25991.

Fluckiger, J.U., Schabel, M.C., DiBella, E.V.R., 2009. Model-based blind estimation of kinetic parameters in dynamic contrast enhanced (DCE)-MRI. Magn. Reson. Med. 62 (6), 1477–1486. https://doi.org/10.1002/mrm.22101.

Fluckiger, J.U., Schabel, M.C., DiBella, E.V.R., 2010. Toward local arterial input functions in dynamic contrast-enhanced MRI. J. Magn. Reson. Imaging 32 (4), 924–934. https://doi.org/10.1002/jmri.22339.

Fung, L.W.-M., Narasimhan, C., Lu, H.-Z., Westerman, M.P., 1989. Reduced water exchange in sickle cell anemia red cells: a membrane abnormality. Biochim. Biophys. Acta Biomembr. 982 (1), 167–172. https://doi.org/10.1016/0005-2736(89)90188-0.

Gaa, T., Neumann, W., Sudarski, S., Attenberger, U.I., Schönberg, S.O., Schad, L.R., Zöllner, F.G., 2017. Comparison of perfusion models for quantitative T1 weighted DCE-MRI of rectal cancer. Sci. Rep. 7 (1). https://doi.org/10.1038/s41598-017-12194-w.

Gaehtgens, P., 1981. Distribution of flow and red cell flux in the microcirculation. Scand. J. Clin. Lab. Invest. 41 (sup156), 83–87. https://doi.org/10.3109/00365518109097437.

Gianolio, E., Ferrauto, G., Gregorio, E.D., Aime, S., 2016. Re-evaluation of the water exchange lifetime value across red blood cell membrane. Biochim. Biophys. Acta Biomembr. 1858 (4), 627–631. https://doi.org/10.1016/j.bbamem.2015.12.029.

Gjedde, A., Wong, D.F., 2022. Four decades of mapping and quantifying neuroreceptors at work in vivo by positron emission tomography. Front. Neurosci. 16. https://doi.org/10.3389/fnins.2022.943512.

Grgac, K., van Zijl, P.C.M., Qin, Q., 2012. Hematocrit and oxygenation dependence of blood ^1H_2O T_1 at 7 Tesla. Magn. Reson. Med. 70 (4), 1153–1159. https://doi.org/10.1002/mrm.24547.

Guo, Y., Lingala, S.G., Bliesener, Y., Lebel, R.M., Zhu, Y., Nayak, K.S., 2017. Joint arterial input function and tracer kinetic parameter estimation from undersampled dynamic contrast-enhanced MRI using a model consistency constraint. Magn. Reson. Med. 79 (5), 2804–2815. https://doi.org/10.1002/mrm.26904.

Gwilliam, M.N., Collins, D.J., Leach, M.O., Orton, M.R., 2021. Quantifying MRI T_1 relaxation in flowing blood: implications for arterial input function measurement in DCE-MRI. Br. J. Radiol. 94 (1119), 20191004. https://doi.org/10.1259/bjr.20191004.

Hindel, S., Söhner, A., Maaß, M., Sauerwein, W., Baba, H.A., Kramer, M., Lüdemann, L., 2017. Validation of interstitial fractional volume quantification by using dynamic contrast-enhanced magnetic resonance imaging in porcine skeletal muscles. Invest. Radiol. 52 (1), 66–73. https://doi.org/10.1097/rli.0000000000000309.

Honig, C.R., Feldstein, M.L., Frierson, J.L., 1977. Capillary lengths, anastomoses, and estimated capillary transit times in skeletal muscle. Am. J. Physiol. Heart Circ. Physiol. 233 (1), H122–H129. https://doi.org/10.1152/ajpheart.1977.233.1.h122.

Jacobs, I., Strijkers, G.J., Keizer, H.M., Janssen, H.M., Nicolay, K., Schabel, M.C., 2016. A novel approach to tracer-kinetic modeling for (macromolecular) dynamic contrast-enhanced MRI. Magn. Reson. Med. 75 (3), 1142–1153. https://doi.org/10.1002/mrm.25704.

Jacquez, J.A., 1985. Compartmental Analysis in Biology and Medicine, second ed. The University of Michigan Press.

Jacquez, J.A., Perry, T., 1990. Parameter estimation: local identifiability of parameters. Am. J. Physiol. Endocrinol. Metab. 258 (4), E727–E736. https://doi.org/10.1152/ajpendo.1990.258.4.e727.

Jafari, R., Chhabra, S., Prince, M.R., Wang, Y., Spincemaille, P., 2017. Vastly accelerated linear least-squares fitting with numerical optimization for dual-input delay-compensated quantitative liver perfusion mapping. Magn. Reson. Med. 79 (4), 2415–2421. https://doi.org/10.1002/mrm.26888.

James, M.L., Gambhir, S.S., 2012. A molecular imaging primer: modalities, imaging agents, and applications. Physiol. Rev. 92 (2), 897–965. https://doi.org/10.1152/physrev.00049.2010.

Jespersen, S.N., Østergaard, L., 2011. The roles of cerebral blood flow, capillary transit time heterogeneity, and oxygen tension in brain oxygenation and metabolism. J. Cereb. Blood Flow Metab. 32 (2), 264–277. https://doi.org/10.1038/jcbfm.2011.153.

Judd, R.M., Reeder, S.B., May-Newman, K., 1999. Effects of water exchange on the measurement of myocardial perfusion using paramagnetic contrast agents. Magn. Reson. Med. 41 (2), 334–342. https://doi.org/10.1002/(sici)1522-2594(199902)41:23.0.co;2-y.

Kallehauge, J.F., Tanderup, K., Duan, C., Haack, S., Pedersen, E.M., Lindegaard, J.C., Fokdal, L.U., Mohamed, S.M.I., Nielsen, T., 2014. Tracer kinetic model selection for dynamic contrast-enhanced magnetic resonance imaging of locally advanced cervical cancer. Acta Oncol. 53 (8), 1064–1072. https://doi.org/10.3109/0284186x.2014.937879.

Kershaw, L.E., Buckley, D.L., 2006. Precision in measurements of perfusion and microvascular permeability with T_1-weighted dynamic contrast-enhanced MRI. Magn. Reson. Med. 56 (5), 986–992. https://doi.org/10.1002/mrm.21040.

Kiser, K., Zhang, J., Das, A.B., Tranos, J.A., Wadghiri, Y.Z., Kim, S.G., 2023. Evaluation of cellular water exchange in a mouse glioma model using dynamic contrast-enhanced MRI with two flip angles. Sci. Rep. 13 (1). https://doi.org/10.1038/s41598-023-29991-1.

Koh, T.S., Cheong, L.H., Hou, Z., Soh, Y.C., 2003. A physiologic model of capillary-tissue exchange for dynamic contrast-enhanced imaging of tumor microcirculation. IEEE Trans. Biomed. Eng. 50 (2), 159–167. https://doi.org/10.1109/tbme.2002.807657.

Kratochvíla, J., Jiřík, R., Bartoš, M., Standara, M., Starčuk, Z., Taxt, T., 2016. Distributed capillary adiabatic tissue homogeneity model in parametric multi-channel blind AIF estimation using DCE-MRI. Magn. Reson. Med. 75 (3), 1355–1365. https://doi.org/10.1002/mrm.25619.

Kuikka, J.T., Bassingthwaighte, J.B., Henrich, M.M., Feinendegen, L.E., 1991. Mathematical modelling in nuclear medicine. Eur. J. Nucl. Med. 18 (5), 351–362. https://doi.org/10.1007/bf02285464.

Kunze, K.P., Rischpler, C., Hayes, C., Ibrahim, T., Laugwitz, K.-L., Haase, A., Schwaiger, M., Nekolla, S.G., 2016. Measurement of extracellular volume and transit time heterogeneity using contrast-enhanced myocardial perfusion MRI in patients after acute myocardial infarction. Magn. Reson. Med. 77 (6), 2320–2330. https://doi.org/10.1002/mrm.26320.

Labadie, C., Lee, J.H., Vetek, G., Springer, C.S., 1994. Relaxographic imaging. J. Magn. Reson. B 105 (2), 99–112. https://doi.org/10.1006/jmrb.1994.1109.

Landis, C.S., Li, X., Telang, F.W., Molina, P.E., Palyka, I., Vetek, G., Springer, C.S., 1999. Equilibrium transcytolemmal water exchange kinetics in skelctal muscle in vivo. Magn. Reson. Med. 42 (3), 467–478. https://doi.org/10.1002/(sici)1522-2594(199909)42:33.0.co;2-0.

Landis, C.S., Li, X., Telang, F.W., Coderre, J.A., Micca, P.L., Rooney, W.D., Latour, L.L., Vétek, G., Pályka, I., Springer Jr., C.S., 2000. Equilibrium transcytolemmal water-exchange kinetics in skeletal muscle in vivo. Magn. Reson. Med. 44 (4), 563–574. https://doi.org/10.1002/1522-2594(200010)44:43.0.CO;2-3.

Larsson, H.B.W., Rosenbaum, S., Fritz-Hansen, T., 2001. Quantification of the effect of water exchange in dynamic contrast MRI perfusion measurements in the brain and heart. Magn. Reson. Med. 46 (2), 272–281. https://doi.org/10.1002/mrm.1188.

Larsson, H.B.W., Vestergaard, M.B., Lindberg, U., Iversen, H.K., Cramer, S.P., 2016. Brain capillary transit time heterogeneity in healthy volunteers measured by dynamic contrast-enhanced T_1-weighted perfusion MRI. J. Magn. Reson. Imaging 45 (6), 1809–1820. https://doi.org/10.1002/jmri.25488.

Li, W., van Zijl, P.C.M., 2020. Quantitative theory for the transverse relaxation time of blood water. NMR Biomed. 33 (5). https://doi.org/10.1002/nbm.4207.

Li, X., Rooney, W.D., Springer, C.S., 2005. A unified magnetic resonance imaging pharmacokinetic theory: intravascular and extracellular contrast reagents. Magn. Reson. Med. 54 (6), 1351–1359. https://doi.org/10.1002/mrm.20684.

Li, W., Grgac, K., Huang, A., Yadav, N., Qin, Q., van Zijl, P.C.M., 2015. Quantitative theory for the longitudinal relaxation time of blood water. Magn. Reson. Med. 76 (1), 270–281. https://doi.org/10.1002/mrm.25875.

Lorthois, S., Duru, P., Billanou, I., Quintard, M., Celsis, P., 2014. Kinetic modeling in the context of cerebral blood flow quantification by $H_2^{15}O$ positron emission tomography: the meaning of the permeability coefficient in Renkin-Crones model revisited at capillary scale. J. Theor. Biol. 353, 157–169. https://doi.org/10.1016/j.jtbi.2014.03.004.

Louie, A.Y., Hüber, M.M., Ahrens, E.T., Rothbächer, U., Moats, R., Jacobs, R.E., Fraser, S.E., Meade, T.J., 2000. In vivo visualization of gene expression using magnetic resonance imaging. Nat. Biotechnol. 18 (3), 321–325. https://doi.org/10.1038/73780.

Luypaert, R., Ingrisch, M., Sourbron, S., de Mey, J., 2012. The Akaike information criterion in DCE-MRI: does it improve the haemodynamic parameter estimates? Phys. Med. Biol. 57 (11), 3609–3628. https://doi.org/10.1088/0031-9155/57/11/3609.

Mazaheri, Y., Kim, N., Lakhman, Y., Jafari, R., Vargas, A., Otazo, R., 2022. Dynamic contrast-enhanced MRI parametric mapping using high spatiotemporal resolution golden-angle radial sparse parallel MRI and iterative joint estimation of the arterial input function and pharmacokinetic parameters. NMR Biomed. 35 (7). https://doi.org/10.1002/nbm.4718.

Murase, K., 2004. Efficient method for calculating kinetic parameters using T_1-weighted dynamic contrast-enhanced magnetic resonance imaging. Magn. Reson. Med. 51 (4), 858–862. https://doi.org/10.1002/mrm.20022.

Muzic, R.F., Cornelius, S., 2001. COMKAT: compartment model kinetic analysis tool. J. Nucl. Med. 42 (4), 636–645.

Nilsson, M., van Westen, D., Ståhlberg, F., Sundgren, P.C., Lätt, J., 2013. The role of tissue microstructure and water exchange in biophysical modelling of diffusion in white matter. Magn. Reson. Mater. Phys. Biol. Med. 26 (4), 345–370. https://doi.org/10.1007/s10334-013-0371-x.

Novikov, D.S., Kiselev, V.G., Jespersen, S.N., 2018. On modeling. Magn. Reson. Med. 79 (6), 3172–3193. https://doi.org/10.1002/mrm.27101.

Parker, G.J.M., Roberts, C., Macdonald, A., Buonaccorsi, G.A., Cheung, S., Buckley, D.L., Jackson, A., Watson, Y., Davies, K., Jayson, G.C., 2006. Experimentally-derived functional form for a population-averaged high-temporal-resolution arterial input function for dynamic contrast-enhanced MRI. Magn. Reson. Med. 56 (5), 993–1000. https://doi.org/10.1002/mrm.21066.

Pedersen, M., Irrera, P., Dastrù, W., Zöllner, F.G., Bennett, K.M., Beeman, S.C., Bretthorst, G.-L., Garbow, J.R., Longo, D.L., 2021. Dynamic contrast enhancement (DCE) MRI-derived renal perfusion and filtration: basic concepts. In: Methods in Molecular Biology, Springer US, pp. 205–227.

Pintaske, J., Martirosian, P., Graf, H., Erb, G., Lodemann, K.-P., Claussen, C.D., Schick, F., 2006. Relaxivity of gadopentetate dimeglumine (Magnevist), gadobutrol (Gadovist), and gadobenate dimeglumine (MultiHance) in human blood plasma at 0.2, 1.5, and 3 Tesla. Invest. Radiol. 41 (3), 213–221. https://doi.org/10.1097/01.rli.0000197668.44926.f7.

Quirk, J.D., Bretthorst, G.L., Duong, T.Q., Snyder, A.Z., Springer, C.S., Ackerman, J.J.H., Neil, J.J., 2003. Equilibrium water exchange between the intra- and extracellular spaces of mammalian brain. Magn. Reson. Med. 50 (3), 493–499. https://doi.org/10.1002/mrm.10565.

Renkin, E.M., 1959. Transport of potassium-42 from blood to tissue in isolated mammalian skeletal muscles. Am. J. Physiol. Legacy Content 197 (6), 1205–1210. https://doi.org/10.1152/ajplegacy.1959.197.6.1205.

Richardson, O.C., Bane, O., Scott, M.L.J., Tanner, S.F., Waterton, J.C., Sourbron, S.P., Carroll, T.J., Buckley, D.L., 2014. Gadofosveset-based biomarker of tissue albumin concentration: technical validation in vitro and feasibility in vivo. Magn. Reson. Med. 73 (1), 244–253. https://doi.org/10.1002/mrm.25128.

Schabel, M.C., 2012. A unified impulse response model for DCE-MRI. Magn. Reson. Med. 68 (5), 1632–1646. https://doi.org/10.1002/mrm.24162.

Schabel, M.C., Morrell, G.R., 2008. Uncertainty in T_1 mapping using the variable flip angle method with two flip angles. Phys. Med. Biol. 54 (1), N1–N8. https://doi.org/10.1088/0031-9155/54/1/n01.

Schabel, M.C., Parker, D.L., 2008. Uncertainty and bias in contrast concentration measurements using spoiled gradient echo pulse sequences. Phys. Med. Biol. 53 (9), 2345–2373. https://doi.org/10.1088/0031-9155/53/9/010.

Schabel, M.C., DiBella, E.V.R., Jensen, R.L., Salzman, K.L., 2010a. A model-constrained Monte Carlo method for blind arterial input function estimation in dynamic contrast-enhanced MRI: II. In vivo results. Phys. Med. Biol. 55 (16), 4807–4823. https://doi.org/10.1088/0031-9155/55/16/012.

Schabel, M.C., Fluckiger, J.U., DiBella, E.V.R., 2010b. A model-constrained Monte Carlo method for blind arterial input function estimation in dynamic contrast-enhanced MRI: I. Simulations. Phys. Med. Biol. 55 (16), 4783–4806. https://doi.org/10.1088/0031-9155/55/16/011.

Shao, X., Zhao, C., Shou, Q., St. Lawrence, K.S., Wang, D.J.J., 2023. Quantification of blood-brain-barrier water exchange and permeability with multidelay diffusion-weighted pseudo-continuous arterial spin labeling. Magn. Reson. Med. 89 (5), 1990–2004. https://doi.org/10.1002/mrm.29581.

Sourbron, S.P., Buckley, D.L., 2011. Tracer kinetic modelling in MRI: estimating perfusion and capillary permeability. Phys. Med. Biol. 57 (2), R1–R33. https://doi.org/10.1088/0031-9155/57/2/r1.

Sourbron, S.P., Buckley, D.L., 2013. Classic models for dynamic contrast-enhanced MRI. NMR Biomed. 26 (8), 1004–1027. https://doi.org/10.1002/nbm.2940.

Spencer, R.G.S., Fishbein, K.W., 2000. Measurement of spin-lattice relaxation times and concentrations in systems with chemical exchange using the one-pulse sequence: breakdown of the Ernst model for partial saturation in nuclear magnetic resonance spectroscopy. J. Magn. Reson. 142 (1), 120–135. https://doi.org/10.1006/jmre.1999.1925.

Springer, C.S., Rooney, W.D., Li, X., 2002. The effects of equilibrium transcytolemmal water exchange on the determination of contrast reagent concentration in vivo. Magn. Reson. Med. 47 (2), 422–424. https://doi.org/10.1002/mrm.10099.

Springer, C.S., Baker, E.M., Li, X., Moloney, B., Pike, M.M., Wilson, G.J., Anderson, V.C., Sammi, M.K., Garzotto, M.G., Kopp, R.P., Coakley, F.V., Rooney, W.D., Maki, J.H., 2022a. Metabolic activity diffusion imaging (MADI): II. Noninvasive, high-resolution human brain mapping of sodium pump flux and cell metrics. NMR Biomed. 36 (1). https://doi.org/10.1002/nbm.4782.

Springer, C.S., Baker, E.M., Li, X., Moloney, B., Wilson, G.J., Pike, M.M., Barbara, T.M., Rooney, W.D., Maki, J.H., 2022b. Metabolic activity diffusion imaging (MADI): I. Metabolic, cytometric modeling and simulations. NMR Biomed. 36 (1). https://doi.org/10.1002/nbm.4781.

St.Lawrence, K.S., Lee, T.-Y., 1998a. An adiabatic approximation to the tissue homogeneity model for water exchange in the brain: I. Theoretical derivation. J. Cereb. Blood Flow Metab. 18 (12), 1365–1377. https://doi.org/10.1097/00004647-199812000-00011.

St. Lawrence, K.S., Lee, T.-Y., 1998b. An adiabatic approximation to the tissue homogeneity model for water exchange in the brain: II. Experimental validation. J. Cereb. Blood Flow Metab. 18 (12), 1378–1385. https://doi.org/10.1097/00004647-199812000-00012.

Strijkers, G.J., Hak, S., Kok, M.B., Springer, C.S., Nicolay, K., 2009. Three-compartment T_1 relaxation model for intracellular paramagnetic contrast agents. Magn. Reson. Med. 61 (5), 1049–1058. https://doi.org/10.1002/mrm.21919.

Sung, K., Daniel, B.L., Hargreaves, B.A., 2013. Transmit B1+ field inhomogeneity and T1 estimation errors in breast DCE-MRI at 3 tesla. J. Magn. Reson. Imaging 38 (2), 454–459. https://doi.org/10.1002/jmri.23996.

Tofts, P.S., 1997. Modeling tracer kinetics in dynamic Gd-DTPA MR imaging. J. Magn. Reson. Imaging 7 (1), 91–101. https://doi.org/10.1002/jmri.1880070113.

Tofts, P.S., Brix, G., Buckley, D.L., Evelhoch, J.L., Henderson, E., Knopp, M.V., Larsson, H.B.W., Lee, T.-Y., Mayr, N.A., Parker, G.J.M., Port, R.E., Taylor, J., Weisskoff, R.M., 1999. Estimating kinetic parameters from dynamic contrast-enhanced T_1-weighted MRI of a diffusable tracer: standardized quantities and symbols. J. Magn. Reson. Imaging 10 (3), 223–232. https://doi.org/10.1002/(sici)1522-2586(199909)10:33.0.co;2-s.

Turco, S., Wijkstra, H., Mischi, M., 2016. Mathematical models of contrast transport kinetics for cancer diagnostic imaging: a review. IEEE Rev. Biomed. Eng. 9, 121–147. https://doi.org/10.1109/rbme.2016.2583541.

Weinmann, H.-J., Ebert, W., Misselwitz, B., Schmitt-Willich, H., 2003. Tissue-specific MR contrast agents. Eur. J. Radiol. 46 (1), 33–44. https://doi.org/10.1016/s0720-048x(02)00332-7.

Wilson, G.J., Woods, M., Springer, C.S., Bastawrous, S., Bhargava, P., Maki, J.H., 2013. Human whole-blood ^1H_2O longitudinal relaxation with normal and high-relaxivity contrast reagents: influence of trans-cell-membrane water exchange. Magn. Reson. Med. 72 (6), 1746–1754. https://doi.org/10.1002/mrm.25064.

Wilson, G.J., Springer, C.S., Bastawrous, S., Maki, J.H., 2016. Human whole blood ^1H_2O transverse relaxation with gadolinium-based contrast reagents: magnetic susceptibility and transmembrane water exchange. Magn. Reson. Med. 77 (5), 2015–2027. https://doi.org/10.1002/mrm.26284.

Yazdani, S., Jaldin-Fincati, J.R., Pereira, R.V.S., Klip, A., 2019. Endothelial cell barriers: transport of molecules between blood and tissues. Traffic 20 (6), 390–403. https://doi.org/10.1111/tra.12645.

Zhang, J., Kim, S., 2013. Uncertainty in MR tracer kinetic parameters and water exchange rates estimated from T_1-weighted dynamic contrast enhanced MRI. Magn. Reson. Med. 72 (2), 534–545. https://doi.org/10.1002/mrm.24927.

Zhang, J., Kim, S.G., 2019. Estimation of cellular-interstitial water exchange in dynamic contrast enhanced MRI using two flip angles. NMR Biomed. 32 (11). https://doi.org/10.1002/nbm.4135.

Zhang, J.L., Rusinek, H., Bokacheva, L., Chen, Q., Storey, P., Lee, V.S., 2009. Use of cardiac output to improve measurement of input function in quantitative dynamic contrast-enhanced MRI. J. Magn. Reson. Imaging 30 (3), 656–665. https://doi.org/10.1002/jmri.21891.

Zhang, Y., Poirier-Quinot, M., Springer, C.S., Balschi, J.A., 2011. Active trans-plasma membrane water cycling in yeast is revealed by NMR. Biophys. J. 101 (11), 2833–2842. https://doi.org/10.1016/j.bpj.2011.10.035.

Acceleration methods for perfusion imaging

11

Li Feng[a] and Nan Wang[b]

[a]*Center for Advanced Imaging Innovation and Research (CAI2R), New York University Grossman School of Medicine, New York, NY, United States*
[b]*Department of Radiology, Stanford University, Stanford, CA, United States*

Acronyms

DCE-MRI	dynamic contrast-enhanced MRI
DSC-MRI	dynamic susceptibility contrast MRI
ASL-MRI	arterial spin-labeling MRI
GBCA	Gadolinium-based contrast agents
FOV	field of view
SNR	signal-to-noise ratio
NSA	number of signal averages
NEX	number of excitations
FFT	fast Fourier transform
NUFFT	nonuniform FFT
SMS	simultaneous multi-slice
RF	radiofrequency
KWIC	k-space weighted image contrast
MRA	MR angiography
POCS	projection onto convex sets
SENSE	SENSitivity Encoding
SMASH	SiMultaneous Acquisition of Spatial Harmonics
GRAPPA	GeneRalized Autocalibrating Partial Parallel Acquisition
ACS	autocalibration signal
SPIRiT	Iterative Self-consistent Parallel Imaging Reconstruction
TSENSE	temporal SENSE
TGRAPPA	temporal GRAPPA
UNFOLD	UNaliasing by Fourier-encoding the OverLaps using the temporal Dimension
k-t BLAST	k-t space Broad-use Linear Acquisition Speed-up Technique
PCA	principal component analysis
ADMM	alternating direction method of multipliers
SVD	singular value decomposition
CNN	conventional neural network

Advances in Magnetic Resonance Technology and Applications, Volume 11, ISSN 2666-9099
https://doi.org/10.1016/B978-0-323-95209-5.00012-X

POMP	Phase-Offset MuliPlanar imaging
CAIPIRINHA	Controlled Aliasing in Parallel Imaging Results in Higher Acceleration
TV	total variation
TGV	total generalized variation
MPPCA	Marchenko-Pastur principal component analysis
GRE	gradient echo
TR	repetition time
TE	echo time
ms	milliseconds
EPI	echo planar imaging

11.1 Introduction

Magnetic Resonance Imaging (MRI) is a powerful and versatile imaging modality for visualizing and measuring perfusion in the human body without radiation exposure. As described in previous chapters, perfusion MRI can be performed using two types of imaging techniques in current clinical practice (Jahng *et al.*, 2014). The first type includes dynamic contrast-enhanced (DCE) MRI and dynamic susceptibility contrast (DSC) MRI, both of which rely on intravenous injection of exogenous gadolinium-based contrast agents (GBCA) to measure contrast-induced signal changes in blood vessels for perfusion quantification. The second type, namely arterial spin-labeling (ASL) MRI, uses magnetically labeled blood as an endogenous tracer for perfusion measurement. One of the major challenges for perfusion MRI, as in many other MRI applications, is the relatively slow imaging speed, which not only leads to long scan times but also imposes a trade-off between achievable spatiotemporal resolution and volumetric coverage (Essig *et al.*, 2013). As a result, fast imaging plays an important role in perfusion MRI for robust, accurate, and reliable clinical implementation.

The purpose of this chapter is to introduce different fast imaging methods that can be used to accelerate perfusion MRI. The chapter has seven main sections. Following the introduction in Section 11.1, Section 11.2 describes the general MRI acquisition and reconstruction process and introduces different ways for accelerating data acquisition. Sections 11.3–11.5 present an overview of state-of-the-art fast imaging methods that can be applied to accelerate perfusion MRI. These methods are classified into three broad categories: image reconstruction from undersampled measurements (Section 11.3), simultaneous multi-slice imaging (Section 11.4), and image denoising to reduce the number of signal averages (Section 11.5). With the knowledge of these fast imaging techniques, Section 11.6 then discusses how one can choose appropriate acceleration methods for different perfusion MRI techniques based on their acquisition schemes with representative examples. This is then followed by Section 11.7 for a summary and a conclusion of the chapter. By the end of this chapter, readers will have learned the basics of fast MRI acquisition and reconstruction, the general principle of different acceleration techniques that can be used in perfusion MRI, and how to choose different acceleration methods for different perfusion MRI techniques.

11.2 General MRI acquisition and reconstruction

This section describes the basic MRI signal equation and general data acquisition and image reconstruction process. Different ways of accelerating data acquisition in MRI are also summarized.

11.2.1 MRI signal equation

Mathematically, MRI acquisition can be described as follows:

$$y(\mathbf{k}) = \int x(\mathbf{r})e^{-i2\pi\mathbf{kr}}d\mathbf{r} + n \tag{11.1}$$

Here, x represents the object to be imaged; y is the acquired MR signal; $\mathbf{r}$ and $\mathbf{k} \in \mathbb{R}^2$ or $\mathbb{R}^3$ indicate 2D or 3D coordinates in image space and k-space, respectively; and n represents Gaussian noise that is acquired together with the MRI signal. With matrix notation, the MRI acquisition process can be represented using the following linear equation by concatenating both $\mathbf{x}$ and $\mathbf{y}$ into vectors:

$$\mathbf{y} = \mathbf{F}\mathbf{x} + \mathbf{n} \tag{11.2}$$

where $\mathbf{F}$ is called an encoding matrix denoting how MRI data are acquired. Eq. (11.2) defines a generalized forward model representing MRI acquisition with a single receiver channel.

Taking a closer look at Eq. (11.1), it can be seen that y is simply the Fourier transform of x. In MRI, data acquisition is performed in the spatial frequency domain known as k-space. The goal of MRI acquisition is to fill out the entire k-space with a predefined field of view (FOV), spatial resolution, and k-space traverse trajectory. MRI offers flexible data acquisition trajectories, including sampling on a line-by-line Cartesian grid or a non-Cartesian grid (*e.g.*, radial), as shown in Fig. 11.1. In some MRI applications with low signal-to-noise ratio (SNR), data acquisition is commonly repeated multiple times to improve the SNR of reconstructed images by averaging all the repeated acquisitions. The number of repetitions is referred to as the number of signal averages (NSA) or the number of excitations (NEX).

With acquired MRI k-space data, an image $\tilde{\mathbf{x}}$ approximating the original image $\mathbf{x}$ (with noise contamination) can be reconstructed as follows:

$$\tilde{\mathbf{x}} = \mathbf{F}^{\mathbf{H}}\mathbf{y} \tag{11.3}$$

Here, $\mathbf{F}^{\mathbf{H}}$ is the Hermitian transpose of $\mathbf{F}$. When the Nyquist sampling criterion is fulfilled (referred to as fully sampled), Eq. (11.2) is a well-defined linear function and $\mathbf{F}$ can be directly inverted. In practice, implementation of Eq. (11.3) needs to take into consideration the acquisition trajectory. For Cartesian sampling, an image can be reconstructed with inverse fast Fourier transform (FFT), while for non-Cartesian sampling, image reconstruction is usually performed using an algorithm called regridding (Block, 2008) or nonuniform FFT (NUFFT) (Fessler and Sutton, 2003).

Cartesian Trajectory

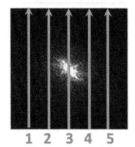

Radial Trajectory

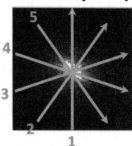

FIG. 11.1

Comparison of 2D Cartesian sampling and 2D radial sampling. Cartesian sampling acquires parallel lines with a fixed frequency encoding direction (arrows), while radial sampling acquires rotating lines that have varying frequency-encoding direction and oversampled k-space center.

Image was reproduced from Figure 1 in J. Magn. Reson. Imaging. 2022 Jul;56(1):45–62 with permission from John Wiley & Sons, Inc.

In dynamic imaging, such as DCE-MRI and DSC-MRI, MRI parameters (T_1, T_2, T_2^*, etc.) change over the course of contrast administration. Under this circumstance, Eq. (11.1) can be extended as follows:

$$y(\mathbf{k}, t) = \int x(\mathbf{r}, t)e^{-i2\pi\mathbf{k}\mathbf{r}} d\mathbf{r} + n \tag{11.4}$$

where x and y both include a time dimension t. Accordingly, $\mathbf{x}$ and $\mathbf{y}$ in Eq. 11.2 can also be extended to represent dynamic k-space data and corresponding images at different time points, which are referred to as dynamic frames.

11.2.2 Acceleration of data acquisition in MRI

In MRI acquisition, only one k-space sample can be measured at a given time point, and all k-space data need to be acquired sequentially. This results in a time-consuming process for MRI acquisition and is the primary cause of slow imaging speed. The total scan time to acquire an MR image can be calculated by multiplying the time to acquire each k-space line (known as repetition time or TR) by the total number of k-space lines needed to form the image, and this time needs to be prolonged proportionally if additional repetitions are needed for signal averaging. The total scan time becomes much longer for acquiring multiple 2D slices or a 3D image. Therefore, accelerated data acquisition plays an essential role in many clinical applications, particularly for dynamic MRI studies. For DCE-MRI and DSC-MRI, fast imaging speed is particularly important since it ensures efficient capture of the quick passage of GBCA in the cardiovascular system with adequate temporal resolution and spatial coverage.

Based on the k-space acquisition process described before, there are multiple ways to accelerate data acquisition in MRI. These methods can roughly be classified into three categories, including: (a) undersampled data acquisition (for both 2D and 3D imaging), (b) simultaneous multi-slice (SMS) acquisition (mostly for multi-slice 2D imaging), and (c) reduced signal averages (for both 2D and 3D imaging). These acceleration methods can be used either to reduce scan time or to increase spatiotemporal resolution and volumetric coverage without prolonging scan time.

Undersampling refers to a sampling strategy that acquires fewer measurements without changing the underlying imaging parameters (*e.g.*, FOV and matrix size). The consequence of undersampling is aliasing image artifacts, as it violates the Nyquist sampling criteria. These artifacts need to be removed using more advanced image reconstruction algorithms. At the same time, MRI provides great flexibility in designing various undersampling schemes that can be tailored to specific image reconstruction methods. Undersampling can be performed for both Cartesian and non-Cartesian acquisitions. It is generally performed only along the phase-encoding dimension, as undersampling the frequency-encoding dimension does not save scan time. Accordingly, non-Cartesian trajectories could provide better undersampling capability, since it does not have a fixed frequency- or phase-encoding direction as shown in Fig. 11.1. In current clinical practice, undersampling is the most widely used acceleration method in MRI, and this will be the main topic of discussion in Section 11.3.

SMS acquisition is an advanced imaging strategy that excites and acquires multiple 2D image slices simultaneously using a specially designed radiofrequency (RF) pulse (Barth *et al.*, 2016). It does not necessarily involve undersampling, but it can be integrated for further improvement in imaging speed. SMS acquisition leads to mix of image content from all excited image slices. Therefore, specific sampling strategies and advanced reconstruction algorithms are needed to resolve the superimposed images from different slice locations. SMS imaging has now become an important fast-imaging technique for multi-slice 2D imaging, and it has seen increasing clinical use. Details about SMS imaging will be summarized in Section 11.4.

Another acceleration method is reducing the number of signal averages for those applications that require repeated acquisitions. Reducing the number of averages does not require modification of sampling scheme and does not generate aliasing artifacts as in undersampled or SMS acquisitions. However, it results in reduced SNR, and this often necessitates additional postprocessing for image denoising. This topic will be discussed in Section 11.5.

11.3 Accelerated MRI through reconstruction of undersampled data

This section focuses on state-of-the-art methods to reconstruct MR images from undersampled k-space data. The ratio between the number of measurements that need to be acquired without acceleration and the number of undersampled measurements is called an undersampling factor or acceleration rate. In case of undersampling, Eq. (11.2) becomes underdetermined. Thus it has infinite solutions and a direct inversion of the encoding matrix creates image artifacts that need to be removed with advanced reconstruction algorithms. Reconstructing an image from undersampled measurements is possible since spatial and/or temporal correlations are present in MR images that can be exploited in image reconstruction. This section classifies different methods for reconstructing undersampled images into two subcategories: linear reconstruction and nonlinear reconstruction.

11.3.1 Linear reconstruction methods

Linear reconstruction methods can be performed by exploiting temporal correlations, spatial correlations, or joint spatiotemporal correlations (Tsao and Kozerke, 2012). For readers who are not familiar with this terminology, linear reconstruction simply implies that the k-space and the to-be-reconstructed image have a linear relationship. Therefore any fluctuations in k-space (*e.g.*, change of noise level) will lead to linearly proportional changes in reconstructed images.

11.3.1.1 Methods exploiting temporal correlations
View sharing

View sharing is one of the earliest methods that exploits temporal correlations or redundancy for accelerated dynamic MRI, and it remains an important imaging technique today in the clinic. In view sharing techniques, some k-space measurements are shared between adjacent dynamic frames. Different versions of view sharing techniques are provided on modern MRI scanners by different vendors, but they can all be traced back to an old view sharing technique called keyhole (Van Vaals *et al.*, 1993). The acquisition scheme for the original keyhole technique is shown in Fig. 11.2A. In the first step, a fully sampled static image (referred to as a reference image) is acquired. In the second step, dynamic images are acquired, and only the k-space center is acquired for each dynamic frame. Keyhole image reconstruction then combines the peripheral k-space measurements from the reference image and the central k-space data in each dynamic frame to form full-resolution dynamic images. In other words, keyhole imaging simply shares the peripheral k-space data in the reference image with all dynamic frames, so that only the k-space center needs to be acquired in dynamic imaging with fast imaging speed. This is possible since for certain MRI applications, such as DCE-MRI, the underlying anatomical structure remains the same and only image contrast is changing, which is mostly determined by the k-space center. Since fully sampled images are generated for all dynamic frames, the final image reconstruction can follow Eq. (11.3).

The limitation of the original keyhole imaging is temporal blurring. This can be improved with different variants (as shown in Fig. 11.2B for an example). The basic idea of these extensions is to acquire the central k-space data more frequently and the peripheral k-space data less frequently, so that the peripheral k-space data can be

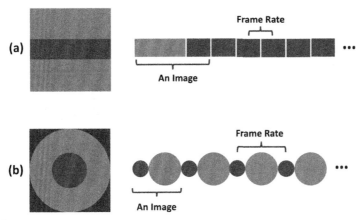

FIG. 11.2

Comparison of standard keyhole dynamic MRI acquisition (A) with an improved variant (B). In (A), the out k-space region (gray) is acquired once, and the k-space center (blue) is updated in dynamic acquisition. An image can be reconstructed combining the two k-space regions. In (B), the out k-space region and k-space center are both acquired in dynamic acquisition to minimize temporal blurring.

Images were adapted from the mriquestions website (https://mriquestions.com/tricks-or-twist.html).

shared in different dynamic frames to reduce temporal blurring (Grist *et al.*, 2012; Hennig *et al.*, 1997; Korosec *et al.*, 1996; Saranathan *et al.*, 2012; Willinek *et al.*, 2008). In addition to Cartesian imaging, view sharing can also be implemented for non-Cartesian imaging. For example, k-space weighted image contrast (KWIC) imaging is a view sharing method for fast radial MRI (Song and Dougherty, 2000), where the sharing of k-space samples increases from the central k-space to outer k-space. In current clinical practice, view sharing MRI methods are mostly used for dynamic contrast-enhanced MR angiography (DCE-MRA) or DCE-MRI.

11.3.1.2 Methods exploiting spatial correlations

In contrast to view sharing methods that exploit temporal correlations, some linear reconstruction techniques exploit spatial correlations to achieve acceleration. Examples include partial Fourier imaging or parallel imaging.

Partial Fourier imaging

Partial Fourier is an old fast-imaging method proposed in the 1980s–1990s (McGibney *et al.*, 1993). It exploits the symmetry property of k-space, so that an MR image can be reconstructed from approximately half of the k-space measurements. Partial Fourier is possible because of the Hermitian symmetry of Fourier transform for a real-valued image:

$$y(-\mathbf{k}) = y^*(\mathbf{k}) \tag{11.5}$$

According to Eq. (11.5), half of the k-space samples are sufficient to reconstruct a real-valued image. Unfortunately, MR images are not truly real valued in practice, and they contain both magnitude and phase due to different factors such as magnetic field inhomogeneity and physiological motion. As a result, Eq. (11.5) cannot be directly used for k-space synthesis. A common solution for addressing this challenge is to acquire a small region of fully sampled k-space center, as shown in Fig. 11.3, from which the phase of the image can be estimated to guide reconstruction of the magnitude image. This is called phase-constrained image reconstruction or simply phase correction (Margosian and Schmitt, 1986; Cuppen and van Est, 1987; Noll et al., 1991). It is possible because the phase of an image typically has a low spatial resolution, and thus a small symmetric region from the k-space center is sufficient for phase estimation. In addition, partial Fourier reconstruction can also be formulated as an iterative phase-constrained optimization problem that can be solved efficiently using the POCS (projection onto convex sets) algorithm (Haacke et al., 1991; Liang et al., 1992). The main limitation of partial Fourier imaging is limited undersampling capability that is less than 2-fold, but it is commonly combined with other image reconstruction methods to achieve more effective acceleration.

Parallel imaging

Parallel imaging, or parallel MRI, first demonstrated for *in vivo* applications in the 1990s, is a disruptive fast-imaging method in the history of MRI (Deshmane et al., 2012). Parallel imaging is flexible, effective, and efficient, and it remains one of the most important and useful imaging techniques in clinical practice today. The general idea behind parallel imaging is that when multicoil arrays are available

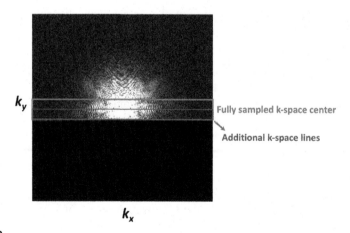

k_y

Fully sampled k-space center

Additional k-space lines

k_x

FIG. 11.3

Partial Fourier k-space sampling. Only half of the k-space is acquired, and additional k-space center lines are acquired to estimate phase from the fully sampled k-space center.

for data acquisition, a linear combination of different coil sensitivities can emulate some gradient-encoding measurements, so that they can be skipped. An image can then be generated through a combination of gradient encoding and coil sensitivity encoding to achieve accelerated data acquisition. Broadly speaking, parallel imaging techniques involve both hardware (design of optimal multicoil arrays) and software (data acquisition and image reconstruction), and this section is primarily focused on the software part.

For multicoil acquisition, Eq. (11.1) can be extended as follows:

$$y_c(\mathbf{k}) = \int m_c(\mathbf{r}) e^{-i2\pi \mathbf{k}\mathbf{r}} d\mathbf{r} + n \tag{11.6}$$

$$m_c(\mathbf{r}) = s_c(\mathbf{r}) x(\mathbf{r}) \tag{11.7}$$

where $c = 1, \ldots, N_c$ indicates different receiving coil elements, m_c is the image to be generated in the c-th coil element, y_c is the acquired k-space data in the c-th coil element, and $s_c(\mathbf{r})$ is the corresponding coil sensitivity. Similar to Eq. (11.2), multicoil acquisition can be represented in the following generalized form:

$$\mathbf{y} = \mathbf{E}\mathbf{x} + \mathbf{n} \tag{11.8}$$

where $\mathbf{E} = \mathbf{FS}$ denotes a multicoil encoding matrix incorporating coil sensitivities from all coil elements. In this case, $\mathbf{y}$ becomes a stack of vectors containing multicoil k-space data and $\mathbf{x}$ is the corresponding multicoil combined image. It was described earlier that Eq. (11.2) becomes underdetermined in case of undersampling, which results in infinite possible solutions. With multicoil coil acquisition, the size of $\mathbf{y}$ is larger than the size of $\mathbf{x}$ in Eq. (11.8) when the undersampling factor is less than the number of coil elements. In this case, Eq. (11.8) becomes overdetermined theoretically with properly designed coil arrays. In this case, parallel imaging aims to find an approximate solution that has the minimum error with the acquired k-space:

$$\tilde{\mathbf{x}} = \underset{x}{\mathrm{argmin}} \|\mathbf{E}\mathbf{x} - \mathbf{y}\|_2^2 \tag{11.9}$$

For an overdetermined linear equation, the best approximated solution can be derived using the Moore–Penrose inversion:

$$\tilde{\mathbf{x}} = \left(\mathbf{E}^H \mathbf{E}\right)^{-1} \mathbf{E}^H \mathbf{y} \tag{11.10}$$

Considering the noise term in Eq. (11.8), this solution can be expressed as the sum of $\tilde{\mathbf{x}}$ with corresponding noise after reconstruction:

$$\tilde{\mathbf{x}} = \mathbf{x} + \left(\mathbf{E}^H \mathbf{E}\right)^{-1} \mathbf{E}^H \mathbf{n} \tag{11.11}$$

Eq. (11.11) indicates that the reconstruction performance of parallel imaging is largely dependent on the encoding matrix $\mathbf{E}$, and a bad conditioning of the linear equation may lead to a strong noise term in Eq. (11.11) causing severe noise amplification.

Since the original introduction of parallel imaging, many reconstruction algorithms have been proposed, which can roughly be classified into two main categories based on how they reconstruct undersampled MR images. The first category aims to remove aliasing artifacts in image space with matrix inversion based on underlying coil sensitivities. This idea, originally conceived in Ra and Rim (1993), was later demonstrated in a work that proposed a practical SNR-optimized implementation called SENSE (SENSitivity Encoding) (Pruessmann et al., 1999). In SENSE reconstruction, a multicoil encoding matrix incorporating preestimated coil sensitivities needs to be constructed. With the encoding matrix, an unaliased image can then be generated following Eq. (11.10). As a result, the SENSE technique might be limited in applications where estimation of coil sensitivities is challenging, and any error in coil sensitivity estimation can be propagated into reconstruction images. SENSE was originally described for Cartesian imaging with regular undersampling (sampling of k-space lines is skipped regularly), which makes it easy to construct the encoding matrix. For non-Cartesian imaging, direct matrix inversion becomes much more challenging, and non-Cartesian SENSE reconstruction has been proposed using an iterative optimization algorithm such as the gradient decent or the conjugate gradient algorithms (Pruessmann et al., 2001; Wright et al., 2014).

The second category of parallel imaging aims to directly estimate the missing k-space data followed by a simple FFT to generate the final image. The first parallel imaging technique to implement this idea was SMASH (SiMultaneous Acquisition of Spatial Harmonics) (Sodickson and Manning, 1997), which aims to approximate low-order harmonics in k-space by a linear combination of coil sensitivities. Specifically, each missing k-space data is reconstructed through a linear combination of the nearest acquired multicoil k-space samples weighted by coil sensitivities. The original SMASH implementation had several limitations, including the need for explicit coil sensitivities and k-space estimation using only the nearest k-space samples. SMASH was later extended to a few improved variants that enable autocalibration and coil-by-coil k-space estimation (Heidemann et al., 2001; Jakob et al., 1998; Griswold et al., 2000), and these innovations ultimately led to a generalized reconstruction approach called GRAPPA (GeneRalized Autocalibrating Partial Parallel Acquisition) (Griswold et al., 2002). For GRAPPA reconstruction, a small region of k-space center is fully sampled, which serves as autocalibration signal (ACS). The ACS is used to estimate weights between different k-space samples (also known as GRAPPA weights or GRAPPA kernel). The estimated weights are then used to estimate missing k-space data from acquired data. The idea of GRAPPA is outlined in Fig. 11.4. In clinical practice, GRAPPA reconstruction is primarily implemented for Cartesian imaging with regular undersampling, which results in shift-invariant GRAPPA weights for simple calibration and fast reconstruction. It can be performed both for 2D imaging or 3D imaging with appropriate adaption (Blaimer et al., 2006). The main advantage of GRAPPA over SENSE-type reconstruction is the elimination of explicit estimation of coil sensitivities. As such, GRAPPA can be more robust in applications where estimation of coil sensitivity maps is difficult. SPIRiT (Iterative Self-consistent Parallel Imaging Reconstruction) is a relatively new method that generalizes GRAPPA to arbitrary undersampling schemes (Lustig and Pauly, 2010).

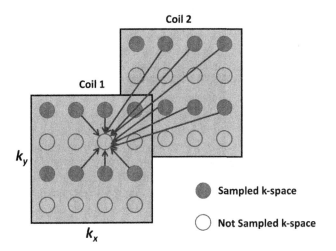

FIG. 11.4

Example of GRAPPA reconstruction with two coil elements. Each nonsampled k-space point can be estimated as a linear combination of k-space neighbors from all coil elements. The weights to combine k-space neighbors can be calculated from fully sampled k-space center.

Same as GRAPPA, SPIRiT also acquires ACS in k-space center and does not require explicit estimation of coil sensitivities. A main advantage of SPIRiT is that it can be easily combined with nonlinear reconstruction methods such as compressed sensing or low-rank reconstruction. However, it requires an iterative reconstruction algorithm that is usually slower compared to standard SENSE or GRAPPA.

A common limitation of all parallel imaging methods is the SNR penalty at high acceleration rates that can lead to noise amplification (Pruessmann *et al.*, 1999), as shown in the following equation.

$$SNR_{acc} = \frac{SNR_{no\text{-}acc}}{g\sqrt{R}} \qquad (11.12)$$

Eq. (11.12) indicates that two factors contribute to the SNR reduction in parallel imaging, one from undersampled k-space measurement ($\sqrt{R}$) and the other one from a spatially dependent term called geometry factor or simply g-factor. In parallel imaging, the g-factor is highly dependent on coil geometry, coil loading, and acquisition orientation. In general, it leads to ill conditioning of the multicoil encoding matrix and, thus, noise amplification. As a result, the achievable acceleration rate using standard parallel imaging is typically limited to 2 to 3 for routine clinical use.

11.3.1.3 Methods exploiting spatiotemporal correlations
Dynamic MRI, such as DCE-MRI of DSC-MRI, provides extensive correlations along the spatial and temporal dimensions, which can both be exploited to improve reconstruction performance toward higher acceleration rates. This section provides an overview of these methods.

Temporal parallel imaging

Soon after the development of standard parallel imaging for accelerated static imaging, temporal parallel imaging methods, mainly temporal SENSE (TSENSE) (Kellman *et al.*, 2001) and temporal GRAPPA (TGRAPPA) (Breuer *et al.*, 2005a), were proposed. In temporal parallel imaging, k-space undersampling is performed using a so-called lattice sampling scheme that employs an interleaved k-t acquisition scheme to regularly shift the undersampled k-space locations along time, as shown in Fig. 11.5. In dynamic imaging, this sampling scheme allows for coil sensitivity estimation from a complete k-space combining multiple adjacent dynamic frames. The estimated coil sensitivity can then be used for reconstructing undersampled images in these dynamic frames. This reconstruction scheme is repeated until all the undersampled dynamic frames are reconstructed. The ability to perform temporal parallel imaging without the need of external ACS stems from the fact that coil sensitivities are generally smooth and change little over time. However, as in standard parallel imaging, excessive acceleration can also lead to severe noise amplification.

Another temporal parallel imaging method that can be applied for accelerated dynamic imaging is called through-time non-Cartesian GRAPPA (Seiberlich *et al.*, 2011). It can be seen from the name that this method was specifically developed for accelerated dynamic imaging using non-Cartesian sampling. The idea of through-time non-Cartesian GRAPPA is to first acquire fully sampled reference dynamic k-space, from which GRAPPA weights for a specific k-space region are estimated along the spatiotemporal dimension. These estimated weights are then used to reconstruct undersampled dynamic k-space acquired with the same acquisition pattern as in the reference. This calibration process needs to be repeated for all k-space regions to estimate all missing k-space data, from which dynamic images can be generated with standard NUFFT. The calibration and reconstruction process for through-time non-Cartesian GRAPPA using radial sampling is shown in Fig. 11.6.

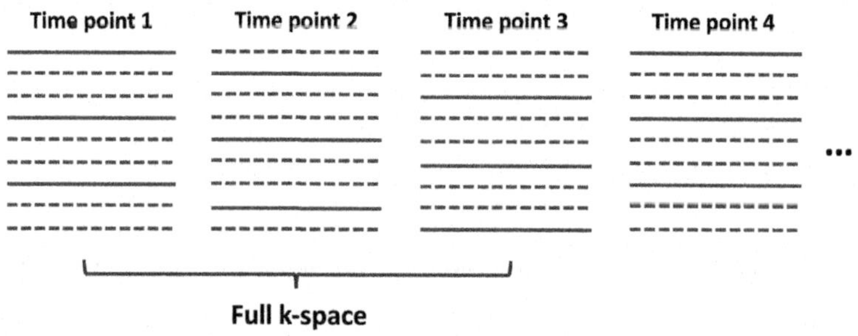

FIG. 11.5

Dynamic lattice k-space sampling (3-fold acceleration) for temporal parallel imaging. K-space sampling is shifted for one line along time, so that connective dynamic frames (*e.g.*, 3 in this example) can be combined to form a fully sampled k-space for coil sensitivity calibration without extra reference data.

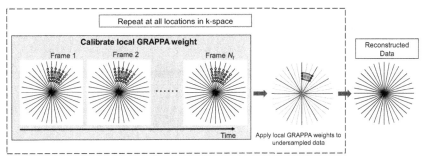

FIG. 11.6

Non-Cartesian radial GRAPPA reconstruction. A fully sampled dynamic k-space data are first acquired for calibration. The weights for each geometry are estimated based on k-space segmentation along time. The calculated weights are applied to reconstruct undersampled dynamic k-space at the same segment. This process is repeated for all k-space regions.

Image was reproduced from Figure 7 in J. Magn. Reson. Imaging 2014;40:1022–1040 with permission from John Wiley & Sons, Inc.

Through-time non-Cartesian GRAPPA enables higher acceleration rates than standard GRAPPA reconstruction because of the extensive information in the reference dynamic data for calibration of GRAPPA weights and the use of non-Cartesian sampling that allows for acceleration along all spatial dimensions and repeated sampling of k-space center. However, through-time non-Cartesian GRAPPA relies on fully sampled dynamic data that need to be acquired with additional scan time, and inconsistency between the acquisition of reference and accelerated k-space may lead to potential reconstruction errors. Recently, the feasibility of self-calibrated through-time GRAPPA reconstruction has been demonstrated, which holds great potential for easier and more robust clinical implementation (Franson *et al.*, 2023).

K-t acceleration methods

K-t acceleration methods are a family of reconstruction techniques to reconstruct undersampled data by exploiting spatiotemporal correlation jointly. Broadly speaking, any accelerated dynamic MRI method that jointly exploits spatiotemporal correlations can be considered a k-t acceleration method, and it can include both linear reconstruction and nonlinear reconstruction techniques. This section describes a few representative examples of k-t acceleration methods that employ linear reconstruction. Similar to parallel imaging, k-t acceleration methods can also be implemented either in the image domain or in k-space domain, and they can be combined with parallel imaging to further improve reconstruction performance.

UNFOLD (UNaliasing by Fourier-encoding the OverLaps using the temporal Dimension) is one of the earliest k-t acceleration methods that exploits spatiotemporal correlations (Madore *et al.*, 1999). The idea of UNFOLD is built on the fact that dynamic information is typically limited to certain region in dynamic imaging, while

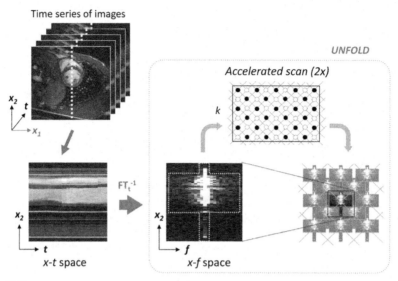

Time series of images

UNFOLD

Accelerated scan (2x)

x_2 t

x_1

k

FT$_t^{-1}$

x_2

x_2

t

f

x-t space

x-f space

FIG. 11.7

Dynamic images can have a sparse representation in the x-f space with lattice k-space sampling. This allows direct separation of image from undersampling artifacts at low acceleration rates.

Image was reproduced from Figure 7 in J. Magn. Reson. Imaging 2012;36:543–560 with permission from John Wiley & Sons, Inc.

the remaining region mostly contains static information. As a result, dynamic images can have a sparse representation when transformed to the so-called x-f space with a temporal FFT, where f represents temporal frequency as shown in Fig. 11.7 for an example. With the lattice sampling scheme as used for temporal parallel imaging, there might be no fold-over artifacts in the x-f space at low acceleration rates. This permits image reconstruction by simply separating the image content and artifacts in the x-f space and transforming back to the original x-t space. The original UNFOLD method, however, requires one to specify a region that includes the true image information in the x-f space, and residual aliasing artifacts can remain at high acceleration rates.

This limitation was later addressed by a method called k-t BLAST (k-t space Broad-use Linear Acquisition Speed-up Technique) (Tsao *et al.*, 2003), which aims to acquire additional training data to estimate the distribution of underlying image in x-f space. This eliminates the need for user interaction, and the use of extra training data to guide the removal of aliasing artifacts in x-f space improves reconstruction performance. As shown in Fig. 11.8, the training data can be acquired directly in central k-space, which can be integrated with undersampled dynamic acquisition. The combination of k-t BLAST with SENSE is known as k-t SENSE (Tsao *et al.*, 2003), which exploits additional coil-encoding capabilities to enable higher acceleration

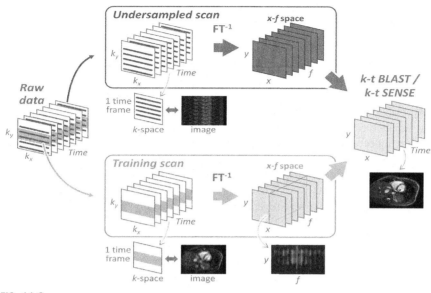

FIG. 11.8

The k-t BLAST and k-t SENSE acquisition and reconstruction framework. The center of k-space is fully sampled, from which low-resolution dynamic images can be generated to guide reconstruction of full-resolution image.

Image was reproduced from Figure 8 in J. Magn. Reson. Imaging 2012;36:543–560 with permission from John Wiley & Sons, Inc.

rates. Instead of performing image reconstruction in the *x-f* space, image reconstruction can also be performed in other spaces. For example, k-t PCA is a technique that performs image reconstruction in *x PC* space, which is obtained through temporal principal component analysis (PCA) (Pedersen *et al.*, 2009). K-t PCA has been demonstrated with better performance than k-t SENSE, because image content in the *x-PC* space can be more compact, thus allowing for better and more effective removal of aliasing artifacts. The main limitation of k-t SENSE and its variants is that the training data typically have a low spatial resolution, which results in partial volume effects. This can lead to underestimation of underlying dynamic image content and can yield spatiotemporal blurring at high acceleration rates.

K-t acceleration methods can be performed directly in k-space to exploit spatiotemporal correlations. For example, k-t GRAPPA is an extension of GRAPPA for accelerated dynamic imaging (Huang *et al.*, 2005). K-t GRAPPA aims to estimate missing k-space data in k-t space, where GRAPPA weights are calibrated from fully sampled spatiotemporal ACS data acquired in central k-space region. The availability of both spatial and temporal correlations in k-t GRAPPA enables better reconstruction quality compared to standard GRAPPA or TGRAPPA. This imaging technique has later been extended to different variants such as k-t^2 GRAPPA (Lai *et al.*, 2008) or PEAK-GRAPPA (Jung *et al.*, 2008) to achieve improved performance.

11.3.2 Nonlinear reconstruction techniques

The main advantages of linear reconstruction methods include fast reconstruction process and easy implementation that facilitates clinical translation. The major limitation, however, is limited acceleration capability. Excessive acceleration tends to yield noise amplification, residual aliasing artifacts, and spatiotemporal blurring. In contrast, nonlinear reconstruction generally allows for higher acceleration rates and improved reconstruction performance. Despite a long history, nonlinear reconstruction did not receive good attention in the field of MRI until the advent of compressed sensing. Indeed, over the past decade, the majority of new reconstruction methods are related to the concept of sparsity, and some of them have seen increasing clinical applications. The rise of artificial intelligence and its wide applications in imaging has further made it easy to translate nonlinear reconstruction into the clinic.

In general, nonlinear reconstruction can be formulated as the following reconstruction problem:

$$\tilde{\mathbf{x}} = \arg\min_{\mathbf{x}} \|\mathbf{F}\mathbf{x} - \mathbf{y}\|_2^2 + \lambda R(\mathbf{x}) \tag{11.13}$$

where R represents a nonlinear function that enforces a constraint or a regularization on the image to be reconstructed. This regularization helps remove image artifacts and/or noise effectively. The left term enforces data consistency, which requires that the reconstructed image should be consistent with acquired k-space data. λ is a weighting parameter to balance data consistency and image regularization. With proper design of an undersampling scheme and selection of the function R, Eq. (11.13) can be adapted to the different nonlinear reconstruction techniques that are summarized in this section. In contrast to linear reconstruction, nonlinear methods typically require an iterative reconstruction algorithm, which often leads to slow reconstruction speed.

11.3.2.1 Compressed sensing

The compressed sensing theory was initially proposed in the 2000s (Candès *et al.*, 2006; Donoho, 2006), and it was soon introduced to MRI as compressed sensing MRI or sparse MRI (Lustig *et al.*, 2007, 2008). The advent of compressed sensing for accelerated MRI has generated great excitement over the past decade, and it is available for routine clinical use today. Implementation of compressed sensing in MRI requires that: (1) an MR image has a sparse representation by itself or in a transform domain, (2) undersampling-induced artifacts are incoherent as added noise, and (3) nonlinear reconstruction is used to promote sparsity and enforce data consistency.

The first requirement in compressed sensing MRI is to choose a suitable sparsifying transform. An MR image may not be sparse by itself, but they can generally find a sparse representation in some transform domains, such as discrete cosine transform, wavelet transform, or spatial finite differences (Lustig *et al.*, 2008). Different sparsifying transforms may lead to different reconstruction performance based on underlying contrast and anatomical structure. As shown in Fig. 11.9 for two

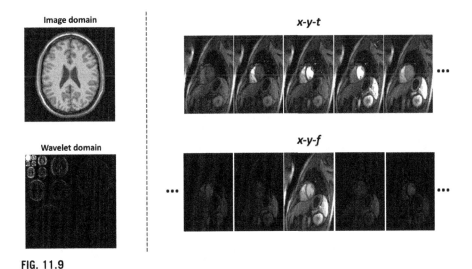

FIG. 11.9

A brain image does not have a sparse representation by itself, but it can have a sparse representation in the wavelet domain, where most wavelet coefficients are close to zero. Dynamic images, such as DCE-MRI, can have a sparse representation after applying a sparsifying transform along the temporal dimension.

examples, a T1-weighted brain image can have a sparse representation in the wavelet transform, while dynamic images, such as DCE-MRI, can have a sparse representation in temporal Fourier domain. Usually, dynamic MRI offers extensive spatiotemporal correlations that can be exploited with a suitable temporal sparsifying transform (Tsao and Kozerke, 2012; Gamper *et al.*, 2008). This permits higher acceleration rates than static imaging. Commonly used temporal sparsifying transforms include temporal FFT, temporal PCA, and temporal finite differences.

The second requirement in compressed sensing MRI is to design an undersampling scheme that produces incoherent aliasing artifacts. For Cartesian imaging, the most commonly used acquisition strategy is random undersampling along the phase-encoding dimension (or both phase- and partition-encoding dimensions for 3D acquisition) (Lustig *et al.*, 2007). Meanwhile, it has been suggested that variable-density random undersampling can better preserve underlying image content by sampling the center of k-space more frequently than the periphery (Lustig *et al.*, 2007, 2008; Moghari *et al.*, 2018), as shown in Fig. 11.10. Since k-space center contains most of the energy and contrast information, acquiring more samples in k-space center can result in better incoherence and thus higher reconstruction quality. Compressed sensing MRI can particularly benefit from non-Cartesian sampling, such as radial and spiral acquisition (Block *et al.*, 2007; Adluru *et al.*, 2009; Feng, 2022a; Feng *et al.*, 2014, 2017, 2016; Li *et al.*, 2022). This is because non-Cartesian sampling inherently permits undersampling along all spatial dimensions, enabling improved undersampling incoherence.

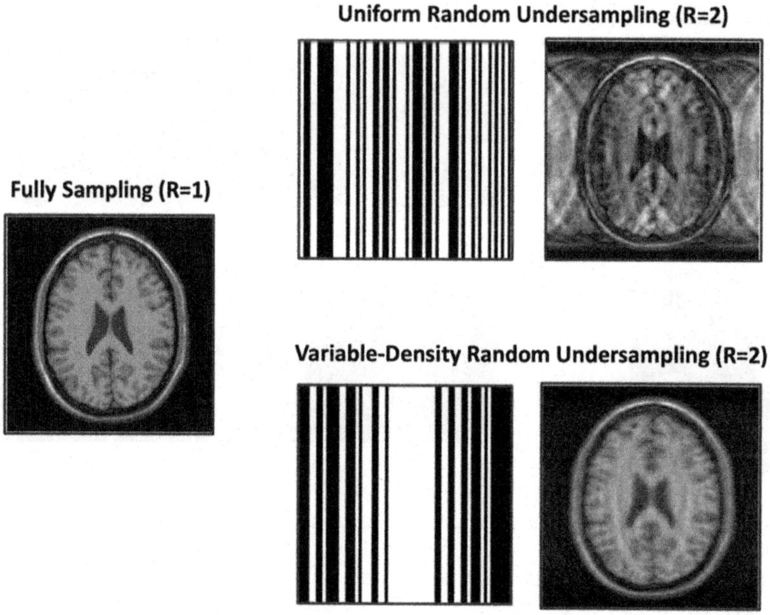

Uniform Random Undersampling (R=2)

Fully Sampling (R=1)

Variable-Density Random Undersampling (R=2)

FIG. 11.10

Comparison of uniform random undersampling and variable-density random undersampling for compressed sensing with Cartesian acquisition. Sampling the k-space center more frequently helps improve incoherence and better preserves image content after undersampling.

The reconstruction of compressed sensing MRI can be formulated as follows:

$$\tilde{\mathbf{x}} = \arg\min_{\mathbf{x}} \|\mathbf{E}\mathbf{x} - \mathbf{y}\|_2^2 + \lambda\|\Psi(\mathbf{x})\|_1 \tag{11.14}$$

Eq. (11.14) is a specific form of Eq. (11.13), in which Ψ indicates a sparsifying transform and the regularization is implemented using an L1 norm, which is given as $\|\mathbf{x}\|_1 = \sum |x_j|$. The multicoil encoding matrix $\mathbf{E}$ represents a combination of compressed sensing with parallel imaging, which has been demonstrated for improved reconstruction performance compared to coil-by-coil compressed sensing reconstruction (Liu *et al.*, 2008; Otazo *et al.*, 2010; Liang *et al.*, 2009). Eq. (11.14) can be solved with different iterative algorithms, such as nonlinear conjugate gradient (Lustig *et al.*, 2007) or alternating direction method of multipliers (ADMM) (Ye, 2019).

After a decade of development and optimization, compressed sensing MRI currently sees a wide variety of clinical applications, and different vendors have provided their compressed sensing MRI products (Feng *et al.*, 2017; Jaspan *et al.*, 2015; Hollingsworth, 2015; Delattre *et al.*, 2020; Yoon *et al.*, 2019). As will be seen in the following techniques, the role of sparsity has gone beyond standard compressed sensing, and it has been an essential component in most modern image reconstruction techniques. General challenges of compressed sensing include slow reconstruction speed, the need for selecting a regularization parameter λ, and potential image artifacts due to overregularization.

11.3.2.2 Low-rank-based reconstruction

In addition to transform sparsity, another feature of dynamic MR that can be exploited for accelerated data acquisition is its low-rank property. Dynamic images can be formed as a matrix $\mathbf{x} \in \mathbb{R}^{N \times T}$, where each column represents a dynamic frame in vector and N equals the total number of pixels in each dynamic frame. Because of temporal correlation, the rank of $\mathbf{x}$, denoted as L, is smaller than the number of dynamic frames T ($e.g., L < T$). In practice, it is expected that $L \ll T$ in most dynamic images, representing reduced degrees of freedom that can be exploited for acceleration.

The low-rank property of dynamic images can be assessed using singular value decomposition (SVD) or PCA. With SVD, for example, $\mathbf{x} \in \mathbb{R}^{N \times T}$ can be decomposed as $\mathbf{x} = \mathbf{U\Sigma V^H}$, with $\mathbf{\Sigma} = \text{diag}([\sigma_i])$ indicating a rectangular diagonal matrix containing all singular values σ_i. If $\mathbf{x}$ has a rank of L, then $\sigma_i = 0$ for $i > L$. Fig. 11.11 shows an example of cardiac perfusion MRI that demonstrates the low-rank representation of dynamic images. The dataset has 40 time points. Singular values, obtained by concatenating each time frame as a column vector followed by SVD, exhibit a rapidly decaying pattern. Performing PCA on the images then results in a compact representation, in which most coefficients are close to zero.

Over the past decade, various low-rank-based reconstruction methods have been developed for fast dynamic MRI. These methods can either exploit the low-rank property of images implicitly or explicitly, as described later. Meanwhile, low-rank-based approaches can also be combined with standard sparsity regularizations for further improvement in reconstruction quality.

Implicit low-rank reconstruction

The low-rank property of dynamic images can be enforced in image reconstruction by putting a constraint on the sum of its nonzero singular values (Candès *et al.*, 2009):

$$\tilde{\mathbf{x}} = \arg \min_{\mathbf{x}} \|\mathbf{Ex} - \mathbf{y}\|_2^2 + \lambda \|\mathbf{x}\|_* \tag{11.15}$$

Here, $\|\bullet\|_*$ denotes the nuclear norm, $\|\mathbf{x}\|_* = \sum_i \sigma_i$. As mentioned before, the low-rank constraint can be combined with sparsity constraints (Lingala *et al.*, 2011; Otazo *et al.*, 2015), and such a combination has been shown to outperform image reconstruction with only sparsity or low-rank constraints.

Explicit low-rank reconstruction

Another way to exploit the low-rank property of dynamic images is to decompose the image matrix into a row basis and a column basis (Liang, 2007). Given an image matrix $\mathbf{x}(\mathbf{r}, t)$ with rank $L < \min(N, T)$, the image can be decomposed as

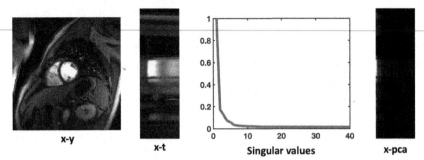

FIG. 11.11

The low-rank representation of a cardiac perfusion MRI dataset. The dataset has 40 time points. Singular values can be obtained by concatenating each time frame as a column vector followed by SVD, and they exhibit a rapidly decaying pattern. Performing PCA on the images results in a compact representation in the x-pca domain, in which most coefficients are close to zero.

$$x(\mathbf{r}, t) = \sum_{l=1}^{L} u_l(\mathbf{r})\phi_l(t) \tag{11.16}$$

or its matrix formation

$$\mathbf{x} = \mathbf{U}\mathbf{\Phi} \tag{11.17}$$

where $\mathbf{U}$ represents the spatial basis and $\mathbf{\Phi}$ represents the temporal basis. The reconstruction of $\mathbf{x}$ can then be implemented by first estimating $\mathbf{\Phi}$ followed by recovery of $\mathbf{U}$.

Various acquisition and reconstruction techniques have been developed for image reconstruction with explicit low-rank decomposition. One approach is to first estimate the temporal basis $\mathbf{\Phi}$ from a subset of fully sampled and low-resolution training data acquired at the center of k-space. In image reconstruction, $\mathbf{\Phi}$ is fixed as known and $\mathbf{U}$ is solved iteratively (Zhao *et al.*, 2012). Since the k-space center contains most of the energy and contrast information, the training data is assumed to contain similar temporal information as the full-resolution data. Specifically, $\mathbf{\Phi}$ can be extracted from the singular vectors of low-resolution dynamic images generated from the training data, and only the dominant components (*e.g.*, the first L major components) are used, which compress the full dynamic images into a low-dimensional image space called subspace. After $\mathbf{\Phi}$ is estimated, image reconstruction can be performed by solving the following cost function:

$$\tilde{\mathbf{U}} = \arg\min_{\mathbf{U}} \|\mathbf{E}\mathbf{U}\mathbf{\Phi} - \mathbf{y}\|_2^2 + \lambda R(\mathbf{U}) \tag{11.18}$$

Compared to the entire dynamic images $\mathbf{x} \in \mathbb{R}^{N \times T}$, $\mathbf{U} \in \mathbb{R}^{N \times L}$ has a much smaller size with reduced degrees of freedom, which then enable improved reconstruction quality.

In addition to basis estimation from additional training data, the temporal basis can also be estimated from the acquired k-space data without additional training data (Feng *et al.*, 2020; Feng, 2022b), estimated from a pharmacokinetic perfusion model (Lingala *et al.*, 2020), or reconstructed jointly in image reconstruction (Ong *et al.*, 2020). However, joint reconstruction of spatial and temporal basis can result in a nonconvex reconstruction problem that requires more advanced algorithm and higher computational burden. The low-rank property can also be exploited in multidimensional dynamic images with improved performance. For example, MR Multitasking is a relatively new technique that extends low-rank subspace reconstruction to multiple temporal dimensions based on low-rank tensor decomposition (Christodoulou *et al.*, 2018). This allows a single acquisition to reconstruct different information (*e.g.*, motion and quantitative parameters) along different temporal dimensions.

11.3.2.3 Model-based reconstruction

Perfusion MRI often involves a quantification step that is performed on reconstructed dynamic images to generate perfusion parameters. Traditionally, quantitative perfusion MRI is performed based on a two-step scheme. In the first step, image reconstruction is performed to reconstruct dynamic images from undersampled k-space. Once the images are generated, perfusion quantification can then be performed based on a pharmacokinetic model in the second step, and perfusion parameter maps can also be generated based on pixel-by-pixel quantification. The pharmacokinetic model can also serve as an explicit constraint for image reconstruction, and it can be incorporated directly into image reconstruction. This represents a novel reconstruction strategy, since (a) the incorporation of the pharmacokinetic model can help improve image reconstruction performance, and (b) image reconstruction can directly generate MR perfusion maps combining the two separate steps that are performed traditionally. Such a reconstruction scheme is referred to as model-based reconstruction.

Mathematically, model-based reconstruction can be performed as follows:

$$\tilde{\mathbf{p}} = \arg\min_{\mathbf{p}} \|\mathbf{E}\mathbf{M}(\mathbf{p}) - \mathbf{y}\|_2^2 + \lambda R(\mathbf{p}) \tag{11.19}$$

Here, $\mathbf{M}$ represents a pharmacokinetic model that is incorporated into image reconstruction, which directly estimates MR perfusion parameters, denoted as $\mathbf{p}$, without producing intermediate dynamic image series. An additional sparsity constraint can also be added as shown in Eq. (11.19). Technically, any perfusion models could be implemented for $\mathbf{M}$. For DCE-MRI, for example, the Patlak model or the Tofts or extended Tofts model can be used (Guo *et al.*, 2017), and corresponding perfusion parameters (*e.g.*, fractional plasma volume $v_p(\mathbf{r})$, extracellular extravascular volume fraction $v_e(\mathbf{r})$, and transfer constant $K^{\text{trans}}(\mathbf{r})$) can be associated with the acquired dynamic k-space $y(\mathbf{k}, t)$ using a function shown as follows:

$$y(\mathbf{k}, t) = f\left(v_p(\mathbf{r}), v_e(\mathbf{r}), K^{\text{trans}}(\mathbf{r}), C_a(\mathbf{r}, t), R_1(\mathbf{r}, 0), M_0(\mathbf{r}), \mathbf{S}\right). \tag{11.20}$$

Here, several variables, including the precontrast relaxation rate $R_1(\mathbf{r},0)$, proton density $M_0(\mathbf{r})$, and arterial input function (AIF) $C_a(\mathbf{r},t)$, can be predetermined, so that the unknown perfusion parameters can be directly estimated based on the following cost function:

$$\left(v_p(\mathbf{r}), v_e(\mathbf{r}), K^{\text{trans}}(\mathbf{r})\right) = \underset{v_p(\mathbf{r}), v_e(\mathbf{r}), K^{\text{trans}}(\mathbf{r})}{\text{argmin}} \left\| y - f\left(v_p(\mathbf{r}), v_e(\mathbf{r}), K^{\text{trans}}(\mathbf{r})\right) \right\|_2^2 \quad (11.21)$$

Adding extra sparsity constraints, Eq. (11.21) becomes:

$$\left(v_p(\mathbf{r}), v_e(\mathbf{r}), K^{\text{trans}}(\mathbf{r})\right) = \underset{v_p(\mathbf{r}), v_e(\mathbf{r}), K^{\text{trans}}(\mathbf{r})}{\text{argmin}} \left\| \mathbf{y} - f\left(v_p(\mathbf{r}), v_e(\mathbf{r}), K^{\text{trans}}(\mathbf{r})\right) \right\|_2^2$$
$$+ \lambda_1 \left\| \Psi v_p(\mathbf{r}) \right\|_1 + \lambda_1 \left\| \Psi v_p(\mathbf{r}) \right\|_1 + \lambda_2 \left\| \Psi K^{\text{trans}}(\mathbf{r}) \right\|_1$$
$$(11.22)$$

Despite an interesting reconstruction strategy, model-based reconstruction has not been widely accepted for accelerated perfusion MRI. This is mainly attributed to the increased computational burden and complexity that are associated with the incorporation of a perfusion model into image reconstruction. Meanwhile, since many pharmacokinetic models are nonlinear, the reconstruction problem in Eq. (11.22) becomes nonconvex, which requires more advanced image reconstruction algorithms. Deep learning represents a promising solution to address this challenge, as will be seen in the following section.

11.3.2.4 Deep learning-based reconstruction

Deep learning-based MRI reconstruction has been the most recent trend in fast MRI. The use of deep learning to reconstruct undersampled MRI data was first demonstrated around 2016 (Wang et al., 2016). Since then, there has been an explosive growth of various deep learning-related fast MRI techniques (Hammernik et al., 2018; Yang et al., 2018; Schlemper et al., 2018; Akçakaya et al., 2019; Mardani et al., 2019; Liu et al., 2019a), and most vendors have now made their deep learning-based imaging reconstruction techniques for clinical use. Indeed, deep learning has opened a new window of opportunity to revolutionize medical imaging research in general, and it has been pushing the translation of accelerated imaging techniques into the clinic. The idea behind deep learning is that a combination of many weighted nonlinear functions, with a multilevel feature model, can be used to learn and represent complex functions. Deep learning utilizes neural networks, such as a conventional neural network (CNN), to learn latent data information and image features. Similar to the neurons in the human brain and the way they work to perform cognitive tasks, a neural network usually has a deep structure consisting of a large number of learning modules with multiple interconnected hidden layers, which enable flexible and effective learning capability.

Most deep learning-based MRI reconstruction methods perform supervised learning or training on a database that consists of fully sampled reference images and paired undersampled images with artifacts. The goal of training is for the neural networks to learn how to remove image artifacts in undersampled images, so that recovered image quality can match that in the references. The training is an iterative process that can be formulated with the following cost function:

$$\tilde{\theta} = \arg \min_{\theta} \left(\lambda_1 \mathbb{E}_{x_u \sim P(x_u)} \|\mathbf{E}(C(\mathbf{x}_u|\theta)) - \mathbf{y}\|_2^2 + \lambda_2 \mathbb{E}_{\mathbf{x}_u \sim P(\mathbf{x}_u)} \left[\|C(\mathbf{x}_u|\theta) - \mathbf{x}\|_p \right] \right)$$

(11.23)

Here $C(\mathbf{x}_u|\theta)$ is an end-to-end mapping function conditioned on a network hyperparameter set θ. Deep learning-based image reconstruction can be treated as a process that aims to learn how to remove artifacts from undersampled images $\mathbf{x}_u$ with the network parameter set. $\|\cdot\|_p$ denotes a p-norm function (*e.g.*, L1-norm or L2-norm) that is used to form a loss function, which enforces that the result of $C(\mathbf{x}_u|\theta)$ should match the reference images $\mathbf{x}$. $\mathbb{E}_{\mathbf{x}_u \sim P(\mathbf{x}_u)}[\,\cdot\,]$ is an expectation operator ensuring that $\mathbf{x}_u$ belongs to the data distribution $P(\mathbf{x}_u)$. It can be seen that Eq. (11.23) has a similar format as Eq. (11.13). Here the right term in Eq. (11.23) can be treated as a data-driven constraint, while the left term ensures data consistency. The main difference, however, is that Eq. (11.13) aims to reconstruct an artifact-free image $\tilde{\mathbf{x}}$, while Eq. (11.23) aims to learn a hyperparameter set θ. Once the training is completed, the learned hyperparameters can then be used to reconstruct new undersampled images efficiently without the need of a fully sampled reference. This process is called inference.

Deep learning-based image reconstruction offers a number of key advantages compared to conventional nonlinear reconstruction methods described before. First, it is a data-driven approach that can be more effective in removing image artifacts and recovering image quality (Wang *et al.*, 2016; Hammernik *et al.*, 2018). Second, although training is a time-consuming process, this does not have to be performed often, and the training can also incorporate different sampling patterns, different types of image contrast (and scan parameters), or even images from different organs to make the learned network more generalizable (Liu *et al.*, 2019a; Johnson *et al.*, 2019, 2021). Once the training is done, the networks can be used to perform new image reconstruction tasks efficiently. In other words, deep learning-based image reconstruction shifts the heavy computational burden to a preprocessing process compared to traditional iterative reconstruction, so that it has great potential for routine clinical use. However, the main challenge of deep learning-based image reconstruction is the need for fully sampled images during training. Although it is relatively easy to collect a large number of static MR datasets for this purpose, it can be much more difficult to collect dynamic MR datasets, and it may not be possible to acquire fully sampled datasets in some applications such as DCE-MRI. As a result, unsupervised or self-supervised deep learning would be highly desired for those applications, and it has become a popular research direction now (Zeng *et al.*, 2021).

In addition to standard image reconstruction, deep learning also holds great potential for model-based reconstruction, so that challenges described in the previous section can be effectively addressed. Similar to Eq. (11.19), model-based deep learning reconstruction can be formulated as the following cost function:

$$\widetilde{\theta} = \arg \min_{\theta} \left(\lambda_1 \mathbb{E}_{\mathbf{x}_u \sim P(\mathbf{x}_u)} \|\mathbf{EM}(C(\mathbf{x}_u|\theta)) - \mathbf{y}\|_2^2 + \lambda_2 \mathbb{E}_{\mathbf{x}_u \sim P(\mathbf{x}_u)} \left[\|C(\mathbf{x}_u|\theta) - \mathbf{p}\|_p \right] \right)$$

$$(11.24)$$

While the mapping function $C(\mathbf{x}_u|\theta)$ aims to generate artifact-free images in Eq. (11.19), it directly generates MR parameters $\mathbf{p}$ in Eq. (11.24). As a result, the loss function to enforce image constraint is formed between the undersampled images and MR parameter maps (the right term), while the loss function to ensure data consistency (the left term) needs to incorporate a perfusion model $\mathbf{M}$ as in conventional model-based reconstruction. This framework has been demonstrated in different MR parameter mapping studies (Liu *et al.*, 2019b, 2020, 2021; Relax-MANTIS, n.d.; Zhou *et al.*, 2018; Feng *et al.*, 2022), and it can be extended for rapid quantitative MRI in general (Feng *et al.*, 2022).

11.4 Accelerated MRI through simultaneous multi-slice imaging

Because of the sequential sampling nature in MRI, acquisition of multiple image slices needs to be performed sequentially. With simultaneous multi-slice (SMS) imaging, it is possible to acquire and reconstruct multiple image slices simultaneously, so that the total scan time can be reduced. In current clinical practice, SMS imaging has become an important fast imaging technique for 2D imaging and has been widely used in many applications such as diffusion MRI or functional MRI. The history of SMS excitation can be traced back a few decades. In 1991 Glover demonstrated the idea of SMS excitation using a technique called POMP (Phase-Offset MultiPlanar imaging). POMP varies the phase in a RF pulse used to excite multiple image slices simultaneously, now known as RF phase encoding or RF encoding, so that each slice can be shifted along the phase-encoding dimension to avoid overlapping. This is based on the time-shifting property of the Fourier Transform—an additional phase in the frequency domain can induce a shift in time (or shift in space in the image domain).

From Eq. (11.1), it can be seen that simultaneously acquired image slices will be superimposed on each other. The original POMP technique was able to separate these images with RF encoding and extended FOV. However, this does not allow for reduction of scan time. SMS imaging without changing FOV became possible with the development of parallel imaging, which enables separation of multiple image slices based on multicoil arrays, as in standard parallel image reconstruction. In 2001, Larkman *et al.* demonstrated the feasibility of exciting multiple image slices simultaneously and then separating them using SENSE reconstruction along the slice dimension (Larkman *et al.*, 2001). However, the performance of this technique is strongly dependent on the geometry of coil array. When the coil sensitivities are similar in the excited slices (*e.g.*, two slices close to each other), this technique suffers from severe g-factor-induced noise amplification.

To overcome these limitations, Breuer *et al.* developed a technique called Controlled Aliasing in Parallel Imaging Results in Higher Acceleration (CAIPIRINHA) (Breuer *et al.*, 2005b). The central concept of CAIPIRINHA is to improve the separation of images at different slice locations by shifting the aliasing patterns from each of them in a controlled way with respect to each other. As shown in Fig. 11.12, this was achieved with RF encoding, which adds distinct phases in the RF pulse when exciting different image slices. This allows each slice to shift a different amount to reduce overlapping and better image separation using parallel imaging. The standard CAIPIRINHA, however, faces extra challenges for implementation in single-shot echo planar imaging (EPI), which has been the cornerstone for many clinical applications such as diffusion MRI or DSC-MRI. For single-shot EPI, all k-space phase-encoding lines within the imaging plane are acquired after a single excitation. Therefore it is not possible to modulate different k-space phase-encoding lines through the phase of corresponding RF pulses. One approach to achieve SMS-EPI is to add a fixed blip in z gradient (Gz) together with the blips in y gradient (Gy) (Nunes *et al.*, 2006). The Gy blips are used to acquire different phase-encoding lines, and the Gz blips are used to introduce the phase difference of the k-space lines acquired across multiple slices. However, because of the finite slice thickness, the accumulation of the phase difference within one slice coming from the frequent Gz blips causes an undesirable "voxel tilting" effect, whereby the readout and slice direction are no longer orthogonal. In this case, the signal of a single voxel is smeared into adjacent voxels.

This challenge was later addressed effectively by a technique called blipped CAIPIRINHA (Setsompop *et al.*, 2012), which applies Gz blips with alternating signs in SMS imaging as shown in Fig. 11.12C. The Gz blips keep the phase difference across different slices, while the alternating signs cancel the accumulation of phase difference within the slice as well as the resulting "voxel tilting" blurring artifacts. This technique has now been widely used for SMS-EPI.

In addition to Cartesian sampling, SMS can also be implemented for 2D non-Cartesian imaging. With non-Cartesian sampling (*e.g.*, spiral or radial acquisition), RF phase encoding does not result in a simple linear shift of the acquired images as in Cartesian imaging, but rather provides more incoherent aliasing artifacts, which allow for separation of multiple image slices using nonlinear reconstruction approaches such as compressed sensing or low-rank-based reconstruction methods (Yang *et al.*, 2019).

11.5 Accelerated MRI through reduced signal averages

As described in Section 11.2, some MRI applications, such as diffusion-weighted imaging and ASL-MRI, suffer from low SNR. Low SNR could prevent accurate clinical diagnosis and also limit the reliability in quantitative MRI. For those applications, repeated data acquisitions are commonly performed in the clinic to improve SNR via signal averaging through repetitions. The main cost of signal averaging

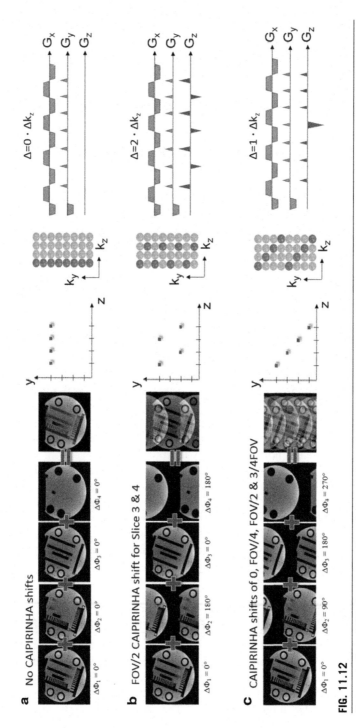

FIG. 11.12

Demonstration of RF phase encoding and/or gradient phase encoding to change aliasing pattern in SMS imaging. (A) Without phase cycling, simultaneously excited 4 image slices are superimposed on each other. (B) With 180° phase cycling, 2 images slices can be shifted by FOV/2. (C) With 90° phase cycling, all the 4 images are shifted by a different amount (0, FOV/4, FOV/2, 3/4FOV).

is prolonged scan time, and as a result, reducing signal averages can be an effective way to accelerate data acquisition if the SNR can be compensated with postprocessing denoising algorithms.

Traditionally, denoising can be performed using various compute vision techniques. For example, the concept of sparsity described in compressed sensing can be used for denoising (Elad and Aharon, 2006). The presence of image correlations provides opportunities for image denoising by exploiting image sparsity based on a total variation (TV) constraint, a total generalized variation (TGV) constraint, or other related sparsity constraints (Beck and Teboulle, 2009; Guo and Chen, 2021; Ma et al., 2018). However, image denoising is generally a challenging task, as it tends to generate image blurring. Dynamic images usually allow for better performance in image denoising due to the extensive spatiotemporal correlations. For example, a novel denoising technique called MPPCA (Marchenko-Pastur principal component analysis) has recently been proposed for diffusion-weighted MR images (Veraart et al., 2016a,b). This technique aims to exploit the spatiotemporal redundancy in diffusion MRI using universal properties of the eigenspectrum of random covariance matrices, so that noise-only principal components can be removed effectively to improve SNR. However, MPPCA exploits temporal redundancy of an image series (e.g., diffusion-weighted images with different b values), and its performance could be limited for static images.

The advent of deep learning for medical image applications brings new opportunities for efficient data-driven image denoising both for static or dynamic images (Tian et al., 2020). Similar to deep learning-based image reconstruction, deep learning-based denoising can be formulated as the following end-to-end mapping function:

$$\tilde{\theta} = \arg \min_{\theta} \left(\mathbb{E}_{\mathbf{x}_n \sim P(\mathbf{x}_n)} \left[\|C(\mathbf{x}_n|\theta) - \mathbf{x}\|_p \right] \right) \tag{11.25}$$

Here, $\mathbf{x}_n$ denotes images with reduced SNR (e.g., with less signal averages) and $\mathbf{x}$ denotes images with good SNR (e.g., with more signal averages). The learning process aims to construct a neural network that can reduce the noise level in $\mathbf{x}_n$ and improve image quality that can match $\mathbf{x}$. When sufficient datasets are available for training, deep learning methods are expected to be more robust and effective than traditional denoising methods. However, as in any deep learning-based applications, the main challenge of deep learning-based denoising is the need for a training dataset with sufficient image pairs.

11.6 Selection of acceleration methods for perfusion MRI

In MRI, fast imaging involves accelerated data acquisition and image reconstruction from acquired k-space data. Therefore selection of a specific acceleration method for perfusion MRI should take into consideration the acquisition scheme for each perfusion MRI technique. This section briefly reviews the acquisition strategies of different perfusion MRI techniques and discusses how to apply and use different accelerated imaging methods to different perfusion MRI techniques with representative examples.

11.6.1 Acceleration methods for DCE-MRI

DCE-MRI exploits the T1 shortening effect associated with GBCA. Therefore, DCE-MRI typically uses a T1-weighted imaging sequence, such as gradient echo (GRE), to measure the signal changes over time caused by the injection of GBCA for perfusion quantification. For most human organs, DCE-MRI acquisition is typically performed using a 3D sequence that provides sufficient SNR and minimizes in-flow effect, and the acquisition is usually repeated for about 5–10 min over the course of contrast enhancement to generate a series of images at different time points. In very demanding applications, such as cardiac perfusion, a 2D acquisition is used. In DCE-MRI, a small repetition time (TR) and a small echo time (TE) less than 5 milliseconds (ms) are usually used to obtain T1-weighted contrast, and a temporal resolution of less than 5 second (s) per volume has been recommended for accurate perfusion quantification (Essig *et al.*, 2013). For DCE-MRI, data acceleration is mostly achieved with undersampled data acquisition, and various image reconstruction algorithms that exploit the spatial and/or temporal correlations in the resulting dynamic images. Both Cartesian and non-Cartesian sampling can be employed in different organs, such as in the brain (Bergamino *et al.*, 2014), the heart (Hamirani and Kramer, 2014), the lung (Chen *et al.*, 2018), the prostate (Mazaheri *et al.*, 2017), the musculoskeletal system (Sujlana *et al.*, 2018), and others (Feng *et al.*, 2017). For 2D cardiac perfusion MRI, SMS acquisition can also be combined with undersampled data acquisition for additional improvement of imaging speed (Yang *et al.*, 2019; Tian *et al.*, 2019; Nazir *et al.*, 2018). In DCE-MRI, the use of GBCA yields increased SNR, so signal averaging is usually not needed in most applications.

11.6.2 Acceleration methods for DSC-MRI

DSC-MRI exploits the increased local susceptibility effect in the vessels caused by the injection of GBCA and measures resulting change of T2* (or T2). As a result, DSC-MRI typically uses a T2*-weighted (sometimes T2-weighted) sequence that has a long TR (1–2 s) and a long TE (greater than 20 ms) to achieve T2*-weighted contrast and minimize T1 contamination or T1 shortening effect due to GBCA extravasation (Boxerman *et al.*, 2020). Similar to DCE-MRI, DSC-MRI acquisition is also repeated over the course of contrast administration to acquire images at different time points, but it generally lasts for a shorter amount of time (∼2 min). The need for a long TR and a long TE necessitates a single-shot sequence for efficient data sampling. In current clinical practice, DSC-MRI acquisition is mostly applied in the brain and often performed using a multi-slice, T2*-weighted, single-shot 2D echo planar imaging (EPI) GRE sequence to enable fast data acquisition with a high temporal resolution (Boxerman *et al.*, 2020). Since DSC-MRI is generally performed using a multi-slice 2D imaging sequence, data acceleration can be achieved with in-plane undersampling and/or SMS acquisition (Han *et al.*, 2021; Chakhoyan *et al.*, 2018). However, due to the need of a relatively long TR in DSC-MRI, acceleration is more often performed to increase spatial resolution and/or slice coverage without sacrificing temporal resolution.

11.6.3 Acceleration methods for ASL-MRI

In contrast to DCE-MRI and DSC-MRI, ASL-MRI does not use exogenous GBCA and relies on magnetically labeled blood as an endogenous tracer to measure perfusion. ASL-MRI involves acquisition of two images, one with labeled blood and the other without labeling. Subtraction of the two images gives the ASL signal for perfusion quantification. An ASL-MRI sequence typically consists of a labeling phase (1.5–2 s), a postlabeling delay for the labeled blood to enter capillary (1.5–2 s), followed by data acquisition (less than 1 s) (Alsop *et al.*, 2015). In ASL-MRI, the labeled inflowing blood signal only accounts for approximately 1% of the total signal in brain tissue, which leads to inherently low ASL signal and poor SNR. Therefore, repeated acquisitions with signal averaging are usually required in ASL-MRI to ensure accurate perfusion quantification. ASL-MRI can be performed both with 2D imaging and 3D imaging, but the current consensus recommends 3D imaging to enable improved SNR (Alsop *et al.*, 2015; Zhu *et al.*, 2018a; Wells *et al.*, 2010; Spann *et al.*, 2017). The need of a long spin labeling process and associated postlabeling delay limit the opportunity to accelerate ASL-MRI, but an effective solution could be to reduce signal averages followed by denoising to recover SNR using either traditional approaches (Zhu *et al.*, 2018b) or deep learning-based approaches (Zhang *et al.*, 2022; Xie *et al.*, 2020; Hales *et al.*, 2020). When ASL-MRI is performed using 2D acquisition, SMS acquisition can also be applied (Wang *et al.*, 2015; Feinberg *et al.*, 2013; Shou *et al.*, 2021; Nanjappa *et al.*, 2021; Cohen *et al.*, 2018).

11.7 Conclusion

This chapter is focused on acceleration methods that can be applied for perfusion MRI. It describes the basics of MRI acquisition and reconstruction process, so that readers can learn how MRI acquisition can be accelerated. The chapter introduces three ways to accelerate MRI acquisition: through undersampling, SMS imaging, or reduced signal averages. To date, undersampling remains the most widely used acceleration method in MRI, which is also the primary topic of discussion in this chapter. This is the main method that can be used in DCE-MRI. For DSC-MRI and ASL-MRI, SMS imaging and reduced signal averages could be more useful, but techniques in these two directions are relatively limited compared to image reconstruction from undersampled data. Effectively, all the methods can also be combined for perfusion MRI to ensure optimal data acquisition and image quality.

Acknowledgment

The authors thank Dr. Lirong Yan from Northwestern University for helpful discussion. The brain MR image used for Figs. 11.9 and 11.10 in this chapter was obtained from the ISMRM Sunrise Course on parallel imaging (http://hansenms.github.io/sunrise/sunrise2013/).

References

Adluru, G., McGann, C., Speier, P., Kholmovski, E.G., Shaaban, A., Dibella, E.V.R., 2009. Acquisition and reconstruction of undersampled radial data for myocardial perfusion magnetic resonance imaging. J. Magn. Reson. Imaging 29, 466–473.

Akçakaya, M., Moeller, S., Weingärtner, S., Uğurbil, K., 2019. Scan-specific robust artificial-neural-networks for k-space interpolation (RAKI) reconstruction: database-free deep learning for fast imaging. Magn. Reson. Med. 81, 439–453.

Alsop, D.C., Detre, J.A., Golay, X., et al., 2015. Recommended implementation of arterial spin-labeled perfusion MRI for clinical applications: a consensus of the ISMRM perfusion study group and the European consortium for ASL in dementia. Magn. Reson. Med. 73, 102–116.

Barth, M., Breuer, F., Koopmans, P.J., Norris, D.G., Poser, B.A., 2016. Simultaneous multi-slice (SMS) imaging techniques. Magn. Reson. Med. 75, 63–81.

Beck, A., Teboulle, M., 2009. Fast gradient-based algorithms for constrained total variation image denoising and deblurring problems. IEEE Trans. Image Process. 18, 2419–2434.

Bergamino, M., Bonzano, L., Levrero, F., Mancardi, G.L., Roccatagliata, L., 2014. A review of technical aspects of T1-weighted dynamic contrast-enhanced magnetic resonance imaging (DCE-MRI) in human brain tumors. Phys. Med. 30, 635–643.

Blaimer, M., Breuer, F.A., Seiberlich, N., et al., 2006. Accelerated volumetric MRI with a SENSE/GRAPPA combination. J. Magn. Reson. Imaging 24, 444–450.

Block, K.T., 2008. Advanced Methods for Radial Data Sampling in Magnetic Resonance Imaging.

Block, K.T., Uecker, M., Frahm, J., 2007. Undersampled radial MRI with multiple coils. Iterative image reconstruction using a total variation constraint. Magn. Reson. Med. 57, 1086–1098.

Boxerman, J.L., Quarles, C.C., Hu, L.S., et al., 2020. Consensus recommendations for a dynamic susceptibility contrast MRI protocol for use in high-grade gliomas. Neuro Oncol. 22, 1262.

Breuer, F.A., Kellman, P., Griswold, M.A., Jakob, P.M., 2005a. Dynamic autocalibrated parallel imaging using temporal GRAPPA (TGRAPPA). Magn. Reson. Med. 53, 981–985.

Breuer, F.A., Blaimer, M., Heidemann, R.M., Mueller, M.F., Griswold, M.A., Jakob, P.M., 2005b. Controlled aliasing in parallel imaging results in higher acceleration (CAIPIRINHA) for multi-slice imaging. Magn. Reson. Med. 53, 684–691.

Candès, E.J., Romberg, J., Tao, T., 2006. Robust uncertainty principles: exact signal reconstruction from highly incomplete frequency information. IEEE Trans. Inf. Theory 52, 489–509.

Candès, E.J., Recht, B., Todd Candès, M.E., Recht, B., 2009. Exact matrix completion via convex optimization. Found. Comput. Math. 9, 717–772. 2009 96.

Chakhoyan, A., Leu, K., Pope, W.B., Cloughesy, T.F., Ellingson, B.M., 2018. Improved spatio-temporal resolution of dynamic susceptibility contrast perfusion MRI in brain tumors using simultaneous multi-slice echo-planar imaging. AJNR Am. J. Neuroradiol. 39, 43–45.

Chen, L., Liu, D., Zhang, J., *et al.*, 2018. Free-breathing dynamic contrast-enhanced MRI for assessment of pulmonary lesions using golden-angle radial sparse parallel imaging. J. Magn. Reson. Imaging 48, 459–468.

Christodoulou, A.G., Shaw, J.L., Nguyen, C., *et al.*, 2018. Magnetic resonance multitasking for motion-resolved quantitative cardiovascular imaging. Nat. Biomed. Eng. 2, 215–226.

Cohen, A.D., Nencka, A.S., Wang, Y., 2018. Multiband multi-echo simultaneous ASL/BOLD for task-induced functional MRI. PloS One 13, e0190427.

Cuppen, J., van Est, A., 1987. Reducing MR imaging time by one-sided reconstruction. Magn. Reson. Imaging 5, 526–527.

Delattre, B.M.A., Boudabbous, S., Hansen, C., Neroladaki, A., Hachulla, A.L., Vargas, M.I., 2020. Compressed sensing MRI of different organs: ready for clinical daily practice? Eur. Radiol. 30, 308–319.

Deshmane, A., Gulani, V., Griswold, M.A., Seiberlich, N., 2012. Parallel MR imaging. J. Magn. Reson. Imaging, 55–72.

Donoho, D.L., 2006. Compressed sensing. IEEE Trans. Inf. Theory 52, 1289–1306.

Elad, M., Aharon, M., 2006. Image denoising via sparse and redundant representations over learned dictionaries. IEEE Trans. Image Process. 15, 3736–3745.

Essig, M., Shiroishi, M.S., Nguyen, T.B., *et al.*, 2013. Perfusion MRI: the five most frequently asked technical questions. AJR Am. J. Roentgenol. 200, 24–34. https://doi.org/10.2214/AJR.12.9543.

Feinberg, D.A., Beckett, A., Chen, L., 2013. Arterial spin labeling with simultaneous multi-slice echo planar imaging. Magn. Reson. Med. 70, 1500.

Feng, L., 2022a. Golden-angle radial MRI: basics, advances, and applications. J. Magn. Reson. Imaging 56, 45–62.

Feng, L., 2022b. 4D Golden-angle radial MRI at subsecond temporal resolution. NMR Biomed. https://doi.org/10.1002/nbm.4844.

Feng, L., Grimm, R., Block, K.T., *et al.*, 2014. Golden-angle radial sparse parallel MRI: combination of compressed sensing, parallel imaging, and golden-angle radial sampling for fast and flexible dynamic volumetric MRI. Magn. Reson. Med. 72, 707–717.

Feng, L., Axel, L., Chandarana, H., Block, K.T., Sodickson, D.K., Otazo, R., 2016. XD-GRASP: golden-angle radial MRI with reconstruction of extra motion-state dimensions using compressed sensing. Magn. Reson. Med. 75, 775–788.

Feng, L., Benkert, T., Block, K.T., Sodickson, D.K., Otazo, R., Chandarana, H., 2017. Compressed sensing for body MRI. J. Magn. Reson. Imaging, 966–987.

Feng, L., Wen, Q., Huang, C., Tong, A., Liu, F., Chandarana, H., 2020. GRASP-Pro: imProving GRASP DCE-MRI through self-calibrating subspace-modeling and contrast phase automation. Magn. Reson. Med. 83, 94–108.

Feng, L., Ma, D., Liu, F., 2022. Rapid MR relaxometry using deep learning: an overview of current techniques and emerging trends. NMR Biomed. 35, e4416.

Fessler, J.A., Sutton, B.P., 2003. Nonuniform fast fourier transforms using min-max interpolation. IEEE Trans. Signal Process. 51, 560–574.

Franson, D., Ahad, J., Liu, Y., *et al.*, 2023. Self-calibrated through-time spiral GRAPPA for real-time, free-breathing evaluation of left ventricular function. Magn. Reson. Med. 89, 536–549.

Gamper, U., Boesiger, P., Kozerke, S., 2008. Compressed sensing in dynamic MRI. Magn. Reson. Med. 59, 365–373.

Grist, T.M., Mistretta, C.A., Strother, C.M., Turski, P.A., 2012. Time-resolved angiography: past, present, and future. J. Magn. Reson. Imaging 36, 1273–1286.

Griswold, M.A., Jakob, P.M., Nittka, M., Goldfarb, J.W., Haase, A., 2000. Partially parallel imaging with localized sensitivities (PILS). Magn. Reson. Med. 44, 602–609.

Griswold, M.A., Jakob, P.M., Heidemann, R.M., *et al.*, 2002. Generalized autocalibrating partially parallel acquisitions (GRAPPA). Magn. Reson. Med. 47, 1202–1210.

Guo, J., Chen, Q., 2021. Image denoising based on nonconvex anisotropic total-variation regularization. Signal Process. 186, 108124.

Guo, Y., Lingala, S.G., Zhu, Y., Lebel, R.M., Nayak, K.S., 2017. Direct estimation of tracer-kinetic parameter maps from highly undersampled brain dynamic contrast enhanced MRI. Magn. Reson. Med. 78, 1566–1578.

Haacke, E.M., Lindskogj, E.D., Lin, W., 1991. A fast, iterative, partial-fourier technique capable of local phase recovery. J. Magn. Reson. 92, 126–145.

Hales, P.W., Pfeuffer, J., Clark, C.A., 2020. Combined denoising and suppression of transient artifacts in arterial spin labeling MRI using deep learning. J. Magn. Reson. Imaging 52, 1413–1426.

Hamirani, Y.S., Kramer, C.M., 2014. Cardiac MRI assessment of myocardial perfusion. Future Cardiol. 10, 349.

Hammernik, K., Klatzer, T., Kobler, E., *et al.*, 2018. Learning a variational network for reconstruction of accelerated MRIData. Magn. Reson. Med. 79, 3055.

Han, M., Yang, B., Fernandez, B., *et al.*, 2021. Simultaneous multi-slice spin- and gradient-echo dynamic susceptibility-contrast perfusion-weighted MRI of gliomas. NMR Biomed. 34.

Heidemann, R.M., Griswold, M.A., Haase, A., Jakob, P.M., 2001. VD-AUTO-SMASH imaging. Magn. Reson. Med. 45, 1066–1074.

Hennig, J., Scheffler, K., Laubenberger, J., Strecker, R., 1997. Time-resolved projection angiography after bolus injection of contrast agent. Magn. Reson. Med. 37, 341–345.

Hollingsworth, K.G., 2015. Reducing acquisition time in clinical MRI by data undersampling and compressed sensing reconstruction. Phys. Med. Biol., R297–R322.

Huang, F., Akao, J., Vijayakumar, S., Duensing, G.R., Limkeman, M., 2005. K-t GRAPPA: a k-space implementation for dynamic MRI with high reduction factor. Magn. Reson. Med. 54, 1172–1184.

Jahng, G.H., Li, K.L., Ostergaard, L., Calamante, F., 2014. Perfusion magnetic resonance imaging: a comprehensive update on principles and techniques. Korean J. Radiol. 15, 554–577.

Jakob, P.M., Grisowld, M.A., Edelman, R.R., Sodickson, D.K., 1998. AUTO-SMASH: a self-calibrating technique for SMASH imaging. SiMultaneous Acquisition of Spatial Harmonics. MAGMA 7, 42–54.

Jaspan, O.N., Fleysher, R., Lipton, M.L., 2015. Compressed sensing MRI: a review of the clinical literature. Br. J. Radiol. 88, 20150487.

Johnson, P.M., Muckley, M.J., Bruno, M., *et al.*, 2019. Joint multi-anatomy training of a variational network for reconstruction of accelerated magnetic resonance image acquisitions. In: Lect Notes Comput Sci (including Subser Lect Notes Artif Intell Lect Notes Bioinformatics). vol. 11905, pp. 71–79. LNCS.

Johnson, P.M., Jeong, G., Hammernik, K., *et al.*, 2021. Evaluation of the robustness of learned MR image reconstruction to systematic deviations between training and test data for the models from the fastMRI challenge. In: Lect Notes Comput Sci (including Subser Lect Notes Artif Intell Lect Notes Bioinformatics). vol. 12964, pp. 25–34. LNCS.

Jung, B., Ullmann, P., Honal, M., Bauer, S., Hennig, J., Markl, M., 2008. Parallel MRI with extended and averaged GRAPPA kernels (PEAK-GRAPPA): optimized spatiotemporal dynamic imaging. J. Magn. Reson. Imaging 28, 1226–1232.

Kellman, P., Epstein, F.H., McVeigh, E.R., 2001. Adaptive sensitivity encoding incorporating temporal filtering (TSENSE). Magn. Reson. Med. 45, 846–852.

Korosec, F.R., Frayne, R., Grist, T.M., Mistretta, C.A., 1996. Time-resolved contrast-enhanced 3D MR angiography. Magn. Reson. Med. 36, 345–351.

Lai, P., Huang, F., Larson, A.C., Li, D., 2008. Fast four-dimensional coronary MR angiography with k-t GRAPPA. J. Magn. Reson. Imaging 27, 659.

Larkman, D.J., Hajnal, J.V., Herlihy, A.H., Coutts, G.A., Young, I.R., Sta Ehnholm, G., 2001. Use of multicoil arrays for separation of signal from multiple slices simultaneously excited. J. Magn. Reson. Imaging 13, 313–317.

Li, Z., Huang, C., Tong, A., Chandarana, H., Feng, L., 2022. Kz-accelerated variable-density stack-of-stars MRI. Magn. Reson. Imaging 97, 56–67.

Liang, Z.P., 2007. Spatiotemporal imaging with partially separable functions. In: 2007 4th IEEE Int Symp Biomed Imaging From Nano to Macro- Proc, pp. 988–991.

Liang, Z.-P., Boada, F., Constable, R., Haacke, E., Lauterbur, P., Smith, M., 1992. Constrained reconstruction methods in MR imaging. Rev. Magn. Reson. Med. 4, 67–185.

Liang, D., Liu, B., Wang, J., Ying, L., 2009. Accelerating SENSE using compressed sensing. Magn. Reson. Med. 62, 1574–1584.

Lingala, S.G., Hu, Y., Dibella, E., Jacob, M., 2011. Accelerated dynamic MRI exploiting sparsity and low-rank structure: k-t SLR. IEEE Trans. Med. Imaging 30, 1042–1054.

Lingala, S.G., Guo, Y., Bliesener, Y., et al., 2020. Tracer kinetic models as temporal constraints during brain tumor DCE-MRI reconstruction. Med. Phys. 47, 37–51.

Liu, B., Zou, Y.M., Ying, L., 2008. Sparsesense: application of compressed sensing in parallel MRI. In: 5th Int Conf Inf Technol Appl Biomed ITAB 2008 Conjunction with 2nd Int Symp Summer Sch Biomed Heal Eng IS3BHE 2008, pp. 127–130.

Liu, F., Samsonov, A., Chen, L., Kijowski, R., Feng, L., 2019a. SANTIS: sampling-augmented neural neTwork with incoherent structure for MR image reconstruction. Magn. Reson. Med. 82, 1890–1904.

Liu, F., Feng, L., Kijowski, R., 2019b. MANTIS: model-augmented neural neTwork with incoherent k-space sampling for efficient MR parameter mapping. Magn. Reson. Med. 82, 174–188.

Liu, F., Kijowski, R., Feng, L., El Fakhri, G., 2020. High-performance rapid MR parameter mapping using model-based deep adversarial learning. Magn. Reson. Imaging 74, 152–160.

Liu, F., Kijowski, R., El Fakhri, G., Feng, L., 2021. Magnetic resonance parameter mapping using model-guided self-supervised deep learning. Magn. Reson. Med. 85, 3211–3226.

Lustig, M., Pauly, J.M., 2010. SPIRiT: iterative self-consistent parallel imaging reconstruction from arbitrary k-space. Magn. Reson. Med. 64, 457–471.

Lustig, M., Donoho, D., Pauly, J.M., 2007. Sparse MRI: the application of compressed sensing for rapid MR imaging. Magn. Reson. Med. 58, 1182–1195.

Lustig, M., Donoho, D.L., Santos, J.M., Pauly, J.M., 2008. Compressed sensing MRI: a look at how CS can improve on current imaging techniques. IEEE Signal Process. Mag., 72–82.

Ma, T.H., Huang, T.Z., Le, Z.X., 2018. Spatially dependent regularization parameter selection for total generalized variation-based image denoising. Comput. Appl. Math. 37, 277–296.

Madore, B., Glover, G.H., Pelc, N.J., 1999. Unaliasing by fourier-encoding the overlaps using the temporal dimension (UNFOLD), applied to cardiac imaging and fMRI. Magn. Reson. Med. 42, 813–828.

Mardani, M., Gong, E., Cheng, J.Y., *et al.*, 2019. Deep generative adversarial neural networks for compressive sensing MRI. IEEE Trans. Med. Imaging 38, 167–179.

Margosian, P., Schmitt, F., 1986. Faster MR imaging methods. Med. Image Process. 0593, 6.

Mazaheri, Y., Akin, O., Hricak, H., 2017. Dynamic contrast enhanced magnetic resonance imaging of prostate cancer: a review of current methods and applications. World J. Radiol. 9, 416.

McGibney, G., Smith, M.R., Nichols, S.T., Crawley, A., 1993. Quantitative evaluation of several partial fourier reconstruction algorithms used in mri. Magn. Reson. Med. 30, 51–59.

Moghari, M.H., Uecker, M., Roujol, S., Sabbagh, M., Geva, T., Powell, A.J., 2018. Accelerated whole-heart MR angiography using a variable-density poisson-disc undersampling pattern and compressed sensing reconstruction. Magn. Reson. Med. 79, 761–769.

Nanjappa, M., Troalen, T., Pfeuffer, J., *et al.*, 2021. Comparison of 2D simultaneous multi-slice and 3D GRASE readout schemes for pseudo-continuous arterial spin labeling of cerebral perfusion at 3 T. Magn. Reson. Mater. Phys. Biol. Med. 34, 437–450.

Nazir, M.S., Neji, R., Speier, P., *et al.*, 2018. Simultaneous multi slice (SMS) balanced steady state free precession first-pass myocardial perfusion cardiovascular magnetic resonance with iterative reconstruction at 1.5 T. J. Cardiovasc. Magn. Reson. 20, 1–11.

Noll, D.C., Nishimura, D.G., Macovski, A., 1991. Homodyne detection in magnetic resonance imaging. IEEE Trans. Med. Imaging 10, 154–163.

Nunes, R., Hajnal, J., Golay, X., Larkman, D., 2006. Simultaneous slice excitation and reconstruction for single shot EPI. Proc. Int. Soc. Magn. Reson. Med. 14, 293.

Ong, F., Zhu, X., Cheng, J.Y., *et al.*, 2020. Extreme MRI: large-scale volumetric dynamic imaging from continuous non-gated acquisitions. Magn. Reson. Med. 84, 1763–1780.

Otazo, R., Kim, D., Axel, L., Sodickson, D.K., 2010. Combination of compressed sensing and parallel imaging for highly accelerated first-pass cardiac perfusion MRI. Magn. Reson. Med. 64, 767–776.

Otazo, R., Candès, E., Sodickson, D.K., 2015. Low-rank plus sparse matrix decomposition for accelerated dynamic MRI with separation of background and dynamic components. Magn. Reson. Med. 73, 1125–1136.

Pedersen, H., Kozerke, S., Ringgaard, S., Nehrke, K., Won, Y.K., 2009. k-t PCA: temporally constrained k-t BLAST reconstruction using principal component analysis. Magn. Reson. Med. 62, 706–716.

Pruessmann, K.P., Weiger, M., Scheidegger, M.B., Boesiger, P., 1999. SENSE: sensitivity encoding for fast MRI. Magn. Reson. Med. 42, 952–962.

Pruessmann, K.P., Weiger, M., Börnert, P., Boesiger, P., 2001. Advances in sensitivity encoding with arbitrary k-space trajectories. Magn. Reson. Med. 46, 638–651.

Ra, J.B., Rim, C.Y., 1993. Fast imaging using subencoding data sets from multiple detectors. Magn. Reson. Med. 30, 142–145.

Relax-MANTIS, n.d. Relax-MANTIS: REference-free LAtent map-eXtracting MANTIS for Efficient MR Parametric Mapping with Unsupervised Deep Learning [https://cds.ismrm. org/protected/19MPresentations/abstracts/1098.html].

Saranathan, M., Rettmann, D.W., Hargreaves, B.A., Clarke, S.E., Vasanawala, S.S., 2012. DIfferential subsampling with Cartesian ordering (DISCO): a high spatio-temporal resolution dixon imaging sequence for multiphasic contrast enhanced abdominal imaging. J. Magn. Reson. Imaging 35, 1484.

Schlemper, J., Caballero, J., Hajnal, J.V., Price, A.N., Rueckert, D., 2018. A deep cascade of convolutional neural networks for dynamic MR image reconstruction. IEEE Trans. Med. Imaging 37, 491–503.

Seiberlich, N., Ehses, P., Duerk, J., Gilkeson, R., Griswold, M., 2011. Improved radial GRAPPA calibration for real-time free-breathing cardiac imaging. Magn. Reson. Med. 65, 492.

Setsompop, K., Gagoski, B.A., Polimeni, J.R., Witzel, T., Wedeen, V.J., Wald, L.L., 2012. Blipped-controlled aliasing in parallel imaging for simultaneous multislice echo planar imaging with reduced g-factor penalty. Magn. Reson. Med. 67, 1210–1224.

Shou, Q., Shao, X., Wang, D.J.J., 2021. Super-resolution arterial spin labeling using slice-dithered enhanced resolution and simultaneous multi-slice acquisition. Front. Neurosci. 15, 1433.

Sodickson, D.K., Manning, W.J., 1997. Simultaneous acquisition of spatial harmonics (SMASH): fast imaging with radiofrequency coil arrays. Magn. Reson. Med. 38, 591–603.

Song, H.K., Dougherty, L., 2000. k-space weighted image contrast (KWIC) for contrast manipulation in projection reconstruction MRI. Magn. Reson. Med. 44, 825–832.

Spann, S.M., Kazimierski, K.S., Aigner, C.S., Kraiger, M., Bredies, K., Stollberger, R., 2017. Spatio-temporal TGV denoising for ASL perfusion imaging. Neuroimage 157, 81–96.

Sujlana, P., Skrok, J., Fayad, L.M., 2018. Review of dynamic contrast-enhanced MRI: technical aspects and applications in the musculoskeletal system. J. Magn. Reson. Imaging 47, 875–890.

Tian, Y., Mendes, J., Pedgaonkar, A., et al., 2019. Feasibility of multiple-view myocardial perfusion MRI using radial simultaneous multi-slice acquisitions. PloS One 14.

Tian, C., Fei, L., Zheng, W., Xu, Y., Zuo, W., Lin, C.W., 2020. Deep learning on image denoising: an overview. Neural Netw. 131, 251–275.

Tsao, J., Kozerke, S., 2012. MRI temporal acceleration techniques. J. Magn. Reson. Imaging 36, 543–560.

Tsao, J., Boesiger, P., Pruessmann, K.P., 2003. k-t BLAST and k-t SENSE: dynamic MRI with high frame rate exploiting spatiotemporal correlations. Magn. Reson. Med. 50, 1031–1042.

Van Vaals, J.J., Brummer, M.E., Thomas Dixon, W., et al., 1993. "Keyhole" method for accelerating imaging of contrast agent uptake. J. Magn. Reson. Imaging 3, 671–675.

Veraart, J., Novikov, D.S., Christiaens, D., Ades-aron, B., Sijbers, J., Fieremans, E., 2016a. Denoising of diffusion MRI using random matrix theory. Neuroimage 142, 394–406.

Veraart, J., Fieremans, E., Novikov, D.S., 2016b. Diffusion MRI noise mapping using random matrix theory. Magn. Reson. Med. 76, 1582–1593.

Wang, Y., Moeller, S., Li, X., et al., 2015. Simultaneous multi-slice Turbo-FLASH imaging with CAIPIRINHA for whole brain distortion-free pseudo-continuous arterial spin labeling at 3 and 7 T. Neuroimage 113, 279–288.

Wang, S., Su, Z., Ying, L., et al., 2016 June. Accelerating magnetic resonance imaging via deep learning. In: Proc—Int Symp Biomed Imaging 2016, pp. 514–517.

Wells, J.A., Thomas, D.L., King, M.D., Connelly, A., Lythgoe, M.F., Calamante, F., 2010. Reduction of errors in ASL cerebral perfusion and arterial transit time maps using image de-noising. Magn. Reson. Med. 64, 715–724.

Willinek, W.A., Hadizadeh, D.R., Von Falkenhausen, M., et al., 2008. 4D time-resolved MR angiography with keyhole (4D-TRAK): more than 60 times accelerated MRA using a combination of CENTRA, keyhole, and SENSE at 3.0T. J. Magn. Reson. Imaging 27, 1455–1460.

Wright, K.L., Hamilton, J.I., Griswold, M.A., Gulani, V., Seiberlich, N., 2014. Non-Cartesian parallel imaging reconstruction. J. Magn. Reson. Imaging 40, 1022–1040.

Xie, D., Li, Y., Yang, H., *et al.*, 2020. Denoising arterial spin labeling perfusion MRI with deep machine learning. Magn. Reson. Imaging 68, 95–105.

Yang, G., Yu, S., Dong, H., *et al.*, 2018. DAGAN: deep de-aliasing generative adversarial networks for fast compressed sensing MRI reconstruction. IEEE Trans. Med. Imaging 37, 1310–1321.

Yang, Y., Meyer, C.H., Epstein, F.H., Kramer, C.M., Salerno, M., 2019. Whole-heart spiral simultaneous multi-slice first-pass myocardial perfusion imaging. Magn. Reson. Med. 81, 852–862.

Ye, J.C., 2019. Compressed sensing MRI: a review from signal processing perspective. BMC Biomed. Eng. 1, 1–17. 2019 11.

Yoon, J.H., Nickel, M.D., Peeters, J.M., Lee, J.M., 2019. Rapid imaging: recent advances in abdominal MRI for reducing acquisition time and its clinical applications. Korean J. Radiol. 20, 1597–1615.

Zeng, G., Guo, Y., Zhan, J., *et al.*, 2021. A review on deep learning MRI reconstruction without fully sampled k-space. BMC Med. Imaging 21, 1–11.

Zhang, L., Xie, D., Li, Y., *et al.*, 2022. Improving sensitivity of arterial spin labeling perfusion MRI in Alzheimer's disease using transfer learning of deep learning-based ASL denoising. J. Magn. Reson. Imaging 55, 1710–1722.

Zhao, B., Haldar, J.P., Christodoulou, A.G., Liang, Z.P., 2012. Image reconstruction from highly undersampled (k, t)-space data with joint partial separability and sparsity constraints. IEEE Trans. Med. Imaging 31, 1809–1820.

Zhou, Z., Zhao, G., Kijowski, R., Liu, F., 2018. Deep convolutional neural network for segmentation of knee joint anatomy. Magn. Reson. Med. 80, 2759–2770.

Zhu, H., He, G., Wang, Z., 2018a. Patch-based local learning method for cerebral blood flow quantification with arterial spin-labeling MRI. Med. Biol. Eng. Comput. 56, 951–956.

Zhu, H., Zhang, J., Wang, Z., 2018b. Arterial spin labeling perfusion MRI signal denoising using robust principal component analysis. J. Neurosci. Methods 295, 10–19.

Further reading

There is a list of review papers that summarize different accelerated MRI methods that the readers can further read.

Barth, M., Breuer, F., Koopmans, P.J., Norris, D.G., Poser, B.A., 2016. Simultaneous multislice (SMS) imaging techniques. Magn. Reson. Med. 75, 63–81.

Deshmane, A., Gulani, V., Griswold, M.A., Seiberlich, N., 2012. Parallel MR imaging. J. Magn. Reson. Imaging, 55–72.

Feng, L., 2022. Golden-angle radial MRI: basics, advances, and applications. J. Magn. Reson. Imaging 56 (1), 45–62.

Feng, L., Benkert, T., Block, K.T., Sodickson, D.K., Otazo, R., Chandarana, H., 2017. Compressed sensing for body MRI. J. Magn. Reson. Imaging, 966–987.

Feng, L., Ma, D., Liu, F., 2022. Rapid MR relaxometry using deep learning: an overview of current techniques and emerging trends. NMR Biomed. 35, e4416.

Lustig, M., Donoho, D.L., Santos, J.M., Pauly, J.M., 2008. Compressed sensing MRI: a look at how CS can improve on current imaging techniques. IEEE Signal Process. Mag., 72–82.

Tsao, J., Kozerke, S., 2012. MRI temporal acceleration techniques. J. Magn. Reson. Imaging 36, 543–560.

Wright, K.L., Hamilton, J.I., Griswold, M.A., Gulani, V., Seiberlich, N., 2014. Non-Cartesian parallel imaging reconstruction. J. Magn. Reson. Imaging 40, 1022–1040.

Ye, J.C., 2019. Compressed sensing MRI: a review from signal processing perspective. BMC Biomed. Eng. 1, 1–17. 2019 11.

Artificial intelligence: The next frontier of perfusion imaging?

12

Cian M. Scannell[a,b], Amedeo Chiribiri[b], and Tim Leiner[c,d]

[a]*Department of Biomedical Engineering, Eindhoven University of Technology, Eindhoven, The Netherlands*
[b]*School of Biomedical Engineering and Imaging Sciences, King's College London, London, United Kingdom*
[c]*Department of Radiology, Mayo Clinic, Minnesota, United States*
[d]*Department of Radiology, University Medical Center Utrecht, Utrecht, The Netherlands*

12.1 Introduction

There is little doubt that artificial intelligence (AI) will have a profound and lasting impact on how medical images are analyzed. Perfusion MRI is not yet realizing the full potential of AI, but the capability of AI to augment the clinical workflow is beginning to be seen in research settings, and it is only a matter of time until AI applications are impacting clinical practice. Indeed, it could be argued that perfusion MRI has more to gain from the use of AI than most applications due to inherently long imaging time, complex postprocessing, and the difficulty of image interpretation. All these challenges can be addressed through the use of AI, and the solution to some or all of these would encourage more routine use, outside of highly specialized centers.

In particular, in a variety of applications, AI has been shown to facilitate faster scanning by allowing the reconstruction of highly undersampled images, also with very quick reconstruction times. It has proved especially adept at segmenting anatomical structures and identifying landmarks on images to automate postprocessing steps, and AI is being developed for classifying diseases and automating reporting. AI could further allow more routine quantification of perfusion MRI by reducing the amount of manual interaction required, leading to precise measurements that are reproducible and enabling more sophisticated diagnoses, the identification of less common diseases, and personalized risk stratification and prognostication.

A major benefit of an AI-based analysis in this instance is that (typically) once a model is trained, it is then fixed and becomes deterministic so that the same AI algorithm applied to the same images will yield the same results with little or no interaction. This is an important consideration in applications such as perfusion MRI, where there is considerable inter- and intraoperator variability and a lack of

Advances in Magnetic Resonance Technology and Applications, Volume 11, ISSN 2666-9099
https://doi.org/10.1016/B978-0-323-95209-5.00002-7

trained operators. For cardiac perfusion MRI, Villa *et al.* showed a strong dependency of the diagnostic accuracy of the modality on the level of training of the operator (Villa *et al.*, 2018). This serves to highlight the need for an objective, user-independent assessment of perfusion. Quantification was proposed as a potential solution to this that could also be used by less experienced operators, but this involved time-consuming manual processing. AI has the potential to add the benefits of speed, automation, and reproducibility to this test.

Perfusion MRI is, however, also likely to be a challenging domain for developing AI. The data are often large 3-dimensional (D) volumes, with an additional temporal component causing both memory and computational time considerations for the use and development of AI. It is also likely that more sophisticated algorithms will be required to leverage the full spatiotemporal information. There is lower signal to noise than purely anatomical MR images, as short acquisition times are required to satisfy the temporal sampling requirements, and the large variations in the appearance of perfusion images between different disease types are a challenge for humans and computers alike. AI algorithms are further hampered by their ability to robustly generalize to images from different domains. In perfusion MRI, this is a significant concern as there is not yet standardization on MR pulse sequences, scan parameters, or acquisition protocols.

12.2 Background

AI has started to significantly impact the field of medical imaging, especially for image processing tasks, such as segmentation and landmark detection. Despite its recent rise to attention, AI is not a new term, and it can refer to a wide variety of methods. It can be loosely defined as any computer program that mimics some aspect of human decision-making, including programs that use a deterministic set of rules and do not learn from data. By this definition, the field, pioneered by Alan Turing, is almost 100 years old.

Machine learning (ML) is the subfield of AI in which rules do not need to be prespecified and explicitly written—they can be learned from data. This can be categorized as either supervised or unsupervised learning. Supervised learning mandates that ground-truth labels exist for the training data and a model is learnt to predict these labels. Unsupervised learning does not require labels but, rather, attempts to learn underlying patterns in the data. Deep learning (DL) is the subfield of ML that has driven the recent advances in medical imaging.

12.2.1 Deep learning

DL is based on (artificial) neural networks (NNs), in which the components of the model (units) loosely approximate neurons, as found in the brain. The most basic type of NN is a fully connected NN, as shown in Fig. 12.1A. This design of fully connected networks is inefficient or infeasible for image processing because of

a

Fully-connected network

b

Convolutional network

c

U-Net

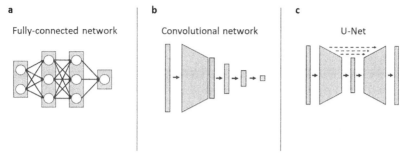

FIG. 12.1

A schematic representation of the basic forms of neural networks that are discussed in this chapter. (A) A fully connected network, (B) a convolutional network, and (C) a U-Net.

the high dimensionality of the inputs (the dimensionality is equal to the number of pixels or voxels in the image). An alternative network design, more suitable for image processing, is the convolutional network (CNN), as visualized in Fig. 12.1B. A CNN combines inputs from only a small region of the previous layer (commonly referred to as the receptive field and designed to loosely resemble the visual cortex). This is implemented as a learnable kernel being convolved with the output of the previous layer. The convolutional kernel is designed to be significantly smaller than the size of the activation it is being convolved with. This enforces sparse interactions between layers and allows the extraction of low-level features, such as edges, without considering the whole image. Parameter sharing is also used such that the same kernel is applied to all inputs to a layer. This aids the training process by greatly reducing the number of parameters to be learned. CNNs also encourage translational invariance in the predictions: as a kernel slides over the whole input, it will detect the same features regardless of their position. This is useful, for example, in classification tasks, when it only matters if an object is in an image and not where it is located (Goodfellow *et al.*, 2016).

The field of image processing was revolutionized when a deep convolutional neural network, AlexNet (Krizhevsky *et al.*, 2012), won the ImageNet Large Scale Visual Recognition Challenge by a large margin in 2012 (Russakovsky *et al.*, 2015). This was the first demonstration that modern computer hardware, such as graphics processing units (GPUs), could be combined with large databases to train deep neural networks to successfully perform computer vision tasks. Since this, deep learning has been widely adopted in the field of medical imaging and has become the *de facto* standard for many processing tasks (Litjens *et al.*, 2017). Trends in perfusion MRI image analysis have followed a similar route where deep learning is now used for many tasks, from reconstruction to detection and segmentation tasks to automating diagnostics and prognostics (Leiner *et al.*, 2019).

The most commonly used network architecture for medical image analysis is now the U-Net (Ronneberger *et al.*, 2015). This is a fully convolutional network in that it only uses convolutional layers and is illustrated in Fig. 12.1C. The architecture has an encoder–decoder structure. The encoder down-samples the input image, using max pooling, to create a low-dimensional embedding. The multistep down-sampling allows the learning of feature representations at different image scales. The decoder takes the low-dimensional embedding and up-samples it to predict the desired output, typically a segmentation map. The network also utilizes skip connections that concatenate the activations of the encoder to the input of the corresponding resolution level of the decoder. This is thought to better allow the recovery of the fine-grained details in the prediction.

12.3 AI in perfusion MRI

AI has the potential to benefit nearly every aspect of perfusion MRI, from the selection of which patients should receive perfusion imaging to the reporting of their images. Fig. 12.2 shows a sample of the steps in the process that could be replaced by a neural network.

The focus of this chapter will be on the MR images, how to form them, how to analyze them, and how to interpret these analyses, but it is important to acknowledge that there is also scope to significantly improve the process before image formation. Sandino *et al.* recently reviewed upstream tasks in radiology and many of their discussions are relevant to perfusion MRI (Sandino *et al.*, 2021).

12.3.1 Acquisition

The first step in an MRI exam is to acquire a series of localizers that are used to identify anatomical structures and plan the subsequent acquisitions. AI would allow a more standardized approach to the planning. A model could be trained that takes the localizers as input and outputs the positions of the desired imaging planes. For applications where 3D volumes are acquired, the accuracy of the planning might be relatively less important as long as the organ of interest is covered, but an AI solution may be faster and thus help to improve patient throughput. A further AI model could allow automatic shimming and together these models could lead to improved B0 magnetic field homogeneity.

One application where the planning may be more difficult and the result more important is cardiac perfusion, where typically there is a 2D acquisition with limited sampling (just three slices) of the region of interest, the left ventricle (LV). Therefore it is crucial that the positioning of these slices is optimized. If the slices begin too low, it is possible to miss a perfusion defect close to the base of the LV. It is also possible to plan slices that are partially above the base or below the apex, thus, missing parts of the ventricle.

The idea of automatically prescribing the acquisition views and shimming has been explored extensively, even before the recent deep learning era, using standard image processing techniques (Frick *et al.*, 2011); now it is seeing renewed attention with the use of AI models (Blansit *et al.*, 2019; Edalati *et al.*, 2022). Edalati *et al.*

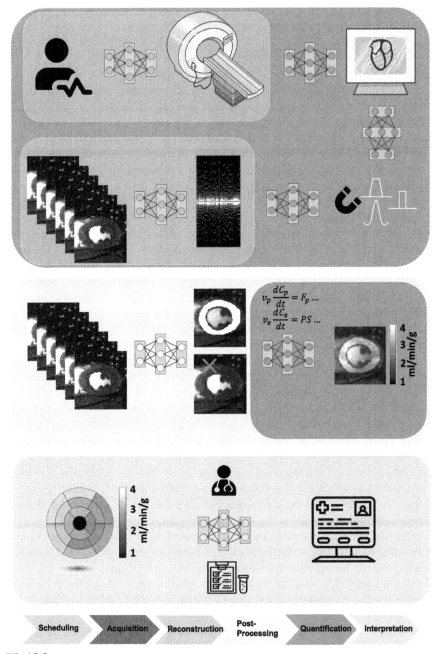

FIG. 12.2

A schematic representation of the steps in perfusion MRI that can be addressed with AI. This could include scheduling the scan and choosing the allocation of scan time, scan planning, selecting the pulse sequences and parameters, image reconstruction, segmentation and landmark detection, quantification, and interpretation and report generation (with interaction from the clinician and other clinical data). Each of these steps will be discussed in detail over the coming sections. The scheme is demonstrated with cardiac images but remains the same for any application.

Parts of the figure were drawn by using pictures from Servier Medical Art. Servier Medical Art by Servier is licensed under a Creative Commons Attribution 3.0 Unported License (https://creativecommons.org/licenses/by/3.0/). It has also used resources from Flaticon.com and modifications of images from references (Scannell et al., 2020b; Tourais et al., 2022) under their license https://creativecommons.org/licenses/by/4.0/.

combined both the problems of scan planning and shimming using two AI models and included a prospective validation of their approach (Edalati *et al.*, 2022). They found less plane angulation errors from planes prescribed using AI as compared to other automated approaches and improved B0 field homogeneity. These works considered prescribing views for cardiac MRI in general, rather than specifically perfusion MRI, but, again, perfusion imaging would have a lot to benefit from the successful deployment of such a system, and it could lead to reduced barriers to access in less specialized centers.

Since an MR scanner is such a flexible system and perfusion imaging is a complex process, there are still many acquisition parameters that can be optimized after choosing the anatomical location to scan. AI can again help to make this work automated and user independent. Work that has already been explored includes, for example, determining the center frequency for bSSFP imaging (Goldfarb *et al.*, 2018), selecting the optimal inversion time (Jiang *et al.*, 2018), and detecting if the current setup will lead to image artifacts (Bahrami *et al.*, 2019). Further possibilities would be to tune the scanner parameters to adapt to the specific acquisition needed or patient requirements, such as image resolution or breath holds.

The final frontier for the use of AI for perfusion MRI acquisition, which has already seen some much more mature developments, is to speed up the acquisition. To satisfy the strict temporal sampling requirements and to allow accurate quantification of perfusion parameters, fast imaging is required. Recent work on compressive sensing-type reconstructions has shown the ability to reconstruct good quality images while sampling below the Nyquist rate. Such reconstructions require a (pseudo-)random undersampling pattern in k-space. Smith *et al.* studied the effect of the randomness in the undersampling pattern on the derived quantitative perfusion values (Smith *et al.*, 2012). They showed that, while the effect is small, there are differences depending on the undersampling. This finding also suggests that since there are differences based on the undersampling there is also an optimal pattern. A recent research direction has focused on the use of AI models to learn the optimal undersampling pattern (van Gorp *et al.*, 2021). Many of these works have also trained the sampling models together with a reconstruction model in an end-to-end manner, so that the sampling is also informed by the reconstruction algorithm in order to improve results (Aggarwal and Jacob, 2020; Bahadir *et al.*, 2020; Huijben *et al.*, 2020; Weiss *et al.*, 2021).

Specific to perfusion MRI, in which the data is spatiotemporal, is the possibility of spatiotemporal (*k-t*) acceleration (Plein *et al.*, 2007; Lebel *et al.*, 2014). *k-t* acceleration can be implemented with dynamically varying random undersampling (Tourais *et al.*, 2022). Learning such a pattern with AI has yet to be explored but could be tackled by adding a temporal component to the models described in the previous paragraph. Recent work by Perlman *et al.* demonstrated some of these capabilities proposing an end-to-end AI pipeline that included a saturation block, a spin dynamics block, and a reconstruction network that are simultaneously trained, while being guided by the MR physics, to discover the optimal acquisition and reconstruct parameter maps (Perlman *et al.*, 2022).

12.3.2 **Reconstruction**

As discussed in the Acquisition section, fast acquisition is required for perfusion MRI. After acquiring fewer data, more advanced image reconstruction or enhancement algorithms are required. In particular, reconstructing images from undersampled k-space data is an ill-posed inverse problem, in that the solution is not unique and is highly sensitive to noise in the input. As such, strong regularization or prior knowledge is required to constrain the optimization problem and reconstruct high-quality images. Classically for compressive sensing reconstructions in MRI, the prior knowledge assumed is that the image will be sparse in some transform (e.g., wavelet) domain (Lustig *et al.*, 2007). The prior knowledge most commonly exploited for perfusion MRI is that the images will be highly correlated in the temporal direction. This regularization can be incorporated by minimizing the temporal total variation (Chen *et al.*, 2014) or enforcing that the reconstructed images should be low rank (Lingala *et al.*, 2011). More advanced reconstruction algorithms can enforce the tracer kinetics of the dynamic contrast enhancement as a temporal constraint (Lingala *et al.*, 2020) or to even directly reconstruct tracer-kinetic parameters from the *k-t* undersampled data (Dikaios *et al.*, 2014; Guo *et al.*, 2017; Correia *et al.*, 2019).

The limitation of all advanced image reconstruction algorithms is that they iteratively solve a nonlinear optimization problem and, therefore, they require long computation times. They also consider the reconstruction of each new image dataset as a completely independent problem from any previously reconstructed images. The use of AI would potentially address parts of these limitations. Instead of solving independent inverse problems for each reconstruction task, a model could be trained to learn the mapping from the raw data to images. This would incorporate information that it has learned during training from different patients and, once training is complete, it would be fast to apply to new data as a single forward pass of a neural network.

The direct mapping of acquired data to images was implemented in AUTOMAP (Zhu *et al.*, 2018), but more recent work has used AI in a less direct manner, incorporating ideas about how the problem should be solved based on traditional iterative image reconstruction algorithms. Hammernik *et al.* (2018) introduced a variational network for MRI reconstruction which 'unfolds' the iterations of the optimization problem into a chain of steps through a neural network and showed improved and more efficient reconstruction of static MR images over compressed sensing. This has inspired several works based on unrolling compressive sensing-type reconstruction in which the regularization functions are replaced with a CNN. The motivation for this stems from the idea that the regularization terms typically employed, e.g., wavelets, are not designed specifically for image reconstruction and that a function that is learned specifically for this task may be more efficient. The idea of unfolding the steps of the iterative algorithm as the stages of the pass through a CNN with learned regularization terms has proven very successful, including as the leading entry for the fast MRI challenge (Pezzotti *et al.*, 2020), albeit for static knee MRI data. Koçanaoğulları *et al.* took this approach to learn a regularized deep image generator network used as a prior to constrain the undersampled reconstruction of kidney DCE-MRI data (Koçanaoğulları *et al.*, 2022), giving improved image quality and temporal consistency that led to better estimation of the tracer-kinetic parameters.

As with the work on AI for acquisition, much of this work has been done with static MRI images, but similar ideas could be applied to quantitative perfusion MRI, possibly using transformers to model temporal dependencies (Shamshad *et al.*, 2022). The MoDL approach again merged the model-based approached with deep learning by incorporating the physics governing the problem to be solved (Aggarwal *et al.*, 2019); this could be extended for perfusion MRI, potentially by adding the tracer-kinetic modeling or quantification as part of the reconstruction. This idea has been addressed in preliminary work by Martín-González *et al.* who proposed a physics-informed self-supervised deep learning framework for reconstructing highly accelerated cardiac perfusion MR data (Martín-González *et al.*, 2021). It could also be considered that the amount of regularization required for reconstructing optimal images is patient specific and could be predicted using AI.

The alternative approach for using AI to reconstruct undersampled perfusion MRI data is to reconstruct low-quality images directly with an inverse Fourier transform and to use AI to improve the quality of these images. This approach may be more straightforward to implement for perfusion MRI as there is not such an emphasis on explicitly accounting for the temporal nature of the data and there are several examples of successful applications of such an idea to cardiac perfusion MRI (Fan *et al.*, 2020; Le *et al.*, 2021; Wang *et al.*, 2022).

12.3.3 Postprocessing

The postprocessing of images is the most developed application of AI in medical imaging, and the same holds true for quantitative perfusion MRI, with a particular focus on automatic segmentation. While in many ways the initial promise of AI in medical imaging has not been fulfilled, there are a number of well-conducted examples of AI in a variety of applications for image segmentation.

For brain perfusion MRI applications, AI has been developed for brain tumor bed detection (Jeong *et al.*, 2021), ischemic stroke lesion segmentation (Pérez Malla *et al.*, 2019), and anatomical segmentation for input into a radiomic analysis (Park *et al.*, 2021). Nalepa *et al.* developed a fully automated pipeline for quantification using AI to extract the tumor region and vascular input function (Nalepa *et al.*, 2020), and there has also been work to directly segment gliomas from multimodel brain MRI (including perfusion images) (Menze *et al.*, 2021). In other applications, it has been used for segmentation in renal perfusion (Haghighi *et al.*, 2018), and Bones *et al.* developed a pipeline for renal perfusion MRI quantification, including cropping before motion correction and automatic renal cortex segmentation (Bones *et al.*, 2022). AI has also been used for tissue segmentation in breast perfusion MRI (Zhang *et al.*, 2019), segmentations have been incorporated to detect breast cancer from screening MRI (Dalmış *et al.*, 2018), Pellicer-Valero *et al.* have segmented multiparametric prostate images (Pellicer-Valero *et al.*, 2022), and Hänsch *et al.* have developed automated CNN-based segmentation of the liver for hepatocellular carcinoma diagnosis (Hänsch *et al.*, 2022).

A noteworthy case study is the segmentation of cardiac perfusion MRI which shows how the spatiotemporal nature of the data and the consideration of the downstream tracer-kinetic quantification lead to choices to be made on what and how to segment. Scannell *et al.* trained a CNN to first identify a time dynamic with good contrast and then input this dynamic into further models to detect a bounding box and to segment the myocardium on this single time dynamic, which was then passed into a pipeline for quantitative perfusion (Scannell *et al.*, 2020b). Alternatively, Yalcinkaya *et al.* trained a model to segment all of the 2D time dynamics (using image patches rather than whole images) (Yalcinkaya *et al.*, 2021), and Sandfort *et al.* used a 3D model with time in the third dimension to improve the segmentation accuracy but, perhaps as expected, this only improves performance if motion correction was also used (Sandfort *et al.*, 2020). Xue *et al.* rather chose to first quantify perfusion on a pixel-wise level and to subsequently segment the quantitative maps (Xue *et al.*, 2020a), this is then combined with a separate model for segmenting a region for the arterial input function (AIF) beforehand (Xue *et al.*, 2020b) so that the quantification is still automated. However, despite the different approaches on how to perform the segmentation, it should be noted that all the models trained in these discussed references are variations of the U-Net model, shown in Fig. 12.1. The U-Net is the *de facto* standard for image segmentation and, despite different input data, consistently performs well for perfusion MRI.

AI has been used additionally for many other postprocessing tasks, including the denoising of perfusion MRI (Xie *et al.*, 2020). For this purpose, models can be trained to take noise-corrupted images as input and to output high-quality images. This is potentially a high-impact area for AI in perfusion MRI due to the physical limits of the acquisition, leading to low SNR images and the increased susceptibility to artifacts in dynamic images. Ulas *et al.* extended upon the idea of AI-based denoising to also constrain the temporal evolution of the images by including the tracer-kinetic model (Ulas *et al.*, 2018a). Ivanovska developed segmentation models that explicitly account for artifacts in breast images (Ivanovska *et al.*, 2019) and Tamada *et al.* worked to reduce artifacts in liver perfusion MRI (Tamada *et al.*, 2020).

Another crucial aspect of any perfusion quantification pipeline is a motion compensation step. This is challenging in perfusion MRI due to the changing contrasts and low SNR. It is also computationally expensive, and thus, there is a large scope for it to benefit from AI. Image registration for motion correction can be addressed in different ways. The most direct approach is to train a model that takes the two images to be aligned as input and predicts the deformation required to align them. A major benefit of this approach is that the model will directly predict the required transformation rather than it being iteratively estimated, as is the case for traditional registration. This means that once a model has been trained, it will be much less computationally expensive to register new images. The other benefit of this approach is that it is not necessary to specify a similarity metric to be optimized, which is particularly advantageous for perfusion imaging, as the assumptions of standard similarity metrics are broken by the dynamic contrast changes. This approach already has several successful applications to perfusion MRI (Lv *et al.*, 2018; Gong *et al.*, 2020; Aprea *et al.*, 2021; Huang *et al.*, 2022). The drawback of this approach is that ground-truth deformations are required for training. These are typically not available, and so deformations are simulated and added to the images for training. Going forward, AI will be used to further augment the data that is acquired, for instance, using super-resolution to increase spatial or temporal resolution but there is currently only very limited applications of this to perfusion MRI (Almansour *et al.*, 2022).

12.3.4 **Quantification**

Many of the image formation, reconstruction, and processing advances discussed hereto have been developed with the aim of improving perfusion quantification, and it should not be a surprise that AI can also play a role in the quantification process. In fact, there is growing evidence that AI can be used to efficiently estimate the kinetic parameters. The estimation of these parameters is an ill-posed inverse problem that is typically solved via nonlinear least-squares fitting. Such problems can converge to local minima; therefore, good initializations or regularization may need to be employed to yield accurate results (Scannell *et al.*, 2020a). As is the case for the similar problem of image reconstruction, AI is beginning to contribute to the solution of these problems in perfusion quantification.

The simplest strategy to quantify perfusion with AI is a direct supervised approach where a model is trained to take in the relevant kinetic information, e.g., the AIF and tissue response curve, and to output the kinetic parameters, such as perfusion. In this case, ground-truth perfusion values are usually not available for training and the neural network is trained using labels estimated by another method. McKinley *et al.* first took this direction with brain perfusion data and compared the performance of a neural network with other ML models (McKinley *et al.*, 2018). In this study, the neural network outperformed simpler linear prediction models, and the neural network approach has since been adopted in several other studies, including in the heart (Scannell *et al.*, 2019), in the brain (Ulas *et al.*, 2018b; Ulas *et al.*, 2019), see Fig. 12.3, in the kidneys (Klepaczko *et al.*, 2020), and in the calf muscle in the context of peripheral artery disease (Zhang *et al.*, 2020).

Van Herten *et al.* took an unsupervised approach using physics-informed neural networks (PINNs) (van Herten *et al.*, 2022). PINNs were introduced by Raissi *et al.* to solve differential equations (Raissi *et al.*, 2019) and have been extended to also fitting model parameters during training. PINNs are based on the idea of the universal approximation theorem, that any function can be well approximated by a neural network, and use the physics of the problem (the differential equations) to train the NN. To be a good approximation of the solution, the residuals of the differential equations with the NN-based approximate solution should be minimized. Therefore the NN can be trained by minimizing these residuals.

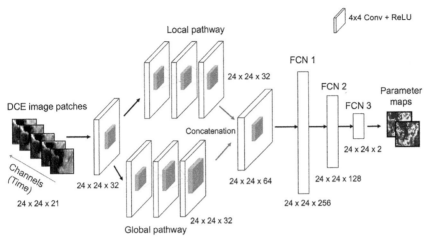

FIG. 12.3

An illustration of a deep learning architecture used for directly estimating kinetic parameters from DCE-MRI data.

Reproduced from Ulas et al. *(2019) under the terms of the Creative Commons Attribution 4.0 International License.*

Since the input and tissue response curves used to estimate the kinetic parameters are time dependent, it may also be beneficial to consider AI models which explicitly account for temporal dependencies. Recurrent neural networks (RNNs) model information over time by using recurrent (or loop) connections in the hidden layers so that the prediction at each time step is based both on the current input and the output from previous time steps. Zou *et al.* took this approach using a long short-term memory (LSTM)-type RNN to quantify perfusion in head and neck DCE-MRI, which improved fitting as compared to the least-squares fitting in synthetic data for undersampled images and reduced the fitting time (Zou *et al.*, 2020). Ottens *et al.* compared different RNN models versus a fully connected network, a CNN, and a U-Net in the setting of DCE-MRI of pancreatic cancer, also with a physics-informed loss (Ottens *et al.*, 2022). This study is noteworthy, as it is the first to systematically assess different network architectures for AI-based perfusion quantification. Based on the trade-off between accuracy and test–retest repeatability, they recommend a gated recurrent unit (GRU) model, a type of RNN. It remains to be seen how generalizable this conclusion will be to different applications.

However, the general approaches are surely transferable and similar ideas are also possible with other data types and models. For example, Kim *et al.* focused on directly generating arterial spin-labeled (ASL) maps from the subtraction images (Kim *et al.*, 2018). In the same field, Luciw *et al.* evaluated CNNs for predicting cerebral blood flow maps from multiple postlabeling delay ASL MRI, with the U-Net giving the best performance (Luciw *et al.*, 2022). Interestingly, they also directly predicted the arterial transit time, which is usually considered separately to the kinetic parameters to stabilize the fitting (Cheong *et al.*, 2003), and they used curriculum learning to improve the performance of the U-Net. Also of note in this work was the quantification of model output uncertainty, through the use of Monte Carlo dropout. This is a relatively straightforward-to-implement measure of uncertainty and would warrant further study, as the uncertainty of the parameters could be important in clinical practice.

Aside from the quantification process itself, the first step in any perfusion quantification pipeline is the identification of the AIF, and in many applications it is not clear how this should be chosen. AI can also help here. There has been work by de la Rosa on estimating the arterial input function (de la Rosa *et al.*, 2021) using deep learning, albeit for CT data in this case, where they directly regress on the AIF through a differentiable singular value decomposition (SVD) deconvolution to allow perfusion quantification (de la Rosa *et al.*, 2020). A similar approach could also be adapted for perfusion MRI. Scannell *et al.* (2022) also recently introduced the AI-AIF to correct signal nonlinearities in the AIF for first-pass myocardial perfusion MRI.

12.3.5 Interpretation

Since much of the work described is recent, now is the time to start to leverage it and to focus on how these analyses can be interpreted for the benefit of the patient. However, since these AI applications have come so recently, especially for less widespread applications, there is yet to be significant inroads into the problems of diagnosis and prognosis prediction.

There have been several applications, for example Park *et al.* that use AI-based image processing to allow the extraction of classical radiomic features for disease classification (predicting treatment response in glioblastoma patients, in this case) (Park *et al.*, 2021). There has also been the landmark multicenter study of Elshafeey *et al.*, who trained a classifier on radiomic features derived from the quantitative perfusion maps to identify pseudoprogression in glioblastoma patients (Elshafeey *et al.*, 2019), and Sudre *et al.* used multiparametric images for ML-based classifications of glioma (Sudre *et al.*, 2020). Going forward, it is likely that AI will allow us to learn even more relevant data-driven features, as compared to the standard radiomic features, to further improve performance, and the ability to handle more complex interactions of features is another potential benefit. Indeed for a similar application, Lee *et al.* were able to distinguish pseudoprogression from true progression in their glioma patients using multiparametric MRI with a LSTM based on a CNN feature extractor; this also outperformed VGG-16 with a single input sequence (Lee *et al.*, 2020). Similar studies have also been performed in other applications, such as breast cancer (Herent *et al.*, 2019), and AI has even allowed more complex questions, such as distinguishing molecular subtypes (Zhu *et al.*, 2019) and assessing collateral flow in stroke patients (Tetteh *et al.*, 2021), to be addressed.

A direction for which the ability of AI to learn complex interactions of high-dimensional data will be hugely beneficial is the interpretation of features in multiparametric images. Indeed, Ho *et al.* extracted deep features using an autoencoder from MR perfusion to classify whether stroke onset time was within 4.5 h, using multiparametric MRI (Ho *et al.*, 2016). Peng *et al.* used multiparametric MRI, including DCE perfusion, diffusion-weighted, and T2-weighted images to improve prostate cancer diagnosis with deep learning (Peng *et al.*, 2021). Antropova *et al.* took this idea further and used deep feature extraction techniques for diagnosis across three breast imaging modalities: mammography, ultrasound, and DCE-MRI (Antropova *et al.*, 2017). The complementary nature of information from completely different imaging modalities will surely be a fruitful direction of research but will also lead to more practical problems to be solved, e.g., what to do if one modality is missing. It will also be useful to consider that there is significant amount of nonimaging data available for each patient. For example, Osama *et al.* proposed a parallel multiparametric feature-embedded Siamese neural network to classify clinical outcomes using the MRI perfusion maps and clinical data in acute ischemic stroke (Osama *et al.*, 2020). Although it must be noted that despite the hype around AI solutions, Anderson *et al.* showed the performance of the radiomics diagnosis to be significantly better than that of a CNN feature extractor in breast cancer (Anderson *et al.*, 2019). It is possible that larger studies and more mature implementations of AI will be required to see the full benefit of the AI approach for diagnosis. An illustration of an AI model for classifying quantitative myocardial perfusion maps is shown in Fig. 12.4.

12.3.6 Next Frontiers

To return to the question in the title: Is AI the next frontier of perfusion MRI? We would argue that the answer is no, that in fact, AI is the current frontier of perfusion MRI. A more relevant question would be: What is the next frontier for AI in perfusion imaging? It is clear that this should be the work required to clinically deploy the

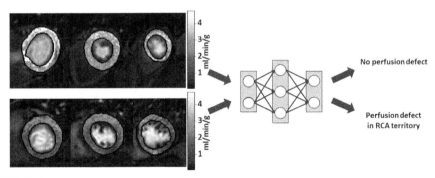

FIG. 12.4

An AI for classifying quantitative myocardial perfusion maps classifying one patient *(top)* as normal and one patient *(bottom)* as having a perfusion defect related to obstructive coronary artery disease in the right coronary artery (RCA).

research discussed in this chapter. A well-documented challenge for AI that is true for all types of medical images is the so-called domain shift. This is a difference between the distribution of the data the model is being applied on and the distribution of the data that it was trained on. It is also well documented that this domain shift can lead to a significant degradation of the performance of an AI model. This needs to be accounted for to allow this clinical deployment.

The domain shift can arise from numerous sources, common examples of these are as follows: the difference in the level of signal, noise, or contrast generated by different scanners, even when using the same pulse sequence; the difference in acquisition parameters leading to different image properties; variability in the scan planning; differences in the anatomy (e.g., size) or amount and appearance of the pathology in the images. Even more subtle differences may also play a role and require careful consideration as they can still impact the use of AI models. This could include how the images are reconstructed and if that includes interpolations, the type and implementation of the interpolation, and if data compression was used during storage.

Campello *et al.* organized an open benchmark challenge for segmentation in multivendor and multidisease cardiac MRI, and it found that data augmentation could bridge some of the gap between domains, but there is still a gap (Campello *et al.*, 2021). This is likely to be an even more complex question in perfusion MRI, as there is a lack of standardization in the acquisition. Even at a single center, with a single scanner, it is possible that the acquisition will vary significantly over time. In reality, it is not yet known how much of an issue the domain shift is in perfusion MRI as very few of the studies referenced in this chapter present an external validation. That is, an analysis of when the model is applied to data much different from the training data, i.e., from a different center or vendor, to test its ability to generalize. Large open-source databases do not yet exist for perfusion MRI but it is one of the aims of the Open Science Initiative for Perfusion Imaging (https://osipi.org/) to collate such resources for benchmarking.

There is much research into training AI models that will be robust to unseen data, such as using domain-adversarial learning (Scannell *et al.*, 2021) and semisupervised learning with a contrastive loss (Hu *et al.*, 2021), but none of this is specific to perfusion and, as such, there is not yet a clear path forward. However, a first step toward this can be to include more diverse training data. To this end, federated learning can facilitate the training of models across centers without transferring the data and has been already demonstrated for multiinstitutional collaborations with MRI data (Guo *et al.*, 2021).

Furthermore, improvements to the data processing workflow are warranted. Many applications that claim to be fully automated are, in fact, not, as the data must be manually extracted and transferred to a dedicated workstation before the relevant parts of the data are identified and the analysis can be started. There has been work to automatically sort the images by the MRI pulse sequence, for further processing, for brain (van der Voort *et al.*, 2021) and cardiac MRI data (Lim *et al.*, 2022), but more work in this direction is required. It is also likely that processing on the scanner or integration with the picture archiving and communication system (PACS) will be required, rather than offline processing, in order to encourage adoption.

12.4 Summary

Significant progress has been made in the past few years toward the augmentation of perfusion MRI using AI. This progress was initially focused on automating image processing tasks, such as anatomical segmentation and landmark detection. These are important steps in automating the workflow to allow deployment in less specialized centers, reducing the workload of radiologists, and increasing patient throughput. More recently, AI has also been used to give efficient solutions to the inverse problems associated with image reconstruction and kinetic parameter estimation; however, these applications are primarily at the proof-of-concept stage and are in need of further validation. Future work is likely to include optimizing the image acquisition, improving image quality, improving in-plane resolution, increasing 3D spatial coverage, reducing scan time, and improving image interpretation with automated reporting. The next generation of research in the field will need an increased focus of developing AI that is more reproducible and generalizable and will need to pay much closer attention to the clinical validation, including testing the AI models in prospective clinical trials. Finally, once these challenges are addressed, AI can begin to really impact daily clinical routine and possibly even expand the usage of perfusion MRI by lowering the barrier to entry and improving the reliability of analyzed perfusion parameters.

References

Aggarwal, H.K., Jacob, M., 2020. J-Modl: joint model-based deep learning for optimized sampling and reconstruction. IEEE J. Sel. Top. Signal Process 14 (6), 1151–1162. https://doi.org/10.1109/JSTSP.2020.3004094.

Aggarwal, H.K., Mani, M.P., Jacob, M., 2019. MoDL: model-based deep learning architecture for inverse problems. IEEE Trans. Med. Imaging 38 (2), 394–405. https://doi.org/10.1109/TMI.2018.2865356.

Almansour, H., et al., 2022. Combined deep learning-based super-resolution and partial Fourier reconstruction for gradient Echo sequences in abdominal MRI at 3 tesla: shortening breath-hold time and improving image sharpness and lesion conspicuity. Acad. Radiol. https://doi.org/10.1016/j.acra.2022.06.003.

Anderson, R., et al., 2019. Evaluating deep learning techniques for dynamic contrast-enhanced MRI in the diagnosis of breast cancer. In: Proc. SPIE. https://doi.org/10.1117/12.2512667.

Antropova, N., Huynh, B.Q., Giger, M.L., 2017. A deep feature fusion methodology for breast cancer diagnosis demonstrated on three imaging modality datasets. Med. Phys. 44 (10), 5162–5171. https://doi.org/10.1002/mp.12453.

Aprea, F., Marrone, S., Sansone, C., 2021. Neural machine registration for motion correction in breast DCE-MRI. In: 2020 25th International Conference on Pattern Recognition (ICPR), pp. 4332–4339, https://doi.org/10.1109/ICPR48806.2021.9412116.

Bahadir, C.D., *et al.*, 2020. Deep-learning-based optimization of the under-sampling pattern in MRI. IEEE Trans. Comput. Imaging 6, 1139–1152. https://doi.org/10.1109/TCI.2020.3006727.

Bahrami, N., *et al.*, 2019. Automated selection of myocardial inversion time with a convolutional neural network: spatial temporal ensemble myocardium inversion network (STEMI-NET). Magn. Reson. Med. 81 (5), 3283–3291. https://doi.org/10.1002/mrm.27680.

Blansit, K., *et al.*, 2019. Deep learning–based prescription of cardiac MRI Planes. Radiol. Artif. Intell. 1 (6), e180069. https://doi.org/10.1148/RYAI.2019180069.

Bones, I.K., *et al.*, 2022. Workflow for automatic renal perfusion quantification using ASL-MRI and machine learning. Magn. Reson. Med. 87 (2), 800–809. https://doi.org/10.1002/MRM.29016.

Campello, V.M., *et al.*, 2021. Multi-centre, multi-vendor and multi-disease cardiac segmentation: the M&Ms challenge. IEEE Trans. Med. Imaging 40 (12), 3543–3554. https://doi.org/10.1109/TMI.2021.3090082.

Chen, C., *et al.*, 2014. Real time dynamic MRI with dynamic total variation. In: *Lecture notes in computer science (including subseries lecture notes in artificial intelligence and lecture notes in bioinformatics)*, 8673 LNCS(PART 1), pp. 138–145, https://doi.org/10.1007/978-3-319-10404-1_18/COVER/.

Cheong, L.H., Koh, T.S., Hou, Z., 2003. An automatic approach for estimating bolus arrival time in dynamic contrast MRI using piecewise continuous regression models. Phys. Med. Biol. 48 (5). https://doi.org/10.1088/0031-9155/48/5/403.

Correia, T., Schneider, T., Chiribiri, A., 2019. Model-based reconstruction for highly accelerated first-pass perfusion cardiac MRI. In: *Lecture notes in computer science (including subseries lecture notes in artificial intelligence and lecture notes in bioinformatics)*, 11765 LNCS, pp. 514–522, https://doi.org/10.1007/978-3-030-32245-8_57/FIGURES/4.

Dalmış, M.U., *et al.*, 2018. Fully automated detection of breast cancer in screening MRI using convolutional neural networks. J. Med. Imaging 5 (1), 1–9. https://doi.org/10.1117/1.JMI.5.1.014502.

de la Rosa, E., *et al.*, 2020. Differentiable deconvolution for improved stroke perfusion analysis. In: Lecture Notes in Computer Science (including subseries Lecture Notes in Artificial Intelligence and Lecture Notes in Bioinformatics). 12267 LNCS, pp. 593–602, https://doi.org/10.1007/978-3-030-59728-3_58/FIGURES/3.

de la Rosa, E., *et al.*, 2021. AIFNet: automatic vascular function estimation for perfusion analysis using deep learning. Med. Image Anal. 74, 102211. https://doi.org/10.1016/J.MEDIA.2021.102211.

Dikaios, N., *et al.*, 2014. Direct parametric reconstruction from undersampled (k, t)-space data in dynamic contrast enhanced MRI. Med. Image Anal. 18 (7), 989–1001. https://doi.org/10.1016/J.MEDIA.2014.05.001.

Edalati, M., *et al.*, 2022. Implementation and prospective clinical validation of AI-based planning and shimming techniques in cardiac MRI. Med. Phys. 49 (1), 129–143. https://doi.org/10.1002/MP.15327.

Elshafeey, N., *et al.*, 2019. Multicenter study demonstrates radiomic features derived from magnetic resonance perfusion images identify pseudoprogression in glioblastoma. Nat. Commun. 10 (1), 3170. https://doi.org/10.1038/s41467-019-11007-0.

Fan, L., *et al.*, 2020. Rapid dealiasing of undersampled, non-Cartesian cardiac perfusion images using U-net. NMR Biomed. 33 (5), e4239. https://doi.org/10.1002/NBM.4239.

Frick, M., *et al.*, 2011. Fully automatic geometry planning for cardiac MR imaging and reproducibility of functional cardiac parameters. J. Magn. Reson. Imaging 34 (2), 457–467. https://doi.org/10.1002/JMRI.22626.

Goldfarb, J.W., Cheng, J., Cao, J.J., 2018. Automatic optimal frequency adjustment for high field cardiac MR imaging via deep learning. In: CMR 2018 – A Joint EuroCMR/SCMR Meeting Abstract Supplement, pp. 437–438.

Gong, Z., *et al.*, 2020. Deep learning of deformable registration for breast DCE-MRI images. In: The Fourth International Symposium on Image Computing and Digital Medicine. Association for Computing Machinery (ISICDM 2020), New York, NY, USA, pp. 229–234, https://doi.org/10.1145/3451421.3451469.

Goodfellow, I., Bengio, Y., Courville, A., 2016. Deep Learning. MIT Press.

Guo, Y., *et al.*, 2017. Direct estimation of tracer-kinetic parameter maps from highly undersampled brain dynamic contrast enhanced MRI. Magn. Reson. Med. 78 (4), 1566–1578. https://doi.org/10.1002/MRM.26540.

Guo, P., *et al.*, 2021. Multi-institutional collaborations for improving deep learning-based magnetic resonance image reconstruction using federated learning. In: Proceedings of the IEEE/CVF Conference on Computer Vision and Pattern Recognition (CVPR), pp. 2423 2432.

Haghighi, M., Warfield, S.K., Kurugol, S., 2018. Automatic renal segmentation in DCE-MRI using convolutional neural networks. In: 2018 IEEE 15th International Symposium on Biomedical Imaging (ISBI 2018), pp. 1534–1537, https://doi.org/10.1109/ISBI.2018.8363865.

Hammernik, K., *et al.*, 2018. Learning a variational network for reconstruction of accelerated MRI data. Magn. Reson. Med. 79 (6), 3055–3071. https://doi.org/10.1002/MRM.26977/ASSET/SUPINFO/MRM26977-SUP-0005-SUPPINFO05.MP4.

Hänsch, A., *et al.*, 2022. Robust liver segmentation with deep learning across DCE-MRI contrast phases. In: Maier-Hein, K., et al. (Eds.), Bildverarbeitung für die Medizin 2022. Springer Fachmedien Wiesbaden, Wiesbaden, pp. 13–18.

Herent, P., *et al.*, 2019. Detection and characterization of MRI breast lesions using deep learning. Diagn. Interv. Imaging 100 (4), 219–225. https://doi.org/10.1016/j.diii.2019.02.008.

Ho, K.C., *et al.*, 2016. A temporal deep learning approach for MR perfusion parameter estimation in stroke. In: 2016 23rd International Conference on Pattern Recognition (ICPR). IEEE, pp. 1315–1320, https://doi.org/10.1109/ICPR.2016.7899819.

Hu, X., *et al.*, 2021. Semi-supervised contrastive learning for label-efficient medical image segmentation. In: de Bruijne, M., et al. (Eds.), Medical Image Computing and Computer Assisted Intervention — MICCAI 2021. Springer International Publishing, Cham, pp. 481–490.

Huang, J., *et al.*, 2022. Deep learning-based deformable registration of dynamic contrast enhanced MR images of the kidney. In: Linte, C.A., Siewerdsen, J.H. (Eds.), Medical Imaging 2022: Image-Guided Procedures, Robotic Interventions, and Modeling. SPIE, pp. 213–222, https://doi.org/10.1117/12.2611768.

Huijben, I.A.M., Veeling, B.S., Van Sloun, R.J.G., 2020. Learning sampling and model-based signal recovery for compressed sensing MRI. In: ICASSP, IEEE International Conference on Acoustics, Speech and Signal Processing — Proceedings, pp. 8906–8910, https://doi.org/10.1109/ICASSP40776.2020.9053331.

Ivanovska, T., *et al.*, 2019. A deep learning framework for efficient analysis of breast volume and fibroglandular tissue using MR data with strong artifacts. Int. J. Comput. Assist. Radiol. Surg. 14 (10), 1627–1633. https://doi.org/10.1007/s11548-019-01928-y.

Jeong, J.J., et al., 2021. Post-op brain tumor bed detection and segmentation using 3D Mask R-CNN for dynamic magnetic resonance perfusion imaging. In: Gimi, B.S., Krol, A. (Eds.), Medical Imaging 2021: Biomedical Applications in Molecular, Structural, and Functional Imaging. SPIE, pp. 431–437, https://doi.org/10.1117/12.2580792.

Jiang, W., et al., 2018. Automatic artifacts detection as operative scan-aided tool in an autonomous MRI environment. In: CMR 2018 – A Joint EuroCMR/SCMR Meeting Abstract Supplement, pp. 1167–1168.

Kim, K.H., Choi, S.H., Park, S.-H., 2018. Improving arterial spin labeling by using deep learning. Radiology 287 (2), 658–666. https://doi.org/10.1148/radiol.2017171154.

Klepaczko, A., et al., 2020. A multi-layer perceptron network for perfusion parameter estimation in DCE-MRI studies of the healthy kidney. Appl. Sci. 10 (16), 5525. https://doi.org/10.3390/APP10165525.

Koçanaoğulları, A., et al., 2022. Learning the regularization in DCE-MR image reconstruction for functional imaging of kidneys. IEEE Access 10, 4102–4111. https://doi.org/10.1109/ACCESS.2021.3139854.

Krizhevsky, A., Sutskever, I., Hinton, G.E., 2012. ImageNet classification with deep convolutional neural networks. In: Proceedings of the 25th International Conference on Neural Information Processing Systems – Volume 1. Curran Associates Inc. (NIPS'12), USA, pp. 1097–1105.

Le, J., et al., 2021. Deep learning for radial SMS myocardial perfusion reconstruction using the 3D residual booster U-net. Magn. Reson. Imaging 83, 178–188. https://doi.org/10.1016/J.MRI.2021.08.007.

Lebel, R.M., et al., 2014. Highly accelerated dynamic contrast enhanced imaging. Magn. Reson. Med. 71 (2), 635–644. https://doi.org/10.1002/MRM.24710.

Lee, J., et al., 2020. Discriminating pseudoprogression and true progression in diffuse infiltrating glioma using multi-parametric MRI data through deep learning. Sci. Rep. 10 (1), 20331. https://doi.org/10.1038/s41598-020-77389-0.

Leiner, T., et al., 2019. Machine learning in cardiovascular magnetic resonance: basic concepts and applications. J. Cardiovasc. Magn. Reson. 61. https://doi.org/10.1186/s12968-019-0575-y.

Lim, R.P., et al., 2022. CardiSort: a convolutional neural network for cross vendor automated sorting of cardiac MR images. Eur. Radiol. 1–14. https://doi.org/10.1007/S00330-022-08724-4/FIGURES/5.

Lingala, S.G., et al., 2011. Accelerated dynamic MRI exploiting sparsity and low-rank structure: K-t SLR. IEEE Trans. Med. Imaging 30 (5), 1042–1054. https://doi.org/10.1109/TMI.2010.2100850.

Lingala, S.G., et al., 2020. Tracer kinetic models as temporal constraints during brain tumor DCE-MRI reconstruction. Med. Phys. 47 (1), 37–51. https://doi.org/10.1002/MP.13885.

Litjens, G., et al., 2017. A survey on deep learning in medical image analysis. Med. Image Anal. 42, 60–88. https://doi.org/10.1016/j.media.2017.07.005.

Luciw, N.J., et al., 2022. Automated generation of cerebral blood flow and arterial transit time maps from multiple delay arterial spin-labeled MRI. Magn. Reson. Med. 88 (1), 406–417. https://doi.org/10.1002/mrm.29193.

Lustig, M., Donoho, D., Pauly, J.M., 2007. Sparse MRI: the application of compressed sensing for rapid MR imaging. Magn. Reson. Med. 58 (6), 1182–1195. https://doi.org/10.1002/MRM.21391.

Lv, J., *et al.*, 2018. Respiratory motion correction for free-breathing 3D abdominal MRI using CNN-based image registration: a feasibility study. Br. J. Radiol. 91 (1083), 20170788. https://doi.org/10.1259/bjr.20170788.

Martín-González, E., *et al.*, 2021. Physics-informed self-supervised deep learning reconstruction for accelerated first-pass perfusion cardiac MRI. In: Lecture Notes in Computer Science (including subseries Lecture Notes in Artificial Intelligence and Lecture Notes in Bioinformatics), 12964 LNCS, pp. 86–95, https://doi.org/10.1007/978-3-030-88552-6_9/FIGURES/5.

McKinley, R., *et al.*, 2018. A machine learning approach to perfusion imaging with dynamic susceptibility contrast MR. Front. Neurol. 9 (SEP), 717. https://doi.org/10.3389/FNEUR.2018.00717.

Menze, B., *et al.*, 2021. Analyzing magnetic resonance imaging data from glioma patients using deep learning. Comput. Med. Imaging Graph. 88, 101828. https://doi.org/10.1016/J.COMPMEDIMAG.2020.101828.

Nalepa, J., *et al.*, 2020. Fully-automated deep learning-powered system for DCE-MRI analysis of brain tumors. Artif. Intell. Med. 102, 101769. https://doi.org/10.1016/j.artmed.2019.101769.

Osama, S., Zafar, K., Sadiq, M.U., 2020. Predicting clinical outcome in acute ischemic stroke using parallel multi-parametric feature embedded Siamese network. Diagnostics. https://doi.org/10.3390/diagnostics10110858.

Ottens, T., *et al.*, 2022. Deep learning DCE-MRI parameter estimation: application in pancreatic cancer. Med. Image Anal. 80, 102512. https://doi.org/10.1016/j.media.2022.102512.

Park, J.E., *et al.*, 2021. Diffusion and perfusion MRI radiomics obtained from deep learning segmentation provides reproducible and comparable diagnostic model to human in post-treatment glioblastoma. Eur. Radiol. 31 (5), 3127–3137. https://doi.org/10.1007/s00330-020-07414-3.

Pellicer-Valero, O.J., *et al.*, 2022. Deep learning for fully automatic detection, segmentation, and Gleason grade estimation of prostate cancer in multiparametric magnetic resonance images. Sci. Rep. 12 (1), 2975. https://doi.org/10.1038/s41598-022-06730-6.

Peng, T., *et al.*, 2021. Can machine learning-based analysis of multiparameter MRI and clinical parameters improve the performance of clinically significant prostate cancer diagnosis? Int. J. Comput. Assist. Radiol. Surg. 16 (12), 2235–2249. https://doi.org/10.1007/s11548-021-02507-w.

Pérez Malla, C.U., *et al.*, 2019. Evaluation of enhanced learning techniques for segmenting ischaemic stroke lesions in brain magnetic resonance perfusion images using a convolutional neural network scheme. Front. Neuroinform. 13 (May), 1–16. https://doi.org/10.3389/fninf.2019.00033.

Perlman, O., *et al.*, 2022. An end-to-end AI-based framework for automated discovery of rapid CEST/MT MRI acquisition protocols and molecular parameter quantification (AutoCEST). Magn. Reson. Med. 87 (6), 2792–2810. https://doi.org/10.1002/MRM.29173.

Pezzotti, N., *et al.*, 2020. An adaptive intelligence algorithm for undersampled knee MRI reconstruction. IEEE Access 8, 204825–204838. https://doi.org/10.1109/ACCESS.2020.3034287.

Plein, S., *et al.*, 2007. Dynamic contrast-enhanced myocardial perfusion MRI accelerated with k-t SENSE. Magn. Reson. Med. 58 (4), 777–785. https://doi.org/10.1002/mrm.21381.

Raissi, M., Perdikaris, P., Karniadakis, G.E., 2019. Physics-informed neural networks: a deep learning framework for solving forward and inverse problems involving nonlinear partial differential equations. J. Comput. Phys. 378, 686–707. https://doi.org/10.1016/j.jcp.2018.10.045.

Ronneberger, O., Fischer, P., Brox, T., 2015. U-net: convolutional networks for biomedical image segmentation. In: Navab, N., et al. (Eds.), Medical Image Computing and Computer-Assisted Intervention -- MICCAI 2015. Springer International Publishing, Cham, pp. 234–241.

Russakovsky, O., et al., 2015. ImageNet large scale visual recognition challenge. Int. J. Comput. Vis. 115 (3), 211–252. https://doi.org/10.1007/s11263-015-0816-y.

Sandfort, V., et al., 2020. Reliable segmentation of 2D cardiac magnetic resonance perfusion image sequences using time as the 3rd dimension. Eur. Radiol. 1–10. https://doi.org/10.1007/s00330-020-07474-5.

Sandino, C.M., et al., 2021. Upstream machine learning in radiology. Radiol. Clin. North Am. 59 (6), 967–985. https://doi.org/10.1016/J.RCL.2021.07.009.

Scannell, C.M., et al., 2019. Deep learning-based prediction of kinetic parameters from myocardial perfusion MRI. *arXiv preprint arXiv:1907.11899*.

Scannell, C.M., Chiribiri, A., et al., 2020a. Hierarchical Bayesian myocardial perfusion quantification. Med. Image Anal. 60, 101611. https://doi.org/10.1016/J.MEDIA.2019.101611.

Scannell, C.M., Veta, M., et al., 2020b. Deep-learning-based preprocessing for quantitative myocardial perfusion MRI. J. Magn. Reson. Imaging 51 (6), 1689–1696. https://doi.org/10.1002/jmri.26983.

Scannell, C.M., Chiribiri, A., Veta, M., 2021. Domain-adversarial learning for multi-Centre, multi-vendor, and multi-disease cardiac MR image segmentation. In: Puyol Anton, E., et al. (Eds.), Statistical Atlases and Computational Models of the Heart. M&Ms and EMIDEC Challenges. Vol. 2020. STACOM, pp. 228–237, https://doi.org/10.1007/978-3-030-68107-4_23.

Scannell, C.M., et al., 2022. AI-AIF: artificial intelligence-based arterial input function for quantitative stress perfusion cardiac magnetic resonance. Eur. Heart J., Digit. Health.

Shamshad, F., et al., 2022. Transformers in Medical Imaging: A Survey. *arXiv preprint arXiv:2201.09873.*, https://doi.org/10.48550/arxiv.2201.09873.

Smith, D.S., et al., 2012. Robustness of quantitative compressive sensing MRI: the effect of random undersampling patterns on derived parameters for DCE- and DSC-MRI. IEEE Trans. Med. Imaging 31 (2), 504–511. https://doi.org/10.1109/TMI.2011.2172216.

Sudre, C.H., et al., 2020. Machine learning assisted DSC-MRI radiomics as a tool for glioma classification by grade and mutation status. BMC Med. Inform. Decis. Mak. 20 (1), 1–14. https://doi.org/10.1186/S12911-020-01163-5/FIGURES/5.

Tamada, D., et al., 2020. Motion artifact reduction using a convolutional neural network for dynamic contrast enhanced MR imaging of the liver. Magn. Reson. Med. Sci. 19 (1), 64–76. https://doi.org/10.2463/mrms.mp.2018-0156.

Tetteh, G., et al., 2021. A Deep Learning Approach to Predicting Collateral Flow in Stroke Patients Using Radiomic Features from Perfusion Images. https://doi.org/10.48550/arxiv.2110.12508.

Tourais, J., et al., 2022. High-resolution free-breathing quantitative first-pass perfusion cardiac MR using Dual-Echo Dixon with Spatio-temporal acceleration. Front. Cardiovasc. Med. 0, 1050. https://doi.org/10.3389/FCVM.2022.884221.

Ulas, C., Tetteh, G., Kaczmarz, S., et al., 2018a. DeepASL: kinetic model incorporated loss for denoising arterial spin labeled MRI via deep residual learning. In: *Lecture Notes in Computer Science (including subseries Lecture Notes in Artificial Intelligence and Lecture Notes in Bioinformatics)*, 11070 LNCS, pp. 30–38, https://doi.org/10.1007/978-3-030-00928-1_4/TABLES/1.

Ulas, C., Tetteh, G., Thrippleton, M.J., *et al.*, 2018b. Direct Estimation of Pharmacokinetic Parameters from DCE-MRI using Deep CNN with Forward Physical Model Loss.

Ulas, C., *et al.*, 2019. Convolutional neural networks for direct inference of pharmacokinetic parameters: application to stroke dynamic contrast-enhanced MRI. Front. Neurol. 9, 1147. https://doi.org/10.3389/fneur.2018.01147.

van der Voort, S.R., Smits, M., Klein, S., 2021. DeepDicomSort: an automatic sorting algorithm for brain magnetic resonance imaging data. Neuroinformatics 19 (1), 159–184. https://doi.org/10.1007/S12021-020-09475-7/FIGURES/22.

van Gorp, H., *et al.*, 2021. Active deep probabilistic subsampling. In: Proceedings of the International Conference on Machine Learning (ICML).

van Herten, R.L.M., *et al.*, 2022. Physics-informed neural networks for myocardial perfusion MRI quantification. Med. Image Anal. 78, 102399. https://doi.org/10.1016/J.MEDIA.2022.102399.

Villa, A.D.M., *et al.*, 2018. Importance of operator training and rest perfusion on the diagnostic accuracy of stress perfusion cardiovascular magnetic resonance. J. Cardiovasc. Magn. Reson. 20 (1), 74. https://doi.org/10.1186/s12968-018-0493-4.

Wang, J., *et al.*, 2022. DEep learning-based rapid spiral image REconstruction (DESIRE) for high-resolution spiral first-pass myocardial perfusion imaging. NMR Biomed. 35 (5). https://doi.org/10.1002/NBM.4661.

Weiss, T., *et al.*, 2021. PILOT: physics-informed learned optimized trajectories for accelerated MRI. J. Mach. Learn. Biomed. Imaging, 6–7. https://doi.org/10.48550/arxiv.1909.05773.

Xie, D., *et al.*, 2020. Denoising arterial spin labeling perfusion MRI with deep machine learning. Magn. Reson. Imaging 68, 95–105. https://doi.org/10.1016/J.MRI.2020.01.005.

Xue, H., Davies, R.H., *et al.*, 2020a. 'Automated inline analysis of myocardial perfusion MRI with deep learning. Radiol. Artif. Intell. 2 (6), e200009. https://doi.org/10.1148/ryai.2020200009.

Xue, H., Tseng, E., *et al.*, 2020b. Automated detection of left ventricle in arterial input function images for inline perfusion mapping using deep learning: a study of 15,000 patients. Magn. Reson. Med. 84 (5), 2788–2800. https://doi.org/10.1002/MRM.28291.

Yalcinkaya, D.M., *et al.*, 2021. Deep learning-based segmentation and uncertainty assessment for automated analysis of myocardial perfusion MRI datasets using patch-level training and advanced data augmentation. In: 2021 43rd Annual International Conference of the IEEE Engineering in Medicine & Biology Society (EMBC), pp. 4072–4078, https://doi.org/10.1109/EMBC46164.2021.9629581.

Zhang, Y., *et al.*, 2019. Automatic breast and Fibroglandular tissue segmentation in breast MRI using deep learning by a fully-convolutional residual neural network U-net. Acad. Radiol. 26 (11), 1526–1535. https://doi.org/10.1016/j.acra.2019.01.012.

Zhang, J.L., *et al.*, 2020. Exercise-induced calf muscle hyperemia: rapid mapping of magnetic resonance imaging using deep learning approach. Physiol. Rep. 8 (16), e14563. https://doi.org/10.14814/phy2.14563.

Zhu, B., *et al.*, 2018. Image reconstruction by domain-transform manifold learning. Nature 555 (7697), 487–492. https://doi.org/10.1038/nature25988.

Zhu, Z., *et al.*, 2019. Deep learning for identifying radiogenomic associations in breast cancer. Comput. Biol. Med. 109, 85–90. https://doi.org/10.1016/j.compbiomed.2019.04.018.

Zou, J., Balter, J.M., Cao, Y., 2020. Estimation of pharmacokinetic parameters from DCE-MRI by extracting long and short time-dependent features using an LSTM network. Med. Phys. 47 (8), 3447–3457. https://doi.org/10.1002/MP.14222.

Applications

Perfusion MRI in the brain: Insights from sickle cell disease and the healthy brain

13

Liza Afzali-Hashemi[a], Koen P.A. Baas[a], John C. Wood[b], and Aart J. Nederveen[a]

[a]*Department of Radiology and Nuclear Medicine, Amsterdam University Medical Center, Location University of Amsterdam, Amsterdam, The Netherlands*
[b]*Division of Hematology, Children's Hospital Los Angeles, Los Angeles, CA, United States*

13.1 Overview

Because of the brain's high energy demand but its lack of intracellular energy storage, it fully relies on the supply of nutrients and oxygen via cerebral blood flow (CBF). Hence, changes in the brain's oxygen supply or demand require fast adaptation through a closely controlled system. Dysfunction of this regulatory system can be detected with Magnetic Resonance Imaging (MRI) using perfusion measurements in the brain and central nervous system (CNS). Such MRI-based measurements can help diagnose numerous diseases that affect cerebrovascular autoregulation as well as provide valuable insights in research settings.

MRI offers several acquisition techniques to evaluate perfusion in the brain and CNS. The most commonly used techniques are dynamic susceptibility contrast (DSC) and arterial spin labeling (ASL) MRI. In DSC-MRI, an exogenous contrast agent is intravenously administered which causes a decrease in blood T2 and T2* (Grandin, 2003; Rosen *et al.*, 1990). The first passage of this contrast agent through the capillaries is captured using a gradient- or spin-echo sequence. From the observed signal loss, CBF can be derived as well as several other hemodynamic parameters. To accurately capture the passage of the contrast agent, the sampling rate of a DSC acquisition is typically below 2 seconds (Wintermark *et al.*, 2005). For each voxel, a time-intensity curve can be measured that allows the reconstruction of parametric maps. The most commonly derived parameters are time to peak (TTP), mean transit time (MTT), cerebral blood volume (CBV), and CBF (Wintermark *et al.*, 2005). CBF and CBV can be quantified by a deconvolution of the time-intensity curves by an arterial input function (AIF), usually derived from a reference voxel within a feeding artery (Ostergaard *et al.*, 1996; Smith *et al.*, 2000a). Dynamic contrast-enhanced (DCE) MRI is similar to DSC but depends on the T1-shortening effect of gadolinium.

https://doi.org/10.1016/B978-0-323-95209-5.00010-6

During a DCE protocol, T1-weighted images are constantly acquired for approximately 5–10 min starting upon gadolinium administration. During this time, gadolinium accumulates within the extracellular space, shortening the T1. Using compartmental modeling, perfusion and capillary permeability can be quantified from the time-intensity curves (Tofts, 2010). Although DCE does provide an estimate for perfusion, it is mostly used to assess the vascular characteristics of tumors due to its sensitivity to leakiness of tumor vasculature (Padhani, 2002).

ASL offers a noninvasive alternative to measure brain and CNS perfusion using water as an endogenous contrast agent. In ASL, water is magnetically labeled using radiofrequency pulses after which the labeled blood flows to the tissue of interest (Williams *et al.*, 1992). After waiting for a certain period of time, referred to as the postlabeling delay (PLD), an image is acquired that contains the signal from the labeled water and static tissue (Alsop and Detre, 1996). Additionally, a control image is acquired without labeling the blood water. Subtraction of the control and label image results in a perfusion-weighted image which can be quantified into CBF using kinetic modeling (Buxton *et al.*, 1998; Wong *et al.*, 1998). Quantification does require the acquisition of a reference image, often referred to as an M0 image. Acquisition of an M0 image typically takes less than 1 min and is even embedded in the ASL sequence of certain MRI vendors. Because a typical voxel in the brain consists of only ~2% labeled blood, ASL has an intrinsically low signal-to-noise ratio (SNR) (Alsop *et al.*, 2015). Therefore a typical ASL acquisition includes >30 repetitions to reach sufficient SNR. Traditionally, ASL acquisitions consisted of only a single PLD but recent advances have enabled multi-PLD acquisitions in a time-efficient manner (Günther, 2007). Having measurements at several inflow times, multi-PLD ASL additionally enables the estimation of the arterial transit time (ATT) and CBV (Chappell *et al.*, 2010). Estimating these parameters leads to more accurate CBF quantification but potentially also offers new biomarkers for cerebrovascular pathology. Although ASL is mostly used to measure perfusion in the brain, it also gets increasingly applied outside the brain including in the spine (Qu *et al.*, 2021; Xing *et al.*, 2015).

Besides ASL and DSC, there are several other, less frequently used MRI techniques for measuring perfusion in the brain and CNS. Intravoxel incoherent motion (IVIM) is a diffusion-weighted MRI technique that can measure, among other parameters, microvascular perfusion (Zhu *et al.*, 2020). IVIM measures translational motion within a given voxel and uses multiple low b-values to uncouple restricted diffusion and pseudo-diffusion effects, which can be used to quantify microperfusion (Zhu *et al.*, 2020; Pavilla *et al.*, 2018). Recently, a technique similar to DSC was introduced to measure CBF using transient hypoxia instead of a contrast agent (Vu *et al.*, 2021a). In this approach, deoxygenated hemoglobin, rather than an exogenous contrast agent, causes a reduction of T2 and T2*. Using a similar quantification strategy, CBF could be quantified from these data. Lastly, phase-contrast MRI can measure total CBF by measuring flow in main feeding arteries (Spilt *et al.*, 2002). Combined with an estimate of measured total brain weight, these measurements can be converted into a global CBF value. Although phase-contrast MRI does not provide regional CBF estimates, it does offer a fast method to estimate global CBF.

Brain and CNS perfusion measurements have increasingly been used to study cerebrovascular hemodynamics in numerous diseases as well as in healthy aging (Toth *et al.*, 2017). For example, perfusion measurements have been used to study type 2 diabetes (Kodl and Seaquist, 2008), Alzheimer's disease (Binnewijzend *et al.*, 2016), moyamoya disease (Zaharchuk *et al.*, 2011), small vessel disease (Shi *et al.*, 2016), sickle cell anemia (Afzali-Hashemi *et al.*, 2021a), stroke (Smith *et al.*, 2000b), and brain tumors (Jarnum *et al.*, 2010). Due to technological advances, spinal cord measurements are increasingly performed during the last decade, for example, to assess perfusion in patients with cervical spondylosis (Ellingson *et al.*, 2019). For an overview of the clinical applications of DSC and ASL measurements in the brain, we refer the reader to reviews by Boxerman *et al.* (2016) and Haller *et al.* (2016).

In the remainder of this chapter, we focus on the application of ASL in patients with sickle cell disease (SCD). Because ASL is noninvasive and easily repeatable, it has become the preferred technique in these patients and used to measure CBF before and after a vasoactive challenge. ASL is increasingly used in these patients to better understand the effects of anemia and hemolysis on the brain and CNS and to evaluate different treatment options. However, there are some disease-specific considerations that have to be accounted for in the acquisition and postprocessing of ASL data. Therefore describing the application of ASL in SCD provides an extensive example of how to tailor brain perfusion measurements for a specific disease, how such measurements can help understand cerebrovascular pathologies, and how to evaluate treatment effects.

13.2 **Sickle cell disease**

SCD is one of the most common genetic blood disorders, affecting over 300,000 newborn infants each year and causing low life expectancy in developed countries (Mburu and Odame, 2019; Piel *et al.*, 2013, 2017). SCD has a high prevalence in malaria-endemic areas, especially in sub-Saharan Africa where malaria is one of the main causes of death (Piel *et al.*, 2017). This is most likely because of natural selection, as sickle cell traits contain alleles that are protective against malaria infection (Karlsson *et al.*, 2014). Individuals with sickle cell trait have one abnormal allele of the β-globin but barely experience any SCD-related complications. Their white blood cells, however, are able to remove malaria infection from the bloodstream (Luzzatto, 2012). Unfortunately, patients with SCD have two abnormal alleles and are highly susceptible to the lethal effects of malaria.

Over the last decades, migration increased the prevalence of SCD in the United States and western and northern Europe, with 3000 affected newborns in the United States and 300 affected newborns in the United Kingdom each year (Thein and Thein, 2016). In patients with SCD, red blood cells tend to polymerize in the deoxygenated state, causing rigid and sticky crescent-shaped cells. These dysfunctional blood cells cause two main problems. First, they cannot bind to oxygen, limiting the optimal oxygen transport to the organs. Second, these highly adhesive cells stick to each other and to the vessel wall, causing obstruction and leading to pain crises in patients with SCD. These complications lead to damage in several organs as well as cerebrovascular complications. SCD patients experience strokes and silent cerebral infarcts (SCIs) leading to reduced cognitive performance that results in unemployment and lower quality of life (DeBaun and Kirkham, 2016; Strouse et al., 2009). SCIs are asymptomatic but are detectable on structural MRI images. Despite the high impact of SCIs on patients, the cause of these infarcts is not fully understood. Previous studies tried to get more insight into this process by studying the cerebral hemodynamics of these patients using ASL perfusion MRI. Studies have found increased CBF, decreased and unaffected cerebrovascular reserve (CVR), and shorter ATTs in patients with severe SCD (genotype HbSS and $HbS\beta^0$). In recent years, altered perfusion was also found in other organs of patients with SCD such as the kidney, liver, and heart, but using ASL in other organs is still in a developing stage. This chapter will discuss the possibilities of ASL in the brain of patients with SCD, provide an overview of previous ASL studies with SCD patients, and highlight disease-specific technical considerations and challenges.

13.2.1 Types of SCD

SCD refers to a category of inherited diseases that consists of several types based on the specific hemoglobin mutation (Pinto et al., 2019). The most common type is HbSS, in which a mutation leading to the production of abnormal hemoglobin, hemoglobin S, is inherited from both parents (Rees et al., 2010). This type of SCD is also called sickle cell anemia. It is the most severe form of SCD, with patients frequently experiencing several symptoms, including episodes of pain, acute chest syndrome, strokes, pulmonary hypertension, and priapism (Ware et al., 2017). The second most prevalent form of SCD is HbSC, where the hemoglobin S mutation is inherited from one parent and hemoglobin C, another type of abnormal hemoglobin, is inherited from the other. Patients with HbSC have similar characteristic clinical features as patients with HbSS, but the symptoms are less severe and less frequent. The two other types of mutations $HbS\beta^+$ and $HbS\beta^0$ occur when the hemoglobin S mutation is coupled with a beta-thalassemia mutation. The production of beta globin is either diminished (β^+) or completely absent (β^0) in beta-thalassemia, resulting in damaged red blood cells. Whereas the disease severity of $HbS\beta^0$ is similar to HbSS patients, patients with $HbS\beta^+$ have a milder form of SCD. Other rare types of SCD are HbSD, HbSO, and HbSE where patients inherit an HbS gene from one parent and another type of abnormal hemoglobin (hemoglobin D, O, and E) from the other. These more uncommon types of SCD range in severity. Individuals with sickle cell trait (HbAS) inherit the HbS gene from one parent and a normal gene from the other. Whereas the red blood cells of these individuals are healthy and disc shaped at rest, the red blood cells can take the sickle cell shape when exposed to conditions that stimulate sickling, including hypoxia and hypothermia (Ashorobi et al., 2022). This might result in sickle cell-related symptoms which can ultimately lead to organ damage when frequently repeated. In a previous study of our group, we divided the genotypes into two groups: severe SCD (genotype HbSS and $HbS\beta^0$) and mild SCD (genotype HbSC and $HbS\beta^+$) and found that CBF in the mild group was comparable to the healthy control group, but significantly lower compared to the severe group (Fig. 13.1).

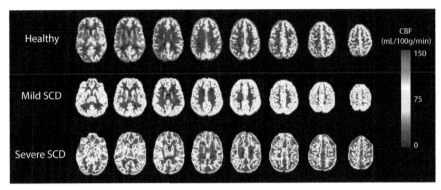

FIG. 13.1

CBF maps of a healthy volunteer, a mild SCD patient, and a severe SCD patient.

13.2.2 **Current treatment options**

Currently, treatment of SCD consists of three types: medication, blood transfusion, and stem cell transplantation. The goal of medication and blood transfusion is to reduce episodes of pain and prevent complications. The most commonly used medication is hydroxyurea, this medication was developed to treat cancer and was used for the first time by sickle cell patients in 1984 (Agrawal *et al.*, 2014). The main advantage of this drug is that it increases the levels of fetal hemoglobin (HbF), which reduces the polymerization of the red blood cells and results in less frequent pain crises and higher quality of life (Canak and Eskazan, 2022). Crizanlizumab and voxelotor are relatively new medications and were approved in 2019 for SCD treatment. Voxelotor increases the oxygen affinity of hemoglobin and reduces hemoglobin polymerization, while crizanlizumab prevents painful vaso-occlusive episodes (Karki *et al.*, 2022). Despite the beneficial effect of these three medications, these drugs have several side effects, including hair loss, nausea, headache, stomach pain, and fever. The severity and frequency of these side effects differ between patients.

Blood transfusion is another type of treatment and can be divided into simple blood transfusion and exchange blood transfusion. Simple transfusion is mostly performed in children with SCD where the blood of the donor is added to the blood of the patients. In adult patients, an exchange transfusion is usually performed where a percentage of the patient's blood is replaced with the donor's blood. Blood transfusion reduces the number of sickled cells, thereby reducing sickle cell-related symptoms. However, the main disadvantage of this treatment is the iron overload that can occur after multiple blood transfusions which can lead to heart and liver failure (Howard, 2016).

Hematopoietic stem cell transplantation (HSCT) is currently the only curative treatment for patients with SCD. During HSCT, the patient's blood cells are destroyed through medication and radiation, and subsequently replaced with stem cells from the donor (Ashorobi and Bhatt, 2022). After a successful HSCT, patients develop normal erythropoiesis and will not experience any SCD-related symptoms. However, HSCT is a high-risk procedure that can cause graft vs host disease that increases the risk of morbidity and mortality, and, therefore, it is mostly performed in SCD patients with severe complications including vaso-occlusions, strokes, and renal failure (Shenoy, 2013). However, HSCT is still in the developing stage for patients with SCD and could be performed in less severely affected patients in the future. The procedure is still evolving, and several medications are currently being investigated to improve the treatment outcome.

13.2.3 ASL in SCD

In SCD patients, resting CBF is elevated to compensate for anemia and maintain oxygen delivery to the brain (Afzali-Hashemi *et al.*, 2021a; Bush *et al.*, 2016; Numaguchi *et al.*, 1990; Oguz *et al.*, 2003). However, this counter mechanism limits the ability to respond to hemodynamic stress, since arterioles and capillaries are already close to maximum vasodilation (Afzali-Hashemi *et al.*, 2021a; Vaclavu *et al.*, 2019). This is potentially dangerous for SCD patients in the case of acute changes in hemoglobin levels or nocturnal desaturation (DeBaun *et al.*, 2012). Currently, ASL is the only noninvasive technique to spatially map the elevated CBF. Moreover, ASL can be repeated which enables measuring before and after a challenge. Thereby, ASL offers the possibility to measure CVR, defined as the capacity of the blood vessels to dilate in response to a challenge such as breath-holding (Macedo-Campos *et al.*, 2018), CO_2 inhalation (Kosinski *et al.*, 2017; Nur *et al.*, 2009; Watchmaker *et al.*, 2018), or administration of acetazolamide (ACZ) (Vaclavu *et al.*, 2019; Kedar *et al.*, 2006). CVR maps from patients with severe and mild SCD as well as from a healthy control are shown in Fig. 13.2. Although CVR can also be measured using phase-contrast MRI (Patrick *et al.*, 1996) and blood oxygenation level-dependent (BOLD) (Davis *et al.*, 1998; Hoge *et al.*, 1999) MRI, ASL has become a popular technique to measure CVR, because it provides quantitative measurements of CBF rather than relative changes (Liu *et al.*, 2019). Given the chronic vasodilatation and many endothelial stressors observed in SCD patients, CVR is believed to be a potential biomarker for cerebrovascular health in these

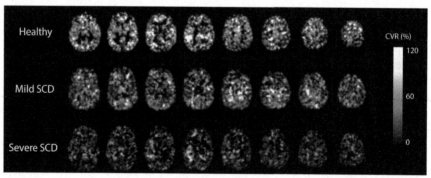

FIG. 13.2

CVR maps of a healthy volunteer, a mild SCD patient, and a severe SCD patient.

patients (Afzali-Hashemi *et al.*, 2021a; Kosinski *et al.*, 2017; Nur *et al.*, 2009; Hebbel *et al.*, 2004; Prohovnik *et al.*, 1989). Although longitudinal studies are currently lacking, CBF and CVR are also thought to be potential biomarkers for SCIs in SCD patients.

ASL has been extensively used in patients with SCD; an overview of ASL studies in patients with SCD and their outcome parameters is presented in Table 13.1. It shows that early studies used continuous and pulsed labeling techniques, whereas pseudo-continuous labeling later became the preferred technique. Moreover, a larger number of patients got included in recent studies and white matter lesions were more frequently assessed. With the introduction of time-encoded ASL, CBF quantification became more accurate because of ATT estimations. Moreover, ATT itself has been studied in SCD and was shown to be reduced in patients compared to healthy controls and in patients with severe genotypes compared to milder genotypes (Afzali-Hashemi *et al.*, 2021a).

Increased attention within the field of SCD research has been placed on the ASL signal found in the draining veins like the superior sagittal sinus (Juttukonda *et al.*, 2019, 2021; Afzali-Hashemi *et al.*, 2021b). This venous ASL signal is found to be hyperintense in SCD patients compared to healthy volunteers, suggesting cerebro-vascular functional shunting. Functional shunting is defined as limited oxygen offloading due to the rapid transit of blood through the capillaries as a result of increased flow (Juttukonda *et al.*, 2019). Studies have shown an inverse relationship between the venous ASL signal and parameters of oxygen metabolism providing evidence for functional shunting (Juttukonda *et al.*, 2021; Afzali-Hashemi *et al.*, 2021b). To further utilize the signal in the draining veins, a dedicated sequence, called water extraction with phase-contrast arterial spin tagging (WEPCAST), was designed to measure blood-brain barrier permeability using ASL (Lin *et al.*, 2018). A previous study using the WEPCAST technique in children with SCD reported a disruption in the blood-brain barrier which was related to hematological abnormalities (Lin *et al.*, 2022). A different approach measuring water transport across the blood-brain barrier relies on the transverse relaxation time (T2) of the ASL label which can be measured using T2 preparation (Liu *et al.*, 2011; Schmid *et al.*, 2015) and multi-echo sequences (Mahroo *et al.*, 2021).

ASL is often used in combination with MRI sequences that offer measurements for venous blood oxygenation, which enable the estimation of oxygen extraction fraction (OEF) by the brain. The most commonly used and extensively validated sequence is called T2-relaxation-under-spin-tagging (TRUST) (Lu and Ge, 2008). Several other techniques like T2-prepared-blood-relaxation-imaging-with-inversion-recovery (T2-TRIR) (Bush *et al.*, 2021; Petersen *et al.*, 2012) and T2-relaxation-under-phase-contrast (TRUPC) (Jiang *et al.*, 2019) have tried to improve these measurements. For an extensive review of the available techniques and their applications, we refer the reader to Jiang and Lu (2022). Using TRUST, previous studies have shown variable results based on the calibration model that is used for converting the T2 relaxation time to venous oxygenation. The bovine model, the first introduced model, reported increased OEF values in patients with SCD compared

Table 13.1 An overview of SCD studies using several ASL techniques with single or multi-PLDs and their outcome parameters[a]

First author	Year	Journal	ASL type	Multi/single PLD	Other imaging techniques	No of patients	Age (year)	Imaging parameters
Oguz et al. (2003)	2003	Radiology	CASL	Single	–	14	8.7 (range: 6–12)	CBF
Strouse et al. (2006)	2006	Blood	CASL	Single	TCD, FLAIR, T2w, DWI, MRA	24	8.5 ± 2.0	CBF, flow velocity, stenosis
Helton et al. (2009)	2009	Pediatric Blood & Cancer	PASL	Single	T1w, T2w, PDw, FLAIR, MRA, DTI, TCD	21	12 (range 5–17)	CBF, lesions, stenosis, FA, ADC, flow velocity
van den Tweel et al. (2009)	2009	Stroke	CASL	Single	FLAIR, MRA	24	13.4 ± 3	CBF, lesions, stenosis
Gevers et al. (2012)	2012	Journal of Magnetic Resonance Imaging	PCASL	Single	FLAIR, MRA	12	14.7 (range 9.3–20.8)	CBF, flow velocity, lesions, stenosis
Arkuszewski et al. (2013)	2013	The Neuroradiology Journal	CASL	Single	FLAIR, MRA	42	8.1 ± 3.3	CBF, lesions, stenosis
Jordan et al. (2016)	2016	BRAIN	PCASL	Single	T1w, FLAIR, T2w, MRA, TRUST	27	27.7 ± 5.0	CBF, Yv, OEF, lesions, stenosis
Juttukonda et al. (2016)	2016	IEEE	PCASL	Single	T1w, MRA, PC	19	Not reported	CBF, flow velocity
Kosinski et al. (2017)	2017	British Journal of Hematology	PASL	Single	BOLD, MRA, FLAIR, T1w	28	No treatment: 14.0 ± 2.5, treatment: mean 12.8 ± 2.2	CBF, CVR (BOLD), lesions
Vaclavu et al. (2016)	2016	American Journal of Neuroradiology	PCASL	Single	T1 blood (IR), MRA, T2w, PC	39	12 ± 2	CBF, T1 blood
Juttukonda et al. (2019)	2019	Journal of Cerebral Blood Flow and Metabolism	PCASL	Single	T1w, FLAIR, MRA, PC	46	26.0 ± 5.9	CBF, OEF, CMRO$_2$, VHS, lesions
Juttukonda et al. (2017)	2017	NMR in Biomedicine	PCASL	Single	T1w, FLAIR, MRA, PC	19	27.5 ± 4.9	CBF, flow velocity, BAT
Bush et al. (2018a)	2018	Magnetic Resonance Imaging	PCASL	Single	T1w, PC	9	18.4 ± 4.4	CBF, flow velocity
Bush et al. (2018b)	2018	Magnetic Resonance in Medicine			PC, TRUST	33	21.8 ± 9.0	CBF, T2 blood, Yv, OEF, CMRO$_2$

Study	Year	Journal	ASL	Single/Multi	Sequences	N	Age	Outcomes
Ford et al. (2018)	2018	Blood	PCASL	Single	T1w, FLAIR	41	10.8 ± 3.9	CBF, lesions
Kawadler et al. (2018)	2018	NMR in Biomedicine	CASL	Multi	T2w	39	20 younger: 11.3 ± 1.4; 19 older: 15.7 ± 2.1	CBF, BAT, SCI
Watchmaker et al. (2018)	2018	Journal of Cerebral Blood Flow and Metabolism	PCASL	Single	FLAIR, TRUST, T1w, MRA	18	27.0 ± 5.0	CBF, OEF, lesions
Whitehead et al. (2018)	2018	American Journal of Neuroradiology	PCASL	Single	T1w, T2w, FLAIR, DWI, MRA	26	Range: 0–20	CBF
Chai et al. (2019)	2019	American Journal of Hematology	PCASL	Single	T1w, FLAIR, MRA	32	Transfused: 19.8 ± 7.2, nontransfused: 23.2 ± 9.1	CBF, OD, lesions
Vaclavu et al. (2019)	2019	Haematologica	PCASL	Single	MRA, FLAIR, T1 blood	36	31.9 ± 11.3	CBF, CVR, lesions, stenosis
Afzali-Hashemi et al. (2021a)	2021	Frontiers in Physiology	PCASL	Multi	T1w, PC	72	Severe: 29 ± 10, mild: 33 ± 12	CBF, ATT, CVR
Juttukonda et al. (2021)	2021	Journal of Cerebral Blood Flow and Metabolism	PCASL	Single	T1w, FLAIR, DWI, MRA, TRUST	69	27 (range: 18.1–40.2)	CBF, OEF, OD, CMRO$_2$, VHS, lesions, GM+WM volume
Afzali-Hashemi et al. (2021b)	2022	Haematologica	PCASL	Single	T2-TRIR, PC, T2w, FLAIR	66	Pediatric: 12.7 ± 2.3, Adult: 32.1 ± 11.2	CBF, VHS, Yv, OEF, flow velocity, lesions
Lin et al. (2022)	2022	Journal of Magnetic Resonance Imaging	—	—	PC, WEPCAST, T1w	21	9.9 ± 1.2	CBF, E, BBB PS
Stotesbury et al. (2022a)	2022	Frontiers in Physiology	PCASL, PASL	Single + multi	FLAIR, T1w, DWI, T2w, MRA	94	16.67 (IQR: 13.32–19.89)	CBF, BAT, lesions, FA
Stotesbury et al. (2022b)	2022	Journal of Cerebral Blood Flow and Metabolism	PCASL, PASL	Single + multi	FLAIR, T1w, MRA	94	16.67 (IQR: 13.32–19.89)	CBF, BAT, VHS, lesions

[a]Age is given as mean ± std., median (range), or median (IQR). Abbreviations: BAT, Bolus arrival time; BBB, bloood-brain barrier; BOLD, blood oxygen level dependent; CASL, continuous arterial spin labeling; CBF, cerebral blood flow; CMRO$_2$, cerebral metabolic rate of oxygen; CVR, cerebrovascular reserve; DTI, diffusion tensor imaging; DWI, diffusion-weighted imaging; E, extraction fraction; FA, fractional anisotropy; FLAIR, fluid-attenuated inversion recovery; GM, gray matter; IR, inversion recovery; MRA, magnetic resonance angiography; OD, oxygen delivery; OEF, oxygen extraction fraction; PASL, pulsed ASL; PC, phase contrast; PCASL, pseudo-continuous arterial spin labeling; PDw, proton density weighted; PS, permeability-surface area product; SCI, silent cerebral infarct; T1w, T1 weighted; T2w, T2 weighted; TCD, transcranial doppler; TRUST, T2 relaxation under spin tagging; VHS, venous hyperintensity signal; WEPCAST, water extraction with phase-contrast arterial spin tagging; WM, white matter; Yv, venous oxygenation.

to healthy controls (Jordan *et al.*, 2016). However, having different red blood cell characteristics, SCD patients are believed to be ill suited to this model. Later, Bush *et al.* presented several calibration models such as the healthy control HbA model and the sickle cell-specific HbS model which was recently updated with the Li-Bush HbS calibration model (Bush *et al.*, 2018b, 2021, 2017). Using the HbS model for SCD patients and HbA model for healthy controls, studies have shown decreased OEF in patients with severe SCD compared to healthy controls (Vaclavu *et al.*, 2020; Vu *et al.*, 2021b). Combined, OEF and CBF measurements offer the calculation of cerebral metabolic rate of oxygen ($CMRO_2$), an estimate for oxygen consumption by the brain. $CMRO_2$ is increasingly being studied in SCD patients. So far, OEF and $CMRO_2$ have been found to be reduced in patients with SCD compared to healthy controls (Vu *et al.*, 2021b) using the sickle cell-specific model for OEF. This is believed to be a result of the capillary shunting that was previously observed, limiting oxygen exchange between blood and tissue and ultimately leading to lower OEF and $CMRO_2$ values (Juttukonda *et al.*, 2021; Afzali-Hashemi *et al.*, 2021b).

13.2.4 Imaging protocol

Over the years the protocols used for perfusion measurements in SCD have matured and, consequently, become more complex. In the following, we list several key elements that should be accounted for.

The higher blood velocity in the neck of SCD patients reduces the labeling efficiency compared to healthy subjects with decreasing B1+ strength (Bush *et al.*, 2018b). Using a population-based B1+, Bush *et al.* (2018b) found a better agreement between PC and pseudo-continuous ASL (PCASL) CBF measurements. The most accurate solution, however, is to acquire a PC scan at the same position as the PCASL labeling plane to measure blood flow velocity at the labeling site. This allows for the calculation of patient-specific labeling efficiency based on simulations (Gevers *et al.*, 2012; Wu *et al.*, 2007). Measuring a subject-specific labeling efficiency is even more important when a CO2 or ACZ challenge is used because these challenges further increase blood velocity.

The longitudinal relaxation time (T1) of blood is longer in patients with SCD (Vaclavu *et al.*, 2016). This is likely to affect the efficiency of the background suppression pulses that are applied in the PCASL sequence. SNR could be increased by adjusting the timing of background suppression pulses based on the longer blood T1. One relatively straightforward possibility is to determine the timing of the background suppression pulses prior to the ASL scan, based on a measured or derived blood T1 value. Alternatively, the timing of the background suppression pulses can be optimized on the fly using a feedback loop (Koolstra *et al.*, 2022). To date, the latter option, however, requires specialized MR software but might be clinically available in the future. Nevertheless, it is recommended to use either method to increase SNR by optimizing the timing of the background suppression pulses.

If available, the use of a multi-PLD sequence is recommended to avoid a mismatch between ATT and PLD. If single-PLD ASL is used, a shorter PLD can be used in patients with SCD compared to healthy volunteers because of the shorter ATT (Afzali-Hashemi *et al.*, 2021a).

Altogether, an imaging protocol for an ASL study could look as follows:

- Survey
- PC survey/angiogram
- Blood T1 measurement
- ASL scan
- M0 scan
- PC at labeling plane
- T1-weighted anatomical scan

For recommendations on ASL acquisition parameters, tailored to specific clinical neuroimaging applications, we refer the reader to recommendations by Lindner *et al.* (2023). In general, a 3D, background suppressed, and acquisition with PCASL labeling is recommended. For most clinical applications, a single PLD of 1800 ms is sufficient, but in research settings and for specific pathologies, a multi-PLD protocol is recommended because of its more accurate CBF quantification and robustness against variations in ATT. Furthermore, a 3–4 mm in-plane resolution and slice thickness is recommended as well as a scan duration of 4–5 min.

13.2.5 **Data analysis protocol**

A standard postprocessing pipeline should include motion correction, coregistration between ASL and M0 image, CBF quantification using one of the available kinetic models, and possible registration to a structural image and/or common template. For a beginner's guide to ASL processing, we refer to Clement *et al.* (2022). However, some disease characteristics need to be taken into account during CBF quantification of ASL data from SCD patients. The longer blood T1 in SCD patients will lead to an overestimation of CBF and needs to be taken into account (Vaclavu *et al.*, 2016). An SCD-specific blood T1 of 1818 ms can be used or blood T1 can be measured using a dedicated scan (Petersen *et al.*, 2012; Li *et al.*, 2017). Both methods resulted in similar agreement with PC MRI (Vaclavu *et al.*, 2016). Interestingly, blood T1 cannot be derived from hematocrit as can be done for healthy volunteers. Second, it is recommended to use the dual-compartment model rather than the single-compartment model (Bush *et al.*, 2018a). This is inconsistent with the consensus recommendation to use a single-compartment model (Alsop *et al.*, 2015), which is again a result of the higher blood T1 in patients with SCD. The single-compartment model only accounts for the T1 of blood. In healthy volunteers, the error due to this assumption is considered acceptable. In SCD patients, however, the difference between blood T1 and tissue T1 is larger; therefore, this assumption results in larger errors. Bush *et al.* (2018a) have shown that this effect was the largest contributor to PCASL underestimation compared to PC CBF. There are several ASL analysis pipelines available. A recent comparison by the open-source initiative for perfusion imaging (OSIPI) lists the different pipelines available (Fan *et al.*, 2023, unpublished). It also summarizes what data the individual pipelines are suited for, what output they produce, and how user friendly they are.

13.3 Case reports

13.3.1 Incidental finding

A 24-year-old woman with SCD genotype HbSS was scanned before and after an ACZ challenge. On the CBF images, a right-left asymmetry was detected, showing hypoperfusion in the right hemisphere (left images Fig. 13.3). On the MOTSA angiogram image, a vascular obstruction in the right middle cerebral artery was seen (right image Fig. 13.3).

13.3.2 Hydroxyurea treatment

A 29-year-old woman with SCD genotype HbSS was scanned 1 month before hydroxyurea treatment and 7 months after (top two rows Fig. 13.4). Gray matter CBF reduced from 105.9 mL/100 g/min before the medication to 72.4 mL/100 g/min after. Gray matter CVR increased from 22.8% before the treatment to 43.4%. The patient felt better after hydroxyurea treatment and the sickle cell-related symptoms appeared less frequently than before the treatment.

13.3.3 Blood transfusion

A 28-year-old man with SCD genotype HbSS was scanned 1 day before and after exchange transfusion. The gray matter CBF before transfusion was 66.8 mL/100 g/min and 64.9 mL/100 g/min after. Gray matter CVR was 47.1% before the transfusion and 49.0% after. Although the CBF and CVR values did not change much, the patient mentioned feeling more energetic after transfusion. The unchanged

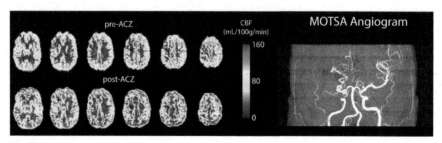

FIG. 13.3

CBF maps of an SCD patient with vascular obstruction in the right middle cerebral artery before and after acetazolamide administration.

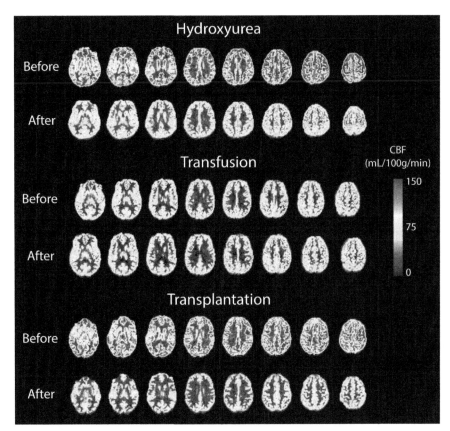

FIG. 13.4

CBF maps from three patients with SCD before and after receiving different treatments: regular intake of hydroxyurea, receiving a blood transfusion, and a successful stem cell transplantation treatment.

CBF and CVR values are most likely because of the strong negative association of CBF with hemoglobin. This patient received exchange transfusion which replaced his hemoglobin with the hemoglobin of the donor. The hemoglobin levels barely increased leading to a small decrease in CBF.

13.3.4 Stem cell transplantation

A 44-year-old woman with SCD genotype HbSS was scanned 3 months before HSCT and 9 months after. Gray matter CBF before HSCT was 92.9 mL/100 g/min and 54.0 mL/100 g/min after. In addition, the gray matter CVR also increased from 29.8% to 78.2%. The patient did not experience any sickle cell-related symptoms.

13.4 Challenges and solutions

13.4.1 Motion

Like other MRI techniques, ASL is susceptible to motion. Subtle head motion between the acquisition of label and control scans can result in a difference signal that can be larger than the perfusion signal itself. This is mostly seen around the edges of the brain. It is, therefore, recommended to always use some form of motion correction before subtracting label and control images. Besides motion between acquisitions, there can also be head motion during the acquisition. This cannot be corrected and can result in severe motion artifacts (Fig. 13.5). In this case, the motion artifacts were observed at one of the delay times of a multi-PLD acquisition. Therefore the motion artifacts were less apparent on the CBF maps obtained after kinetic modeling. Nevertheless, this scan had to be excluded. This example also illustrates that the image quality of individual PWIs needs to be evaluated separately. Although healthy controls can generally lie still for a longer time than patients, there are no SCD-specific characteristics that need to be taken into account regarding motion. The susceptibility of ASL to motion highlights the need to keep the scan duration as short as possible. This, however, is not easy given the poor SNR intrinsic to the technique. If one still wants to keep the scan time as short as possible in light of possible motion artifacts or patient comfort, we would recommend either increasing the voxel size or tuning the labeling block such that less acoustic noise is generated (van der Meer et al., 2014).

13.4.2 SNR in white matter

Whereas several articles demonstrated the accuracy of ASL for gray matter, there is skepticism about the reliability of ASL in white matter because of its lower CBF and longer ATT. However, patients with SCD have longer blood T1, shorter ATT, and higher CBF, which increase the temporal signal-to-noise ratio (tSNR) in the white

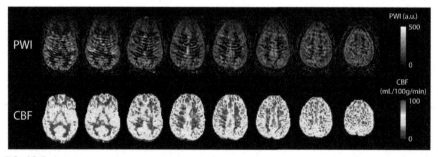

FIG. 13.5

ASL scan of a 25-year-old male SCD patient. One of the PWIs that was part of a multi-PLD acquisition showed severe motion artifact. After kinetic modeling, motion artifacts were less apparent on the CBF maps.

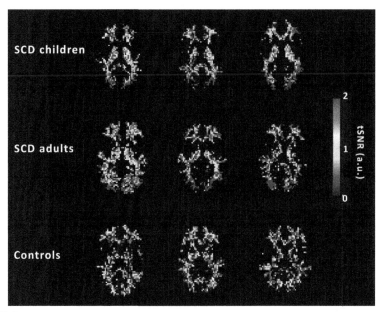

FIG. 13.6

Example of temporal signal-to-noise ratio (tSNR) maps from 35 averages from three representative children (*first row*) and adults (*second row*) with sickle cell disease (SCD), as well as for healthy controls (*third row*).

matter of these patients. Using a single-PLD PCASL with 35 control-label pairs, the tSNR was significantly higher in children and adult patients with SCD compared to controls (Fig. 13.6). This could mean that the measurements of WM ASL might be more accurate in patients with SCD compared to healthy participants.

13.4.3 Quantification challenges

A major advantage of ASL is that it offers quantification of CBF rather than only qualitative measures. However, this also comes with challenges, some of which are specific to SCD. Research laboratories often use in-house built postprocessing pipelines but without providing detailed descriptions of individual processing steps or proper quality control. This hampers the reproducibility and interpretation of ASL studies. The OSIPI task force therefore aims to "create open access resources for perfusion imaging research in order to eliminate the practice of duplicate development, improve the reproducibility of perfusion imaging research, and speed up the translation into tools for discovery science, drug development, and clinical practice" (osipi.org). Quantification of ASL data from SCD patients should be done with extra care. Online available postprocessing pipelines or CBF quantification software on the MRI system itself are always configured for healthy volunteers and do not take into account the SCD-specific characteristics that were discussed in this chapter. When quantifying the CBF of SCD patients, one must ensure that these characteristics, of which most importantly longer blood T1 and shorter ATT, are taken into account.

13.5 Learning and knowledge outcomes

Perfusion MRI is a valuable tool for assessing cerebrovascular hemodynamics in both healthy individuals and those with various diseases affecting the brain and CNS. In particular, ASL has significantly advanced our understanding of SCD pathophysiology. Despite these advancements, there is still much potential for quantitative MRI to be further explored. Challenges exist, but there is promise in the acceleration of MRI acquisitions and the application of artificial intelligence (AI) in the near future. These developments could enable scan protocols to contain accurate measurements of spatially resolved hemodynamic parameters in addition to conventional perfusion measurements.

References

Afzali-Hashemi, L., Baas, K.P.A., Schrantee, A., Coolen, B.F., van Osch, M.J.P., Spann, S.M., et al., 2021a. Impairment of cerebrovascular hemodynamics in patients with severe and milder forms of sickle cell disease. Front. Physiol. 12, 645205.

Afzali-Hashemi, L., Vaclavu, L., Wood, J.C., Nederveen, A.J., Mutsaerts, H.J.M.M., Schrantee, A., et al., 2021b. Assessment of functional shunting in patients with sickle cell disease. Blood 138.

Agrawal, R.K., Patel, R.K., Shah, V., Nainiwal, L., Trivedi, B., 2014. Hydroxyurea in sickle cell disease: drug review. Indian J. Hematol. Blood Transfus. 30 (2), 91–96.

Alsop, D.C., Detre, J.A., 1996. Reduced transit-time sensitivity in noninvasive magnetic resonance imaging of human cerebral blood flow. J. Cereb. Blood Flow Metab. 16 (6), 1236–1249.

Alsop, D.C., Detre, J.A., Golay, X., Gunther, M., Hendrikse, J., Hernandez-Garcia, L., et al., 2015. Recommended implementation of arterial spin-labeled perfusion MRI for clinical applications: a consensus of the ISMRM perfusion study group and the European consortium for ASL in dementia. Magn. Reson. Med. 73 (1), 102–116.

Arkuszewski, M., Krejza, J., Chen, R., Melhem, E.R., 2013. Sickle cell anemia: reference values of cerebral blood flow determined by continuous arterial spin labeling MRI. Neuroradiol. J. 26 (2), 191–200.

Ashorobi, D., Bhatt, R., 2022. Bone Marrow Transplantation in Sickle Cell Disease. StatPearls, Treasure Island (FL).

Ashorobi, D., Ramsey, A., Yarrarapu, S.N.S., Bhatt, R., 2022. Sickle Cell Trait. StatPearls, Treasure Island (FL).

Binnewijzend, M.A., Benedictus, M.R., Kuijer, J.P., van der Flier, W.M., Teunissen, C.E., Prins, N.D., et al., 2016. Cerebral perfusion in the predementia stages of Alzheimer's disease. Eur. Radiol. 26 (2), 506–514.

Boxerman, J.L., Shiroishi, M.S., Ellingson, B.M., Pope, W.B., 2016. Dynamic susceptibility contrast MR imaging in glioma: review of current clinical practice. Magn. Reson. Imaging Clin. N. Am. 24 (4), 649–670.

Bush, A.M., Borzage, M.T., Choi, S., Vaclavu, L., Tamrazi, B., Nederveen, A.J., et al., 2016. Determinants of resting cerebral blood flow in sickle cell disease. Am. J. Hematol. 91 (9), 912–917.

Bush, A., Borzage, M., Detterich, J., Kato, R.M., Meiselman, H.J., Coates, T., et al., 2017. Empirical model of human blood transverse relaxation at 3 T improves MRI T2 oximetry. Magn. Reson. Med. 77 (6), 2364–2371.

Bush, A., Chai, Y.Q., Choi, S.Y., Vaclavu, L., Holland, S., Nederveen, A., et al., 2018a. Pseudo continuous arterial spin labeling quantification in anemic subjects with hyperemic cerebral blood flow. Magn. Reson. Imaging 47, 137–146.

Bush, A.M., Coates, T.D., Wood, J.C., 2018b. Diminished cerebral oxygen extraction and metabolic rate in sickle cell disease using T2 relaxation under spin tagging MRI. Magn. Reson. Med. 80 (1), 294–303.

Bush, A., Vu, C., Choi, S., Borzage, M., Miao, X., Li, W., et al., 2021. Calibration of T2 oximetry MRI for subjects with sickle cell disease. Magn. Reson. Med. 86 (2), 1019–1028.

Buxton, R.B., Frank, L.R., Wong, E.C., Siewert, B., Warach, S., Edelman, R.R., 1998. A general kinetic model for quantitative perfusion imaging with arterial spin labeling. Magn. Reson. Med. 40 (3), 383–396.

Canak, B., Eskazan, A.E., 2022. Spotlight commentary–voxelotor: a new kid on the block in the treatment of sickle cell disease. Br. J. Clin. Pharmacol. 88 (6), 2564–2565.

Chai, Y.Q., Bush, A.M., Coloigner, J., Nederveen, A.J., Tamrazi, B., Vu, C., et al., 2019. White matter has impaired resting oxygen delivery in sickle cell patients. Am. J. Hematol. 94 (4), 467–474.

Chappell, M.A., MacIntosh, B.J., Donahue, M.J., Gunther, M., Jezzard, P., Woolrich, M.W., 2010. Separation of macrovascular signal in multi-inversion time arterial spin labelling MRI. Magn. Reson. Med. 63 (5), 1357–1365.

Clement, P., Castellaro, M., Okell, T.W., Thomas, D.L., Vandemaele, P., Elgayar, S., et al., 2022. ASL-BIDS, the brain imaging data structure extension for arterial spin labeling. Sci. Data 9 (1).

Davis, T.L., Kwong, K.K., Weisskoff, R.M., Rosen, B.R., 1998. Calibrated functional MRI: mapping the dynamics of oxidative metabolism. Proc. Natl. Acad. Sci. U. S. A. 95 (4), 1834–1839.

DeBaun, M.R., Kirkham, F.J., 2016. Central nervous system complications and management in sickle cell disease. Blood 127 (7), 829–838.

DeBaun, M.R., Sarnaik, S.A., Rodeghier, M.J., Minniti, C.P., Howard, T.H., Iyer, R.V., et al., 2012. Associated risk factors for silent cerebral infarcts in sickle cell anemia: low baseline hemoglobin, sex, and relative high systolic blood pressure. Blood 119 (16), 3684–3690.

Ellingson, B.M., Woodworth, D.C., Leu, K., Salamon, N., Holly, L.T., 2019. Spinal cord perfusion MR imaging implicates both ischemia and hypoxia in the pathogenesis of cervical spondylosis. World Neurosurg. 128, E773–E81.

Fan, H., Mutsaerts, H.J.M.M., Anazodo, U., Arteaga, D., Baas, K.P.A., Buchanan, C., 2023. The Open Source Initiative for Perfusion Imaging (OSIPI): ASL Pipeline Inventory. Manuscript submitted for publication.

Ford, A.L., Ragan, D.K., Fellah, S., Binkley, M.M., Fields, M.E., Guilliams, K.P., et al., 2018. Silent infarcts in sickle cell disease occur in the border zone region and are associated with low cerebral blood flow. Blood 132 (6), 1714–1723.

Gevers, S., Nederveen, A.J., Fijnvandraat, K., van den Berg, S.M., van Ooij, P., Heijtel, D.F., et al., 2012. Arterial spin labeling measurement of cerebral perfusion in children with sickle cell disease. J. Magn. Reson. Imaging 35 (4), 779–787.

Grandin, C.B., 2003. Assessment of brain perfusion with MRI: methodology and application to acute stroke. Neuroradiology 45 (11), 755–766.

Günther, M., 2007. Highly efficient accelerated acquisition of perfusion inflow series by cycled arterial spin labeling. In: Proceedings of the 15th Annual Meeting of ISMRM, p. 380.

Haller, S., Zaharchuk, G., Thomas, D.L., Lovblad, K.O., Barkhof, F., Golay, X., 2016. Arterial spin labeling perfusion of the brain: emerging clinical applications. Radiology 281 (2), 337–356.

Hebbel, R.P., Osarogiagbon, R., Kaul, D., 2004. The endothelial biology of sickle cell disease: inflammation and a chronic vasculopathy. Microcirculation 11 (2), 129–151.

Helton, K.J., Paydar, A., Glass, J., Weirich, E.A., Hankins, J., Li, C.S., et al., 2009. Arterial spin-labeled perfusion combined with segmentation techniques to evaluate cerebral blood flow in white and gray matter of children with sickle cell anemia. Pediatr. Blood Cancer 52 (1), 85–91.

Hoge, R.D., Atkinson, J., Gill, B., Crelier, G.R., Marrett, S., Pike, G.B., 1999. Investigation of BOLD signal dependence on cerebral blood flow and oxygen consumption: the deoxyhemoglobin dilution model. Magn. Reson. Med. 42 (5), 849–863.

Howard, J., 2016. Sickle cell disease: when and how to transfuse. Hematol. Am. Soc. Hemat., 625–631.

Jarnum, H., Steffensen, E.G., Knutsson, L., Frund, E.T., Simonsen, C.W., Lundbye-Christensen, S., et al., 2010. Perfusion MRI of brain tumours: a comparative study of pseudo-continuous arterial spin labelling and dynamic susceptibility contrast imaging. Neuroradiology 52 (4), 307–317.

Jiang, D., Lu, H., 2022. Cerebral oxygen extraction fraction MRI: techniques and applications. Magn. Reson. Med. 88 (2), 575–600.

Jiang, D.R., Lu, H.Z., Parkinson, C., Su, P., Wei, Z.L., Pan, L., et al., 2019. Vessel-specific quantification of neonatal cerebral venous oxygenation. Magn. Reson. Med. 82 (3), 1129–1139.

Jordan, L.C., Gindville, M.C., Scott, A.O., Juttukonda, M.R., Strother, M.K., Kassim, A.A., et al., 2016. Non-invasive imaging of oxygen extraction fraction in adults with sickle cell anaemia. Brain 139 (Pt. 3), 738–750.

Juttukonda, M.R., Jordan, L.C., Gindville, M.C., Pruthi, S., Donahue, M.J., 2016. Quantitation of arterial spin labeling MRI labeling efficiency in high cervical velocity conditions using phase contrast angiography. IEEE.

Juttukonda, M.R., Jordan, L.C., Gindville, M.C., Davis, L.T., Watchmaker, J.M., Pruthi, S., et al., 2017. Cerebral hemodynamics and pseudo-continuous arterial spin labeling considerations in adults with sickle cell anemia. NMR Biomed. 30 (2).

Juttukonda, M.R., Donahue, M.J., Davis, L.T., Gindville, M.C., Lee, C.A., Patel, N.J., et al., 2019. Preliminary evidence for cerebral capillary shunting in adults with sickle cell anemia. J. Cereb. Blood Flow Metab. 39 (6), 1099–1110.

Juttukonda, M.R., Donahue, M.J., Waddle, S.L., Davis, L.T., Lee, C.A., Patel, N.J., et al., 2021. Reduced oxygen extraction efficiency in sickle cell anemia patients with evidence of cerebral capillary shunting. J. Cereb. Blood Flow Metab. 41 (3), 546–560.

Karki, N.R., Saunders, K., Kutlar, A., 2022. A critical evaluation of crizanlizumab for the treatment of sickle cell disease. Expert. Rev. Hematol. 15 (1), 5–13.

Karlsson, E.K., Kwiatkowski, D.P., Sabeti, P.C., 2014. Natural selection and infectious disease in human populations. Nat. Rev. Genet. 15 (6), 379–393.

Kawadler, J.M., Hales, P.W., Barker, S., Cox, T.C.S., Kirkham, F.J., Clark, C.A., 2018. Cerebral perfusion characteristics show differences in younger versus older children with sickle cell anaemia: results from a multiple-inflow-time arterial spin labelling study. NMR Biomed. 31 (6).

Kedar, A., Drane, W.E., Shaeffer, D., Nicole, M., Adams, C., 2006. Measurement of cerebrovascular flow reserve in pediatric patients with sickle cell disease. Pediatr. Blood Cancer 46 (2), 234–238.

Kodl, C.T., Seaquist, E.R., 2008. Cognitive dysfunction and diabetes mellitus. Endocr. Rev. 29 (4), 494–511.

Koolstra, K., Staring, M., de Bruin, P., van Osch, M.J.P., 2022. Subject-specific optimization of background suppression for arterial spin labeling magnetic resonance imaging using a feedback loop on the scanner. NMR Biomed. 35 (9).

Kosinski, P.D., Croal, P.L., Leung, J., Williams, S., Odame, I., Hare, G.M., et al., 2017. The severity of anaemia depletes cerebrovascular dilatory reserve in children with sickle cell disease: a quantitative magnetic resonance imaging study. Br. J. Haematol. 176 (2), 280–287.

Li, W., Liu, P., Lu, H., Strouse, J.J., van Zijl, P.C.M., Qin, Q., 2017. Fast measurement of blood T1 in the human carotid artery at 3T: accuracy, precision, and reproducibility. Magn. Reson. Med. 77 (6), 2296–2302.

Lin, Z.X., Li, Y., Su, P., Mao, D., Wei, Z.L., Pillai, J.J., et al., 2018. Non-contrast MR imaging of blood-brain barrier permeability to water. Magn. Reson. Med. 80 (4), 1507–1520.

Lin, Z.X., Lance, E., McIntyre, T., Li, Y., Liu, P.Y., Lim, C., et al., 2022. Imaging blood-brain barrier permeability through MRI in pediatric sickle cell disease: a feasibility study. J. Magn. Reson. Imaging 55 (5), 1551–1558.

Lindner, T., Bolar, D.S., Achten, E., Barkhof, F., Bastos-Leite, A.J., Detre, J.A., et al., 2023. Current state and guidance on arterial spin labeling perfusion MRI in clinical neuroimaging. Magn. Reson. Med. 89 (5), 2024–2047.

Liu, P., Uh, J., Lu, H., 2011. Determination of spin compartment in arterial spin labeling MRI. Magn. Reson. Med. 65 (1), 120–127.

Liu, P.Y., De Vis, J.B., Lu, H.Z., 2019. Cerebrovascular reactivity (CVR) MRI with CO2 challenge: a technical review. NeuroImage 187, 104–115.

Lu, H.Z., Ge, Y.L., 2008. Quantitative evaluation of oxygenation in venous vessels using T2-relaxation-under-spin-tagging MRI. Magn. Reson. Med. 60 (2), 357–363.

Luzzatto, L., 2012. Sickle cell anaemia and malaria. Mediterr. J. Hematol. Infect. Dis. 4 (1).

Macedo-Campos, R.D., Adegoke, S.A., Figueiredo, M.S., Braga, J.A.P., Silva, G.S., 2018. Cerebral vasoreactivity in children with sickle cell disease: A transcranial doppler study. J. Stroke Cerebrovasc. Dis. 27 (10), 2703–2706.

Mahroo, A., Buck, M.A., Huber, J., Breutigam, N.J., Mutsaerts, H.J.M.M., Craig, M., et al., 2021. Robust multi-TE ASL-based blood-brain barrier integrity measurements. Front. Neurosci., 15.

Mburu, J., Odame, I., 2019. Sickle cell disease: reducing the global disease burden. Int. J. Lab. Hematol. 41, 82–88.

Numaguchi, Y., Haller, J.S., Humbert, J.R., Robinson, A.E., Lindstrom, W.W., Gruenauer, L.M., et al., 1990. Cerebral blood flow mapping using stable xenon-enhanced CT in sickle cell cerebrovascular disease. Neuroradiology 31 (4), 289–295.

Nur, E., Kim, Y.S., Truijen, J., van Beers, E.J., Davis, S.C.A.T., Brandjes, D.P., et al., 2009. Cerebrovascular reserve capacity is impaired in patients with sickle cell disease. Blood 114 (16), 3473–3478.

Oguz, K.K., Golay, X., Pizzini, F.B., Freer, C.A., Winrow, N., Ichord, R., et al., 2003. Sickle cell disease: continuous arterial spin-labeling perfusion MR imaging in children. Radiology 227 (2), 567–574.

Ostergaard, L., Weisskoff, R.M., Chesler, D.A., Gyldensted, C., Rosen, B.R., 1996. High resolution measurement of cerebral blood flow using intravascular tracer bolus passages. Part I: mathematical approach and statistical analysis. Magn. Reson. Med. 36 (5), 715–725.

Padhani, A.R., 2002. Dynamic contrast-enhanced MRI in clinical oncology: current status and future directions. J. Magn. Reson. Imaging 16 (4), 407–422.

Patrick, J.T., Fritz, J.V., Adamo, J.M., Dandonna, P., 1996. Phase-contrast magnetic resonance angiography for the determination of cerebrovascular reserve. J. Neuroimaging 6 (3), 137–143.

Pavilla, A., Arrigo, A., Mejdoubi, M., Duvauferrier, R., Gambarota, G., Saint-Jalmes, H., 2018. Measuring cerebral hypoperfusion induced by hyperventilation challenge with intravoxel incoherent motion magnetic resonance imaging in healthy volunteers. J. Comput. Assist. Tomogr. 42 (1), 85–91.

Petersen, E.T.D.V.J., Alderliesten, T., Kersbergen, K.J., Benders, M., Hendrikse, J., van den Berg, C.A.T., 2012. Simultaneous OEF and haematocrit assessment using T2 prepared blood relaxation imaging with inversion recovery. In: International Society of Magnetic Resonance in Medicine. International Society of Magnetic Resonance in Medicine, Melbourne.

Piel, F.B., Hay, S.I., Gupta, S., Weatherall, D.J., Williams, T.N., 2013. Global burden of sickle cell anaemia in children under five, 2010-2050: modelling based on demographics, excess mortality, and interventions. PLoS Med. 10 (7).

Piel, F.B., Steinberg, M.H., Rees, D.C., 2017. Sickle cell disease. N. Engl. J. Med. 376 (16), 1561–1573.

Pinto, V.M., Balocco, M., Quintino, S., Forni, G.L., 2019. Sickle cell disease: a review for the internist. Intern. Emerg. Med. 14 (7), 1051–1064.

Prohovnik, I., Pavlakis, S.G., Piomelli, S., Bello, J., Mohr, J.P., Hilal, S., et al., 1989. Cerebral hyperemia, stroke, and transfusion in sickle-cell disease. Neurology 39 (3), 344–348.

Qu, J., Kong, Q., Guo, Y., Kuehn, B., Sun, Y., Zhu, J., 2021. Recent progress in ASL outside the brain. Chin. J. Acad. Radiol. 4 (4), 220–228.

Rees, D.C., Williams, T.N., Gladwin, M.T., 2010. Sickle-cell disease. Lancet 376 (9757), 2018–2031.

Rosen, B.R., Belliveau, J.W., Vevea, J.M., Brady, T.J., 1990. Perfusion imaging with NMR contrast agents. Magn. Reson. Med. 14 (2), 249–265.

Schmid, S., Teeuwisse, W.M., Lu, H., van Osch, M.J., 2015. Time-efficient determination of spin compartments by time-encoded pCASL T2-relaxation-under-spin-tagging and its application in hemodynamic characterization of the cerebral border zones. NeuroImage 123, 72–79.

Shenoy, S., 2013. Hematopoietic stem-cell transplantation for sickle cell disease: current evidence and opinions. Ther. Adv. Hematol. 4 (5), 335–344.

Shi, Y.L., Thrippleton, M.J., Makin, S.D., Marshall, I., Geerlings, M.I., de Craen, A.J.M., et al., 2016. Cerebral blood flow in small vessel disease: a systematic review and meta-analysis. J. Cereb. Blood Flow Metab. 36 (10), 1653–1667.

Smith, A.M., Grandin, C.B., Duprez, T., Mataigne, F., Cosnard, G., 2000a. Whole brain quantitative CBF and CBV measurements using MRI bolus tracking: comparison of methodologies. Magn. Reson. Med. 43 (4), 559–564.

Smith, A.M., Grandin, C.B., Duprez, T., Mataigne, F., Cosnard, G., 2000b. Whole brain quantitative CBF, CBV, and MTT measurements using MRI bolus tracking: implementation and application to data acquired from hyperacute stroke patients. J. Magn. Reson. Imaging 12 (3), 400–410.

Spilt, A., Box, F.M.A., van der Geest, R.J., Reiber, J.H.C., Kunz, P., Kamper, A.M., *et al.*, 2002. Reproducibility of total cerebral blood flow measurements using phase contrast magnetic resonance imaging. J. Magn. Reson. Imaging 16 (1), 1–5.

Stotesbury, H., Hales, P.W., Hood, A.M., Koelbel, M., Kawadler, J.M., Saunders, D.E., *et al.*, 2022a. Individual watershed areas in sickle cell anemia: an arterial spin labeling study. Front. Physiol. 13.

Stotesbury, H., Hales, P.W., Koelbel, M., Hood, A.M., Kawadler, J.M., Saunders, D.E., *et al.*, 2022b. Venous cerebral blood flow quantification and cognition in patients with sickle cell anemia. J. Cereb. Blood Flow Metab. 42 (6), 1061–1077.

Strouse, J.J., Cox, C.S., Melhem, E.R., Lu, H.Z., Kraut, M.A., Razumovsky, A., *et al.*, 2006. Inverse correlation between cerebral blood flow measured by continuous arterial spin-labeling (CASL) MRI and neurocognitive function in children with sickle cell anemia (SCA). Blood 108 (1), 379–381.

Strouse, J.J., Jordan, L.C., Lanzkron, S., Casella, J.F., 2009. The excess burden of stroke in hospitalized adults with sickle cell disease. Am. J. Hematol. 84 (9), 548–552.

Thein, M.S., Thein, S.L., 2016. World sickle cell day 2016: a time for appraisal. Indian J. Med. Res. 143, 678–681.

Tofts, P.S., 2010. T1-weighted DCE imaging concepts: modelling, acquisition and analysis. MAGNETOM Flash 3, 30–35.

Toth, P., Tarantini, S., Csiszar, A., Ungvari, Z., 2017. Functional vascular contributions to cognitive impairment and dementia: mechanisms and consequences of cerebral autoregulatory dysfunction, endothelial impairment, and neurovascular uncoupling in aging. Am. J. Physiol. Heart Circ. Physiol. 312 (1), H1–H20.

Vaclavu, L., van der Land, V., Heijtel, D.F.R., van Osch, M.J.P., Cnossen, M.H., Majoie, C.B. L.M., *et al.*, 2016. In vivo T1 of blood measurements in children with sickle cell disease improve cerebral blood flow quantification from arterial spin-labeling MRI. Am. J. Neuroradiol. 37 (9), 1727–1732.

Vaclavu, L., Meynart, B.N., Mutsaerts, H.J.M.M., Petersen, E.T., Majoie, C.B.L.M., VanBavel, E.T., *et al.*, 2019. Hemodynamic provocation with acetazolamide shows impaired cerebrovascular reserve in adults with sickle cell disease. Haematologica 104 (4), 690–699.

Vaclavu, L., Petr, J., Petersen, E.T., Mutsaerts, H., Majoie, C.B.L., Wood, J.C., *et al.*, 2020. Cerebral oxygen metabolism in adults with sickle cell disease. Am. J. Hematol. 95 (4), 401–412.

van den Tweel, X.W., Nederveen, A.J., Majoie, C.B.L.M., van der Lee, J.H., Wagener-Schimmel, L., van Walderveen, M.A.A., *et al.*, 2009. Cerebral blood flow measurement in children with sickle cell disease using continuous arterial spin labeling at 3.0-tesla MRI. Stroke 40 (3), 795–800.

van der Meer, J.N., Heijtel, D.F.R., van Hest, G., Plattel, G.J., van Osch, M.J.P., van Someren, E.J.W., *et al.*, 2014. Acoustic noise reduction in pseudo-continuous arterial spin labeling (pCASL). Magn. Reson. Mater. Phys. Biol. Med. 27 (3), 269–276.

Vu, C., Chai, Y.Q., Coloigner, J., Nederveen, A.J., Borzage, M., Bush, A., et al., 2021a. Quantitative perfusion mapping with induced transient hypoxia using BOLD MRI. Magn. Reson. Med. 85 (1), 182–195.

Vu, C., Bush, A., Choi, S., Borzage, M., Miao, X., Nederveen, A.J., et al., 2021b. Reduced global cerebral oxygen metabolic rate in sickle cell disease and chronic anemias. Am. J. Hematol. 96 (8), 901–913.

Ware, R.E., de Montalembert, M., Tshilolo, L., Abboud, M.R., 2017. Sickle cell disease. Lancet 390 (10091), 311–323.

Watchmaker, J.M., Juttukonda, M.R., Davis, L.T., Scott, A.O., Faraco, C.C., Gindville, M.C., et al., 2018. Hemodynamic mechanisms underlying elevated oxygen extraction fraction (OEF) in moyamoya and sickle cell anemia patients. J. Cereb. Blood Flow Metab. 38 (9), 1618–1630.

Whitehead, M.T., Smitthimedhin, A., Webb, J., Mahdi, E.S., Khademian, Z.P., Carpenter, J.L., et al., 2018. Cerebral blood flow and marrow diffusion alterations in children with sickle cell anemia after bone marrow transplantation and transfusion. Am. J. Neuroradiol. 39 (11), 2132–2139.

Williams, D.S., Detre, J.A., Leigh, J.S., Koretsky, A.P., 1992. Magnetic resonance imaging of perfusion using spin inversion of arterial water. Proc. Natl. Acad. Sci. U. S. A. 89 (1), 212–216.

Wintermark, M., Sesay, M., Barbier, E., Borbely, K., Dillon, W.P., Eastwood, J.D., et al., 2005. Comparative overview of brain perfusion imaging techniques. Stroke 36 (9), e83–e99.

Wong, E.C., Buxton, R.B., Frank, L.R., 1998. Quantitative imaging of perfusion using a single subtraction (QUIPSS and QUIPSS II). Magn. Reson. Med. 39 (5), 702–708.

Wu, W.C., Fernandez-Seara, M., Detre, J.A., Wehrli, F.W., Wang, J., 2007. A theoretical and experimental investigation of the tagging efficiency of pseudocontinuous arterial spin labeling. Magn. Reson. Med. 58 (5), 1020–1027.

Xing, D., Zha, Y., Yan, L., Wang, K., Gong, W., Lin, H., 2015. Feasibility of ASL spinal bone marrow perfusion imaging with optimized inversion time. J. Magn. Reson. Imaging 42 (5), 1314–1320.

Zaharchuk, G., Do, H.M., Marks, M.P., Rosenberg, J., Moseley, M.E., Steinberg, G.K., 2011. Arterial spin-labeling MRI can identify the presence and intensity of collateral perfusion in patients with Moyamoya disease. Stroke 42 (9), 2485–U183.

Zhu, G.M., Federau, C., Wintermark, M., Chen, H., Marcellus, D.G., Martin, B.W., et al., 2020. Comparison of MRI IVIM and MR perfusion imaging in acute ischemic stroke due to large vessel occlusion. Int. J. Stroke 15 (3), 332–342.

Perfusion MRI in the heart: Arterial spin labeling

14

Verónica Aramendía-Vidaurreta[a] and Frank Kober[b]

[a]*Department of Radiology, Clínica Universidad de Navarra, Pamplona, Spain*
[b]*Aix-Marseille Univ, CNRS, CRMBM, Marseille, France*

14.1 Overview

14.1.1 Motivation and targeted applications

Perfusion assessment in the heart is an important addition to standard clinical protocols, not only for diagnosing pathologies that are specifically vascular, but also for those that may indirectly involve circulatory aspects such as microvascular function. Myocardial perfusion MRI can mean many things ranging from visual delineation of impaired vascular supply after myocardial infarction to full absolute quantitative mapping of capillary myocardial blood flow (MBF) under rest and stress conditions. While first-pass dynamic contrast-enhanced (DCE) MRI is the only way currently available on scanners to do such measurements, a full quantitative assessment is rather complicated and time consuming in clinical routine examinations. Currently, clinical MR scanners do not supply all-in-one analysis packages, although community-developed tools for MBF quantification from first-pass measurements show strong progress and are more widely distributed. Instead, various compromises in the form of "semiquantitative" markers have been elaborated to obtain information going beyond a simple visual detection of hypoperfused areas. Among them, the most frequently used are upslope integral and time to peak after bolus injection. These markers, if used with care, can help comparing values related to MBF among subjects, but there is a risk of bias depending on various factors. First pass offers excellent contrast changes due to perfusion, but the translation of the contrast changes into accurate and reproducible physiological measures is still not straightforward. ASL, on the other hand, offers a much simpler quantification, which, in theory, is also more robust. However, the contrast changes between labeling and control acquisitions are very small compared with those occurring in first-pass acquisitions, which gives ASL a poor sensitivity despite higher tissue blood flow in the myocardium than in the brain.

In contrast to human clinical applications, sensitivity is of little concern when studying rodent animal models of human pathologies, because rodents feature much higher capillary tissue perfusion in general. For the same reasons, first-pass MRI measurements in rodents are more complicated than in humans, which is why rodent myocardial ASL has become a robust routine method readily employed in several application studies.

As in other organs, the major motivation for using ASL is its ability to provide a repeatable, quantitative microvascular assessment without using exogenous contrast agents.

14.1.2 Promises and benefits

First-pass gadolinium-enhanced DCE perfusion imaging is now a well-established technique in the heart, consensus on quantitative MBF mapping acquisition and postprocessing methods is becoming real, and gadolinium-based contrast agent safety in general is now very well managed in MRI. Nonetheless, there has been growing interest in contrast-free cardiac MRI methods. On the one hand, the cardiac patient population is particularly affected by contrast agent intolerance due to a higher proportion of patients also presenting with chronic kidney failure and, therefore, with contraindication to contrast agents. For instance, about 30% of heart failure patients also have kidney failure (Ahmed and Campbell, 2008), excluding them from contrast agent injection. The most common cause of heart failure is coronary artery disease (CAD) (Purek *et al.*, 2006), where myocardial perfusion is an important diagnostic marker. Beyond inaccessibility for a (small) subgroup of patients, the injection represents discomfort for patients and additional time spent by technologists, causing a growing general reluctance to using them in certain imaging centers. Recently, potential retention of gadolinium contrast agents in tissue such as brain, liver, skin, and bone has been pointed out (Guo *et al.*, 2018). On the other hand, when it comes to quantitative MBF measurements, ASL promises better reproducibility than first pass due to less dependence on experimental parameters such as the arterial input function (AIF), whose influence has to be taken into account in first-pass postprocessing. Such better reproducibility has, however, never been shown by direct comparison in human subjects. A major strength of ASL MBF is that it is immediately quantitative: the signal is directly proportional to capillary blood flow. As in other organs, an ASL acquisition does not require particular preparations and can be run from the console as any other imaging sequence. It can also be freely repeated under varying physiologic conditions, for instance, during pharmacologic or physiologic stress, which is particularly relevant in cardiac MRI to explore vascular functional responses. Nevertheless, to date, myocardial ASL has not reached a clinical routine application status, and there are no vendor-supplied acquisition or postprocessing methods available. Myocardial ASL is, therefore, still a topic of ongoing method research.

14.1.3 Preclinical vs clinical use

ASL has been used in both large (swine) and small (rats, mice) animals, not only for studying animal models of human disease, but also for setting up and evaluating new ASL approaches under precisely controlled physiologic conditions. Dimensions and physiologic parameters of the swine heart are of the same order of magnitude as those of the human heart, such that the same MRI methods as those for humans can be applied. Due to major differences in rodent cardiovascular and hemodynamic values, however, the conditions for perfusion imaging in small-animal preclinical MRI are very different from clinical MRI. Heart rates and breath rates are more than five times higher and so is capillary blood flow (approximately 7 ml/g/min in mice at rest vs 1 ml/g/min in humans). Velocities in major vessels are similar to those humans, although these vessels are much smaller. Finally, the time for blood to recirculate in the entire body is only a few seconds in mice as opposed to half a minute in humans. All these facts, but particularly the high capillary blood flow values, are in favor of ASL for measuring myocardial perfusion in rodents. Nonetheless, there are implications for optimizing MRI sequences in both situations.

14.1.4 Specifics of myocardial ASL

14.1.4.1 Signal detection efficiency in the heart

The promise of good sensitivity due to higher tissue blood flow at rest in the human heart (roughly 1 ml/g/min) than in the brain (gray matter: roughly 0.5 ml/g/min) is compromised by several factors. (i) The detection sensitivity is reduced to about one-third due to the larger volume that has to be covered by the receive radio-frequency antenna(e) compared with the brain. (ii) Due to cardiac motion, only one-third of the cardiac cycle is actually usable for signal detection, reducing the acquisition efficiency further. (iii) Breathing motion further limits acquisition efficiency, since it either imposes a limit on the maximum acquisition duration (breath-hold) or requires parts of the data to be discarded. (iv) Highly efficient sequences such as turbo-spin echo or EPI are more difficult to use due to cardiac motion on the one hand and to a more heterogeneous B_0 field distribution on the other. Balanced steady-state free-precession (bSSFP) sequences, however, also provide good signal efficiency, and they have become common practice in the heart.

14.1.4.2 Motion

Motion management during a cardiac MRI study is crucial for the quantification of myocardial perfusion with ASL. The low signal-to-noise ratio (SNR) of the perfusion-weighted signal makes the technique more vulnerable to patient breathing, heart wall movement, and involuntary motion artifacts. Thus several acquisitions are required to obtain reliable perfusion measurements. Motion during this time can cause the structures of the heart to change in size and position with movements in and out of the selected image plane, causing both in-plane and through-plane motion, which can be even higher in basal or apical slices in comparison to the mid-ventricular slice. Since ASL is a subtraction technique, motion between control and label images, as well as within the temporal series, can significantly affect perfusion quantification. This is especially relevant in myocardial ASL due to the high signal intensity of the blood pool that is surrounded by the (relatively thin) myo-cardium and the low resolution of the acquired images, which causes partial volume effects in the presence of small misalignments. This can create signal outliers within the myocardium both with large positive and negative perfusion values.

14.1.4.3 Blood flow patterns in larger and smaller vessels, timing

Blood supply to the myocardium via the coronary arteries is pulsatile. This is due to an increase of pressure in the aortic root after aortic valve closure, but changes in vascular resistance due to tissue compaction in systole also play an important role. Dynamic measurements via flow probes have shown that this pulsatility varies significantly across different arteries. Flow velocity in the main coronary arteries reaches 80 cm/s in diastole and drops to 30 cm/s in systole (Hadjiloizou et al., 2008). Whereas the left anterior descending artery shows flow variations of 50% at rest (Fig. 14.1), the right tree has a much smoother time course.

Due to the impedance of the coronary tree, flow pulsatility, however, shallows out rapidly after two bifurcations. These circumstances still show that for maintaining a constant labeling efficiency in any acquisition scheme, the labeling should occur in a distinct cardiac phase that has to be the same for every heartbeat. Prospective gating is, therefore, necessary for myocardial ASL acquisitions even if more and more cardiac MRI techniques move to retrospective or self-gated variants. In pulsed ASL, labeling is done instantaneously and upstream of the imaging plane, and due to the flow variations, the optimal moment of pulsed labeling within the cardiac cycle has to be determined. Flow-driven inversions, as those used in continuous or pseudo-continuous ASL, would suffer from the pulsatility but also from the tortuosity of the arterial tree adding significant uncertainty to the labeling efficiency.

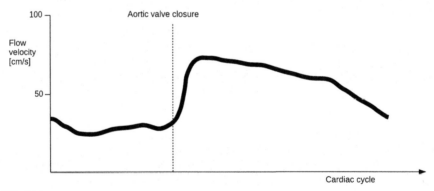

FIG. 14.1

Coronary flow in the left main coronary branch across the cardiac cycle based on intravascular flow data. Flow velocity at the beginning of the diastole is roughly 50% higher than in end-diastole.

Schematic representation of data from Hadjiloizou, N., Davies, J.E., Malik, I.S., Aguado-Sierra, J., Willson, K., Foale, R.A., Parker, K.H., Hughes, A.D., Francis, D.P., Mayet, J., 2008. Differences in cardiac microcirculatory wave patterns between the proximal left mainstem and proximal right coronary artery. Am. J. Physiol. Heart Circ. Physiol. 295, H1198–H1205.

14.1.5 **History of development and current status of methods**

14.1.5.1 *Clinical*

Inspired from the first successful results using ASL in the brain (Kim, 1995; Williams *et al.*, 1992), the feasibility of a double-gated flow-sensitive alternating inversion-recovery (FAIR) echo-planar imaging (EPI) method in the human and in swine heart was shown shortly after (Poncelet *et al.*, 1999). It became also clear that, at rest, MBF measurements in the human heart were highly variable, whereas the ASL signal under pharmacologic stress was sufficient to provide more reliable measurements. In a second study, carried out using a Look–Locker FAIR snapshot fast low-angle shot (FLASH) method, ASL was able to distinguish between CAD patients and healthy volunteers, although the absolute MBF values obtained were greatly overestimated compared with literature values obtained with other imaging techniques such as positron emission tomography (Wacker *et al.*, 2003). Since these first myocardial ASL studies in humans, scanner hardware has improved in many ways, progressively allowing multichannel detection, parallel imaging acceleration (Do *et al.*, 2014), reliable bSSFP readouts (Zun *et al.*, 2009), and improved ECG triggering reliability (Do *et al.*, 2017). Postprocessing has contributed to improved robustness to motion and enabled free-breathing acquisitions (Aramendía-Vidaurreta *et al.*, 2020). Compressed sensing acceleration has been used to reduce the readout window such that FAIR-bSSFP acquisitions could alternatively be done in end-systole rather than in end-diastole (Henningsson *et al.*, 2021), where a larger in-plane area of myocardium is available for evaluation. Various improvements in acquisition and labeling strategy have been proposed, such as steady-pulsed ASL (spASL) and velocity-selective ASL (VS-ASL). To date, most published studies have remained technical in nature, but there are also several successful demonstrations where ASL was sensitive enough to show perfusion defects in patients.

Typical acquisition times for FAIR-ASL and spASL in a single slice are between 2 and 5 min, depending on the selected strategy. With breath-holding and a FAIR labeling strategy, six breath-holds (around 12-s duration and 15-s delay between breath-holds) are typically performed, with the acquisition of one label and one control image each (Zun *et al.*, 2009). With FAIR labeling and synchronized breathing or free breathing, 62 images are typically acquired (Aramendía-Vidaurreta *et al.*, 2022). With spASL labeling, a series of 128 label and 128 control images were acquired under free breathing (Capron *et al.*, 2015) resulting in about 4-min acquisition time.

14.1.5.2 *Preclinical*

In rodents, the effect of perfusion on Look–Locker T1 measurements was studied in various theoretical and fundamental papers (Bauer *et al.*, 1996, 1997) before attempting to quantify MBF via T1 measurements directly using the same method for the first time (Belle *et al.*, 1998). Most small-animal myocardial ASL studies, including the first one, used variants of the Look–Locker FAIR (LLFAIR) technique. This gradient-echo-based technique was later refined by different retrospective respiratory gating approaches and higher spatial resolution (Kober *et al.*, 2005,

2004; Vandsburger *et al.*, 2010) and validated against microspheres (Jacquier *et al.*, 2011). Since typical acquisition times remained relatively long (around 25 min), segmented k-space acquisition has been used to acquire a single-slice MBF map with reduced acquisition time, sacrificing some of the effective resolution due to motion (Campbell-Washburn *et al.*, 2013a). A multi-slice ASL approach was proposed, allowing acquisition of three slices within 15 min (Campbell-Washburn *et al.*, 2013b), although MBF values seemed higher compared to the reference single-slice technique. The Look–Locker ASL approach was also directly compared with accelerated first-pass perfusion measurements, showing an advantage of ASL in high blood flow conditions, whereas first pass was found advantageous with low blood flow values (Naresh *et al.*, 2015). Potential sources of error in the LLFAIR measurement were later addressed in a detailed analysis (Kampf *et al.*, 2014). A fully retrospectively gated LLFAIR approach was finally introduced, allowing reduction of the acquisition time and reconstruction of MBF maps in different cardiac phases (Gutjahr *et al.*, 2015). Other than these Look–Locker approaches and similarly to most techniques used in humans, Abeykoon *et al.* (Abeykoon *et al.*, 2012) proposed a signal-intensity-based FAIR-ASL method.

To improve efficiency via a steady-pulsed ASL scheme using a cine readout, the cine-ASL method was introduced by inserting one labeling pulse in the cine-readout scheme of every cardiac cycle (Troalen *et al.*, 2013). This allowed shorter acquisition times at equal spatial resolution and robustness. The sensitivity of cine-ASL was indeed shown roughly three times larger than that of LLFAIR, whereas MBF values were comparable. As a cine imaging technique, cine-ASL also provides several perfusion maps across the cardiac cycle (Troalen *et al.*, 2014). A remaining drawback of cine-ASL is that a separate T1 measurement is necessary for absolute quantification.

14.2 Imaging protocol

At the time this chapter was written, no vendor-supplied myocardial ASL techniques were available. However, for both preclinical and clinical implementations, the techniques may be obtained via user-to-user interactions. Certain preclinical and clinical methods for acquisition and postprocessing can be obtained upon reasonable request from the chapter authors. This section reviews and discusses existing strategies from a fundamental and practical point of view and gives underlying literature references.

14.2.1 Labeling strategies

14.2.1.1 FAIR ASL

Flow-sensitive alternating inversion recovery (FAIR) labeling, invented by Kim *et al.* in 1995 (Kim, 1995), has been used in most myocardial ASL studies due to its ability to label blood flowing in all directions. A nonselective pulse is used for the label readout to invert a large volume, considering blood flowing above and below the image plane and static tissue, and a slice-selective pulse is used for the control readout to invert a small volume, considering that blood within this region will flow out of the slice during the inversion time and only static tissue will remain inverted. To assure that the slice-selective inversion pulse inverts the image plane, despite labeling pulse imperfections and the presence of motion, an increased slice-selective volume is used that includes equal sized gaps above and below the image plane.

The lower row of Fig. 14.2 shows an illustration of blood flow through the heart in a schematic three-chamber view for FAIR labeling over the course of two cardiac cycles.

Short TRs are desirable to reduce scan time, but a minimum repetition time of approximately 6 s has been mostly used to allow an almost complete recovery of the label magnetization (97% recovery considering a T1 of 1.664 s at 3T and an RR (time interval between consecutive R waves on an ECG) interval of 1 s), which is inverted every two repetition times. The use of presaturation pulses in the sequence

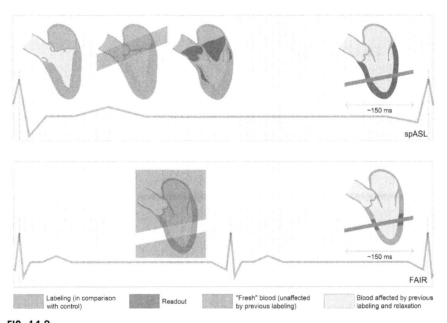

FIG. 14.2

Illustration of blood flow through the heart during the cardiac cycle for FAIR and spASL labeling strategies. Upper row: spASL labeling represented during one ECG cycle. Blood in the aortic root is labeled at every end-systolic phase and flows through the coronaries to the imaging slice being read out in late diastole. Labeled blood is fully replaced every cardiac cycle. Blood labeled in the left atrium exchanges with left-ventricular blood in diastole and is then also fully replaced. Lower row: FAIR labeling and readout occurring in mid-diastole of two consecutive cardiac cycles. The labeling slab thickness is larger than the imaging slice thickness to assure inversion of all spins in the imaging slice.

can help to minimize static tissue variations. Simulations of arterial blood magnetization after the use of presaturation and inversion pulses with different repetition times showed similar signal patterns, indicating that low perfusion values are caused by saturation of arterial blood outside the imaging slice and that shortening the repetition time to four cardiac cycles provided comparable perfusion values to those obtained with a longer repetition time of six cardiac cycles (Aramendía-Vidaurreta et al., 2019). An example MBF map acquired with a FAIR-bSSFP sequence during free breathing along with values quantified in segmental regions is shown in Fig. 14.3. The acquired pixel size was 2.3×2.3 mm^2 and the slice thickness 10 mm.

In preclinical applications, the FAIR labeling method has been used in conjunction with a Look–Locker readout strategy (Belle et al., 1998), where magnetization recovery after global or slice-selective inversion is sampled at every heartbeat using ECG-gated acquisitions. The readouts were either single gradient echoes or larger segments of k-space with small flip angles. The Look–Locker sampling strategy is particularly advantageous at the high heart rates encountered in rodents, but FAIR-Look–Locker acquisitions have also been used in two human studies (Fidler et al., 2004; Keith et al., 2017).

14.2.1.2 Steady-pulsed ASL and cine-ASL

Steady-pulsed ASL (Capron et al., 2013, 2015) has been proposed to improve FAIR's relatively poor data collection efficiency (signal acquisitions per unit time) caused by its long repetition times while preserving the pulsed character of the labeling. Relying on the fact that blood in the aorta is (i) completely exchanged with every heartbeat and (ii) almost at rest after closure of the aortic valve, the labeling pulse targets the aortic root and blood in the coronaries downstream at every heartbeat. Readouts are done at the end of each diastole, ensuring a relatively quiescent position of the myocardium. The control scan is acquired with the inversion slab geometry mirrored by the imaging slice, as for the EPISTAR (Edelman et al., 1994) labeling scheme known from brain ASL.

The upper row of Fig. 14.2 shows an illustration of blood flow through the heart in a three-chamber view for spASL labeling over the course of a cardiac cycle.

The high flow velocities in the coronaries in diastole should ensure small transit times from the labeled volume to the imaging slice. spASL is, therefore, a pulsed labeling technique where the entire blood volume entering the myocardium is labeled periodically while allowing image acquisitions in every heartbeat.

There is, however, an uncertainty concerning the final blood magnetization arriving in the arteriole in the imaging slice due to dispersion of the created "inversion bolus." First, blood spins undergo inversions not only once in the aortic root, but also in the left atrium one heartbeat before arriving in the aorta. Therefore, the inversion efficiency is somewhat uncertain, given that there are potentially several spin inversion and relaxation delays in sequence. According to an analysis in a human study (Capron et al., 2015), blood magnetization in the aortic root after several repetitions was approximately zero rather than -M$_0$. This fact was later used in a pig study, where the inversion pulses were replaced by saturation pulses to

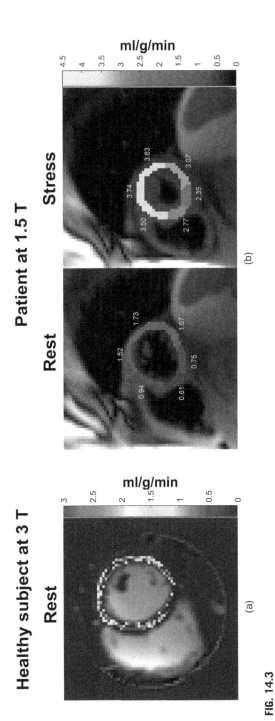

FIG. 14.3

Myocardial blood flow (MBF) maps obtained with a FAIR-bSSFP sequence. (A) FAIR MBF map acquired during free breathing from a healthy volunteer at 3T after pairwise motion correction. (B) FAIR MBF map measurements obtained per segment of a patient at 1.5T under rest (middle) and under continuous infusion of adenosine (right).

Data: V. Aramendia-Vidaurreta, Clínica Universidad de Navarra. Pamplona, Spain.

reduce residual magnetization transfer effects (Javed *et al.*, 2019) and showed equal performance. Globally, the signal per scan produced with spASL was shown smaller than with FAIR, which is compensated by the larger number of averages per unit time.

For preclinical applications, the spASL strategy has been employed in a specific implementation known under the name cine-ASL (Troalen *et al.*, 2013), where one gradient echo of a cine-readout train was replaced by a labeling pulse in every cardiac cycle.

Fig. 14.4 shows examples of MBF maps acquired with a GRAPPA-2 accelerated spASL-bSSFP sequence from a healthy volunteer (acquired pixel size 1.7×1.7 mm^2, slice thickness 5 mm) and with cine-ASL from a mouse under rest and adenosine stress (acquired pixel size 195×391 µm^2, slice thickness 1 mm).

14.2.1.3 *Velocity-selective labeling*

To overcome transit time effects occurring with the two geometric labeling strategies described before, velocity-selective labeling (VS-ASL) has been proposed first in the brain (Wong *et al.*, 2006). Inversion pulses can be designed in combination with bipolar gradients such as to modify only those spins whose displacement velocity is above a defined value of velocities along one direction. This way, coronary blood is being targeted by its characteristic flow velocity instead of its upstream location. More recently, feasibility of velocity-selective labeling has also been studied in the heart (Jao and Nayak, 2018) with positive results regarding feasibility, but also pointing out two main challenges: (i) the velocity selectivity can only be applied in a single direction, raising questions on its efficacy in the tortuous vessel geometry in the heart, and (ii) the cutoff velocity must be sufficiently high to avoid targeting

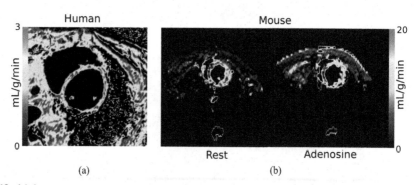

FIG. 14.4

(A) spASL MBF map acquired during one breath-hold from a healthy volunteer at 3T. The difference between chest muscle on the left and myocardium is visible despite a significant noise level. Artifacts are visible in the inferior wall. (B) Cine-ASL MBF maps of a mouse acquired within 2.5 min at 7T under rest (middle) and under continuous infusion of adenosine via the tail vein (right).

Data: F. Kober, A. Tonson, CRMBM, Marseille, France.

cardiac tissue motion and still be sensitive to flow in the coronary branches, requiring a compromise. Optimization of pulse profiles with sharper velocity profiles and still sufficient robustness to motion and B_0 and B_1 field inhomogeneities have improved the technique later, reaching labeling performances comparable with those of FAIR (Landes *et al.*, 2020).

14.2.2 Motion strategies used during acquisition

Table 14.1 summarizes different cardiac and respiratory motion management strategies for human acquisitions that have been used in studies published so far. They are explained in more detail in the following sections. Of note, the combined cardiac and respiratory strategies themselves can also have an effect on the degree of expected motion during a cardiac study. Studies have reported both nonsignificant (Raper *et al.*, 1967) and significant (Fang *et al.*, 2008) differences in heart rate variability between controlled breathing, such as breath-holding or synchronized breathing, and spontaneous breathing, the latter being better suited to minimize this variability. For reasons explained in a dedicated section later, the following strategies mainly apply to the human heart.

14.2.2.1 Respiratory motion

In human studies, the effects of respiration on the position of the heart have been well studied by MRI (Wang *et al.*, 1995). As explained in more detail later, different strategies exist to deal with respiratory motion. In brief, these are breath-holding (Zun *et al.*, 2009), synchronized breathing (Aramendía-Vidaurreta *et al.*, 2022), respiratory- or navigator-gating (Wang *et al.*, 2010), and free breathing (Capron *et al.*, 2015) in combination with retrospective motion correction or segmentation algorithms.

Patient breathing can be monitored with a respiratory bellow placed on the abdomen. Breath-holding instructions (breath-in, breath-out, breath-hold) are given in general by the scanner sending voice recordings at the beginning of the ASL sequence to minimize motion. In this manner, subjects are able to hold their breath during the acquisition of a pair of label and control images, which takes an approximate period of 12 s, depending on the patient's heart rate. Thus multiple breath-holds are required to acquire sufficient image pairs for signal averaging.

Subjects can also be instructed to synchronize their breathing with the sequence by learning to recognize the MRI sounds of both the labeling and readout pulses of the sequence. This increases scan efficiency by avoiding the use of breath-holding instructions but also requires cooperation of the subject.

Free-breathing strategies are desirable in clinical practice. Respiratory triggering or navigator-gating techniques allow subjects to breathe normally, but prolong scan time considerably. Alternatively, retrospective nonrigid registration algorithms allow minimizing in-plane motion while reducing scan time.

All these strategies present various challenges. Using multiple voluntary breath-holds limits the number of acquired image pairs, and attention is needed to ensure that

Table 14.1 Strategies for managing cardiac and respiratory motion in ASL at the time of data acquisition.

Motion	Strategy	Advantages	Disadvantages
Respiratory	Breath-holding	– Is common in most cardiac MR sequences – Ensures relatively good stability	– Requires patient cooperation to follow instructions – Multiple breath-holds are required to acquire sufficient image pairs – Changes in heart rate might occur in relation to the breath-hold
	Synchronized breathing	– Allows shorter scan times because no breath-holding instructions nor respiratory triggering is needed – May be perceived as less stressful by certain subjects compared with breath-holding	– Requires patient cooperation to synchronize breathing – Changes in heart rate might occur in relation to the breath-hold
	Respiratory triggering or navigator gating	– Allows subjects to breathe freely	– Slow breathing rates or irregular breathing decreases scan efficiency
	Free breathing and retrospective in-plane motion correction	– Allows shorter scan times because no breath-holding instructions nor respiratory triggering is needed – Allows subjects to breathe freely	– Images can be affected by through-plane motion caused by respiration
Cardiac	Single ECG triggering (e.g., FAIR)	– Constant inversion time and simpler MBF quantification – Potentially simpler software implementation	– Motion can occur in the presence of heart rate variations – There is a need to acquire additional images at short and long inversion times for image quantification
	Double ECG triggering (e.g., FAIR)	– Ensures better motion stability across heartbeats	– Timing of the inversion pulse varies in the presence of heart rate variability
	Gating at end-systole	– Depicts a thicker myocardial area in the imaging slice, allowing better separation from borders and more pixels to be included in evaluated ROIs	– The end-systolic period is shorter than the mid-diastole, which requires the use of faster readouts to prevent the effects of motion
	Gating at mid-diastole	– Longer quiescent phase of the heart, reducing motion effects during readout – Acquisition window may be longer	– During mid-diastole the number of pixels within the myocardium is reduced in comparison to the systolic cardiac phase – Diastolic period is shorter during stress if heart rate is higher

Advantages and disadvantages related to each approach are listed.

the heart returns to the same position to avoid misalignment of the myocardium between image pairs, where the displacement of the heart can be as large as 8.3 mm (Liu *et al.*, 1993). The success of synchronized breathing is highly dependent on the subject's cooperation, which can also be demanding in clinical practice. The application of existing motion correction or segmentation techniques is not straightforward, and their specific parameters should be validated in the context of myocardial ASL.

14.2.2.2 Cardiac motion

Prospective cardiac triggering in myocardial ASL studies can be performed with the use of the electrocardiogram (ECG) signal by prospectively detecting the R-wave peak of the cardiac cycle. To circumvent cardiac motion, both labeling and readout are synchronized to the same cardiac phase. Typically, the window used for signal acquisition in human myocardial ASL studies has a duration of 150 ms and is positioned in mid-diastole, where the heart remains relaxed for a longer time. However, heart rate variability can be a confounder, especially during pharmacological stress or in patients with arrhythmia. The optimal trigger delay should be adjusted manually by visual inspection of the ECG or on a short-axis cine scan acquired beforehand.

Cardiac motion is especially problematic in FAIR implementations because the optimal inversion time (TI) is longer than one cardiac cycle, and both labeling and readout should be acquired in the same cardiac phase. In this context, both single- and double-ECG gating have been used. In both cases, the labeling pulse is triggered by ECG after an adjustable delay (TD) set before acquisition (Poncelet et al., 1999; Zun et al., 2009). Single-gated ASL then uses a fixed inversion time after which the image is acquired. Double-ECG gating triggers both labeling and readout using the same adjustable delay, making it more robust to heart rate variations, but at the expense of inversion time variations across image pairs depending on the subject's heart rate (Do et al., 2017). These variations need to be recorded and accounted for in the quantification process. Additional images at short and long TIs are acquired to fit inversion recovery curves, and the interpolated signal difference is then measured at the average TI.

Systolic imaging has been shown feasible with shortened acquisition windows via the use of compressed sensing techniques. This method showed similar temporal signal-to-noise ratios relative to diastolic imaging but offered a higher number of pixels available for quantification within the myocardium (Henningsson et al., 2021).

The timing of sequence elements for cardiac FAIR and spASL acquisitions over several cardiac cycles is shown in Fig. 14.5 with the use of ECG triggering, illustrating the small timing windows available for acquisition over the total acquisition time, explaining parts of the low efficiency of myocardial ASL.

14.2.2.3 Preclinical vs clinical

In preclinical studies, for obvious reasons, the subjects cannot be instructed to hold their breath. In rodents, high heart and breath rates hamper the use of navigator-based acquisitions. Cardiac and respiratory motion handling and sequence design have, therefore, evolved along a different path than in humans. Instead of using a single acquisition after each inversion, the FAIR labeling technique can be combined with ECG-gated Look–Locker readouts, allowing a significant increase in acquisition efficiency (Look–Locker-FAIR-Gradient Echo (LLFAIRGE), (Belle et al., 1998; Campbell-Washburn et al., 2013a; Kober et al., 2004)). Using ECG gating, the inversion recovery is then efficiently sampled at every heartbeat with RR intervals in the range of 100–300 ms. In addition, these acquisitions simultaneously provide T1 maps that can be used in the analysis process. Owing to the high heart rates in

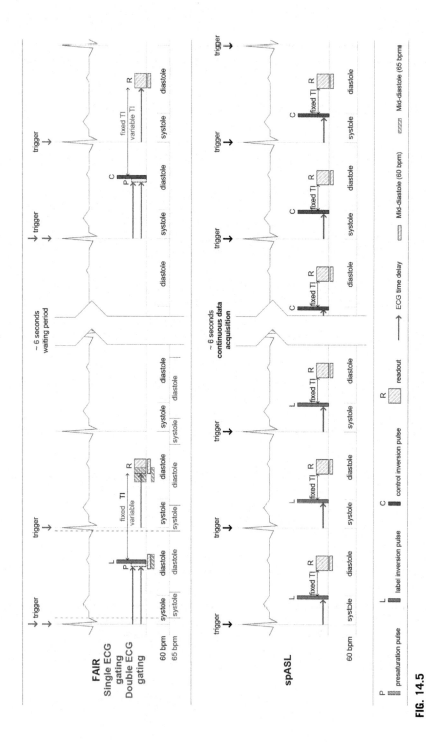

FIG. 14.5

Timing diagram for FAIR and spASL sequences for human acquisitions. Nonselective or label and slice-selective or control inversion pulses are followed by the imaging readouts. In FAIR, the use of additional presaturation pulses minimizes static tissue variations, and the use of single (blue (light gray in print version)) and double (green (gray in print version)) ECG gating allows positioning label and readout modules in the same portion of the cardiac cycle, considering heartbeats of 60 and 65 bpm. The former triggers both labeling and readout in the same cardiac phase at the cost of TI variability between image pairs depending on the subject's heart rate. The latter triggers both labeling and readout in the same cardiac phase at the cost of TI variability between image pairs depending on the subject's heart rate. A minimum repetition time of ~6s is required to allow for T1 recovery of the label before the next acquisition. In spASL, labeling occurs at end-systole and imaging at mid-diastole at every heartbeat. Data is acquired continuously, resulting in much higher scan efficiency than FAIR despite lower labeling efficiency due to the label distance.

small animals, labeling and acquisitions can be done in a segmented fashion over many cardiac cycles, allowing visualization of perfusion maps in several cardiac phases (Troalen *et al.*, 2014). Only in a subset of the studies cited before, respiratory motion has been addressed by a retrospective exclusion of images acquired during respiratory periods with strong motion, which led to a slightly better image quality. In general, however, respiratory motion in rodent studies has received less attention than in human studies. This is partly due to the large number of signals acquired over many cardiac and several respiratory cycles, but also to the fact that respiratory cycles in anesthetized rodents feature relatively long quiescent respiratory phases and comparatively short intervals with stronger motion. Rodent ASL acquisitions are, therefore, generally more robust to respiratory motion than those in humans, and they benefit not only from the higher capillary and arterial blood flow, but also from the much higher heart rates in rodents.

14.2.3 **Practical considerations**

14.2.3.1 Clinical

Previous research has shown the feasibility of myocardial ASL at both 1.5T and 3T (Aramendía-Vidaurreta *et al.*, 2020), but most studies have been performed at 3T due to the higher field strength that increases both the SNR and the duration of the labeling magnetization (longer relaxation time of blood). This field strength is also advantageous for first-pass and late gadolinium enhancement sequences (Bartoli *et al.*, 2022; Wang *et al.*, 2021) at the cost of more susceptibility artifacts.

As for any cardiovascular MRI exam, a thorax array in combination with a spine coil can be used for signal reception and the body coil for transmission. The most typical ASL sequence parameters used in a myocardial ASL study are specified in Table 14.2. Within clinical cardiac MRI protocols, ASL must be performed prior to the injection of gadolinium contrast agents that shorten blood T1. Therefore, rest–stress ASL images are acquired before stress–rest first pass. ECG gating is required and heart rate and blood pressure are recorded for every subject before and after the pharmacologic stress agents. While studies including stress agents are not free of risks and are, therefore, limited to patients requiring such an exam, milder forms of stress such as handgrip exercise, leg elevation, MR-compatible treadmill, or cold pressor test are compatible with volunteer studies.

14.2.3.2 Preclinical

Myocardial ASL has been implemented for rats and mice on different scanner hardware ranging from 4.7 to 9.4 T. For FAIR and FAIR Look–Locker sequences, a volume transmit RF coil covering at least the torso and the abdomen is necessary. For the cine-ASL sequence, no particular coil setup is needed as long as the entire heart is excited by the transmit coil. The system needs to be equipped with ECG and optional respiratory monitoring and gating capabilities. As for other cardiac MRI protocols, for rat imaging, a decoupled multichannel receive RF coil is recommended. Table 14.2 presents typical parameters for both rats and mice. Since the performance of preclinical scanner hardware significantly varies across sites, sequence

Table 14.2 Commonly used human myocardial ASL sequence parameters.

Parameters		Values	Description
Labeling	Slab in FAIR	Nonselective (NS): dependent on transmit coil length (~390 mm) Slice selective (SS): 24–40 mm	NS inversion: The labeled blood volume of the nonselective inversion pulse is dependent on the transmit coil size. The volume should be centered at the image slice SS inversion: Thickness depends on the slice thickness adding two gaps of equal size above and below it to assure the image slice is inverted despite the presence of heart movements
	Slab in spASL	Labeling slab: 60 mm	In the label image, the inversion slab covers the aortic root and is placed in the basal heart. In the control image, the inversion slab is placed symmetrically to the imaging slice
	Pulse	Hyperbolic secant	Adiabatic inversion pulses are commonly used
Readout	Sequence	bSSFP	Most of the recent studies used a snapshot bSSFP sequence although EPI has been used successfully
	Acceleration	GRAPPA or SENSE factor 2	Parallel imaging acceleration has been shown useful for reducing the time window in the cardiac cycle
	Preparation		Fat suppression has been used in some studies
	Slice thickness and positioning	Short-axis slice of 8–10 mm	A short-axis slice of the heart is typically selected. For this purpose, sagittal, axial, and coronal localizers are acquired followed by long- and short-axis scans of the left ventricle
	Number of slices	1–3	Typically, a single mid-ventricular slice is acquired with a bSSFP readout. Three slices (basal, mid-ventricular, and apical) have been acquired with an optimized EPI readout (Javed and Nayak, 2020)
	Pixel size	2×2 to 3×3 mm^2 for bSSFP readouts	Pixel size values range from 2×2 mm^2 to 3×3 mm^2 for bSSFP readouts

Table 14.3 Recommended parameters for preclinical myocardial ASL in mice and rats. If different, values for rats are given in parenthesis.

Parameters		Look–Locker FAIR-Gradient-Echo	Cine-ASL	Comments
Labeling	Slab thickness	2 (4) mm	2 (4) mm	
	Pulse duration	4–8 ms	4–8 ms	Minimize duration, use adiabatic pulses
	Slab position	Same as imaging slice	Cover entire aortic root	Parallel to imaging slice
	Pulse timing	On QRS trigger pulse	In end-systole	
Readout	Field of view	25×25 (40×40) mm^2	25×25 (40×40) mm^2	
	Matrix	128×64 or 128×128	128×64 or 128×128	
	Acceleration	GRAPPA-2	GRAPPA-2	Can be used with array receive coils provided that SNR is sufficient
	Readout type	Gradient echo	Gradient echo	Minimize all gradient durations including excitation pulse
	Echo time	Minimize (1–1.5 ms)	Minimize (1–1.5 ms)	Use partial Fourier readout 30%
	Repetition time	Minimize	Minimize	
	Readout gating	One echo per cardiac cycle	One cine readout per cardiac cycle	Most acquisitions use a single phase encoding step per cardiac cycle, but segmentation is possible
	Number of time points	35 (25) echoes after inversion, should last 5 s	35 (25) cine blocks used for averaging	
	Flip angle	15°	8°	
	Slice thickness	1 (2) mm	1 (2) mm	Orientation: left-ventricular short axis

timings may also vary, and sequences should be optimized as recommended in Table 14.3. Values for rats are shown in parentheses.

For minimizing the influence of anesthesia on hemodynamics and vasodilation, isoflurane is preferred. Furthermore, since even isoflurane has hemodynamic and vasodilatory effects, it should be delivered at the minimum necessary concentration. Temperature should be monitored and maintained at physiologic levels.

14.3 **Data analysis protocol**

14.3.1 **Outline**

Image processing in myocardial ASL generally employs the following steps:

- Reordering of label and control images
- Retrospective motion management: (Optional): Discarding image outliers and/or image registration (pairwise or groupwise approaches, see later)

- If applicable: averaging along label and control image series
- Calculation of relative difference map between label and control
    OR:
- Fitting of an exponential model to calculate T1 maps from Look–Locker data
- Myocardial segmentation: Segmentation to extract myocardial (global or regional) MBF values
- Quantification of MBF: Application of the corresponding model to obtain a quantitative MBF map

14.3.2 **Retrospective motion management**

In humans, typical breath rates are comparatively low on the scale of the experiment time, making the management of motion in myocardial ASL necessary. Due to the large signal differences in the left-ventricular blood pool between label and control images, the subtraction between the two can produce errors far larger than the expected signal difference due to perfusion if the myocardium is misaligned due to motion (Fig. 14.6).

To minimize these errors, one alternative is to discard image outliers. In free breathing myocardial ASL studies, several control and label images are averaged to obtain a quantitative MBF map. Image outliers with excessive motion can contaminate blood flow quantification and are often discarded. Visual inspection can be time consuming, and a number of strategies have been suggested for their identification and removal.

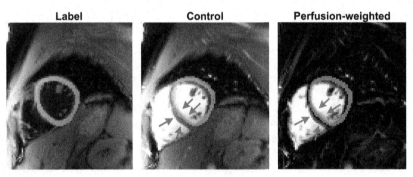

| Label | Control | Perfusion-weighted |

FIG. 14.6

Representative example of a motion-corrupted ASL pair. From left to the right: label, control and perfusion-weighted (control minus label) images. The green (gray in print version) region of interest has been delineated over the label image, representing the myocardial tissue signal that will be used for quantification. However, when the same region is applied to the control and perfusion-weighted image, the presence of motion contaminates the signal due to the high intensity of the blood pool. The red (dark gray in print version) arrows indicate the segments with higher motion between control and label images.

Extreme motion outliers in the perfusion-weighted signal were identified as those exceeding the mean signal by two standard deviations (Aramendía-Vidaurreta *et al.*, 2020, 2019, 2022; Wang *et al.*, 2010). Contour-based algorithms in combination with the cross-correlation coefficient of a contour within the myocardium allowed rejection of the most poorly correlated image pairs (Capron *et al.*, 2015). Evaluation of the intensity variation of a voxel located in the anterior myocardial segment identified motion in the superio-inferior direction due to the high intensity contrast between lungs and myocardium (Aramendía-Vidaurreta *et al.*, 2022).

Another alternative is image registration. Existing research recognizes the critical role played by registration algorithms in cardiac MRI studies to allow image acquisition in free-breathing studies and ease its clinical application. Retrospectively aligning images with respect to each other can minimize motion and facilitate image segmentation.

Rigid registrations have been used assuming that the heart is a rigid structure, which can be valid in certain scenarios but may compromise the accuracy of the results. Nonrigid registrations are able to deal with the fact that the heart is a nonrigid organ that gets deformed during the cardiac cycle and due to respiration (Makela *et al.*, 2002).

Previous research has registered first-pass myocardial perfusion images using both rigid and nonrigid approaches in the long- and short-axis planes of the heart (Kellman *et al.*, 2005). Nonrigid schemes showed a significantly better performance than rigid registrations in areas such as the right ventricle. Rigid registration failed to register those subjects with high septal wall movement of the myocardium. Through-plane motion was more evident in the long-axis images of the heart and could not be entirely corrected during registration.

A key issue for their application in myocardial ASL is the different signal intensity that exists between the interleaved control and label images, which can be understood by the algorithms as motion. Working independently with these images has facilitated registration (Aramendía-Vidaurreta *et al.*, 2022; Javed *et al.*, 2015). Another challenge is the low signal-to-noise ratio of the perfusion-weighted signal, which makes image registration delicate. In this context, both pairwise and groupwise algorithms have been compared (Aramendía-Vidaurreta *et al.*, 2022). In the former, a reference image was selected to which all other images were aligned. In the latter, the entire dataset was registered within a single optimization procedure. In this particular context, pairwise registration showed better performance. Registration of myocardial ASL images has also reduced the need of manual segmentation, thus facilitating the ASL analysis pipeline (Javed *et al.*, 2015).

As outlined before, in preclinical studies, respiratory motion has less impact on the ASL experiment in general. However, discarding motion-affected images (Kober *et al.*, 2004) or rebinning (Vandsburger *et al.*, 2010) of k-space data has been shown to improve image quality.

14.3.3 **Myocardial segmentation**

Segmentation of the left ventricular myocardium is required to obtain averaged perfusion signal values globally, per coronary artery territory and per segment, as defined by the American Heart Association guidelines (Cerqueira *et al.*, 2002).

These regions of interest can be drawn in all individual perfusion-weighted images to minimize motion artifacts, but the process can be time consuming, especially if the number of images is large. In addition, care must be taken when delineating these regions to prevent partial volume effects due to the low ASL image resolution and the proximity of the ventricular blood pool and the epicardial fat.

In the context of myocardial ASL, most segmentations have been performed manually (Zun *et al.*, 2009). Semiautomatic segmentations based on the use of Canny edge filters and morphological operators have also been suggested (Aramendía-Vidaurreta *et al.*, 2020).

More recently, a considerable literature has emerged around deep learning automatic segmentation of the myocardium in MRI (Bartoli *et al.*, 2022; Wang *et al.*, 2021). In this context, segmentation of left ventricular myocardium on short- and long-axis cine images has been achieved, as well as in other regions such as the papillary muscles or blood pool. However, the cine-bSSFP images used for segmentation are advantageous for automatic methods due to the high blood–tissue contrast and spatial resolution. Direct application of these deep learning techniques in myocardial ASL images remains challenging. Therefore, methods for the assessment of model uncertainty had been evaluated in myocardial ASL studies as a measure of image segmentation quality and a specific false-positive vs false-negative trade-off to prevent myocardial oversegmentation. This demonstrated high correlations between deep learning and manual segmentations (Do *et al.*, 2020; Gordaliza *et al.*, 2019).

14.3.4 Quantification of myocardial blood flow: Models

For quantification of MBF, the myocardial perfusion-weighted signal needs to be converted to absolute units of milliliters of blood per gram of tissue per minute (ml/g/min). For this purpose, both single- and two-compartment models have been used in the myocardial ASL literature.

For the FAIR technique, the single-compartment model was first proposed by Kim *et al.* in 1995 (Kim, 1995) and uses label and control difference images as a measure of perfusion. Considering equal T1 values for blood and myocardium, quantification reduces to the following equation:

$$\text{MBF} = \frac{\Delta M \lambda}{\beta M_0 \text{TI} \exp^{\text{TI}/\text{T1}}}, \tag{14.1}$$

where λ is the blood–tissue partition coefficient, ΔM is the mean signal difference between control and label images, β is the labeling efficiency (assumed to be 1), and M_0 is the equilibrium magnetization derived from a separate baseline image, TI is the inversion time, and T1 is the longitudinal relaxation time of blood in minutes. In double-ECG triggering sequences, variations in the TI exist. Thus, prior to quantification, label and control images need to be separately fitted to an inversion recovery model with three parameters (M_0, β, and T1). Subsequently, the mean signal difference between control and label images is obtained at the point of the average TI.

The two-compartment model was first proposed by Detre *et al.* in 1992 (Detre *et al.*, 1992) and later adapted in an isolated rat heart study by Bauer *et al.* (Bauer *et al.*, 1997) using a Look–Locker inversion-recovery sequence for determining maps or regional mean values of T1 observed after selective or nonselective inversion.

$$\mathrm{MBF} = \frac{\lambda}{\mathrm{T1_{bl}}} \left(\frac{\mathrm{T1_{ns}}}{\mathrm{T1_{ss}}} - 1 \right), \tag{14.2}$$

where λ is the blood–tissue partition coefficient, $\mathrm{T1_{bl}}$ is the longitudinal relaxation time of blood in minutes, and $\mathrm{T1_{ns}}$ and $\mathrm{T1_{ss}}$ are the longitudinal relaxation time of the myocardium after the nonselective and slice-selective inversions, respectively. In humans, this model has been used with Modified Look–Locker (MOLLI) (Keith *et al.*, 2017) and earlier with Look–Locker snapshot-FLASH or bSSFP acquisitions (Fidler *et al.*, 2004; Northrup *et al.*, 2008; Wacker *et al.*, 2003).

For the spASL technique, a specific model was developed based on the modified Bloch equations (Capron *et al.*, 2013) leading to the following relation once a steady state of magnetization is reached in both situations label and control when using FLASH readouts:

$$\mathrm{MBF} = \frac{\Delta M_\infty \lambda}{2\beta M_0 \mathrm{T1}^{app*}}, \tag{14.3}$$

where ΔM_∞ is the magnetization difference in the steady state, $\mathrm{T1}^{app*}$ is the saturation-corrected longitudinal relaxation time of tissue, and β is the average labeling efficiency ($\beta = 1$ for a complete inversion). The model was later extended to bSSFP readouts (Capron *et al.*, 2015).

λ values reported in literature ranged from 0.8 to 1 ml/g (Poncelet *et al.*, 1999; Wang *et al.*, 2010; Zun *et al.*, 2009). All quantification models generally rely on the fast exchange assumption, which is generally considered to be valid for ASL in the myocardium (*e.g.*, Bauer *et al.*, 1997).

14.4 Applications
14.4.1 Clinical

Measuring perfusion under rest and stress conditions is of high importance in the context of CAD. The narrowing of coronary vessels is detected sooner during hyperemic conditions because of their lower perfusion increase in comparison to normal vessels (Gould and Lipscomb, 1974). Thus myocardial perfusion reserve, defined as the ratio of stress to rest perfusion, might be altered in case of coronary stenosis. In healthy volunteers, the effects of pharmacological stress have been replicated with mild forms of stress, such as handgrip exercise, leg elevation, MR-compatible treadmill, or cold pressor test. One study also reported the feasibility of ASL to quantify perfusion with adenosine pharmacological vasodilator in healthy subjects (Yoon *et al.*, 2017).

To date, studies evaluating ASL in clinical situations were focused on coronary artery disease (CAD) only, but beyond potential clinical usefulness, they have shown that rest–stress comparisons can be done in an efficient way using ASL, and that under pharmacologic stress, ASL showed promise for CAD diagnosis. Briefly, Wacker *et al.* published in 2003 the first study at 2T comparing perfusion reserve within the anterior and posterior myocardium in patients with coronary stenosis (Wacker *et al.*, 2003). After that, studies have used a segmental analysis of the myocardium to differentiate between normally perfused and abnormally perfused segments. In patients with suspected CAD, pharmacological stress agents (both adenosine and regadenoson) are commonly used in the clinical practice. One ASL study, however, employed isometric handgrip exercise in these patients as an endothelial stressor to quantify perfusion defects (Javed *et al.*, 2020).

Nevertheless, ASL in humans currently lacks validation against gold standards such as PET imaging. Proper quantitative comparisons with first-pass MRI have not been done by the time this chapter was written. One study has, however, compared ASL perfusion reserve against that measured with semiquantitative first-pass MRI (Aramendía-Vidaurreta *et al.*, 2020) and shown a moderate correlation in a group of CAD patients. A summary of clinically oriented rest–stress evaluation studies in healthy volunteers and clinical studies in patients with CAD is given in Table 14.4.

14.4.2 Preclinical

Myocardial ASL in small animals today appears as a validated and robust technique for myocardial perfusion mapping. Table 14.5 lists selected preclinical application studies in which myocardial ASL played an important role. The vast majority of studies employed rodent models. One acute infarction study was done on a porcine model. In most studies, ASL perfusion mapping was used along with other MRI and non-MRI techniques to characterize animal models of human disease. In several cases, the evaluation was done longitudinally to monitor disease progression or the effect of treatment.

14.5 Safety considerations

There are no specific safety considerations applicable for myocardial ASL other than those applying for standard cardiac MRI. Pharmacologic stress exams are submitted to specific limitations of applicability, precautions, and authorizations.

14.6 Challenges and solutions

A major challenge for future studies is to provide validation of myocardial ASL in humans against gold-standard techniques such as PET. Yet, the lack of vendor-supplied acquisition methods has hampered wider use of myocardial ASL even in clinical research studies. Successful cardiac rest–stress comparisons can greatly

Table 14.4 Application studies in healthy volunteers to evaluate various forms of cardiac stress and clinical application studies in patients with suspected CAD where MBF was measured under both rest and stress.

Application	ASL technique	Field strength	Stress	References
Stress/healthy subjects	FAIR-bSSFP	3T	Passive leg elevation and isometric handgrip	Zun *et al.* (2009)
Stress/healthy subjects	FAIR-bSSFP	3T	Stress balls at 0.5 Hz	Wang *et al.* (2010)
Stress/healthy subjects	FAIR-bSSFP	3T	Cold pressor test	Capron *et al.* (2015)
Stress/healthy subjects	FAIR-bSSFP	3T	Adenosine infusion	Yoon *et al.* (2017)
Stress/healthy subjects	FAIR-bSSFP	3T	Passive leg elevation	Aramendía-Vidaurreta *et al.* (2019)
CAD patients (*n* = 13)	FAIR-FLASH	2T	Adenosine (0.56 mg/kg) over 6 min	Wacker *et al.* (2003)
CAD patients (*n* = 13)	FAIR-bSSFP	3T	Adenosine (0.14 mg/kg/min) over 6 min	Zun *et al.* (2011b)
CAD patients (*n* = 29)	FAIR-bSSFP	3T	Adenosine (0.14 mg/kg/min)	Zun *et al.* (2011a)
CAD patients (*n* = 11)	FAIR-bSSFP	3T	Isometric handgrip over 5 min	Javed *et al.* (2020)
CAD patients (*n* = 16)	FAIR-bSSFP	1.5T	Regadenoson (0.4 mg) over 5 min	Aramendía-Vidaurreta *et al.* (2022)
Hemodialysis patients (n = 12)	FAIR-bSFFP	3T	Dialysis treatment	Buchanan et al. (2017)

contribute to building confidence in this technique, but because pharmacologic stress tests are not completely free of risks, they are reserved to clinical protocols in patients requiring a stress test. MR method development research centers, therefore, cannot easily perform them. An extension of the application field from CAD diagnosis to pathologies with more subtle, microvascular alterations could help highlight the potential advantages in absolute quantification that ASL has compared with other techniques. Collaboration of MR researchers with clinicians is, therefore, of major importance, particularly in the field of myocardial ASL.

The feasibility of several new acquisition method approaches (VSASL, spASL) has been demonstrated, but these have not been used in applications so far. Future validation studies may pick up these newer methods as well to make them useful in applications.

Table 14.5 Preclinical application studies in which myocardial ASL played an important role by application type.

Disease/Application	Model	ASL technique	References
Type 2 diabetes	Rat	LLFAIRGE	Iltis *et al.* (2005b)
Type 2 diabetes/Nutrition/ Treatment	Mouse	Cine-ASL	Abdesselam *et al.* (2015)
Hypertension/Type 1 diabetes	Rat	LLFAIRGE	Iltis *et al.* (2005a)
Hypertension	Rat	LLFAIRGE	Caudron *et al.* (2013)
Hypertrophy	Rat	LLFAIR-FLASH	Waller *et al.* (2008)
Hypertrophy	Mouse	Cine-ASL	Ku *et al.* (2021)
Aortic stenosis	Mouse	Cine-ASL	Quast *et al.* (2022)
Infarction	Rat	LLFAIR-FLASH	Waller *et al.* (2001)
Infarction	Rat	LLFAIR-FLASH	Nahrendorf *et al.* (2002)
Infarction	Mouse	LLFAIRGE-Multi-slice	Dongworth *et al.* (2017)
Infarction/cell transplant	Rat	LLFAIR-FLASH	Zhang *et al.* (2012)
Heart failure	Rat	LLFAIRGE	Merabet *et al.* (2012)
Heart failure	Rat	LLFAIRGE	Desrois *et al.* (2014)
Stress mechanisms / Adenosine	Mouse	Cine-ASL	Shah *et al.* (2021)
Infarction	Swine	FAIR-EPI	Do *et al.* (2016), Do *et al.* (2016)

Finally, the availability of ready-to-use and robust motion correction and segmentation methods may be improved in the future, opening the myocardial ASL method to research groups not actively developing methods themselves.

In contrast, preclinical myocardial ASL methods are now relatively mature, and the agreement of literature studies regarding normal and abnormal perfusion values in rodents is good. However, as for human myocardial ASL, no vendor-supplied preclinical methods are currently distributed.

The need for ECG electrodes in preclinical studies is an additional burden for certain routine protocols, where self-gating approaches are now more commonly used for standard cine-MRI. Future research on preclinical myocardial ASL methods may, therefore, explore how self-gating may be incorporated in myocardial ASL techniques despite the pulsating blood flow in larger vessels.

14.7 **Learning and knowledge outcomes**

This chapter gives an overview of the current status of myocardial ASL methods and applications in humans and rodents. The specific challenges involved in applying ASL to the heart should become clear to the reader as well as the major motivation for using this technique. The differences between clinical and preclinical implementations and applications should become clear. Underlying theory, existing acquisition- and postprocessing methods are discussed, and practical recommendations are given that should help the reader implement the technique. A brief literature review of applications showed the current status of this technique in the clinical and preclinical worlds.

References

Abdesselam, I., Pepino, P., Troalen, T., Macia, M., Ancel, P., Masi, B., Fourny, N., Gaborit, B., Giannesini, B., Kober, F., Dutour, A., Bernard, M., 2015. Time course of cardiometabolic alterations in a high fat high sucrose diet mice model and improvement after GLP-1 analog treatment using multimodal cardiovascular magnetic resonance. J. Cardiovasc. Magn. Reson. 17, 95.

Abeykoon, S., Sargent, M., Wansapura, J.P., 2012. Quantitative myocardial perfusion in mice based on the signal intensity of flow sensitized CMR. J. Cardiovasc. Magn. Reson. 14, 73.

Ahmed, A., Campbell, R.C., 2008. Epidemiology of chronic kidney disease in heart failure. Heart Fail. Clin. 4, 387–399.

Aramendía-Vidaurreta, V., García-Osés, A., Vidorreta, M., Bastarrika, G., Fernández-Seara, M.A., 2019. Optimal repetition time for free breathing myocardial arterial spin labeling. NMR Biomed., e4077.

Aramendía-Vidaurreta, V., Echeverría-Chasco, R., Vidorreta, M., Bastarrika, G., Fernández-Seara, M.A., 2020. Quantification of myocardial perfusion with vasodilation using arterial spin labeling at 1.5T. J. Magn. Reson. Imaging 53, 777–788.

Aramendía-Vidaurreta, V., Gordaliza, P.M., Vidorreta, M., Echeverría-Chasco, R., Bastarrika, G., Muñoz-Barrutia, A., Fernández-Seara, M.A., 2022. Reduction of motion effects in myocardial arterial spin labeling. Magn. Reson. Med. 87, 1261–1275.

Bartoli, A., Fournel, J., Ait-Yahia, L., Cadour, F., Tradi, F., Ghattas, B., Cortaredona, S., Million, M., Lasbleiz, A., Dutour, A., Gaborit, B., Jacquier, A., 2022. Automatic deep-learning segmentation of Epicardial adipose tissue from low-dose chest CT and prognosis impact on COVID-19. Cell 11, 1034.

Bauer, W., Hiller, K., Roder, F., Rommel, E., Ertl, G., Haase, A., 1996. Magnetization exchange in capillaries by microcirculation affects diffusion-controlled spin-relaxation: a model which describes the effect of perfusion on relaxation enhancement by intravascular contrast agents. Magn. Reson. Med. 35, 43–55.

Bauer, W.R., Roder, F., Hiller, K.H., Han, H., Fröhlich, S., Rommel, E., Haase, A., Ertl, G., 1997. The effect of perfusion on T1 after slice-selective spin inversion in the isolated cardioplegic rat heart: measurement of a lower bound of intracapillary-extravascular water proton exchange rate. Magn. Reson. Med. 38, 917–923.

Belle, V., Kahler, E., Waller, C., Rommel, E., Voll, S., Hiller, K., Bauer, W., Haase, A., 1998. In vivo quantitative mapping of cardiac perfusion in rats using a noninvasive MR spin-labeling method. J. Magn. Reson. Imaging 8, 1240–1245.

Buchanan, C., Mohammed, A., Cox, E., Köhler, K., Canaud, B., Taal, M.W., Selby, N.M., Francis, S., McIntyre, C.W., 2017. Intradialytic cardiac magnetic resonance imaging to assess cardiovascular responses in a short-term trial of hemodiafiltration and hemodialysis. J. Am. Soc. Nephrol. 28, 1269–1277.

Campbell-Washburn, A.E., Price, A.N., Wells, J.A., Thomas, D.L., Ordidge, R.J., Lythgoe, M.F., 2013a. Cardiac arterial spin labeling using segmented ECG-gated look-locker FAIR: variability and repeatability in preclinical studies. Magn. Reson. Med. 69, 238–247.

Campbell-Washburn, A.E., Zhang, H., Siow, B.M., Price, A.N., Lythgoe, M.F., Ordidge, R.J., Thomas, D.L., 2013b. Multislice cardiac arterial spin labeling using improved myocardial perfusion quantification with simultaneously measured blood pool input function. Magn. Reson. Med. 70, 1125–1136.

Capron, T., Troalen, T., Cozzone, P.J., Bernard, M., Kober, F., 2013. Cine-ASL: a steady-pulsed arterial spin labeling method for myocardial perfusion mapping in mice. Part II. Theoretical model and sensitivity optimization. Magn. Reson. Med. 70, 1399–1408.

Capron, T., Troalen, T., Robert, B., Jacquier, A., Bernard, M., Kober, F., 2015. Myocardial perfusion assessment in humans using steady-pulsed arterial spin labeling. Magn. Reson. Med. 74, 990–998.

Caudron, J., Mulder, P., Nicol, L., Richard, V., Thuillez, C., Dacher, J.-N., 2013. MR relaxometry and perfusion of the myocardium in spontaneously hypertensive rat: correlation with histopathology and effect of anti-hypertensive therapy. Eur. Radiol. 23, 1871–1881.

Cerqueira, M.D., Weissman, N.J., Dilsizian, V., Jacobs, A.K., Kaul, S., Laskey, W.K., Pennell, D.J., Rumberger, J.A., Ryan, T.J., Verani, M.S., 2002. Standardized myocardial segmentation and nomenclature for tomographic imaging of the heart. J. Cardiovasc. Magn. Reson. 4, 203–210.

Desrois, M., Kober, F., Lan, C., Dalmasso, C., Cole, M., Clarke, K., Cozzone, P.J., Bernard, M., 2014. Effect of isoproterenol on myocardial perfusion, function, energy metabolism and nitric oxide pathway in the rat heart—a longitudinal MR study. NMR Biomed. 27, 529–538.

Detre, J.A., Leigh, J.S., Williams, D.S., Koretsky, A.P., 1992. Perfusion imaging. Magn. Reson. Med. 23, 37–45.

Do, H.P., Jao, T.R., Nayak, K.S., 2014. Myocardial arterial spin labeling perfusion imaging with improved sensitivity. J. Cardiovasc. Magn. Reson. 16, 15.

Do, H.P., Ramanan, V., Jao, T.R., Wright, G.A., Nayak, K.S., Ghugre, N.R., 2016. Non-contrast myocardial perfusion assessment in porcine acute myocardial infarction using arterial spin labeled CMR. In: J. Cardiovasc. Magn. Reson. 18 S1 SCMR 19th Annu. Sci. Sess. January 27–30, Los Angeles, CA, USA, p. O7.

Do, H.P., Yoon, A.J., Fong, M.W., Saremi, F., Barr, M.L., Nayak, K.S., 2017. Double-gated myocardial ASL perfusion imaging is robust to heart rate variation. Magn. Reson. Med. 77, 1975–1980.

Do, H.P., Guo, Y., Yoon, A.J., Nayak, K.S., 2020. Accuracy, uncertainty, and adaptability of automatic myocardial ASL segmentation using deep CNN. Magn. Reson. Med. 83, 1863–1874.

Dongworth, R.K., Campbell-Washburn, A.E., Cabrera-Fuentes, H.A., Bulluck, H., Roberts, T., Price, A.N., Hernández-Reséndiz, S., Ordidge, R.J., Thomas, D.L., Yellon, D.M., Lythgoe, M.F., Hausenloy, D.J., 2017. Quantifying the area-at-risk of myocardial infarction in-vivo using arterial spin labeling cardiac magnetic resonance. Sci. Rep. 7, 2271.

Edelman, R.R., Siewert, B., Darby, D.G., Thangaraj, V., Nobre, A.C., Mesulam, M.M., Warach, S., 1994. Qualitative mapping of cerebral blood flow and functional localization with echo-planar MR imaging and signal targeting with alternating radio frequency. Radiology 192, 513–520.

Fang, Y., Sun, J.-T., Li, C., Poon, C.-S., Wu, G.-Q., 2008. Effect of different breathing patterns on nonlinearity of heart rate variability. In: Annu. Int. Conf. IEEE Eng. Med. Biol. Soc. IEEE Eng. Med. Biol. Soc. Annu. Int. Conf. 2008, pp. 3220–3223.

Fidler, F., Wacker, C.M., Dueren, C., Weigel, M., Jakob, P.M., Bauer, W.R., Haase, A., 2004. Myocardial perfusion measurements by spin-labeling under different vasodynamic states. J. Cardiovasc. Magn. Reson. 6, 509–516.

Gordaliza, P.M., Aramendia-Vidaurreta, V., Vaquero, J.J., Bastarrika, G., Fernández-Seara, M.A., Muñoz-Barrutia, M.A., 2019. Automatic segmentation of the myocardium in cardiac arterial spin labelling images using a deep learning model facilitates myocardial blood flow quantification. In: Proc Intl Soc Mag Reson Med 27 Annu. Meet. Int. Soc. Magn. Reson. Med. 11–17 May, Montréal, QC, Canada, p. 4794.

Gould, K.L., Lipscomb, K., 1974. Effects of coronary stenoses on coronary flow reserve and resistance. Am. J. Cardiol. 34, 48–55.

Guo, B.J., Yang, Z.L., Zhang, L.J., 2018. Gadolinium deposition in brain: current scientific evidence and future perspectives. Front. Mol. Neurosci. 11.

Gutjahr, F.T., Kampf, T., Winter, P., Meyer, C.B., Williams, T., Jakob, P.M., Bauer, W.R., Ziener, C.H., Helluy, X., 2015. Quantification of perfusion in murine myocardium: a retrospectively triggered T1 -based ASL method using model-based reconstruction. Magn. Reson. Med. 74, 1705–1715.

Hadjiloizou, N., Davies, J.E., Malik, I.S., Aguado-Sierra, J., Willson, K., Foale, R.A., Parker, K.H., Hughes, A.D., Francis, D.P., Mayet, J., 2008. Differences in cardiac microcirculatory wave patterns between the proximal left mainstem and proximal right coronary artery. Am. J. Physiol. Heart Circ. Physiol. 295, H1198–H1205.

Henningsson, M., Carlhäll, C.-J., Kihlberg, J., 2021. Myocardial arterial spin labeling in systole and diastole using flow-sensitive alternating inversion recovery with parallel imaging and compressed sensing. NMR Biomed. 34, e4436.

Iltis, I., Kober, F., Dalmasso, C., Cozzone, P.J., Bernard, M., 2005a. Noninvasive characterization of myocardial blood flow in diabetic, hypertensive, and diabetic-hypertensive rats using spin-labeling MRI. Microcirculation 12, 607–614.

Iltis, I., Kober, F., Desrois, M., Dalmasso, C., Lan, C., Portha, B., Cozzone, P.J., Bernard, M., 2005b. Defective myocardial blood flow and altered function of the left ventricle in type 2 diabetic rats: a noninvasive in vivo study using perfusion and cine magnetic resonance imaging. Invest. Radiol. 40, 19–26.

Jacquier, A., Kober, F., Bun, S., Giorgi, R., Cozzone, P.J., Bernard, M., 2011. Quantification of myocardial blood flow and flow reserve in rats using arterial spin labeling MRI: comparison with a fluorescent microsphere technique. NMR Biomed. 24, 1047–1053.

Jao, T.R., Nayak, K.S., 2018. Demonstration of velocity selective myocardial arterial spin labeling perfusion imaging in humans. Magn. Reson. Med. 80, 272–278.

Javed, A., Nayak, K.S., 2020. Single-shot EPI for ASL-CMR. Magn. Reson. Med. 84, 738–750.

Javed, A., Jao, T.R., Nayak, K.S., 2015. Motion correction facilitates the automation of cardiac ASL perfusion imaging. In: J. Cardiovasc. Magn. Reson. 17 S1 18th Annu. SCMR Sci. Sess. 4–7 Feb, Nice, France, p. P51.

Javed, A., Lee, N.G., Do, H.P., Ghugre, N., Wright, G., Wong, E., Nayak, K.S., 2019. Optimization of steady-pulsed arterial spin labeling for myocardial perfusion imaging. In: Proc Intl Soc Mag Reson Med 27 Annu. Meet. Int. Soc. Magn. Reson. Med. 11–17 May, Montréal, QC, Canada, p. 2210.

Javed, A., Yoon, A., Cen, S., Nayak, K.S., Garg, P., 2020. Feasibility of coronary endothelial function assessment using arterial spin labeled CMR. NMR Biomed. 33, e4183.

Kampf, T., Helluy, X., Gutjahr, F.T., Winter, P., Meyer, C.B., Jakob, P.M., Bauer, W.R., Ziener, C.H., 2014. Myocardial perfusion quantification using the T1-based FAIR-ASL method: the influence of heart anatomy, cardiopulmonary blood flow and look-locker readout. Magn. Reson. Med. 71, 1784–1797.

Keith, G.A., Rodgers, C.T., Chappell, M.A., Robson, M.D., 2017. A look-locker acquisition scheme for quantitative myocardial perfusion imaging with FAIR arterial spin labeling in humans at 3 tesla. Magn. Reson. Med. 78, 541–549.

Kellman, P., Larson, A.C., Hsu, L.-Y., Chung, Y.-C., Simonetti, O.P., McVeigh, E.R., Arai, A.E., 2005. Motion-corrected free-breathing delayed enhancement imaging of myocardial infarction. Magn. Reson. Med. 53, 194–200.

Kim, S.G., 1995. Quantification of relative cerebral blood flow change by flow-sensitive alternating inversion recovery (FAIR) technique: application to functional mapping. Magn. Reson. Med. 34, 293–301.

Kober, F., Iltis, I., Izquierdo, M., Desrois, M., Ibarrola, D., Cozzone, P.J., Bernard, M., 2004. High-resolution myocardial perfusion mapping in small animals in vivo by spin-labeling gradient-echo imaging. Magn. Reson. Med. 51, 62–67.

Kober, F., Iltis, I., Cozzone, P.J., Bernard, M., 2005. Myocardial blood flow mapping in mice using high-resolution spin labeling magnetic resonance imaging: influence of ketamine/xylazine and isoflurane anesthesia. Magn. Reson. Med. 53, 601–606.

Ku, M.-C., Kober, F., Lai, Y.-C., Pohlmann, A., Qadri, F., Bader, M., Carrier, L., Niendorf, T., 2021. Cardiovascular magnetic resonance detects microvascular dysfunction in a mouse model of hypertrophic cardiomyopathy. J. Cardiovasc. Magn. Reson. 23, 63.

Landes, V., Javed, A., Jao, T., Qin, Q., Nayak, K., 2020. Improved velocity-selective labeling pulses for myocardial ASL. Magn. Reson. Med. 84, 1909–1918.

Liu, Y.L., Riederer, S.J., Rossman, P.J., Grim, R.C., Debbins, J.P., Ehman, R.L., 1993. A monitoring, feedback, and triggering system for reproducible breath-hold MR imaging. Magn. Reson. Med. 30, 507–511.

Makela, T., Clarysse, P., Sipila, O., Pauna, N., Pham, Q.C., Katila, T., Magnin, I.E., 2002. A review of cardiac image registration methods. IEEE Trans. Med. Imaging 21, 1011–1021.

Merabet, N., Bellien, J., Glevarec, E., Nicol, L., Lucas, D., Remy-Jouet, I., Bounoure, F., Dreano, Y., Wecker, D., Thuillez, C., Mulder, P., 2012. Soluble epoxide hydrolase inhibition improves myocardial perfusion and function in experimental heart failure. J. Mol. Cell. Cardiol. 52, 660–666.

Nahrendorf, M., Hiller, K.-H., Theisen, D., Hu, K., Waller, C., Kaiser, R., Haase, A., Ertl, G., Brinkmann, R., Bauer, W.R., 2002. Effect of transmyocardial laser revascularization on myocardial perfusion and left ventricular remodeling after myocardial infarction in rats. Radiology 225, 487–493.

Naresh, N.K., Chen, X., Moran, E., Tian, Y., French, B.A., Epstein, F.H., 2015. Repeatability and variability of myocardial perfusion imaging techniques in mice: Comparison of arterial spin labeling and first-pass contrast-enhanced MRI. Magn. Reson. Med. 75, 2394–2405.

Northrup, B.E., McCommis, K.S., Zhang, H., Ray, S., Woodard, P.K., Gropler, R.J., Zheng, J., 2008. Resting myocardial perfusion quantification with CMR arterial spin labeling at 1.5 T and 3.0 T. J. Cardiovasc. Magn. Reson. 10, 53.

Poncelet, B.P., Koelling, T.M., Schmidt, C.J., Kwong, K.K., Reese, T.G., Ledden, P., Kantor, H.L., Brady, T.J., Weisskoff, R.M., 1999. Measurement of human myocardial perfusion by double-gated flow alternating inversion recovery EPI. Magn. Reson. Med. 41, 510–519.

Purek, L., Laule-Kilian, K., Christ, A., Klima, T., Pfisterer, M.E., Perruchoud, A.P., Mueller, C., 2006. Coronary artery disease and outcome in acute congestive heart failure. Heart Br. Card. Soc. 92, 598–602.

Quast, C., Kober, F., Becker, K., Zweck, E., Hoffe, J., Jacoby, C., Flocke, V., Gyamfi-Poku, I., Keyser, F., Piayda, K., Erkens, R., Niepmann, S., Adam, M., Baldus, S., Zimmer, S., Nickenig, G., Grandoch, M., Bönner, F., Kelm, M., Flögel, U., 2022. Multiparametric MRI identifies subtle adaptations for demarcation of disease transition in murine aortic valve stenosis. Basic Res. Cardiol. 117, 29.

Raper, A.J., Richardson, D.W., Kontos, H.A., Patterson, J.L., 1967. Circulatory responses to breath holding in man. J. Appl. Physiol. 22, 201–206.

Shah, S.A., Reagan, C.E., French, B.A., Epstein, F.H., 2021. Molecular mechanisms of adenosine stress T1 mapping. Circ. Cardiovasc. Imaging 14, e011774.

Troalen, T., Capron, T., Cozzone, P.J., Bernard, M., Kober, F., 2013. Cine-ASL: A steady-pulsed arterial spin labeling method for myocardial perfusion mapping in mice. Part I. Experimental study. Magn. Reson. Med. 70, 1389–1398.

Troalen, T., Capron, T., Bernard, M., Kober, F., 2014. In vivo characterization of rodent cyclic myocardial perfusion variation at rest and during adenosine-induced stress using cine-ASL cardiovascular magnetic resonance. J. Cardiovasc. Magn. Reson. 16, 18.

Vandsburger, M.H., Janiczek, R.L., Xu, Y., French, B.A., Meyer, C.H., Kramer, C.M., Epstein, F.H., 2010. Improved arterial spin labeling after myocardial infarction in mice using cardiac and respiratory gated look-locker imaging with fuzzy C-means clustering. Magn. Reson. Med. 63, 648–657.

Wacker, C.M., Fidler, F., Dueren, C., Hirn, S., Jakob, P.M., Ertl, G., Haase, A., Bauer, W.R., 2003. Quantitative assessment of myocardial perfusion with a spin-labeling technique: preliminary results in patients with coronary artery disease. J. Magn. Reson. Imaging 18, 555–560.

Waller, C., Hiller, K.H., Kahler, E., Hu, K., Nahrendorf, M., Voll, S., Haase, A., Ertl, G., Bauer, W.R., 2001. Serial magnetic resonance imaging of microvascular remodeling in the infarcted rat heart. Circulation 103, 1564–1569.

Waller, C., Hiller, K.-H., Pfaff, D., Gattenlöhner, S., Ertl, G., Bauer, W.R., 2008. Functional mechanisms of myocardial microcirculation in left ventricular hypertrophy: a hypothetical model of capillary remodeling post myocardial infarction. Microvasc. Res. 75, 104–111.

Wang, Y., Riederer, S.J., Ehman, R.L., 1995. Respiratory motion of the heart: kinematics and the implications for the spatial resolution in coronary imaging. Magn. Reson. Med. 33, 713–719.

Wang, D.J., Bi, X., Avants, B.B., Meng, T., Zuehlsdorff, S., Detre, J.A., 2010. Estimation of perfusion and arterial transit time in myocardium using free-breathing myocardial arterial spin labeling with navigator-echo. Magn. Reson. Med. 64, 1289–1295.

Wang, Y., Zhang, Y., Wen, Z., Tian, B., Kao, E., Liu, X., Xuan, W., Ordovas, K., Saloner, D., Liu, J., 2021. Deep learning based fully automatic segmentation of the left ventricular endocardium and epicardium from cardiac cine MRI. Quant. Imaging Med. Surg. 11, 1600–1612.

Williams, D.S., Detre, J.A., Leigh, J.S., Koretsky, A.P., 1992. Magnetic resonance imaging of perfusion using spin inversion of arterial water. Proc. Natl. Acad. Sci. U. S. A. 89, 212–216.

Wong, E.C., Cronin, M., Wu, W.-C., Inglis, B., Frank, L.R., Liu, T.T., 2006. Velocity-selective arterial spin labeling. Magn. Reson. Med. 55, 1334–1341.

Yoon, A.J., Do, H.P., Cen, S., Fong, M.W., Saremi, F., Barr, M.L., Nayak, K.S., 2017. Assessment of segmental myocardial blood flow and myocardial perfusion reserve by adenosine-stress myocardial arterial spin labeling perfusion imaging. J. Magn. Reson. Imaging 46, 413–420.

Zhang, H., Qiao, H., Frank, R.S., Huang, B., Propert, K.J., Margulies, S., Ferrari, V.A., Epstein, J.A., Zhou, R., 2012. Spin-labeling magnetic resonance imaging detects increased myocardial blood flow after endothelial cell transplantation in the infarcted heart. Circ. Cardiovasc. Imaging 5, 210–217.

Zun, Z., Wong, E.C., Nayak, K.S., 2009. Assessment of myocardial blood flow (MBF) in humans using arterial spin labeling (ASL): feasibility and noise analysis. Magn. Reson. Med. 62, 975–983.

Zun, Z., Jao, T., Varadarajan, P., Pai, R.G., Wong, E.C., Nayak, K.S., 2011a. Myocardial ASL perfusion reserve test detects angiographic CAD in initial cohort of 29 patients. In: Proc Intl Soc Mag Reson Med 19 Annu. Meet. Int. Soc. Magn. Reson. Med. 7–13 May, Montréal, QC, Canada, p. 221.

Zun, Z., Varadarajan, P., Pai, R.G., Wong, E.C., Nayak, K.S., 2011b. Arterial spin labeled CMR detects clinically relevant increase in myocardial blood flow with vasodilation. JACC Cardiovasc. Imaging 4, 1253–1261.

Perfusion MRI in the heart: First-pass perfusion

15

N. Sharrack[a], J.D. Biglands[a,b], S. Plein[a], and D.A. Broadbent[a,b]

[a]*Leeds Institute of Cardiovascular and Metabolic Medicine, University of Leeds, Leeds, United Kingdom*

[b]*Medical Physics and Engineering, Leeds Teaching Hospitals NHS Trust, Leeds, United Kingdom*

Abbreviations

AIF	arterial input function
ATP	adenosine triphosphate
BAT	bolus arriving time
bSSFP	balanced steady-state free precession
CAD	coronary artery disease
CMD	coronary microvascular dysfunction
CNR	contrast-to-noise ratio
CTO	chronic total occlusion
DCE	dynamic contrast enhanced
DCM	dilated cardiomyopathy
EPI	echo-planar imaging
GRE	gradient echo
HCM	hypertrophic cardiomyopathy
MACE	major adverse coronary events
MBF	myocardial blood flow
MPR	myocardial perfusion reserve
PCI	percutaneous coronary intervention
PDw	protein density weighting
PET	positron emission tomography
RF	radiofrequency
RPP	rate-pressure product
SCD	sudden cardiac death
SNR	signal-to-noise ratio
TE	echo time

Advances in Magnetic Resonance Technology and Applications, Volume 11, ISSN 2666-9099
https://doi.org/10.1016/B978-0-323-95209-5.00019-2

15.1 **Overview**

First-pass myocardial perfusion MRI is well established as an accurate technique to assess myocardial ischemia, without ionizing radiation (Heitner *et al.*, 2019; Kwong *et al.*, 2019; Nagel *et al.*, 2019). It is an essential part of the diagnostic and prognostic assessment of cardiovascular disease with high levels of evidence in major clinical cardiology guidelines (Gulati *et al.*, 2021). The presence of myocardial ischemia indicates an increased risk of cardiac events and adverse cardiovascular outcomes (Pezel *et al.*, 2021; Vincenti *et al.*, 2017). Visual interpretation of perfusion datasets relies on experienced reporters to identify regional perfusion defects. However, the acquired data can be used to derive quantitative maps, displaying absolute values of myocardial blood flow (MBF). The potential advantages of quantitation over visual reading include removal of operator dependence, simpler and faster analysis, the ability to detect disease with global rather than regional reduction of MBF, such as coronary microvascular dysfunction (CMD), and the provision of an objective and serially evaluable endpoint.

Myocardial perfusion MRI requires fast data acquisition to account for the rapid dynamic signal changes during the passage of contrast agent through the heart and the need for cardiac and respiratory motion compensation. If quantitative perfusion values are required from the images, then further constraints are placed on the acquisition and substantial postprocessing of the data is required. This chapter gives a concise overview of these challenges and constraints and describes the various approaches of addressing them, before reviewing the evidence supporting myocardial perfusion MRI in clinical practice and the way these images are evaluated.

First, the myocardial perfusion imaging sequence will be discussed. This needs to be performed in such a way as to deal with both the contractile motion of the beating heart and breathing motion. The spatial resolution needs to be adequate to visualize small perfusion defects while maintaining adequate image quality. Simultaneously, the temporal resolution must be sufficient to characterize the contrast bolus passing through both the blood pool and the myocardium.

The process of generating quantitative myocardial blood flow estimates will then be discussed. This involves localizing the blood pool to generate an arterial input function (AIF) as well as extracting relevant myocardial signal intensity profiles (on a pixel-wise or segmental basis). The nonlinear signal response to contrast agent concentration must then be accounted for before the data are analyzed with an appropriate perfusion model to obtain estimates of myocardial perfusion (and potentially additional physiological parameters). Finally, the results must be presented in a way that is appropriate to the disease in question.

Third, the clinical evaluation of myocardial perfusion datasets will be discussed. The three main clinical indications for myocardial perfusion MRI are coronary artery disease (CAD), coronary microvascular disease (CMD), and cardiomyopathy. Evidence supporting the use of myocardial perfusion MRI in the diagnosis and management of these conditions will be described before describing how visual and quantitative analysis of myocardial perfusion MRI datasets is integrated into clinical reporting.

Finally, the current challenges and future directions for myocardial perfusion MRI will be discussed.

15.2 Imaging protocol

Several factors need to be considered when defining a cardiac dynamic contrast-enhanced (DCE) imaging protocol for quantitative analysis, to ensure suitability of the acquired image data. The images must be sensitive to contrast agent presence (across the expected range of concentrations) while providing adequate image quality. They must have sufficient temporal resolution for accurate quantification, and sufficient spatial resolution to allow isolation of the myocardium from the blood pool to avoid biasing influence from partial volume effects. The overall acquisition duration must also be sufficient for accurate quantification of the relevant tracer kinetic parameters. Within these restrictions, the protocol should aim to provide the optimal anatomical coverage to address the clinical questions. A Society for Cardiovascular Magnetic Resonance task force statement (Kramer *et al.*, 2020) provides details of recommended methods for acquiring multiple cardiovascular MRI sequences including perfusion imaging, although it should be noted that those recommendations are not specifically for quantitative analysis and further adaptation or optimization may be required when imaging for that purpose.

15.2.1 Cardiac synchronization and coverage

For myocardial DCE-MRI, several factors influence the choice of sequence. The combination of physiological motion and rapidly changing tracer concentrations imposes tighter requirements on the required speed and temporal resolution of imaging than for many other organs.

To acquire images at the same phase of the cardiac cycle, acquisition is triggered using a marker of cardiac rhythm. Most commonly, electrocardiography (ECG) is used, although pulse monitoring is an alternative. This is necessary to ensure the cardiac tissue is in the same shape and position for each dynamic acquisition.

The time available for imaging is determined by the patient's heart rate and the need to avoid artifacts arising from motion during data acquisition. This imposes a requirement to acquire data with a very short temporal footprint within each heartbeat. Acquiring 3D volumes is therefore challenging, so acquisition of multiple 2D slices is the clinical standard, although this does result in different slices being acquired at different cardiac phases.

The need for rapid imaging also limits the achievable spatial resolution; so, perfusion assessment has typically focused on the thick-walled left ventricle rather than

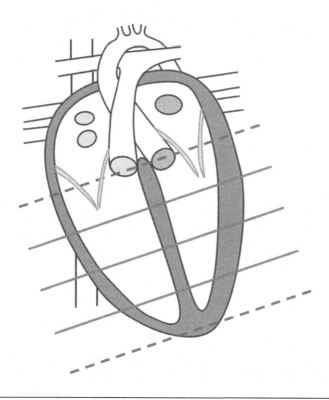

FIG. 15.1

"Three-of-five" approach for reproducible planning of myocardial perfusion MRI imaging.
Initially five equally spaced slices are prescribed with the outer slices placed over the
mitral valve plane and the apex of the left ventricle. The outer two (*dashed*) are then discarded
prior to acquisition of the inner three.

Figure adapted with permission from Broadbent, D.A., 2015. Quantitative Dynamic Contrast Enhanced
Magnetic Resonance Imaging for Evaluation of the Myocardium in Ischaemic Heart Disease (PhD Thesis).
University of Leeds.

the thinner-walled right ventricle or atria. Typically, 3 short axis slices (of around
8–10 mm thickness) are acquired over the left ventricle at basal, mid-cavity, and apical levels (see Fig. 15.1), allowing observed perfusion to be mapped to segments
1–16 of the commonly used American Heart Association (AHA) 17 segment model
(Cerqueira *et al.*, 2002). Some centers also choose to acquire additional perpendicular (long axis) slices to include the apical cap (segment 17), although if there is not
sufficient time to allow imaging of additional slices per cardiac cycle this may
require reducing the temporal resolution by imaging each slice only on alternate
heartbeats.

An alternative strategy is to acquire 3D datasets covering the whole heart
(or at least the ventricles). However, this increases the temporal footprint of the
acquisition making it challenging to perform with sufficient spatial resolution.

In clinical practice, 2D acquisitions remain the standard, and more widespread adoption of 3D methods will rely on advances in acceleration techniques.

15.2.2 Sequence type

Given the tight time constraints for myocardial perfusion imaging, a rapid sequence type with a very short repetition time is required. Sequences that can be used for this purpose are spoiled gradient echo (GRE), balanced steady-state free precession (bSSFP), or hybrid echo planar imaging (h-EPI). Conventional EPI, despite being extremely rapid, is not typically used due to its high sensitivity to artifacts (both from motion and susceptibility effects). Spoiled GRE, bSSFP, and h-EPI, typically in combination with relatively modest parallel imaging factors, are all capable of acquiring single slices with sufficient in-plane spatial resolution (around 3 mm or smaller voxel size) in under 150 ms, allowing three slices to be acquired per cardiac cycle (as described before) for heart rates up to approximately 130 bpm.

Each sequence type has benefits and disadvantages, and the appropriate choice may depend on the scanner and field strength. Spoiled GRE is the most robust sequence to frequency shifts (due to off-resonance or magnetic susceptibility effects) and is, therefore, widely used at higher field strengths (3 T or greater) where these pose greater challenges. The bSSFP sequence generates greater signal-to-noise ratio (SNR) and contrast-to-noise ratio (CNR) than spoiled GRE and is typically faster, although due to the sensitivity to frequency effects its use is often restricted to field strengths up to 1.5 T. Similarly, h-EPI is highly sensitive to susceptibility artifacts and can only be used with very short echo trains at higher field strengths. Its key benefit is speed, being the fastest of these three sequences, and it may, therefore, find application to increase spatial resolution or coverage, or if imaging at very high heart rates.

15.2.3 Magnetization preparation

To enhance T1 contrast weighting, the image readout methods described before are typically combined with magnetization preparation for myocardial perfusion imaging. This is generally magnetization saturation (although early versions used inversion), followed by a delay prior to image acquisition to allow partial recovery of longitudinal magnetization (as shown in Fig. 15.2). To achieve optimal CNR in the myocardium, this delay (measured from the center of the preparation pulse to the acquisition of the central line of k-space) is typically around 70–100 ms.

Signal contrast is then dominated by the T1 dependent recovery during this delay, rather than the inherent contrast weighting of the readout phase. This, therefore, allows strong T1 weighting to be generated even with low flip angle readouts (as shown in Fig. 15.3).

Successful saturation preparation eradicates dependence of the signal strength on previous acquisitions. This is beneficial as the prior magnetization evolution can be difficult to model for nonsaturation recovery sequences, where each acquisition

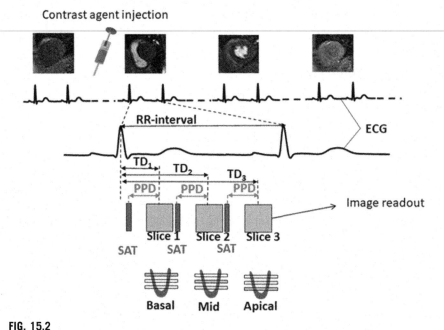

FIG. 15.2

Graphical representation of a typical three slice saturation recovery perfusion sequence. The center of k-space for each slice is acquired at a different cardiac phase (the corresponding trigger delay, TD) after the R wave trigger but at the same time (prepulse delay, PPD) after the preceding magnetization preparation pulse (SAT).

Figure reproduced with permission from Broadbent, D., Kidambi, A., Biglands, J., 2015.
EACVI CMR Pocket Guides: Physics for Clinicians.

cannot be considered in isolation—especially in a scenario where both the T1 (due to varying contrast agent concentration) and timings (due to cardiac cycle length variation and potentially missed triggers) are constantly changing. By nulling magnetization, the signal intensity for a given acquisition is made to depend only on what has happened since the preparation pulse. As a result, the signal behavior can be more readily modeled, although it should be noted that even relatively small imperfections in saturation performance can have potentially significant biasing effects on signal, particularly in the baseline (Broadbent *et al.*, 2016).

15.2.3.1 Temporal resolution

As discussed in the previous section, there is a need to synchronize data acquisition to the cardiac cycle and to acquire data within a short temporal footprint in each cardiac cycle. A high temporal resolution is also required for myocardial DCE as more rapid changes in contrast agent concentrations are observed in the heart than in other organs. These rapid changes in contrast arise as the myocardium is the most proximal tissue perfused by the left side of the heart, and so the contrast agent bolus has

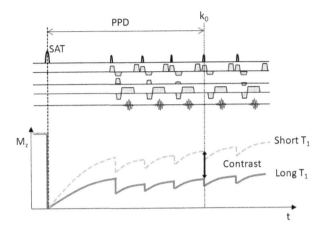

FIG. 15.3

Evolution of longitudinal magnetization (M_z) to generate T1 contrast in a saturation recovery spoiled gradient echo sequence. The difference in Mz at the time the readout pulse for the center of k-space (k_0) is applied depends primarily on T1 and PPD, but is also partly dependent on the number, flip angle, and timing of the readout pulses before it. However, with effective magnetization saturation, the value of Mz prior to the saturation pulse is irrelevant, as the magnetization is nulled by it. The signal strength will also depend on receiver gain, echo time, and T2*.

Figure adapted with permission from Broadbent, D., Kidambi, A., Biglands, J., 2015.
EACVI CMR Pocket Guides: Physics for Clinicians.

undergone less intravascular dispersion than for more distal tissues. Indeed, the AIF is measured directly from the left ventricular cavity; so, for the first pass of the bolus, the contrast agent will only have passed through the venous return from the injection site and the pulmonary vasculature and not through any systemic arteries as would be the case for DCE-MRI of other organs.

As for most quantitative DCE applications, it is the need to accurately characterize the AIF, rather than the perfused tissue, that dictates the required temporal resolution. However, high temporal resolution is required even for nonquantitative myocardial perfusion imaging where the AIF is not measured, and the same technique of acquiring images every heartbeat is typically used even for visual assessment. This is because even resting myocardium is very highly perfused, so rates of contrast enhancement after arrival are much more rapid than in other organs.

Due to this requirement, myocardial DCE-MRI data is not typically segmented across multiple heartbeats but all data for each image is generally acquired in a single shot, within a single cardiac cycle. Due to the gaps between acquisitions though, the temporal resolution is not defined by the duration of the image acquisition (as it is for DCE applications where data is acquired contiguously). Instead, it is determined by the duration of the cardiac cycle, and whether new images for each slice are acquired every heartbeat, or less frequently to allow extra slices to be acquired. It therefore typically ranges from 0.5 s (for a high, 120 bpm, heart rate with imaging repeated every beat) to 3 s (for a low, 40 bpm, heart rate with imaging repeated every other beat).

15.2.3.2 Imaging duration

Myocardial DCE is commonly referred to clinically as "first-pass perfusion imaging," as visual reporting of perfusion generally focuses on the first pass of the contrast agent (before recirculation). Quantitative assessment of perfusion alone may also be performed with such a short acquisition (indeed some tracer kinetic models are not suited to analyzing longer datasets unless extracellular contrast agents are used).

Acquisition duration is often therefore as short as around 30–40 s, and sometimes visual assessment is based purely on data acquired during a breath-hold covering the first pass to limit respiratory motion. However, analysis using detailed tracer kinetic models which account for leakage of contrast agent to the interstitium (and allow measurement of additional physiological parameters beyond perfusion) requires a longer acquisition (1–2 min). Note that in practice most vendor sequences require the number of measurements to be specified, rather than the imaging duration in time units.

15.2.4 Adaptions for the nonlinear signal response to contrast agent

The purpose of contrast agents in DCE-MRI is to transiently shorten the longitudinal magnetization recovery time, T1 (or, equivalently, to increase the longitudinal magnetization recovery rate, R1 = 1/T1), of protons close to the contrast agent molecules. R1 increases in proportion to the contrast agent concentration with a gradient defined as the longitudinal relaxivity (r1), while T1 decreases with a nonlinear relationship.

MR signal intensities in most sequences increase with decreasing T1 (increasing R1). However, the relationship between MRI signal intensity and both T1 and R1 is also nonlinear, and so is the relationship between signal intensity and contrast agent concentration as shown in Fig. 15.4. As can be seen, the signal enhancement response gets increasingly less steep at higher concentrations and will eventually reach a plateau once R1 is fast enough to allow full recovery between each RF pulse. In fact, signal intensity will eventually start to decrease, as the contrast agent also has the effect of accelerating transverse magnetization decay (shortening T2 and T2*), and at high concentrations the signal loss due to this effect will exceed the signal gain due to increased longitudinal recovery.

While MR signal intensity can be used as a marker of contrast agent concentration, this nonlinearity must be accounted for in quantitative analysis. Data analysis and acquisition are inherently linked, and steps can be taken during the acquisition to generate data more amenable to accurate analysis. A key requirement is to ensure that the plateau of the curve in Fig. 15.4 is not reached, i.e., that the signal response

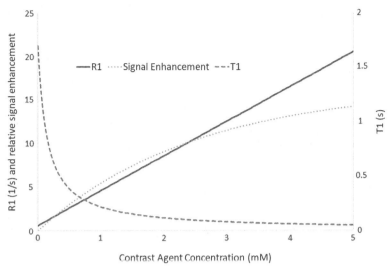

FIG. 15.4

Representative responses of R1, T1, and signal enhancement to contrast agent concentration for a saturation recovery sequence. Note that while R1 exhibits a linear response, neither T1 nor signal enhancement do, with the gradient of both becoming less steep as concentration increases. The gradient of the R1 line is the contrast agent relaxivity (r1).

is not saturated. To do this, an appropriate balance between the sequence parameters and the anticipated maximal contrast agent concentration must be used.

Alternatively, data can be acquired in such a way that the signal response can be considered approximately linear. In this case, signal enhancement can be used as a direct surrogate for contrast agent concentration in the quantification process (although some bias will always remain). To approach this linear regime, the contrast agent dose and/or T1 sensitivity of the sequence must be kept low (the latter typically by using short saturation time values in the sequence). While achievable, quantifying such data is very challenging due to the different ranges of contrast agent concentrations encountered in the blood pool (used for the AIF) and the myocardium itself (typically there is around an order of magnitude difference in peak concentrations). For any acquisition designed to constrain the AIF to this approximately linear regime, the signal response, and so the contrast-to-noise ratio, of the myocardial data will be very low. To circumvent this, strategies involving dual-acquisition methods (where the AIF and myocardial data are acquired separately) have been developed as described in the following section.

15.2.4.1 Dual-acquisition methods

By acquiring the AIF and myocardial data separately, both can be acquired from within the approximately linear signal response range, and both with acceptable CNR. This can be achieved in two distinct ways.

Firstly, the same image acquisition method can be used but with separate contrast agent administrations (Christian *et al.*, 2004; Hsu *et al.*, 2006; Kostler *et al.*, 2004). A small bolus is injected initially, which is used solely to sample the AIF. By using a small bolus, the peak concentration in the blood is minimized. Sequence parameters that generate sufficient CNR in the myocardium with a standard bolus can therefore also be used to measure the AIF with this small bolus, without significant signal nonlinearity effects. Prior to analysis, the measured AIF signal enhancement is scaled up by the ratio of the contrast agent doses (typically around a 1:10 ratio is used).

While this dual-bolus technique has been used successfully, it does come with some practical difficulties. It is essential that the dose scaling is accurate, not only in terms of total dose but also in terms of injection profile. To generate an AIF that is simply scaled in amplitude and not changed temporally, the contrast agent should be diluted, or the injection rate reduced, rather than just using a smaller volume. Further to this, by acquiring the data at different time points, even if only after a short delay, there may be physiological changes between acquisitions (such as changes in heart rate or cardiac output). This may mean that the myocardial response observed is not necessarily representative of that from the observed AIF. Dual-bolus protocols are also demanding on staff with multiple potential sources of human error and, therefore, not easily incorporated into clinical protocols.

An alternative strategy is to use the same contrast agent injection but with independent imaging acquisitions for the AIF and myocardial data (Gatehouse *et al.*, 2004). In this technique, commonly referred to as dual-sequence imaging, an additional interleaved slice with reduced sensitivity is introduced for the AIF acquisition alongside the standard (normally three) slices for the myocardial data. The reduced sensitivity is achieved in magnetization prepared sequences by using a reduced delay time (Fig. 15.5).

In comparison to the dual-bolus method, the dual-sequence method circumvents many of the practical difficulties of administering two independent boluses with carefully controlled dose ratios and is much easier incorporated into clinical routine. Further challenges are, however, introduced. First, the addition of an extra slice purely for AIF measurement adds to the overall timing challenges, especially at high heart rates. This is somewhat mitigated by the fact that the left ventricular blood pool (in which the AIF is measured) is relatively large, so a low spatial resolution can be used for this extra slice. Coupled with the short magnetization preparation delay, this means the temporal duration of the extra slot can be much shorter than that of the individual myocardial slices where higher spatial resolution is required. A second challenge is that while this approach can keep both the AIF and myocardial data in the approximately linear regime, the constant of proportionality between signal enhancement and contrast agent concentration will differ between the acquisitions. As discussed later, this sensitivity difference must be accounted for in the quantitative analysis even if signal linearity is assumed.

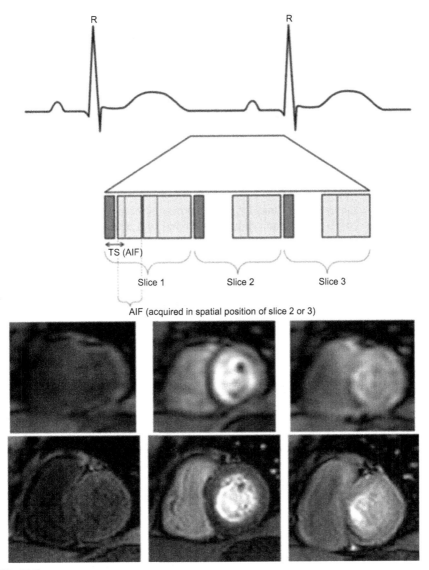

R   R

TS (AIF)

Slice 1   Slice 2   Slice 3

AIF (acquired in spatial position of slice 2 or 3)

FIG. 15.5

Schematic and example images demonstrating one method for interleaving an additional low-resolution image acquisition for the AIF into a conventional 3 slice perfusion series (i.e., the dual-sequence method). In this case, the additional slice shares a nonslice selective preparation pulse with slice 1 but is colocated with slice 2 or 3 (to avoid interference between the low and full resolution slices), although alternative strategies exist, including use of an additional preparation pulse for the extra AIF slice.

Figure reproduced with permission from Broadbent, D.A., 2015. Quantitative Dynamic Contrast Enhanced Magnetic Resonance Imaging for Evaluation of the Myocardium in Ischaemic Heart Disease (PhD thesis). University of Leeds.

15.2.4.2 Transverse decay effects

While T1 shortening is the predominant contrast agent effect exploited in DCE imaging, the agents also accelerate the rate of transverse decay and, therefore, shorten T2 and T2*. This can reduce signal intensities compared to those predicted from longitudinal recovery effects alone and can cause signal intensity to decrease with increasing contrast agent at high concentrations (leading to an undesirable nonmonotonic relationship between signal intensity and concentration).

In practice, transverse decay effects are minimized by using short echo time (TE) acquisitions, and it is not expected that concentrations beyond those corresponding to peak signal enhancement would be encountered with typical imaging parameters. However, at the highest concentrations encountered, there is potential for the effect to have a biasing influence on quantification if unaccounted for. To account for them a dual-echo acquisition (Kellman *et al.*, 2006) may be used to allow calculation of signal intensities to those from a (practically infeasible) zero-TE acquisition which would be unaffected by these transverse decay effects. Such effects are only expected to be practically relevant in the blood pool (peak concentrations are much lower in the myocardium) and may not always be significant there. For the dual-sequence method described before, a dual-echo acquisition would typically only be used with the interleaved AIF scan (if used at all), as there are drawbacks from this method, including an increase in the imaging temporal footprint.

15.3 Data analysis protocol

15.3.1 Localizing the myocardium and blood pool

All quantitative analyses of DCE-MRI data require the AIF to be localized in the image. If quantitative maps are being generated, the tissue of interest (in this case the myocardium) must also be localized. Despite the many challenges of myocardial DCE-MRI, the presence of the left ventricular blood pool in the image makes it relatively easy to obtain an AIF free from partial volume effects. Similarly, the myocardium, at least on the central slice of the heart, is large enough to obtain a myocardial signal curve with good CNR. On the other hand, a problem peculiar to myocardial perfusion measurement is the fact that the myocardium is adjacent to the source of the AIF. Great care needs to be taken not to let the endocardial border encroach on the left-ventricular blood pool, when localizing the myocardium, lest the myocardial curves be contaminated with AIF voxels, resulting in perfusion overestimates (Biglands *et al.*, 2011).

An example of localization is shown in Fig. 15.6, which shows regions of interest isolating the myocardium and a region within the blood pool, clear from any surrounding structures. Historically, these regions have been drawn manually using some form of dedicated software. There are a range of automated approaches for carrying out these tasks that are achieving growing reliability. However, they can fail, particularly if the image quality is suboptimal due to artifacts or poor image SNR.

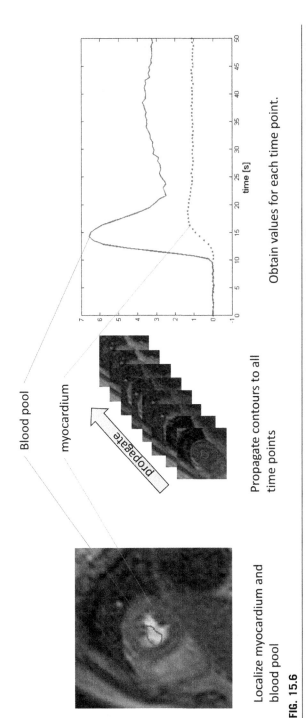

Localize myocardium and blood pool

Propagate contours to all time points

Obtain values for each time point.

FIG. 15.6

Localization of the myocardium and blood pool to generate signal intensity vs time curves. Region of interest must be identified within the blood pool and the myocardium, typically using the time point where the first-pass bolus arrives in the blood pool, which provides maximum contrast between the myocardium, blood pool, and surroundings. These ROIs must then be propagated to the full time series, either through image registration or by manual adjustment. Mean pixel values within the ROI are then obtained at each time point to provide a mean value for each time point in the curve. Note: In perfusion mapping, averaging is not performed for the myocardium and there will be a myocardial curve for every pixel. However, the myocardial region of interest is often still used to exclude voxels from surrounding tissues.

If motion has been dealt with adequately during acquisition (e.g., by breath-holding), then the AIF and myocardial regions can simply be copied to each time point in the series and used to generate signal vs time curves. Otherwise, motion is compensated for postacquisition using an image registration algorithm that spatially deforms and translates consecutive images so that the motion is removed from the time series. The patient is not required to hold their breath, improving patient comfort through the scan, the need for correcting contours over the time series is removed, and the perfusion measurement technique does not need to be limited to the first pass. However, automated motion correction can fail and so a manual check of the data is advisable.

Cardiac motion does not only take place within the axial image plane. Through-plane motion (i.e., the heart moving perpendicular to the image plane) can take place either through breathing or due to the contractile motion of the heart if the ECG triggering is imperfect. No motion correction method, automated or otherwise, can correct for myocardial tissue shifting outside of the imaging slice. The effects are ameliorated by using relatively thick slice thickness so that through-plane motion is small compared to the slice thickness. However, this problem can only be fully addressed using a 3D acquisition, where motion correction algorithms can fully characterize the motion and obtain a constant position in the heart across the data series.

15.3.2 Surface coil inhomogeneity correction

In general, most cardiac imaging is performed using combinations of surface radiofrequency (RF) receive coils (typically an anterior coil on the patient's chest in conjunction with a spine coil). These coils give good SNR and allow parallel imaging, but these benefits come at the expense of poor intrinsic signal uniformity. Consequently, anterior parts of the heart can exhibit higher signal intensity than more posterior parts, even if the magnetic properties are the same. For accurate quantitative analysis of DCE data, it is necessary to correct for this spatial inhomogeneity. This can be achieved using signal normalization techniques built into the sequence on the scanner or, if necessary, in postprocessing. Such correction methods are typically based on fitting curved surfaces to baseline images (ideally proton density weighted phases, if acquired) and then normalizing all phases of dynamic data using this sensitivity map (Hsu et al., 2003; Kremers et al., 2010; Nielles-Vallespin et al., 2015).

Note that if a model-based signal nonlinearity correction (see next section) is performed on a voxel-wise basis, independent surface coil inhomogeneity correction may not be necessary, as the local sensitivity variations are effectively incorporated into the nonlinearity correction.

15.3.3 Signal nonlinearity correction

As discussed in the Imaging protocol section, MRI signal intensity does not typically exhibit a linear response to contrast agent concentration. However, for accurate quantification, the inputs to the deconvolution process should be equal to, or at least be proportional to, the concentrations in the tissue and blood. Two approaches can be taken during analysis, dependent on the acquisition method.

15.3.3.1 Assuming linearity

It is not generally practical to keep both the AIF and myocardial data within the approximately linear range simultaneously (without the CNR of the myocardial data being too low); so, the assumption of linearity is typically only made in conjunction with dual-acquisition methods, as discussed before. For relative quantification, for example calculation of myocardial perfusion reserve (the ratio between blood flow at stress and rest), this can be done directly. However, for absolute quantification, the different scaling of the AIF and myocardial data must be accounted for.

For the dual-bolus approach this is relatively straightforward: the AIF can simply be scaled by the ratio of contrast agent doses. For the dual-sequence approach, however, the sensitivity difference between the two sequences must be accounted for using a model-based approach (Cernicanu and Axel, 2006).

15.3.3.2 Model-based correction

Although the signal response to contrast agent concentration is nonlinear, it is predictable. The signal intensity will depend on a combination of known sequence parameters, variable tissue/blood properties that are affected by the contrast agent (typically assumed to be only T1), and other constant but unknown factors such as coil sensitivity. Together, these can form a mathematical model describing the predicted signal response. For saturation recovery sequences, the simplest form of this model (Eq. 15.1) is to assume perfect saturation and to neglect the effect of the readout pulses on magnetization evolution and that of the contrast agent on transverse decay rates. Under these assumptions, the signal evolution can be described by simple exponential recovery toward the equilibrium value S_0:

$$S = S_0 \left(1 - e^{\frac{-TS}{T1}} \right) \tag{15.1}$$

More detailed signal models can be derived that include the effect of the train of readout pulses. These will typically describe the magnetization evolution to the acquisition of the central line of k-space, as this contributes most strongly to image contrast. The exact form of the equation will depend on the sequence type and detailed acquisition parameters to accurately reflect the full magnetization evolution arising from each RF pulse and recovery period. Relaxing the assumption of perfect saturation is more challenging, as the signal model for a given image then also becomes dependent on the magnetization evolution history during previous image acquisitions rather than just the current one (Broadbent et al., 2016).

From this signal model it is possible to convert the observed signal intensity at each dynamic phase to a longitudinal relaxation time (T1). Subsequently, the change of R1 from baseline (ΔR1), which is directly proportional to contrast agent concentration, can be calculated. However, before this can be done, the value of S_0 must be determined. This is achieved by acquiring extra data immediately prior to the dynamic scan, so that the blood and tissue magnetic properties are the same as during the DCE baseline. The determined value of S_0 at baseline is then assumed to remain constant throughout the remainder of the DCE measurement.

A common approach is to obtain proton density weighted (PDw) images at the start of the baseline by omitting the saturation pulse from some initial phases. To a first approximation this is equivalent to sampling with infinite TS, and so under the simplified model given before the PDw signal intensity directly equals S_0. However, in practice, this simple modeling is insufficient, as the effect of the readout RF pulses is not negligible and more detailed modeling is required. With this approach, the two baseline measurements (SR DCE and PDw) can be used to form an estimate of baseline T1 (Cernicanu and Axel, 2006), which can then be used to determine S_0 for the DCE sequence at baseline.

Alternatively, relaxometry techniques can be used to independently measure the baseline T1 directly. However, this approach can be highly sensitive to bias due to imperfect saturation pulse performance, whereas the proton density-weighted approach, or use of an additional postcontrast T1 measurement, can be more resilient to such effects (Broadbent *et al.*, 2016).

15.3.3.3 Transverse decay effects

The modeling presented before disregards the fact that contrast agents accelerate transverse magnetization decay as well as longitudinal recovery. Due to the short echo times used and due to the shorter value of native T2* compared to native T1, the signal loss due to these effects is generally small in comparison to the desirable signal gain caused by T1 shortening. However, if they become significant, then neglecting to account for them may cause errors in physiological parameter estimates.

As described previously, one approach to address this is to perform a dual-echo acquisition (Kellman *et al.*, 2006), allowing estimation of T2* and extrapolation of signal intensities back to a zero-TE value. This can be performed at each dynamic phase using a monoexponential decay model (Eq. 15.2), prior to independent conversion of the estimated TE $= 0$ signal intensities to ΔR1 using the aforementioned methodology.

$$S(TE) = S_o e^{\frac{-TE}{T2^*}} \tag{15.2}$$

15.3.4 **Temporal segmentation**

To perform quantitative DCE analysis, some segmentation of the data may be needed in the temporal domain. First, any integrated additional scans for nonlinearity correction (such as initial PDw phases) need separating, although as the number of phases of this is known a priori this task is generally trivial.

Secondly, the point at which contrast agent first arrives needs to be identified to allow separating the baseline, for example, for model-based signal nonlinearity correction. The timing of this will depend on patient-specific physiological characteristics as well as when the injection was initiated in relation to the start of scanning. This may be performed manually, or automatic methods may be implemented (Jacobs *et al.*, 2016). In either case, when this is performed to identify a true baseline signal, it is important that no points are affected by contrast agent presence, so the selection of the end of the baseline should be performed conservatively early to avoid bias from the initial contrast agent arrival.

Thirdly, the deconvolution process assumes the contrast agent arrival in the AIF is coincident with that in the tissue. However, there is some delay (the bolus arrival time, BAT) due to the time taken for the contrast agent to pass from the LV cavity out of the heart and through the coronary arteries. Again, this may be corrected for by translating the AIF and myocardial curves manually, or it may be automated (Cheong *et al.*, 2003; Natsume *et al.*, 2015). Alternatively, the BAT may be incorporated as an additional free parameter within the deconvolution (Kershaw and Buckley, 2006).

Finally, if an analysis method is used that is only suitable for the first pass (e.g., Fermi deconvolution), then the end of that period needs identification. As per the other time points, this may be done either manually by identifying the local minimum that follows the first pass peak in the AIF or automatically (Biglands *et al.*, 2011).

For the dual-bolus technique, the relevant steps of the earlier segmentation need to be done independently for each bolus; if the data is acquired during a single long acquisition, the two curves need temporal shifting to be coincident. In this technique a single temporal shift is applied which accounts for both the relatively long delay between the two boluses, and the much shorter BAT (described above).

15.3.5 Temporal interpolation

In many DCE-MRI applications, the temporal sampling rate is determined solely by scan parameters, with subsequent dynamic phases typically acquired contiguously (or at least regularly). In cardiac DCE, however, timing is also dependent on the subject's cardiac rhythm, as the acquisition is physiologically triggered. Consequently, the temporal resolution will vary depending on heart rate. Furthermore, even within acquisitions there will be some variability in cardiac cycle duration throughout the scan; so, the temporal spacing of the measured data will not be perfectly regular.

Data is, therefore, typically linearly interpolated to a regular temporal spacing, prior to tracer kinetic analysis. Interpolation may be done to the same (average) temporal spacing as in the measured data, although some studies suggest that upsampling the data to a higher temporal resolution may improve quantitation accuracy (Milidonis *et al.*, 2022).

15.3.6 Tracer kinetic analysis

15.3.6.1 Choosing a model

Once time series for the blood pool and the myocardium have been obtained and converted to contrast agent concentrations, a method is needed to calculate the myocardial blood flow. The earliest approaches simply measured certain properties of the first-pass myocardial curve, such as the upslope gradient or the area under the curve.

These semiquantitative perfusion indices correlated with myocardial perfusion and had some utility in diagnosis (Al-Saadi *et al.*, 2000; Huber *et al.*, 2012; Nagel *et al.*, 2003; Plein *et al.*, 2005), but they were not derived from any underlying model, and so are not considered quantitative perfusion measurements.

Quantitative models for measuring perfusion from DCE-MRI curves are based on the central-volume principle. Mathematically, this relates the AIF concentration vs time curve, $C_{AIF}(t)$, to the tissue curve, $C_{myo}(t)$, as in Eq. (15.3), i.e., as a convolution operation (denoted $\otimes$) with the impulse response function, $R(t)$, multiplied by the flow rate, Q:

$$C_{myo}(t) = Q \cdot R(t) \otimes C_{AIF}(t) \qquad (15.3)$$

$R(t)$ represents the fraction of contrast agent in the tissue at time t. Therefore the MBF can be measured from the initial maximum value of $Q \cdot R(t)$, where $R(t) = 1$. However, an analytical solution to Eq. (15.3) by deconvolution is mathematically unstable and produces physiologically unrealistic impulse response functions. Therefore $R(t)$ needs to be constrained in some way.

Some of the earliest methods to quantitative perfusion approached this by simply applying some nonphysiological constraints on the impulse response function. A popular method was to constrain $R(t)$ to be a Fermi function (Jerosch-Herold *et al.*, 1998). There was no theoretical reason to use this function to constrain the previous equation other than that it gave a good approximation to the shape of $R(t)$ and allowed the deconvolution to proceed. However, it was only a good representation to the first-pass peak and so the curves had to be limited to the first pass in order to use this method. Other model-free methods were less restrictive and required only that $R(t)$ be a smooth, monotonic function, but typically a smoothness parameter was required to be set for the solution to proceed (Jerosch-Herold *et al.*, 2002; Pack *et al.*, 2008).

In contrast to these methods, parametric deconvolution methods attempt to derive an analytic form for $R(t)$ based on tracer-kinetic modeling. All such models must necessarily make assumptions about the tissue's structure in terms of the number of spaces involved (interstitial and vascular) and how contrast agent is allowed to mix between those spaces. With increasing complexity, more parameters may be extracted from the model, such as the fractional sizes of the interstitial and vascular spaces and the rates of transfer between them. However, it is important to understand that such models may be inappropriate if the rate of transfer of contrast agent between the interstitial and vascular spaces is too fast or too slow. If the rate is so fast that two spaces act as a single space (flow limited), then it will not be possible to measure the size of the individual spaces or the flow rate between them. Conversely, if the flow rate is so slow that there is no measurable leakage into the interstitial space during the measurement (permeability limited), then those parameters will be similarly inaccessible. In such cases, the model is overparametrized and may generate incorrect estimates.

This issue can cause confusion with models that use the K^{trans} parameter (Sourbron and Buckley, 2012, 2011). K^{trans} is defined as the plasma flow rate multiplied by the extraction fraction (E), which is the proportion of tracer that is extracted into the capillary bed after one pass of the tracer through the tissue. In a flow-limited system, where E = 0, K^{trans} can be interpreted simply as the plasma flow. In a permeability-limited system, where E = 1, K^{trans} should be interpreted as the product of the capillary permeability and the surface area (PS). The evidence suggests that contrast agent transfer in the heart is typically in an intermediate state between these two limits, but care should be taken interpreting such models because transfer can become flow limited at lower flow rates (Broadbent *et al.*, 2013).

Deciding which quantitative perfusion method is the best for clinical use is not straightforward. Most methods have been compared with other measurements, such as animal models using microspheres as a reference standard (Christian *et al.*, 2004; Kraitchman *et al.*, 1996) or human perfusion using alternative methods (Fritz-Hansen *et al.*, 2008; Ritter *et al.*, 2006), but all such comparisons have their own sets of limitations. While microspheres provide a useful tool for validation against an absolute reference in preclinical studies their application is invasive, and there are no gold-standard methods for validating absolute myocardial blood flow quantification suitable for clinical research. Another approach is to validate measurements in terms of their ability to diagnose disease (Biglands *et al.*, 2015; Hsu *et al.*, 2018; Mordini *et al.*, 2014). In the context of ischemic heart disease, these comparisons have shown that MBF values are useful in diagnosis but they have not provided persuasive evidence to favor one method above others.

15.3.7 Water exchange

In healthy tissue, the extravascular, extracellular cellular contrast agents used in DCE-MRI should reach the interstitial space (between the cells) but not leak into the cells themselves. However, contrast agents increase image signal intensity indirectly, by shortening the T1 (and T2) of water within the tissue. Although contrast agents cannot leak inside the cell, water—whose T1 has been modified by those contrast agents—can. The extent to which this may be a problem for making measurements from DCE-MRI data depends on the cellular–interstitial water exchange rate.

Analogous to the previous argument regarding the vascular–interstitial transfer rate, if the cellular–interstitial rate is very fast then the interstitial and cellular spaces will act as one well-mixed compartment and water exchange will not affect experiments. If it is very slow then the signal response becomes biexponential (corresponding to the two signal decay rates within and outside of the cells), which can be accounted for easily. However, if there is an intermediate exchange rate then the simple two-compartment models described before may be inadequate and yield erroneous measurements. This water exchange effect is described more fully elsewhere in this book, refer to Chapter 10 for further detail.

Models that take into account the effect of water exchange have been developed (Buckley, 2018; Landis *et al.*, 1999). In the context of extracellular volume made via

T1 maps, these models have even been used to measure the residence time of water within the cells of mice and humans (Coelho-Filho, Mongeon, et al., 2013a, Coelho-Filho, Shah, et al., 2013b; Goldfarb and Zhao, 2016). However, in order to obtain these measurements, much higher doses than would be used clinically were employed (0.15 to 2 mmol/kg, 3 to 4 times the normal dose). A normal clinical dose of 0.05 mmol/kg is much less sensitive to water-exchange effects. Experiments at this more realistic dosage have shown DCE-MRI data to be water exchange insensitive and it is not routinely taken into account in quantitative myocardial perfusion (Buckley et al., 2008; Lundin et al., 2016).

15.4 Applications

The interpretation of quantitative perfusion results can depend on the disease of interest. When systemic perfusion changes across the whole myocardium are expected, such as in some cardiomyopathies, e.g., dilated cardiomyopathy or diabetic cardiomyopathy, generating a single perfusion value for the whole myocardium may be sensible because it maximizes the CNR in the data and produces a single number for the global myocardial perfusion.

However, if localized perfusion defects are expected (for instance in ischemic heart disease), then the question arises as to how best represent the perfusion values spatially. Historically this has been done on a region-based analysis. In the context of ischemic heart disease, the American Heart association (AHA) (Cerqueira et al., 2002) recommends a model for mapping the myocardial territories to coronary arteries. The three myocardial slices are subdivided radially into 6 segments (4 for the apical slice), and quantitative perfusion is performed for each segment, producing 16 perfusion values (17 if a fourth apex measurement is included). These are conventionally represented in a bull's-eye plot that allows cardiologists to visualize the quantitative perfusion data (see Fig. 15.7). The model has been used widely as a way of maximizing the contrast-to-noise ratio of perfusion curves, while maintaining spatial information that can be related to localized perfusion defects. However, such a direct mapping of myocardial region to individual coronary artery is oversimplistic and disregards one of the key advantages MRI offers to perfusion imaging: much higher in-plane spatial resolution than other imaging modalities.

More recent perfusion analyses have therefore generated perfusion values for each individual voxel in the map (Hsu et al., 2012; Kellman et al., 2017; Xue et al., 2019; Zarinabad et al., 2012). Such methods rely on high-quality image registration and model fitting which must be robust against noise in the enhancement curves. However, if perfusion maps can be generated reliably, they are preferable because they maintain the spatial resolution of the original images and negate the need for information reduction schemes such as the AHA mapping.

A further consideration in the reporting and interpretation of quantitative perfusion results is the units used. Deconvolution directly produces results in units of blood flow per unit volume of myocardium (e.g., MBF in ml/min/ml). However, it is also common to convert this to blood flow per unit mass (ml/min/g) using an assumed myocardial density (typically of around 1.05 g/ml).

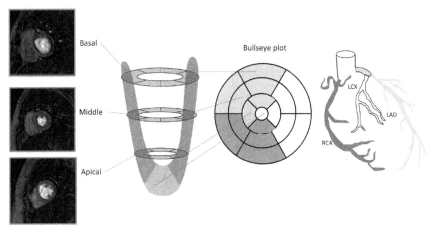

FIG. 15.7

The American Heart Association model recommends three axial slices through the myocardium. The myocardium is subdivided radially into six segments (four in the apical slice), and perfusion is assessed in each segment. These segments are associated with different coronary arteries, although it is understood that such mapping is oversimplistic and coronary territories will differ between individuals.

15.4.1 Stress perfusion MRI

The coronary arterioles act as the principal resistance vessels and modulate coronary blood flow. Coronary autoregulation describes the capacity of the heart to maintain steady myocardial perfusion across a range of perfusion pressures. This mechanism protects the heart from myocardial ischemia at rest. For this reason, it is necessary to perform perfusion measurements under stress conditions to reveal myocardial ischemia. With investigations such as MRI, where exercise-induced stress is difficult due to the constraints of the imaging apparatus, stress is typically induced pharmacologically using an infusion of a vasodilator like adenosine. Alternatively, a positive inotropic agent such as dobutamine may be used, but the associated increase in heart rate and the risk of inducing significant arrhythmia make these stressors a rare choice in clinical myocardial perfusion MRI.

Stress testing is an essential part of the diagnostic and prognostic assessment of cardiovascular disease featuring in various international practice guidelines. The recent 2021 AHA chest pain guidelines make several Class 1 recommendations for stress perfusion MRI in the assessment of acute and stable chest pain (Gulati *et al.*, 2021). As well as a diagnostic and prognostic aid for patients with known or suspected CAD, stress perfusion MRI has the potential to improve patient outcomes and reduce healthcare costs. The role of stress perfusion MRI extends to other forms of cardiac disease (e.g., cardiomyopathies), where it may provide new pathophysiological insights. These different indications will be discussed in more detail later.

Diagnostic studies using quantitative MBF or semiquantitative perfusion estimates often express their results in terms of the myocardial perfusion reserve (MPR), the ratio of the stress and resting myocardial blood flows (Eq. 15.4).

$$MPR = \frac{stress\ MBF}{rest\ MBF} \qquad (15.4)$$

Assuming that the patient is in maximal vasodilation when the stress measurement is taken, the ratio of the stress and rest measurements can be taken as a measure of the ability of the system to maintain flow in the face of a change in myocardial demand, i.e., its reserve of possible flow increase before maximal vasodilation occurs.

When available, rest and stress myocardial perfusion data are analyzed separately to derive both rest and stress MBF. The MPR can be displayed as a further polar plot. Further outputs from the analysis may include the AIF, myocardial signal intensity profiles, and other data that can be used for quality assurance.

A reduction in stress MBF or MPR may imply pathology such as CAD or CMD but can also be caused by inadequate stress. Verification of adequate stress is typically achieved by reviewing the patient's symptoms (flushing, breathlessness, and chest tightness for vasodilators), heart rate response (rise of ≥ 10 bpm) during the study, and splenic switch-off on the acquired stress images (Manisty *et al.*, 2015). Systolic blood pressure (fall of >10 mmHg) is often listed as an indicator of adequate stress, but in practice is inconsistently observed and should not be used as a reliable marker. It is important to note that the splenic switch-off sign only occurs with adenosine and is not present when using other stress agents such as regadenoson. However, these signs should not be used in isolation, and certain patient groups, particularly heart failure patients, may have a blunted hemodynamic response.

15.4.1.1 Clinical evaluation of quantitative myocardial perfusion MRI

Cardiac positron emission tomography (PET) is currently considered to be the clinical reference method for quantitative assessment of MBF, while microspheres serve as the experimental standard (Herzog *et al.*, 2009; Shah *et al.*, 2013). Quantification of MBF by MRI has been validated against microspheres, PET, and fractional flow reserve in a few small studies (Jerosch-Herold *et al.*, 2002; Kellman *et al.*, 2017; Lockie *et al.*, 2011; Wilke *et al.*, 1997). A recent study comparing 13N-NH3 PET and MRI has shown good correlation between myocardial perfusion quantified by PET and MRI and suggested myocardial perfusion MRI as a useful alternative in clinical practice (Engblom *et al.*, 2017).

The three main clinical indications for quantitative myocardial perfusion MRI include CAD, CMD, and cardiomyopathy. These will be discussed in further detail later.

15.4.1.2 Coronary artery disease

Myocardial stress perfusion MRI is well established as an indispensable tool in the assessment of patients with CAD. In the recent 2021 AHA chest pain guidelines, myocardial stress perfusion MRI has a class I indication for intermediate risk patients without known CAD, and a 2a recommendation for sequential testing after an inconclusive coronary computed tomography angiographic study (Gulati *et al.*, 2021). Myocardial stress perfusion MRI is also recommended for the diagnosis of ischemia in patients with no obstructive CAD or for coronary microvascular dysfunction (CMD). The 2016 European Society Cardiology (ESC) heart failure guidelines suggest the use of myocardial stress perfusion MRI for the assessment of myocardial ischemia and viability to guide revascularization decisions in patients with coronary artery disease and heart failure (Ponikowski *et al.*, 2016).

Although the important role of myocardial stress perfusion MRI in the assessment of patients with CAD is well established, quantitative analysis has several advantages over visual read in the assessment of patients with CAD.

First, visual, or qualitative assessment, is less accurate compared to quantitative assessment. This was confirmed in a study where visual assessment of first-pass perfusion was found to underestimate ischemic burden in multivessel CAD (Kotecha *et al.*, 2019), where quantitation was better at identifying three-vessel disease (87% vs 40%) and two-vessel disease (71% vs 48%). In another study of 67 patients undergoing ischemia assessment and invasive coronary angiography, the diagnostic accuracy of quantitative assessment was found to be higher than that of semiquantitative (92% vs 75–82%) (Mordini *et al.*, 2014). Visual read may, therefore, underestimate ischemic burden, as well as potentially missing cases of balanced triple vessel disease.

Second, the ability of quantitative DCE-MRI to accurately quantify ischemic burden may allow clinicians to assess treatment efficacy by serial imaging. In a study looking at patients with total coronary artery occlusion (CTO), MPR in the CTO regions improved significantly after revascularization (Bucciarelli-Ducci *et al.*, 2016). In a randomized myocardial stress perfusion MRI study comparing percutaneous coronary imaging (PCI) with CABG in multivessel CAD, there was a greater improvement in MPR when revascularization was undertaken using PCI, compared to CABG (Arnold *et al.*, 2013).

Third, quantitative DCE-MRI may offer further prognostic information, over visual assessment. In a study of 395 patients with suspected CAD, quantitative ischemic burden performed better than visual assessment and quantitative analysis provided incremental prognostic value over visual read (Sammut *et al.*, 2018). In the largest quantitative myocardial DCE-MRI study to date involving 1049 patients with known or suspected CAD, reduced MBF and MPR measured automatically using in-line artificial intelligence generated perfusion mapping, which was a strong, independent prognostic marker of outcomes (Knott *et al.*, 2020). Furthermore, of importance, in patients without regional perfusion defects on visual read, both stress MBF and MPR remained independently associated with MACE, with MPR being independently associated with death and MACE. This demonstrates the additional utility of perfusion quantification.

Fourth, an optimal stress MBF threshold of $<1.94\,ml/min/g$ achieved excellent accuracy (AUC 0.90) and was shown to correlate well with FFR positive lesions on invasive coronary angiography, with FFR positive lesions showing significantly lower stress MBF and MPR compared to FFR negative lesions (Kotecha *et al.*, 2019). In some patients, this may absolve the need for invasive coronary angiography, which is associated with complications. This was similar to an earlier study of patients with suspected CAD undergoing invasive coronary angiography with FFR, where an MPR (corrected for rate-pressure product, (RPP)) cut off of 2.04 achieved 93% sensitivity and 57% specificity for identifying significant CAD

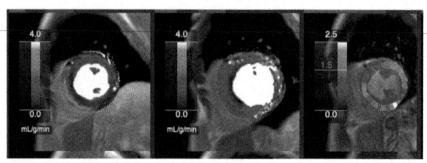

FIG. 15.8

An example of a stress myocardial perfusion map in a patient with an LAD infarct, who has significant peri-infarct ischemia. Color pixels correspond to MBF in units per ml/g/min. Endocardial, epicardial borders and RV insertion points were automatically detected, using CVI42 (Circle Cardiovascular Imaging, Calgary, Canada).

(Costa *et al.*, 2007). It is important to note that these studies have all been relatively small and used a variety of acquisition methods and models to derive thresholds.

Additionally, the high spatial resolution of MRI allows the assessment of transmural differences in MBF, which improves sensitivity of detecting CAD. In healthy patients, there are higher flows in the subendocardium, compared to the subepicardium. However, this is reversed at stress and in patients with CAD or CMD, where perfusion to the subendocardium is reduced (Fairbairn *et al.*, 2014). This transmural perfusion gradient may be increased by functionally significant epicardial stenosis.

MRI is often not used in the assessment of acute coronary syndromes. However, one study that assessed 64 STEMI patients acutely and 6 months after PCI found that acute MBF was found to predict both functional recovery and infarct size reduction (Borlotti *et al.*, 2019).

Fig. 15.8 shows an example of a quantitative myocardial stress perfusion map of a patient who has an LAD infarct with significant peri-infarct ischemia.

15.4.2 Coronary microvascular disease

CMD affects a large proportion of patients with angina and nonobstructive CAD. It is defined by impaired flow augmentation in response to pharmacological vasodilatation in the presence of nonobstructive CAD (Rahman *et al.*, 2019). Recently, it has been shown that CMD heralds an increased risk of major adverse cardiovascular outcomes (MACE) (Taqueti *et al.*, 2017). A number of noninvasive techniques have been proposed for diagnosing CMD. Myocardial perfusion quantification has an important role to play in the diagnosis and evaluation of these patients. One of the main challenges when there is a global reduction in stress MBF and MPR is the differentiation between three-vessel disease and CMD. In a study of 75 patients with angina and nonobstructive CAD, perfusion quantification outperformed visual assessment for the detection of CMD (AUC 0.88 vs 0.60, respectively) (Rahman *et al.*, 2021).

Interestingly, MPR significantly outperformed stress MBF (AUC 0.64), illustrating the potential utility of acquiring rest perfusion. In another study looking at 50 patients with stable angina who underwent quantitative myocardial DCE-MRI and invasive coronary physiology testing, in patients without a regional perfusion defect, a threshold of 1.82 ml/min discriminated between three-vessel disease and CMD (AUC 0.94, $P < 0.001$), with a lower MBF value in patients with three-vessel disease (Kotecha et al., 2019). In another study of 85 patients with chest pain and nonobstructive epicardial coronary disease, investigators invasively measured CFR and microvascular resistance with a Doppler wire at rest, during cycling and after pharmacological hyperemia with intravenous adenosine and compared these results to quantitative myocardial DCE-MRI (Rahman et al., 2021). In the 55 patients who completed the entire protocol, CMD (defined as a CFR <2.5) was diagnosed in 69% of patients. Around 80% of patients with low coronary flow reserve (CFR) had inducible ischemia defined as an endocardial-to-epicardial perfusion ratio of <1.0 during myocardial stress perfusion MRI. The 45 patients with CMD had a lower MPR (2.01 ± 0.11 vs 2.68 ± 0.49), driven by higher resting MBF (1.37 ± 0.37 ml/min/g vs 1.13 ± 0.20 ml/min/g) and a lower endocardial-to-epicardial perfusion ratio (0.93 ± 0.08 vs 1.05 ± 0.11).

15.4.2.1 Cardiomyopathies

Quantitative myocardial DCE-MRI has several emerging roles in the assessment of cardiomyopathy, namely in the assessment and management of hypertrophic cardiomyopathy (HCM), dilated cardiomyopathy (DCM), diabetic cardiomyopathy, and others.

Perfusion defects have long been appreciated in HCM patients and can be independent of epicardial coronary disease, conventional risk factors, or patient symptoms (Maron et al., 2009). In patients with hypertrophic cardiomyopathy, myocardial perfusion and perfusion reserve have shown prognostic value, and these parameters may become an integral part of clinical risk stratification (Cecchi et al., 2003). Quantitative perfusion using PET was investigated in 23 patients with HCM, where it was found that HCM patients had impaired perfusion reserve compared to a comparator cohort, even in nonhypertrophied myocardial segments (Camici et al., 1991). Petersen et al. undertook a quantitative myocardial DCE-MRI study in 35 patients with HCM and found that MBF was reduced in HCM, particularly in the subendocardium and in proportion to the degree of hypertrophy (Petersen et al., 2007). In the largest quantitative myocardial DCE-MRI study to date in HCM, it has been suggested that coronary microvascular dysfunction may occur early in phenotype development of HCM, with abnormal perfusion even in remote myocardium (Camaioni et al., 2020). It was shown that 78% of HCM patients had perfusion defects on clinical read, with low global MBF throughout the population, lowest in the most hypertrophied and fibrotic segments. Interestingly, in over one-fifth of patients, stress MBF fell below rest MBF values in at least one myocardial segment giving an MPR <1. The authors gave several explanations for this, including microscopic steal, altered myocardial vasomotor response to adenosine in HCM, and altered myocardial mechanics causing perfusion defects secondary to vasodilatation-induced tachycardia. A recent abstract presented showed that

100% of patients with apical HCM had perfusion defects. They showed that apical microvascular ischemia was the hallmark feature of apical HCM, occurring even in cases where apical hypertrophy does not reach conventional diagnostic criteria (Hughes *et al.*, 2021).

Sudden cardiac death (SCD) is a devastating complication of HCM and may occur as the initial disease presentation, frequently in asymptomatic or mildly symptomatic young people (Cecchi *et al.*, 2003). Perfusion quantification may help in the potential use of ischemia as a testable risk factor for sudden cardiac death (SCD) and as a potential prognostic and therapeutic target in HCM patients.

A recent quantitative myocardial DCE-MRI study has shown that patients with DCM exhibit CMD, the severity of which is associated with the degree of LV impairment (Gulati *et al.*, 2019). In DCM, DCE-MRI demonstrated reduced stress MBF (3.07 ± 1.02 ml/min/g vs 3.53 ± 0.79 ml/min/g, $n = 65$), with greater impairment in patients with more severe LV systolic dysfunction.

Quantitative myocardial DCE-MRI has also had an emerging role in the assessment of patients with diabetes mellitus. Diabetic cardiomyopathy is defined by the existence of abnormal myocardial structure and performance in the absence of other cardiac risk factors, such as CAD, hypertension, and significant valvular disease, in individuals with diabetes mellitus.

Using both MRI (Larghat *et al.*, 2014) and PET (Murthy *et al.*, 2012), myocardial perfusion during vasodilator stress has been shown to be impaired in patients with diabetes in the absence of epicardial CAD. This is important since CMD is associated with a 2.5% annual major adverse event rate, including cardiovascular mortality, nonfatal MI, nonfatal stroke, and congestive HF (Rubinshtein *et al.*, 2010). Larghat *et al.* assessed MPR in 65 patients with no significant coronary artery stenosis on angiography and showed that MPR was lower in diabetics than nondiabetics (2.10 ± 0.76 vs 2.84 ± 1.25, respectively, $P = 0.01$) (Larghat *et al.*, 2014).

Furthermore, in a study of 48 cardiac transplant recipients followed up for 7.1 years, quantitative myocardial DCE-MRI outperformed invasive coronary angiography for the detection of allograft vasculopathy (Miller *et al.*, 2014).

15.4.3 Integration of quantitative perfusion into clinical reporting

There are no current guidelines on how visual and quantitative perfusion data should be combined in clinical practice. In many institutions, visual analysis remains the primary analysis, which is followed by review of the quantitative perfusion maps for confirmation of the findings as well as for the detection of global perfusion abnormalities that may go undetected by visual interpretation alone. However, in experienced centers, primary review of the quantitative perfusion maps is already becoming the primary or sole analysis method. Lack of evidence for these strategies against the wealth of evidence for visual read may be a hindrance for this approach until larger studies have been published, but this approach may be particularly useful for less experienced sites. Fig. 15.9 suggests how quantitative myocardial DCE-MRI may be integrated into clinical assessment.

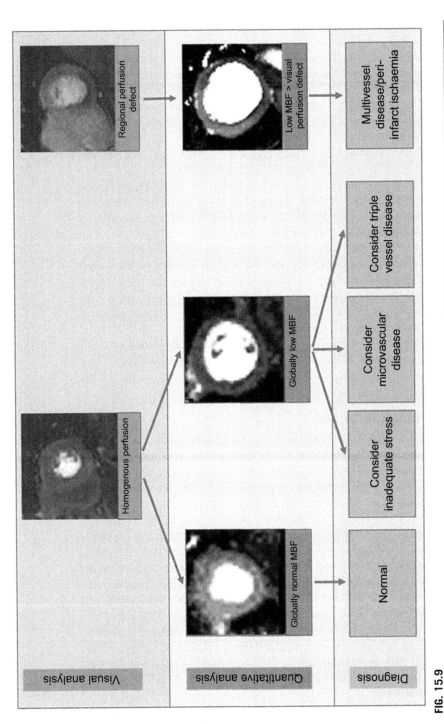

FIG. 15.9

Potential scenarios when quantitative perfusion analysis may be integrated into clinical practice. In a patient with visually homogenous perfusion, normal stress MBF/MPR by quantitative myocardial DCE-MRI can reaffirm the diagnosis of a normal perfusion study. Conversely, visually homogenous perfusion with low-stress MBF/MPR suggests either inadequate vasodilatory response (check splenic switch-off and hemodynamic response), coronary microvascular disease (which may show an endo to epicardial perfusion gradient), or severe triple vessel disease (where the pattern of perfusion is typically heterogeneous). In a patient with a regional perfusion defect on visual analysis, quantitative myocardial DCE-MRI may help identify the extent of disease (visual analysis is relative and may underestimate disease extent) and peri-infarct ischemia. MBF, myocardial blood flow.

Figure reproduced from Sharrack, N., Chiribiri, A., Schwitter, J., Plein, S., 2022. How to do quantitative myocardial perfusion cardiovascular magnetic resonance. Eur. Heart J. Cardiovasc. Imaging 23, 315–318. doi:10.1093/EHJCI/JEAB193.

Quantitative myocardial perfusion maps may supplement visual interpretation in several ways:

Confirmation of visual read: Successive or simultaneous visual and quantitative analysis may enhance diagnostic certainty where the two strategies agree and, when results are discrepant, may alert the reader to the presence of artifacts.

Adequacy of hemodynamic response: When inadequate hemodynamic response is suspected based on a lack of clinical response or absence of splenic switch-off, review of quantitative myocardial perfusion maps can help confirm (low-stress MBF and/or MPR) or refute (high-stress MBF and/or MPR) this suspicion.

Suspected CMD: Visual interpretation of myocardial perfusion CMR has limited ability to detect CMD. In patients with no regional visual perfusion defects and adequate hemodynamic response, but for low-stress MBF or MPR on quantitative myocardial perfusion mapping, CMD is a likely diagnosis.

Disease extent: Visual read of myocardial perfusion CMR compares signal changes between different myocardial regions and is, thus, adjusted for the lowest perfused area in an image, potentially masking less severe defects elsewhere. Quantitative analysis provides objective absolute blood flow values for each region. This may be advantageous in multivessel CAD, where quantitative myocardial perfusion CMR may better identify disease extent than visual read.

Follow-up studies and research: Quantitative myocardial perfusion CMR provides absolute numbers of MBF, which can help assess treatment effects.

15.5 Safety considerations

Myocardial perfusion MRI uses gadolinium-based contrast agents, and the same safety considerations should apply to the use of these agents as in scanning other parts of the body. A specific safety consideration related to myocardial perfusion MRI relates to the use of pharmacological stress agents. The most commonly used agents are the vasodilators adenosine, regadenoson, dipyridamole, or adenosine triphosphate (ATP). These agents have an excellent safety profile, and severe side effects are rarely encountered (Bruder *et al.*, 2009). Adenosine, ATP, and regadenoson commonly cause symptoms such as flushing, chest pain, palpitations, and breathlessness. More severe, but rare side effects include transient heart block, transient hypotension, or bronchospasm. It is important to carefully screen patients who may be at an increased risk of these complications such as those with preexisting AV block or severe asthma, prior to undertaking a stress perfusion MRI scan. Continuous monitoring of symptoms, heart rate, and blood pressure during the administration of the stress agent is also recommended and the relevant safety equipment and medications to deal with any complications should be easily available in the MRI department. It is also recommended that emergency responses should be regularly rehearsed by the MRI team.

15.6 **Challenges and solutions**

Although myocardial stress perfusion MRI has formed the basis of several important research studies such as MR-INFORM (Nagel *et al.*, 2019) and CE-MARC (Greenwood *et al.*, 2016), and the prognostic significance of perfusion quantification has been demonstrated in a large study (Knott *et al.*, 2020), large multicenter, randomized trials using quantitative myocardial perfusion are lacking. Further prospective, multicenter randomized trials using quantitative myocardial DCE-MRI to noninvasively guide revascularization decisions could provide the evidence needed to change clinical guidelines.

Registration, segmentation, and other errors may lead to erroneous MBF maps and automated and manual quality checks should be a routine part of the interpretation of quantitative myocardial perfusion imaging. Dark rim artifacts affect the diagnostic accuracy of myocardial stress perfusion MRI and can mimic subendocardial perfusion defects leading to false positive diagnosis of CAD. When using quantitative myocardial DCE-MRI, MBF is generally lower in true perfusion defects compared to dark rim artifacts but may remain a source of error.

Furthermore, as described in the preceding sections, cardiac DCE-MRI is typically performed using multi-slice acquisitions with a limited number of slices (often 3) across the left ventricle. Such acquisitions do not, therefore, provide whole heart (or even whole ventricle) coverage. Exploitation of advanced acceleration methods and of improved intrinsic signal-to-noise ratios of modern hardware may allow these limitations to be resolved. Techniques being explored to allow this include use of simultaneous multi-slice (SMS) acquisitions and compressed sensing to allow more complete coverage with 2D acquisitions (McElroy *et al.*, 2022) or of non-Cartesian sampling strategies to allow 3D acquisitions (Mendes *et al.*, 2020).

Current limitations include the lack of consensus on optimal acquisition and analysis techniques, which ideally should yield MBF/MPR estimates consistent with and interchangeable across studies and preferably also with PET. There is also no intervendor standardization. This remains an important hindrance for the implementation of quantitative myocardial DCE-MRI into routine clinical practice. Furthermore, without normal MBF and MPR reference ranges, one cannot confidently diagnose pathology such as CMD or follow up patients between centers or scanners.

Finally, improvements in the availability of facilities that perform stress CMR, and the continued training of cardiologists, radiographers, and radiologists with advanced experience in cardiac MRI, would facilitate translation of quantitative stress myocardial perfusion imaging from the research domain into routine clinical practice.

Key future objectives include better standardization of acquisition protocols, model selection and reporting, and the development of robust thresholds for differentiation of health and disease—objectives which are best achieved through multicenter, collaborative research.

15.7 **Learning and knowledge outcomes**

1. ECG triggered saturation recovery sequences are commonly used in cardiac DCE, to produce consistent image contrast despite the motion of the heart and the dependence of pulse sequence timings on heart rate.
2. Due to the large difference in peak contrast agent concentrations in the blood pool and the myocardium, further measures (such as the use of dual-sequence imaging) are required to allow accurate perfusion quantification.
3. Myocardial blood flow measurements require an arterial input function to be measured, typically in the left ventricular blood pool.
4. A range of tracer kinetic models can be used to measure myocardial blood flow values from dynamic contrast-enhanced cardiac MR images.
5. To avoid incorrect diagnosis and management, it is important to always check your patient has been adequately stressed when performing stress perfusion MRI.
6. Quantitative myocardial DCE-MRI can help provide important clinical information, above and beyond that derived from stress perfusion MRI, particularly in the detailed assessment of CAD, CMD, and cardiomyopathies.
7. Current limitations of the implementation of quantitative myocardial DCE-MRI into routine clinical practice include the lack of consensus on optimal acquisition and analysis techniques and lack of intervendor standardization for normal MBF and MPR thresholds.

Declarations

The authors have nothing to declare.

References

Al-Saadi, N., Nagel, E., Gross, M., Bornstedt, A., Schnackenburg, B., Klein, C., Klimek, W., Oswald, H., Fleck, E., 2000. Noninvasive detection of myocardial ischemia from perfusion reserve based on cardiovascular magnetic resonance. Circulation 101, 1379–1383.

Arnold, J.R., Karamitsos, T.D., van Gaal, W.J., Testa, L., Francis, J.M., Bhamra-Ariza, P., Ali, A., Selvanayagam, J.B., Westaby, S., Sayeed, R., Jerosch-Herold, M., Neubauer, S., Banning, A.P., 2013. Residual ischemia after revascularization in multivessel coronary artery disease: insights from measurement of absolute myocardial blood flow using magnetic resonance imaging compared with angiographic assessment. Circ. Cardiovasc. Interv. 6, 237–245. https://doi.org/10.1161/CIRCINTERVENTIONS.112.000064.

Biglands, J., Magee, D., Boyle, R., Larghat, A., Plein, S., Radjenovic, A., 2011. Evaluation of the effect of myocardial segmentation errors on myocardial blood flow estimates from DCE-MRI. Phys. Med. Biol. 56, 2423–2443. https://doi.org/10.1088/0031-9155/56/8/007.

Biglands, J., Magee, D., Sourbron, S., Plein, S., Greenwood, J., Radjenovic, A., 2015. A comparison of the diagnostic performance of four quantitative myocardial perfusion estimation methods used in cardiac magnetic resonance imaging: a CE-MARC sub-study. Radiology 275, 393–402.

Borlotti, A., Jerosch-Herold, M., Liu, D., Viliani, D., Bracco, A., Alkhalil, M., de Maria, G.L., Ox, A.M.I.S.I, Channon, K.M., Banning, A.P., Choudhury, R.P., Neubauer, S., Kharbanda, R.K., Dall'Armellina, E., 2019. Acute microvascular impairment post-reperfused STEMI is reversible and has additional clinical predictive value: a CMR OxAMI study. JACC Cardiovasc. Imaging 12, 1783–1793. https://doi.org/10.1016/j.jcmg.2018.10.028.

Broadbent, D.A., Biglands, J.D., Larghat, A., Sourbron, S.P., Radjenovic, A., Greenwood, J.P., Plein, S., Buckley, D.L., 2013. Myocardial blood flow at rest and stress measured with dynamic contrast-enhanced MRI: comparison of a distributed parameter model with a fermi function model. Magn. Reson. Med. 70, 1591–1597. https://doi.org/10.1002/MRM.24611.

Broadbent, D.A., Biglands, J.D., Ripley, D.P., Higgins, D.M., Greenwood, J.P., Plein, S., Buckley, D.L., 2016. Sensitivity of quantitative myocardial dynamic contrast-enhanced MRI to saturation pulse efficiency, noise and t1 measurement error: comparison of non-linearity correction methods. Magn. Reson. Med. 75, 1290–1300. https://doi.org/10.1002/MRM.25726.

Bruder, O., Schneider, S., Nothnagel, D., Dill, T., Hombach, V., Schulz-Menger, Eike Nagel, J., Lombardi, M., van Rossum, A.C., Wagner, A., Schwitter, J., Senges, J., Sabin, G.V., Sechtem, U., Heiko Mahrholdt, H., 2009. EuroCMR (European Cardiovascular Magnetic Resonance) registry: results of the German pilot phase. J. Am. Coll. Cardiol. 54, 1457–1466. https://doi.org/10.1016/j.jacc.2009.07.003.

Bucciarelli-Ducci, C., Auger, D., di Mario, C., Locca, D., Petryka, J., O'Hanlon, R., Grasso, A., Wright, C., Symmonds, K., Wage, R., Asimacopoulos, E., del Furia, F., Lyne, J.C., Gatehouse, P.D., Fox, K.M., Pennell, D.J., 2016. CMR guidance for recanalization of coronary chronic total occlusion. JACC Cardiovasc. Imaging 9, 547–556. https://doi.org/10.1016/j.jcmg.2015.10.025.

Buckley, D.L., 2018. Shutter-speed dynamic contrast-enhanced MRI: is it fit for purpose? Magn. Reson. Med., 1–13. https://doi.org/10.1002/mrm.27456.

Buckley, D.L., Kershaw, L.E., Stanisz, G.J., 2008. Cellular-interstitial water exchange and its effect on the determination of contrast agent concentration in vivo: dynamic contrast-enhanced MRI of human internal obturator muscle. Magn. Reson. Med. 60, 1011–1019.

Camaioni, C., Knott, K.D., Augusto, J.B., Seraphim, A., Rosmini, S., Ricci, F., Boubertakh, R., Xue, H., Hughes, R., Captur, G., Lopes, L.R., Brown, L.A.E., Manisty, C., Petersen, S.E., Plein, S., Kellman, P., Mohiddin, S.A., Moon, J.C., 2020. Inline perfusion mapping provides insights into the disease mechanism in hypertrophic cardiomyopathy. Heart 106, 824–829. https://doi.org/10.1136/heartjnl-2019-315848.

Camici, P., Chiriatti, G., Lorenzoni, R., Bellina, R.C., Gistri, R., Italiani, G., Parodi, O., Salvadori, P.A., Nista, N., Papi, L., et al., 1991. Coronary vasodilation is impaired in both hypertrophied and nonhypertrophied myocardium of patients with hypertrophic cardiomyopathy: a study with nitrogen-13 ammonia and positron emission tomography. J. Am. Coll. Cardiol. 17, 879–886. https://doi.org/10.1016/0735-1097(91)90869-b.

Cecchi, F., Olivotto, I., Gistri, R., Lorenzoni, R., Chiriatti, G., Camici, P.G., 2003. Coronary microvascular dysfunction and prognosis in hypertrophic cardiomyopathy. N. Engl. J. Med. 349, 1027–1035. https://doi.org/10.1056/NEJMoa025050.

Cernicanu, A., Axel, L., 2006. Theory-based signal calibration with single-point T1 measurements for first-pass perfusion quantitative perfusion MRI studies. Acad. Radiol. 13, 686–693. https://doi.org/10.1016/j.acra.2006.02.040.

Cerqueira, M.D., Weissman, N.J., Dilsizian, V., Jacobs, A.K., Kaul, S., Laskey, W.K., Pennell, D.J., Rumberger, J.A., Ryan, T., Verani, M.S., 2002. Standardized myocardial segmentation and nomenclature for tomographic imaging of the heart. Circulation 105, 539–542. https://doi.org/10.1161/hc0402.102975.

Cheong, L.H., Koh, T.S., Hou, Z., 2003. An automatic approach for estimating bolus arrival time in dynamic contrast MRI using piecewise continuous regression models. Phys. Med. Biol. 48, N83–N88. https://doi.org/PiiS0031-9155(03)52601-X https://doi.org/10.1088/0031-9155/48/5/403.

Christian, T.F., Rettmann, D.W., Aletras, A.H., Liao, S.L., Taylor, J.L., Balaban, R.S., Arai, A.E., 2004. Absolute myocardial perfusion in canines measured by using dual-bolus first-pass MR imaging. Radiology 232, 677–684. https://doi.org/10.1148/radiol.2323030573.

Coelho-Filho, O.R., Mongeon, F.P., Mitchell, R., Moreno, H., Nadruz, W., Kwong, R., Jerosch-Herold, M., 2013a. Role of transcytolemmal water-exchange in magnetic resonance measurements of diffuse myocardial fibrosis in hypertensive heart disease. Circ. Cardiovasc. Imaging 6, 134–141.

Coelho-Filho, O.R., Shah, R.V., Mitchell, R., Neilan, T.G., Moreno, H., Simonson, B., Kwong, R., Rosenzweig, A., Das, S., Jerosch-Herold, M., 2013b. Quantification of cardiomyocyte hypertrophy by cardiac magnetic resonance: implications for early cardiac remodeling. Circulation 128, 1225–1233.

Costa, M.A., Shoemaker, S., Futamatsu, H., Klassen, C., Angiolillo, D.J., Nguyen, M., Siuciak, A., Gilmore, P., Zenni, M.M., Guzman, L., Bass, T.A., Wilke, N., 2007. Quantitative magnetic resonance perfusion imaging detects anatomic and physiologic coronary artery disease as measured by coronary angiography and fractional flow reserve. J. Am. Coll. Cardiol. 50, 514–522. https://doi.org/10.1016/j.jacc.2007.04.053.

Engblom, H., Xue, H., Akil, S., Carlsson, M., Hindorf, C., Oddstig, J., Hedeer, F., Hansen, M.S., Aletras, A.H., Kellman, P., Arheden, H., 2017. Fully quantitative cardiovascular magnetic resonance myocardial perfusion ready for clinical use: a comparison between cardiovascular magnetic resonance imaging and positron emission tomography. J. Cardiovasc. Magn. Reson. 19, 78. https://doi.org/10.1186/s12968-017-0388-9.

Fairbairn, T.A., Motwani, M., Mather, A.N., Biglands, J.D., Larghat, A.M., Radjenovic, A., Greenwood, J.P., Plein, S., 2014. Cardiac MR imaging to measure myocardial blood flow response to the cold pressor test in healthy smokers and nonsmokers. Radiology 270. https://doi.org/10.1148/radiol.13122345.

Fritz-Hansen, T., Hove, J.D., Kofoed, K.F., Kelbaek, H., Larsson, H.B.W., 2008. Quantification of MRI measured myocardial perfusion reserve in healthy humans: a comparison with positron emission tomography. J. Magn. Reson. Imaging 27, 818–824. https://doi.org/10.1002/jmri.21306.

Gatehouse, P.D., Elkington, A.G., Ablitt, N.A., Yang, G.Z., Pennell, D.J., Firmin, D.N., 2004. Accurate assessment of the arterial input function during high-dose myocardial perfusion cardiovascular magnetic resonance. J. Magn. Reson. Imaging 20, 39–45. https://doi.org/10.1002/jmri.20054.

Goldfarb, J.W., Zhao, W., 2016. Effects of transcytolemmal water exchange on the assessment of myocardial extracellular volume with cardiovascular MRI. NMR Biomed. 29, 499–506.

Greenwood, J.P., Ripley, D.P., Berry, C., McCann, G.P., Plein, S., Bucciarelli-Ducci, C., Armellina, E.D., Prasad, A., Bijsterveld, P., Foley, J.R., Mangion, K., Sculpher, M., Walker, S., Everett, C.C., Cairns, D.A., Sharples, L.D., Brown, J.M., 2016. Effect of care guided by cardiovascular magnetic resonance, myocardial perfusion scintigraphy, or

NICE guidelines on subsequent unnecessary angiography rates: the CE-MARC 2 randomized clinical trial. JAMA 316, 1051–1060. https://doi.org/10.1001/JAMA.2016.12680.

Gulati, A., Ismail, T.F., Ali, A., Hsu, L.Y., Goncalves, C., Ismail, N.A., Krishnathasan, K., Davendralingam, N., Ferreira, P., Halliday, B.P., Jones, D.A., Wage, R., Newsome, S., Gatehouse, P., Firmin, D., Jabbour, A., Assomull, R.G., Mathur, A., Pennell, D.J., Arai, A.E., Prasad, S.K., 2019. Microvascular dysfunction in dilated cardiomyopathy: a quantitative stress perfusion cardiovascular magnetic resonance study. JACC Cardiovasc. Imaging 12, 1699–1708. https://doi.org/10.1016/j.jcmg.2018.10.032.

Gulati, M., Levy, P.D., Mukherjee, D., Amsterdam, E., Bhatt, D.L., Birtcher, K.K., Blankstein, R., Boyd, J., Bullock-Palmer, R.P., Conejo, T., Diercks, D.B., Gentile, F., Greenwood, J.P., Hess, E.P., Hollenberg, S.M., Jaber, W.A., Jneid, H., Joglar, J.A., Morrow, D.A., O'Connor, R.E., Ross, M.A., Shaw, L.J., 2021. 2021 AHA/ACC/ASE/CHEST/SAEM/SCCT/SCMR guideline for the evaluation and diagnosis of chest pain: Executive summary: a report of the American College of Cardiology/American Heart Association joint committee on clinical practice guidelines. Circulation 144, e368–e454. https://doi.org/10.1161/CIR.0000000000001030.

Heitner, J.F., Kim, R.J., Kim, H.W., Klem, I., Shah, D.J., Debs, D., Farzaneh-Far, A., Polsani, V., Kim, J., Weinsaft, J., Shenoy, C., Hughes, A., Cargile, P., Ho, J., Bonow, R.O., Jenista, E., Parker, M., Judd, R.M., 2019. Prognostic value of vasodilator stress cardiac magnetic resonance imaging: a multicenter study with 48 000 patient-years of follow-up. JAMA Cardiol. 4, 256–264. https://doi.org/10.1001/jamacardio.2019.0035.

Herzog, B.A., Husmann, L., Valenta, I., Gaemperli, O., Siegrist, P.T., Tay, F.M., Burkhard, N., Wyss, C.A., Kaufmann, P.A., 2009. Long-term prognostic value of 13N-Ammonia myocardial perfusion positron emission tomography. Added value of coronary flow reserve. J. Am. Coll. Cardiol. 54, 150–156. https://doi.org/10.1016/j.jacc.2009.02.069.

Hsu, L.-Y., Rhoads, K.L., Aletras, A.H., Arai, A.E., 2003. LNCS 2879—Surface coil intensity correction and non-linear intensity normalization improve pixel-resolution parametric maps of myocardial MRI perfusion. LNCS 2879, 975–976.

Hsu, L.Y., Rhoads, K.L., Holly, J.E., Kellman, P., Aletras, A.H., Arai, A.E., 2006. Quantitative myocardial perfusion analysis with a dual-bolus contrast-enhanced first-pass MRI technique in humans. J. Magn. Reson. Imaging 23, 315–322. https://doi.org/10.1002/jmri.20502.

Hsu, L.-Y., Groves, D.W., Aletras, A.H., Kellman, P., Arai, A.E., 2012. A quantitative pixel-wise measurement of myocardial blood flow by contrast-enhanced first-pass CMR perfusion imaging: microsphere validation in dogs and feasibility study in humans. JACC Cardiovasc. Imaging 5, 154–166. https://doi.org/10.1016/j.jcmg.2011.07.013.

Hsu, L.-Y., Jacobs, M., Benovoy, M., Ta, A.D., Conn, H.M., Winkler, S., Greve, A.M., Chen, M.Y., Shanbhag, S.M., Bandettini, W.P., Arai, A.E., 2018. Diagnostic performance of fully automated pixel-wise quantitative myocardial perfusion imaging by cardiovascular magnetic resonance. JACC Cardiovasc. Imaging, 1–11. https://doi.org/10.1016/j.jcmg.2018.01.005.

Huber, A., Sourbron, S., Klauss, V., Schaefer, J., Bauner, K.U., Schweyer, M., Reiser, M., Rummeny, E., Rieber, J., 2012. Magnetic resonance perfusion of the myocardium: semiquantitative and quantitative evaluation in comparison with coronary angiography and fractional flow reserve. Invest. Radiol. 47, 332–338.

Hughes, R.K., Augusto, J.B., Knott, K., Seraphim, A., Joy, G., Mohiddin, S., Captur, G., Lopes, L.R., Kellman, P., Moon, J.C., 2021. 20 Apical Ischaemia Is Ubiquitous in Apical Hypertrophic Cardiomyopathy and Occurs before Overt Hypertrophy.

Jacobs, M., Benovoy, M., Chang, L.C., Arai, A.E., Hsu, L.Y., 2016. Evaluation of an automated method for arterial input function detection for first-pass myocardial perfusion cardiovascular magnetic resonance. J. Cardiovasc. Magn. Reson. 18, 1–11. https://doi.org/10.1186/S12968-016-0239-0/TABLES/1.

Jerosch-Herold, M., Wilke, N., Stillman, A.E., 1998. Magnetic resonance quantification of the myocardial perfusion reserve with a Fermi function model for constrained deconvolution. Med. Phys. 25, 73–84.

Jerosch-Herold, M., Swingen, C., Seethamraju, R.T., 2002. Myocardial blood flow quantification with MRI by model-independent deconvolution. Med. Phys. 29, 886–897. https://doi.org/10.1118/1.1473135.

Kellman, P., Aletras, A.H., Hsu, L.Y., McVeigh, E.R., Arai, A.E., 2006. T2* measurement during first-pass contrast-enhanced cardiac perfusion imaging. Magn. Reson. Med. 56, 1132–1134. https://doi.org/10.1002/mrm.21061.

Kellman, P., Hansen, M.S., Nielles-Vallespin, S., Nickander, J., Themudo, R., Ugander, M., Xue, H., 2017. Myocardial perfusion cardiovascular magnetic resonance: optimized dual sequence and reconstruction for quantification. J. Cardiovasc. Magn. Reson. 19, 43. https://doi.org/10.1186/s12968-017-0355-5.

Kershaw, L.E., Buckley, D.L., 2006. Precision in measurements of perfusion and microvascular permeability with T1-weighted dynamic contrast-enhanced MRI. Magn. Reson. Med. 56, 986–992. https://doi.org/10.1002/mrm.21040.

Knott, K.D., Seraphim, A., Augusto, J.B., Xue, H., Chacko, L., Aung, N., Petersen, S.E., Cooper, J.A., Manisty, C., Bhuva, A.N., Kotecha, T., Bourantas, C.V., Davies, R.H., Brown, L.A.E., Plein, S., Fontana, M., Kellman, P., Moon, J.C., 2020. The prognostic significance of quantitative myocardial perfusion: an artificial intelligence-based approach using perfusion mapping. Circulation 141, 1282–1291. https://doi.org/10.1161/CIRCULATIONAHA.119.044666.

Kostler, H., Ritter, C., Lipp, M., Beer, M., Hahn, D., Sandstede, J., 2004. Prebolus quantitative MR heart perfusion imaging. Magn. Reson. Med. 52, 296–299. https://doi.org/10.1002/mrm.20160.

Kotecha, T., Martinez-Naharro, A., Boldrini, M., Knight, D., Hawkins, P., Kalra, S., Patel, D., Coghlan, G., Moon, J., Plein, S., Lockie, T., Rakhit, R., Patel, N., Xue, H., Kellman, P., Fontana, M., 2019. Automated pixel-wise quantitative myocardial perfusion mapping by CMR to detect obstructive coronary artery disease and coronary microvascular dysfunction: validation against invasive coronary physiology. JACC Cardiovasc. Imaging 12, 1958–1969. https://doi.org/10.1016/j.jcmg.2018.12.022.

Kraitchman, D.L., Wilke, N., Hexeberg, E., JeroschHerold, M., Wang, Y., Parrish, T.B., Chang, C.N., Zhang, Y., Bache, R.J., Axel, L., 1996. Myocardial perfusion and function in dogs with moderate coronary stenosis. Magn. Reson. Med. 35, 771–780.

Kramer, C.M., Barkhausen, J., Bucciarelli-Ducci, C., Flamm, S.D., Kim, R.J., Nagel, E., 2020. Standardized cardiovascular magnetic resonance imaging (CMR) protocols: 2020 update. J. Cardiovasc. Magn. Reson. 22, 1–18. https://doi.org/10.1186/s12968-020-00607-1.

Kremers, F.P., Hofman, M.B., Groothuis, J.G., Jerosch-Herold, M., Beek, A.M., Zuehlsdorff, S., Nielles-Vallespin, S., van Rossum, A.C., Heethaar, R.M., 2010. Improved correction of spatial inhomogeneities of surface coils in quantitative analysis of first-pass myocardial perfusion imaging. J. Magn. Reson. Imaging 31, 227–233. https://doi.org/10.1002/jmri.21998.

Kwong, R.Y., Ge, Y., Steel, K., Bingham, S., Abdullah, S., Fujikura, K., Wang, W., Pandya, A., Chen, Y.Y., Mikolich, J.R., Boland, S., Arai, A.E., Bandettini, W.P.,

Shanbhag, S.M., Patel, A.R., Narang, A., Farzaneh-Far, A., Romer, B., Heitner, J.F., Ho, J.Y., Singh, J., Shenoy, C., Hughes, A., Leung, S.W., Marji, M., Gonzalez, J.A., Mehta, S., Shah, D.J., Debs, D., Raman, S.V., Guha, A., Ferrari, V.A., Schulz-Menger, J., Hachamovitch, R., Stuber, M., Simonetti, O.P., 2019. Cardiac magnetic resonance stress perfusion imaging for evaluation of patients with chest pain. J. Am. Coll. Cardiol. 74, 1741–1755. https://doi.org/10.1016/j.jacc.2019.07.074.

Landis, C.S., Li, X., Telang, F.W., Molina, P.E., Palyka, I., Vetek, G., Springer, C.S., 1999. Equilibrium transcytolemmal water-exchange kinetics in skeletal muscle in vivo. Magn. Reson. Med. 42, 467–478.

Larghat, A.M., Swoboda, P.P., Biglands, J.D., Kearney, M.T., Greenwood, J.P., Plein, S., 2014. The microvascular effects of insulin resistance and diabetes on cardiac structure, function, and perfusion: a cardiovascular magnetic resonance study. Eur. Heart J. Cardiovasc. Imaging 15, 1368–1376.

Lockie, T., Ishida, M., Perera, D., Chiribiri, A., de Silva, K., Kozerke, S., Marber, M., Nagel, E., Rezavi, R., Redwood, S., Plein, S., 2011. High-resolution magnetic resonance myocardial perfusion imaging at 3.0-Tesla to detect hemodynamically significant coronary stenoses as determined by fractional flow reserve. J. Am. Coll. Cardiol. 57, 70–75. https://doi.org/10.1016/j.jacc.2010.09.019.

Lundin, M., Sörensson, P., Kellman, P., Sigfridsson, A., Ugander, M., 2016. Measurement of the intracellular lifetime of water to estimate myocardial cell size is not feasible in humans using clinical contrast agent doses at 1.5T. J. Cardiovasc. Magn. Reson. 18, 1–2. https://doi.org/10.1186/1532-429X-18-S1-P237.

Manisty, C., Ripley, D.P., Herrey, A.S., Captur, G., Wong, T.C., Petersen, S.E., Plein, S., Peebles, C., Schelbert, E.B., Greenwood, J.P., Moon, J.C., 2015. Splenic switch-off: a tool to assess stress adequacy in adenosine perfusion cardiac MR imaging. Radiology 276, 732–740. https://doi.org/10.1148/radiol.2015142059.

Maron, M.S., Olivotto, I., Maron, B.J., Prasad, S.K., Cecchi, F., Udelson, J.E., Camici, P.G., 2009. The case for myocardial ischemia in hypertrophic cardiomyopathy. J. Am. Coll. Cardiol. 54, 866–875. https://doi.org/10.1016/j.jacc.2009.04.072.

McElroy, S., Ferrazzi, G., Nazir, M.S., Evans, C., Ferreira, J., Bosio, F., Mughal, N., Kunze, K.P., Neji, R., Speier, P., Stab, D., Ismail, T.F., Masci, P.G., Villa, A.D.M., Razavi, R., Chiribiri, A., Roujol, S., 2022. Simultaneous multislice steady-state free precession myocardial perfusion with full left ventricular coverage and high resolution at 1.5 T. Magn. Reson. Med. https://doi.org/10.1002/mrm.29229.

Mendes, J.K., Adluru, G., Likhite, D., Fair, M.J., Gatehouse, P.D., Tian, Y., Pedgaonkar, A., Wilson, B., DiBella, E.V.R., 2020. Quantitative 3D myocardial perfusion with an efficient arterial input function. Magn. Reson. Med. 83, 1949–1963. https://doi.org/10.1002/mrm.28050.

Milidonis, X., Nazir, M.S., Chiribiri, A., 2022. Impact of temporal resolution and methods for correction on cardiac magnetic resonance perfusion quantification. J. Magn. Reson. Imaging. https://doi.org/10.1002/JMRI.28180.

Miller, C.A., Naish, J.H., Shaw, S.M., Yonan, N., Williams, S.G., Clark, D., Bishop, P.W., Ainslie, M.P., Borg, A., Coutts, G., Parker, G.J., Ray, S.G., Schmitt, M., 2014. Multiparametric cardiovascular magnetic resonance surveillance of acute cardiac allograft rejection and characterisation of transplantation-associated myocardial injury: a pilot study. J. Cardiovasc. Magn. Reson. 16, 52. https://doi.org/10.1186/s12968-014-0052-6.

Mordini, F.E., Haddad, T., Hsu, L.Y., Kellman, P., Lowrey, T.B., Aletras, A.H., Bandettini, W.P., Arai, A.E., 2014. Diagnostic accuracy of stress perfusion CMR in comparison with quantitative coronary angiography: fully quantitative, semiquantitative, and qualitative assessment. JACC Cardiovasc. Imaging 7, 14–22. https://doi.org/10.1016/j.jcmg.2013.08.014.

Murthy, V.L., Naya, M., Foster, C.R., Gaber, M., Hainer, J., Klein, J., Dorbala, S., Blankstein, R., di Carli, M.F., 2012. Association between coronary vascular dysfunction and cardiac mortality in patients with and without diabetes mellitus. Circulation 126, 1858–1868. https://doi.org/10.1161/CIRCULATIONAHA.112.120402.

Nagel, E., Klein, C., Paetsch, I., Hettwer, S., Schnackenburg, B., Wegscheider, K., Fleck, E., 2003. Magnetic resonance perfusion measurements for the noninvasive detection of coronary artery disease. Circulation 108, 432–437. https://doi.org/10.1161/01.CIR.0000080915.35024.A9.

Nagel, E., Greenwood, J.P., McCann, G.P., Bettencourt, N., Shah, A.M., Hussain, S.T., Perera, D., Plein, S., Bucciarelli-Ducci, C., Paul, M., Westwood, M.A., Marber, M., Richter, W.-S., Puntmann, V.O., Schwenke, C., Schulz-Menger, J., Das, R., Wong, J., Hausenloy, D.J., Steen, H., Berry, C., 2019. Magnetic resonance perfusion or fractional flow reserve in coronary disease. N. Engl. J. Med. 380, 2418–2428. https://doi.org/10.1056/nejmoa1716734.

Natsume, T., Ishida, M., Kitagawa, K., Nagata, M., Sakuma, H., Ichihara, T., 2015. Theoretical considerations in measurement of time discrepancies between input and myocardial time-signal intensity curves in estimates of regional myocardial perfusion with first-pass contrast-enhanced MRI. Magn. Reson. Imaging 33, 1059–1065. https://doi.org/10.1016/J.MRI.2015.06.015.

Nielles-Vallespin, S., Kellman, P., Hsu, L.-Y., Arai, A.E., 2015. FLASH proton density imaging for improved surface coil intensity correction in quantitative and semi-quantitative SSFP perfusion cardiovascular magnetic resonance. J. Cardiovasc. Magn. Reson. 17, 16. https://doi.org/10.1186/s12968-015-0120-6.

Pack, N.A., DiBella, E.V.R., Rust, T.C., Kadrmas, D.J., McGann, C.J., Butterfield, R., Christian, P.E., Hoffman, J.M., 2008. Estimating myocardial perfusion from dynamic contrast-enhanced CMR with a model-independent deconvolution method. J. Cardiovasc. Magn. Reson. 10, 1–15. https://doi.org/10.1186/1532-429X-10-52/FIGURES/9.

Petersen, S.E., Jerosch-Herold, M., Hudsmith, L.E., Robson, M.D., Francis, J.M., Doll, H.A., Selvanayagam, J.B., Neubauer, S., Watkins, H., 2007. Evidence for microvascular dysfunction in hypertrophic cardiomyopathy: new insights from multiparametric magnetic resonance imaging. Circulation 115, 2418–2425. https://doi.org/10.1161/CIRCULATIONAHA.106.657023.

Pezel, T., Hovasse, T., Kinnel, M., Unterseeh, T., Champagne, S., Toupin, S., Garot, P., Sanguineti, F., Garot, J., 2021. Prognostic value of stress cardiovascular magnetic resonance in asymptomatic patients with known coronary artery disease. J. Cardiovasc. Magn. Reson. 23, 19. https://doi.org/10.1186/s12968-021-00721-8.

Plein, S., Radjenovic, A., Ridgway, J.P., Barmby, D., Greenwood, J.P., Ball, S.G., Sivananthan, M.U., 2005. Coronary artery disease: myocardial perfusion MR imaging with sensitivity encoding versus conventional angiography. Radiology 235, 423–430. https://doi.org/10.1148/radiol.2352040454.

Ponikowski, P., Voors, A.A., Anker, S.D., Bueno, H., Cleland, J.G.F., Coats, A.J.S., Falk, V., Gonzalez-Juanatey, J.R., Harjola, V.P., Jankowska, E.A., Jessup, M., Linde, C., Nihoyannopoulos, P., Parissis, J.T., Pieske, B., Riley, J.P., Rosano, G.M.C., Ruilope, L.M., Ruschitzka, F., Rutten, F.H., van der Meer, P., Group, E.S.C.S.D, 2016. 2016 ESC Guidelines for the diagnosis and treatment of acute and chronic heart failure: the Task Force for the diagnosis and treatment of acute and chronic heart failure of the European Society of Cardiology (ESC)Developed with the special contribution of the Heart Failure Association (HFA) of the ESC. Eur. Heart J. 37, 2129–2200. https://doi.org/10.1093/eurheartj/ehw128.

Rahman, H., Ryan, M., Lumley, M., Modi, B., McConkey, H., Ellis, H., Scannell, C., Clapp, B., Marber, M., Webb, A., Chiribiri, A., Perera, D., 2019. Coronary microvascular dysfunction is associated with myocardial ischemia and abnormal coronary perfusion during exercise. Circulation 140, 1805–1816. https://doi.org/10.1161/CIRCULATIONAHA.119.041595.

Rahman, H., Scannell, C.M., Demir, O.M., Ryan, M., McConkey, H., Ellis, H., Masci, P.G., Perera, D., Chiribiri, A., 2021. High-resolution cardiac magnetic resonance imaging techniques for the identification of coronary microvascular dysfunction. JACC Cardiovasc. Imaging 14, 978–986. https://doi.org/10.1016/j.jcmg.2020.10.015.

Ritter, C., Brackertz, A., Sandstede, J., Beer, M., Hahn, D., Kostler, H., 2006. Absolute quantification of myocardial perfusion under adenosine stress. Magn. Reson. Med. 56, 844–849. https://doi.org/10.1002/mrm.21020.

Rubinshtein, R., Yang, E.H., Rihal, C.S., Prasad, A., Lennon, R.J., Best, P.J., Lerman, L.O., Lerman, A., 2010. Coronary microcirculatory vasodilator function in relation to risk factors among patients without obstructive coronary disease and low to intermediate Framingham score. Eur. Heart J. 31, 936–942. https://doi.org/10.1093/eurheartj/ehp459.

Sammut, E.C., Villa, A.D.M., di Giovine, G., Dancy, L., Bosio, F., Gibbs, T., Jeyabraba, S., Schwenke, S., Williams, S.E., Marber, M., Alfakih, K., Ismail, T.F., Razavi, R., Chiribiri, A., 2018. Prognostic value of quantitative stress perfusion cardiac magnetic resonance. JACC Cardiovasc. Imaging 11, 686–694. https://doi.org/10.1016/j.jcmg.2017.07.022.

Shah, R., Heydari, B., Coelho-Filho, O., Murthy, V.L., Abbasi, S., Feng, J.H., Pencina, M., Neilan, T.G., Meadows, J.L., Francis, S., Blankstein, R., Steigner, M., di Carli, M., Jerosch-Herold, M., Kwong, R.Y., 2013. Stress cardiac magnetic resonance imaging provides effective cardiac risk reclassification in patients with known or suspected stable coronary artery disease. Circulation 128, 605–614. https://doi.org/10.1161/CIRCULATIONAHA.113.001430.

Sourbron, S.P., Buckley, D.L., 2011. On the scope and interpretation of the Tofts models for DCE-MRI. Magn. Reson. Med. 66, 735–745. https://doi.org/10.1002/mrm.22861.

Sourbron, S.P., Buckley, D.L., 2012. Tracer kinetic modelling in MRI: estimating perfusion and capillary permeability. Phys. Med. Biol. 57, R1–33. https://doi.org/10.1088/0031-9155/57/2/R1.

Taqueti, V.R., Shaw, L.J., Cook, N.R., Murthy, V.L., Shah, N.R., Foster, C.R., Hainer, J., Blankstein, R., Dorbala, S., di Carli, M.F., 2017. Excess cardiovascular risk in women relative to men referred for coronary angiography is associated with severely impaired coronary flow reserve, not obstructive disease. Circulation 135, 566–577. https://doi.org/10.1161/CIRCULATIONAHA.116.023266.

Vincenti, G., Masci, P.G., Monney, P., Rutz, T., Hugelshofer, S., Gaxherri, M., Muller, O., Iglesias, J.F., Eeckhout, E., Lorenzoni, V., Pellaton, C., Sierro, C., Schwitter, J., 2017. Stress perfusion CMR in patients with known and suspected CAD: prognostic value and optimal ischemic threshold for revascularization. JACC Cardiovasc. Imaging 10, 526–537. https://doi.org/10.1016/J.JCMG.2017.02.006.

Wilke, N., Jerosch-Herold, M., Wang, Y., Huang, Y., Christensen, B.V., Stillman, A.E., Ugurbil, K., McDonald, K., Wilson, R.F., 1997. Myocardial perfusion reserve: assessment with multisection, quantitative, first-pass MR imaging. Radiology 204, 373–384. https://doi.org/10.1148/radiology.204.2.9240523.

Xue, H., Brown, L.A.E., Nielles-Vallespin, S., Plein, S., Kellmann, P., 2019. Automatic in-line quantitative myocardial perfusion mapping: processing algorithm and implementation. Magn. Reson. Med., 1–49. https://doi.org/10.1002/mrm.27954.

Zarinabad, N., Chiribiri, A., Hautvast, G.L.T.F., Ishida, M., Schuster, A., Cvetkovic, Z., Batchelor, P.G., Nagel, E., 2012. Voxel-wise quantification of myocardial perfusion by cardiac magnetic resonance. Feasibility and methods comparison. Magn. Reson. Med. 68, 1994–2004. https://doi.org/10.1002/mrm.24195.

Perfusion MRI of the lungs 16

Giles Santyr

Peter Gilgan Centre for Research and Learning, The Hospital for Sick Children, Department of Medical Biophysics, University of Toronto, Toronto, ON, Canada

16.1 Overview

Respiratory diseases, including obstructive (*e.g.*, chronic obstructive pulmonary disease, cystic fibrosis, and asthma), restrictive (*e.g.*, idiopathic pulmonary fibrosis, IPF), and vascular disease (*e.g.*, pulmonary hypertension) are significant contributors to morbidity and mortality and represent a major healthcare burden to society. While symptoms can be managed, few cures exist due to a lack of tools for early detection and monitoring of novel treatments. Clinical assessment and management of lung diseases typically involves the use of pulmonary function tests (PFTs) such as spirometry and blood gas analysis. PFTs are whole-lung measurements and, while sensitive to disease, cannot identify regional deficiencies in lung function. The inability to pinpoint the site of dysfunction limits the role of PFTs in diagnosis of early lung disease, perhaps most notably in children. Lung imaging is typically performed using planar X-ray and computed tomography (CT), which can reveal changes in anatomy (*e.g.*, lung density) but usually not before clinical symptoms manifest. X-ray imaging methods are generally insensitive to functional changes such as deficits in gas exchange, perfusion and ventilation, which may well precede measurable structural changes. For this reason, nuclear medicine imaging techniques such as single-photon emission tomography (SPECT) and positron emission tomography (PET) can be used when functional lung information such as ventilation and perfusion is required. However, nuclear medicine imaging is limited in spatial resolution, lacks specificity, and, similar to CT, results in additional radiation dose to the patient.

In contrast to X-ray and nuclear medicine imaging methods, magnetic resonance imaging (MRI) provides an approach for probing lung anatomy and function without the use of ionizing radiation, which may be particularly well suited for tracking the progression and regression of lung disease. However, the lung contains mainly air with only approximately 20% soft tissue, making proton MRI challenging due to low signal. This low signal is further diminished by the presence of microscopic magnetic field nonuniformities arising from susceptibility differences at the gas–tissue interface.

Advances in Magnetic Resonance Technology and Applications, Volume 11, ISSN 2666-9099
https://doi.org/10.1016/B978-0-323-95209-5.00021-0

Furthermore, respiratory motion can be problematic leading to image blurring and other artifacts. These challenges have been addressed in recent years, enabling MRI to provide anatomical images of the lung that rival X-ray CT. This improvement in MR image quality has enabled the use of both endogenous and exogenous contrast mechanisms to detect and quantitate lung perfusion changes associated with a variety of diseases, including COPD, CF, and IPF.

Dynamic contrast-enhanced MRI (DCE-MRI) is considered the mainstay of lung perfusion MRI. This enables quantitative perfusion measures by tracking the uptake by the lung of an intravenous bolus of a gadolinium-based agent (*e.g.*, Gd-DTPA). However, injection of a gadolinium-based agent may not be desirable, especially in pediatric applications. Two promising approaches for lung perfusion MRI that do not involve the injection of an exogenous contrast agent are arterial spin labeling (ASL) and Fourier Decomposition (FD) MRI. More recently, hyperpolarized (HP) gas MRI has been developed for imaging the lung, providing a further palette of novel micro-anatomical and functional lung measures. Specifically, MRI of the dissolved phases of HP ^{129}Xe has been demonstrated to be sensitive to perfusion in the lung. HP MRI of the lung will be discussed in a later section.

The purpose of this chapter is to introduce the reader to the basic methodologies and application of lung perfusion MRI. To begin, the major technical advances leading to effective use of MRI in the lung are described. Next, the main methods for lung perfusion MRI using both exogenous and endogenous contrast are explained, including emergent hyperpolarized gas MRI techniques. The application of these methods to various lung diseases is described. Finally, current challenges for lung MRI are presented along with potential solutions and future directions.

16.2 Technical advances in lung MRI

The lung has always been a very desirable organ to image with MRI, but only recently have hardware and software advances positioned lung MRI for practical application (Voskrebenzev and Vogel-Claussen, 2021). Specifically, the use of radiofrequency (RF) multicoil arrays designed for the thorax has improved lung MR image quality as well as permitted acceleration by parallel imaging approaches. Imaging speed has been further improved by compressed sensing as well as improved gradient performance and pulse sequences. These improvements in speed have enabled complete volumetric imaging within a reasonable breath-hold duration to mitigate blurring and artifacts due to respiratory motion. For longer image acquisitions, prospective respiratory gating or retrospective motion correction algorithms have significantly improved lung MR image quality during free breathing.

While MRI has generally benefitted from the increased signal associated with the use of higher magnetic field strengths (*e.g.*, 3T), the benefits to lung imaging are less clear. This has to do with the increase in magnetic susceptibility associated the air–tissue interface, which scales proportionately with static field strength, leading to reduced apparent transverse relaxation times, principally T2*, of a few ms at clinical field strengths. Reduced T2* means reduced signal in gradient echo imaging, the mainstay of lung MRI. Furthermore, increased susceptibility causes higher distortion of the static field, leading to increased artifacts (*i.e.*, blurring, geometric distortion). Nevertheless, improvements in lung signal due to advances in coils and reduction of respiratory motion artifacts have outweighed susceptibility disadvantages, making lung MRI feasible at 3T. Furthermore, pulse sequences incorporating ultra-short TE (UTE), which can help ameliorate the effects of susceptibility, hold promise for further improvement of lung image quality at high static field strengths.

16.2.1 Respiratory motion compensation

When lung MR image acquisition times exceed a reasonable breath-hold duration (>10–20 s) or when functional information related to ventilation/perfusion is desired, respiratory/cardiac gating strategies can be helpful. Traditional "hard gating" approaches have relied on an external signal provided by a device that directly monitors respiratory motion (*e.g.*, bellows) and/or cardiac motion (*e.g.*, ECG monitor). This signal triggers MRI data acquisition only when a repeated position is obtained (*e.g.*, inspiration) or synchronizes the data acquisition such that the effects of motion can be removed or characterized in the reconstruction process. However, hard gating relies on a robust external signal which can be difficult to obtain reliably, especially in subjects who do not execute periodic breathing. More commonly, respiratory/cardiac gating is performed using "self-gating" approaches which extract the motion directly from the acquired MR data. Self-gated approaches typically employ a special MR signal, such as a navigator echo, reflecting the periodic motion of the diaphragm, which is then used to prospectively trigger image acquisition synchronized to the motion in real time. Self-gating can also be used for retrospective motion correction, when the acquired motion-encoded signal is exploited in the reconstruction process to select particular views to minimize the deleterious effects of the motion (*e.g.*, ghosting artifacts). This has the added advantage of capturing and potentially characterizing the entire periodic motion directly, important for functional lung imaging.

16.2.2 Ultra-short echo time (UTE)

One of the most significant recent advances in lung MRI has been the development of UTE techniques that can provide echo times as short as 10–50 μs to capture rapidly decaying lung signal, particularly at high static magnetic field strength (Voskrebenzev and Vogel-Claussen, 2021; Torres *et al.*, 2019; Månsson *et al.*, 2003). UTE acquisitions are typically performed using "center-out" non-Cartesian k-space trajectories (*e.g.*, radial) to facilitate short echo times (Johnson *et al.*, 2013; Herrmann *et al.*, 2016). A further benefit of such trajectories is the inherent oversampling of the central k-space frequency (k_0) signal during each acquisition. Due to respiratory/cardiac motion, the phase and/or magnitude of this signal varies between acquisitions, allowing for tracking of motion as described before (*i.e.*, self-gating) without the need for a navigator echo or external triggering signals. Furthermore, any inadvertent gross motion by the subject will also be detectable in the initial phase/magnitude, allowing for the exclusion of unwanted data which may corrupt overall image quality (Zhu *et al.*, 2020). The most commonly used non-Cartesian k-space trajectory for UTE MRI is 3D radial (Herrmann *et al.*, 2016). However, this trajectory can be inefficient for high isotropic resolution, requiring large numbers of projections leading to increased scan times. To alleviate this, 3D UTE imaging using spiral-based trajectories such as cones (Gurney *et al.*, 2006) or Fermat looped-orthogonally encoded trajectories (FLORET) have also been demonstrated for lung (Willmering *et al.*, 2019). These trajectories sample k-space more efficiently, reducing the number of acquisitions needed and decreasing overall scan time. FLORET in particular has emerged as an efficient acquisition strategy with favorable properties for lung imaging (Willmering *et al.*, 2019).

16.3 Exogenous contrast-enhanced approaches

Notwithstanding the intrinsic T1, T2, and T2* contrast between vascular and non-vascular lung tissues allowing for direct measurement of pulmonary blood volume as described in Section 16.4, lung perfusion MRI typically relies on the use of an exogenous contrast agent to specifically enhance the signal from blood. This is generally achieved by rapid 3D time-resolved, gradient-recalled echo (GRE) imaging of the lung following bolus injection and first passage of an intravenous gadolinium-based agent. More recently, gaseous contrast agents (*i.e.*, oxygen) have shown potential for revealing lung perfusion changes without the need for gadolinium injection.

16.3.1 Intravenous agents (DCE-MRI)

Dynamic contrast-enhanced MRI (DCE-MRI) approaches are considered the gold standard for lung perfusion MRI, especially when quantitative information (*e.g.*, pulmonary blood flow and volume) is sought. There are two main DCE-MRI approaches depending on the type of contrast and/or intravenous agent employed: (i) paramagnetic-based (T1) and (ii) susceptibility-based (T2*). These two concepts have previously been introduced in Chapters 2 and 3, respectively, so will not be described in detail here. Instead, the following two sections will focus on key considerations when using these two approaches in the lung.

16.3.1.1 Paramagnetic-based methods

The main effect of a paramagnetic agent such as gadolinium (Gd) is to reduce the T1 relaxation time of blood and surrounding tissues, leading to GRE signal enhancement in well-perfused regions, relative to poorly perfused regions of the lung. Most lung perfusion MRI methods use a bolus injection of a gadolinium chelate (*e.g.*, Gd-DTPA) at a concentration of 0.05–0.1 mM/kg bodyweight and a bolus injection rate of 3–5 ml/s (Fink *et al.*, 2006). The relatively low concentration of the agent ensures sufficient T1 shortening in the small pulmonary vessels without excessive T2* shortening, which would reduce the already low signal in the lung (Saunders *et al.*, 2022). Images are acquired at baseline and dynamically during signal enhancement using 3D GRE imaging, with flip angles ranging between 20 and 40°, to achieve good signal contrast-to-noise ratio in the lung. Images are acquired at a temporal resolution below the transit time of the bolus in the lungs (3–4 s), as shown in Fig. 16.1. This rapid temporal resolution (<one 3D volume per second) is achieved by minimizing TE and TR, often in concert with parallel imaging and view sharing (Ley and Ley-Zaporozhan, 2012).

DCE-MRI images can be analyzed qualitatively (*e.g.*, subtraction), semiquantitatively by descriptive parameters (*e.g.*, slope of enhancement curve), or quantitatively using analytical models describing the kinetics of the contrast agent (Ley and Ley-Zaporozhan, 2012). Quantitation of perfusion in the lung follows indicator dilution theory, generally assuming a linear relationship between signal enhancement and concentration of the contrast agent as well as minimal extravasation during first pass (Meier and Zierler, 1954). With an estimation of the arterial input function (AIF) obtained from an ROI placed on the pulmonary artery, perfusion parameters such as mean transit time (MTT), pulmonary blood flow (PBF), and pulmonary blood volume (PBV) can be extracted regionally which agree well with radioisotope imaging (Ohno *et al.*, 2007). Nevertheless, these pulmonary perfusion metrics must be interpreted with caution as they have been shown to depend on lung inflation, with differences seen during breath-holds compared to free breathing (Saunders *et al.*, 2022). Furthermore, the relatively large size of the lung and the multiple branching structure of the pulmonary arteries often lead to an overestimation of pulmonary perfusion, particularly in central regions heavily weighted by the arteries. A dual-bolus approach (Risse *et al.*, 2006) or reference ROIs placed on both the pulmonary artery and vein along with cross-correlation analysis can help suppress intrapulmonary vessels leading to improved estimates of PBF and PBV (Risse *et al.*, 2009).

While only the first-pass effects of Gd-DTPA have thus far been considered, it is worth pointing out that the later pharmacokinetics of DCE-MRI allows for detection and measurement of blood perfusion in lung tumors. Of particular interest to the lung are identification and characterization of pulmonary nodules, which can be challenging to detect using X-ray methods such as chest CT. Using DCE-MRI, malignant pulmonary neoplasms have been shown to have increased microvasculature compared to benign nodules, consistent with angiogenesis as confirmed by biopsy (Du *et al.*, 2022).

16.3.1.2 Susceptibility-based methods

In addition to the effect on T1, gadolinium-based contrast agents also increase the susceptibility of blood thereby reducing T2* and providing dynamic susceptibility contrast (DSC) that can potentially be exploited for imaging lung perfusion (Chapter 3). Unfortunately, the intrinsic T2* of the lung is already very short (a few ms) due to air–tissue interfaces, making it challenging to measure T2*

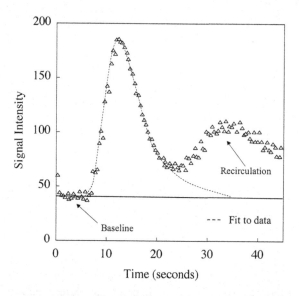

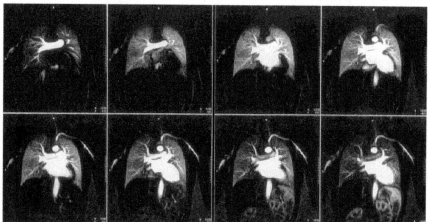

FIG. 16.1

Dynamic contrast-enhanced (DCE) MRI of the lung. (Top) Time dependence of the MRI signal from the pulmonary tissue following bolus injection of Gd-DTPA, with characteristic peak indicating the first pass of the exogenous agent, used to perform quantitative perfusion imaging. (Bottom) Time-resolved coronal images of the lungs of a healthy subject showing signal enhancement from the major pulmonary vessels and blush from the lung parenchyma.

From Hatabu, H., Tadamura, E., Levin, D.L., Chen, Q., Li, W., Kim, D., et al. 1999a. Quantitative assessment of pulmonary perfusion with dynamic contrast-enhanced MRI. Magn. Reson. Med. 42 (6), 1033–8, and as reprinted in Hopkins, S.R., Prisk, G.K. 2010. Lung perfusion measured using magnetic resonance imaging: new tools for physiological insights into the pulmonary circulation. J. Magn. Reson. Imaging. 32 (6), 1287–301, used with permission.

reductions due to gadolinium in the lung with this approach. One noteworthy exception is the use of iron-based blood pool contrast agents such as superparamagnetic iron oxide nanoparticles (SPIONs), which substantially increase the susceptibility of blood (Gharagouzloo *et al.*, 2015). When injected, SPIONs in concert with inhaled gases can be sensitive to lung perfusion as described further in Section 16.3.2.2.

16.3.2 **Gaseous agents**

While the intrinsic proton T2* and T1 relaxation times of lung tissue are known to depend on alveolar size and alveolar oxygen concentration, respectively (Futura Publishing Company Inc, 1996), the effects of blood flow on lung relaxation are more subtle and less well understood. Application of an exogenous gaseous agent (*e.g.*, air enriched in oxygen) can help shed light on pulmonary perfusion. Furthermore, nonproton gas MRI methods (*e.g.*, ^{3}He, ^{129}Xe, and ^{19}F) offer new tools for the investigation of pulmonary blood flow. The combination of gas MRI with the aforementioned DCE-MRI methods provides a powerful approach for the acquisition of ventilation and perfusion maps, respectively, the ratio (*i.e.*, V/Q) of which provides insight into diseases which alter the matching of these parameters (*e.g.*, pulmonary embolism).

16.3.2.1 *Oxygen-enhanced lung MRI*

It is well established that the paramagnetic property of molecular oxygen, leading to T1 shortening, can be exploited to reflect regional alveolar oxygenation partial pressure (p_AO_2) (Edelman *et al.*, 1996). In this approach, termed oxygen-enhanced (OE) MRI, lung images are obtained under conditions of breathing normal room air alternately with breathing pure oxygen. To first order, a difference image shows regions of decreased ventilation associated with disease (*e.g.*, obstruction). Deeper analysis of OE reveals that signal changes with p_AO_2 directly reflect pulmonary blood oxygen content, which can be attributed to a combination of ventilation/perfusion ratio and diffusing capacity of the lung, collectively called the oxygen transfer function (OTF) (Jakob *et al.*, 2004). The OTF can be measured as the slope of 1/T1 vs applied oxygen concentration in the breathing gas mixture and, to the extent that it is not limited by diffusion, provides a map of ventilation/perfusion mismatch, shown to be useful for identifying pulmonary embolism (Edelman *et al.*, 1996).

16.3.2.2 *Hyperpolarized and perfluorinated gases*

Using spin-exchange optical pumping (and other methods), both ^{3}He and ^{129}Xe gases can be hyperpolarized (HP) to a level some 4–5 orders of magnitude larger than thermal equilibrium, making them detectable using conventional MRI techniques. The inert nature of these gases, coupled with this very high polarization and the absence of any background signal, provides a very sensitive approach for interrogating the lung environment in the vicinity of blood vessels, including perfusion. Perfusion imaging of healthy pig lungs has been demonstrated using the susceptibility effect of a bolus injection of Gd-DTPA on T2* of HP ^{3}He gas (Dimitrov *et al.*, 2005). SPIONs injected intravenously have also been shown to deplete the HP ^{3}He gas signal in the lung, allowing for perfusion mapping in an animal model of pulmonary embolism (Viallon *et al.*, 2000). Compared to ^{3}He gas, ^{129}Xe gas is advantageous for perfusion imaging as xenon is relatively soluble in the bloodstream and has a distinctive chemical shift when bound to hemoglobin, which also depends on oxygenation state. This potentially powerful approach for mapping blood perfusion and oxygenation is described in more detail in Section 16.5.1.

Despite the promise of HP gas MRI, the cost, infrastructure/personnel requirements, and current lack of availability of this technology may hamper its clinical translation. Therefore, alternative ^{19}F gas lung MRI approaches have been pursued in recent years, notably using perfluorinated gases such as sulfur hexafluoride (SF$_6$) and perfluoropropane (PFP) (Couch *et al.*, 2019). Sufficient signal in the lung is achievable with these gases without the need for hyperpolarization since these gases have several chemically equivalent ^{19}F atoms per molecule, relatively short T1 relaxation times enabling signal averaging, and ^{19}F is 100% isotopically abundant. Perhaps most importantly, perfluorinated gases can be mixed with oxygen without signal loss, permitting repeated and continuous use (*i.e.*, multiple breaths, washin/washout studies) without anoxia: an important advantage compared to the anoxic mixtures required for HP gas MRI. ^{19}F MRI using PFP gas has been shown to provide measures of pulmonary perfusion based on T2* shortening following an intravenous injection of Gd-based contrast agent in mice (Neal *et al.*, 2020). The sensitivity of ^{19}F T1 on the partial pressure of SF$_6$ gas in the lungs has been shown to permit measurement of ventilation/perfusion ratio in mice (Adolphi and Kuethe, 2008).

16.4 Endogenous contrast approaches

The value of a noncontrast approach for quantitation of lung perfusion cannot be overstated given potential health concerns, such as nephrogenic systemic fibrosis (NSF) associated with the use of certain gadolinium-based agents (notably linear chelates). While NSF occurs primarily in patients with compromised renal function, long-term effects of gadolinium are not well understood and there is a general trend toward reducing the reliance of MRI on injectable agents, especially in vulnerable populations such as children. By resolving the lung proton T1 recovery into two components corresponding to the intra and extravascular signals, respectively, it is possible to estimate the fractional pulmonary blood volume (fPBV) regionally without the need for exogenous contrast agents (Gaass *et al.*, 2012). fPBV maps have been shown to demonstrate the effect of gravity on the distribution of pulmonary blood and agree qualitatively with DCE-MRI and contrast-enhanced CT (Gaass *et al.*, 2012). It has also been shown that lung T2* decay can be similarly resolved into intra/extravascular contributions (Triphan *et al.*, 2015), offering a viable approach for fPBV mapping, particularly with the advent of UTE sequences as described before (Hatabu *et al.*, 1999).

In addition to using relaxation effects to separate MRI signals arising from the vascular pool relative to the nonvascular compartments, flow effects can also be exploited. Endogenous lung perfusion MRI techniques such as arterial spin labeling (ASL) or Fourier Decomposition (FD) take advantage of the inflow of blood into the imaging volume rather than the more subtle endogenous relaxation differences between blood and other lung tissues and may have significant potential in the lung as will be described in the following sections.

16.4.1 Arterial spin labeling

Arterial spin labeling (ASL) takes advantage of the ability of MRI to magnetically "tag" (*i.e.*, label) proton spins associated with water in blood then subsequently readout signal from those tagged spins at a later point in time when they have entered a slice of interest via blood flow. In this way, water serves as a freely diffusible tracer of blood flow without the need for an exogenous agent (Chapter 4). Compared to other organs, this approach is particularly advantageous in the highly perfused lungs, which receive the entire cardiac output. Typically for lung ASL imaging, two ECG-gated images (a "tagged" and a "control" image) are acquired 5–8 s apart during a single breath-hold using either SSFP or fast (or turbo) spin echo pulse sequences. These sequences are well suited for the lung, because they reduce artifacts caused by respiratory and cardiac motion and employ an echo to mitigate signal loss due to susceptibility (Bolar *et al.*, 2006). To avoid incidental magnetization transfer (MT) effects, a common ASL approach in the lung employs flow-sensitive alternating inversion recovery (FAIR or FAIRER) that first applies a nonselective inversion preparatory pulse to the entire torso during diastole in order to differentially tag inflowing (*i.e.*, arterial) spins located outside the imaged slice compared to stationary spins located within the slice (tagged image). A second image is then acquired with the preparatory tagging pulse applied only to the slice of interest, leaving inflowing spins unperturbed (control image). Subtraction of the control image from the tagged image results in elimination of signal from stationary tissue, which is unchanged between the two acquisitions (Mai *et al.*, 1999), as shown in Fig. 16.2. The spatially centered application of the tagging pulses in the two acquisitions results in effective MT compensation (Voskrebenzev and Vogel-Claussen, 2021). The time between the preparatory pulse and the imaging read-out pulse (or inversion time, TI) will determine the ASL signal, proportional to the amount of pulmonary arterial blood delivered during the previous heart cycle. For the lungs, TI is typically chosen to be ~80% of the cardiac R-R interval derived from the ECG signal (600–800 ms) to allow a complete systolic ejection period to pass to capture all the stroke volume during a period of low flow and avoid pulsatility effects (Hopkins and Prisk, 2010).

Shortly following initial inception for angiography (Edelman *et al.*, 1994), ASL using FAIR was demonstrated for pulmonary perfusion in humans (Mai and Berr, 1999) and subsequently shown to be able to detect pulmonary embolism

Control Image: Selective Inversion   Tag Image: Nonselective Inversion   Difference Image (with lung mask)

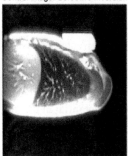 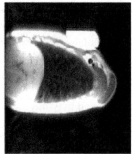 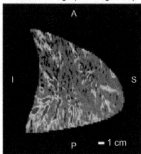

FIG. 16.2

Arterial spin labeling (ASL) MRI of the right lung of a healthy subject seen in the sagittal plane. First panel (left): acquired (control) image following a slice-selective inversion pulse "tag." Note that the bright signal in the lung of the control image corresponds to vessels where blood flowing into the slice of interest is not affected by the tag. Second panel (center): acquired image of same plane as shown at left but following a nonselective inversion pulse, effectively tagging blood throughout the entire volume. Third panel (right): subtraction of the tag image (center) from control image (left), revealing the signal predominately from the pulmonary vessels.

From Hopkins, S.R., Prisk, G.K. 2010. Lung perfusion measured using magnetic resonance imaging: new tools for physiological insights into the pulmonary circulation. J. Magn. Reson. Imaging. 32 (6), 1287–301, used with permission.

(Mai *et al.*, 2000) as well as map gravitational and lung inflation effects (Fan *et al.*, 2010). More recently, FAIR has been used to measure pulmonary perfusion at 3T and has been shown to compare favorably with both phase-contrast MR angiography and SPECT perfusion imaging in healthy subjects (Greer *et al.*, 2018). Other investigations of ASL in the lung have demonstrated the effects of pulsatility and tracer saturation effects on quantitative pulmonary blood flow measurements, emphasizing the importance of the choice of the acquired plane (Bolar *et al.*, 2006). Specifically, use of the coronal plane, which contains portions of both the left and right pulmonary arteries, can lead to underestimation of PBF due to the presence of significant in-plane flow. For this reason, the axial and, more commonly, the sagittal planes are typically chosen (Hopkins and Prisk, 2010).

Since ASL is inherently a planar imaging technique, its clinical application in the lung has been limited by the inability to obtain 3D information critical for diagnosis (Ohno *et al.*, 2022). One further limitation of ASL in the lung is that the available signal from the tagged blood is diminished by T1 relaxation occurring during TI, thereby requiring subtraction to detect the relatively weak signal. Therefore any misregistration between the two breath-hold images obtained at potentially different respiratory levels can lead to artifacts, especially near the blood vessels and diaphragm. In order to address this limitation, single-shot ASL techniques, which completely suppress stationary tissues without the need for subtraction, have been developed (Fischer *et al.*, 2008). A particularly promising ASL approach for reduction of misregistration artifacts involves the combination of FAIR and UTE, where the center-out acquisition of the latter helps reduce respiratory motion effects (Tibiletti *et al.*, 2016).

16.4.2 Free-breathing MRI and Fourier decomposition

There has been considerable interest in exploiting the natural periodic changes in intrinsic lung proton MRI signal associated with blood flow to image lung perfusion during quiescent free breathing. These techniques exploit the detectable changes in the magnitude of the lung vascular signal during the cardiac cycle to generate perfusion-weighted images, primarily due to different inflow (*i.e.*, time-of-flight) effects in a particular slice (Bauman *et al.*, 2009). Similarly, the signal changes corresponding to respiratory motion can be used to separate and quantitate the ventilation contribution to the signal, but will not be described in this chapter.

The most straightforward free-breathing (FB) method, referred to as Fourier Decomposition (FD), involves application of a pixel-wise, 1D Fourier transform along the time domain of the series of coregistered (usually using nonrigid registration), time-resolved lung images, yielding a map of the temporal frequency distribution of the signal. The signals corresponding to the ventilation and cardiac frequencies provide ventilation- and perfusion-weighted images, respectively (Bauman *et al.*, 2009). As such, FD methods rely on high temporal resolution in order to satisfy the Nyquist sampling criterion ($\sim$3 image/s for typical cardiac frequencies) (Voskrebenzev and Vogel-Claussen, 2021). Significant advancements on the FD approach include the use of a nonuniform Fourier Transform to mitigate variations in cardiac frequency (Bondesson *et al.*, 2019), and the Matrix Pencil (MP) method for more robust estimation of the spectral components (Bauman and Bieri, 2017). Since the FD method relies on the magnitude of the signal from the pulmonary vessels, it is generally performed using rapid 2D bSSFP acquisitions, which accentuate signal from fluids such as blood. However, bSSFP can be challenging in the lung at higher static magnetic field strength due to increased off-resonance effects (*e.g.*, banding artifacts). More recently, transient spoiled gradient-recalled echo (SPGR) has been demonstrated for lung perfusion imaging at 1.5T and 3T using the MP approach (Bauman *et al.*, 2019).

To aid clinical translation at 3T, FB approaches that are amenable to conventional fast gradient-recalled echo (FGRE) are desirable. These include SELf-gated Non-contrast-Enhanced FUnctional Lung (SENCEFUL) MRI and Phase-Resolved Functional Lung (PREFUL) MRI. In SENCEFUL, a self-gating signal is used to sort quasi-randomly acquired data according to cardiac/respiratory phase (Fischer *et al.*, 2014), following which a Fourier Transform is applied to yield the spectral components and corresponding FD images as described before. SENCEFUL has been shown to be superior to conventional FD at 3T and provides perfusion-weighted images that agree well with DCE-MRI following gadolinium injection (Fischer *et al.*, 2014).

PREFUL MRI rapidly acquires 2D FGRE (or FLASH) images during free breathing that are coregistered and sorted in the time domain to generate synthesized pixel-wise respiratory and cardiac cycle maps (Voskrebenzev *et al.*, 2018). Fig. 16.3

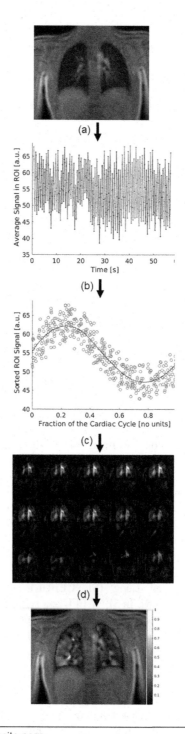

(a)

(b)

(c)

(d)

FIG. 16.3

See figure legend on opposite page.

shows a generalized image postprocessing pipeline for lung perfusion mapping with PREFUL. Briefly, a region of interest is chosen in a major vessel (*e.g.*, pulmonary artery) to establish a cardiac waveform in the time domain. This is then used to fit lung signals to create a pixel-wise map of waveforms associated with the lung parenchyma, yielding a set of perfusion-weighted images along a single synthesized cardiac cycle. These time-domain images can be used in a number of ways. For example, this time series may be collapsed along the temporal dimension to produce a perfusion-weighted maximum intensity projection (MIP) and then analyzed for defects. Signal changes that are out of phase with the cardiac cycle (*i.e.*, either occurring earlier or later in the cycle) may be examined by calculating the time to peak (TTP) on a voxel-wise basis (Voskrebenzev *et al.*, 2018; Pohler *et al.*, 2020). Perfusion quantification including normalized perfusion (Q_N) and quantitative perfusion (Q_{Quant}) can be estimated from exchange fraction and blood fraction and more accurately quantify perfusion as compared to FD MRI (Glandorf *et al.*, 2021). Furthermore, the pulmonary pulse wave transit time (pPTT) can be calculated to measure time delays associated with arterial blood flow from the pulmonary arteries into the lung microvasculature (Pohler *et al.*, 2020) and has been validated with echocardiography (Pohler *et al.*, 2021).

The insufficient temporal resolution of 3D FB MRI acquisitions typically means that it is applied in a single-slice fashion, which can miss disease in the rest of the lung. Several 2D acquisitions acquired serially to cover the lung could alleviate this concern, but this leads to lengthy scans (up to 10 min) and increased sensitivity to motion artifacts and misregistration between acquired slices. More recently, UTE sequences (*e.g.*, radial) have been applied to FB MRI in order to increase signal and speedup data acquisition, enabling 3D isotropic resolution for ventilation imaging with SENCEFUL (Mendes Pereira *et al.*, 2019) and PREFUL (Klimes *et al.*, 2021). Due to the use of a nonselective RF pulse insensitive to inflow, the application of FB methods to 3D perfusion imaging remains an unsolved challenge (Klimeš *et al.*, 2021).

16.5 **Other approaches**

16.5.1 **Hyperpolarized ^{129}Xe gas**

As described in Section 16.3.2.2, hyperpolarized (HP) gas MRI is a very promising tool for the investigation of lung ventilation. Due to its larger gyromagnetic ratio, ^{3}He provides the strongest signal, but it is an exceedingly rare by-product of the

FIG. 16.3　Perfusion-weighted PREFUL MRI processing pipeline. (A) The average signal within a region of interest (ROI) enclosing the main blood vessel is fit to a piecewise sine function (red (black in print version) and blue (gray in print version)). (B) The phase of each image is determined and used to construct a synthesized cardiac cycle. (C) Following nonparametric regression, 15 equidistant cardiac phases are interpolated (D) and a normalized perfusion-weighted map is determined from the peak cardiac phase (Fischer *et al.*, 2014).

Figure courtesy of Samal Munidasa.

radioactive decay of tritium and, therefore, limited in availability. ^{129}Xe is substantially more abundant, relatively inexpensive, and although lower in signal compared to ^{3}He, can be polarized more rapidly and potentially to higher levels (Nikolaou et al., 2013). Perhaps the most unique and exciting feature of hyperpolarized ^{129}Xe is the potential to probe the various dissolved phases in the lung. Chemical shift differences permit detection of ^{129}Xe in the gas phase as well as in several dissolved-phase compartments, including the lung parenchymal tissue and plasma, red blood cells (RBC), and adipose tissue (Månsson et al., 2003). In particular, the ability to probe the RBCs provides an opportunity to measure regional differences in the size, flow, and oxygenation of the pulmonary blood pool. Fig. 16.4A shows a

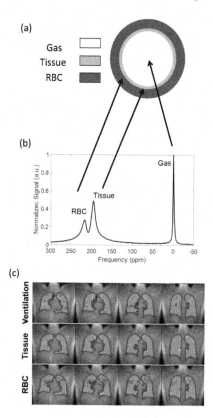

FIG. 16.4

(A) Conceptual diagram of a single alveolus with xenon dissolving from the airspace into the pulmonary tissue and blood. (B) Representative human in vivo dissolved-phase ^{129}Xe spectrum. (C) Representative dissolved-phase ^{129}Xe imaging using a 1-pt Dixon approach to decompose gas, tissue, and RBC signal (Rankine et al., 2020).

Figure courtesy of Dr. Brandon Zanette.

conceptual picture of a representative alveolus indicating the lung compartment. Fig. 16.4B shows a representative ^{129}Xe spectrum obtained from a human subject, demonstrating the distinct peaks associated with the compartments shown in Fig. 16.4A. Dynamic ^{129}Xe spectroscopic approaches such as chemical shift saturation recovery (CSSR) (Månsson *et al.*, 2003) allow a direct measure of the transfer of xenon atoms between the gas phase and the dissolved phases. By resolving the tissue and RBC dissolved phases directly as a function of a characteristic delay time, CSSR is sensitive to gas exchange and perfusion (Zanette *et al.*, 2018).

The relatively large chemical shift differences between the gas and dissolved peaks of ^{129}Xe signal ($\sim$200 ppm) can be used to efficiently separate these compartments spatially by exploiting the phase evolution of the gradient echo signal as a function of echo time to encode and reconstruct images corresponding to each peak separately (Fig. 16.4C) (Mugler *et al.*, 2010). Non-Cartesian spatial encoding strategies (*i.e.*, spiral) have also been shown to provide a further improvement in temporal resolution, allowing the regional mapping of gas exchange curves (Zanette *et al.*, 2018; Doganay *et al.*, 2016), which can be used to extract regional microanatomical and functional tissue information such as barrier thickness and RBC compartment size and blood flow changes using appropriate theoretical models (Zanette and Santyr, 2019).

The distinctive chemical shift of ^{129}Xe in the RBCs is a consequence of the binding of xenon to the hemoglobin (Hb) molecule. This chemical shift depends on the conformation of Hb affected by oxygenation state, providing a potential means of measuring blood oxygenation saturation (S_aO_2) *in vivo* (Norquay *et al.*, 2017). S_aO_2 measured in the lung with ^{129}Xe correlates with pulse oximetry and reflects hyperoxia in ventilated rodents (Friedlander *et al.*, 2021). The cardiogenic dependence of the lung RBC signal amplitude and frequency, due to changes in S_aO_2 through the cardiac cycle, has been demonstrated in humans (Bier *et al.*, 2019).

16.5.2 Intravoxel incoherent motion (IVIM)

The value of diffusion-weighted MRI (DWI) for the study of tissue microstructure has received considerable attention in recent years, particularly for the characterization of pulmonary nodules (Mori *et al.*, 2008). As an extension of this approach, intravoxel incoherent motion (IVIM) associated with microscopic blood flow through the quasi-random capillary network of pulmonary lesions can also be characterized (Le Bihan *et al.*, 1988). In particular, measurement of the early decay component of the DWI signal vs diffusion sensitization factor (b) curve allows for the extraction of the tissue blood volume fraction. The combination of DWI, IVIM, and diffusion kurtosis imaging has been shown to differentiate malignant and benign lung lesions (Zheng *et al.*, 2021). Blood volume fractions measured in the lung with IVIM compare well with DCE-MRI, without the need for contrast agents (Carinci *et al.*, 2015). Using IVIM, the maturation of the pulmonary vasculature during gestation has been demonstrated in the human fetus (Ercolani *et al.*, 2021).

16.6 Clinical applications

16.6.1 Obstructive lung diseases

Chronic obstructive pulmonary disease (COPD) is a leading cause of global morbidity and mortality and represents a significant healthcare burden to our society. While the loss of lung ventilatory function (*e.g.*, dyspnea) from emphysema and small airways destruction is one of the hallmarks of COPD, vascular changes also underlie the disease. Atherosclerosis accompanying COPD is known to affect pulmonary blood flow and is detectable using lung perfusion MRI, typically with DCE-MRI methods. Perfusion defect percentage (QDP) measured with DCE-MRI has been shown to be abnormal in COPD patients, consistent with CT and PFTs (Schiwek *et al.*, 2022). Ventilation/perfusion ratio (V/Q) assessed with the combination of DCE-MRI and HP gas MRI has demonstrated improvement following treatment in COPD subjects (Singh *et al.*, 2022). Deep learning approaches have been used to automate characterization of COPD subjects based on perfusion metrics extracted from DCE-MRI (Winther *et al.*, 2020). PREFUL has demonstrated significant promise in adults with COPD, where reductions in perfusion could be detected compared to healthy individuals (Voskrebenzev *et al.*, 2018).

Cystic Fibrosis (CF) remains the most common, incurable autosomal recessive disorder in Caucasian populations (Elborn, 2016). In the lung, CF causes obstruction due to the buildup of sticky mucous resulting in repeated infection and irreversible remodeling and dilation of the airways (*i.e.*, bronchiectasis) ultimately leading to organ failure. In CF lung disease, signal enhancement and time-to-peak differences of Gd-enhanced imaging have been used to identify regions of abnormal lung parenchyma compared to normal tissue (Amaxopoulou *et al.*, 2018). Perfusion values measured with ASL have been shown to be reduced in the upper lobes of CF patients compared to healthy control subjects and correlate well with FEV_1 (Schraml *et al.*, 2012). ASL has revealed reductions in pulmonary perfusion in young patients with CF, which correlate well with PFTs, specifically forced expiratory volume in 1 s (FEV_1) (Schraml *et al.*, 2012). PREFUL MRI has demonstrated significant promise in adults with CF where perfusion abnormalities could be detected compared to healthy individuals (Voskrebenzev *et al.*, 2018; Kunz *et al.*, 2021). QDP derived from DCE-MRI has been shown to correlate strongly with PREFUL MRI in both CF (Behrendt *et al.*, 2022) and COPD (Behrendt *et al.*, 2020). MP MRI has been used to measure improvements in lung perfusion following inhalation therapy with salbutamol in children with stable CF (Kieninger *et al.*, 2022).

16.6.2 Interstitial lung disease, idiopathic pulmonary fibrosis, and lung infection

Interstitial lung diseases (ILDs) describe a large group of disorders most of which cause progressive scarring (or fibrosis) of the parenchymal tissue between the airspaces in the lungs. ILDs, most notably IPF, have poor prognosis in part due to difficulties in stratification for treatment, including lung transplant. DCE-MRI has been used to investigate lung perfusion abnormalities in IPF (Torres *et al.*, 2022), revealing capillary perfusion deficits that correlate with physiological gas exchange measures (Weatherley *et al.*, 2021). By employing quantitative modeling, DCE-MRI has been shown to detect alterations in permeability, perfusion, and extravascular volume in IPF (Montesi *et al.*, 2021). Perhaps most promising, RBC and barrier images obtained with [129]Xe MRI (similar to that shown in Fig. 16.4) have been shown to be predictive of mortality and lung transplant in IPF subjects (Rankine *et al.*, 2020) and COVID-19-infected subjects (Grist *et al.*, 2021), not revealed by other methods, including CT. Defects in pulmonary ventilation and perfusion due to COVID-19 infection have also been shown using PREFUL (Lévy *et al.*, 2022). Fig. 16.5 shows defects (indicated by arrows) in ventilation and perfusion maps, along with segmented defect maps, obtained from a pediatric subject at 3 months post-COVID-19 infection.

16.6.3 Pulmonary hypertension

Pulmonary hypertension (PH) is a heterogeneous medical condition affecting the heart and lungs and associated with a number of diseases, requiring complex functional and anatomical assessment. The increase in pulmonary vascular resistance accompanying PH reduces cardiac output, leading to abnormal blood flow in the pulmonary arteries entering the lungs, among other consequences. MRI is a promising nonionizing imaging tool for quantitative assessment of PH, particularly the combination of DCE-MRI (Lechartier *et al.*, 2022), FB MRI (Rengier *et al.*, 2019), and exogenous gas MRI (Saunders *et al.*, 2022). ASL MRI has revealed lower blood flow in patients with pulmonary arterial hypertension (PAH) compared to healthy volunteers (Hopkins *et al.*, 2021). OE-MRI has been shown to correlate with V/Q scintigraphy and RV end diastolic volume in subjects with PH (Saunders *et al.*, 2022). FB MRI with PREFUL has demonstrated significant promise in adults with

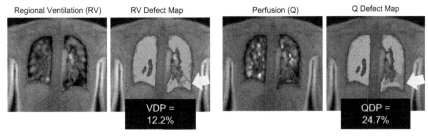

FIG. 16.5

PREFUL MRI-derived regional ventilation (RV) and corresponding ventilation defect percentage (VDP) map (left) and perfusion (Q) and corresponding perfusion defect percentage (QDP) map (right) for a pediatric COVID-19 survivor. Yellow (gray in print version) arrows indicate ventilation and perfusion defects in the bottom left lung.

Figure courtesy of Samal Munidasa.

chronic thromboembolic hypertension (CTEPH) where perfusion abnormalities could be detected compared to healthy individuals (Voskrebenzev *et al.*, 2018), consistent with DCE-MRI and dual-energy CT (Moher Alsady *et al.*, 2021). Both the amplitude and frequency shift of the RBC [129]Xe MRI signal have been shown to provide a distinct signature of disease, permitting the differentiation of pulmonary hypertension and left heart failure from COPD and idiopathic pulmonary fibrosis (Wang *et al.*, 2019).

16.6.4 Pediatric lung diseases

Radiation risk-free approaches such as MRI are desirable for imaging the lungs of children, particularly to repeatedly monitor disease progression or response to treatment. Congenital lung malformations can be characterized using DCE-MRI (Kellenberger *et al.*, 2020) and FB MRI where perfusion deficits have been shown to be consistent with LCI (Willers *et al.*, 2022). Fig. 16.6 shows the feasibility of PREFUL MRI for perfusion imaging in neonates and infants and demonstrating detection of perfusion defects in a neonate consistent with structural UTE imaging. Lung perfusion reductions due to primary ciliary dyskinesia (PCD), an inherited disorder difficult to diagnose, have been shown using FB MRI (Nyilas *et al.*, 2018). DCE-MRI reveals pulmonary blood flow impairment in children with congenital

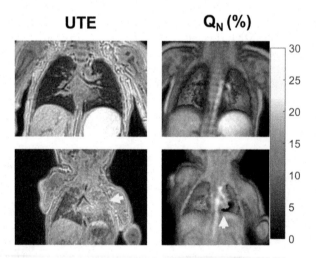

FIG. 16.6

Representative structural (UTE) images and normalized perfusion (Q_N) images obtained with PREFUL from a healthy 8-month-old female recruited from the general population (top) and a 4-day-old female neonate recruited from the NICU with clinical history of respiratory distress at birth and neonatal encephalopathy (bottom). Yellow (gray in print version) arrows highlight regions of structural and perfusion abnormalities in the left lung (Zanette *et al.*, 2022).

Figure courtesy of Dr. Brandon Zanette.

diaphragmatic hernia (CDH) which correlates with PFTs (Gross *et al.*, 2022). Chronic lung disease of prematurity (CLDP), including bronchopulmonary dysplasia (BPD), is a growing concern as almost 1 in 10 children are born prematurely, and almost half those babies require ventilator support. BPD is characterized by pulmonary vascular stunting and associated with obstructive lung disease later in life. ^{129}Xe MRI has been shown to reveal changes in RBC signals in a hyperoxic rat model of BPD compared to controls (Fliss *et al.*, 2021).

16.7 Challenges and solutions

Despite the considerable promise of perfusion MRI of the lung, several challenges remain to be surmounted. Exogenous contrast-enhanced approaches, while quantitative, require an injection of paramagnetic agent which may be contraindicated. This will require improved endogenous approaches (*e.g.*, ASL and free-breathing MRI) which will need to be demonstrated to cover the entire lung, be slice orientation independent and have sufficient spatial and temporal resolution to provide quantitative perfusion maps. Such improvements in speed and image quality will likely be facilitated in future by non-Cartesian k-space trajectories and ultra-short echo time strategies as well as the use of compressed sensing. Hyperpolarized ^{129}Xe and other inert gas approaches for lung perfusion MRI offer exciting potential, especially for elucidating vascular and interstitial lung disease, but are not widely available as yet. Implementation on a wide range of clinical systems will be required to validate and confirm clinical potential of these novel approaches in future. Finally, robust image analysis tools for quantitative lung perfusion MRI remain to be developed and validated, including automated image segmentation and feature extraction. In future, it is likely that automated image processing algorithms, such as deep learning, will play an increasingly important role as experience with lung perfusion MRI continues to grow.

16.8 Learning and knowledge outcomes

Technology advances in recent years have contributed to significant improvements in image quality, positioning perfusion MRI to make important contributions to diagnosis of obstructive pulmonary diseases, hypertension, pulmonary fibrosis, and other lung diseases. Especially for monitoring of disease progression and regression following treatment, lung perfusion MRI may be well suited compared to X-ray imaging since it involves no ionizing radiation, an important consideration for vulnerable populations such as children. While dynamic imaging following intravenous injection of a paramagnetic agent, such as Gd-DTPA, will continue to provide a mainstay for quantitative perfusion MRI of the lung, gaseous agents such as oxygen, perfluorinated gases, and hyperpolarized ^{129}Xe gas are poised to provide complementary information, based on differences in perfusion reflected by changes in gas transfer. Novel forms of endogenous contrast in the lung due to variations in blood inflow during and between image acquisition, such as free-breathing (FB) MRI and arterial spin labeling (ASL), are promising approaches for perfusion imaging of the lung without the need for intravenous agents or ventilated gases. These endogenous approaches are expected to play an increasing role in perfusion MRI of the lung as they are more translatable and can be deployed more broadly in nonresearch settings, especially when exogenous approaches are contraindicated or unavailable. Furthermore, these methods typically require minimal patient compliance, which may prove essential for very young or very sick patients.

Acknowledgments

Thanks to Dr. Brandon Zanette and Samal Munidasa for important contributions and assistance with figures and proofreading the manuscript.

References

Adolphi, N.L., Kuethe, D.O., 2008. Quantitative mapping of ventilation-perfusion ratios in lungs by 19F MR imaging of T1 of inert fluorinated gases. Magn. Reson. Med. 59 (4), 739–746.

Amaxopoulou, C., Gnannt, R., Higashigaito, K., Jung, A., Kellenberger, C.J., 2018. Structural and perfusion magnetic resonance imaging of the lung in cystic fibrosis. Pediatr. Radiol. 48 (2), 165–175.

Bauman, G., Bieri, O., 2017. Matrix pencil decomposition of time-resolved proton MRI for robust and improved assessment of pulmonary ventilation and perfusion. Magn. Reson. Med. 77 (1), 336–342.

Bauman, G., Puderbach, M., Deimling, M., Jellus, V., Chefd'hotel, C., Dinkel, J., et al., 2009. Non-contrast-enhanced perfusion and ventilation assessment of the human lung by means of Fourier decomposition in proton MRI. Magn. Reson. Med. 62 (3), 656–664.

Bauman, G., Pusterla, O., Bieri, O., 2019. Functional lung imaging with transient spoiled gradient echo. Magn. Reson. Med. 81 (3), 1915–1923.

Behrendt, L., Voskrebenzev, A., Klimeš, F., Gutberlet, M., Winther, H.B., Kaireit, T.F., et al., 2020. Validation of automated perfusion-weighted phase-resolved functional lung (PREFUL)-MRI in patients with pulmonary diseases. J. Magn. Reson. Imaging 52 (1), 103–114.

Behrendt, L., Smith, L.J., Voskrebenzev, A., Klimeš, F., Kaireit, T.F., Pöhler, G.H., et al., 2022. A dual center and dual vendor comparison study of automated perfusion-weighted phase-resolved functional lung magnetic resonance imaging with dynamic contrast-enhanced magnetic resonance imaging in patients with cystic fibrosis. Pulm Circ. 12 (2), e12054.

Bier, E.A., Robertson, S.H., Schrank, G.M., Rackley, C., Mammarappallil, J.G., Rajagopal, S., et al., 2019. A protocol for quantifying cardiogenic oscillations in dynamic. NMR Biomed. 32 (1), e4029.

Bolar, D.S., Levin, D.L., Hopkins, S.R., Frank, L.F., Liu, T.T., Wong, E.C., et al., 2006. Quantification of regional pulmonary blood flow using ASL-FAIRER. Magn. Reson. Med. 55 (6), 1308–1317.

Bondesson, D., Schneider, M.J., Gaass, T., Kuhn, B., Bauman, G., Dietrich, O., *et al.*, 2019. Nonuniform Fourier-decomposition MRI for ventilation- and perfusion-weighted imaging of the lung. Magn. Reson. Med. 82 (4), 1312–1321.

Carinci, F., Meyer, C., Phys, D., Breuer, F.A., Triphan, S., Choli, M., *et al.*, 2015. Blood volume fraction imaging of the human lung using intravoxel incoherent motion. J. Magn. Reson. Imaging 41 (5), 1454–1464.

Couch, M.J., Ball, I.K., Li, T., Fox, M.S., Biman, B., Albert, M.S., 2019. 19 F MRI of the lungs using inert fluorinated gases: challenges and new developments. J. Magn. Reson. Imaging 49 (2), 343–354.

Dimitrov, I.E., Insko, E., Rizi, R., Leigh, J.S., 2005. Indirect detection of lung perfusion using susceptibility-based hyperpolarized gas imaging. J. Magn. Reson. Imaging 21 (2), 149–155.

Doganay, O., Stirrat, E., McKenzie, C., Schulte, R.F., Santyr, G.E., 2016. Quantification of regional early stage gas exchange changes using hyperpolarized (129)Xe MRI in a rat model of radiation-induced lung injury. Med. Phys. 43 (5), 2410.

Du, Y., Zhang, S., Liang, T., Shang, J., Guo, C., Lian, J., *et al.*, 2022. Dynamic contrast-enhanced MRI perfusion parameters are imaging biomarkers for angiogenesis in lung cancer. Acta Radiol. 64, 572–580. https://doi.org/10.1177/02841851221088581.

Edelman, R.R., Siewert, B., Adamis, M., Gaa, J., Laub, G., Wielopolski, P., 1994. Signal targeting with alternating radiofrequency (STAR) sequences: application to MR angiography. Magn. Reson. Med. 31 (2), 233–238.

Edelman, R.R., Hatabu, H., Tadamura, E., Li, W., Prasad, P.V., 1996. Noninvasive assessment of regional ventilation in the human lung using oxygen-enhanced magnetic resonance imaging. Nat. Med. 2 (11), 1236–1239.

Elborn, J.S., 2016. Cystic fibrosis. Lancet 388 (10059), 2519–2531.

Ercolani, G., Capuani, S., Antonelli, A., Camilli, A., Ciulla, S., Petrillo, R., *et al.*, 2021. IntraVoxel incoherent motion (IVIM) MRI of fetal lung and kidney: can the perfusion fraction be a marker of normal pulmonary and renal maturation? Eur. J. Radiol. 139, 109726.

Fan, L., Liu, S.Y., Xiao, X.S., Sun, F., 2010. Demonstration of pulmonary perfusion heterogeneity induced by gravity and lung inflation using arterial spin labeling. Eur. J. Radiol. 73 (2), 249–254.

Fink, C., Risse, F., Semmler, W., Schoenberg, S.O., Kauczor, H.U., Reiser, M.F., 2006. MRI of pulmonary perfusion. Radiologe 46 (4), 290. 2-6, 8-9.

Fischer, A., Pracht, E.D., Arnold, J.F., Kotas, M., Flentje, M., Jakob, P.M., 2008. Assessment of pulmonary perfusion in a single shot using SEEPAGE. J. Magn. Reson. Imaging 27 (1), 63–70.

Fischer, A., Weick, S., Ritter, C.O., Beer, M., Wirth, C., Hebestreit, H., *et al.*, 2014. SElf-gated non-contrast-enhanced FUnctional lung imaging (SENCEFUL) using a quasi-random fast low-angle shot (FLASH) sequence and proton MRI. NMR Biomed. 27 (8), 907–917.

Fliss, J.D., Zanette, B., Friedlander, Y., Sadanand, S., Lindenmaier, A.A., Stirrat, E., *et al.*, 2021. Hyperpolarized 129Xe magnetic resonance spectroscopy in a rat model of bronchopulmonary dysplasia. Am. J. Physiol. Lung Cell. Mol. Physiol. 321 (3), L507–L517.

Friedlander, Y., Zanette, B., Lindenmaier, A.A., Fliss, J., Li, D., Emami, K., *et al.*, 2021. Effect of inhaled oxygen concentration on (129) Xe chemical shift of red blood cells in rat lungs. Magn. Reson. Med. 86 (3), 1187–1193.

Futura Publishing Company, Inc, 1996. Application of Magnetic Resonance to the Study of Lung. Futura Publishing Company, Inc., New York.

Gaass, T., Dinkel, J., Bauman, G., Zaiss, M., Hintze, C., Haase, A., et al., 2012. Non-contrast-enhanced MRI of the pulmonary blood volume using two-compartment-modeled T1-relaxation. J. Magn. Reson. Imaging 36 (2), 397–404.

Gharagouzloo, C.A., McMahon, P.N., Sridhar, S., 2015. Quantitative contrast-enhanced MRI with superparamagnetic nanoparticles using ultrashort time-to-echo pulse sequences. Magn. Reson. Med. 74 (2), 431–441.

Glandorf, J., Klimes, F., Behrendt, L., Voskrebenzev, A., Kaireit, T.F., Gutberlet, M., et al., 2021. Perfusion quantification using voxel-wise proton density and median signal decay in PREFUL MRI. Magn. Reson. Med. 86 (3), 1482–1493.

Greer, J.S., Maroules, C.D., Oz, O.K., Abbara, S., Peshock, R.M., Pedrosa, I., et al., 2018. Non-contrast quantitative pulmonary perfusion using flow alternating inversion recovery at 3T: a preliminary study. Magn. Reson. Imaging 46, 106–113.

Grist, J.T., Chen, M., Collier, G.J., Raman, B., Abueid, G., McIntyre, A., et al., 2021. Hyperpolarized. Radiology 301 (1), E353–E360.

Gross, V., Zahn, K., Maurer, K., Wessel, L., Schaible, T., Schoenberg, S.O., et al., 2022. MR lung perfusion measurements in adolescents after congenital diaphragmatic hernia: correlation with spirometric lung function tests. Eur. Radiol. 32 (4), 2572–2580.

Gurney, P.T., Hargreaves, B.A., Nishimura, D.G., 2006. Design and analysis of a practical 3D cones trajectory. Magn. Reson. Med. 55 (3), 575–582.

Hatabu, H., Alsop, D.C., Listerud, J., Bonnet, M., Gefter, W.B., 1999. T2* and proton density measurement of normal human lung parenchyma using submillisecond echo time gradient echo magnetic resonance imaging. Eur. J. Radiol. 29 (3), 245–252.

Herrmann, K.H., Krämer, M., Reichenbach, J.R., 2016. Time efficient 3D radial UTE sampling with fully automatic delay compensation on a clinical 3T MR scanner. PloS One 11 (3), 1–16.

Hopkins, S.R., Prisk, G.K., 2010. Lung perfusion measured using magnetic resonance imaging: new tools for physiological insights into the pulmonary circulation. J. Magn. Reson. Imaging 32 (6), 1287–1301.

Hopkins, S.R., Sa, R.C., Prisk, G.K., Elliott, A.R., Kim, N.H., Pazar, B.J., et al., 2021. Abnormal pulmonary perfusion heterogeneity in patients with Fontan circulation and pulmonary arterial hypertension. J. Physiol. 599 (1), 343–356.

Jakob, P.M., Wang, T., Schultz, G., Hebestreit, H., Hebestreit, A., Hahn, D., 2004. Assessment of human pulmonary function using oxygen-enhanced T(1) imaging in patients with cystic fibrosis. Magn. Reson. Med. 51 (5), 1009–1016.

Johnson, K.M., Fain, S.B., Schiebler, M.L., Nagle, S., 2013. Optimized 3D ultrashort echo time pulmonary MRI. Magn. Reson. Med. 70 (5), 1241–1250.

Kellenberger, C.J., Amaxopoulou, C., Moehrlen, U., Bode, P.K., Jung, A., Geiger, J., 2020. Structural and perfusion magnetic resonance imaging of congenital lung malformations. Pediatr. Radiol. 50 (8), 1083–1094.

Kieninger, E., Willers, C., Röthlisberger, K., Yammine, S., Pusterla, O., Bauman, G., et al., 2022. Effect of salbutamol on lung ventilation in children with cystic fibrosis: comprehensive assessment using spirometry, multiple-breath washout, and functional lung magnetic resonance imaging. Respiration 101 (3), 281–290.

Klimes, F., Voskrebenzev, A., Gutberlet, M., Obert, A.J., Pohler, G.H., Grimm, R., et al., 2021. Repeatability of dynamic 3D phase-resolved functional lung (PREFUL) ventilation MR imaging in patients with chronic obstructive pulmonary disease and healthy volunteers. J. Magn. Reson. Imaging 54 (2), 618–629.

Klimeš, F., Voskrebenzev, A., Gutberlet, M., Kern, A.L., Behrendt, L., Grimm, R., *et al.*, 2021. 3D phase-resolved functional lung ventilation MR imaging in healthy volunteers and patients with chronic pulmonary disease. Magn. Reson. Med. 85 (2), 912–925.

Kunz, A.S., Weng, A.M., Wech, T., Knapp, J., Petritsch, B., Hebestreit, H., *et al.*, 2021. Non-contrast pulmonary perfusion MRI in patients with cystic fibrosis. Eur. J. Radiol. 139, 109653.

Le Bihan, D., Breton, E., Lallemand, D., Aubin, M.L., Vignaud, J., Laval-Jeantet, M., 1988. Separation of diffusion and perfusion in intravoxel incoherent motion MR imaging. Radiology 168 (2), 497–505.

Lechartier, B., Chaouat, A., Aubert, J.D., Schwitter, J., 2022. Swiss Society for Pulmonary H. magnetic resonance imaging in pulmonary hypertension: an overview of current applications and future perspectives. Swiss Med. Wkly. 152, w30055.

Lévy, S., Heiss, R., Grimm, R., Grodzki, D., Hadler, D., Voskrebenzev, A., *et al.*, 2022. Free-breathing low-field MRI of the lungs detects functional alterations associated with persistent symptoms after COVID-19 infection. Invest. Radiol. 57 (11), 742–751.

Ley, S., Ley-Zaporozhan, J., 2012. Pulmonary perfusion imaging using MRI: clinical application. Insights Imaging 3 (1), 61–71.

Mai, V.M., Berr, S.S., 1999. MR perfusion imaging of pulmonary parenchyma using pulsed arterial spin labeling techniques: FAIRER and FAIR. J. Magn. Reson. Imaging 9 (3), 483–487.

Mai, V.M., Hagspiel, K.D., Christopher, J.M., Do, H.M., Altes, T., Knight-Scott, J., *et al.*, 1999. Perfusion imaging of the human lung using flow-sensitive alternating inversion recovery with an extra radiofrequency pulse (FAIRER). Magn. Reson. Imaging 17 (3), 355–361.

Mai, V.M., Hagspiel, K.D., Altes, T., Goode, A.R., Williams, M.B., Berr, S.S., 2000. Detection of regional pulmonary perfusion deficit of the occluded lung using arterial spin labeling in magnetic resonance imaging. J. Magn. Reson. Imaging 11 (2), 97–102.

Månsson, S., Wolber, J., Driehuys, B., Wollmer, P., Golman, K., 2003. Characterization of diffusing capacity and perfusion of the rat lung in a lipopolysaccharide disease model using hyperpolarized 129Xe. Magn. Reson. Med. 50 (6), 1170–1179.

Meier, P., Zierler, K.L., 1954. On the theory of the indicator-dilution method for measurement of blood flow and volume. J. Appl. Physiol. 6 (12), 731–744.

Mendes Pereira, L., Wech, T., Weng, A.M., Kestler, C., Veldhoen, S., Bley, T.A., *et al.*, 2019. UTE-SENCEFUL: first results for 3D high-resolution lung ventilation imaging. Magn. Reson. Med. 81 (4), 2464–2473.

Moher Alsady, T., Kaireit, T.F., Behrendt, L., Winther, H.B., Olsson, K.M., Wacker, F., *et al.*, 2021. Comparison of dual-energy computer tomography and dynamic contrast-enhanced MRI for evaluating lung perfusion defects in chronic thromboembolic pulmonary hypertension. PloS One 16 (6), e0251740.

Montesi, S.B., Zhou, I.Y., Liang, L.L., Digumarthy, S.R., Mercaldo, S., Mercaldo, N., *et al.*, 2021. Dynamic contrast-enhanced magnetic resonance imaging of the lung reveals important pathobiology in idiopathic pulmonary fibrosis. ERJ Open Res. 7 (4).

Mori, T., Nomori, H., Ikeda, K., Kawanaka, K., Shiraishi, S., Katahira, K., *et al.*, 2008. Diffusion-weighted magnetic resonance imaging for diagnosing malignant pulmonary nodules/masses: comparison with positron emission tomography. J. Thorac. Oncol. 3 (4), 358–364.

Mugler, J.P., Altes, T.A., Ruset, I.C., Dregely, I.M., Mata, J.F., Miller, G.W., *et al.*, 2010. Simultaneous magnetic resonance imaging of ventilation distribution and gas uptake in the human lung using hyperpolarized xenon-129. Proc. Natl. Acad. Sci. U. S. A. 107 (50), 21707–21712.

Neal, M.A., Pippard, B.J., Simpson, A.J., Thelwall, P.E., 2020. Dynamic susceptibility contrast (19) F-MRI of inhaled perfluoropropane: a novel approach to combined pulmonary ventilation and perfusion imaging. Magn. Reson. Med. 83 (2), 452–461.

Nikolaou, P., Coffey, A.M., Walkup, L.L., Gust, B.M., Whiting, N., Newton, H., et al., 2013. Near-unity nuclear polarization with an open-source 129Xe hyperpolarizer for NMR and MRI. Proc. Natl. Acad. Sci. U. S. A. 110 (35), 14150–14155 (Research Support, N.I.H., Extramural Research Support, Non-U.S. Gov't Research Support, U.S. Gov't, Non-P.H.S.).

Norquay, G., Leung, G., Stewart, N.J., Wolber, J., Wild, J.M., 2017. Xe chemical shift in human blood and pulmonary blood oxygenation measurement in humans using hyperpolarized. Magn. Reson. Med. 77 (4), 1399–1408.

Nyilas, S., Bauman, G., Pusterla, O., Sommer, G., Singer, F., Stranzinger, E., et al., 2018. Structural and functional lung impairment in primary ciliary dyskinesia. Assessment with magnetic resonance imaging and multiple breath washout in comparison to spirometry. Ann. Am. Thorac. Soc. 15 (12), 1434–1442.

Ohno, Y., Murase, K., Higashino, T., Nogami, M., Koyama, H., Takenaka, D., et al., 2007. Assessment of bolus injection protocol with appropriate concentration for quantitative assessment of pulmonary perfusion by dynamic contrast-enhanced MR imaging. J. Magn. Reson. Imaging 25 (1), 55–65.

Ohno, Y., Hanamatsu, S., Obama, Y., Ueda, T., Ikeda, H., Hattori, H., et al., 2022. Overview of MRI for pulmonary functional imaging. Br. J. Radiol. 95 (1132), 20201053.

Pohler, G.H., Klimes, F., Voskrebenzev, A., Behrendt, L., Czerner, C., Gutberlet, M., et al., 2020. Chronic thromboembolic pulmonary hypertension perioperative monitoring using phase-resolved functional lung (PREFUL)-MRI. J. Magn. Reson. Imaging 52 (2), 610–619.

Pohler, G.H., Loffler, F., Klimes, F., Behrendt, L., Voskrebenzev, A., Gonzalez, C.C., et al., 2021. Validation of phase-resolved functional lung (PREFUL) magnetic resonance imaging pulse wave transit time compared to echocardiography in chronic obstructive pulmonary disease. J. Magn. Reson. Imaging 56, 605–615.

Rankine, L.J., Wang, Z., Wang, J.M., He, M., McAdams, H.P., Mammarappallil, J., et al., 2020. Xenon gas exchange magnetic resonance imaging as a potential prognostic marker for progression of idiopathic pulmonary fibrosis. Ann. Am. Thorac. Soc. 17 (1), 121–125.

Rengier, F., Melzig, C., Derlin, T., Marra, A.M., Vogel-Claussen, J., 2019. Advanced imaging in pulmonary hypertension: emerging techniques and applications. Int. J. Cardiovasc. Imaging 35 (8), 1407–1420.

Risse, F., Semmler, W., Kauczor, H.U., Fink, C., 2006. Dual-bolus approach to quantitative measurement of pulmonary perfusion by contrast-enhanced MRI. J. Magn. Reson. Imaging 24 (6), 1284–1290.

Risse, F., Kuder, T.A., Kauczor, H.U., Semmler, W., Fink, C., 2009. Suppression of pulmonary vasculature in lung perfusion MRI using correlation analysis. Eur. Radiol. 19 (11), 2569–2575.

Saunders, L.C., Hughes, P.J.C., Alabed, S., Capener, D.J., Marshall, H., Vogel-Claussen, J., et al., 2022. Integrated cardiopulmonary MRI assessment of pulmonary hypertension. J. Magn. Reson. Imaging 55 (3), 633–652.

Schiwek, M., Triphan, S.M.F., Biederer, J., Weinheimer, O., Eichinger, M., Vogelmeier, C.F., et al., 2022. Quantification of pulmonary perfusion abnormalities using DCE-MRI in COPD: comparison with quantitative CT and pulmonary function. Eur. Radiol. 32 (3), 1879–1890.

Schraml, C., Schwenzer, N.F., Martirosian, P., Boss, A., Schick, F., Schafer, S., *et al.*, 2012. Non-invasive pulmonary perfusion assessment in young patients with cystic fibrosis using an arterial spin labeling MR technique at 1.5 T. MAGMA 25 (2), 155–162.

Singh, D., Wild, J.M., Saralaya, D., Lawson, R., Marshall, H., Goldin, J., *et al.*, 2022. Effect of indacaterol/glycopyrronium on ventilation and perfusion in COPD: a randomized trial. Respir. Res. 23 (1), 26.

Tibiletti, M., Bianchi, A., Stiller, D., Rasche, V., 2016. Pulmonary perfusion quantification with flow-sensitive inversion recovery (FAIR) UTE MRI in small animal imaging. NMR Biomed. 29 (12), 1791–1799.

Torres, L., Kammerman, J., Hahn, A.D., Zha, W., Nagle, S.K., Johnson, K., *et al.*, 2019. Structure-function imaging of lung disease using ultrashort Echo time MRI. Acad. Radiol. 26 (3), 431–441.

Torres, L.A., Lee, K.E., Barton, G.P., Hahn, A.D., Sandbo, N., Schiebler, M.L., *et al.*, 2022. Dynamic contrast enhanced MRI for the evaluation of lung perfusion in idiopathic pulmonary fibrosis. Eur. Respir. J. 60 (4), 2102058. https://doi.org/10.1183/13993003.02058-2021.

Triphan, S.M., Jobst, B.J., Breuer, F.A., Wielputz, M.O., Kauczor, H.U., Biederer, J., *et al.*, 2015. Echo time dependence of observed T1 in the human lung. J. Magn. Reson. Imaging 42 (3), 610–616.

Viallon, M., Berthezene, Y., Decorps, M., Wiart, M., Callot, V., Bourgeois, M., *et al.*, 2000. Laser-polarized (3)He as a probe for dynamic regional measurements of lung perfusion and ventilation using magnetic resonance imaging. Magn. Reson. Med. 44 (1), 1–4.

Voskrebenzev, A., Vogel-Claussen, J., 2021. Proton MRI of the lung: how to tame scarce protons and fast signal decay. J. Magn. Reson. Imaging 53 (5), 1344–1357.

Voskrebenzev, A., Gutberlet, M., Klimeš, F., Kaireit, T.F., Schönfeld, C., Rotärmel, A., *et al.*, 2018. Feasibility of quantitative regional ventilation and perfusion mapping with phase-resolved functional lung (PREFUL) MRI in healthy volunteers and COPD, CTEPH, and CF patients. Magn. Reson. Med. 79 (4), 2306–2314.

Wang, Z., Bier, E.A., Swaminathan, A., Parikh, K., Nouls, J., He, M., *et al.*, 2019. Diverse cardiopulmonary diseases are associated with distinct xenon magnetic resonance imaging signatures. Eur. Respir. J. 54 (6).

Weatherley, N.D., Eaden, J.A., Hughes, P.J.C., Austin, M., Smith, L., Bray, J., *et al.*, 2021. Quantification of pulmonary perfusion in idiopathic pulmonary fibrosis with first pass dynamic contrast-enhanced perfusion MRI. Thorax 76 (2), 144–151.

Willers, C., Maager, L., Bauman, G., Cholewa, D., Stranzinger, E., Raio, L., *et al.*, 2022. School-age structural and functional MRI and lung function in children following lung resection for congenital lung malformation in infancy. Pediatr. Radiol. 52, 1255–1265.

Willmering, M.M., Robison, R.K., Wang, H., Pipe, J.G., Woods, J.C., 2019. Implementation of the FLORET UTE sequence for lung imaging. Magn. Reson. Med. 82 (3), 1091–1100.

Winther, H.B., Gutberlet, M., Hundt, C., Kaireit, T.F., Alsady, T.M., Schmidt, B., *et al.*, 2020. Deep semantic lung segmentation for tracking potential pulmonary perfusion biomarkers in chronic obstructive pulmonary disease (COPD): the multi-ethnic study of atherosclerosis COPD study. J. Magn. Reson. Imaging 51 (2), 571–579.

Zanette, B., Santyr, G., 2019. Accelerated interleaved spiral-IDEAL imaging of hyperpolarized. Magn. Reson. Med. 82 (3), 1113–1119.

Zanette, B., Stirrat, E., Jelveh, S., Hope, A., Santyr, G., 2018. Physiological gas exchange mapping of hyperpolarized. Med. Phys. 45 (2), 803–816.

Zanette, B., Schrauben, E.M., Munidasa, S., Stirrat, E., Couch, M.J., Grimm, R., Voskrebenzev, A., Vogel-Klaussen, J., Seethamraju, R., Macgowan, C.K., Greer, M.C., Tam, E.W.Y., Santyr, G., 2022. Clinical feasibility of structural and functional MRI in free-breathing neonates and infants. J. Magn. Reson. Imaging 55 (6), 1696–1707.

Zheng, Y., Li, J., Chen, K., Zhang, X., Sun, H., Li, S., *et al.*, 2021. Comparison of conventional DWI, intravoxel incoherent motion imaging, and diffusion kurtosis imaging in differentiating lung lesions. Front. Oncol. 11, 815967.

Zhu, X., Chan, M., Lustig, M., Johnson, K.M., Larson, P.E.Z., 2020. Iterative motion-compensation reconstruction ultra-short TE (iMoCo UTE) for high-resolution free-breathing pulmonary MRI. Magn. Reson. Med. 83 (4), 1208–1221.

Applications of quantitative perfusion MRI in the liver

17

Maxime Ronot[a,b,c], Philippe Garteiser[b,c], and Bernard E. Van Beers[a,b,c]

[a]*Department of Radiology, APHP, University Hospitals Paris Nord Val de Seine, Beaujon, Clichy, Hauts-de-Seine, France*
[b]*University Paris Cité, Paris, France*
[c]*Laboratory of Imaging Biomarkers, INSERM U1149, Centre for Research on Inflammation, Paris, France*

17.1 Overview

The liver is a large, richly vascularized, solid, and mobile abdominal organ. The liver has a dual blood supply, with approximately 25% from arterial blood inflow (hepatic artery) and 75% from venous inflow (portal vein). On a microscopic scale, the vascular network is made not of continuous capillaries but of fenestrated sinusoids where the two afferent vascular systems communicate through trans-sinusoidal and transvasal connections, as well as peribiliary plexi. Around the sinusoids, the perisinusoidal space of Disse connects with the vascular pole of the hepatocytes. Additionally, adaptive mechanisms exist where an increase in the arterial blood supply compensates for a decrease in the portal venous supply. This arterioportal balance is called the "hepatic buffer response" (Itai and Matsui, 1997) and directly affects how one models the arterial and venous components of the hepatic perfusion and vascular permeability.

Perfusion imaging and quantification in the liver can be done with various MR imaging methods. Most applications rely on the injection of a gadolinium chelate contrast medium and on data acquisition by rapid temporal sampling of signal-time curves providing information on the variations in contrast agent concentrations over time. The resulting curves can be analyzed with descriptive semiquantitative approaches to extract simple parameters such as the time to peak, the blood flow or volume, or the mean transit time. Mathematical perfusion models can also be used, with increasing complexity levels depending on the number of vascular inputs and compartments considered. The most frequently used models include two compartments that allow extraction of the transfer constant between plasma and extravascular extracellular space (K^{trans}, ml min^{-1} 100 g^{-1}) and the extravascular extracellular fractional volume (v_e, %).

Advances in Magnetic Resonance Technology and Applications, Volume 11, ISSN 2666-9099
https://doi.org/10.1016/B978-0-323-95209-5.00004-0

Other MRI methods include evaluating liver perfusion with arterial spin labeling (Martirosian *et al.*, 2019) or with diffusion MRI, using the intravoxel incoherent motion (IVIM) model that assumes that part of the restriction in diffusion is related to changes in hepatic microperfusion (Luciani *et al.*, 2008).

Perfusion quantification in the liver faces specific challenges because of its size, significant deformations with breathing motion, dual blood supply, and changing sinusoidal permeability with disease. Perfusion quantification in the liver has been studied in two main contexts:

(1) Assessment of liver tumors, particularly malignant tumors including hepatocellular carcinomas (HCC) and liver metastases, to improve tumor detection and characterization, treatment response assessment, and patient follow-up.

(2) Investigating chronic liver diseases to monitor disease severity and refine patient prognostication.

The purpose of this chapter is to describe the technical challenges and clinical applications of MRI perfusion quantification in the liver.

17.2 Imaging protocol

The optimal liver perfusion analysis requires high spatial resolution to examine small lesions or focal areas, a high temporal resolution to correctly resolve the rapid signal changes caused by the contrast agent, adequate modeling of liver perfusion, and compatibility with anatomical imaging methods for combined assessment of perfusion and conventional metrics. The objective is to quantify both the whole liver and focal lesions accurately. Notably, the need for extensive coverage and high temporal and spatial resolution increases data collection time.

17.2.1 Dynamic contrast-enhanced MRI

Classically, dynamic contrast-enhanced MRI (DCE-MRI) is performed with a T_1-weighted fat-saturated three-dimensional imaging sequence and Cartesian k-space sampling in a breath-hold. The imaging sequence lasts 10–20 s and is repeated during the arterial phase (about 30–40 s after diffusible contrast agent injection), portal venous phase (60–70 s after injection), and delayed phase (3–5 min postinjection). With this method, the arterial input function variations cannot be sampled adequately. Moreover, the method is susceptible to respiratory motion, resulting in suboptimal image quality in patients who cannot correctly hold their breath. For perfusion imaging, correct sampling of the concentration vs time curves is essential. A temporal resolution <3 s is usually recommended, but a resolution of 1 s may be needed to visualize aortic peak enhancement.

Several methods have been developed to obtain time-resolved MRI of the liver. In patients with diffuse liver disease, single-section time-resolved perfusion MRI was obtained by continuously scanning with a T_1-weighted fast spoiled gradient-echo sequence, preceded by a non-section-selective 90° pulse and a spoiler gradient. To limit motion artifacts, cardiac triggering and respiratory tracking were used. The time resolution was <1 s (Annet *et al.*, 2003). More recently, a three-dimensional (3D) perfusion MRI has been performed with a T_1-weighted spoiled gradient-echo sequence in the coronal plane and a time resolution of 2–5 s (Hagiwara *et al.*, 2008).

The perfusion methods can be further accelerated by combining non-Cartesian *k*-space sampling, typically using radial or spiral trajectories, with spatial and temporal undersampling (Chen *et al.*, 2015; Chandarana *et al.*, 2013). Images can be reconstructed from this accelerated data by exploiting data redundancies using parallel imaging and/or compressed sensing methods (Wright *et al.*, 2014). Free-breathing 3D perfusion MRI of the liver with a time resolution of 2 s is feasible with these advanced methods, including golden-angle radial sparse parallel (GRASP) imaging (Feng *et al.*, 2016) and undersampled spiral acquisition with 3D through-time generalized autocalibrating partially parallel acquisition (GRAPPA) (Hagiwara *et al.*, 2008). GRASP not only speeds up the scan time by using compressed sensing and parallel imaging concurrently, but it is also less sensitive to motion than Cartesian sampling owing to its radial acquisition scheme. It has recently been reported in patients with liver fibrosis and HCC that a single MRI examination with free-breathing GRASP can provide both images with a sufficient spatial resolution for anatomic evaluation and a high temporal resolution for pharmacokinetic modeling of liver and tumors (Weiss *et al.*, 2019; Yoon *et al.*, 2022).

17.2.2 Other methods

Dynamic susceptibility contrast-MRI (DSC-MRI) has not been often used in the liver. Indeed, for susceptibility imaging, the contrast agent needs to be compartmentalized within the intravascular space. This compartmentalization occurs in the normal brain when small-molecular-weight contrast agents are injected because of an intact blood-organ barrier. In contrast, small-molecular-weight contrast agents rapidly diffuse in the hepatic extravascular space of Disse because the sinusoids are fenestrated in the normal liver.

Another approach to evaluate liver perfusion uses sequences that measure molecular diffusion (diffusion-weighted MRI) with the intravoxel incoherent motion (IVIM) model. IVIM assumes that local perfusion influences the diffusion of water molecules at a microscopic level. This method was initially developed by Le Bihan *et al.* (1986), who developed the concept that water circulating in randomly oriented capillaries (at a voxel level) imitates random motion "pseudo-diffusion." This motion attenuates the diffusion MRI signal, which depends upon the speed of the circulating blood and the vascular architecture. Like true molecular diffusion, the effect of pseudo-diffusion on signal attenuation depends on the *b* value. However, the amount of signal attenuation from pseudo-diffusion is larger than that of molecular tissue diffusion. Therefore the relative contribution of pseudo-diffusion to the diffusion-weighted signal is only significant at very low *b* values, allowing differentiating between diffusion and perfusion components. The IVIM parameters include the true diffusion coefficient (D or D_{slow}), related to perfusion-free molecular diffusion restriction in tissue (the slow component of diffusion); pseudo-diffusion or perfusion-related diffusion (D* or D_{fast}), which is linked to perfusion; and the perfusion fraction (f), representing the fractional volume occupied by flowing spins in the voxel. Finally, liver perfusion assessment without contrast material injection can also be performed with arterial spin labeling (Martirosian *et al.*, 2019), but this approach is limited to research studies and is not used in the clinic.

17.3 **Data analysis protocol**

Two main approaches are used to quantify liver perfusion with DCE-MRI: a model-free approach based on the analysis of the shape of the signal-time curves and a pharmacokinetic approach based on various perfusion models.

17.3.1 **Model-free methods**

The analysis of tissue signal-time curves is not based on any underlying physiological model or assumption (model-free approaches). Instead, hepatic signal enhancement curves are analyzed to produce parametric maps of curve parameters, including maximum signal enhancement, time to peak, the enhancement slope, or area under the curve (Roberts *et al.*, 2006). Moreover, computers can be used to automate the detection of hyperenhancing areas.

The model-free evaluation is confounded by the vascular input of the contrast agent, which may differ significantly among patients depending on the injection protocol (volume, concentration, injection rate) and physiological conditions (respiration, cardiac output). These effects can be accounted for to some extent by normalizing descriptive parameters in the liver to those of a reference tissue, such as the muscle or a nondiseased portion of the liver. Correction is performed by calculating ratios to obtain relative parameters. However, these simple approaches may be very sensitive to the quality of the acquisitions. Model-free methods have been mostly studied and applied in the early phase of the clinical applications of perfusion quantification of the liver. While it is still used by several teams, it has progressively been replaced by compartmental pharmacokinetic models both in research studies and in the clinic.

17.3.2 **Compartmental pharmacokinetic models**

Compartmental pharmacokinetic models use a top-down vision, i.e., these models aim to represent the liver as one or several functional well-mixed compartments to reproduce the MRI signal measurements. The models incorporate specific liver perfusion characteristics and can consider either the contribution of the arterial input (e.g., in tumors) or the arterial and portal venous input (e.g., in chronic liver disease) obtained from the signal enhancement curves within the afferent blood vessels.

The details of models have been described in Sections 1 and 2, but the peculiarities of hepatic models are worth discussing. A dual-input single-compartment model is often used (Materne *et al.*, 2002). In this model, the dual vascular supply is taken into consideration. The signal from the aorta and the portal vein is measured with two different inflow rate constants. The liver, comprising the liver sinusoids, extravascular and extracellular space (the space of Disse), is modeled as a single compartment assuming that lateral diffusional equilibration of small-molecular-weight contrast agents (such as extracellular gadolinium-based contrast agents) occurs virtually instantaneously because the sinusoids are fenestrated with pores of a diameter of 50–200 nm, as discussed before (Goresky and Rose, 1977). The mathematical equation for this dual-input single-compartment model is as follows:

$$\frac{dC_L(t)}{dt} = k_{1a}C_a(t - \tau_a) + k_{1p}C_p(t - \tau_p) - k_2C_L(t)$$

where $C_L(t)$, $C_a(t)$, and $C_p(t)$ (μM) represent the contrast agent concentration in the liver tissue, aorta, and portal vein compartments, respectively; k_{1a} represents the aortic inflow rate constant, k_{1p} represents the portal venous inflow rate constant, and k_2 represents the outflow rate constant in $ml\,min^{-1}\,100\,g^{-1}$, and τ_a and τ_b (min) represent the transit time from the aorta and portal vein to the liver.

Several additional features parameters can be derived from these parameters, such as the arterial fraction (AF, also called hepatic perfusion index, HPI (%)), the total liver perfusion (F, $ml\,min^{-1}\,100\,g^{-1}$), the mean transit time (MTT, min), or distribution volume (DV, %), as follows:

$$AF = HPI = \frac{k_{1a}}{k_{1a} + k_{1p}}$$

$$F = k_{1a} + k_{1p}$$

$$DV = \frac{k_{1a} + k_{1p}}{k_2}$$

$$MTT = \frac{1}{k_2}$$

To compute these perfusion parameters, the vascular inputs are extracted from regions of interest (ROI) drawn in the abdominal aorta and the portal vein. The delays for both inputs are usually included in the model to consider the temporal offset between the actual input to the compartment and the signals measured in the ROIs. Of course, adding these two fitting parameters increases the degrees of freedom of the models, with a risk of convergence to local minima or overfitting. A simpler version uses fixed delays. Here, the arterial delay is estimated from a curve analysis because of the sharp signal intensity increase, but the estimation of the portal venous delay can be more challenging. A simple approach assumes that the arterial-portal venous delays are equal (Materne *et al.*, 2002) or that the portal venous delay can be fixed to zero (Sourbron *et al.*, 2012). Another method is to include delays in the model and use advanced optimization procedures such as finding the optimum estimation from multiple initial conditions (Leporq *et al.*, 2012). The main limitation of these approaches is the increase in computing time because of the repetition of the fitting procedure.

The dual-input model is suited to model liver perfusion. However, hepatic tumors do not have a portal venous input, except at a very early stage, which can be disregarded (Liu and Matsui, 2007). Consequently, a simpler single-input model can be used. Goodness-of-fit techniques have shown that single- and dual-input models should be used in hepatic tumors and diffuse liver diseases, respectively (Banerji *et al.*, 2012). Several single-input models have been described to assess focal liver lesions. The most frequently used models (e.g., the Kety or Tofts model) include two compartments: the plasma and the extravascular extracellular space. In these models, the transfer constant between these two compartments (names K^{trans}, ml min^{-1} 100 g^{-1}) and the extravascular extracellular fractional volume (named ν_e, %) are assessed. An extended version of the Kety or Tofts model that considers the tracer's contribution to the vascular space has been described. K^{trans}, ν_e, and the fractional plasma volume (v_p, %) are assessed here. More complex models exist, such as the St. Lawrence and Lee model, which is a distributed-parameter model with four free parameters: the plasma perfusion (F, ml min^{-1} 100 g^{-1}), the extraction fraction (E, %), the extravascular volume (v_e, %), and the capillary mean transit time (MTT, min). Additional parameters such as the capillary permeability-surface (PS, ml min^{-1} 100 g^{-1}) or the fractional plasma volume (v_p, %) can be derived from this model (Michoux *et al.*, 2008).

There is some debate about the best tumor perfusion model to use. The St. Lawrence and Lee model is more physiologically relevant, because it separates perfusion from permeability. However, it is significantly less precise than the simpler compartmental models due to the significant interdependency of the free parameters and sensitivity to the initial values. The extended Tofts model is an excellent compromise to estimate tumor perfusion parameters, because supervised fitting or data interpolation is unnecessary (Leach *et al.*, 2003). K^{trans} and ν_e have been recommended as primary and secondary endpoints of perfusion measurements (Leach *et al.*, 2003). Both variables should be obtained, especially because the reproducibility of ν_e is better than that of K^{trans} and because v_e can be used as a surrogate for vascular permeability by determining the volume accessible to contrast agents with different molecular weights (Michoux *et al.*, 2008; Galbraith *et al.*, 2002; Padhani *et al.*, 2002).

Authors have suggested that a similar model may be used for assessing both the hepatic parenchyma (i.e., diffuse liver diseases) and focal lesions (i.e., tumors) instead of using a dual-input single-compartment model in the former and a single-input two-compartment model in the latter. For this purpose, a dual-input single-compartment model (Chen *et al.*, 2015; Pahwa *et al.*, 2018) or a dual-input two-compartment distributed model has been proposed (Koh *et al.*, 2008, 2011;

Ghodasara *et al.*, 2017). With the dual-input two-compartment model, the vascular and the extravascular/extracellular spaces are considered two different compartments. This assumption may be particularly relevant in advanced chronic liver disease (advanced fibrosis or cirrhosis) and in malignancies such as hepatocellular carcinomas in which sinusoids progressively transform into continuous capillaries (i.e., fenestrae are occluded) (Van Beers *et al.*, 2003). Overall, the diagnostic relevance of the various pharmacokinetic models must be assessed in context by comparing their goodness of fit, repeatability, and concordance with known perfusion parameter measurements (Shukla-Dave *et al.*, 2019).

17.4 Challenges and solutions

Several challenges in accurate perfusion quantification of hepatic perfusion are direct consequences of the use of perfusion MRI approaches: the nonlinear relationship between signal intensity and tracer concentration, as well as high sensitivity to signal intensity saturation and inflow artifacts. These limitations are discussed in Sections 1 and 2, and will not be discussed further here. Other challenges are more specific to the liver as an organ, including motion correction and the consequences of the large organ volume.

17.4.1 Correcting for liver movement

Radiographic studies using fixed landmarks, radioscopy, and CT have shown that the liver is an organ that moves considerably during breathing, shifting approximately 20 mm along its head–feet axis, 10 mm along its anteroposterior axis, and 5 mm laterally (Langen and Jones, 2001). However, because perfusion MRI usually lasts 2 to 3 min, breath-holding acquisitions cannot be obtained. Moreover, the liver deforms during the respiration cycle. Therefore precise image registration is important to ensure correct pixel scale matching and to maintain data quality for quantitative pharmacokinetic models.

Non-co-registered datasets of 2D images collected over time suffer from temporal noise and misregistration in both inputs and tissue functions because of respiratory motion. This results in a significant reduction of the robustness of perfusion measurements. This is particularly problematic for the portal venous input function because the main portal vein follows an oblique orientation relative to the head–feet and anteroposterior axes. Furthermore, this complex orientation is subject to significant intersubject variability. To cope with this, coronal views may be preferred. Another possibility is to include a registration-based motion correction method in the postprocessing chain. Many approaches have been described, from the most straightforward combination of rotation and translation to deep learning (Liu *et al.*, 2018). Some methods use segmented liver as region of interest to continuously register perfusion images despite intensity variations (Zhang *et al.*, 2020).

The registration methods usually include image deformation, computation of an objective function, and iterative minimization. The deformation and minimization are common issues in many other fields. However, finding the best objective function remains challenging in the liver as the MRI signal varies over time because of contrast agent transport. Image similarity (e.g., gradient, intensity, or entropy) is typically used, but modality-independent neighborhood descriptors (Towards Realtime Multimodal Fusion for Image-Guided Interventions Using Self-Similarities, 2020) show low sensitivity to contrast agent-induced intensity variation. Instead of registering one image onto another, some methods work with complete time series, i.e., time-resolved 2D images or time-resolved 3D volumes, at once. For example, principal component analysis-based group-wise registration (Huizinga *et al.*, 2016) is robust to contrast changes in time-resolved 3D quantitative MRI datasets.

17.4.1.1 *The liver is a large organ*
Imaging dynamic processes such as perfusion usually involve a compromise between spatial coverage and spatiotemporal resolution. As a result, the spatial resolution and scanned volume are typically reduced to improve the temporal resolution. This is not an issue for most focal lesions but can be problematic for small lesions (i.e., *circa* the spatial resolution) or heterogeneous diffuse diseases.

Partial k-space updating methods during 4D acquisition can also be used. These sequences combine keyhole acquisition and optimized k-space filling. Keyhole acquisition assumes that dynamic information is bandlimited in k-space. The keyhole methods update data in the center more frequently than in other parts of the k-space. For each dynamic step, only a central elliptical cylinder of the k-space (low frequencies including dynamic information) is refreshed with a random readout. Dynamic sequences are followed or preceded by the readout of the periphery of k-space (high frequencies including static information). This method drastically decreases the acquisition time. The acquisitions may be combined with parallel and partial Fourier imaging to decrease the acquisition time further. The time gained may be used to add slices and image the entire liver with adequate spatial and temporal resolution. The k-space filling scheme in these sequences also reduces motion sensitivity, an essential issue for free-breathing perfusion imaging.

17.4.2 **Reproducibility of liver perfusion measurements**
The perfusion parameters derived from perfusion MRI are subject to variation, depending on the acquisition strategy, the reconstruction methods, and the software (Goh *et al.*, 2007; Heye *et al.*, 2013a,b). For instance, Ng *et al.* have shown that the intrapatient coefficients of variation of K^{trans} and the initial area under the curve (AUC) for liver lesions were 9% and 10% (Ng *et al.*, 2010). In that study, a 20% reduction of a perfusion parameter after treatment resulted in confidence estimates changes ranging from 71% to 87%. To ensure variability is as limited as possible, several strategies can be applied. First, acquisition protocols should be standardized as much as possible in multicentric studies, and central reviewing should be favored

(Leach *et al.*, 2012). Second, hepatic perfusion measurements should be performed in fasting patients because the portal venous flow increases after a meal, thus impacting the arterial–portal venous balance discussed before (Jajamovich *et al.*, 2014). Third, it is recommended to perform measurement repeatability analysis before initiating clinical trials or studies involving perfusion quantification. This is especially important when perfusion MRI is used in longitudinal studies, for example, to assess response to treatment. In this case, the treatment effect on perfusion measurements (critical % change) should be higher than the repeatability index (%), which corresponds to $2.77 \times \text{wCV}$, where wCV is the within-subject coefficient of variation. For instance, K^{trans} has a wCV of 15%–20% based on a literature review. This suggests that a change of approximately 40%–55% is required in a patient to be confident that the detected change is not caused by the repeatability error (https://qibawiki.rsna.org/images/1/12/DCE-MRI_Quantification_Profile_v1.0.pdf, 2012). Of note, the reproducibility of D* in IVIM has also been shown to be poor, limiting its use to assess perfusion in routine clinical practice (Kakite *et al.*, 2015; Andreou *et al.*, 2013).

17.5 Case studies of applications

The two main clinical applications of liver perfusion are the assessment of the severity and prognostic of chronic advanced liver diseases and liver oncology. Because of its inherent complexity, perfusion is not meant to compete with morphological cross-sectional imaging methods. Instead, it is more clinically relevant to determine the potential benefits of adding perfusion mapping to conventional imaging and how such measurements may influence the management of patients.

17.5.1 Chronic and advanced liver diseases

Regardless of the cause, chronic liver injuries lead to inflammatory macrophage infiltration of the hepatic parenchyma and increased fibrogenic mediator release. The hepatic stellate cells in the perisinusoidal space of Disse undergo phenotypical activation into myofibroblasts, secreting large amounts of extracellular matrix proteins, primarily collagen, that accumulate in the extravascular space and lead to sinusoidal capillarization, i.e., loss of the sinusoidal fenestrations. The enlargement of the extravascular space, the capillarization of the sinusoids, and the active contraction of the activated stellate cells decrease the volume of the sinusoidal space and increase the intrahepatic vascular resistance leading to portal hypertension and portal venous inflow decrease (Friedman, 2003; Tacke, 2017). This perfusion decrease is only partly compensated by increased arterial perfusion (the buffer response). Therefore total liver perfusion decreases, and liver atrophies.

The first studies assessing perfusion changes in chronic liver diseases were published in the 1990s with perfusion CT. Miles *et al.* and Blomley *et al.* described the hepatic buffer response with a slope model (Miles *et al.*, 1993; Blomley *et al.*, 1995).

Using a dual-input single-compartment model, we confirmed the decrease of the whole liver perfusion and the increase in arterial fraction and mean transit time in patients with cirrhosis compared to normal controls and patients with various degrees of liver fibrosis deposition (Van Beers *et al.*, 2001). Our research team also showed that perfusion CT could identify the early stages of liver fibrosis. However, the significant overlap in the perfusion parameters between patient groups limited the clinical usefulness of the method in patients with mild and moderate fibrosis (Ronot *et al.*, 2010).

Initial MRI perfusion studies were performed in animal models of chronic liver disease and confirmed the previous results (Materne *et al.*, 2002; Van Beers *et al.*, 2003; Zhou *et al.*, 2013). Only subsequently were liver MRI perfusion studies performed in humans, which reported that in cirrhosis, a reduction of the portal venous perfusion and an increase of both the arterial perfusion and the mean transit time were seen (Annet *et al.*, 2003; Hagiwara *et al.*, 2008; Baxter *et al.*, 2009). Perfusion changes were also observed in intermediate stages of liver fibrosis (Annet *et al.*, 2003; Hagiwara *et al.*, 2008) but were more marked in cirrhotic patients, where the perfusion changes were shown to correlate with the severity of liver dysfunction and portal hypertension (Annet *et al.*, 2003). Although these results call for confirmation in larger studies, they suggest that imaging perfusion quantification could help assess cirrhosis severity. Similar results were reported in a recent study using a dual-input two-compartment model. In this study, the total hepatic perfusion, the fractional volume of the vascular space, and the permeability–surface area product decreased, whereas the fractional volume of the extravascular extracellular space increased in patients with cirrhosis compared to normal controls (Pahwa *et al.*, 2018; Ghodasara *et al.*, 2017) (Fig. 17.1).

With IVIM, we showed that rats with liver fibrosis had reduced in vivo ADC (apparent diffusion coefficient) values compared to controls, although this difference disappeared ex vivo, underscoring the role of decreased perfusion in ADC decrease related to liver fibrosis (Annet *et al.*, 2007). Furthermore, an IVIM study published by Luciani *et al.* confirmed the decrease in ADC in patients with liver cirrhosis compared to controls and showed that the restriction in diffusion was caused mainly by changes in perfusion and, to a lesser extent, by the reduction in pure hepatic diffusion (Luciani *et al.*, 2008). These results were confirmed by Yoon *et al.* in patients with chronic liver disease [111] and by Chow *et al.* and Zhang *et al.* in rodents (Chow *et al.*, 2012; Zhang *et al.*, 2013).

Several teams have more recently investigated the IVIM model for assessing steatohepatitis with contradictory results. A study by Lee *et al.* did not show any significant correlation between the hepatic fat fraction and IVIM-assessed perfusion parameters (Guiu *et al.*, 2012), while Guiu *et al.* (Guiu *et al.*, 2012) found decreased parenchymal perfusion in liver steatosis with IVIM diffusion MRI. In a study in rabbits by Joo *et al.*, the perfusion fraction was found to be significantly lower in rabbits with nonalcoholic fatty liver disease (NAFLD) than in controls with a progressive perfusion decrease as the severity of NAFLD increased (Joo *et al.*, 2014).

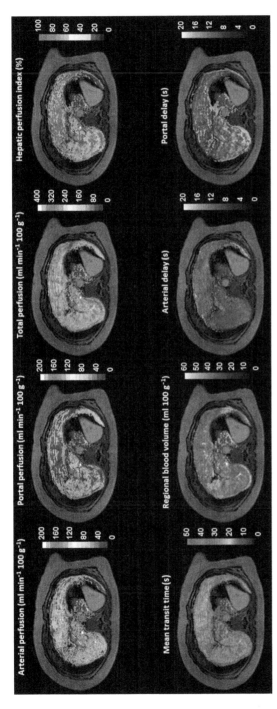

FIG. 17.1

Examples of parametric perfusion maps (arterial, portal, and total perfusion; hepatic perfusion index; mean transit time; and regional blood volume) obtained from a patient with cirrhosis. These maps were computed from images with a 4D keyhole acquisition with a 1.7-s temporal resolution. Gd-DOTA was used as a tracer. Perfusion parameters were extracted using a nonlinear least square fit on a dual-input one-compartment model, including delays (arterial and portal). These maps well illustrate the adaptive response in the liver, aiming to counterbalance the portal perfusion drop by increasing arterial perfusion to keep the total perfusion. This explains the significant rise in the hepatic perfusion index. Mean transit time increased, illustrating the restriction of the low-molecular-weight tracer molecules in the Disse due to collagen deposition in liver cirrhosis. Mean perfusion parameter values were arterial, portal, and total perfusion: 72, 54, and 126 mL min^{-1} 100 g^{-1}, respectively; hepatic perfusion index: 57.1%; mean transit time: 27.2 s; and regional blood volume: 57.1 mL 100 g^{-1}.

Finally, arterial spin labeling (ASL) has been marginally investigated (Martirosian *et al.*, 2019). This method has observed changes in liver perfusion after a meal challenge, and differences in response to the meal challenge between controls and patients with liver cirrhosis have been detected (Cox *et al.*, 2019). In general, however, ASL is challenging in the liver due to organ motion, dual vascular input, and a small hepatic artery that is not amenable to targeting.

17.5.2 Hepatic tumors

Conventional MRI acquisition sequences for assessing liver tumors include non-contrast-enhanced T_2-weighted, T_1-weighted, and diffusion-weighted imaging and DCE T_1-weighted imaging acquired during the arterial, portal venous, and delayed phases. MRI is accurate for detecting, characterizing, and monitoring liver tumors. Hepatic tumors' signal intensity and enhancement patterns have been extensively described (https://www.acr.org/-/media/ACR/Files/RADS/LI-RADS/LI-RADS-2018-Core.pdf, 2023). However, in clinical practice, images are analyzed visually, and quantitative signal intensity enhancement or perfusion quantification is not routinely performed.

Imaging is also central in assessing antitumoral locoregional or systemic treatments. In most cases, purely visual and morphological (i.e., nonquantitative) imaging criteria are used. The "response evaluation criteria in solid tumors" (RECIST) is the most validated response classification. It is used worldwide, based on the measurement and monitoring of the tumors' largest diameters. These criteria are approved as the standard method to evaluate tumor response in clinical trials and are considered a surrogate endpoint to predict survival in patients with solid tumors (Therasse *et al.*, 2000; Eisenhauer *et al.*, 2009). However, for liver tumors, especially hepatocellular carcinomas (HCC), the RECIST criteria fail to grasp the oncological benefits of treatments that induce intratumoral changes, including necrosis but do not necessarily translate into tumor shrinkage.

Several other imaging criteria have been proposed to assess the effects of treatment on hepatic tumors. With these criteria, the tumor viability and vascularization are assessed based on changes in signal enhancement after contrast material injection (Bruix *et al.*, 2001; Lencioni and Llovet, 2010; Choi *et al.*, 2007). These criteria include the modified version of the RECIST criteria (mRECIST) (Lencioni and Llovet, 2010) and the European Association for the Study of the Liver (EASL) criteria in which the viable portion of the tumors (i.e., areas showing contrast enhancement on the arterial phase) is measured. Additionally, with the Choi criteria (Choi *et al.*, 2007), the tumor diameter and signal intensity enhancement are measured on the CT scans.

Morphological changes, e.g., tumor diameter decrease, may appear weeks after treatment initiation. Patients could benefit from better detection of early modifications to modulate their treatment. Quantitative perfusion methods can potentially discriminate responders from nonresponders at early time points, because they capture subtle vascularization changes that occur early (Kamel *et al.*, 2009). Moreover, quantitative perfusion might help predict patient response before treatment (Nakamura *et al.*, 2018; Luo *et al.*, 2019).

17.5.3 **Detection and characterization of focal liver lesions**

MRI is very accurate for the detection of hepatic tumors (Takahashi *et al.*, 2003). Studies have shown that parameters derived from model-free visual perfusion approaches differ between benign tumors and malignancies (Alicioglu *et al.*, 2013; Chen *et al.*, 2018). Liver malignancies can be characterized using either a dual-input single-compartment model or a dual-input two-compartment model (e.g., metastases vs HCC) (Pahwa *et al.*, 2018; Ghodasara *et al.*, 2017). It has also been reported that hypoenhancing and hyperenhancing neuroendocrine liver metastases that correspond to different tumor biology can be differentiated based on differences in perfusion and distribution volume (Koh *et al.*, 2011).

Studies in patients with liver cirrhosis have assessed the value of perfusion MRI for differentiating between HCC precursors (dysplastic nodules) and overt tumors. The arterial hepatic blood flow and the arterial fraction progressively increase toward HCC compared to the surrounding hepatic parenchyma, while the distribution volume and the portal venous blood flow significantly decrease (Taouli *et al.*, 2013). Similar results have been published with perfusion CT (Ippolito *et al.*, 2008a,b). Moreover, the perfusion parameters evaluated with perfusion CT were correlated with tumor differentiation. Well-differentiated HCC harbor significantly higher perfusion values than less differentiated tumors (Sahani *et al.*, 2007).

Overt hepatic metastases result in significant hemodynamic changes in the liver, including an increase in arterial perfusion and a decrease in portal venous perfusion. These liver hemodynamic changes were first described with scintigraphy and CT (Miles and Kelley, 1997; Leggett *et al.*, 1997; Meijerink *et al.*, 2008) and confirmed with MRI (Totman *et al.*, 2005). However, from a clinical point of view, detecting hepatic hemodynamic changes in a metastatic liver is of limited value because the tumors can already be detected with morphological imaging sequences. Studies showing similar results in small animals with occult liver metastases (with either Doppler or perfusion-CT) (Totman *et al.*, 2005; Leen *et al.*, 1996; Tsushima *et al.*, 2001; Cuenod *et al.*, 2001) are more interesting, although the results have not yet been confirmed in humans.

Some results with IVIM have been reported in patients with liver tumors. We observed that the IVIM parameters did not improve the determination of malignancy and the characterization of hepatic tumor type compared to the ADC (Doblas *et al.*, 2013). The limited usefulness of IVIM in assessing liver diseases and tumors might be partially explained by the poor reproducibility of the perfusion-related diffusion parameters (Kakite *et al.*, 2015).

17.5.4 **Assessing response to treatment—Locoregional therapies**

Assessing liver tumor response and tumor recurrence after percutaneous ablation (mainly after radiofrequency or microwave ablation) is based on the depiction of complete tumor destruction (referred to as complete ablation) and on the reappearance of viable tumor tissue, respectively. This has been extensively described with cross-sectional imaging and proved robust (Chopra *et al.*, 2001; Forner *et al.*, 2009).

Therefore perfusion quantification is usually not needed. However, it has been suggested in a perfusion CT study that the quantification of blood volume may help detect local tumor recurrence (Meijerink et al., 2009), but similar results have not been reported with MRI to date.

Perfusion CT and MR imaging have been assessed in patients undergoing intraarterial treatments, especially chemoembolization (TACE). Significant changes in tumor and peri-tumor perfusion parameters occur after chemoembolization because of a combination of ischemia and drug-related modifications. Choi et al. prospectively evaluated the feasibility of using perfusion CT for the follow-up of TACE in rabbits and showed that arterial perfusion decrease changes could be observed as early as 1 week after treatment. The authors concluded that perfusion CT could help assess the early response to treatment (Choi et al., 2010). However, very few clinical studies have been conducted. Ippolito et al. used DCE-MRI in tumor residues after TACE. They found differences in relative arterial enhancement, relative venous enhancement, relative late enhancement, maximum enhancement, and time to peak between completely and incompletely treated tumors similar to those of Choi et al. in animals (Ippolito et al., 2016).

DCE-MRI to quantify tumor perfusion after embolization has been assessed in a rabbit study (Wang et al., 2007). The authors showed that MRI could depict serial reductions in liver tumor perfusion during transarterial embolization. A group from Chicago has also reported significant changes in perfusion during embolization using a model-free approach (Larson et al., 2008; Gaba et al., 2008; Wang et al., 2010). After the treatments, they reported a significant decrease in the AUC of time–intensity curves. Similar results were reported by Taouli et al., who observed higher arterial fraction and lower portal venous hepatic blood flow in untreated HCCs compared to those treated with TACE (Taouli et al., 2013).

The use of perfusion models in depicting early changes related to tumor necrosis or predicting good or poor responders has been reported. Braren et al. showed in a rodent model that v_e, the extravascular extracellular volume fraction, correlated with residual tumor. Its quantification 1 day after embolization was promising for predicting tumor necrosis (Braren et al., 2011). Michielsen et al. showed that perfusion parameters before TACE were predictive of progression-free survival, independent of tumor size and the number of lesions (Michielsen et al., 2011).

Selective internal radiation therapy (SIRT) is a more recent locoregional treatment for hepatic malignancies (Memon et al., 2011). Glass or resin microspheres labeled with a radionuclide (usually 90-yttrium) are injected into the tumor feeders. Like other intraarterial treatments, SIRT is usually performed in nonresectable or diffuse, infiltrating, and/or multifocal tumors. Tumor response assessment after SIRT is challenging because most changes occur 3 to 6 months after treatment. Morsbach et al. recently published preliminary results with perfusion CT in patients with colorectal liver metastases treated with SIRT (Morsbach et al., 2013). The authors showed that a significant difference in arterial perfusion was found in pretreatment CT perfusion between responders and nonresponders and that 1-year survival was significantly higher in patients with high pretreatment arterial perfusion.

17.5.5 **Assessing response to treatment—Systemic chemotherapy and targeted therapies**

Systemic treatment of liver metastases, especially those of colorectal origin, is based on cytotoxic agents that can be associated with targeted therapy. Several systemic treatments have improved overall survival in advanced HCC, including targeted therapies and immunomodulating agents. However, conventional cytotoxic agents cause tumor fibrosis and shrinkage without necrosis (Rubbia-Brandt et al., 2007), while targeted therapies lead to sustained tumor stabilization but reduce tumor vascularization and necrosis. Therefore the computation of perfusion changes after targeted therapy could be of major interest (Fig. 17.2). Although quantitative perfusion MRI is rarely used in oncologic trials, some studies have included this method in their design, aside from conventional tumor evaluation criteria with the RECIST.

Hsu et al. performed MRI-based perfusion quantification before and after 14 days of treatment in a phase II study evaluating sorafenib combined with tegafur/uracil as first-line therapy for advanced HCC (Hsu et al., 2011). They selected the tumor region with the highest signal enhancement and measured the volume transfer constant, K^{trans}. The baseline K^{trans} was significantly higher in patients with limited disease than those with progressive disease and significantly decreased after treatment. A vascular response, defined as $\geq 40\%$ decrease in K^{trans} after treatment, was associated with improved progression-free and overall survival. These results were confirmed in more recent studies by the same research group showing that the semiquantitative perfusion parameter "HCC peak enhancement" before systemic therapy and high peak reduction 1 week after systemic therapy were markers of good overall survival (Chen et al., 2016, 2017). Finally, another study evaluating sunitinib in treating advanced HCC found a significant decrease in K^{trans} in all patients and a larger decrease in patients with controlled disease than in those with progressive disease (Zhu et al., 2009). Recently, Onuoha et al. investigated whether the early perfusion change in HCC predicts the long-term therapeutic response to atezolizumab plus bevacizumab (Onuoha et al., 2022). The tumor-to-liver signal ratio in the arterial phase was used to estimate the tumor perfusion. The authors showed that the HCC perfusion changes of the responders were significantly higher than that of the nonresponders.

Most published studies evaluating liver metastases have focused on colorectal metastases treated with targeted therapies in combination with conventional cytotoxic agents, and similar results have been reported with K^{trans} and other perfusion parameters. Coenegrachts et al. showed that the baseline k_{ep} (k_{ep} is K^{trans}/v_e, i.e., the rate constant between the extravascular extracellular space and blood plasma) was significantly higher in responders than in nonresponders. They also showed that the responders had a significant decrease in k_{ep} after 6 weeks of treatment (Coenegrachts et al., 2012). De Bruyne et al. reported progression-free survival improvement in patients with $>40\%$ reduction in K^{trans} after treatment (De Bruyne et al., 2012). These results were supported by those from Hirashima et al. in lesions treated with bevacizumab and FOLFIRI chemotherapy, showing a

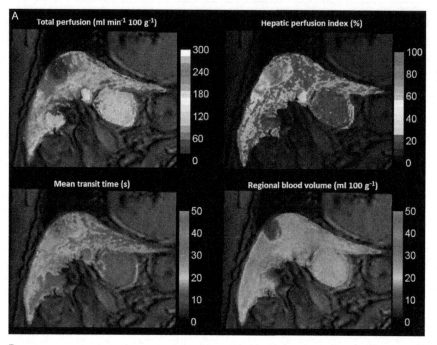

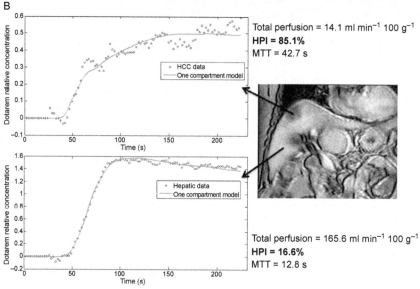

Total perfusion = 14.1 ml min^{-1} 100 g^{-1}
HPI = 85.1%
MTT = 42.7 s

Total perfusion = 165.6 ml min^{-1} 100 g^{-1}
HPI = 16.6%
MTT = 12.8 s

FIG. 17.2

Perfusion maps (A—arterial, portal venous, total perfusion; hepatic perfusion index; mean transit time; and regional blood volume) obtained from a patient with hepatocellular carcinoma 7 days after initiating sorafenib treatment. These maps are computed from images with a 3D FLASH acquisition with 2.0-s temporal resolution. Gadoteric acid is used as contrast agent. Perfusion parameters are extracted using a nonlinear least square fit on the dual-input single-compartment model, including two delays (arterial and portal venous). (B) The lesion is in the dome of the liver. Comparison between the lesion (*upper part*) and the surrounding liver (*lower part*). The lesion shows a substantial total perfusion decrease with a substantial hepatic perfusion index (HPI) and mean transit time (MTT) increase, consistent with tumor response to treatment (B).

variation in both K^{trans} and k_{ep} as early as 1 week after treatment to help predict response to chemotherapy (Hirashima *et al.*, 2012). There are few studies regarding other liver metastases, such as those from neuroendocrine tumors. However, existing results are consistent with those observed in colorectal liver metastases, with a lower pretreatment distribution volume and high arterial flow fraction being associated with a better response to treatment (Miyazaki *et al.*, 2012). Finally, in a mouse model of liver metastases, arterial spin labeling has been used to assess the acute hemodynamic changes caused by a targeted vascular disrupting agent (Johnson *et al.*, 2016).

17.6 Learning and knowledge outcomes

The current situation of perfusion MRI in the liver is somewhat paradoxical. Numerous studies reporting promising results in various clinical settings fail to translate into applications in clinical practice. While perfusion MRI could theoretically be used in combination with morphological imaging for the assessment of tumors and hepatic disease and could help solve unmet needs in various clinical scenarios, the expected application of these results into routine clinical practice has not occurred.

There are many possible explanations for this. First, quantitative imaging is difficult in the liver because it is a mobile, blood-filled, flexible organ with dual vascular input and fenestrated sinusoids. Data collection at relevant spatiotemporal resolution and 3D coverage are, therefore, very challenging. This requires advanced image acquisition methods or registration techniques. Second, patients have different morphotypes, breath-holding capacities, hepatic iron concentrations, and fasting states. Third, perfusion MRI packages are often trademarked, making direct comparison difficult. Furthermore, the MRI signal is affected by the magnetic field strength and the overall architecture of machines and antennas. Moreover, most published studies are single center and retrospective, without standardized acquisition parameters, postprocessing methods, or predicted outcomes. This results in significant intra- and intervendor, software, reader, and patient variability, as discussed before. Observed statistical differences in populations are, therefore, rarely applicable to individuals. There are also still concerns about the need for additional scanning time and the complexity of imaging interpretation.

Overcoming these limitations will require a significant collective effort. More clinical validation studies of liver perfusion mapping are needed. Importantly, the medical imaging community has never had more powerful tools to succeed, as illustrated by initiatives such as the Quantitative Imaging Biomarkers Alliance, the Radiological Society of North America (https://www.rsna.org/en/research/quantitative-imaging-biomarkers-alliance, 2023), the Biomarker Inventory, or the European Society of Radiology (https://www.myesr.org/research/biomarkers-inventory, 2022).

References

Alicioglu, B., Guler, O., Bulakbasi, N., Akpinar, S., Tosun, O., Comunoglu, C., 2013. Utility of semi-quantitative parameters to differentiate benign and malignant focal hepatic lesions. Clin. Imaging 37 (4), 692–696.

American College of Radiology, 2023. CT/MRI Liver Imaging Reporting and Data System v2018 Core. https://www.acr.org/-/media/ACR/Files/RADS/LI-RADS/LI-RADS-2018-Core.pdf. Accessed January 2023.

Andreou, A., Koh, D.M., Collins, D.J., et al., 2013. Measurement reproducibility of perfusion fraction and pseudodiffusion coefficient derived by intravoxel incoherent motion diffusion-weighted MR imaging in normal liver and metastases. Eur. Radiol. 23, 428–434.

Annet, L., Materne, R., Danse, E., Jamart, J., Horsmans, Y., Van Beers, B.E., 2003. Hepatic flow parameters measured with MR imaging and Doppler US: correlations with degree of cirrhosis and portal hypertension. Radiology 229 (2), 409–414.

Annet, L., Peeters, F., Abarca-Quinones, J., Leclercq, I., Moulin, P., Van Beers, B.E., 2007. Assessment of diffusion-weighted MR imaging in liver fibrosis. J. Magn. Reson. Imaging 25 (1), 122–128.

Banerji, A., Naish, J.H., Watson, Y., Jayson, G.C., Buonaccorsi, G.A., Parker, G.J., 2012. DCE-MRI model selection for investigating disruption of microvascular function in livers with metastatic disease. J. Magn. Reson. Imaging 35 (1), 196–203.

Baxter, S., Wang, Z.J., Joe, B.N., Qayyum, A., Taouli, B., Yeh, B.M., 2009. Timing bolus dynamic contrast-enhanced (DCE) MRI assessment of hepatic perfusion: initial experience. J. Magn. Reson. Imaging 29 (6), 1317–1322.

Blomley, M.J., Coulden, R., Dawson, P., et al., 1995. Liver perfusion studied with ultrafast CT. J. Comput. Assist. Tomogr. 19 (3), 424–433.

Braren, R., Altomonte, J., Settles, M., et al., 2011. Validation of preclinical multiparametric imaging for prediction of necrosis in hepatocellular carcinoma after embolization. J. Hepatol. 55 (5), 1034–1040.

Bruix, J., Sherman, M., Llovet, J.M., et al., 2001. Clinical management of hepatocellular carcinoma. Conclusions of the Barcelona-2000 EASL conference. European Association for the Study of the Liver. J. Hepatol. 35 (3), 421–430.

Chandarana, H., Feng, L., Block, T.K., et al., 2013. Free-breathing contrast-enhanced multiphase MRI of the liver using a combination of compressed sensing, parallel imaging, and golden-angle radial sampling. Invest. Radiol. 48 (1), 10–16.

Chen, Y., Lee, G.R., Wright, K.L., et al., 2015. Free-breathing liver perfusion imaging using 3-dimensional through-time spiral generalized autocalibrating partially parallel acquisition acceleration. Invest. Radiol. 50 (6), 367–375.

Chen, B.B., Hsu, C.Y., Yu, C.W., et al., 2016. Dynamic contrast-enhanced MR imaging of advanced hepatocellular carcinoma: comparison with the liver parenchyma and correlation with the survival of patients receiving systemic therapy. Radiology 281 (2), 454–464.

Chen, B.B., Hsu, C.Y., Yu, C.W., et al., 2017. Early perfusion changes within 1 week of systemic treatment measured by dynamic contrast-enhanced MRI may predict survival in patients with advanced hepatocellular carcinoma. Eur. Radiol. 27 (7), 3069–3079.

Chen, H.J., Roy, T.L., Wright, G.A., 2018. Perfusion measures for symptom severity and differential outcome of revascularization in limb ischemia: preliminary results with arterial spin labeling reactive hyperemia. J. Magn. Reson. Imaging 47 (6), 1578–1588.

Choi, H., Charnsangavej, C., Faria, S.C., et al., 2007. Correlation of computed tomography and positron emission tomography in patients with metastatic gastrointestinal stromal tumor treated at a single institution with imatinib mesylate: proposal of new computed tomography response criteria. J. Clin. Oncol. 25 (13), 1753–1759.

Choi, S.H., Chung, J.W., Kim, H.C., et al., 2010. The role of perfusion CT as a follow-up modality after transcatheter arterial chemoembolization: an experimental study in a rabbit model. Invest. Radiol. 45 (7), 427–436.

Chopra, S., Dodd 3rd, G.D., Chintapalli, K.N., Leyendecker, J.R., Karahan, O.I., Rhim, H., 2001. Tumor recurrence after radiofrequency thermal ablation of hepatic tumors: spectrum of findings on dual-phase contrast-enhanced CT. AJR Am. J. Roentgenol. 177 (2), 381–387.

Chow, A.M., Gao, D.S., Fan, S.J., et al., 2012. Liver fibrosis: an intravoxel incoherent motion (IVIM) study. J. Magn. Reson. Imaging 36 (1), 159–167.

Coenegrachts, K., Bols, A., Haspeslagh, M., Rigauts, H., 2012. Prediction and monitoring of treatment effect using T1-weighted dynamic contrast-enhanced magnetic resonance imaging in colorectal liver metastases: potential of whole tumour ROI and selective ROI analysis. Eur. J. Radiol. 81 (12), 3870–3876.

Cox, E.F., Palaniyappan, N., Aithal, G.P., Guha, I.N., Francis, S.T., 2019. Using MRI to study the alterations in liver blood flow, perfusion, and oxygenation in response to physiological stress challenges: meal, hyperoxia, and hypercapnia. J. Magn. Reson. Imaging 49 (6), 1577–1586.

Cuenod, C., Leconte, I., Siauve, N., et al., 2001. Early changes in liver perfusion caused by occult metastases in rats: detection with quantitative CT. Radiology 218 (2), 556–561.

De Bruyne, S., Van Damme, N., Smeets, P., et al., 2012. Value of DCE-MRI and FDG-PET/CT in the prediction of response to preoperative chemotherapy with bevacizumab for colorectal liver metastases. Br. J. Cancer 106 (12), 1926–1933.

Doblas, S., Wagner, M., Leitao, H.S., et al., 2013. Determination of malignancy and characterization of hepatic tumor type with diffusion-weighted magnetic resonance imaging: comparison of apparent diffusion coefficient and intravoxel incoherent motion-derived measurements. Invest. Radiol. 48 (10), 722–728.

Eisenhauer, E.A., Therasse, P., Bogaerts, J., et al., 2009. New response evaluation criteria in solid tumours: revised RECIST guideline (version 1.1). Eur. J. Cancer 45 (2), 228–247.

Feng, L., Axel, L., Chandarana, H., Block, K.T., Sodickson, D.K., Otazo, R., 2016. XD-GRASP: golden-angle radial MRI with reconstruction of extra motion-state dimensions using compressed sensing. Magn. Reson. Med. 75 (2), 775–788.

Forner, A., Ayuso, C., Varela, M., et al., 2009. Evaluation of tumor response after locoregional therapies in hepatocellular carcinoma: are response evaluation criteria in solid tumors reliable? Cancer 115 (3), 616–623.

Friedman, S.L., 2003. Liver fibrosis—from bench to bedside. J. Hepatol. 38 (Suppl 1), S38–S53.

Gaba, R.C., Wang, D., Lewandowski, R.J., et al., 2008. Four-dimensional transcatheter intraarterial perfusion MR imaging for monitoring chemoembolization of hepatocellular carcinoma: preliminary results. J. Vasc. Interv. Radiol. 19 (11), 1589–1595.

Galbraith, S.M., Lodge, M.A., Taylor, N.J., et al., 2002. Reproducibility of dynamic contrast-enhanced MRI in human muscle and tumours: comparison of quantitative and semi-quantitative analysis. NMR Biomed. 15 (2), 132–142.

Ghodasara, S., Pahwa, S., Dastmalchian, S., Gulani, V., Chen, Y., 2017. Free-breathing 3D liver perfusion quantification using a dual-input two-compartment model. Sci. Rep. 7 (1), 17502.

Goh, V., Halligan, S., Bartram, C.I., 2007. Quantitative tumor perfusion assessment with multidetector CT: are measurements from two commercial software packages interchangeable? Radiology 242 (3), 777–782.

Goresky, C.A., Rose, C.P., 1977. Blood-tissue exchange in liver and heart: the influence of heterogeneity of capillary transit times. Fed. Proc. 36 (12), 2629–2634.

Guiu, B., Petit, J.M., Capitan, V., et al., 2012. Intravoxel incoherent motion diffusion-weighted imaging in nonalcoholic fatty liver disease: a 3.0-T MR study. Radiology 265 (1), 96–103.

Hagiwara, M., Rusinek, H., Lee, V.S., et al., 2008. Advanced liver fibrosis: diagnosis with 3D whole-liver perfusion MR imaging—initial experience. Radiology 246 (3), 926–934.

Heye, T., Davenport, M.S., Horvath, J.J., et al., 2013a. Reproducibility of dynamic contrast-enhanced MR imaging. Part I. Perfusion characteristics in the female pelvis by using multiple computer-aided diagnosis perfusion analysis solutions. Radiology 266 (3), 801–811.

Heye, T., Merkle, E.M., Reiner, C.S., et al., 2013b. Reproducibility of dynamic contrast-enhanced MR imaging. Part II. Comparison of intra- and interobserver variability with manual region of interest placement versus semiautomatic lesion segmentation and histogram analysis. Radiology 266 (3), 812–821.

Hirashima, Y., Yamada, Y., Tateishi, U., et al., 2012. Pharmacokinetic parameters from 3-tesla DCE-MRI 103 surrogate biomarkers of antitumor effects of bevacizumab plus FOLFIRI in colorectal cancer with liver metastasis. Int. J. Cancer 130 (10), 2359–2365.

Hsu, C.-Y., Shen, Y.-C., Yu, C.-W., et al., 2011. Dynamic contrast-enhanced magnetic resonance imaging biomarkers predict survival and response in hepatocellular carcinoma patients treated with sorafenib and metronomic tegafur/uracil. J. Hepatol. 55 (4), 858–865.

https://qibawiki.rsna.org/images/1/12/DCE-MRI_Quantification_Profile_v1.0.pdf, 2012. Accessed January 5, 2023.

https://www.myesr.org/research/biomarkers-inventory, 2022. Accessed January 6, 2023.

Huizinga, W., Poot, D.H., Guyader, J.M., et al., 2016. PCA-based groupwise image registration for quantitative MRI. Med. Image Anal. 29, 65–78.

Ippolito, D., Sironi, S., Pozzi, M., et al., 2008a. Perfusion computed tomographic assessment of early hepatocellular carcinoma in cirrhotic liver disease: initial observations. J. Comput. Assist. Tomogr. 32 (6), 855–858.

Ippolito, D., Sironi, S., Pozzi, M., et al., 2008b. Hepatocellular carcinoma in cirrhotic liver disease: functional computed tomography with perfusion imaging in the assessment of tumor vascularization. Acad. Radiol. 15 (7), 919–927.

Ippolito, D., Trattenero, C., Talei Franzesi, C., et al., 2016. Dynamic contrast-enhanced magnetic resonance imaging with gadolinium ethoxybenzyl diethylenetriamine pentaacetic acid for quantitative assessment of vascular effects on hepatocellular-carcinoma lesions treated by transarterial chemoembolization or radiofrequency ablation. J. Comput. Assist. Tomogr. 40 (5), 692–700.

Itai, Y., Matsui, O., 1997. Blood flow and liver imaging. Radiology 202 (2), 306–314.

Jajamovich, G.H., Dyvorne, H., Donnerhack, C., Taouli, B., 2014. Quantitative liver MRI combining phase contrast imaging, elastography, and DWI: assessment of reproducibility and postprandial effect at 3.0 T. PLoS One 9 (5), e97355.

Johnson, S.P., Ramasawmy, R., Campbell-Washburn, A.E., et al., 2016. Acute changes in liver tumour perfusion measured non-invasively with arterial spin labelling. Br. J. Cancer 114 (8), 897–904.

Joo, I., Lee, J.M., Yoon, J.H., Jang, J.J., Han, J.K., Choi, B.I., 2014. Nonalcoholic fatty liver disease: intravoxel incoherent motion diffusion-weighted MR imaging-an experimental study in a rabbit model. Radiology 270 (1), 131–140.

Kakite, S., Dyvorne, H., Besa, C., *et al.*, 2015. Hepatocellular carcinoma: short-term reproducibility of apparent diffusion coefficient and intravoxel incoherent motion parameters at 3.0T. J. Magn. Reson. Imaging 41 (1), 149–156.

Kamel, I.R., Liapi, E., Reyes, D.K., Zahurak, M., Bluemke, D.A., Geschwind, J.F., 2009. Unresectable hepatocellular carcinoma: serial early vascular and cellular changes after transarterial chemoembolization as detected with MR imaging. Radiology 250 (2), 466–473.

Koh, T.S., Thng, C.H., Lee, P.S., *et al.*, 2008. Hepatic metastases: in vivo assessment of perfusion parameters at dynamic contrast-enhanced MR imaging with dual-input two-compartment tracer kinetics model. Radiology 249 (1), 307–320.

Koh, T.S., Thng, C.H., Hartono, S., *et al.*, 2011. Dynamic contrast-enhanced MRI of neuroendocrine hepatic metastases: a feasibility study using a dual-input two-compartment model. Magn. Reson. Med. 65 (1), 250–260.

Langen, K.M., Jones, D.T., 2001. Organ motion and its management. Int. J. Radiat. Oncol. Biol. Phys. 50 (1), 265–278.

Larson, A.C., Wang, D., Atassi, B., *et al.*, 2008. Transcatheter intraarterial perfusion: MR monitoring of chemoembolization for hepatocellular carcinoma—feasibility of initial clinical translation. Radiology 246 (3), 964–971.

Le Bihan, D., Breton, E., Lallemand, D., Grenier, P., Cabanis, E., Laval-Jeantet, M., 1986. MR imaging of intravoxel incoherent motions: application to diffusion and perfusion in neurologic disorders. Radiology 161, 401–407.

Leach, M.O., Brindle, K.M., Evelhoch, J.L., *et al.*, 2003. Assessment of antiangiogenic and antivascular therapeutics using MRI: recommendations for appropriate methodology for clinical trials. Br. J. Radiol. 76 Spec No 1, S87–S91.

Leach, M.O., Morgan, B., Tofts, P.S., *et al.*, 2012. Imaging vascular function for early stage clinical trials using dynamic contrast-enhanced magnetic resonance imaging. Eur. Radiol. 22 (7), 1451–1464.

Leen, E., Angerson, W.G., Cooke, T.G., McArdle, C.S., 1996. Prognostic power of Doppler perfusion index in colorectal cancer. Correlation with survival. Ann. Surg. 223 (2), 199–203.

Leggett, D.A., Kelley, B.B., Bunce, I.H., Miles, K.A., 1997. Colorectal cancer: diagnostic potential of CT measurements of hepatic perfusion and implications for contrast enhancement protocols. Radiology 205 (3), 716–720.

Lencioni, R., Llovet, J.M., 2010. Modified RECIST (mRECIST) assessment for hepatocellular carcinoma. Semin. Liver Dis. 30 (1), 52–60.

Leporq, B., Dumortier, J., Pilleul, F., Beuf, O., 2012. 3D-liver perfusion MRI with the MS-325 blood pool agent: a noninvasive protocol to asses liver fibrosis. J. Magn. Reson. Imaging 35 (6), 1380–1387.

Liu, Y., Matsui, O., 2007. Changes of intratumoral microvessels and blood perfusion during establishment of hepatic metastases in mice. Radiology 243 (2), 386–395.

Liu, J., Pan, Y., Li, M., *et al.*, 2018. Applications of deep learning to MRI images: a survey. Big Data Mining Anal. 1 (1), 1–18.

Luciani, A., Vignaud, A., Cavet, M., *et al.*, 2008. Liver cirrhosis: intravoxel incoherent motion MR imaging—pilot study. Radiology 249 (3), 891–899.

Luo, Y., Pandey, A., Ghasabeh, M.A., *et al.*, 2019. Prognostic value of baseline volumetric multiparametric MR imaging in neuroendocrine liver metastases treated with transarterial chemoembolization. Eur. Radiol. 29, 5160–5171.

Martirosian, P., Pohmann, R., Schraml, C., *et al.*, 2019. Spatial-temporal perfusion patterns of the human liver assessed by pseudo-continuous arterial spin labeling MRI. Z. Med. Phys. 29 (2), 173–183.

Materne, R., Smith, A.M., Peeters, F., *et al.*, 2002. Assessment of hepatic perfusion parameters with dynamic MRI. Magn. Reson. Med. 47 (1), 135–142.

Meijerink, M.R., van Waesberghe, J.H., van der Weide, L., van den Tol, P., Meijer, S., van Kuijk, C., 2008. Total-liver-volume perfusion CT using 3-D image fusion to improve detection and characterization of liver metastases. Eur. Radiol. 18 (10), 2345–2354.

Meijerink, M.R., van Waesberghe, J.H., van der Weide, L., *et al.*, 2009. Early detection of local RFA site recurrence using total liver volume perfusion CT initial experience. Acad. Radiol. 16 (10), 1215–1222.

Memon, K., Lewandowski, R.J., Kulik, L., Riaz, A., Mulcahy, M.F., Salem, R., 2011. Radioembolization for primary and metastatic liver cancer. Semin. Radiat. Oncol. 21 (4), 294–302.

Michielsen, K.D.K.F., Verslype, C., Dymarkowski, S., van Malenstein, H., Oyen, R., Maleux, G., Vandecaveye, V., 2011. Pretreatment DCE-MRI for prediction of PFS in patients with inoperable HCC treated with TACE. Cancer Imaging 11 (Spec No A:S114).

Michoux, N., Huwart, L., Abarca-Quinones, J., *et al.*, 2008. Transvascular and interstitial transport in rat hepatocellular carcinomas: dynamic contrast-enhanced MRI assessment with low- and high-molecular weight agents. J. Magn. Reson. Imaging 28 (4), 906–914.

Miles, K.A., Kelley, B.B., 1997. Altered perfusion adjacent to hepatic metastases. Clin. Radiol. 52 (2), 162–163.

Miles, K.A., Hayball, M.P., Dixon, A.K., 1993. Functional images of hepatic perfusion obtained with dynamic CT. Radiology 188 (2), 405–411.

Miyazaki, K., Orton, M.R., Davidson, R.L., *et al.*, 2012. Neuroendocrine tumor liver metastases: use of dynamic contrast-enhanced MR imaging to monitor and predict radiolabeled octreotide therapy response. Radiology 263 (1), 139–148.

Morsbach, F., Pfammatter, T., Reiner, C.S., *et al.*, 2013. Computed tomographic perfusion imaging for the prediction of response and survival to transarterial radioembolization of liver metastases. Invest. Radiol. 48 (11), 787–794.

Nakamura, Y., Kawaoka, T., Higaki, T., *et al.*, 2018. Hepatocellular carcinoma treated with sorafenib: arterial tumor perfusion in dynamic contrast-enhanced CT as early imaging biomarkers for survival. Eur. J. Radiol. 98, 41–49.

Ng, C.S., Raunig, D.L., Jackson, E.F., *et al.*, 2010. Reproducibility of perfusion parameters in dynamic contrast-enhanced MRI of lung and liver tumors: effect on estimates of patient sample size in clinical trials and on individual patient responses. AJR Am. J. Roentgenol. 19, W134–W140.

Onuoha, E., Smith, A.D., Cannon, R., Khushman, M., Kim, H., 2022. Perfusion change of hepatocellular carcinoma during atezolizumab plus bevacizumab treatment: a pilot study. J. Gastrointest. Cancer. https://doi.org/10.1007/s12029-022-00858-4. Online ahead of print.

Padhani, A.R., Hayes, C., Landau, S., Leach, M.O., 2002. Reproducibility of quantitative dynamic MRI of normal human tissues. NMR Biomed. 15 (2), 143–153.

Pahwa, S., Liu, H., Chen, Y., *et al.*, 2018. Quantitative perfusion imaging of neoplastic liver lesions: a multi-institution study. Sci. Rep. 8 (1), 4990.

Quantitative Imaging Biomarkers Alliance, 2023. Rsna.org Website. https://www.rsna.org/en/research/quantitative-imaging-biomarkers-alliance. Accessed Janary 6, 2023.

Roberts, C., Issa, B., Stone, A., Jackson, A., Waterton, J.C., Parker, G.J., 2006. Comparative study into the robustness of compartmental modeling and model-free analysis in DCE-MRI studies. J. Magn. Reson. Imaging 23 (4), 554–563.

Ronot, M., Asselah, T., Paradis, V., et al., 2010. Liver fibrosis in chronic hepatitis C virus infection: differentiating minimal from intermediate fibrosis with perfusion CT. Radiology 256 (1), 135–142.

Rubbia-Brandt, L., Giostra, E., Brezault, C., et al., 2007. Importance of histological tumor response assessment in predicting the outcome in patients with colorectal liver metastases treated with neo-adjuvant chemotherapy followed by liver surgery. Ann. Oncol. 18 (2), 299–304.

Sahani, D.V., Holalkere, N.-S., Mueller, P.R., Zhu, A.X., 2007. Advanced hepatocellular carcinoma: CT perfusion of liver and tumor tissue—initial experience. Radiology 243 (3), 736–743.

Shukla-Dave, A., Obuchowski, N.A., Chenevert, T.L., et al., 2019. Quantitative imaging biomarkers alliance (QIBA) recommendations for improved precision of DWI and DCE-MRI derived biomarkers in multicenter oncology trials. J. Magn. Reson. Imaging 49 (7), e101–e121.

Sourbron, S., Sommer, W.H., Reiser, M.F., Zech, C.J., 2012. Combined quantification of liver perfusion and function with dynamic gadoxetic acid-enhanced MR imaging. Radiology 263 (3), 874–883.

Tacke, F., 2017. Targeting hepatic macrophages to treat liver diseases. J. Hepatol. 66 (6), 1300–1312.

Takahashi, N., Yoshioka, H., Yamaguchi, M., Saida, Y., Itai, Y., 2003. Accelerated dynamic MR imaging with a parallel imaging technique for hypervascular hepatocellular carcinomas: usefulness of a test bolus in examination and subtraction imaging. J. Magn. Reson. Imaging 18 (1), 80–89.

Taouli, B., Johnson, R.S., Hajdu, C.H., et al., 2013. Hepatocellular carcinoma: perfusion quantification with dynamic contrast-enhanced MRI. AJR Am. J. Roentgenol. 201 (4), 795–800.

Therasse, P., Arbuck, S.G., Eisenhauer, E.A., et al., 2000. New guidelines to evaluate the response to treatment in solid tumors. European Organization for Research and Treatment of Cancer, National Cancer Institute of the United States, National Cancer Institute of Canada. J. Natl. Cancer Inst. 92 (3), 205–216.

Totman, J.J., O'Gorman, R.L., Kane, P.A., Karani, J.B., 2005. Comparison of the hepatic perfusion index measured with gadolinium-enhanced volumetric MRI in controls and in patients with colorectal cancer. Br. J. Radiol. 78 (926), 105–109.

SpringerLink (Ed.), 2020. Towards Realtime Multimodal Fusion for Image-Guided Interventions Using Self-Similarities.

Tsushima, Y., Blomley, M.J., Yokoyama, H., Kusano, S., Endo, K., 2001. Does the presence of distant and local malignancy alter parenchymal perfusion in apparently disease-free areas of the liver? Dig. Dis. Sci. 46 (10), 2113–2119.

Van Beers, B.E., Leconte, I., Materne, R., Smith, A.M., Jamart, J., Horsmans, Y., 2001. Hepatic perfusion parameters in chronic liver disease: dynamic CT measurements correlated with disease severity. AJR Am. J. Roentgenol. 176 (3), 667–673.

Van Beers, B.E., Materne, R., Annet, L., et al., 2003. Capillarization of the sinusoids in liver fibrosis: noninvasive assessment with contrast-enhanced MRI in the rabbit. Magn. Reson. Med. 49 (4), 692–699.

Wang, D., Bangash, A.K., Rhee, T.K., *et al.*, 2007. Liver tumors: monitoring embolization in rabbits with VX2 tumors—transcatheter intraarterial first-pass perfusion MR imaging. Radiology 245 (1), 130–139.

Wang, D., Jin, B., Lewandowski, R.J., *et al.*, 2010. Quantitative 4D transcatheter intraarterial perfusion MRI for monitoring chemoembolization of hepatocellular carcinoma. J. Magn. Reson. Imaging 31 (5), 1106–1116.

Weiss, J., Ruff, C., Grosse, U., Grözinger, G., Horger, M., Nikolaou, K., Gatidis, S., 2019. Assessment of hepatic perfusion using GRASP MRI: bringing liver MRI on a new level. Invest. Radiol. 54, 737–743.

Wright, K.L., Hamilton, J.I., Griswold, M.A., Gulani, V., Seiberlich, N., 2014. Non-Cartesian parallel imaging reconstruction. J. Magn. Reson. Imaging 40 (5), 1022–1040.

Yoon, J.H., Lee, J.M., Yu, M.H., Hur, B.Y., Grimm, R., Sourbron, S., Chandarana, H., Son, Y., Basak, S., Lee, K.B., Yi, N.J., Lee, K.W., Suh, K.S., 2022. Simultaneous evaluation of perfusion and morphology using GRASP MRI in hepatic fibrosis. Eur. Radiol. 32, 34–45.

Zhang, Y., Jin, N., Deng, J., *et al.*, 2013. Intra-voxel incoherent motion MRI in rodent model of diethylnitrosamine-induced liver fibrosis. Magn. Reson. Imaging 31 (6), 1017–1021.

Zhang, T., Li, Z., *et al.*, 2020. Improved registration of DCE-MR images of the liver using a prior segmentation of the region of interest. In: Paper Presented at the Medical Imaging 2016: Image Processing.

Zhou, L., Chen, T.W., Zhang, X.M., *et al.*, 2013. Liver dynamic contrast-enhanced MRI for staging liver fibrosis in a piglet model. J. Magn. Reson. Imaging.

Zhu, A.X., Sahani, D.V., Duda, D.G., *et al.*, 2009. Efficacy, safety, and potential biomarkers of sunitinib monotherapy in advanced hepatocellular carcinoma: a phase II study. J. Clin. Oncol. 27 (18), 3027–3035.

Perfusion MRI in the kidneys: Arterial spin labeling

18

Maria A. Fernández-Seara[a,b] and Rebeca Echeverria-Chasco[a,b]

[a]*Department of Radiology, Clínica Universidad de Navarra, Pamplona, Spain*
[b]*IdiSNA, Instituto de Investigación Sanitaria de Navarra, Pamplona, Spain*

18.1 Overview

The kidneys are highly perfused organs that receive approximately 20% of the cardiac output, through the renal arteries. They are critical organs whose main function is to filter blood. This process takes place in filtering units, called nephrons. Nephrons are located in the renal cortex, the outer portion of the kidney, which is approximately 1 cm thick and receives a high percentage of the renal blood flow (RBF). In contrast, flow to the inner portion, called the renal medulla, is only a small percentage (less than 10%). Each nephron contains a filter, the glomerulus, which consists of a cluster of arterial capillaries. Blood flows into the glomerulus through afferent arterioles. In the glomerulus, filtration is driven by high glomerular blood hydrostatic pressure. Filtered blood leaves the glomerulus by the efferent arteriole, which circulates parallel to the tubule, reabsorbing filtered water, minerals, and nutrients. Renal perfusion, thus, provides the driving force for glomerular filtration, and it influences the glomerular filtration rate (GFR), which is considered the gold standard measurement of kidney function. Clinically, GFR measurements are complex to realize; so, generally GFR is estimated from serum creatinine.

Kidney perfusion is measured in physiological units of mL of blood per minute per 100 g of tissue, with normal cortical values being about $300 \, \text{mL} \, \text{min}^{-1} \, (100 \, \text{g})^{-1}$ in healthy human subjects. Renal perfusion is an alternative indicator of kidney function. It can be evaluated by using nuclear imaging techniques, such as renal scintigraphy, which utilize a radioactive tracer to provide RBF measurements (Peters, 1998). Doppler ultrasonography can also be used to measure the Doppler resistive index, a useful parameter for quantifying alterations in RBF (Tublin *et al.*, 2003).

MRI provides several alternative methods to the imaging techniques mentioned before, with the advantages of being noninvasive and quantitative. Phase-contrast MRI sequences can measure blood flow in the renal arteries (de Boer *et al.*, 2022) but are not able to assess renal parenchymal perfusion. A variant of diffusion-weighted MRI, called intravoxel incoherent motion (IVIM), uses a biexponential model to estimate voxel-wise the flowing fraction (f), a parameter related to

Advances in Magnetic Resonance Technology and Applications, Volume 11, ISSN 2666-9099
https://doi.org/10.1016/B978-0-323-95209-5.00013-1

perfusion (Ljimani *et al.*, 2020); however, computing quantitative perfusion values from f is not straightforward. Two other MRI techniques can provide voxel-wise perfusion measurements in the kidney: arterial spin labeling (ASL) and dynamic contrast-enhanced MRI (DCE-MRI).

DCE-MRI requires intravenous injection of a gadolinium-based contrast agent. Image intensity changes during the first pass of the contrast agent are sampled using a fast imaging sequence and used to estimate tissue perfusion by tracer kinetic analysis. ASL, on the other hand, uses magnetically labeled arterial blood as an endogenous tracer. The effects of the label are evaluated by comparison with a control condition. Differences in signal intensity between label and control images can be converted to perfusion using an appropriate mathematical model. This technique as it is applied to the kidneys will be described in detail in the following sections.

18.2 Imaging protocol

ASL in the brain has been the main focus of research for years due to the difficulties in imaging other organs, although the feasibility of ASL in the kidneys was reported in 1995 (Roberts *et al.*, 1995), being one of the first studies performed outside the brain. However, renal ASL presented several difficulties that limited its application. The main challenges in renal ASL are kidney motion and field inhomogeneities, which can affect the B_0 and B_1 fields and, thus, can degrade the quality of the perfusion measurements. Because of these issues, the configuration for brain ASL cannot be directly applied to the kidneys. Nevertheless, renal ASL has gained relevance in the last decade, in part fostered by technical developments in abdominal imaging and motivated by the interest in avoiding the use of gadolinium-based contrast agents for renal patients. Recently, a panel of experts working under the framework of the PARENCHIMA project, funded by a European Cooperation in Science and Technology (COST) action, has published technical recommendations on renal ASL with the aim of providing a set of guidelines for image acquisition, facilitating the standardization of methods across sites and vendors (Nery *et al.*, 2020).

In general, ASL sequences consist of two main components that are practically independent: (i) the preparation module in which the inflowing arterial blood is labeled and (ii) the readout module that generates the paired images of the kidney tissue under control and label conditions. Other strategies can be employed to improve the quality of the renal perfusion signal, such as pre- and postsaturation pulses or background-suppression pulses.

Renal ASL can be performed either on 1.5 or 3.0 T systems as it has been demonstrated in published studies, although at 3.0 T higher signal-to-noise ratio (SNR) is expected due to the increase in field strength and the longer relaxation times (blood and tissue T1s), whereas at 1.5 T, lower B_0 inhomogeneities and reduced susceptibility effects should be found.

18.2.1 **Labeling schemes**

ASL labeling is performed by employing RF pulses and gradients, and the manner in which these pulses are combined results in the different schemes that are available to label blood. Labeling efficiency is one of the most important factors in ASL, which directly affects the quality of the perfusion images and the accuracy of perfusion quantification, and, thus, needs to be maximized for renal ASL. Two ASL variants, flow-sensitive alternating inversion recovery (FAIR) and pseudo-continuous ASL (PCASL), are the most employed labeling schemes in the kidneys. Despite PCASL being the recommended labeling method to measure brain perfusion according to a consensus document (Alsop *et al.*, 2015), because of its intrinsic higher SNR and a more controlled duration of the inverted bolus compared to pulsed ASL (PASL) methods, for renal ASL both PCASL and FAIR strategies are recommended (Nery *et al.*, 2020).

Note that after inversion the longitudinal magnetization of the labeled blood recovers to its initial equilibrium state, a process which is governed by the relaxation time constant T1; therefore, the duration of the labeling is only a few seconds, depending on the blood T1 and on the efficiency of the labeling pulses. Thus the time delay between the labeling and the readout, named postlabeling delay (PLD) in PCASL and inversion time (TI) in FAIR, needs to be large enough to allow for all the labeled blood to arrive at the renal parenchyma before image acquisition but not too long to minimize T1 relaxation.

18.2.1.1 *PCASL implementation*

The first human renal ASL study was carried out using continuous ASL (CASL). In CASL, a long RF pulse is applied in combination with a long gradient to label the arterial blood spins as they cross a plane, known as the labeling plane or inversion plane. The labeling plane is located approximately perpendicular to the suprarenal abdominal aorta, between 8 and 10 cm above the center of the kidneys, as shown in Fig. 18.1. CASL constant RF pulse and gradient generate a flow-driven adiabatic inversion of the inflowing blood magnetization, because blood spins follow the effective field in the rotating frame. Although the feasibility of CASL was demonstrated in the kidney, CASL implementation is limited by the RF transmission, which requires a continuous mode that is not readily available on most MRI scanners.

PCASL (Dai *et al.*, 2008; Wu *et al.*, 2007) was introduced as an alternative implementation to CASL, in which a similar inversion is achieved by dividing the long RF pulse and gradient used in CASL into multiple short, high-power RF pulses and gradients separated by a gap, which can be optimized to reduce both magnetization transfer (MT) effects and power deposition. PCASL implementation is more compatible with clinical MRI scanners. A sequence diagram is shown in Fig. 18.2.

Theoretically, PCASL efficiency increases with the B_1 average (i.e., the average B_1 value over time). However, this value is restricted by specific absorption rate (SAR) limits. A B_1 of 1.6 µT should provide high efficiency while keeping SAR

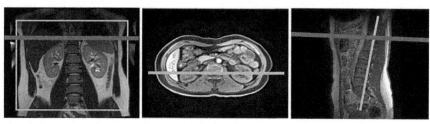

FIG. 18.1

Location of PCASL planes: the labeling plane (in *purple*) and imaging plane (in *orange*) shown in coronal, axial, and sagittal views over anatomical images.

Reproduced with permission from the open access article Nery, F., Buchanan, C.E., Harteveld, A.A., Odudu, A., Bane, O., Cox, E.F., et al., 2020. Consensus-based technical recommendations for clinical translation of renal ASL MRI. Magn. Reson. Mater. Phys. Biol. Med. 33 (1), 141–161, under the creative commons license (http://creativecommons.org/licenses/by/4.0/).

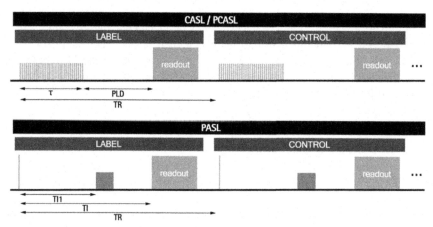

FIG. 18.2

CASL/PCASL and FAIR sequence diagrams for label and control configurations, with labeling pulses in *dark purple*, control pulses in *light purple*, and the *orange block* represents the readout. TR is the repetition time. For CASL/PCASL: τ, labeling duration; *PLD*, postlabeling delay. For FAIR: the *green block* represents the QUIPSSII/Q2TIPS pulses, TI_1: the bolus duration, TI: inversion time.

Reproduced with permission from the open access article Nery, F., Buchanan, C.E., Harteveld, A.A., Odudu, A., Bane, O., Cox, E.F., et al., 2020. Consensus-based technical recommendations for clinical translation of renal ASL MRI. Magn. Reson. Mater. Phys. Biol. Med. 33 (1):141–161, under the creative commons license (http://creativecommons.org/licenses/by/4.0/).

values within acceptable limits. For PCASL (PCASL RF pulses and gradients are depicted in Fig. 18.3), the Hann-shaped RF pulse is widely employed for its spatial selectivity, with a typical pulse duration $\delta = 500$ µs. The time between RF pulses (T) should be short (1000 µs or shorter if it is allowed by the scanner) to improve the response to off-resonance frequencies.

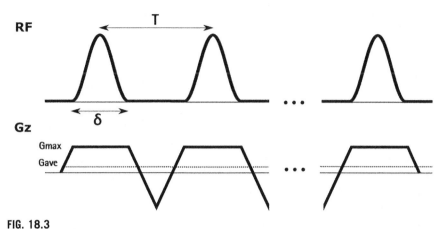

FIG. 18.3

PCASL labeling pulses and gradients. Hann-shaped pulses are typically employed with a duration δ and a period of T. G_{max} is the slice selection gradient amplitude and G_{ave} is the net average gradient.

Reproduced with permission from the open access article Nery, F., Buchanan, C.E., Harteveld, A.A., Odudu, A., Bane, O., Cox, E.F., et al., 2020. Consensus-based technical recommendations for clinical translation of renal ASL MRI. Magn. Reson. Mater. Phys. Biol. Med. 33 (1):141–161, under the creative commons license (http://creativecommons.org/licenses/by/4.0/).

PCASL can be implemented according to two schemes: unbalanced or balanced. In the former, the average gradient (G_{ave}) is present during labeling, but during the control, the slice-selection gradient is entirely refocused resulting in zero average gradient, whereas in the latter the same gradient waveform is employed for both labeling and control. Although both configurations have been employed in renal ASL, the unbalanced PCASL configuration is preferred, as it is inherently less sensitive to off-resonance frequencies. Nonetheless, even for unbalanced PCASL, field inhomogeneities can deteriorate the labeling efficiency, and these can be large near the inversion plane as a consequence of the different nearby interfaces (i.e., air-bone, air-tissue, tissue-bone). A difference of a few centimeters in the location of the labeling plane can result in large variations in the frequency offset (Echeverria-Chasco *et al.*, 2021; Greer *et al.*, 2019). In addition, PCASL inversion efficiency depends on flow velocities, which can be high and variable in the aorta due to age and disease (Garcia *et al.*, 2018) and present a pulsatile pattern. A careful choice of PCASL parameters can improve the unbalanced PCASL performance, increasing its robustness in the presence of off-resonance frequencies and pulsatile blood flow profiles (Echeverria-Chasco *et al.*, 2021). A robust PCASL configuration requires a G_{max}/G_{ave} ratio of 6–7 and a G_{ave} value that depends on the aortic blood flow velocities, being 0.4 mT/m appropriate for young subjects and 0.5 mT/m for patients and elderly subjects.

Labeling duration (i.e., the duration of the RF train of pulses, - τ -) is typically between 1.5 and 1.8 s, which provides adequate SNR without significantly increasing the repetition time and the power deposition. Usually, for single-PLD configurations, the PLD is in the range of 1.2–1.5 s while for multi-PLD acquisitions, a minimum of three PLDs has been employed in published studies (e.g., Kim *et al.*, 2017; Shimizu *et al.*, 2017), ranging from 0.5 to 2.0 s.

18.2.1.2 FAIR implementation

In PASL, a nearly instantaneous RF pulse is used to invert the magnetization in a large volume of blood. High efficiencies on the order of 97% are reached in the kidneys, and labeling is largely insensitive to variations in flow velocity. Among all the labeling variants in PASL, the FAIR scheme is the most frequently used.

In FAIR, two types of inversion are employed: nonselective and selective. In the nonselective inversion (label condition), a large volume (centered in the imaging plane) is inverted. The selective inversion (control condition) inverts a slab with a thickness equal to the sum of the imaging plane thickness and a gap of 10–20mm at each side of the plane, to allow for the movement of the kidneys. The selective-inversion slab should be carefully positioned avoiding the aorta to maximize blood signal difference between control and label images. For performing the inversions, the adiabatic frequency offset corrected inversion (FOCI) pulse is widely employed because of its improved inversion profile. As the gradient configuration (selective or nonselective) is the only difference between FAIR acquisitions, this scheme is not sensitive to MT effects. A FAIR sequence diagram is shown in Fig. 18.2.

In FAIR the duration of the labeled blood bolus is unknown. Thus QUIPSSII or Q2TIPs strategies (Wong *et al.*, 1998) are employed to destroy the longitudinal blood magnetization after a time known as TI_1 and, therefore, accurately define the duration of the labeled bolus, which is needed for RBF quantification. These pulses need to be applied to the region where labeled blood is flowing (i.e., aorta, renal arteries) but leaving a small gap between the saturation slab and the imaging plane, in order not to saturate the in-plane signal. An example of the selective- and non-selective-inversion slabs and the QUIPSSII/Q2TIPs slab location can be found in Fig. 18.4.

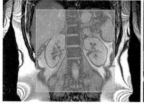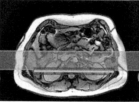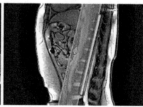

FIG. 18.4

FAIR planes location for the imaging plane (in *orange*), the selective-inversion slab (in *purple*), and QUIPSSII/Q2TIPS saturation slab (in *green*) shown in coronal, axial, and sagittal views over anatomical images.

Reproduced with permission from the open access article Nery, F., Buchanan, C.E., Harteveld, A.A., Ödüdü, A., Bane, O., Cox, E.F., et al., 2020. Consensus-based technical recommendations for clinical translation of renal ASL MRI. Magn. Reson. Mater. Phys. Biol. Med. 33 (1):141–161, under the creative commons license (http://creativecommons.org/licenses/by/4.0/).

As the TI in single-delay acquisitions is in the range of 1.8–2.0 s, the bolus duration, TI_1, is typically around 1.0–1.2 s. In multi-delay acquisitions several TIs are acquired ranging from 0.1 to 2.7 s, according to published studies (Cutajar et al., 2012, 2014; Li et al., 2018; Cai et al., 2017; Conlin et al., 2018).

18.2.2 Other sequence details

In order to improve the quality of perfusion data, different strategies have been developed and implemented in renal ASL sequences.

18.2.2.1 Pre- and postsaturation pulses

Presaturation pulses can be employed in PCASL and FAIR acquisitions to increase the accuracy of perfusion signal quantification as static tissue signal variations between label and control conditions are reduced. Presaturation pulses are applied to a selective imaging slab, of a thickness equal to the imaging plane thickness plus a gap of 10–20 mm at each side of the plane to account for the kidney movement, at the beginning of each TR to null the magnetization of the static tissue in the imaging plane. A common scheme employed in published studies uses WET saturation pulses (Ogg et al., 1994).

Postsaturation pulses can be used in FAIR to reduce the effects of imperfect inversion profiles between label and control acquisitions. A single *sinc*-shaped saturation pulse is generally employed immediately after the selective or nonselective inversion (Buchanan et al., 2018, 2020).

18.2.2.2 Background-suppression pulses

Background-suppression (BS) pulses were originally proposed to attenuate the signal from the static tissue, reducing nonperfusion artifacts and improving the perfusion signal temporal SNR (tSNR), as the percentage of ASL signal is very low compared to the static tissue signal (on the order of 4%–5% in healthy kidneys and 2%–3% in patients). Largely, BS is implemented by employing adiabatic inversion pulses such as FOCI or hyperbolic secant (HS) that are selectively applied before the labeling and nonselectively after the labeling pulses. The number of pulses and their timing depend on the desired suppression level (e.g., a tissue suppression of 90% of its value in the equilibrium state), the labeling scheme, labeling duration, and PLD/TI. BS timings are calculated by employing nonlinear optimization for a range of T1s (e.g., for the kidneys the range of T1s = [800–2500 ms]) and considering the presaturation pulses, if they have been employed.

BS is recommended based on published results in renal ASL that demonstrated BS was feasible and could improve the signal quality and the tSNR (Robson et al., 2009; Gardener and Francis, 2010; Bones et al., 2019). In addition, it has been reported that under free-breathing acquisitions, perfusion measurements were robust when BS was employed, and although retrospective motion correction could improve image quality it could not replace BS (Taso et al., 2019a). Different

BS schemes have been proposed in the kidneys for PCASL (Echeverria-Chasco et al., 2021; Cutajar et al., 2012; Robson et al., 2009, 2016; Taso et al., 2019a) and FAIR (Cutajar et al., 2012; Gardener and Francis, 2010; Nery et al., 2019; Shirvani et al., 2019).

18.2.2.3 Strategies to minimize motion effects during acquisition

The motion of the kidneys during the respiratory cycle presents a significant challenge for data acquisition and can degrade the quality of ASL data, since perfusion-weighted images are obtained on a voxel-wise basis by subtracting label from control images acquired at different time stamps. Native kidneys that are very close to the lungs are especially prone to respiratory motion artifacts. Motion leads to blurring, because multiple images must be subtracted and averaged to produce the final perfusion-weighted image, and motion artifacts can be on the order of perfusion signal intensity, which can result in errors in the ASL images. Different techniques have been used in renal ASL to minimize respiratory motion effects during image acquisition which can be employed individually or in combination.

Respiratory triggering employs bellows to monitor the respiratory motion and by setting a threshold on the bellows signal, at a given percentage of the inspiration peak that depends on the labeling scheme, PLD/TI, and sequence configuration (pre/postsaturation or BS pulses), the sequence is triggered to acquire all the label and control images in a similar phase of the respiratory cycle (Echeverria-Chasco et al., 2021; Cutajar et al., 2012). Even though it can be easily implemented, it increases the acquisition time and respiratory movement is not expected to be compensated completely. Another strategy is to perform breath-holding acquisitions (Shimizu et al., 2017; Li et al., 2018); however, these are long as usually a pair of control and label images is acquired in a single breath-hold, and they need to be repeated several times, which can be difficult and demanding for patients. A third strategy known as synchronized breathing requires the patient to coordinate breathing with the pulse sequence sounds by performing repeated patterns of breath-in/breath-out/breath-holding (Robson et al., 2016; Artz et al., 2011a). This strategy can be combined with respiratory triggering to improve data quality. However, it requires high levels of cooperation, and some patients cannot tolerate this pattern over long periods of time.

18.2.2.4 M_0 image acquisition

Perfusion-weighted images are generated in arbitrary units and converted to physiological RBF units during quantification (see Section 18.3.2. Quantification). To that end, the signal of the tissue in the equilibrium state needs to be known; so, for such purpose, a proton density image (also known as M_0 image) is acquired. This image is acquired no matter the labeling scheme employed. For acquiring the M_0 image, no BS pulses, preparation pulses, or labeling pulses can be employed, and the TR should be long (7–8 s) to allow the measurement of the proton density. The readout should be identical to the one used to acquire the label and control images. The M_0 image can be acquired in the same sequence as control and label images, or separately but under the same conditions (transmitter/receiver gain and shimming parameters) as those employed for the ASL acquisition.

18.2.3 **Imaging readouts**

The readout scheme is also an important element of the ASL sequence. Readout should be rapid to minimize the ASL signal decay across slices. In addition, as ASL presents inherently low SNR, imaging schemes that provide the highest image SNR and reduce signal loss and distortions are desired. Thus short echo times (TE) are preferred.

Readouts can be configured to acquire a single slice or several slices (multi-slice). Single-slice acquisitions are faster and easier to plan, while being adequate to image perfusion in renal disease as most pathologies affect the kidney uniformly. Multi-slice acquisitions have the advantage of covering more kidney tissue (which is especially helpful for some renal pathologies, such as unilateral or focal lesions); however, locating the imaging slab can be more difficult (especially in FAIR, as the selective inversion must avoid the aorta) and BS cannot be optimized for all the slices (BS is optimized for a single time point), resulting in different levels of tissue suppression across the slices, which leads to more nonperfusion artifacts to appear in the final acquired slices. In addition, in multi-slice configurations, imaging sequences should be fast to acquire all the slices before the decay of the ASL signal. Thus the single-slice acquisition is considered sufficient in most kidney pathologies as indicated in the renal ASL recommendations paper (Nery *et al.*, 2020), although multiple slices have been acquired in a number of studies published in the literature.

For 2D acquisitions, several readout schemes have been employed in renal studies, such as balanced steady-state free precession (bSSFP), gradient-echo echo-planar imaging (GE-EPI), spin-echo echo-planar imaging (SE-EPI), or single-shot rapid acquisition with relaxation enhancement (RARE). In a study performed by Buchanan *et al.* (2018) in which four PASL sequences with different imaging schemes (bSSFP, GE-EPI, SE-EPI and RARE) were compared, the SE-EPI readout showed the highest tSNR, repeatability, and the lowest signal variation across slices. According to Nery *et al.* (2020), GE-EPI readouts are not encouraged for renal imaging because of their susceptibility to magnetic field inhomogeneities; bSSFP sequences provide high SNR with a very short TE, but acquisition time is long because of SAR limitations, which can limit the number of slices in multi-slice acquisitions and they are also affected by field inhomogeneities; RARE schemes provide high SNR but require long TE and SAR is high; EPI sequences might need long TE for high-resolution acquisitions; and SE-EPI are susceptible to chemical shift artifacts. Therefore the final recommendation is the use of the SE-EPI sequence, although RARE and bSSFP schemes are also adequate for renal perfusion imaging.

3D readouts have not been frequently employed because of the underlying challenges in volumetric acquisitions in abdominal organs due to motion. However, in the last few years more studies employing 3D readouts have been reported, with readouts based on the gradient and spin echo GRASE (Cutajar *et al.*, 2012, 2014; Cai *et al.*, 2017) or 3D RARE (Robson *et al.*, 2016; Taso *et al.*, 2019b) pulse sequences, demonstrating the feasibility and viability to quantify perfusion employing these readouts.

Slice thickness should be 4–8 mm for 2D acquisitions and 3–6 mm for 3D acquisitions, with an in-plane resolution of 2–4 mm. The recommended TR, which includes all the pulse sequence events, should be 4–6 s (Nery *et al.*, 2020). Because of the low intrinsic SNR of ASL, a minimum of 20 ASL pairs should be acquired in single-delay acquisitions, to achieve sufficient signal in the perfusion-weighted image.

Other strategies such as fat suppression combined with the readout can help to minimize fat-displacement artifacts, especially in EPI-based readouts. Partial-Fourier or parallel imaging can also be helpful to accelerate the acquisition, reducing the TE and shortening the TR and the acquisition time.

18.2.4 Special considerations for transplanted kidneys

Transplanted kidneys are placed in the lower abdomen near the iliac fossa, and the renal artery and vein are clamped to iliac artery and vein, respectively. Kidney location and orientation depend on anatomical constraints of the patient and vary across patients (which can receive a second or third transplanted kidney) (see Fig. 18.5).

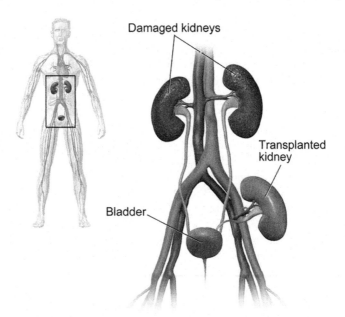

Kidney Transplant

FIG. 18.5

Renal allograft location and vascular anatomy.

Adapted from https://commons.wikimedia.org/wiki/File:KidneyTransplant.png. Under the Creative Commons Attribution-Share Alike 4.0 International license.

Because of the location of the transplanted kidney, respiratory movement has a lesser impact on transplants than on native kidneys, enabling acquisitions to be performed in free breathing. Image orientation tends to be similar to native kidneys, in coronal orientation, although because of the location of the allograft, the orientation might change to sagittal/coronal or oblique/coronal. Regarding imaging readouts, the same sequences used for native kidneys can be employed. Both PCASL and FAIR labeling schemes have been successfully employed in transplanted kidneys; however, a few considerations should be taken into account.

For PCASL, the labeling plane is typically placed in the lower abdominal aorta, between 8 and 10 cm above the center of the kidney and perpendicular to the artery (see Fig. 18.6). PCASL parameters similar to those used for native kidneys can be employed in transplanted kidneys; however, due to the lower velocity of the blood at the labeling plane location, the average gradient should be increased ($G_{ave} = 0.5$–0.6 mT/m).

For FAIR, the selective slab positioning can be a challenge as the orientation of the kidney can make difficult positioning the slab while avoiding the aorta or the iliac artery; thus, special care should be taken (see Fig. 18.7). The other sequence parameters are identical to the ones suggested for native kidneys.

18.2.5 **Special considerations for animal imaging**

ASL studies in animals have been carried out in high-field scanners at 3.0, 7.0, or 11.5 T, in which both FAIR and PCASL approaches have been successfully employed in animals such as rat, mouse, or swine. A study comparing both labeling approaches reported that PCASL provided higher sensitivity than FAIR (Duhamel *et al.*, 2014).

Regarding PCASL and FAIR configurations, sequences can be configured as for human kidneys but taking into consideration the anatomy of the animal. For example, the labeling plane in PCASL should be located a few centimeters above the kidneys, or the selective-slab thickness should encompass the animal kidneys and a few millimeters extra to compensate for kidney movement. An example of FAIR planes location is shown in Fig. 18.8.

Regarding the imaging scheme, similar considerations as in humans must be taken into account, such as a rapid sequence, with short TE and high SNR. However,

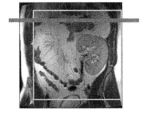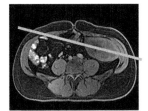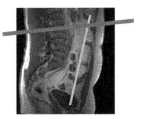

FIG. 18.6

Location of PCASL planes for the renal allograft: the labeling plane (in *purple*) and imaging plane (in *orange*) shown in coronal, axial, and sagittal views over anatomical images.

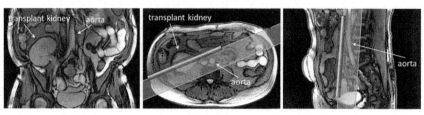

FIG. 18.7

FAIR planes location in the transplanted kidney for the imaging plane (in *orange*), the selective-inversion slab (in *purple*), and QUIPSSII/Q2TIPS saturation slab (in *green*) shown in coronal, axial, and sagittal views over anatomical images.

Reproduced with permission from the open access article Nery, F., Buchanan, C.E., Harteveld, A.A., Odudu, A., Bane, O., Cox, E.F., et al., 2020. Consensus-based technical recommendations for clinical translation of renal ASL MRI. Magn. Reson. Mater. Phys. Biol. Med. 33 (1):141–161, under the creative commons license (http://creativecommons.org/licenses/by/4.0/).

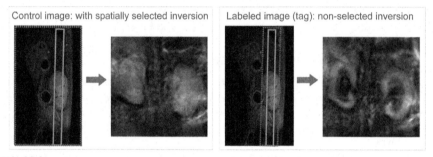

FIG. 18.8

Example of FAIR plane positioning in a mouse. Imaging plane is shown in *yellow* and the inversion in *red*. In the control image, nonselective inversion is performed (NS slab is represented in *red*) whereas in the label image, a selective inversion is performed (slab shown in *red*).

Reproduced with permission from the open access book chapter Ku, M.-C., Fernández-Seara, M.A., Kober, F., Niendorf, T., 2021. Noninvasive renal perfusion measurement using arterial spin labeling (ASL) MRI: basic concept. In: Preclinical MRI of the Kidney, pp. 229–239, under the creative commons license (http://creativecommons.org/licenses/by/4.0/).

because of the smaller anatomy of the animals, slice thickness, FOV, or in-plane resolution should be adjusted accordingly.

18.3 **Data analysis protocol**

Data processing allows improving data quality for ASL quantification. Some sequences can compute ASL maps at the scanner console by simply subtracting the control and label images and averaging the ASL pairs; however, several artifacts due to motion can be expected if no motion correction is performed. Although it can be helpful to have the ASL map on-line, most renal scientists perform the data analysis off-line.

To date, the entire pipeline for renal ASL data analysis is not available in any commercial software. Instead, the majority of ASL studies have been analyzed by employing custom scripts coded in MATLAB (MathWorks, Natick, MA) or Python (Python Software Foundation). Generally, the basic steps in the data analysis are image registration to correct for motion, RBF quantification, and, finally, kidney segmentation.

18.3.1 Image registration

Retrospective motion correction is needed to reduce motion artifacts and partial volume effects that can appear in the control and label image subtraction. Motion correction must be performed if the acquisition was in free breathing and is strongly advised for respiratory triggering, breath-holding, and synchronized breathing. Note that the M_0 image, which is required for quantification, should be also aligned with the set of ASL data. As the right and left kidneys can move independently, they should be registered separately.

Motion correction is carried out by employing image registration techniques, which realign the kidneys based on a reference image (i.e., all the images are registered to a fixed image), or in a groupwise manner, in which the complete set of ASL images are registered together. The latter has been reported to be more robust to signal intensity changes in the images as it takes into consideration the temporal information. An approach to perform groupwise registration based on a principal components analysis (PCA) metric as similarity measure has been successfully used to register ASL images (Echeverria-Chasco et al., 2021). Both rigid transformations and nonrigid transformations have been employed in the kidneys, although the latter are more favorable as they enable image deformation. Among the software packages employed for registering the images are Elastix (Echeverria-Chasco et al., 2021; Klein et al., 2010), ANTs (Avants et al., 2008; Taso et al., 2019a), and MATLAB (MathWorks, Natick, MA) (Cox et al., 2017).

18.3.2 Quantification

The perfusion map is calculated by subtracting voxel by voxel the signal from the control and label images and, because a minimum of 20 ASL pairs is acquired, the resulting 20 difference images are averaged to obtain a single perfusion map in arbitrary units. The averaged perfusion-weighted image (PWI [%]) reflects the percentage of ASL signal (i.e., perfusion signal is divided voxel-wise by the M_0 signal). To improve the quality of the perfusion maps, outliers should be detected and removed before averaging. Different outlier rejection methods have been reported, which can be categorized as manual (Cox et al., 2017) or automatic methods. Automatic approaches employ distinct strategies such as the information of the respiratory bellows (Robson *et al.*, 2009) or metrics such as the standard deviation to define a threshold (Echeverria-Chasco *et al.*, 2021).

A quantification model is employed to convert the perfusion signal to RBF in physiological units of mL min^{-1} (100 g)$^{-1}$ of tissue. The quantification is key to facilitating comparisons across subjects and scanners.

Among all the quantification methods for renal perfusion, with their different assumptions and complexities, the single-compartment model is the most utilized. The model assumes that by the time the image is acquired the labeled spins remain in the blood, thus neglecting the signal relaxation differences between blood and tissue, simplifying the model. The advantages of this model are that it allows the quantification of RBF by employing a single PLD, provided that the PLD is longer than the arterial transit time, which facilitates the acquisition and reduces the acquisition time, and the model can be easily applied to the perfusion map without requiring complex procedures such as optimizers or fitting algorithms. The single-compartment Buxton models (Buxton *et al.*, 1998) are shown in Eq. (18.1) for PCASL, and in Eq. (18.2) for FAIR with QUIPSSII. This model was recommended by the panel of renal ASL experts.

$$\text{RBF}\left(\text{mL min}^{-1} \ (100\text{g})^{-1}\right) = \text{PWI} \frac{6000\lambda}{2\alpha T1_b} \frac{e^{\frac{PLD}{T1_b}}}{1 - e^{\frac{-\tau}{T1_b}}} \tag{18.1}$$

$$\text{RBF}\left(\text{mL min}^{-1}(100\text{g})^{-1}\right) = \text{PWI} \frac{6000\lambda}{2\alpha TI_1} e^{\frac{TI}{T1_b}} \tag{18.2}$$

where α is the labeling efficiency (assumed 0.85 for PCASL and 0.95 for FAIR), if BS pulses have been employed, a factor of 0.93 for each BS pulse should be added. λ is the tissue-blood water partition coefficient (assumed 0.9 mL/g), $T1_b$ is the arterial blood longitudinal relaxation time (assumed as 1.65 s at 3.0 T, and 1.48 s at 1.5 T). For PCASL, PLD is the postlabeling delay and τ is the labeling duration. In FAIR, TI is the inversion time and TI_1 is the bolus duration. Note that in multi-slice acquisitions, PLD and TIs should be modified for each slice according to their real PLD/TI.

For multi-delay acquisitions, the model includes the different perfusion-weighted maps acquired at different time points. These models have the advantage of measuring the arterial transit time (ATT) and the bolus duration, which in patients with suspected arterial blood delay can be beneficial. However, additional scan time is required, and because fewer ASL pairs are acquired SNR and tSNR are reduced at each individual PLD/TI and motion artifacts are accentuated due to the longer acquisition. In addition, model fitting is more complicated than for single-delay acquisitions. Examples of multi-delay studies in PCASL and FAIR can be found in the literature (Kim *et al.*, 2017; Shimizu *et al.*, 2017; Cutajar *et al.*, 2012, 2014; Li *et al.*, 2018; Cai *et al.*, 2017; Conlin *et al.*, 2018).

18.3.3 Kidney segmentation

Kidney segmentation is necessary to estimate the quantitative values of perfusion and to compute statistics in the regions of interest (ROIs). As mentioned, the kidney is composed of two main structures: the cortex and the medulla, which play different roles. Thus nephrologists have strong interest in having the perfusion values measured in both regions. Typically, vascular structures, collecting systems, and cysts that can also appear in the image are excluded from the ROIs. However, although typically reported, RBF values measured in the medulla are less reliable as they present lower SNR and medullary segmentation is less accurate and prone to errors. On the contrary, cortical perfusion values are more reliable due to the higher blood flow and they are less affected by nonperfusion artifacts.

To date, the majority of studies have segmented the kidneys manually by defining ROIs in the cortex and medulla in the M_0 image or anatomical images that are aligned with the perfusion maps, whereas others employ the perfusion-weighted image or the RBF maps to directly draw the ROIs. However, performing the ROIs in the PWI or RBF maps is not recommended, as bias can be introduced and hyperintense signals due to vessels can be baffled and included in the ROIs. Thus ROIs should be drawn on anatomic images such as the M_0 or other anatomical images with high resolution if they are registered to the ASL images (Nery *et al.*, 2020). As manual segmentation is tedious and prone to errors in case of nonexperienced operators, semiautomatic methods have gained relevance. These approaches require the intervention of the operator to initialize the algorithm, by drawing the kidney contour or by providing a seed. Interest in fully automatic approaches is growing in recent years due to the development of machine learning and deep learning methods in medical image processing. Deep learning has been employed in ASL in combination with other imaging techniques such as T2 mapping (Daniel *et al.*, 2021) or T2* information to perform the segmentation between the cortex and the medulla (Lauersen *et al.*, 2021). Other methods that automatically segment the kidney and define cortical and medullary ROIs rely on T1 mapping information (Cox *et al.*, 2017). However, all these methods need further validation. To date, no studies with a fully automated pipeline for kidney segmentation exist.

18.4 **Case studies of applications**

Renal ASL has been applied to study kidney pathophysiology in several kidney diseases, including chronic kidney disease (CKD), acute kidney injury (AKI), kidney transplantation, renal masses, lupus nephritis, metabolic syndrome, hypertension, diabetes, heart failure, and renovascular disease. Studies on some these pathologies are described in more detail below.

18.4.1 **Chronic kidney disease**

In CKD, glomerular pathology and tubular atrophy converge leading to progressive and diffuse renal destruction (Schelling, 2016). To assess renal parenchymal damage, blood flow must be monitored. Several studies have reported perfusion measurements obtained with renal ASL in CKD patients (Echeverria-Chasco *et al.*, 2021; Li *et al.*, 2018; Cai *et al.*, 2017; Michaely *et al.*, 2004; Rossi *et al.*, 2012; Gillis *et al.*, 2016; Mora-Gutiérrez *et al.*, 2017; Emilien *et al.*, 1999; Lu *et al.*, 2021).

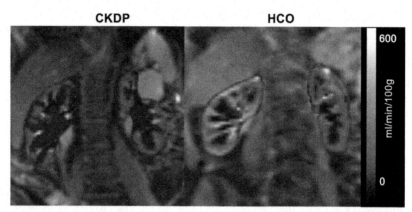

FIG. 18.9

Renal blood flow maps (in $mL^{-1} min^{-1} (100g)^{-1}$) of two representative subjects: a CKD patient (CKDP) and a healthy control (HCO).

Reproduced from Echeverria-Chasco, R., Vidorreta, M., Aramendía-Vidaurreta, V., Cano, D., Escalada, J.,
Garcia-Fernandez, N., et al., 2021. Optimization of pseudo-continuous arterial spin labeling for renal
perfusion imaging. Magn. Reson. Med. 85 (3), 1507–1521 with permission.

In these studies, cortical RBF has been consistently found to be lower in CKD patients than healthy controls (Fig. 18.9) and to correlate with eGFR, regardless of the underlying etiology. Those studies that differentiated patients across CKD stages (Mora-Gutiérrez *et al.*, 2017; Lu *et al.*, 2021) have reported reduced cortical RBF as CKD progressed (Fig. 18.10).

The medulla is particularly susceptible to the development of hypoxia and perfusion injury. However, ASL perfusion measurements in the medulla have only been reported in a small group of studies, due to the technical challenges encountered when trying to measure RBF in the medulla, which include low perfusion SNR and partial volume effects. Nonetheless, a recent study has evaluated RBF in the outer medulla (Lu *et al.*, 2021), where blood flow is not as low as in the inner medulla, finding significantly lower values in CKD patients than healthy controls.

18.4.2 Acute kidney injury

AKI is a relatively frequent event, characterized by sudden renal function loss, that can lead to the development of CKD. Renal ischemia is thought to be important in the pathogenesis of acute renal failure (Schrier and Wang, 2004) and renal perfusion alterations appear to be a crucial factor in AKI occurrence. Although renal perfusion status in AKI has been studied with ultrasound, ASL studies on AKI are scarce. A pilot study using ASL FAIR to evaluate RBF in a small group of patients found lower RBF in the cortex and the medulla, in patients compared with healthy controls (Dong *et al.*, 2013). Another study on an animal model of AKI showed, using ASL

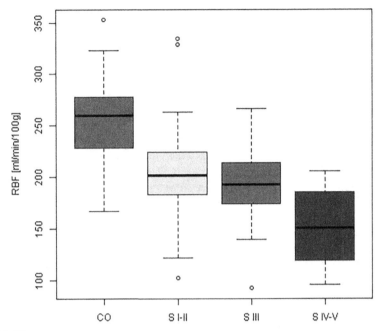

FIG. 18.10

Cortical RBF values measured using renal ASL in a healthy control group (CO) and CKD patient groups, separated by CKD stage. Significant differences were found between the CO group and each of the patient groups, and between patient groups (S I–II and S IV–V).

Reproduced from Mora-Gutiérrez, J.M., Garcia-Fernandez, N., Slon Roblero, M.F., Páramo, J.A., Escalada, F.J., Wang, D.J.J., et al., 2017. Arterial spin labeling MRI is able to detect early hemodynamic changes in diabetic nephropathy. J. Magn. Reson. Imaging. 46 (6), with permission.

FAIR, significantly reduced perfusion values in the mouse kidney after moderate and severe kidney injury (Hueper *et al.*, 2013).

18.4.3 Kidney transplantation

Kidney transplantation is the therapy of choice in end-stage renal disease, improving survival and quality of life for patients, in comparison with dialysis. Short-term survival of the allograft has been significantly improved in the last decade; however, there is a need to increase long-term allograft survival (Wang *et al.*, 2016), which could be achieved with new methods to monitor allograft function after transplantation, since biopsy monitoring is invasive and suffers from sampling bias.

Several studies using ASL in real transplant have demonstrated its good reproducibility (Artz *et al.*, 2011b; Bane *et al.*, 2020) and highlighted the translational potential of the technique in this area (Heusch *et al.*, 2014; Cutajar *et al.*, 2015; Niles *et al.*, 2016; Ren *et al.*, 2016; Ahn *et al.*, 2020; Wang *et al.*, 2021;

Yu *et al.*, 2021). ASL has shown promise for detecting subclinical pathology even in allografts with stable function (Wang *et al.*, 2021). ASL-derived RBF has been correlated with allograft fibrosis degree, assessed with biopsy; moreover, it has shown potential to predict long-term allograft dysfunction (Yu *et al.*, 2021). ASL is also very attractive for repeated longitudinal evaluation of allograft function in transplant recipients and donors (Niles *et al.*, 2016).

18.4.4 Renal masses

Angiogenesis is a feature of the formation and growth of tumoral masses. Tumor vascularity and angiogenic activity can be assessed by measuring blood flow. Thus one of the earliest clinical applications of renal ASL to be explored was its use in the evaluation of renal cancer (De Bazelaire *et al.*, 2008). Since this initial work, the potential of renal ASL in the clinical management of renal cancer has been illustrated in several studies. ASL-derived perfusion levels have been correlated with histopathologic findings (Lanzman *et al.*, 2012). ASL shows promise as an early predictor of response to antiangiogenic therapy and may help to differentiate therapy responders from nonresponders (Tsai *et al.*, 2021). Finally, the value of ASL in the assessment of tumor vascularity has been compared with DCE-MRI (Zhang *et al.*, 2016), showing a moderate correlation between measures derived from the two techniques.

18.4.5 ASL measurements in animal models of kidney disease

Renal ASL has been applied in animal models of AKI and CKD. In a mouse model of AKI, ASL at 7 T allowed evaluation of serial changes in cortical renal perfusion with ischemia-induced moderate and severe acute renal failure (Hueper *et al.*, 2013), which correlated with kidney volume loss and tubular injury. Longitudinal changes in renal perfusion due to acute and chronic renal allograft rejection have also been examined in mouse models of kidney transplantation (Hueper *et al.*, 2017), where the perfusion impairment was closely related to histological damage.

In CKD, animal models have allowed investigators to probe specific aspects of the disease. Ischemic injury induced by renal artery stenosis (RAS) and the effects of various treatments have been investigated in mouse studies, where ASL showed decreased perfusion with RAS that recovered with the treatments (Jiang *et al.*, 2017; Kashyap *et al.*, 2018; Zou *et al.*, 2018a,b). In a rat model of glomerular fibrosis (Conlin *et al.*, 2018), renal cortical perfusion measured using ASL at 3 T decreased with fibrosis severity, suggesting that it can be a sensitive marker for characterizing glomerular fibrosis before it develops into severe renal injury.

ASL has also been used to study ischemia-reperfusion injury in animal models, which is a major cause of delayed graft function in kidney transplantation, showing decreased cortical perfusion in kidneys with severe injury (Greite *et al.*, 2018).

18.5 Safety considerations

The noncontrast nature of ASL is particularly attractive in renal applications due to the concerns with nephrogenic systemic fibrosis in patients with renal impairments. Regarding the data acquisition using ASL sequences, the safety considerations are those that apply to any MRI examination. Since generally ASL sequences have high SAR, the rules for scanning patients with medical devices should be carefully considered (Jabehdar Maralani *et al.*, 2020).

18.6 **Challenges and solutions**

Susceptibility effects and motion have hampered the application of ASL in abdominal organs. Due to the averaging required to increase SNR, ASL sequences cannot be run in a single breath-hold. Therefore other strategies to compensate for or mitigate respiratory motion need to be employed, particularly when imaging native kidneys. Multiple breath-hold acquisitions have been attempted (Martirosian *et al.*, 2004), but motion artifacts are still present since the location of the kidneys is not necessarily the same for each breath-hold. Respiratory triggering and navigator gating can be used to acquire the data during the expiratory phase of the breathing cycle; however, in practice these strategies lengthen the scan time considerably (Gardener and Francis, 2010; Song *et al.*, 2010; Tan *et al.*, 2014). Synchronized breathing, a technique in which the subject is taught to synchronize the breath with the sequence sounds, can work fairly well but requires the subject's cooperation and it can be more difficult to perform for patients (Robson *et al.*, 2009). Finally, free breathing is the most comfortable alternative for patients, but it needs to be combined with background suppression (Taso *et al.*, 2019a) and effective motion correction strategies during image postprocessing.

ASL measurements in the medulla are challenging due to lower SNR than in the cortex, caused by its lower perfusion, partial volume effects between cortex and medulla and the uncertainty in the kinetics of labeled water, which could exchange with the surrounding tissue before penetrating deep into the medulla. Because of these reasons blood flow values in the medulla are not considered reliable with current measurement approaches (Nery *et al.*, 2020). Nonetheless, if medullary perfusion is of interest, ROIs should be carefully selected to include a sufficient number of voxels within the medulla to minimize partial volume effects and SNR could be increased by acquiring and averaging a larger number of images. As alterations in medullary perfusion play a role in certain pathologies, future technical developments are needed to improve the measurement reproducibility.

To promote the widespread implementation and clinical translation of renal ASL, efforts are being made by renal MRI scientists to harmonize the implementation of the technique across MRI vendors (Buchanan *et al.*, 2022) and to standardize the acquisition parameters. In 2020 a group of ASL experts followed a modified Delphi process to reach consensus on an approach that was considered robust and reproducible (Nery *et al.*, 2020) and that can be implemented across scanner vendors. As a default protocol, the panel recommended PCASL or FAIR labeling combined with a single-slice 2D readout, with background suppression and a simple but robust quantification model based on single-delay data. These efforts at harmonization should increase the technique's reproducibility and facilitate the realization of multicenter clinical studies, paving the way for its use in clinical practice.

Although a single-delay ASL acquisition was a feature of the recommended strategy, due to simplicity of acquisition and data analysis, multi-time point methods could be beneficial if ATT is expected to be abnormally long in a given patient group. Analysis of multi-delay data can provide an estimate of ATT, facilitating a more accurate quantification of renal perfusion (Kim *et al.*, 2017; Shimizu *et al.*, 2017; Shirvani *et al.*, 2019). Reduced sensitivity to transit time can also be achieved using flow-based labeling strategies, such as velocity or acceleration selective ASL (Franklin *et al.*, 2021); however, current implementations of these techniques carry an SNR penalty in comparison with the conventional spatially selective approaches.

Finally, although most of the work in renal ASL has been done at field strengths of 1.5 or 3 T, imaging at higher magnetic field ($\geq$7 T) could potentially yield higher SNR, but challenges like transmit field inhomogeneity and high SAR need to be overcome (Li *et al.*, 2018).

18.7 Learning and knowledge outcomes

This chapter presented detailed technical recommendations for the implementation of renal ASL sequences, data acquisition, and data analysis. An overview of the use of renal ASL in clinical applications was also provided. Finally, remaining challenges to the widespread use of the technique were highlighted and potential solutions were outlined.

References

Ahn, H.S., Yu, H.C., Kwak, H.S., Park, S.H., 2020. Assessment of renal perfusion in transplanted kidney patients using pseudo-continuous arterial spin labeling with multiple post-labeling delays. Eur. J. Radiol. 130 (July), 109200.

Alsop, D.C., Detre, J.A., Golay, X., Gunther, M., Hendrikse, J., Hernandez-Garcia, L., *et al.*, 2015. Recommended implementation of arterial spin-labeled perfusion MRI for clinical applications: a consensus of the ISMRM perfusion study group and the European consortium for ASL in dementia. Magn. Reson. Med. 73 (1), 102–116.

Artz, N.S., Sadowski, E.A., Wentland, A.L., Grist, T.M., Seo, S., Djamali, A., *et al.*, 2011a. Arterial spin labeling MRI for assessment of perfusion in native and transplanted kidneys. Magn. Reson. Imaging 29 (1), 74–82.

Artz, N.S., Stawski, R.S., Almeida, D.M., Lachman, M.E., Rosnick, C.B., 2011b. Reproducibility of renal perfusion MR imaging in native and transplanted kidneys using non-contrast arterial spin labeling. J. Magn. Reson. Imaging 33 (6), 1414–1421.

Avants, B.B., Epstein, C.L., Grossman, M., Gee, J.C., 2008. Symmetric diffeomorphic image registration with cross-correlation: evaluating automated labeling of elderly and neurodegenerative brain. Med. Image Anal. 12 (1), 26–41.

Bane, O., Hectors, S., Gordic, S., Kennedy, P., Wagner, M., Weiss, A., *et al.*, 2020. Multiparametric magnetic resonance imaging shows promising results to assess renal transplant dysfunction with fibrosis. Physiol. Behav. 97 (2), 414–420.

Bones, I.K., Harteveld, A.A., Franklin, S.L., van Osch, M.J.P., Hendrikse, J., Moonen, C.T.W., *et al.*, 2019. Enabling free-breathing background suppressed renal pCASL using fat imaging and retrospective motion correction. Magn. Reson. Med. 82 (1), 276–288.

Buchanan, C., Cox, E., Francis, S., 2018 Jun. Evaluation of 2D imaging schemes for pulsed arterial spin labeling of the human kidney cortex. Diagnostics 8 (3), 43.

Buchanan, C.E., Mahmoud, H., Cox, E.F., McCulloch, T., Prestwich, B.L., Taal, M.W., *et al.*, 2020. Quantitative assessment of renal structural and functional changes in chronic kidney disease using multi-parametric magnetic resonance imaging. Nephrol. Dial. Transplant. 35 (6), 955–964.

Buchanan, C.E., Li, H., Morris, D.M., Daniel, A.J., Sousa, J., Sourbron, S., Thomas, D.L., Priest STF, A.N., 2022. A travelling kidney study using a harmonised multiparametric renal MRI protocol. In: Proceedings of the 27th annual meeting of ISMRM, Montreal, Canada, p. 0482.

Buxton, R.B., Frank, L.R., Wong, E.C., Siewert, B., Warach, S., Edelman, R.R., 1998 Sep. A general kinetic model for quantitative perfusion imaging with arterial spin labeling. Magn. Reson. Med. 40 (3), 383–396.

Cai, Y.Z., Li, Z.C., Zuo, P.L., Pfeuffer, J., Li, Y.M., Liu, F., *et al.*, 2017. Diagnostic value of renal perfusion in patients with chronic kidney disease using 3D arterial spin labeling. J. Magn. Reson. Imaging 46 (2), 589–594.

Conlin, C.C., Huang, Y., Gordon, B.A.J., Zhang, J.L., 2018. Quantitative characterization of glomerular fibrosis with magnetic resonance imaging: a feasibility study in a rat glomerulonephritis model. Am. J. Physiol. Ren. Physiol. 314 (5), F747–F752.

Cox, E.F., Buchanan, C.E., Bradley, C.R., Prestwich, B., Mahmoud, H., Taal, M., *et al.*, 2017 Sep. Multiparametric renal magnetic resonance imaging: validation, interventions, and alterations in chronic kidney disease. Front. Physiol. 8 (SEP), 696.

Cutajar, M., Thomas, D.L., Banks, T., Clark, C.A., Golay, X., Gordon, I., 2012 Apr. Repeatability of renal arterial spin labelling MRI in healthy subjects. Magn. Reson. Mater. Phys. Biol. Med. 25 (2), 145–153.

Cutajar, M., Thomas, D.L., Hales, P.W., Banks, T., Clark, C.A., Gordon, I., 2014. Comparison of ASL and DCE MRI for the non-invasive measurement of renal blood flow: quantification and reproducibility. Eur. Radiol. 24 (6), 1300–1308.

Cutajar, M., Hilton, R., Olsburgh, J., Marks, S.D., Thomas, D.L., Banks, T., *et al.*, 2015. Renal blood flow using arterial spin labelling MRI and calculated filtration fraction in healthy adult kidney donors pre-nephrectomy and post-nephrectomy. Eur. Radiol. 25 (8), 2390–2396.

Dai, W., Garcia, D., de Bazelaire, C., Alsop, D.C., 2008. Continuous flow-driven inversion for arterial spin labeling using pulsed radio frequency and gradient fields. Magn. Reson. Med. 60 (6), 1488–1497.

Daniel, A.J., Buchanan, C.E., Allcock, T., Scerri, D., Cox, E.F., Prestwich, B.L., *et al.*, 2021. Automated renal segmentation in healthy and chronic kidney disease subjects using a convolutional neural network. Magn. Reson. Med. 86 (2), 1125–1136.

De Bazelaire, C., Alsop, D.C., George, D., Pedrosa, I., Wang, Y., Michaelson, M.D., *et al.*, 2008 Sep 1. Magnetic resonance imaging—measured blood flow change after antiangiogenic therapy with PTK787/ZK 222584 correlates with clinical outcome in metastatic renal cell carcinoma. Clin. Cancer Res. 14 (17), 5548–5554.

de Boer, A., Villa, G., Bane, O., Bock, M., Cox, E.F., Dekkers, I.A., *et al.*, 2022. Consensus-based technical recommendations for clinical translation of renal phase contrast MRI. J. Magn. Reson. Imaging 55 (2), 323–335.

Dong, J., Yang, L., Su, T., Yang, X.D., Chen, B., Zhang, J., et al., 2013. Quantitative assessment of acute kidney injury by noninvasive arterial spin labeling perfusion MRI: a pilot study. Sci. China Life Sci. 56 (8), 745–750.

Duhamel, G., Prevost, V., Girard, O.M., Callot, V., Cozzone, P.J., 2014. High-resolution mouse kidney perfusion imaging by pseudo-continuous arterial spin labeling at 11.75 T. Magn. Reson. Med. 71 (3), 1186–1196.

Echeverria-Chasco, R., Vidorreta, M., Aramendía-Vidaurreta, V., Cano, D., Escalada, J., Garcia-Fernandez, N., et al., 2021 Oct. Optimization of pseudo-continuous arterial spin labeling for renal perfusion imaging. Magn. Reson. Med. 85 (3), 1507–1521.

Emilien, G., Maloteaux, J.M., Geurts, M., Hoogenberg, K., Cragg, S., 1999. Dopamine receptors-physiological understanding to therapeutic intervention potential. Pharmacol. Ther. 84 (2), 133–156.

Franklin, S.L., Bones, I.K., Harteveld, A.A., Hirschler, L., van Stralen, M., Qin, Q., et al., 2021. Multi-organ comparison of flow-based arterial spin labeling techniques: spatially non-selective labeling for cerebral and renal perfusion imaging. Magn. Reson. Med. 85 (5), 2580–2594.

Garcia, J., van der Palen, R.L.F., Bollache, E., Jarvis, K., Rose, M.J., Barker, A.J., et al., 2018. Distribution of blood flow velocity in the normal aorta: effect of age and gender. J. Magn. Reson. Imaging 47 (2), 487–498.

Gardener, A.G., Francis, S.T., 2010. Multislice perfusion of the kidneys using parallel imaging: image acquisition and analysis strategies. Magn. Reson. Med. 63 (6), 1627–1636.

Gillis, K.A., McComb, C., Patel, R.K., Stevens, K.K., Schneider, M.P., Radjenovic, A., et al., 2016. Non-contrast renal magnetic resonance imaging to assess perfusion and corticomedullary differentiation in health and chronic kidney disease. Nephron 133 (3), 183–192.

Greer, J., Wang, Y., Pedrosa, I., Madhuranthakam, A.J., 2019. Pseudo-continuous arterial spin labeled renal perfusion imaging at 3T with improved robustness to off-resonance. In: Proceedings of the 27th Scientific Meeting. International Society for Magnetic Resonance in Medicine, Montreal, Canada, p. 4959.

Greite, R., Thorenz, A., Chen, R., Jang, M.S., Rong, S., Brownstein, M.J., et al., 2018. Renal ischemia-reperfusion injury causes hypertension and renal perfusion impairment in the CD1 mice which promotes progressive renal fibrosis. Am. J. Physiol. Ren. Physiol. 314 (5), F881–F892.

Heusch, P., Wittsack, H.J., Blondin, D., Ljimani, A., Nguyen-Quang, M., Martirosian, P., et al., 2014. Functional evaluation of transplanted kidneys using arterial spin labeling MRI. J. Magn. Reson. Imaging 40 (1), 84–89.

Hueper, K., Gutberlet, M., Rong, S., Hartung, D., Mengel, M., Lu, X., et al., 2013. Acute kidney injury: arterial spin labeling to monitor renal perfusion impairment in mice—comparison with histopathologic results and renal function. Radiology 270 (1), 130367.

Hueper, K., Schmidbauer, M., Thorenz, A., Bräsen, J.H., Gutberlet, M., Mengel, M., et al., 2017. Longitudinal evaluation of perfusion changes in acute and chronic renal allograft rejection using arterial spin labeling in translational mouse models. J. Magn. Reson. Imaging 46 (6), 1664–1672.

Jabehdar Maralani, P., Schieda, N., Hecht, E.M., Litt, H., Hindman, N., Heyn, C., et al., 2020. MRI safety and devices: an update and expert consensus. J. Magn. Reson. Imaging 51 (3), 657–674.

Jiang, K., Ferguson, C.M., Ebrahimi, B., Tang, H., Kline, T.L., Burningham, T.A., et al., 2017. Noninvasive assessment of renal fibrosis with magnetization transfer MR imaging: validation and evaluation in murine renal artery stenosis. Radiology 283 (1), 77–86.

Kashyap, S., Osman, M., Ferguson, C.M., Nath, M.C., Roy, B., Lien, K.R., *et al.*, 2018. Ccl2 deficiency protects against chronic renal injury in murine renovascular hypertension. Sci. Rep. 8 (1), 1–12.

Kim, D.W., Shim, W.H., Yoon, S.K., Oh, J.Y., Kim, J.K., Jung, H., *et al.*, 2017. Measurement of arterial transit time and renal blood flow using pseudocontinuous ASL MRI with multiple post-labeling delays: feasibility, reproducibility, and variation. J. Magn. Reson. Imaging 46 (3), 813–819.

Klein, S., Staring, M., Murphy, K., Viergever, M.A., Pluim, J.P.W., 2010. Elastix: a toolbox for intensity-based medical image registration. IEEE Trans. Med. Imaging 29 (1), 196–205.

Lanzman, R.S., Robson, P.M., Sun, M.R., Patel, A.D., Mentore, K., Wagner, A.A., *et al.*, 2012. Arterial spin-labeling MR imaging of renal masses: correlation with histopathologic findings. Radiology 265 (3), 799–808.

Lauersen, M.O., Koylu, B., Haddock, B., Sorensen, J.A., 2021. Kidney segmentation for quantitative analysis applying MaskRCNN architecture. In: 2021 IEEE Symposium Series on Computational Intelligence, SSCI 2021 - Proceedings, pp. 1–6.

Li, X., Auerbach, E.J., Van de Moortele, P.F., Ugurbil, K., Metzger, G.J., 2018. Quantitative single breath-hold renal arterial spin labeling imaging at 7T. Magn. Reson. Med. 79 (2), 815–825.

Ljimani, A., Caroli, A., Laustsen, C., Francis, S., Mendichovszky, I.A., Bane, O., *et al.*, 2020. Consensus-based technical recommendations for clinical translation of renal diffusion-weighted MRI. Magn. Reson. Mater. Phys. Biol. Med. 33 (1), 177–195.

Lu, F., Yang, J., Yang, S., Bernd, K., Fu, C., Yang, C., *et al.*, 2021. Use of three-dimensional arterial spin labeling to evaluate renal perfusion in patients with chronic kidney disease. J. Magn. Reson. Imaging 54 (4), 1152–1163.

Martirosian, P., Klose, U., Mader, I., Schick, F., 2004. FAIR true-FISP perfusion imaging of the kidneys. Magn. Reson. Med. 51 (2), 353–361.

Michaely, H.J., Schoenberg, S.O., Ittrich, C., Dikow, R., Bock, M., Guenther, M., 2004. Renal disease: value of functional magnetic resonance imaging with flow and perfusion measurements. Investig. Radiol. 39 (11), 698–705.

Mora-Gutiérrez, J.M., Garcia-Fernandez, N., Slon Roblero, M.F., Páramo, J.A., Escalada, F.J., Wang, D.J.J., *et al.*, 2017. Arterial spin labeling MRI is able to detect early hemodynamic changes in diabetic nephropathy. J. Magn. Reson. Imaging 46 (6), 1810–1817.

Nery, F., De Vita, E., Clark, C.A., Gordon, I., Thomas, D.L., 2019. Robust kidney perfusion mapping in pediatric chronic kidney disease using single-shot 3D-GRASE ASL with optimized retrospective motion correction. Magn. Reson. Med. 81 (5), 2972–2984.

Nery, F., Buchanan, C.E., Harteveld, A.A., Odudu, A., Bane, O., Cox, E.F., *et al.*, 2020. Consensus-based technical recommendations for clinical translation of renal ASL MRI. Magn. Reson. Mater. Phys. Biol. Med. 33 (1), 141–161.

Niles, D.J., Artz, N.S., Djamali, A., Sadowski, E.A., Grist, T.M., Fain, S.B., 2016. Longitudinal assessment of renal perfusion and oxygenation in transplant donor-recipient pairs using arterial spin labeling and blood oxygen level-dependent magnetic resonance imaging. Investig. Radiol. 51 (2), 113–120.

Ogg, R.J., Kingsley, R.B., Taylor, J.S., 1994. WET, a T1-and B1-insensitive water-suppression method for in vivo localized 1H NMR spectroscopy. J. Magn. Reson. Ser. B 104 (1), 1–10.

Peters, A.M., 1998. Non-invasive imaging of renal structure and function scintigraphic imaging of renal function. Exp. Nephrol. 6, 391–397.

Ren, T., Wen, C.-L., Chen, L.-H., Xie, S.-S., Cheng, Y., Fu, Y.-X., et al., 2016. Evaluation of renal allografts function early after transplantation using intravoxel incoherent motion and arterial spin labeling MRI. Magn. Reson. Imaging 34 (7), 908–914.

Roberts, D.A., Detre, J.A., Bolinger, L., Insko, E.K., Lenkinski, R.E., Pentecost, M.J., et al., 1995. Renal perfusion in humans: MR imaging with spin tagging of arterial water. Radiology 196, 281–286.

Robson, P.M., Madhuranthakam, A.J., Dai, W., Pedrosa, I., Rofsky, N.M., Alsop, D.C., et al., 2009 Jun. Strategies for reducing respiratory motion artifacts in renal perfusion imaging with arterial spin labeling. Magn. Reson. Med. 61 (6), 1374–1387.

Robson, P.M., Madhuranthakam, A.J., Smith, M.P., Sun, M.R.M., Dai, W., Rofsky, N.M., et al., 2016 Feb. Volumetric arterial spin-labeled perfusion imaging of the kidneys with a three-dimensional fast spin echo acquisition. Acad. Radiol. 23 (2), 144–154.

Rossi, C., Artunc, F., Martirosian, P., Schlemmer, H.P., Schick, F., Boss, A., 2012. Histogram analysis of renal arterial spin labeling perfusion data reveals differences between volunteers and patients with mild chronic kidney disease. Investig. Radiol. 47 (8), 490–496.

Schelling, J.R., 2016. Tubular atrophy in the pathogenesis of chronic kidney disease progression. Pediatr. Nephrol. 31 (5), 693–706.

Schrier, R.W., Wang, W., 2004. Acute renal failure and Sepsis. N. Engl. J. Med. 351 (2), 159–169.

Shimizu, K., Kosaka, N., Fujiwara, Y., Matsuda, T., Yamamoto, T., Tsuchida, T., et al., 2017. Arterial transit time-corrected renal blood flow measurement with pulsed continuous arterial spin labeling MR imaging. Magn. Reson. Med. Sci. 16 (1), 38–44.

Shirvani, S., Tokarczuk, P., Statton, B., Quinlan, M., Berry, A., Tomlinson, J., et al., 2019. Motion-corrected multiparametric renal arterial spin labelling at 3 T: reproducibility and effect of vasodilator challenge. Eur. Radiol. 29 (1), 232–240.

Song, R., Loeffler, R.B., Hillenbrand, C.M., 2010. Improved renal perfusion measurement with a dual navigator-gated Q2TIPS fair technique. Magn. Reson. Med. 64 (5), 1352–1359.

Tan, H., Koktzoglou, I., Prasad, P.V., 2014. Renal perfusion imaging with two-dimensional navigator gated arterial spin labeling. Magn. Reson. Med. 71 (2), 570–579.

Taso, M., Guidon, A., Alsop, D.C., 2019a. Influence of background suppression and retrospective realignment on free-breathing renal perfusion measurement using pseudo-continuous ASL. Magn. Reson. Med. 81 (4), 2439–2449.

Taso, M., Zhao, L., Guidon, A., Litwiller, D.V., Alsop, D.C., 2019 Aug. Volumetric abdominal perfusion measurement using a pseudo-randomly sampled 3D fast-spin-echo (FSE) arterial spin labeling (ASL) sequence and compressed sensing reconstruction. Magn. Reson. Med. 82 (2), 680–692.

Tsai, L.L., Bhatt, R.S., Strob, M.F., Jegede, O.A., Sun, M.R.M., Alsop, D.C., et al., 2021 Feb 1. Arterial spin labeled perfusion MRI for the evaluation of response to tyrosine kinase inhibition therapy in metastatic renal cell carcinoma. Radiology 298 (2), 332–340.

Tublin, M.E., Bude, R.O., Platt, J.F., 2003. The resistive index in renal Doppler sonography: where do we stand? Am. J. Roentgenol. 180 (4), 885–892.

Wang, J.H., Skeans, M.A., Israni, A.K., 2016. Current status of kidney transplant outcomes: dying to survive. Adv. Chronic Kidney Dis. 23 (5), 281–286.

Wang, W., Yu, Y., Li, X., Chen, J., Zhang, Y., Zhang, L., et al., 2021. Early detection of subclinical pathology in patients with stable kidney graft function by arterial spin labeling. Eur. Radiol. 31 (5), 2687–2695.

Wong, E.C., Buxton, R.B., Frank, L.R., 1998. Quantitative imaging of perfusion using a single subtraction (QUIPSS and QUIPSS II). Magn. Reson. Med. 39 (5), 702–708.

Wu, W.C., Fernandez-Seara, M., Detre, J.A., Wehrli, F.W., Wang, J., 2007. A theoretical and experimental investigation of the tagging efficiency of pseudocontinuous arterial spin labeling. Magn. Reson. Med. 58 (5), 1020–1027.

Yu, Y.M., Wang, W., Wen, J., Zhang, Y., Lu, G.M., Zhang, L.J., 2021. Detection of renal allograft fibrosis with MRI: arterial spin labeling outperforms reduced field-of-view IVIM. Eur. Radiol. 31 (9), 6696–6707.

Zhang, Y., Kapur, P., Yuan, Q., Xi, Y., Carvo, I., Signoretti, S., et al., 2016 Feb 1. Tumor vascularity in renal masses: correlation of arterial spin-labeled and dynamic contrast-enhanced magnetic resonance imaging assessments. Clin. Genitourin. Cancer 14 (1), e25–e36.

Zou, X., Kwon, S.H., Jiang, K., Ferguson, C.M., Puranik, A.S., Zhu, X., et al., 2018a. Renal scattered tubular-like cells confer protective effects in the stenotic murine kidney mediated by release of extracellular vesicles. Sci. Rep. 8 (1), 1–12.

Zou, X., Jiang, K., Puranik, A.S., Jordan, K.L., Tang, H., Zhu, X., et al., 2018b. Targeting murine mesenchymal stem cells to kidney injury molecule-1 improves their therapeutic efficacy in chronic ischemic kidney injury. Stem Cells Transl. Med. 7 (5), 394–403.

DCE-MRI in the kidneys

19

Dario Livio Longo[a] and Walter Dastrù[b]

[a]*Institute of Biostructures and Bioimaging (IBB), National Research Council of Italy (CNR), Torino, Italy*
[b]*Department of Molecular Biotechnologies and Health Sciences, University of Turin, Torino, Italy*

19.1 Overview

Renal function evaluation is important and needed in radiological examinations for assessing a variety of renal diseases, including renovascular diseases, renal transplants, and progression of acute kidney injury to chronic kidney disease (Sato *et al.*, 2020). Functional impairment usually precedes anatomical and morphological alterations; therefore, early detection could potentially delay or pause disease progression by offering more therapeutic options. In clinical practice, several biomarkers have been developed and are currently used to assess kidney function, such as serum creatinine, urine output, or renal scintigraphy. However, all these methods have serious and well-known limitations: dependence on age and body mass, insensitivity to early kidney dysfunction, radiation exposure, and limited anatomical information (Cottam *et al.*, 2022).

In the last decades, magnetic resonance imaging (MRI) has been shown to provide complementary functional information in addition to superb anatomical images. In particular, MRI can noninvasively assess several aspects of renal function from filtration to perfusion, oxygenation, microstructure, and acid–base balance (Zhang and Lee, 2020; Selby *et al.*, 2018; Caroli *et al.*, 2018; Grenier *et al.*, 2016; Mahmoud *et al.*, 2016; Longo *et al.*, 2013, 2017, 2021). Renal hemodynamic parameters, including renal blood flow (RBF) and glomerular filtration rate (GFR), can be obtained with dynamic contrast-enhanced (DCE) MRI, also termed magnetic resonance renography (MRR), by monitoring the transit of a gadolinium-based contrast agent through kidney regions (Notohamiprodjo *et al.*, 2010; Bokacheva *et al.*, 2008; Pedersen *et al.*, 2021). DCE-MRI holds multiple advantages over renal scintigraphy, offering superior spatial resolution, full three-dimensional coverage of the kidneys, and no radiation exposure. Moreover, DCE-MRI can distinguish the signal of the cortex from the medulla and the collecting systems. Although DCE-MRI can provide unique measurements of relevance to the pathophysiology of the

Advances in Magnetic Resonance Technology and Applications, Volume 11, ISSN 2666-9099
https://doi.org/10.1016/B978-0-323-95209-5.00020-9

kidneys, lack of accepted protocols for GFR evaluation and complicated analysis procedures still hamper wide clinical adoption.

This chapter introduces the reader to the application of DCE-MRI for functional renal imaging. Typical imaging protocols and fast T1-weighted (T1w) sequences are discussed, followed by different approaches for image analysis to calculate physiological parameters, and concluding with examples of applications to various kidney diseases at both the clinical and preclinical level.

19.2 Imaging protocol

In a conventional DCE-MRI protocol, repeated T1w images are acquired both before (baseline) and after Gadolinium-based contrast agent injection. Under conditions of heavy T1 weighting, the MR signal intensity can be considered directly proportional to the molar concentration of the contrast agent (CA). Therefore the time course of the MR signal (also called renogram) can be considered a good estimate of the time course of the molar concentration of the CA.

At sufficient temporal resolution, a normal DCE-MRI curve following CA injection can be described by three phases: vascular, parenchymal, and excretory (Fig. 19.1). The first segment of the renographic curve is the vascular phase that corresponds to the first pass of the CA, reflected by a steep linear rise. The parenchymal phase is characterized by a slow increase up to a second peak that corresponds to the recirculation of the CA in the parenchyma. The excretory phase is characterized by a slow decay of the signal that corresponds to the release of the contrast material into the calices of the collecting system.

19.2.1 Precontrast T1 measurement

Usually, a precontrast injection T1 measurement is performed to allow the conversion of signal intensity time curves to Gadolinium (Gd) concentration curves. A variety of methods can be used to determine the longitudinal relaxation time T1: inversion recovery (IR), saturation recovery or variable TR (VTR) method, the variable flip angle (VFA) method, or the Look–Locker modified inversion recovery. Comprehensive description of the most common pulse sequence implementations for each T1 measurement method and their acquisition protocols have been described by Garteiser *et al.* (2021).

- *Inversion Recovery (IR)*: The standard reference method for measuring the T1 is the inversion recovery method which makes use of an inversion of the magnetization M_0 obtained by applying a 180° RF pulse. Following that pulse, the system is allowed to recover over an interval called the inversion time (TI). After the TI time interval, the magnetization is sampled with a readout technique. When magnetization has fully recovered after a long and constant repetition time (TR), IR is repeated with a different value of TI.

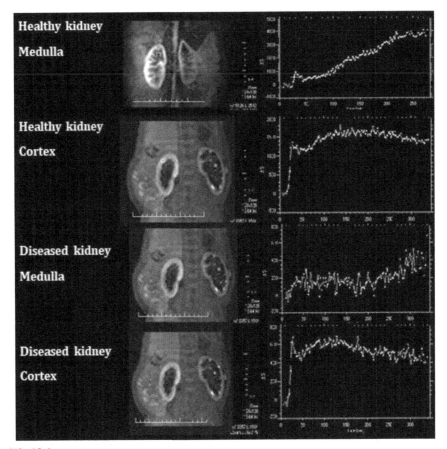

FIG. 19.1

Example of MR renogram, the result of DCE-MRI of pig kidneys with healthy (*left*) and diseased (*right*) kidneys. The yellow ROIs are shown to illustrate where the dynamic curves were drawn following an intravenous injection of 0.05 ml kg^{-1} of Gd-DTPA-BMA performed as a single bolus administered by hand 10 s after start of a dynamic gradient-echo sequence, with a single-phase acquisition time of 2.2 s. A total of 700 dynamic phases were acquired during 7 min.

Reproduced with permission from Pedersen, M., et al., 2021. Dynamic contrast enhancement (DCE) MRI-derived renal perfusion and filtration: basic concepts. Methods Mol. Biol. 2216, 205–227.

To produce highly accurate T1 measurements, the length of TI should range from 0 to 5 or even 7 times the longest T1 value expected in the sample. Furthermore, a good number of different TI values must be used. However, for in vivo studies, where the timing requirements are more stringent, a smaller range of TI values (from 0 to 3 times the longest expected T1) and a limited number of different TI values (down to as few as a couple of values) may be the only feasible choice.

When the acquisition is finished and a collection of images is obtained, T1 values can be obtained by fitting the pixel intensity vs TI curve against the equation:

$$M(t) = M(0)\left(1 - 2e^{-TI/T_1}\right) \tag{19.1}$$

The IR acquisition approach produces T1 maps with the best accuracy and dynamic range, but it has the drawback of being time consuming. In fact, even with very fast EPI readouts, it takes no less than 15–20 min, which often makes it unsuitable for in vivo applications (Hueper et al., 2014).

- *Saturation Recovery (SR)*: With this method, a series of images are acquired by applying 90° saturation pulses followed by a readout technique. The TR is changed ranging from short values to values that reach multiples of the T1 expected in the sample. When the TR is short with respect to T1, the system cannot fully recover its magnetization up to its equilibrium value and is said to be saturated. The series of images that is acquired by varying the TR will show different amount of saturation in different locations according to their T1.

The same considerations that have been described for choosing TI values in IR can be repeated here for TR. The image series is analyzed by fitting each pixel intensity vs TR curve against the equation:

$$M(t) = M(0)\left(1 - e^{-TR/T_1}\right) \tag{19.2}$$

The saturation recovery method is faster with respect to IR, but it has the drawback of a limited dynamic range and, hence, a limited T1 range.

- *Variable Flip Angle (VFA)*: Alternatively, it is possible to acquire a series of T1 weighted images without applying any saturation or inversion pulses but by varying the flip angle used for excitation. The TR must be kept constant. The pixel intensity vs flip angle curve can then be fitted against equation:

$$M(\theta) = M(0)\left(\frac{1 - e^{-TR/T_1}}{1 - \cos\theta \, e^{-TR/T_1}}\right) \sin\theta \tag{19.3}$$

to calculate the T1 of the tissue (Stikov et al., 2015; Fennessy et al., 2012; Schabel and Parker, 2008).

- *Look–Locker (LL)*: The Look–Locker method consists of sampling the signal repeatedly while the magnetization recovers from an inversion pulse. It offers the advantage of a decreased acquisition time with respect to the IR method, but its use is limited because of the reduced signal-to-noise ratio (SNR) that is available for an accurate T1 estimation (Messroghli et al., 2004; Nkongchu and Santyr, 2005).

The signal is analyzed according to the following equations:

$$M(t) = A - Bexp\left(-\frac{t}{T_1^*}\right)$$

(19.4)

with:

$$T_1 = T_1^*\left(\frac{B}{A} - 1\right)$$

(19.5)

19.2.2 Dynamic contrast measurement

After the assessment of precontrast T1, the dynamic acquisition can be started. The standard scheme of acquisition for a DCE-MRI protocol consists of a dynamic T1-weighted gradient-echo sequence while a Gd-based contrast agent is injected. The standard dose of contrast agent is 0.1 mmol Gd kg^{-1}, although a lower concentration can be used in patients with decreased renal function to reduce susceptibility effects upon accumulation in the collecting systems (Taton *et al.*, 2019). A few baseline precontrast images are acquired before injection to measure the MR signal at baseline, when no CA is present in the bloodstream and tissues. The injection of the contrast agent solution starts a short time after the dynamic sequence begins. A dynamic signal enhancement will develop initially in the main blood vessels (vascular phase) then in the most perfused tissues. After some time, the signal will begin decreasing indicating the start of a washout phase. The entire experiment typically lasts less than 5 min.

The MR signal in a T1-weighted image is generally not linear with 1/T1 value, which in turn changes during the passage of the CA reflecting its concentration in the blood/tissues and being dependent with the chosen acquisition sequence. Therefore equations describing the longitudinal magnetization as a function of T1 and the sequence parameters (such as Eqs. (19.3)–(19.5) are needed to convert signal into T1 values and then concentrations. Crucial to the success of the dynamic acquisition is the choice of the time resolution of the experiment, i.e., the time between the acquisition of successive images. A high temporal resolution, i.e., a short interval between images (usually 1 to 4 s per image), is of particular importance for quantitative studies of the kidney perfusion and GFR. This fact stems from the need to accurately measure the vascular phase of the CA and is particularly crucial for accurate sampling of the arterial input function (AIF) during the initial period after CA injection (ca. 30–40 s).

The T1-weighted image, in order to be rapidly acquired, most of the time exploits one of the following sampling techniques: GRE with Cartesian or radial sampling. Among these main groups of MRI acquisition schemes, a number of subtechniques exist that differ slightly with respect to the approach to sampling k-space (such as parallel imaging or compressed sensing) (Han and Cho, 2016; Michaely *et al.*, 2007; Heacock *et al.*, 2017).

- *Gradient recalled echo* (*GRE*): Techniques of this kind are the most common sequences for DCE-MRI studies. Such techniques use a single RF pulse for excitation combined with a reversal of the readout gradient. Usually, a short flip angle is used by GRE imaging to achieve very short TR, therefore allowing fast MRI acquisitions and increasing the contrast between enhanced and not enhanced kidney tissue. When the TR is set to a value that is shorter than both T1 and T2, a steady-state magnetization is reached, hence GRE sequences can be distinguished in unbalanced, RF-spoiled, and balanced steady-state free precession (SSFP) (Scheffler, 1999; Denolin et al., 2005; Scheffler and Lehnhardt, 2003). Both three-dimensional (3D) and two-dimensional (2D) acquisitions can be acquired, with the first providing whole kidney coverage, but requiring longer acquisition times, whereas the second one can be acquired with higher spatial and temporal resolution but sampling only one slice (Takahashi et al., 2021). In addition, 3D acquisitions enable the use of spatial coregistration to reduce motion artifacts (see Section 19.3.2).

- *Radial sampling*: With radial acquisition, the center of k-space is oversampled by individual lines or groups of lines (blades) that repeatedly pass through this region by several rotations. The major benefit of radial sampling is relative insensitivity to motion artifacts that allows fast DCE-MRI acquisitions during free breathing (Stemkens et al., 2019). Moreover, 3D acquisitions with high spatial and temporal resolution can be achieved (Pandey et al., 2017; Kurugol et al., 2020).

19.2.3 Contrast agents

Typical injection protocols in DCE-MRI use standard doses of Gd-based contrast agent (0.1 mmol Gd kg^{-1} body weight) within 5–10 s with automatic power injectors. On the other hand, several studies have exploited lower contrast doses, owing to the much higher blood perfusion in kidneys, showing similar results in the accuracy of the calculated renal functional parameters (Rusinek et al., 2001). Additionally, even ultralow doses (up to one-tenth of standard dose) provided consistent results by reducing susceptibility effects due to high gadolinium concentrations in the renal medulla and collecting systems (Taylor et al., 1997; Lee et al., 2001).

In principle, any small molecular weight contrast agent can be used, including linear and macrocyclic compounds. However, each contrast agent is endowed with different pharmacokinetic properties, hence blood pool clearance rates, which can affect the arterial input function time curve. Moreover, agents that possess a mixed renal and liver clearance (e.g., gadoxetate disodium/eovist and gadobenate dimeglumine/multihance), arising from moderate serum albumin binding properties, although endowing enhanced relaxivity (Avedano et al., 2013, 2007), are not optimal for DCE-MRI, because of a quite different blood clearance kinetics.

Even though limited to preclinical investigations, alternatives to gadolinium contrast agents are currently explored, based on Mn(II) complexes (Gale et al., 2018; Leone et al., 2022) or on iodinated contrast media that are visualized by exploiting CEST (Chemical Exchange Saturation Transfer) MRI contrast, showing comparable results to that achievable with conventional Gd-based CAs (Anemone et al., 2017; Longo et al., 2016; Irrera et al., 2020).

19.3 **Data analysis protocol**

Quantification of DCE-MRI data is an important step to measuring several functional or physiological properties at a voxel or Region of Interest (ROI) level. Outputs obtained from image analysis are expected to deliver accurate, reliable, reproducible, and biologically relevant values that reflect only the patient biology and are ideally not affected by the scanner or by the acquisition protocol. However, very few software packages with the MRI instrumentation or commercial ones allow DCE-MRI analysis. None of them provides a complete solution, and most of them are not optimized for reliable quantitative analysis, requiring proper validation by using synthetic image data or digital reference objects (Semmineh et al., 2017). To overcome these limitations, several open-source packages for renal perfusion and filtration quantification have been developed by different groups (Zollner et al., 2013; Barnes et al., 2015; Ortuno et al., 2013; Debus et al., 2019). Consequently, significant variations can exist in calculated perfusion parameters, hampering data comparison in multiinstitutional clinical trials (Heye et al., 2013; Huang et al., 2014). In spite of these challenges, several strategies can be adopted to minimize the various sources of variability and improve the quantification of DCE-MRI measurements (Kim, 2018a).

DCE-MRI analysis is typically carried out on a pixel level to produce spatially resolved parametric maps or on individual ROIs. Several steps are involved in DCE-MRI analysis, including segmentation of the kidneys, coregistration of the time series, identification of the arterial input function (AIF), conversion of signal intensity (or enhancement) time curves to concentration time curves, and pharmacokinetic analysis by applying a proper model to calculate perfusion and filtration estimates.

19.3.1 **Kidney segmentation**

Accurate kidney segmentation is needed for tracer kinetic analysis of DCE-MRI data, with more complex kinetic models requiring delineation of renal cortex and medulla regions. Inaccurate segmentation can lower GFR accuracy because of partial volume effects (Gutierrez et al., 2010). Segmentation of renal compartments can be carried out manually, as is done in today's clinical routine, by selecting 3D volumes at different time points and delineating the contours of each structure according to the observer's knowledge of the underlying anatomy. However, manual segmentation is time consuming and results in inter- and even intra-observer variability. Alternatively, semiautomatic or fully automatic segmentation methods can be exploited to alleviate these problems (Yang et al., 2016). Among these methods, 3D active contours, region growing-based techniques, spatial correlation, and graph-cut methods have been most investigated (Li et al., 2011; Rusinek et al., 2007; Zollner et al., 2009; Khrichenko and Darge, 2010). More recently, advances in convolutional neural networks (CNN) and availability of software for their implementations have attracted the exploitation of deep learning approaches for segmentation in medical imaging (Lundervold and Lundervold, 2019). New CNN approaches are emerging for segmentation of the kidneys in DCE-MR images for both healthy and diseased kidneys (Klepaczko *et al.*, 2021; Kline *et al.*, 2017). However, challenges remain in segmenting internal renal structures, particularly in diseased kidneys, because of different intensities and contrast changes in each compartment during perfusion. Interestingly, some studies have shown that manual segmentation does not provide superior performance when compared to fully automatic segmentation processes that are less demanding and less user dependent (Hanson *et al.*, 2017).

19.3.2 Image coregistration

During the period of image acquisition (for up to several minutes) when gadolinium contrast agent molecules pass throughout the kidney tissue, motion related to pulsation, breathing, or peristalsis can reduce image quality and introduce artifacts in the MRI data. Several strategies have been proposed to overcome motion artifacts, including image acquisition techniques, postprocessing approaches, or a combination of both (Zollner *et al.*, 2020). Among the first group are breath-hold strategies, including breath-hold during first pass of the contrast agent (Melbourne *et al.*, 2011), retrospective gating by using navigator echoes (Attenberger *et al.*, 2010), or radial readout schemes (Eikefjord *et al.*, 2015). Recently, a combination of multiple approaches for free-breathing renal DCE-MRI based on radial k-space sampling, parallel imaging, and compressed sensing, dubbed GRASP, was proposed to obtain both morphological and quantitative functional information within a single acquisition (Riffel *et al.*, 2016).

In the second group, image postprocessing registration techniques are dependent on the geometric transformation model that is used to model the expected geometric change(s) of the kidneys during breathing or cardiac motion (Viergever *et al.*, 2016). Displacements during normal breathing can be up to 20 mm and are usually modeled using rigid transformation (Positano *et al.*, 2013; de Senneville *et al.*, 2008; Buonaccorsi *et al.*, 2006) or deformable models (Fig. 19.2) (Zollner *et al.*, 2009; Tokuda *et al.*, 2011). Another confounder is the passage of the contrast agent, which produces large signal intensity variations that can transform the shape of the kidneys on MRI.

Further developments have been proposed by combining simultaneous registration and segmentation of the whole kidney to enforce time-course similarity of voxels for improving GFR estimation (Hodneland *et al.*, 2014).

A new category of deep learning techniques exploited for image registration is also emerging in renal DCE-MRI, demonstrating high potential for medical registration (Mazurowski *et al.*, 2019; Lv *et al.*, 2018).

19.3.3 Arterial input function

Knowledge of the arterial plasma concentration of the Gd-based agent (commonly dubbed arterial input function, or AIF) from the whole blood signal is needed for pharmacokinetic analysis, independently of the chosen model. A more

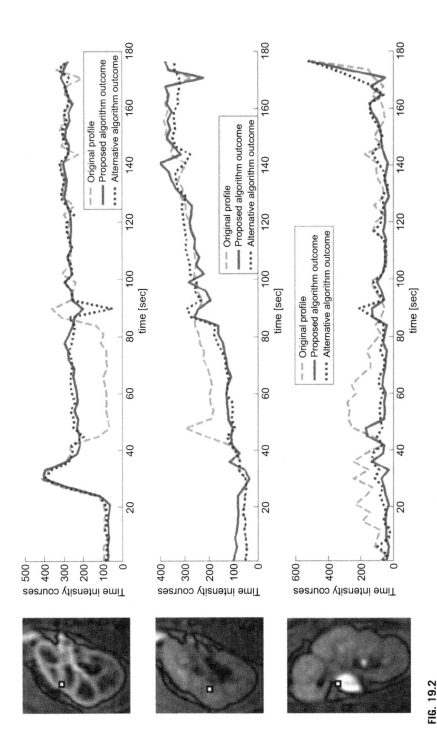

FIG. 19.2

Evaluation of the registration algorithm based on the comparison between the mean intensity time courses within a ROI before registration and after registration. ROI selection is depicted to the left of the time courses. The selected time frames are those where the kidney compartment is maximally enhanced: ∼30s, 100s, and 170s for the cortex, medulla, and pelvis, respectively.

Reproduced with permission from Zöllner, F.G., et al., 2009. Assessment of 3D DCE-MRI of the kidneys using non-rigid image registration and segmentation of voxel time courses. Comput. Med. Imaging Graph. 33, 171–181.

detailed description about the AIF can be found in Chapter 8 of this book: "Arterial input function: a friend or a foe?"

The AIF can be selected and measured for each subject, usually defined in the abdominal aorta, thus taking into account within-subject variation, but at the cost of being operator dependent (Zhang et al., 2015a; Kleppesto et al., 2022), or a population-based AIF can be used, with the limitation of ignoring variability between patients (Fedorov et al., 2014; Georgiou et al., 2019). For individual AIF, ROIs are usually positioned immediately above the origin of the renal arteries and in the slice position through the middle of the aorta to minimize partial volume effects in images acquired in the oblique-coronal plane to reduce the effect of respiratory motion.

Moreover, it has been shown that the size and the location of the aortic ROI can affect the estimation of renal perfusion and filtration values and, hence, the reproducibility of these parameters (Mendichovszky et al., 2009; Cutajar et al., 2010).

In many studies, it is not possible to obtain reliable AIF measurements because of low temporal/spatial resolution or lack of a suitable artery within the imaging field of view. Therefore an assumed form of the AIF can be used. This population average AIF can be generated either using mathematical equations that allow to reproduce different temporal resolution values or obtained from direct measurements (Parker et al., 2006). Alternative approaches are emerging based on automatic calculation of the individual AIF by principal component analysis (Sanz-Requena et al., 2015; Kim, 2018b) or on blind deconvolution approaches, where the AIF is estimated from the tissue tracer concentration derived from one or multiple ROIs (Kratochvila et al., 2016). However, these approaches are more complex since the need for high SNR and for high temporal resolution to capture the fast vascular distribution phase of the bolus (Fluckiger et al., 2009; Bartos et al., 2014).

19.3.4 Quantification of contrast agent concentration

The first step in DCE-MRI processing for quantitative analysis is the calculation of Gd concentration in the tissue of interest from the MRI signal enhancement. This means calculating the concentrations along time from the calculated or approximated longitudinal relaxation rates that are in turn obtained from the measured signals. For T1-weighted images, tracer concentration is dependent on.

$$c_{Gd} = \frac{\left(\frac{1}{T_1} - \frac{1}{T_1^0}\right)}{r_1} \tag{19.6}$$

where c_{Gd} is the concentration of the contrast agent, T1 is the longitudinal relaxation time of the tissue that is reduced from its native value $T1^0$ by the presence of a concentration c_{Gd} of the contrast agent, and r_1 is the relaxivity of the contrast agent that is specific for each contrast agent and is dependent on the magnetic field strength (Caravan et al., 1999; Laurent et al., 2006). Since relaxivities are difficult to measure in living tissue, in vitro determined values are commonly used, although the relaxivity can be slightly different in vivo (Shuter et al., 1996).

Native T1 values can be measured by several methods, as described previously in Section 19.2.1 of this chapter, comprising faster methods as the variable flip angle or the newer MP2RAGE (magnetization prepared rapid two gradient echo sequence) approach or by more accurate but less rapid inversion recovery-based approaches (Bergen *et al.*, 2020; Bharadwaj Das *et al.*, 2022; Dekkers *et al.*, 2019; Bane *et al.*, 2018).

In addition, the relation between the increase of the MR signal and corresponding decrease of T1 is dependent on the specific MR sequence used (Bokacheva *et al.*, 2007). Considering a simple spoiled gradient echo (usually dubbed FLASH or GRE) that is the most common sequence used for DCE-MRI studies, the signal S can be expressed as follows:

$$S = S_0 \frac{\left(1 - e^{-TR/T_1}\right) \sin \theta}{1 - e^{-TR/T_1} \cos \theta} \tag{19.7}$$

where TR is the repetition time, S_0 is the fully relaxed signal (TR $>>$ T1) measured before the contrast agent injection, and θ is the flip angle (FA). Therefore knowing r_1, T_1^0, and FA, a clear relationship exists between S and changes in T1 (that reflects the contrast agent concentration c_{Gd}). Note that an implicit assumption in both Eqs. (19.1) and (19.2) is a fast water exchange regime between tissue compartments, although differences in concentrations can be expected between different compartments (blood plasma, extravascular extracellular space, and intracellular space) due to limited water exchange rates across them, potentially affecting DCE-MRI analysis (Wiart *et al.*, 2006; Buckley, 2002; Zhang and Kim, 2019; Buckley *et al.*, 2008).

Quantitative analysis with pharmacokinetic modeling requires finding the arterial plasma concentration, that is, the concentration in the blood of an artery feeding the tissue of interest. The AIF should therefore be measured close to the tissue and in large vessel, since small arteries can introduce partial volume errors (Calamante *et al.*, 2003; Hansen *et al.*, 2009). By using Eqs. (19.1) and (19.2) the blood concentration (c_b) can be obtained from the blood MR signal. Since contrast agents are extracellular, the plasma concentration (c_p) from the whole blood must be corrected by the patient hematocrit (H) with the following equation:

$$c_p = \frac{c_b}{1 - H} \tag{19.8}$$

A standard value of $H = 0.37$–0.45 can be used, although H can vary considerably between patients or during treatments (Sahoo *et al.*, 2016; Just *et al.*, 2011). Alternatively, the relationship between T1 and the MR signal intensity can be obtained by using a calibration curve calculated with water phantoms doped with gadolinium at several concentrations. However, the signal intensity is not only proportional to the tissue T1, but is also dependent on system gain, coil, proton density, and FA, thereby leading to less accurate quantification (Bokacheva *et al.*, 2007).

19.3.5 Contrast agent kinetic analysis

The second step relies on pharmacokinetic analysis of the Gd concentration as a function of time to model the distribution of the contrast agent inside the tissue (body), reflecting physiological parameters. Several processing methods have been implemented, that can be grouped in direct, deconvolution, compartment model, and semiparametric ones. In this chapter we will briefly describe only few of them, the reader can refer to several reviews for an in-depth description of all these methods and for new ones (Sourbron, 2010; Khalifa *et al.*, 2014). Of note, there is no consensus yet about the best model for analyzing kidney DCE-MRI studies.

19.3.5.1 Upslope method

This method is based on the assumption of the flow of tracer inside the kidney vasculature that is valid until the contrast agent starts to flow into the renal tubules. The regional blood flow can be calculated by the maximum slope of the kidney concentration curve divided by the maximum arterial concentration:

$$F_P = \frac{slope\ (c_t(t))}{(c_A(t))} \tag{19.9}$$

where F_p is the tissue plasma flow, c_t is the concentration curve of the tracer in the kidney, and $c_A(t)$ is the concentration of the contrast agent inside the artery (Montet *et al.*, 2003; Wah *et al.*, 2018). Although the upslope method is quite simple, the measurement of the maximum of the concentration inside the artery is quite challenging. Additionally, only images acquired right after the bolus injection can be exploited for analysis, making it demanding for high temporal resolution acquisition protocols and high SNR.

A similar approach has been proposed by Baumann and Rudin for the calculation of the glomerular filtration rate (GFR), by considering the flow of the contrast agent from the cortex to the medulla, ignoring the outflow of the contrast from the medulla and without the need for the AIF (Baumann and Rudin, 2000). In this model, the medullar concentration C_m is related to the cortical tracer concentration C_c according to:

$$\frac{dc_m}{dt} = k_{cl}c_c(t) \tag{19.10}$$

where K_{cl} is an estimate of the clearance rate in the initial phase after tracer administration upon the contrast agent transport from cortex to medulla.

A further extension of this approach for GFR calculation is the Patlak–Rutland method, by considering unilateral tracer flow between two compartments, the vascular space and the nephron space (Hackstein *et al.*, 2003). However, such method strongly depends on the choice of the time interval over which a slope is calculated, hence affecting GFR estimates (Hackstein *et al.*, 2005).

19.3.5.2 Parametric deconvolution

This approach is based on the assumption that the contrast agent concentration curve has some known analytical form, but without any physiological relationship. Examples of these functions are the Fermi model, gamma variate functions, and polynomial representations (Jerosch-Herold *et al.*, 1998; Mouridsen *et al.*, 2006; Larsson *et al.*, 2008). This approach is more robust to noisy data, providing higher accuracy in the plasma flow estimate and in other calculated parameters, including time to peak, mean transit time, maximum signal intensity, although with little physiological interpretation (Michaely *et al.*, 2006).

19.3.5.3 Compartment models

These models provide a well-defined relationship between a parametric representation of the contrast agent concentration time curves and the underlying physiological parameters, including exchange rates between tissue compartments. The concept of a compartment is that of a space where the gadolinium contrast agent is dissolved and well mixed, where the concentration is spatially uniform. In tissue, the Gd-based contrast agent can distribute across several compartments in which the outlets of one compartment form the inlets of another, with the principle of mass conservation.

With this perspective, a two-compartment model has been adopted by several groups, with the compartments representing vascular and tubular, and GFR defined as the flow of the contrast agent from the vascular to the tubular compartment. Annet et al. implemented this two-compartment model by introducing the delay and dispersion of the bolus between aorta and renal vasculature, allowing the calculation of renal filtration:

$$\frac{dC_k(t)}{dt} = k_{21}C_a(t) - k_{12}C_k(t) \tag{19.11}$$

where C_a represents the concentration of the contrast agent in the plasma of the renal cortex, C_k is the concentration of the contrast agent in the renal tubules, k_{21} is the GFR, and k_{12} is the output of the renal tubules (Annet *et al.*, 2004).

By solving for $C_k(t)$:

$$C_k(t) = k_{21} \int_0^t C_a(\tau)e^{-k_{12}(t-\tau)}d\tau \tag{19.12}$$

An extension of this model was proposed by Sourbron et al., for measuring both GFR and renal perfusion (Sourbron *et al.*, 2008), although with a slight overestimation of GFR values (Buckley *et al.*, 2006). Other models have been proposed by combining more compartments for providing functional information regarding cortex and medulla subregions (Koh *et al.*, 2006). These three-compartment models exploit separate cortical and medullary DCE-MRI time curves with a shared vascular and tubular compartment, allowing calculation of GFR values in good correlation with renal scintigraphy measurements (Lee *et al.*, 2007; Zhang *et al.*, 2008). However, models with more compartments require higher data quality to extract meaningful and reliable additional parameters from the acquired images. Therefore model selection should be based also on a careful statistical evaluation (to assess whether additional free parameters improve the fitting of the data) or by exploiting automatic methods (Ewing and Bagher-Ebadian, 2013; Li *et al.*, 2012; Donaldson *et al.*, 2010; Duan *et al.*, 2017).

19.3.5.4 Shape or semiquantitative analysis

A simple and heuristic approach based on calculation of parameters such as time to peak, slope, initial area under the curve, maximum enhancement, from the shape of the signal intensity time curve (Irrera *et al.*, 2020; Zollner *et al.*, 2016). This approach does not require information regarding native T1 map, AIF, and acquisition protocols (such as FA); hence, it is simple and fast (no curve fitting is required). However, variations in native T1 can strongly affect the calculated parameters. The main limitations are difficulties in comparing results between MRI scanners, and the calculated parameters do not have any clear relationship with tissue physiological properties.

19.4 Case studies of applications

Since DCE-MRI can provide measurements of renal perfusion and filtration, it can be used in several pathological conditions that can lead to acute or chronic kidney disease, including renal artery stenosis, urinary obstructions, renal transplant, and renal masses. We will provide few representative examples for each of them and look at recent applications, including preclinical studies with animal models. Additional studies and applications have been described in previous review articles to which the reader may refer (Bokacheva *et al.*, 2008; Grenier *et al.*, 2003; Prasad, 2006; Zhang *et al.*, 2014; Caroli *et al.*, 2021).

19.4.1 Renal artery stenosis

Arterial narrowing upon either fibromuscular dysplasia or atherosclerosis can lead to renal artery stenosis (RAS), a potential but curable source of renovascular hypertension and renal impairment. Unfortunately, based on the amount of renal dysfunction, patients may be asymptomatic since a partially or normally functioning contralateral kidney may compensate for the damaged kidney. Noninvasive DCE-MRI can potentially identify patients with RAS by assessing single kidney filtration (Fig. 19.3). In patients with atherosclerotic renovascular disease, DCE-MRI was investigated for assessing single kidney glomerular filtration rate (SK-GFR) and compared with standard radioisotope measurements (Buckley *et al.*, 2006; Chrysochou *et al.*, 2008). Both SK-GFR estimates obtained by using a Patlak–Rutland plot and a compartmental model correlated well with the radioisotope standard (Spearman's rho of 0.81 and of 0.71 for the Patlak–Rutland and the compartmental model, respectively). DCE-MRI can also be applied to evaluate the functional outcome in patients with atherosclerotic RAS following renal artery revascularization. SK-GFR was measured by using a two-region filtration model at baseline and 4 months after revascularization and compared with radioisotope measures (Lim *et al.*, 2013). A good agreement was observed between SK-GFR values from DCE-MRI and radioisotopes, suggesting that DCE-MRI can potentially replace radioisotope measurements. More accurate estimations of GFR can be obtained in hypertensive patients when

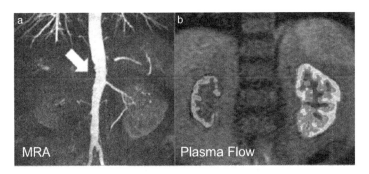

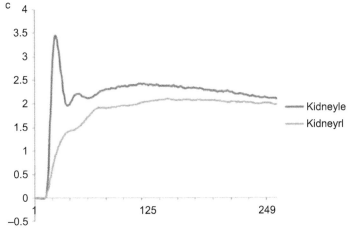

FIG. 19.3

(A) Maximum intensity projection of a high-grade stenosis of the right renal artery. The right kidney is atrophic and shows delayed and reduced enhancement. (B) Plasma flow map: plasma flow is considerably reduced in the right kidney (*right*: 49.3 ml/(100 ml × min)) vs *left*: 192.1 ml/(100 ml × min)). (C) Fitted DCE-MRI curves of the right (*red curve*) and the left (*blue curve*) kidney. The curve of the left kidney exhibits a typical shape: an early peak corresponds to the first pass of the contrast agent. A slow ascend characterizes the recirculation of the contrast agent and the glomerular–tubular transfer. A slow decay of signal intensity corresponds with the excretory function of the kidney. The right kidney shows loss of the early arterial peak and a delayed passage of the contrast media.

Reproduced with permission from Notohamiprodjo, M., Reiser, M.F., Sourbron, S.P., 2010. Diffusion and perfusion of the kidney. Eur. J. Radiol. 76 (3), 337–347.

applying novel postprocessing procedures performing automated image registration using fat-only images from a modified Dixon dual-echo acquisition (de Boer *et al.*, 2018). GFR values obtained from the fat-registered images by applying a two-compartment model correlated well with creatinine-based GFR (Spearman's correlation coefficient = 0.68, $P = 0.04$).

19.4.2 Urinary obstruction

Obstructive anomalies of the urinary tract or of the kidneys can lead to renal failure if not surgically corrected. The gold standard for assessing loss of renal function is dynamic renal scintigraphy for measuring split or differential renal function. Several approaches based on DCE-MRI have also emerged, providing accurate evaluation of split renal function and discriminating normal and obstructive kidneys.

In patients with ureteropelvic junction obstruction, single kidney GFR estimated by applying the Baumann–Rudin model to DCE-MRI data provided excellent correlation to nuclear renography measure of the split renal function ratio and accurately discriminated functional from mechanical obstruction (Krepkin *et al.*, 2014). Even in patients with chronic urinary obstruction, split renal function values assessed with the Patlak–Rutland method from DCE-MRI curves were considered to provide equivalent value to that calculated by renal scintigraphy in moderately dilated kidneys (Claudon *et al.*, 2014). However, in severely dilated kidneys, the split renal function obtained by DCE-MRI was slightly underestimated, making questionable the substitution. Exploitation of DCE-MRI in children with congenital anomalies of the kidney and urinary tract demonstrated no significant differences in the calculated split renal function as well as in the classification of drainage curves between DCE-MRI with the Patlak–Rutland method and dynamic renal scintigraphy (Fig. 19.4) (Damasio *et al.*, 2019). In patients with obstructive hydronephrosis, DCE-MRI single kidney GFR is comparable to GFR obtained from renal scintigraphy, with good reliability for both nonhydronephrotic and hydronephrotic kidneys (Wang *et al.*, 2019).

19.4.3 Kidney transplantation

The possibility of detecting early complications after kidney transplantation is critical for the survival and long-term outcome of renal grafts. Major complications are characterized by an ischemic injury to the tubules, leading to tubular dysfunction, decrease in GFR, and reduction of renal blood flow. DCE-MRI can provide measurements of single kidney GFR that is an important indicator of allograft renal function.

Sade et al. calculated K^{trans} by applying a quantitative pharmacokinetic model to evaluate renal perfusion in native kidneys before and after preemptive transplantation (Sade *et al.*, 2017). They observed that native kidneys were still viable and functioning at 6 months after transplantation, with a decreased perfusion. Prospective comparison of the DCE-MRI approach to assess GFR for determination of allograft renal function provided contradictory results. GFR determined by a modified two-compartment model after renal transplantation provided higher accuracy and precision than GFR calculated from ^{99m}Tc-DTPA-based single-photon emission computed tomography (SPECT) technique (Wan-Li *et al.*, 2019). In another study, renographies acquired with low-dose DCE-MRI and analyzed by a two-compartment pharmacokinetic model provided GFR estimates with large variability when compared to those obtained with ^{51}Cr-EDTA in kidney transplant recipients (Taton *et al.*, 2019).

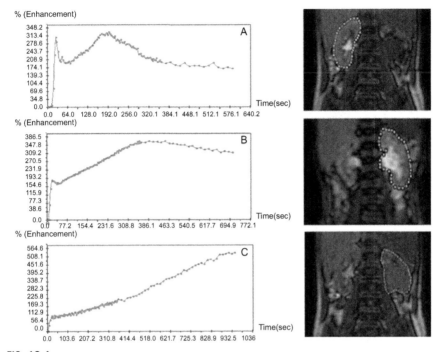

FIG. 19.4

DCE-MRI urography drainage curves classification. Curves are classified under a normal, borderline, and accumulation patterns. Normal curve shows a decreasing of the signal intensity after the filtration peak (A); borderline curve remains stable after the filtration peak, drawing a plateau (B); accumulation curve shows an ever-increasing signal intensity (C).

Reprinted with permission from Damasio, M.B., et al., 2019. Comparative study between functional MR urography and renal scintigraphy to evaluate drainage curves and split renal function in children with congenital anomalies of kidney and urinary tract (CAKUT). Front. Pediatr. 7, 527.

19.4.4 **Renal tumors**

DCE-MRI is commonly exploited in oncology for the noninvasive assessment of tumor microvasculature properties, as extravasation of the low molecular weight Gd-based contrast agent allows assessment of vascular volume and permeability (Jackson *et al.*, 2007). Moreover, DCE-MRI is emerging as a potential imaging biomarker for measuring the efficacy of antiangiogenic treatments (Fig. 19.5) (Hylton, 2006; O'Connor *et al.*, 2007). In particular, pharmacokinetic analysis of the concentration time curve allows to estimate K^{trans}, a parameter that reflects a combination of permeability and blood flow (Brix *et al.*, 2010; Consolino *et al.*, 2017, 2016).

Differentiation of renal tumor subtype is important in clinical practice for treatment planning. DCE-MRI has shown promising results for renal cell carcinoma subtype characterization, both exploiting calculated K^{trans} estimates or by a simpler

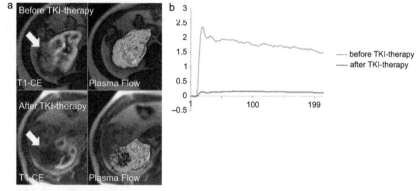

FIG. 19.5

Sarcomatoid transformed clear cell renal cell carcinoma of the right kidney before and 4 weeks after targeted TKI-therapy with Sunitinib (A). Clinically the patient showed therapy response. Before therapy a relatively high perfused mass (plasma flow 144.4 ml 100ml^{-1} min^{-1}) infiltrating the kidney can be seen (B). After TKI therapy the mass shows slight shrinkage. The mass is now hypovascularized (B) and only minimal plasma flow (21.0 ml 100ml^{-1} min^{-1}) can be determined.

Reprinted with permission from Notohamiprodjo, M., Reiser, M.F., Sourbron, S.P., 2010. Diffusion and perfusion of the kidney. Eur. J. Radiol. 76 (3): 337–347.

contrast enhancement ratios approach to distinguish between clear cell and non-clear-cell renal carcinoma with high sensitivity and specificity (Zeng *et al.*, 2015; Yamamoto *et al.*, 2017).

Owing to the high spatial resolution of MR images, spatial variation in structure and function can be quantified inside tumors and new approaches are evaluating the benefits of this intratumor heterogeneity as an additional biomarker (O'Connor *et al.*, 2015; Longo *et al.*, 2015). Recently, statistical clustering of K^{trans} and k$_{ep}$ parametric maps (obtained from an extended Tofts model) into three tumor subregions allowed differentiation of high- from low-grade clear cell renal cell carcinomas with high accuracy and sensitivity (Xi *et al.*, 2018).

In patients with primary renal cell carcinoma (RCC) and metastases, the plasma flow parameter obtained from a two-compartment exchange model showed a significant reduction for patients treated with sunitinib in comparison to untreated ones (Braunagel *et al.*, 2015). In another study in patients with metastatic renal cell cancer receiving sunitinib, higher baseline K^{trans} values calculated from the Tofts model were significantly associated with a longer progression-free survival (Hudson *et al.*, 2018).

19.4.5 **Other diseases**

Since altered renal perfusion is common across a wide range of diseases, DCE-MRI has also been investigated in patient with liver cirrhosis. GFR values calculated from low-dose DCE-MRI using a gamma variate function were significantly more accurate than creatinine-based GFR estimations in patients with cirrhosis

(Vivier *et al.*, 2011). In patients with chronic liver disease, DCE-MRI GFR by applying a three-compartment model significantly correlated with the estimated GFR from serum creatinine (Spearman's $r = 0.49$) (Bane *et al.*, 2016).

19.4.6 DCE-MRI in animal models of kidney diseases

Preclinical studies using animal models are important for characterizing kidney diseases associated with an impaired renal function. Animal models provide a controlled experimental design and allow validation of new DCE-MRI acquisition and analysis methods.

In a murine model of unilateral renal artery stenosis, GFR and renal perfusion were calculated from a modified two-compartment model 2 weeks after surgery. Both single kidney GFR and renal perfusion were comparable to that obtained with the inulin clearance (Spearman's $= 0.94$) (Jiang *et al.*, 2018). Implementation of a fast multi-slice DCE-MRI approach successfully differentiated control from stenotic mouse kidneys based on GFR and renal perfusion estimates (Jiang *et al.*, 2019).

Adriamycin-induced proteinuric renal injury in rodents is an established model for chronic kidney disease, characterized by glomerulosclerosis and fibrosis (Lee and Harris, 2011). In adriamycin-challenged rats a consistent and progressive decrease of kidney perfusion was detected as early as 6 days after exposure, with the calculated washout rate of the contrast agent significantly reduced at all time points after adriamycin (Egger *et al.*, 2015).

Ischemia-reperfusion injury is a major cause of acute kidney injury (AKI) and of delayed graft function and loss in kidney transplantation. DCE-MRI has been extensively used for detecting changes in GFR between ischemic and healthy kidneys in rats in the unilateral AKI model, showing significantly decreased GFR values using a two-compartment model (Zollner *et al.*, 2015, 2014). Additionally, the renal perfusion estimates, RBF (renal blood flow) and RBV (renal blood volume), calculated from a deconvolution analysis of the DCE-MRI time curves showed significantly reduced renal filtration between clamped and contralateral kidneys of the same mouse in the unilateral ischemia-reperfusion injury model (Fig. 19.6) (Irrera *et al.*, 2020).

In other AKI models, DCE-MRI showed earlier detection following cisplatin-induced AKI than the commonly used serum biomarkers (creatinine, blood urea nitrogen) (Privratsky *et al.*, 2019). Furthermore, the unilateral renal clearance rate derived from DCE-MRI can monitor the efficacy of mitochondria-targeted antioxidant treatment in reducing the severity of renal ischemia-reperfusion injury in rats (Liu *et al.*, 2018).

19.5 Safety considerations

The nephrogenic systemic fibrosis (NSF) is a rare and potentially fatal condition that was observed in patients with acute kidney injury or with chronic kidney disease (stage 4 or 5) with an estimated GFR (eGFR) of less than $30\,\mathrm{ml\,min^{-1}}\,1.73\,\mathrm{m^{-2}}$ upon administration of some exogenous Gd-based contrast agents (Agarwal *et al.*, 2009).

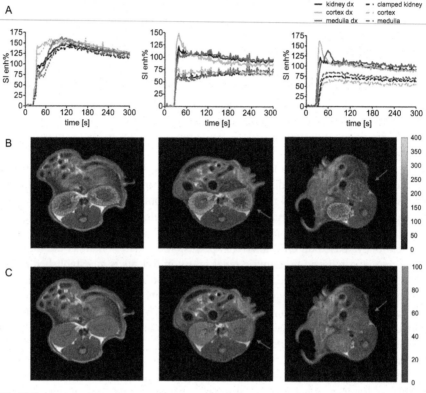

FIG. 19.6

DCE-MRI enhancement (SI enh%) time curves following gadoteridol injection (A) for whole kidneys and the cortex and medulla regions. No differences in time curve enhancements were observed for the kidneys of the healthy group (first column). After unilateral ischemia-reperfusion injury in the post-3-day (second column) and post-1-week (third column) groups, a marked decrease in DCE-MRI enhancement time curves was observed for all regions of the clamped kidneys, compared with the contralateral ones. Representative images of the calculated parametric maps RBF [ml 100 ml^{-1} min^{-1}] (B) and RBV [ml 100 ml^{-1}] (C) obtained from the deconvolution analysis of DCE-MRI acquisitions upon gadoteridol injection at different time points following AKI damage (first column, baseline; second column, after 3 days; third column, after 1 week). The arrow shows the clamped kidney.

Adapted with permission from Irrera, P., et al., 2020. Dual assessment of kidney perfusion and pH by exploiting a dynamic CEST-MRI approach in an acute kidney ischemia–reperfusion injury murine model. NMR Biomed. 33(6): e4287.

Later, it was recognized that not all the Gd-based contrast agents had the same risk, with the macrocyclic agents or linear agents with partial hepatobiliary excretion being those with lower risk (Behzadi *et al.*, 2018). In addition, several studies that exploited lower doses still achieved similar accuracy in assessing renal perfusion and filtration (Lee *et al.*, 2001; Wright *et al.*, 2014).

Owing to the association between gadolinium administration and retention and NSF, a consensus has not yet been reached regarding to which extent DCE-MRI should be applied in patients with impaired renal function (Hellman, 2011; Mathur *et al.*, 2020). Consequently, DCE-MRI studies are limited to patients with less severe reductions in GFR and even more in those with more advanced chronic kidney diseases, thus limiting its widespread applicability (Bauerle *et al.*, 2021). On the other hand, it is noteworthy that recent findings show that the low risk of retaining the most stable Gd-based complexes may outweigh the risk of NSF in this population (Woolen *et al.*, 2020).

19.6 **Challenges and solutions**

Several challenges remain for bringing functional DCE-MRI of the kidneys to the clinical environment. Some of them are related to technical barriers: new and faster sequences for T1 mapping, parallel acquisition, or non-Cartesian sampling available only in few research centers; need for highly specialized technicians; in-house implementation of sequences; and postprocessing (Lietzmann *et al.*, 2012; Coll-Font *et al.*, 2021; Zhang *et al.*, 2015b). Potential solutions are dependent on progressive harmonization on both acquisition and analysis protocols across MRI platforms and between centers (Claudon *et al.*, 2014). Recently, the European-funded COST action initiative "PARENCHIMA" has addressed this issue by providing an open access protocol collection and wide-ranging recommendations for preclinical renal MRI used in translational research for several renal functional imaging techniques, including DCE-MRI, to improve reproducibility and comparability of studies (Pohlmann *et al.*, 2021; Irrera *et al.*, 2021; Zollner *et al.*, 2021; Mendichovszky *et al.*, 2020).

Accurate measurements of the AIF are needed to improve the quality of the calculated gadolinium concentrations, hence of the derived perfusion estimates. Coronal acquisition can help reduce AIF inflow artifacts.

Significant improvements in SNR and spatial resolution or acquisition time are still possible by moving renal imaging from 3 T to higher field strengths of 7 T, requiring close collaboration with MRI manufacturers to overcome specific challenges regarding SAR limitations and RF inhomogeneity (de Boer *et al.*, 2016).

No robust evidence has been provided so far regarding the utility of renal MRI to improve diagnosis and for assessing treatment response in patients with different types of renal disease, because most of the studies are small in size. Therefore implementation of large, multicenter, and clinically led studies with prospective follow-up is needed, including evaluation of repeatability studies for assessing the accuracy of multiparametric renal MRI approaches (Claudon *et al.*, 2014; de Boer *et al.*, 2021). More comparative studies are also needed to assess the accuracy of the DCE-MRI-derived estimates relative to other methods or techniques, although reference standard methods are not available (Taton *et al.*, 2019; Alhummiany *et al.*, 2022).

New Machine Learning and Deep Learning approaches have been recently adopted for several steps related to image acquisition, reconstruction, segmentation, and postprocessing, with potential significant improvements in kidney function quantifications (Klepaczko *et al.*, 2021).

19.7 Learning and knowledge outcomes

This chapter addressed recent advances in renal MRI showing that several measurements of renal structure and function are feasible and accurate, leading to additional insights into the pathophysiology of different kidney diseases both at preclinical and clinical level. Moreover, a further step can be achieved by implementing MRI-based multiparametric imaging for the simultaneous assessment of structure, function, and perfusion as a promising tool for diagnosis, prognosis, and treatment monitoring in kidney disease. However, protocol standardization and large-scale studies are still needed to enable a broader adoption at the clinical level, which is still limited because of concerns about potential adverse effects of gadolinium-based contrast agents in patients with severe renal insufficiency.

References

Agarwal, R., *et al.*, 2009. Gadolinium-based contrast agents and nephrogenic systemic fibrosis: a systematic review and meta-analysis. Nephrol. Dial. Transplant. 24 (3), 856–863.

Alhummiany, B.A., *et al.*, 2022. Bias and precision in magnetic resonance imaging-based estimates of renal blood flow: assessment by triangulation. J. Magn. Reson. Imaging 55 (4), 1241–1250.

Anemone, A., Consolino, L., Longo, D.L., 2017. MRI-CEST assessment of tumor perfusion using X-ray iodinated agents: comparison with a conventional Gd-based agent. Eur. Radiol. 27 (5), 2170–2179.

Annet, L., *et al.*, 2004. Glomerular filtration rate: assessment with dynamic contrast-enhanced MRI and a cortical-compartment model in the rabbit kidney. J. Magn. Reson. Imaging 20 (5), 843–849.

Attenberger, U.I., *et al.*, 2010. Retrospective respiratory triggering renal perfusion MRI. Acta Radiol. 51 (10), 1163–1171.

Avedano, S., *et al.*, 2007. Maximizing the relaxivity of HSA-bound gadolinium complexes by simultaneous optimization of rotation and water exchange. Chem. Commun. (Camb.) 45, 4726–4728.

Avedano, S., *et al.*, 2013. Coupling fast water exchange to slow molecular tumbling in Gd3+ chelates: why faster is not always better. Inorg. Chem. 52 (15), 8436–8450.

Bane, O., *et al.*, 2016. Assessment of renal function using intravoxel incoherent motion diffusion-weighted imaging and dynamic contrast-enhanced MRI. J. Magn. Reson. Imaging 44 (2), 317–326.

Bane, O., *et al.*, 2018. Accuracy, repeatability, and interplatform reproducibility of T1 quantification methods used for DCE-MRI: results from a multicenter phantom study. Magn. Reson. Med. 79 (5), 2564–2575.

Barnes, S.R., *et al.*, 2015. ROCKETSHIP: a flexible and modular software tool for the planning, processing and analysis of dynamic MRI studies. BMC Med. Imaging 15, 19.

Bartos, M., *et al.*, 2014. The precision of DCE-MRI using the tissue homogeneity model with continuous formulation of the perfusion parameters. Magn. Reson. Imaging 32 (5), 505–513.

Bauerle, T., Saake, M., Uder, M., 2021. Gadolinium-based contrast agents: what we learned from acute adverse events, nephrogenic systemic fibrosis and brain retention. Rofo 193 (9), 1010–1018.

Baumann, D., Rudin, M., 2000. Quantitative assessment of rat kidney function by measuring the clearance of the contrast agent Gd(DOTA) using dynamic MRI. Magn. Reson. Imaging 18 (5), 587–595.

Behzadi, A.H., *et al.*, 2018. Immediate allergic reactions to gadolinium-based contrast agents: a systematic review and meta-analysis. Radiology 286 (2), 471–482.

Bergen, R.V., Ryner, L., Essig, M., 2020. Comparison of DCE-MRI parametric mapping using MP2RAGE and variable flip angle T1 mapping. Magn. Reson. Imaging.

Bharadwaj Das, A., *et al.*, 2022. Estimation of contrast agent concentration in DCE-MRI using 2 flip angles. Invest. Radiol. 57 (5), 343–351.

Bokacheva, L., *et al.*, 2007. Quantitative determination of Gd-DTPA concentration in T1-weighted MR renography studies. Magn. Reson. Med. 57 (6), 1012–1018.

Bokacheva, L., *et al.*, 2008. Assessment of renal function with dynamic contrast-enhanced MR imaging. Magn. Reson. Imaging Clin. N. Am. 16 (4), 597–611. viii.

Braunagel, M., *et al.*, 2015. Dynamic contrast-enhanced magnetic resonance imaging measurements in renal cell carcinoma: effect of region of interest size and positioning on interobserver and intraobserver variability. Invest. Radiol. 50 (1), 57–66.

Brix, G., *et al.*, 2010. Tracer kinetic modeling of tumor angiogenesis based on dynamic contrast-enhanced CT and MRI measurements. Eur. J. Nucl. Med. Mol. Imaging 37 (Suppl 1), S30–S51.

Buckley, D.L., 2002. Transcytolemmal water exchange and its effect on the determination of contrast agent concentration in vivo. Magn. Reson. Med. 47 (2), 420–424.

Buckley, D.L., *et al.*, 2006. Measurement of single kidney function using dynamic contrast-enhanced MRI: comparison of two models in human subjects. J. Magn. Reson. Imaging 24 (5), 1117–1123.

Buckley, D.L., Kershaw, L.E., Stanisz, G.J., 2008. Cellular-interstitial water exchange and its effect on the determination of contrast agent concentration in vivo: dynamic contrast-enhanced MRI of human internal obturator muscle. Magn. Reson. Med. 60 (5), 1011–1019.

Buonaccorsi, G.A., *et al.*, 2006. Comparison of the performance of tracer kinetic model-driven registration for dynamic contrast enhanced MRI using different models of contrast enhancement. Acad. Radiol. 13 (9), 1112–1123.

Calamante, F., Yim, P.J., Cebral, J.R., 2003. Estimation of bolus dispersion effects in perfusion MRI using image-based computational fluid dynamics. Neuroimage 19 (2 Pt 1), 341–353.

Caravan, P., *et al.*, 1999. Gadolinium(III) chelates as MRI contrast agents: structure, dynamics, and applications. Chem. Rev. 99 (9), 2293–2352.

Caroli, A., *et al.*, 2018. Diffusion-weighted magnetic resonance imaging to assess diffuse renal pathology: a systematic review and statement paper. Nephrol. Dial. Transplant. 33 (Suppl 2), ii29–ii40.

Caroli, A., Remuzzi, A., Lerman, L.O., 2021. Basic principles and new advances in kidney imaging. Kidney Int. 100 (5), 1001–1011.

Chrysochou, C., Buckley, D.L., Kalra, P.A., 2008. Magnetic resonance imaging: advances in the investigation of atheromatous renovascular disease. J. Nephrol. 21 (4), 468–477.

Claudon, M., *et al.*, 2014. Chronic urinary obstruction: evaluation of dynamic contrast-enhanced MR urography for measurement of split renal function. Radiology 273 (3), 801–812.

Coll-Font, J., *et al.*, 2021. Modeling dynamic radial contrast enhanced MRI with linear time invariant systems for motion correction in quantitative assessment of kidney function. Med. Image Anal. 67, 101880.

Consolino, L., *et al.*, 2016. Functional imaging of the angiogenic switch in a transgenic mouse model of human breast cancer by dynamic contrast enhanced magnetic resonance imaging. Int. J. Cancer 139 (2), 404–413.

Consolino, L., *et al.*, 2017. Assessing tumor vascularization as a potential biomarker of imatinib resistance in gastrointestinal stromal tumors by dynamic contrast-enhanced magnetic resonance imaging. Gastric Cancer 20 (4), 629–639.

Cottam, D., Azzopardi, G., Forni, L.G., 2022. Biomarkers for early detection and predicting outcomes in acute kidney injury. Br. J. Hosp. Med. (Lond.) 83 (8), 1–11.

Cutajar, M., *et al.*, 2010. The importance of AIF ROI selection in DCE-MRI renography: reproducibility and variability of renal perfusion and filtration. Eur. J. Radiol. 74 (3), e154–e160.

Damasio, M.B., *et al.*, 2019. Comparative study between functional MR urography and renal scintigraphy to evaluate drainage curves and split renal function in children with congenital anomalies of kidney and urinary tract (CAKUT). Front. Pediatr. 7, 527.

de Boer, A., *et al.*, 2016. 7 T renal MRI: challenges and promises. MAGMA 29 (3), 417–433.

de Boer, A., *et al.*, 2018. Modified dixon-based renal dynamic contrast-enhanced MRI facilitates automated registration and perfusion analysis. Magn. Reson. Med. 80 (1), 66–76.

de Boer, A., *et al.*, 2021. Multiparametric renal MRI: an intrasubject test–retest repeatability study. J. Magn. Reson. Imaging 53 (3), 859–873.

de Senneville, B.D., *et al.*, 2008. Improvement of MRI-functional measurement with automatic movement correction in native and transplanted kidneys. J. Magn. Reson. Imaging 28 (4), 970–978.

Debus, C., *et al.*, 2019. MITK-ModelFit: a generic open-source framework for model fits and their exploration in medical imaging—design, implementation and application on the example of DCE-MRI. BMC Bioinform. 20 (1), 31.

Dekkers, I.A., *et al.*, 2019. Reproducibility of native T1 mapping for renal tissue characterization at 3 T. J. Magn. Reson. Imaging 49 (2), 588–596.

Denolin, V., Azizieh, C., Metens, T., 2005. New insights into the mechanisms of signal formation in RF-spoiled gradient echo sequences. Magn. Reson. Med. 54 (4), 937–954.

Donaldson, S.B., *et al.*, 2010. A comparison of tracer kinetic models for T1-weighted dynamic contrast-enhanced MRI: application in carcinoma of the cervix. Magn. Reson. Med. 63 (3), 691–700.

Duan, C., *et al.*, 2017. Are complex DCE-MRI models supported by clinical data? Magn. Reson. Med. 77 (3), 1329–1339.

Egger, C., *et al.*, 2015. Adriamycin-induced nephropathy in rats: functional and cellular effects characterized by MRI. J. Magn. Reson. Imaging 41 (3), 829–840.

Eikefjord, E., *et al.*, 2015. Use of 3D DCE-MRI for the estimation of renal perfusion and glomerular filtration rate: an intrasubject comparison of FLASH and KWIC with a comprehensive framework for evaluation. AJR Am. J. Roentgenol. 204 (3), W273–W281.

Ewing, J.R., Bagher-Ebadian, H., 2013. Model selection in measures of vascular parameters using dynamic contrast-enhanced MRI: experimental and clinical applications. NMR Biomed. 26 (8), 1028–1041.

Fedorov, A., *et al.*, 2014. A comparison of two methods for estimating DCE-MRI parameters via individual and cohort based AIFs in prostate cancer: a step towards practical implementation. Magn. Reson. Imaging 32 (4), 321–329.

Fennessy, F.M., *et al.*, 2012. Practical considerations in T1 mapping of prostate for dynamic contrast enhancement pharmacokinetic analyses. Magn. Reson. Imaging 30 (9), 1224–1233.

Fluckiger, J.U., Schabel, M.C., Dibella, E.V., 2009. Model-based blind estimation of kinetic parameters in dynamic contrast enhanced (DCE)-MRI. Magn. Reson. Med. 62 (6), 1477–1486.

Gale, E.M., *et al.*, 2018. A Manganese-based alternative to gadolinium: contrast-enhanced MR angiography, excretion, pharmacokinetics, and metabolism. Radiology 286 (3), 865–872.

Garteiser, P., *et al.*, 2021. Experimental protocols for MRI mapping of renal T1. Methods Mol. Biol. 2216, 383–402.

Georgiou, L., *et al.*, 2019. A functional form for a representative individual arterial input function measured from a population using high temporal resolution DCE MRI. Magn. Reson. Med. 81 (3), 1955–1963.

Grenier, N., *et al.*, 2003. Functional MRI of the kidney. Abdom. Imaging 28 (2), 164–175.

Grenier, N., Merville, P., Combe, C., 2016. Radiologic imaging of the renal parenchyma structure and function. Nat. Rev. Nephrol. 12 (6), 348–359.

Gutierrez, D.R., *et al.*, 2010. Partial volume effects in dynamic contrast magnetic resonance renal studies. Eur. J. Radiol. 75 (2), 221–229.

Hackstein, N., Heckrodt, J., Rau, W.S., 2003. Measurement of single-kidney glomerular filtration rate using a contrast-enhanced dynamic gradient-echo sequence and the Rutland-Patlak plot technique. J. Magn. Reson. Imaging 18 (6), 714–725.

Hackstein, N., *et al.*, 2005. Glomerular filtration rate measured using the Patlak plot technique and contrast-enhanced dynamic MRI with different amounts of gadolinium-DTPA. J. Magn. Reson. Imaging 22 (3), 406–414.

Han, S., Cho, H., 2016. Temporal resolution improvement of calibration-free dynamic contrast-enhanced MRI with compressed sensing optimized turbo spin echo: the effects of replacing turbo factor with compressed sensing accelerations. J. Magn. Reson. Imaging 44 (1), 138–147.

Hansen, A.E., *et al.*, 2009. Partial volume effect (PVE) on the arterial input function (AIF) in T1-weighted perfusion imaging and limitations of the multiplicative rescaling approach. Magn. Reson. Med. 62 (4), 1055–1059.

Hanson, E., *et al.*, 2017. Workflow sensitivity of post-processing methods in renal DCE-MRI. Magn. Reson. Imaging 42, 60–68.

Heacock, L., *et al.*, 2017. Comparison of conventional DCE-MRI and a novel golden-angle radial multicoil compressed sensing method for the evaluation of breast lesion conspicuity. J. Magn. Reson. Imaging 45 (6), 1746–1752.

Hellman, R.N., 2011. Gadolinium-induced nephrogenic systemic fibrosis. Semin. Nephrol. 31 (3), 310–316.

Heye, T., et al., 2013. Reproducibility of dynamic contrast-enhanced MR imaging. Part I. Perfusion characteristics in the female pelvis by using multiple computer-aided diagnosis perfusion analysis solutions. Radiology 266 (3), 801–811.

Hodneland, E., et al., 2014. Segmentation-driven image registration- application to 4D DCE-MRI recordings of the moving kidneys. IEEE Trans. Image Process. 23 (5), 2392–2404.

Huang, W., et al., 2014. Variations of dynamic contrast-enhanced magnetic resonance imaging in evaluation of breast cancer therapy response: a multicenter data analysis challenge. Transl. Oncol. 7 (1), 153–166.

Hudson, J.M., et al., 2018. The prognostic and predictive value of vascular response parameters measured by dynamic contrast-enhanced-CT, -MRI and -US in patients with metastatic renal cell carcinoma receiving sunitinib. Eur. Radiol. 28 (6), 2281–2290.

Hueper, K., et al., 2014. T1-mapping for assessment of ischemia-induced acute kidney injury and prediction of chronic kidney disease in mice. Eur. Radiol. 24 (9), 2252–2260.

Hylton, N., 2006. Dynamic contrast-enhanced magnetic resonance imaging as an imaging biomarker. J. Clin. Oncol. 24 (20), 3293–3298.

Irrera, P., et al., 2020. Dual assessment of kidney perfusion and pH by exploiting a dynamic CEST-MRI approach in an acute kidney ischemia–reperfusion injury murine model. NMR Biomed. 33 (6), e4287.

Irrera, P., et al., 2021. Dynamic contrast enhanced (DCE) MRI-derived renal perfusion and filtration: experimental protocol. Methods Mol. Biol. 2216, 429–441.

Jackson, A., et al., 2007. Imaging tumor vascular heterogeneity and angiogenesis using dynamic contrast-enhanced magnetic resonance imaging. Clin. Cancer Res. 13 (12), 3449–3459.

Jerosch-Herold, M., Wilke, N., Stillman, A.E., 1998. Magnetic resonance quantification of the myocardial perfusion reserve with a Fermi function model for constrained deconvolution. Med. Phys. 25 (1), 73–84.

Jiang, K., et al., 2018. Measurement of murine single-kidney glomerular filtration rate using dynamic contrast-enhanced MRI. Magn. Reson. Med. 79 (6), 2935–2943.

Jiang, K., et al., 2019. Measurement of murine kidney functional biomarkers using DCE-MRI: a multi-slice TRICKS technique and semi-automated image processing algorithm. Magn. Reson. Imaging 63, 226–234.

Just, N., et al., 2011. Assessment of the effect of hematocrit-dependent arterial input functions on the accuracy of pharmacokinetic parameters in dynamic contrast-enhanced MRI. NMR Biomed. 24 (7), 902–915.

Khalifa, F., et al., 2014. Models and methods for analyzing DCE-MRI: a review. Med. Phys. 41 (12), 124301.

Khrichenko, D., Darge, K., 2010. Functional analysis in MR urography—made simple. Pediatr. Radiol. 40 (2), 182–199.

Kim, H., 2018a. Variability in quantitative DCE-MRI: sources and solutions. J. Nat. Sci. 4 (1).

Kim, H., 2018b. Modification of population based arterial input function to incorporate individual variation. Magn. Reson. Imaging 45, 66–71.

Klepaczko, A., Eikefjord, E., Lundervold, A., 2021. Healthy kidney segmentation in the Dce-Mr images using a convolutional neural network and temporal signal characteristics. Sensors (Basel) 21 (20).

Kleppesto, M., *et al.*, 2022. Operator dependency of arterial input function in dynamic contrast-enhanced MRI. MAGMA 35 (1), 105–112.

Kline, T.L., *et al.*, 2017. Performance of an artificial multi-observer deep neural network for fully automated segmentation of polycystic kidneys. J. Digit. Imaging 30 (4), 442–448.

Koh, T.S., *et al.*, 2006. A biphasic parameter estimation method for quantitative analysis of dynamic renal scintigraphic data. Phys. Med. Biol. 51 (11), 2857–2870.

Kratochvila, J., *et al.*, 2016. Distributed capillary adiabatic tissue homogeneity model in parametric multi-channel blind AIF estimation using DCE-MRI. Magn. Reson. Med. 75 (3), 1355–1365.

Krepkin, K., *et al.*, 2014. Dynamic contrast-enhanced MR renography for renal function evaluation in ureteropelvic junction obstruction: feasibility study. AJR Am. J. Roentgenol. 202 (4), 778–783.

Kurugol, S., *et al.*, 2020. Prospective pediatric study comparing glomerular filtration rate estimates based on motion-robust dynamic contrast-enhanced magnetic resonance imaging and serum creatinine (eGFR) to (99m)Tc DTPA. Pediatr. Radiol. 50 (5), 698–705.

Larsson, H.B., *et al.*, 2008. Dynamic contrast-enhanced quantitative perfusion measurement of the brain using T1-weighted MRI at 3 T. J. Magn. Reson. Imaging 27 (4), 754–762.

Laurent, S., Elst, L.V., Muller, R.N., 2006. Comparative study of the physicochemical properties of six clinical low molecular weight gadolinium contrast agents. Contrast Media Mol. Imaging 1 (3), 128–137.

Lee, V.W., Harris, D.C., 2011. Adriamycin nephropathy: a model of focal segmental glomerulosclerosis. Nephrology (Carlton) 16 (1), 30–38.

Lee, V.S., *et al.*, 2001. MR renography with low-dose gadopentetate dimeglumine: feasibility. Radiology 221 (2), 371–379.

Lee, V.S., *et al.*, 2007. Renal function measurements from MR renography and a simplified multicompartmental model. Am. J. Physiol. Renal Physiol. 292 (5), F1548–F1559.

Leone, L., *et al.*, 2022. A neutral and stable macrocyclic Mn(II) complex for MRI tumor visualization. ChemMedChem, e202200508.

Li, X., *et al.*, 2011. Renal cortex segmentation using optimal surface search with novel graph construction. Med. Image Comput. Comput. Assist. Interv. 14 (Pt 3), 387–394.

Li, X., *et al.*, 2012. Statistical comparison of dynamic contrast-enhanced MRI pharmacokinetic models in human breast cancer. Magn. Reson. Med. 68 (1), 261–271.

Lietzmann, F., *et al.*, 2012. DCE-MRI of the human kidney using BLADE: a feasibility study in healthy volunteers. J. Magn. Reson. Imaging 35 (4), 868–874.

Lim, S.W., *et al.*, 2013. Prediction and assessment of responses to renal artery revascularization with dynamic contrast-enhanced magnetic resonance imaging: a pilot study. Am. J. Physiol. Renal Physiol. 305 (5), F672–F678.

Liu, X., *et al.*, 2018. Mitochondria-targeted antioxidant MitoQ reduced renal damage caused by ischemia–reperfusion injury in rodent kidneys: longitudinal observations of T2-weighted imaging and dynamic contrast-enhanced MRI. Magn. Reson. Med. 79 (3), 1559–1567.

Longo, D.L., *et al.*, 2013. Imaging the pH evolution of an acute kidney injury model by means of iopamidol, a MRI-CEST pH-responsive contrast agent. Magn. Reson. Med. 70 (3), 859–864.

Longo, D.L., *et al.*, 2015. Cluster analysis of quantitative parametric maps from DCE-MRI: application in evaluating heterogeneity of tumor response to antiangiogenic treatment. Magn. Reson. Imaging 33 (6), 725–736.

Longo, D.L., *et al.*, 2016. In vitro and in vivo assessment of nonionic iodinated radiographic molecules as chemical exchange saturation transfer magnetic resonance imaging tumor perfusion agents. Invest. Radiol. 51 (3), 155–162.

Longo, D.L., *et al.*, 2017. Noninvasive evaluation of renal pH homeostasis after ischemia reperfusion injury by CEST-MRI. NMR Biomed. 30 (7).

Longo, D.L., *et al.*, 2021. Renal pH imaging using chemical exchange saturation transfer (CEST) MRI: basic concept. Methods Mol. Biol. 2216, 241–256.

Lundervold, A.S., Lundervold, A., 2019. An overview of deep learning in medical imaging focusing on MRI. Z. Med. Phys. 29 (2), 102–127.

Lv, J., *et al.*, 2018. Respiratory motion correction for free-breathing 3D abdominal MRI using CNN-based image registration: a feasibility study. Br. J. Radiol. 91 (1083), 20170788.

Mahmoud, H., *et al.*, 2016. Imaging the kidney using magnetic resonance techniques: structure to function. Curr. Opin. Nephrol. Hypertens. 25 (6), 487–493.

Mathur, M., Jones, J.R., Weinreb, J.C., 2020. Gadolinium deposition and nephrogenic systemic fibrosis: a radiologist's primer. Radiographics 40 (1), 153–162.

Mazurowski, M.A., *et al.*, 2019. Deep learning in radiology: an overview of the concepts and a survey of the state of the art with focus on MRI. J. Magn. Reson. Imaging 49 (4), 939–954.

Melbourne, A., *et al.*, 2011. The effect of motion correction on pharmacokinetic parameter estimation in dynamic-contrast-enhanced MRI. Phys. Med. Biol. 56 (24), 7693–7708.

Mendichovszky, I.A., Cutajar, M., Gordon, I., 2009. Reproducibility of the aortic input function (AIF) derived from dynamic contrast-enhanced magnetic resonance imaging (DCE-MRI) of the kidneys in a volunteer study. Eur. J. Radiol. 71 (3), 576–581.

Mendichovszky, I., *et al.*, 2020. Technical recommendations for clinical translation of renal MRI: a consensus project of the Cooperation in Science and Technology Action PARENCHIMA. MAGMA 33 (1), 131–140.

Messroghli, D.R., *et al.*, 2004. Modified Look-Locker inversion recovery (MOLLI) for high-resolution T1 mapping of the heart. Magn. Reson. Med. 52 (1), 141–146.

Michaely, H.J., *et al.*, 2006. Renal artery stenosis: functional assessment with dynamic MR perfusion measurements—feasibility study. Radiology 238 (2), 586–596.

Michaely, H.J., *et al.*, 2007. Semiquantitative assessment of first-pass renal perfusion at 1.5 T: comparison of 2D saturation recovery sequences with and without parallel imaging. AJR Am. J. Roentgenol. 188 (4), 919–926.

Montet, X., *et al.*, 2003. Noninvasive measurement of absolute renal perfusion by contrast medium-enhanced magnetic resonance imaging. Invest. Radiol. 38 (9), 584–592.

Mouridsen, K., *et al.*, 2006. Bayesian estimation of cerebral perfusion using a physiological model of microvasculature. Neuroimage 33 (2), 570–579.

Nkongchu, K., Santyr, G., 2005. An improved 3-D Look- -Locker imaging method for T(1) parameter estimation. Magn. Reson. Imaging 23 (7), 801–807.

Notohamiprodjo, M., Reiser, M.F., Sourbron, S.P., 2010. Diffusion and perfusion of the kidney. Eur. J. Radiol. 76 (3), 337–347.

O'Connor, J.P., *et al.*, 2007. DCE-MRI biomarkers in the clinical evaluation of antiangiogenic and vascular disrupting agents. Br. J. Cancer 96 (2), 189–195.

O'Connor, J.P., *et al.*, 2015. Imaging intratumor heterogeneity: role in therapy response, resistance, and clinical outcome. Clin. Cancer Res. 21 (2), 249–257.

Ortuno, J.E., *et al.*, 2013. DCE@urLAB: a dynamic contrast-enhanced MRI pharmacokinetic analysis tool for preclinical data. BMC Bioinform. 14, 316.

Pandey, A., *et al.*, 2017. Multiresolution imaging using golden angle stack-of-stars and compressed sensing for dynamic MR urography. J. Magn. Reson. Imaging 46 (1), 303–311.

Parker, G.J., *et al.*, 2006. Experimentally-derived functional form for a population-averaged high-temporal-resolution arterial input function for dynamic contrast-enhanced MRI. Magn. Reson. Med. 56 (5), 993–1000.

Pedersen, M., *et al.*, 2021. Dynamic contrast enhancement (DCE) MRI-derived renal perfusion and filtration: basic concepts. Methods Mol. Biol. 2216, 205–227.

Pohlmann, A., *et al.*, 2021. Recommendations for preclinical renal MRI: a comprehensive open-access protocol collection to improve training, reproducibility, and comparability of studies. Methods Mol. Biol. 2216, 3–23.

Positano, V., *et al.*, 2013. Automatic 2D registration of renal perfusion image sequences by mutual information and adaptive prediction. MAGMA 26 (3), 325–335.

Prasad, P.V., 2006. Functional MRI of the kidney: tools for translational studies of pathophysiology of renal disease. Am. J. Physiol. Renal Physiol. 290 (5), F958–F974.

Privratsky, J.R., *et al.*, 2019. Dynamic contrast-enhanced MRI promotes early detection of toxin-induced acute kidney injury. Am. J. Physiol. Renal Physiol. 316 (2), F351–F359.

Riffel, P., *et al.*, 2016. "One-stop shop": free-breathing dynamic contrast-enhanced magnetic resonance imaging of the kidney using iterative reconstruction and continuous golden-angle radial sampling. Invest. Radiol. 51 (11), 714–719.

Rusinek, H., Lee, V.S., Johnson, G., 2001. Optimal dose of Gd-DTPA in dynamic MR studies. Magn. Reson. Med. 46 (2), 312–316.

Rusinek, H., *et al.*, 2007. Performance of an automated segmentation algorithm for 3D MR renography. Magn. Reson. Med. 57 (6), 1159–1167.

Sade, R., *et al.*, 2017. Value of dynamic MRI using the Ktrans technique for assessment of native kidneys in pre-emptive renal transplantation. Acta Radiol. 58 (8), 1005–1011.

Sahoo, P., *et al.*, 2016. Comparison of actual with default hematocrit value in dynamic contrast enhanced MR perfusion quantification in grading of human glioma. Magn. Reson. Imaging 34 (8), 1071–1077.

Sanz-Requena, R., *et al.*, 2015. Automatic individual arterial input functions calculated from PCA outperform manual and population-averaged approaches for the pharmacokinetic modeling of DCE-MR images. J. Magn. Reson. Imaging 42 (2), 477–487.

Sato, Y., Takahashi, M., Yanagita, M., 2020. Pathophysiology of AKI to CKD progression. Semin. Nephrol. 40 (2), 206–215.

Schabel, M.C., Parker, D.L., 2008. Uncertainty and bias in contrast concentration measurements using spoiled gradient echo pulse sequences. Phys. Med. Biol. 53 (9), 2345–2373.

Scheffler, K., 1999. A pictorial description of steady-states in rapid magnetic resonance imaging. Concepts Magn. Reson. 11 (5), 291–304.

Scheffler, K., Lehnhardt, S., 2003. Principles and applications of balanced SSFP techniques. Eur. Radiol. 13 (11), 2409–2418.

Selby, N.M., *et al.*, 2018. Magnetic resonance imaging biomarkers for chronic kidney disease: a position paper from the European Cooperation in Science and Technology Action PARENCHIMA. Nephrol. Dial. Transplant. 33 (Suppl 2), ii4–ii14.

Semmineh, N.B., *et al.*, 2017. A population-based digital reference object (DRO) for optimizing dynamic susceptibility contrast (DSC)-MRI methods for clinical trials. Tomography 3 (1), 41–49.

Shuter, B., *et al.*, 1996. The relaxivity of Gd-EOB-DTPA and Gd-DTPA in liver and kidney of the wistar rat. Magn. Reson. Imaging 14 (3), 243–253.

Sourbron, S., 2010. Technical aspects of MR perfusion. Eur. J. Radiol. 76 (3), 304–313.

Sourbron, S.P., *et al.*, 2008. MRI-measurement of perfusion and glomerular filtration in the human kidney with a separable compartment model. Invest. Radiol. 43 (1), 40–48.

Stemkens, B., *et al.*, 2019. A dual-purpose MRI acquisition to combine 4D-MRI and dynamic contrast enhanced imaging for abdominal radiotherapy planning. Phys. Med. Biol. 64 (6), 06NT02.

Stikov, N., *et al.*, 2015. On the accuracy of T1 mapping: searching for common ground. Magn. Reson. Med. 73 (2), 514–522.

Takahashi, H., *et al.*, 2021. MR characteristics of mucinous tubular and spindle cell carcinoma (MTSCC) of the kidney: comparison with clear cell and papillary subtypes of renal cell carcinoma. Abdom. Radiol. (NY) 46 (11), 5250–5259.

Taton, B., *et al.*, 2019. A prospective comparison of dynamic contrast-enhanced MRI and (51) Cr-EDTA clearance for glomerular filtration rate measurement in 42 kidney transplant recipients. Eur. J. Radiol. 117, 209–215.

Taylor, J., *et al.*, 1997. Magnetic resonance renography: optimisation of pulse sequence parameters and Gd-DTPA dose, and comparison with radionuclide renography. Magn. Reson. Imaging 15 (6), 637–649.

Tokuda, J., *et al.*, 2011. Impact of nonrigid motion correction technique on pixel-wise pharmacokinetic analysis of free-breathing pulmonary dynamic contrast-enhanced MR imaging. J. Magn. Reson. Imaging 33 (4), 968–973.

Viergever, M.A., *et al.*, 2016. A survey of medical image registration—under review. Med. Image Anal. 33, 140–144.

Vivier, P.H., *et al.*, 2011. Kidney function: glomerular filtration rate measurement with MR renography in patients with cirrhosis. Radiology 259 (2), 462–470.

Wah, T.M., *et al.*, 2018. Renal cell carcinoma perfusion before and after radiofrequency ablation measured with dynamic contrast enhanced MRI: a pilot study. Diagnostics (Basel) 8 (1).

Wang, Y.C., *et al.*, 2019. The accuracy of renal function measurements in obstructive hydro-nephrosis using dynamic contrast-enhanced MR renography. AJR Am. J. Roentgenol. 213 (4), 859–866.

Wan-Li, Z., *et al.*, 2019. Prospective comparison between DCE-MRR and (99 m) Tc-DTPA-based SPECT for determination of allograft renal function. J. Magn. Reson. Imaging 49 (1), 262–269.

Wiart, M., *et al.*, 2006. In vivo quantification of regional myocardial blood flow: validity of the fast-exchange approximation for intravascular T1 contrast agent and long inversion time. Magn. Reson. Med. 56 (2), 340–347.

Woolen, S.A., *et al.*, 2020. Risk of nephrogenic systemic fibrosis in patients with stage 4 or 5 chronic kidney disease receiving a group II gadolinium-based contrast agent: a systematic review and meta-analysis. JAMA Intern. Med. 180 (2), 223–230.

Wright, K.L., *et al.*, 2014. Quantitative high-resolution renal perfusion imaging using 3-dimensional through-time radial generalized autocalibrating partially parallel acquisi-tion. Invest. Radiol. 49 (10), 666–674.

Xi, Y., *et al.*, 2018. Statistical clustering of parametric maps from dynamic contrast enhanced MRI and an associated decision tree model for non-invasive tumor grading of T1b solid clear cell renal cell carcinoma. Eur. Radiol. 28 (1), 124–132.

Yamamoto, A., *et al.*, 2017. Differentiation of subtypes of renal cell carcinoma: dynamic contrast-enhanced magnetic resonance imaging versus diffusion-weighted magnetic resonance imaging. Clin. Imaging 41, 53–58.

Yang, X., *et al.*, 2016. Renal compartment segmentation in DCE-MRI images. Med. Image Anal. 32, 269–280.

Zeng, M., Cheng, Y., Zhao, B., 2015. Measurement of single-kidney glomerular filtration function from magnetic resonance perfusion renography. Eur. J. Radiol. 84 (8), 1419–1423.

Zhang, J., Kim, S.G., 2019. Estimation of cellular-interstitial water exchange in dynamic contrast enhanced MRI using two flip angles. NMR Biomed. 32 (11), e4135.

Zhang, J.L., Lee, V.S., 2020. Renal perfusion imaging by MRI. J. Magn. Reson. Imaging 52 (2), 369–379.

Zhang, J.L., *et al.*, 2008. Functional assessment of the kidney from magnetic resonance and computed tomography renography: impulse retention approach to a multicompartment model. Magn. Reson. Med. 59 (2), 278–288.

Zhang, J.L., *et al.*, 2014. New magnetic resonance imaging methods in nephrology. Kidney Int. 85 (4), 768–778.

Zhang, J., *et al.*, 2015a. Effect of T2* correction on contrast kinetic model analysis using a reference tissue arterial input function at 7 T. MAGMA 28 (6), 555–563.

Zhang, Y.D., *et al.*, 2015b. Feasibility study of high-resolution DCE-MRI for glomerular filtration rate (GFR) measurement in a routine clinical modal. Magn. Reson. Imaging 33 (8), 978–983.

Zollner, F.G., *et al.*, 2009. Assessment of 3D DCE-MRI of the kidneys using non-rigid image registration and segmentation of voxel time courses. Comput. Med. Imaging Graph. 33 (3), 171–181.

Zollner, F.G., *et al.*, 2013. UMMPerfusion: an open source software tool towards quantitative MRI perfusion analysis in clinical routine. J. Digit. Imaging 26 (2), 344–352.

Zollner, F.G., *et al.*, 2014. Renal perfusion in acute kidney injury with DCE-MRI: deconvolution analysis versus two-compartment filtration model. Magn. Reson. Imaging 32 (6), 781–785.

Zollner, F.G., *et al.*, 2015. Functional imaging of acute kidney injury at 3 Tesla: investigating multiple parameters using DCE-MRI and a two-compartment filtration model. Z. Med. Phys. 25 (1), 58–65.

Zollner, F.G., *et al.*, 2016. An open source software for analysis of dynamic contrast enhanced magnetic resonance images: UMM Perfusion revisited. BMC Med. Imaging 16, 7.

Zollner, F.G., *et al.*, 2020. Image registration in dynamic renal MRI-current status and prospects. MAGMA 33 (1), 33–48.

Zollner, F.G., *et al.*, 2021. Analysis protocol for dynamic contrast enhanced (DCE) MRI of renal perfusion and filtration. Methods Mol. Biol. 2216, 637–653.

MRI of skeletal muscle perfusion

20

Fatemeh Adelnia[a], Donnie Cameron[b], and David A. Reiter[c]

[a]*Department of Radiology and Radiological Sciences, Vanderbilt University Medical Center, Vanderbilt University Institute of Imaging Science, Nashville, TN, United States*
[b]*C.J. Gorter MRI Center, Department of Radiology, Leiden University Medical Center, Leiden, The Netherlands*
[c]*Department of Radiology and Imaging Sciences, Biomedical Engineering, and Orthopedics, Emory University School of Medicine, Atlanta, GA, United States*

20.1 Overview

Perfusion in skeletal muscle plays a critical role in supporting muscle function and cellular homeostasis through the delivery of oxygen and nutrients and the removal of metabolic waste products. The primary function of skeletal muscle is the production of force through energetically demanding muscle contractions. At rest, muscle metabolism is relatively low, supported by a resting blood flow of $<5\,ml\,min^{-1}\,100\,g^{-1}$ tissue. Muscle microvasculature can produce the largest change in blood volume of any organ in the body, provoked by increases in local oxygen metabolism as great as 100-fold from rest to peak exercise, resulting in up to a 20-fold increase in blood flow (Heinonen *et al.*, 2015; Joyner and Casey, 2015). In addition to performing muscle contractions, skeletal muscle is the largest organ by mass, comprising 40%–45% of the total body weight, and serves as the body's largest and most dynamic reservoir of protein (Poole *et al.*, 2013; Wang *et al.*, 2022). Skeletal muscle mass is highly regulated through the dynamic balance of synthesis (anabolism) and breakdown (catabolism) of muscle protein. This balance is disrupted in acute and chronic disease, leading to muscle loss and increased morbidity and mortality. Multiple MRI approaches have been established for measuring perfusion in skeletal muscle as it relates to muscle function and homeostasis. Perfusion MRI is emerging as a tool for improved assessment of muscle pathophysiology in a variety of cardiometabolic and musculoskeletal diseases, with the potential for providing biomarkers useful for developing and monitoring treatment strategies.

While a detailed review of muscle microvascular anatomy and physiology is beyond the scope of this chapter, there are some important characteristics that help inform the application of MRI perfusion measurements. Skeletal muscle blood flow is supplied by the largest capillary bed in the body (Poole *et al.*, 2013).

Advances in Magnetic Resonance Technology and Applications, Volume 11, ISSN 2666-9099
https://doi.org/10.1016/B978-0-323-95209-5.00003-9

Muscle microvascular blood flow delivery is provided by one or more feed arteries, supplying a hierarchical network of arterioles, all of which can actively regulate blood flow. These arterioles exhibit decreasing luminal diameter with each branch point, feeding smaller arterioles ending with the terminal arterioles. Terminal arterioles are the smallest vessel capable of controlling blood flow, and supply blood directly to a "capillary unit" consisting of up to 20 capillary vessels (6 μm average diameter in healthy individuals), which drain into a single postcapillary venule (Fig. 20.1). In contrast to other organs like the brain, liver, lung, and kidney, the capillary unit in skeletal muscle has a columnar architecture with capillaries positioned between muscle fibers and oriented longitudinally along fibers. This distribution of capillaries around cells minimizes the diffusion distance across cells, and thus provides efficient transfer of circulating oxygen and nutrients to the muscle, as well as removal of waste. The capillary unit in humans is 500–1000 μm in length, 200–500 μm in width, and 100–200 μm in depth (Murrant *et al.*, 2017). One terminal arteriole typically supplies blood to two capillary units flowing in opposing directions along muscle fibers—this is termed the capillary fascicle (Mendelson *et al.*, 2021). Capillary fascicles are distributed among each other within muscle groups

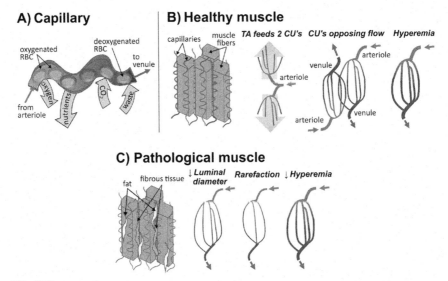

FIG. 20.1

Illustration of capillaries in skeletal muscle. (A) Capillaries support the most direct delivery of oxygen and nutrients to skeletal muscle cells and removal of metabolic waste. (B) Capillary units (CUs) exhibit a columnar architecture with capillaries positioned between muscle fibers and oriented longitudinally along fibers. One terminal arteriole (TA) supplies blood to two CUs that are oriented in opposite directions along the muscle. Capillary units are commonly distributed among each other and in opposing directions. After stimulus, arterioles increase blood flow to affected CUs via vasodilation. (C) Pathological changes to the CU include reduced capillary diameter, loss of capillary count (rarefaction), and reduced hyperemia in response to stress.

and between motor units, and oriented in opposing directions, providing redundancy in the delivery of perfusion to the tissue from multiple terminal arterioles (Poole *et al.*, 2013).

Perfusion impairments are involved in aging and an array of acute and chronic diseases, having substantial impact on clinical outcomes. Impairments can be both direct and indirect consequences of disease, which is often of fundamental interest in research and clinical settings and in the development and monitoring of treatments. Architecturally, capillary density decreases through loss of vessels (rarefaction) with pathology, limiting perfusion capacity (Bradley *et al.*, 1990; Sakkas *et al.*, 2003; Prior *et al.*, 2016). Additionally, delivery of oxygenated blood to the capillary bed can be impaired with reductions in capillary luminal diameter (Morifuji *et al.*, 2012; Padilla *et al.*, 2006), where vessels become too small for efficient transport of oxygen-carrying red blood cells (McClatchey *et al.*, 2017).

In addition to architectural derangements, microvascular dysfunction—the dysregulation of blood flow—leads to perfusion impairments impacting muscle function and cardiovascular health (Dvoretskiy *et al.*, 2020; Houben *et al.*, 2017; Hudlicka, 2011). Blood flow to the capillary unit is regulated by complex interactions between systemic (hormonal and neurovascular) and local (metabolic, myogenic, and endothelial) mechanisms (Hudlicka, 2011). This demand-driven regulation of blood flow is impaired in aging and disease, motivating the combination of MRI perfusion measurements with stress studies. Two types of stresses are routinely used in combination with perfusion MRI: muscle contraction (functional hyperemia) and temporary restriction of blood flow (reactive hyperemia). These two paradigms have been widely used in the study of muscle and microvascular physiology and their implementations vary in terms of intensity (*i.e.*, muscle contraction level or pneumatic cuff pressure) and duration (*i.e.*, $\sim$1 s up to several minutes) of the stress. The intricacies of these stresses have been described elsewhere (Limberg *et al.*, 2020) and are beyond the scope of this chapter, but will influence the selection of the perfusion technique and specific implementation in terms of required temporal resolution.

It is important to have prior knowledge of the anticipated microvascular changes when selecting quantitative MRI approaches, understanding that there are several abnormalities that could be initiating, relevant, or key to perfusion impairment. Impairments could be structural and/or functional and might be observed at rest or in response to stress. Four primary MRI modalities have been used to characterize skeletal muscle perfusion, each affording advantages and disadvantages for a given application: dynamic contrast-enhanced MRI (DCE-MRI), arterial spin labeling (ASL), intravoxel incoherent motion (IVIM), and blood oxygenation level-dependent (BOLD) MRI. The remainder of this section provides an overview of each method as it relates to some of the specific attributes of underlying microvascular pathology described before. Since previous chapters have described the theory behind DCE-MRI (Chapter 2), ASL (Chapter 4), and BOLD (Chapter 5), we will only provide technical details as they relate to applications in skeletal muscle. Since IVIM is not described in previous chapters, we will include a brief overview of the theory here.

20.1.1 **DCE-MRI**

Dynamic contrast-enhanced MRI is the only method discussed here that provides unambiguous tracking and monitoring of tracer concentration changes in circulation over time. This allows for description of physiologic quantities of the microvascular system like perfusion (in units of ml of blood min^{-1} $100 g^{-1}$ of tissue), and microvessel permeability and surface area (Tofts *et al.*, 1999). Resting perfusion measures can be obtained with a rapid T_1-weighted sequence acquired dynamically, starting before the injection of a T_1-shortening tracer, and continuing during washout. Because DCE-MRI uses a rapid T_1-weighted sequence, it can capture larger imaging volumes than other MRI techniques, which is beneficial in conditions where perfusion is heterogeneous. The most common tracers for skeletal muscle perfusion studies are FDA-approved gadolinium-based contrast agents (GBCAs) such as gadolinium diethylenetriamine penta-acetic acid (Gd-DTPA). These agents are small enough in size (less than 0.5 kDa) to allow rapid extravasation from the plasma capillary space to the interstitial space of the tissue. Delivery of GBCAs is usually flow limited, making them suitable for studying vascular permeability and damage to the microvascular barrier as a direct or indirect pathologic consequence of diseases or conditions such as aging and muscle injury. The primary limitation of DCE-MRI is the safety risks associated with GBCAs, which are contraindicated for patients with renal insufficiency, a condition often occurring in patients with microvascular impairments. Since perfusion measurements using DCE-MRI require bolus tracking, and these GBCAs have a half-life of ~90 min, they are less practical for repeated measurements within one session. Because of these limitations, DCE-MRI of skeletal muscle is used less frequently than other MRI modalities in clinical and research studies when contrast is not already clinically indicated.

20.1.2 **ASL**

Details on the fundamentals of ASL have been described in Chapter 4. ASL, like DCE-MRI, permits the measurement of skeletal muscle perfusion in physiologic units, but without the need for exogenous contrast. Baseline perfusion in skeletal muscle—less than 5 ml min^{-1} $100 g^{-1}$—is too time consuming to reliably measure with ASL, due to the technique's inherently low signal-to-noise ratio (SNR). For this reason, a vasoactive stimulus is required to induce sufficient perfusion in muscle, allowing for dynamic quantification of blood flow magnitude and time course. ASL techniques are capable of high-temporal-resolution acquisitions, sufficient to capture blood flow dynamics during hyperemia. Further, because no tracer is required, ASL does not impose the same limitations as DCE-MRI on the number of perfusion measurements per scan session.

20.1.3 **IVIM**

IVIM was initially introduced to provide perfusion contrast in the brain for functional imaging studies (Le Bihan, 1988). It has since been broadly applied to multiple tissue systems, including skeletal muscle. IVIM imaging uses a diffusion-weighted sequence and has most widely been implemented using pulsed gradient spin echo (PGSE). The PGSE sequence consists of a pair of bipolar gradients where the first gradient lobe imposes a linearly varying phase shift to the transverse magnetization. After a delay, the second lobe reverses this phase shift. If translational molecular motion occurs during the delay interval, as is the case with water diffusion in muscle fibers and perfusion of blood water in the microvasculature, the incoherent destructive phase accumulation caused by the bipolar gradients results in a net signal loss (Le Bihan, 1988).

Sensitization of the signal attenuation to the rate of molecular motion is controlled by the diffusion gradient strength (G), gradient duration (δ), and the separation of the two gradient lobes (Δ). This is frequently expressed as a single diffusion-weighting parameter (b) with units of time per distance squared (*i.e.*, s/mm (Joyner and Casey, 2015)), as follows for the PGSE sequence:

$$b = \gamma^2 G^2 \delta^2 \left(\Delta - \frac{\delta}{3} \right) \tag{20.1}$$

where γ is the gyromagnetic ratio (Stejskal and Tanner, 1965). The b-value is inversely proportional to the rate of molecular motion and so smaller values (*i.e.*, $b < 250\,\text{mm}^2/\text{s}$) are sensitive to more rapid motions of blood perfusion and larger values are sensitive to slower tissue–water diffusion outside the vascular compartment.

The perfusion and diffusion signals are superimposed within each voxel, but due to the large difference in their respective motional rates (perfusion is at least an order of magnitude larger than tissue–water diffusion), these two phenomena can be separately measured with a sufficient b-value range and using appropriate signal modeling. The most common IVIM signal model assumes two separate compartments represented using a biexponential function:

$$S(b) = S(0) \cdot \left(f \cdot exp^{(-bD^*)} + (1 - f) \cdot exp^{(-bD)} \right) \tag{20.2}$$

where $S(0)$ is the signal intensity in the absence of diffusion gradients, f is the fraction of signal attributed to perfusion, D^* is the rate of blood water molecular motions in the microvasculature (*i.e.*, pseudo-diffusion coefficient), and D is the apparent diffusion coefficient of water in the extravascular space.

IVIM parameters, unlike DCE-MRI and ASL, do not provide perfusion properties in units of blood flow but rather units of diffusion (*i.e.*, mm^2/s). The quantity $f \times D^*$ is thought to be IVIM's closest representation of blood flow, showing a correlation with ASL-derived perfusion before and after exercise (Ohno *et al.*, 2020), while f alone provides a microvascular structure metric.

IVIM can be applied to measure muscle perfusion at rest and after stress. Because it relies on multiple b-values and gradient directions, and the biexponential model is SNR challenged, IVIM has a much lower time resolution compared with DCE-MRI, ASL, and BOLD, making it less ideal for capturing rapid changes in perfusion. However, IVIM studies of hyperemia have been used to measure slower dynamic changes in perfusion parameters on the order of minutes (Adelnia *et al.*, 2019a; Filli *et al.*, 2015) rather than seconds (Englund *et al.*, 2015), supporting its use in functional muscle studies.

20.1.4 **BOLD**

The BOLD signal in muscle mainly reflects venous oxygenation, but is also dependent on blood volume, perfusion, and orientation of vascular architecture relative to the main magnetic field. Specifically, as oxygen-saturated arterial red blood cells release oxygen across concentration gradients to the surrounding tissue (Fig. 20.1A), their magnetic susceptibility transitions from diamagnetic to paramagnetic, resulting in a net reduction in tissue T_2^* values (Lebon et al., 1998). Because BOLD signal contrast relies on substantive fluctuations in blood oxygenation, it is used in combination with stimuli in muscle studies (Caroca et al., 2021). Although skeletal muscle cells contribute to susceptibility-related signal decay, both numerical simulations and experimental studies suggest that the contribution of the extravascular BOLD effect to skeletal muscle T_2^* is too small to be practically important (Sanchez et al., 2010).

BOLD is the only perfusion MRI approach that provides information specific to oxygen delivery and utilization. While DCE-MRI, ASL, and IVIM are frequently used as proxies for studying blood flow changes in response to increased metabolism, they reflect blood water signal alone. This is particularly important when assessing functional perfusion in conditions like diabetes, where decreases in capillary luminal diameter can lead to uncoupling of circulating red blood cells from plasma at the capillary level (McClatchey et al., 2017). Thus BOLD might provide a more sensitive biomarker of impairment in functional perfusion as compared with other measures.

20.2 **Imaging protocol**
20.2.1 **Participant setup**

The anatomy of interest should be centered on the scanning bed as much as possible, to mitigate B_0 and B_1 inhomogeneities. For exercise protocols, the muscle(s) of interest should be isolated using pads, sandbags, and straps to minimize motions during exercise and limit gross movement artifacts. Ideally, a flexible, lightweight RF coil should be used for improved sensitivity. Receive coils should not be fastened too tightly, to avoid influencing perfusion, and cable lengths and positioning of ancillary coil hardware should be considered when exercise or inflator-cuff apparatus is built or purchased.

20.2.2 **Vasoactive stimulus**

Two types of stresses are routinely used with muscle perfusion MRI: (i) reactive hyperemia methods (Fig. 20.2) and (ii) functional hyperemia methods (Fig. 20.3). Vasoactive stimulus paradigms are currently essential for skeletal muscle ASL and BOLD studies and have also been incorporated into DCE-MRI and IVIM studies.

Reactive hyperemia methods use a pneumatic cuff to occlude the arterial supply proximal to the imaging volume during data acquisition. These methods show

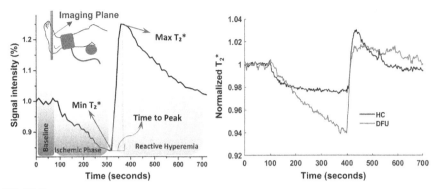

FIG. 20.2

Example of reactive hyperemia experiment with BOLD MRI. *Left*: Time course illustrating BOLD T_2^* signal changes during resting phase or baseline for a 1 min (*blue*; dark gray in print version), a 3-min ischemic period provoked by air cuff inflation (*green*; gray in print version), and a 6.5-min reactive hyperemia phase (*red*; light gray in print version). *Right*: Comparison of reactive hyperemia time course T_2^* signal changes from the medial plantar forefoot in a healthy individual (HC, *black*) and a diabetic patient with a medial plantar foot ulcer (DFU, *red*; light gray in print version). The diabetic patient demonstrates greater reduction of T_2^* during the ischemia phase (from 100 to 400s) and blunted peak T_2^* during hyperemia.

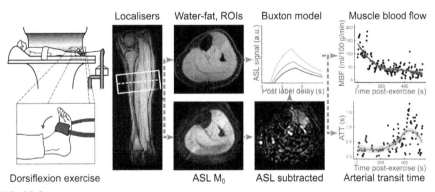

FIG. 20.3

Example functional ischemia experiment with pulsed arterial spin labeling (PASL). From left to right: an example of an in-scanner dorsiflexion exercise where a healthy volunteer lifts a standardized weight, fastened over the dorsum of the foot, in time with a visual/audio guide; sagittal localizer showing positioning of PASL and water–fat-separation imaging stacks over the thickest part of the lower leg; a region of interest (ROI) is drawn in the tibialis anterior on axial water–fat images, and used to estimate M_0 and ASL subtraction signals at multiple postlabel delays for input into the Buxton general kinetic model; and examples of postexercise time courses for muscle blood flow and arterial transit times. Note how blood flow decreases and arterial transit time increases after hyperemia.

several advantages and are often the preferred stimulus for skeletal muscle studies. Commercial cuffs provide rapid and reproducible inflation pressures, making them subject-effort independent and relatively simple to set up. Based on muscle blood flow (MBF), reactive hyperemia demonstrates greater reproducibility compared with functional hyperemia, with intraclass correlation coefficients of 0.98 versus 0.87, respectively (Lopez *et al.*, 2015). These protocols typically include 1 min of baseline data prior to cuff inflation and are succeeded by 2 min of postarterial-occlusion data acquisition (Englund *et al.*, 2016). The duration of ischemia and cuff inflation pressure should be carefully determined, as both have an influence on the hyperemic response. However, a maximum duration of 5 min is suggested for maximal blood flow response, with diminishing increases for longer durations (Hillestad, 1963). Cuff pressures from 30 to 50 mmHg above the individual's brachial systolic blood pressure are commonly used to induce venous and arterial occlusion, though these pressures do not always result in complete arrest of arterial blood flow. If complete arterial occlusion is needed, arterial blood flow measurements distal to the cuff should be used to verify arrested flow per individual for the selected cuff pressure. Cuff-occlusion procedures are largely motion free, allowing for dynamic BOLD (T_2^*) measurements during ischemia such as time to maximum ischemia and percent decrease in T_2^* at maximum ischemia. These measures reflect oxygen availability and utilization. Postischemia response characteristics include peak perfusion amplitude and time-to-peak amplitude, which are markers of vascular reserve and function, respectively (Fig. 20.2).

Functional hyperemia uses an exercise-based stimulus to study the blood flow response during hyperemia (Fig. 20.3). The intensity and duration of the exercise has a strong influence on the perfusion parameters (Frank *et al.*, 1999), with a higher workload leading to greater MBF and a shorter arterial transit time (ATT) (Conlin *et al.*, 2019). Thus standardization of exercise is important to allow for meaningful comparisons between individuals. Exercise paradigms range from repeated short bouts (Frank *et al.*, 1999), to longer durations until exhaustion (Pollak *et al.*, 2012). The selection of exercise paradigm should also consider the role of central and peripheral cardiovascular effects on blood flow regulation. For example, small-muscle-mass exercise (*e.g.*, knee extension, plantar flexion) is used if specificity to peripheral microvascular function is desired (Esposito *et al.*, 2015; Haykowsky *et al.*, 1985). The exercise setup depends strongly on the selected MRI method, research question, and anatomy of interest. For example, in-magnet exercise is used when rapid transient blood flow dynamics are of interest (Towse *et al.*, 2016), while out-of-magnet exercise imposes fewer constraints on exercise devices, but applications are limited to prolonged blood flow response (Federau *et al.*, 2020; Nguyen *et al.*, 2016, 2017). Inexpensive MRI-compatible, custom-built devices can provide reproducible measurements while commercial ergometers can provide workload and force standardization, often recording exercise output and providing real-time feedback during exercise. Biofeedback can be provided to participants inside the magnet using a projector or screen to encourage exercise compliance and help limit motion to periods of "dead time" in the scanning sequence.

20.2.3 **General protocol considerations**

Good-quality localizers should be used to identify landmarks in the anatomy of interest to allow the placement of perfusion imaging stacks. High-resolution T_1-weighted or water–fat separation images can be acquired for muscle-specific or region-of-interest selection during the analysis phase. These anatomical images should be colocalized with the perfusion volume to aid analysis workflow. Participants should lie on the MRI table for five minutes before perfusion imaging, to allow for stabilization of transient venous filling and functional vascular adaptations (Duteil *et al.*, 2006).

Perfusion imaging typically requires image capture of rapid tracer concentration changes or blood flow dynamics; thus, a fast readout is generally preferred. Appropriate sequences for DCE-MRI include a T_1-weighted gradient-echo sequence with parallel imaging. Single-shot echo-planar imaging (EPI) is also commonly used to accelerate ASL, BOLD, and IVIM approaches, but can lead to susceptibility-related distortions and fat-shift artifacts, so should be optimized accordingly. Perfusion acquisition parameters should balance adequate temporal resolution with reasonable spatial resolution and SNR. Typical FOV and matrix sizes yield in-plane voxel sizes of 1.5–3 mm and slice thicknesses of 5–10 mm. Perfusion approaches with limited intrinsic SNR, like ASL and BOLD, sometimes use larger in-plane voxel sizes of 4–6 mm.

The abundance of fat in skeletal muscle and bone marrow requires efficient fat suppression for IVIM, ASL, and BOLD methods. Unsuppressed fat signals can lead to parameter estimation errors with the IVIM model (Adelnia *et al.*, 2019b; Cameron *et al.*, 2017), for example, as lipids exhibit very slow diffusion (*i.e.*, $\sim 1 \times 10^{-6}$ mm^2/s) (Brandejsky *et al.*, 2012). They can also lead to subtraction errors in ASL in the event of small shifts in anatomy. Further, when fat suppression is imperfect, in regions of inhomogeneous B_0 for example, fat-shift artifacts can appear when EPI readouts are used. Fat suppression via spectral fat saturation or water-specific excitation should therefore be used, in combination with advanced B_0 shimming, such as image-based shimming. When significant residual fat signal remains, phase encoding bandwidth should be minimized to reduce the chemical shift displacement of macroscopic fat into muscles of interest. This can be partially achieved using a smaller matrix size, parallel imaging methods like SENSE and GRAPPA (Pipe, 2009), or multishot EPI readouts—at the expense of increased echo times and longer acquisition times. Because muscle has a relatively short T_2 ($\sim$30–45 ms at 3 T), the TE should be minimized, when possible, via asymmetric echo-train sampling, for example. This is also important for ASL, to avoid more BOLD-like T_2* weighting of images.

In addition to B_0 shimming, B_1 shimming may be beneficial in MSK perfusion imaging. B_1 inhomogeneities at 3 T and greater can produce regions of low SNR leading to parameter estimation errors. For ASL, where high-flip-angle labeling pulses are used, B_1 variation across and within the extremities (Brink *et al.*, 2016) can lead to inflow of unlabeled spins and inconsistent labeling (Schepers *et al.*, 2005), hampering comparisons between limbs.

20.2.4 DCE-MRI imaging protocol

An appropriate T_1-weighted sequence for DCE-MRI should balance the trade-offs between capture volume, spatial resolution, and temporal resolution as they relate to the anatomy/pathology (*i.e.*, diffuse vs. focal), physiologic conditions (rest vs. stress), and analysis approach (semiquantitative vs. quantitative). For example, evaluation of soft tissue tumors based on kinetic modeling of resting perfusion can be reliably performed at temporal resolutions as low as 16 s, allowing for multi-slice coverage (Huang *et al.*, 2016). Semiquantitative analysis of signal intensity dynamics during hyperemia in muscle can be accurately assessed with temporal resolution of ~4 s, though quantitative modeling that relies on accurate measurement of the arterial input function (AIF) requires a sampling rate of less than 400 ms (Lutz *et al.*, 2004). Thus single-slice acquisitions have typically been used for quantitative hyperemia studies to allow adequate temporal resolution. Both 2D and 3D spoiled gradient-echo sequences are used, with TRs ranging from 2 to 10 ms, TEs ranging from 0.5 to 2 ms, and flip angles 10–60 degrees, to provide T_1 weighting with minimal T_2^* weighting (Sujlana *et al.*, 2018). Saturation recovery-prepared gradient-echo sequences have also been optimized and used for improved T_1 sensitivity (Conlin *et al.*, 2019; Lutz *et al.*, 2004; Thompson *et al.*, 2005; Zhang *et al.*, 2019, 2016). As with all DCE-MRI procedures, images are acquired continuously, capturing several images prior to contrast injection, and continuing 3–6 min afterward to permit extraction of time–intensity perfusion curves, using a GBCA concentration of 0.1 mmol/kg injected at a rate of 2–5 ml/s, followed by a 20-ml saline flush at the same rate (Bayer *et al.*, 2018; El Rafei *et al.*, 2018; Galbraith *et al.*, 2002). The DCE-MRI sequence should be applied in the latter half of the protocol to avoid T_1-shortening effects on other acquisitions. Like applications in other tissues, for quantitative perfusion modeling, a T_1-mapping sequence should be collected prior to DCE-MRI using the same sequence parameters as the DCE-MRI sequence (El Rafei *et al.*, 2018), allowing DCE-MRI signal intensity curves to be converted into units of contrast agent concentration.

For reactive hyperemia protocols, it is common to inject contrast when the cuff is released, though this approach creates challenges for AIF quantification, which requires a difficult-to-achieve temporal resolution (Lutz *et al.*, 2004). Alternatively, if contrast is injected at the beginning of cuff occlusion, allowing adequate time to mix with circulating blood, a step-input AIF can be assumed upon cuff release, simplifying quantification (Thompson *et al.*, 2005). When combined with a functional hyperemia protocol, contrast is often injected moments before the end of exercise. Recent work has demonstrated high test–retest reproducibility, within 5%, for quantitative perfusion during functional hyperemia (Zhang *et al.*, 2019).

20.2.5 ASL imaging protocol

The optimal skeletal muscle ASL sequence depends on the application of interest. In reactive hyperemia studies where hyperperfusion and normalization occur on a short timescale (Raynaud *et al.*, 2001), a pulsed ASL (PASL) protocol may be preferred, particularly when quantification of peak blood flow amplitude and time to peak are of interest. PASL uses a single RF pulse for near-instantaneous inversion (Kwong et al., 1995; Kim and Tsekos, 1997), and so is capable of measuring perfusion with a temporal resolution of up to 2 s (Raynaud et al., 2001). In contrast, continuous and pseudo-continuous ASL (CASL and pCASL) use a series of pulses to achieve inversion, reducing temporal resolution to a minimum of 6–8 s (Wu et al., 2009; Groezinger et al., 2014). CASL and pCASL do, however, offer a theoretical improvement in SNR, because labeled blood spins accumulate in tissue throughout an extended labeling duration (Wong et al., 1998a). If MBF alone is of interest, CASL and pCASL may be suitable (Englund et al., 2016; Wu et al., 2008), particularly for longer timescale functional hyperemia applications. However, their lower temporal resolution could mean a higher chance of missing or miscalculating peak perfusion in reactive hyperemia studies (Raynaud et al., 2001), although work by Englund et al. has shown similar perfusion time-course metrics between PASL and pCASL. Furthermore, with CASL and pCASL techniques, blood must travel from a proximal labeling plane to the slice of interest. This can lead to significant T_1 decay of the label, and loss of SNR, for larger multi-slice stacks and slice gaps (Jezzard et al., 2018). The longer labeling duration required for CASL and pCASL also leads to increased motion sensitivity, limiting applicability during exercise (Baligand et al., 2021). Lastly, CASL techniques are difficult to apply due to hardware requirements: a factor that has limited their widespread use.

Buxton's general kinetic model is the most widely used for ASL (Buxton et al., 1998), permitting MBF calculation with one or more postlabel delays (PLDs). Single-PLD acquisitions are the most reported and are more widely supported by scanner manufacturers; however, recent work has shown the benefit of specialized multi-PLD acquisitions (Conlin et al., 2019; Veeger et al., 2022), which do not depend on ATT. This is particularly important, as ATT is expected to vary between subjects and between slices (Veeger et al., 2022), as well as in pathologic conditions, which may lead to errors in MBF estimates. To accurately measure both MBF and ATT, PLDs should be sufficient in number, range, and spacing to adequately sample the Buxton curve; however, these parameters will typically vary depending on the anatomy and arterial supply. To acquire several PLDs within a single TR interval, a Look–Locker approach can be used to repeatedly sample the magnetization after labeling (Look and Locker, 1970). In recent work, Veeger et al. used 10 PLDs, ranging from 600 to 2400 ms in steps of 200 ms, though a shorter first PLD is recommended to capture more inflow of labeled blood (Veeger et al., 2022).

For PASL acquisitions, a blood bolus of a well-defined duration is required for quantitative perfusion parameter estimates. The temporal width of this bolus can be set using a Quantitative Imaging of Perfusion Using a Single Subtraction (QUIPSS) module; the labeling pulse defines the beginning of the bolus, while the QUIPSS module determines its end. This decreases the influence of transit delays on perfusion parameters (Wong et al., 1998b). The bolus width can then be input into the Buxton fit for pulsed ASL.

Background suppression is an important inclusion, reducing physiologic noise and fat signals in muscle (Baligand et al., 2021). This can include a water excitation technique (WET) module (Ogg *et al.*, 1994), both pre- and postsaturation, to mitigate B_1 inhomogeneities and variations in T_1 across the anatomy of interest.

20.2.6 IVIM imaging protocol

PGSE with fat suppression and single-shot EPI readout is the predominant sequence used for IVIM studies in the extremities. Though capillaries are primarily oriented parallel with muscle fibers, most applications use diffusion encoding in three orthogonal directions with trace-weighted images providing a directional average of IVIM perfusion parameters. Initial work supports orientation dependence of the IVIM signal in muscle (Karampinos *et al.*, 2010), with additional work needed to establish these models as they relate to morphologic and functional changes. In cases where dynamic measures are desired, obtaining one diffusion-sensitizing direction oriented parallel to the muscle's long axis can reduce the acquisition time and increase the temporal resolution (Adelnia *et al.*, 2019b; Riexinger *et al.*, 2021; Yao and Sinha, 2000).

There is currently no consensus on the optimal *b*-value sampling for IVIM measurements in muscle. For any application, *b*-value sampling must include small values (between 0 and $200 \, \text{s/mm}^2$) for perfusion and larger values (between 200 and $1000 \, \text{s/mm}^2$) to define D. Most protocols use 12 *b*-values evenly split between perfusion and diffusion ranges. Selection of *b*-values should consider the application (Englund *et al.*, 2022). For example, when measuring perfusion changes during a transient state, such as after exercise, fewer *b*-values might be used to permit increased temporal resolution, though capillary anisotropy effects could bias the absolute value measured (Filli *et al.*, 2015). Resting perfusion benefits from a larger number of *b*-values to improve the ability to detect the relatively small blood pool (Adelnia *et al.*, 2019b).

Because IVIM imaging requires relatively lengthy imaging times for acquisition of multiple diffusion-weighted images, it has not been used as widely for characterizing rapid temporal changes in perfusion, *i.e.*, blood flow response from short-lived muscle contractions (Towse *et al.*, 2016). For more strenuous exertions, IVIM has shown sensitivity to differences in postexercise hyperemia in aging and musculoskeletal disease (Federau *et al.*, 2020; Adelnia *et al.*, 2019b), supporting its use with vasoactive stress.

20.2.7 BOLD imaging protocol

A T_2^*-weighted spoiled GRE with a single-shot EPI readout is the predominant sequence used for BOLD imaging in muscle. This is typically combined with fat suppression to mitigate the deleterious effect of chemically shifted signals from subcutaneous and bone marrow fat. Most implementations use an axial 2D multi-slice acquisition in the muscle/extremity of interest, balancing number of slices per TR with the desired temporal resolution. A large slice thickness of $\sim 1 \, \text{cm}$ is often used to increase SNR since BOLD signal amplitude changes can be relatively small. Multi-echo sequences permit dynamic quantitative T_2^* mapping, affording improved sensitivity and specificity to the BOLD effect, with a minor impact on temporal resolution (Damon *et al.*, 2007).

20.3 **Data analysis protocol**

20.3.1 **Registration**

Image registration is the first preprocessing step recommended for longer dynamic scans, to correct for undesired subject motion, which may affect signal processing and fitting. Given that most perfusion protocols require a motion-associated stimulus, with motion arising from either exercise or cuff inflation and deflation (Englund *et al.*, 2016), it is beneficial to internally register all acquired images to a "rest" scan from the beginning of the protocol, to account for residual motion or small shifts in anatomy.

20.3.2 **Segmentation and masking**

Given the relatively low SNR of most perfusion techniques in skeletal muscle, ROI-based analyses are often used over voxel-wise analyses. Ideally, ROIs should be drawn on high-resolution anatomical images; however, this requires the anatomical images to be registered to the perfusion MRI data, with the transforms being applied to the ROIs. When drawing ROIs, the observer should draw within the fascia and exclude intramuscular tendons, to avoiding adding noise from low-signal regions.

For ASL, a blood vessel mask should be applied after segmentation to limit the influence of large vessels on perfusion estimates. It is assumed that voxels with very high ASL signals, more than two standard deviations above the slice mean (Gao *et al.*, 2014), for example, represent such vessels.

20.3.3 **DCE-MRI**

DCE-MRI muscle data are evaluated using two approaches: (a) semiquantitative and (b) quantitative.

Semiquantitative analysis uses the raw T_1-weighted signal intensity. This approach is attractive as it obviates rigorous requirements on data acquisition and provides robust parameters that include the arrival time of the tracer (AT), signal enhancement or initial slope (IS), time to peak (TTP), and area under the contrast-agent curve (AUC). However, these parameters represent a mixture of microcirculatory and tissue properties with a limited direct connection to quantitative physiologic correlates.

Quantitative analysis relies on signal enhancement of T_1-weighted images and the effective concentration of the tracer in tissue and blood plasma. A two-compartment kinetic model is often used to relate the blood plasma and tissue tracer concentrations to quantitative parameters (k^{trans}, k_{ep}, v_e) (Tofts *et al.*, 1999; Patlak *et al.*, 1983; Tofts, 1997). The size of the GBCA strongly influences the concentration time course in the extravascular extracellular space (EES) and vascular space (Jaspers *et al.*, 2009, 2010); therefore, the model assumptions should be modified for different GBCAs to allow accurate perfusion parameter estimation (Jaspers *et al.*, 2009, 2010). Bias in perfusion parameters may also arise due to model assumptions that neglected the influence of water exchange, the proton exchange rate across exchange sites being much faster than the difference in relaxation rates between the compartments. Buckley *et al.* proposed a rapid spoiled GRE sequence with a short TR (3.4 ms), 30 degree flip angle, and a standard dose of GBCA (0.1 mmol/kg) to minimize the influence of cellular–interstitial water exchange on perfusion parameter estimation (Buckley *et al.*, 2008).

Robust measurement of DCE-MRI perfusion parameters in skeletal muscle is challenging and depends on several factors, including, but not limited to, a good scanning protocol, reproducible injection technique, and appropriate model assumptions and AIF estimation. These factors may further vary between different scanners or centers, leading to poor reproducibility in DCE-MRI parameters. For example, interscan coefficient of variation is reported as high as 51% for influx rate in patients (Versluis *et al.*, 2011). In another study, semiquantitative parameters showed relatively lower within-patient coefficient of variation (*i.e.*, 12% for enhancement and 17% for AUC) compared with quantitative parameters (*i.e.*, 49% for K_{ep}, 16% for v_e) (Galbraith *et al.*, 2002). Although semiquantitative parameters seem to be more robust in muscle studies, the comparison of patients using these parameters is more difficult as these parameters have limited connections to physiologic and pathophysiologic attributes. Thus protocols should balance the specificity of quantitative DCE-MRI analysis with the reproducibility of semiquantitative analysis in muscle studies.

20.3.4 **ASL**

Due to the temporal spacing of tag and control images, BOLD effects can limit the sensitivity of the ASL analysis, since the oxygenation of inflowing blood may differ from tag to control. This inconsistency can be corrected for using surround subtraction (Lu *et al.*, 2006). Briefly, for an interleaved ASL acquisition of the form "Tag[0], Control[0], Tag[1], Control[1], …" surround subtraction is performed as per the following scheme: "Control [0] – (Tag[0] + Tag[1])/2, (Control[0] + Control[1])/2 – Tag[1], …" This also has the added benefit of increasing the SNR.

When multi-PLD data are acquired, flip-angle-sweep correction is needed to account for saturation effects in the Look–Locker sampling of the magnetization. Such sampling typically acquires multiple images at intervals much shorter than T_1, which can lead to substantial saturation of the longitudinal magnetization. This can be ameliorated during acquisition using a variable-flip-angle approach, but additional T_1 correction will restore the ASL data to a form conducive to Buxton fitting.

After preprocessing, the resulting ASL signal can be fitted to an appropriate model. When only one PLD is available, the single-compartment models for PASL and pCASL are used (Raynaud *et al.*, 2001; Alsop *et al.*, 2015). However, when multiple PLDs are available, the ASL data should be fitted to the Buxton model (Buxton *et al.*, 1998), to isolate the independent effects of MBF and ATT on the ASL signal. This can be performed using the *basil* toolbox (Chappell *et al.*, 2009), for example, from FSL (FMRIB Software Library, Oxford, UK; fsl.fmrib.ox.ac.uk).

20.3.5 **IVIM model fitting**

The standard signal model for IVIM analysis is a biexponential (Eq. 20.2) which is typically fit using a nonlinear least squares approach either in a single step or in two steps: fitting first the diffusion parameters to b-values $>200\,s/mm^2$ and then fitting the perfusion parameters using all b-values (Englund et al., 2022; Wurnig et al., 2015; Callot et al., 2003; Pekar et al., 1992; Sigmund et al., 2012). Some vendor implementations of PGSE do not include the influence of imaging gradients in b-value calculations which, if uncorrected, can impart bias and inconsistencies in estimated IVIM parameters between scanners. Consistent with the known difficulties in fitting biexponential models, the more rapid rate-constant parameter, D^*, is the most difficult parameter to fit reliably (Adelnia et al., 2019a; Cameron et al., 2017). The stretched exponential signal model has been proposed as more numerically stable, providing comparable results to the biexponential model (Yao et al., 2021), as well as estimation of microvascular volume fractions that are more aligned with histologic measurements of capillary volume fraction (Reiter et al., 2021).

Fitting of the biexponential model to noisy data is subject to parameter estimation errors, which are beyond the scope of this chapter but have been reviewed elsewhere (Adelnia et al., 2019a; Cameron et al., 2017; Wurnig et al., 2015; Reiter et al., 2021). Additionally, beyond nonnegative least squares, several fitting methods such as Bayesian and neural network approaches aimed at improving estimation accuracy and precision have also been described and are the topic of ongoing research (Gurney-Champion et al., 2018; Jerome et al., 2014; Koopman et al., 2021). Bayesian methods show potential for improving repeatability and reliability of IVIM parameters.

20.3.6 **BOLD**

The BOLD response to vasoactive stimulus is typically calculated using the average T_2^*-weighted signal within an ROI, producing a time-trace signal before, during, and after the active paradigm (Li et al., 2017). In the case of multi-echo EPI, T_2^* values are calculated from exponential fitting of the averaged signal decay curve over an ROI. Note that the BOLD response is often normalized to the baseline for simple comparison over all subjects. Fig. 20.2 shows the BOLD signal acquired with reactive hyperemia. The ischemic phase is typically described by the minimum ischemic value (T_2^* min), which reflects the minimal BOLD signal during the ischemic phase relative to the baseline. Reactive hyperemia is described with the peak value (T_2^*max) referring to the height of the BOLD signal maximum during hyperemia relative to baseline, and the TTP referring to the time interval between cuff deflation and T_2^* max. T_2^* max and TTP values are shown to be the most reproducible BOLD parameters having both physiologic and pathophysiologic correlates. Despite the widespread use of BOLD MRI with ischemia–reperfusion, and consistency between different studies with coefficients of variation close to 10%, none of those studies describe in detail how the BOLD parameters such as T_2^* max and TTP were obtained from the signal time courses. To further improve BOLD measurements and minimize operator bias, Schewzow and colleagues recently proposed a method to numerically determine the peak value, TTP, and slope of the tangent by fitting the BOLD signal to a sigmoid function (Schewzow et al., 2013). Studies have shown that this method improves the quantitative analysis of BOLD data and represents an important step toward standardization of skeletal muscle fMRI quantification and its clinical applications (Schewzow et al., 2013; Schmid et al., 2014); however, further validation is essential.

20.4 Case studies of applications

Muscle perfusion MRI techniques have been applied in healthy human studies and the study of a variety of conditions for diagnosis and assessing therapies. There are also examples of preclinical applications (Baligand et al., 2011; Bertoldi et al., 2008), including in a mouse model of muscular dystrophy (Latroche et al., 2015). Clinical studies have largely focused on aging, peripheral arterial disease (PAD), and diabetes, as detailed later.

20.4.1 Aging

Age-associated changes in perfusion have been investigated using multiple MRI approaches. In a multiparametric MRI study, resting biexponential IVIM parameters did not indicate any age effect on perfusion, though DCE-MRI revealed increases in the extravascular–extracellular volume fraction, v_e (Yoon et al., 2018). In contrast, using stretched exponential IVIM parameters, Adelnia et al. observed an age-associated decline in resting perfusion attributed to reduced muscle oxidative capacity (Adelnia et al., 2019a). Postexercise IVIM perfusion fraction has also shown a greater increase in young versus older healthy adults (Adelnia et al., 2019b). ASL has shown lower peak perfusion and postexercise hyperemia in older versus younger adults (Wray et al., 2009a), and increased perfusion was observed in older but not younger adults after an antioxidant intervention (Wray et al., 2009b). BOLD perfusion indices have been seen to correlate with age (Tonson et al., 2017), and older adults also exhibited reduced BOLD peak amplitude and slower TTP during reactive hyperemia. A lower T_2^* maximum is interpreted to reflect an increase in vessel wall rigidity, predilation of the vasculature, and rarefaction with aging (Schulte et al., 2008). BOLD MRI with a functional hyperemia stimulus has demonstrated sensitivity to an aerobic exercise intervention in older adults, showing increased peak post-contraction BOLD response by 33% in the exercise group compared with a control group (Hurley et al., 2019).

20.4.2 Peripheral arterial disease

Functional assessment of PAD using MRI could be a valuable diagnostic tool to determine disease severity and evaluate therapeutic efficacy (Thompson et al., 2005; Versluis et al., 2011, 2012). Despite previously unsuccessful studies

(Versluis *et al.*, 2011), Versluis *et al.* used DCE-MRI with a cuff paradigm to differentiate PAD patients from healthy controls (Versluis *et al.*, 2012). They showed a reduced plasma volume fraction, ν_p, in patients (3.2%–4.8% lower) versus healthy controls, for all muscles in the calf, possibly due to impaired microvascular dilatative capacity as a result of endothelial dysfunction. The rate constant k and AUC were also lower ($P < 0.01$) in patients compared to healthy controls, indicating lower vascular reactivity in these patients in line with the results of another study (Versluis *et al.*, 2012).

ASL has been able to differentiate PAD patients from controls using both cuff occlusion and exercise hyperemia procedures (Lopez *et al.*, 2015). ASL parameters are associated with disease severity in PAD (Wu *et al.*, 2009; Englund *et al.*, 2015): where the TTP increases with increasing severity, the hyperemic response decreases, and the duration of the hyperemic phase becomes longer. These parameters improved after percutaneous transluminal angioplasty, though not in all muscles (Groezinger *et al.*, 2014).

BOLD measures of reactive hyperemia have shown a significantly lower relative increase in peak T_2^* and delayed TTP in PAD patients (Fig. 20.2), with TTP correlating with ankle-brachial index (Englund *et al.*, 2015; Ledermann *et al.*, 2006). This reduced peak T_2^* is thought to be influenced by impaired blood flow and longer contact time between capillary blood and surrounding tissue, which leads to more rapid deoxygenation of hemoglobin. Increased vessel wall rigidity or preexisting vasomotor relaxation in PAD can lead to lower TTP and peak amplitudes. BOLD parameters also show improvements with successful revascularization, supporting their use for evaluating how treatment strategies influence oxygen delivery at the tissue level (Bajwa *et al.*, 2016; Huegli *et al.*, 2009).

20.4.3 Diabetes

DCE-MRI has been applied in patients with diabetes to quantify the capillary permeability of the lower extremity muscles. Wang *et al.* reported a reduction in the transfer constant, K^{trans}, of the tibialis anterior muscle in patients with Type 2 diabetes ($K^{trans} = 0.03\,\text{min}^{-1}$) compared with healthy controls ($K^{trans} = 0.1\,\text{min}^{-1}$) (Wang *et al.*, 2011). After percutaneous angioplasty, K^{trans} in patients increased to $0.06\,\text{min}^{-1}$, but was still less than that in healthy controls, highlighting the ability of DCE-MRI to elucidate disease-related vessel wall alterations despite restoration of blood flow.

A small number of ASL studies in diabetes have reported reduced perfusion in the leg (Zheng *et al.*, 2014) and foot (Zheng *et al.*, 2015; Edalati *et al.*, 2019), showing regions of ischemia in the foot adjacent to diabetic ulcers (Edalati *et al.*, 2019).

20.4.4 Myopathy

Myopathies are rare heterogeneous systemic disorders that can be difficult to differentiate due to overlapping symptoms of muscle weakness, pain, and dysfunction. Preliminary work using IVIM has shown reduced D^* in the quadriceps muscles of dermatomyositis patients compared with healthy individuals (Sigmund *et al.*, 2012). Significantly lower perfusion fraction and higher tissue diffusivity have been reported in thigh muscle from autoimmune myositis compared with muscular dystrophy, though these parameters showed limited sensitivity and specificity, ranging from 57% to 99% and 0.8% to 75%, respectively (Ran *et al.*, 2021).

20.4.5 Muscle injury

DCE-MRI has been used to estimate tissue inflammation and microvascular permeability during rehabilitation after acute muscle strain injury (Bayer *et al.*, 2018). This work showed significantly greater initial rate of enhancement and maximal enhancement in the injured muscle compared to uninjured reference muscle, with this difference persisting six months after injury, and suggests a link between the magnitude of the trauma, the inflammatory response, and muscle atrophy.

20.5 Safety considerations

Administration of GBCAs is necessary in DCE-MRI and carries the risk of toxicity and nephrogenic systemic fibrosis (NSF). Apart from NSF, which is most common in individuals with impaired kidney function, there is evidence that gadolinium can accumulate in brain, bone, and kidney tissue in individuals with normal kidney function, leading to toxicity. These risks should be considered when studying individuals in whom contrast is not clinically indicated. The renal function of participants should be tested before DCE-MRI; an estimated glomerular filtration rate (eGFR) less than $30\,\mathrm{ml\,min^{-1}\,1.73\,m^{-2}}$ is commonly considered a contraindication for administration of GBCAs.

The use of vasoactive stimuli carries varying levels of risk depending on the individual. Arterial occlusion paradigms have been proposed in the literature with increasing amplitude of cuff pressure and duration of occlusion potentially leading to greater discomfort in study subjects. Care must be taken with arterial occlusion paradigms, as occlusion may lead to worsening of pain in patients with critical ischemia or ulceration. Exercise workloads should likewise be curtailed in patients with exercise limitations and joint pain.

20.6 Challenges and solutions

While applications of perfusion MRI methods demonstrate utility for characterizing a variety of perfusion properties useful in distinguishing degradation of the microvasculature, some challenges remain to improve these methods in terms of applicability/feasibility, descriptiveness, sensitivity, and suitability for widespread use.

For screening, diagnostic applications, and therapeutic monitoring, perfusion methods should ideally provide noninvasive, rapid, robust, and sensitive measures.

Providing full coverage of the body region is particularly important, as many pathologies present with focal or heterogeneous perfusion impairments. Further, recent work demonstrates differences in perfusion parameters and oxidative capacity along the muscle length supporting the need for multi-slice coverage (Veeger *et al.*, 2022; Boss *et al.*, 2018; Heskamp *et al.*, 2021). DCE-MRI shows potential for providing quantitative perfusion measures with extensive volume coverage, although further work is needed to account for the AIF (Galanakis *et al.*, 2020). The AIF in DCE-MRI muscle studies is commonly obtained via an elliptical ROI drawn within major vessels such as the femoral artery (El Rafei *et al.*, 2018) or the tibial-fibular trunk (Versluis *et al.*, 2011, 2012). This method relies on the presence of large vessels in the imaging volume, and errors due to inflow and partial volume must be considered carefully. The distal aspect of the arteries within the volume needs to be selected to avoid inflow artifacts, and voxels exhibiting partial volume effects should be excluded. The R_1 time curves from the remaining voxels are then averaged together, and the mean time course is used as an AIF (Buckley *et al.*, 2008; Wang *et al.*, 2011). Some studies have also tested dynamic susceptibility contrast MRI (DSC-MRI) in skeletal muscle (Zhang *et al.*, 2016; Goyault *et al.*, 2012), potentially obtaining large-volume measures of perfusion without the influence of capillary permeability. Both DCE-MRI and DSC-MRI rely on GBCAs, making them less suitable for routine use and screening. However, recent work demonstrates the feasibility of low-dose GBCA (0.05 mmol/kg) acquisitions, potentially tolerated by individuals with impaired kidney function (Zhang *et al.*, 2008) and permitting more than one DCE-MRI injection in a single scan session (Zhang *et al.*, 2019). Newly emerging ASL techniques, like a split-label PASL design (Baligand *et al.*, 2021), are under development to overcome perfusion confounding technical factors in multi-slice implementations related to slice positioning. IVIM also affords multi-slice coverage demonstrating potential for discriminating differences in perfusion at rest (Adelnia *et al.*, 2019a), though further work is needed to standardize IVIM protocols and validate IVIM perfusion measures to physiologic measures of blood flow.

Descriptive measurements of microvascular function that necessitate the use of vasoactive stress exams are currently limited by image acquisition speed and spatial coverage. Muscle blood flow dynamics can be rapid, particularly when reactive hyperemia stimulus paradigms are used, and so high-temporal-resolution sampling strategies are required, particularly for quantitative modeling. Single-slice DCE-MRI and ASL acquisitions currently provide sufficient temporal resolution for quantitative blood flow measurements. BOLD imaging can provide adequate temporal resolution and volumetric coverage to capture semiquantitative oxygen-utilization and hyperemia indices, though it lacks the ability to quantify blood flow. As many perfusion measures rely on image subtraction or serial pixel-wise comparisons, they are vulnerable to motion, making them difficult to use during exercise. New methods allowing measurements during exercise would permit more detailed assessment of stress response. Currently, this is addressed using devices that can synchronize image capture with momentary pauses in exercise (Richardson *et al.*, 1985), though techniques using retrospective reconstruction with motion correction might permit measurements during exercise. Recent work shows improved functional assessment of perfusion with simultaneous dynamic measurements of perfusion and tissue-level oxygen utilization in single-slice acquisitions (Englund *et al.*, 2015; Duteil *et al.*, 2004). More work is needed to extend these approaches to multi-slice imaging.

Microvascular architecture has been an elusive measure because of definitive limits on MRI resolution. IVIM models demonstrate mixed results on changes in perfusion volume fraction with pathology, likely due to variability in acquisition and quantification methods (Englund *et al.*, 2015; Adelnia *et al.*, 2019b; Cameron *et al.*, 2017; Reiter *et al.*, 2021). Initial work has demonstrated the potential for extending IVIM models to characterize additional microstructural parameters like capillary anisotropy (Karampinos *et al.*, 2010). More work is needed to evaluate the utility of these approaches in pathologic conditions.

Perfusion MRI holds untapped potential for studying dysregulation of skeletal muscle mass. DCE-MRI provides measures of vessel permeability, permeable surface area, and extravascular volume fraction, which could be critical in understanding conditions like sarcopenia and muscle wasting, which are thought to involve diminished uptake of amino acids necessary for protein synthesis (Poole *et al.*, 2013; Wang *et al.*, 2022). For existing methods to be more widely adopted, there is a need for standardization of acquisition and analysis protocols and multisite reproducibility studies. Model fitting can be influenced by temporal resolution, acquisition length, SNR, and AIF selection, among other factors. In general, MR protocols with higher temporal resolution, more signal averaging, and postprocessing software allowing automatic AIF selection could help reduce perfusion parameter variability in normal muscle (El Rafei *et al.*, 2018), increasing their sensitivity to pathologic changes. Initial work provides guidelines for analysis procedures, demonstrating reliability measures for ROI and pixel-wise analysis as it relates to SNR (El Rafei *et al.*, 2018; Galbraith *et al.*, 2002). Model assumptions are also still being scrutinized for various methods such as the assumption of fast exchange on K^{trans} the with Tofts model (Buckley *et al.*, 2008).

Despite the widespread use of vasoactive stress protocols, clear guidance is lacking on robust ways to compute time-course parameters such as HPV and TTP. Most studies rely on manual determination of these parameters, which is time consuming and operator dependent and could contribute to the variations observed between studies. Standardized approaches for modeling these dynamics could lead to improved reliability and repeatability of functional perfusion measures (Schewzow *et al.*, 2013, 2015; Schmid *et al.*, 2014).

20.7 Learning and knowledge outcomes

DCE-MRI, ASL, IVIM, and BOLD MRI approaches have been established for measuring perfusion characteristics in skeletal muscle.

Perfusion MRI is emerging as a tool for improved assessment of muscle pathophysiology in a variety of diseases, with the potential for providing biomarkers for pharmaceutical interventions.

Vasoactive stimuli including arterial occlusion and exercise paradigms provide functional measures of the microvascular system, are currently essential for ASL and BOLD acquisitions, and can be used in DCE-MRI and IVIM studies.

Pulsed ASL methods are the most used ASL techniques for measuring non-contrast perfusion in muscle, because of their finer temporal resolution and near-instantaneous labeling, which make quantitative and dynamic measurements of blood flow response feasible.

DCE-MRI, ASL, and IVIM are used as proxies for studying blood flow changes in response to increased metabolism, but they reflect blood water signal alone, while BOLD is the only approach providing information specific to oxygen delivery and utilization.

Preliminary studies in patients have shown sensitivity of semiquantitative and quantitative perfusion parameters to impaired skeletal muscle perfusion. Future work is needed to establish robust protocols for widespread use.

Acknowledgments

The authors acknowledge support from the following institutes of the National Institutes of Health (NIH): National Institute on Aging (1K25AG076864-01), and the National Institute of Diabetes and Digestive and Kidney Diseases Diabetic Complications Consortium (RRID:SCR_001415, grants U24DK076169, U24DK115255).

References

Adelnia, F., Cameron, D., Bergeron, C.M., *et al.*, 2019a. The role of muscle perfusion in the age-associated decline of mitochondrial function in healthy individuals. Front. Physiol. 10, 427.

Adelnia, F., Shardell, M., Bergeron, C.M., *et al.*, 2019b. Diffusion-weighted MRI with intravoxel incoherent motion modeling for assessment of muscle perfusion in the thigh during post-exercise hyperemia in younger and older adults. NMR Biomed. 32 (5), e4072.

Alsop, D.C., Detre, J.A., Golay, X., *et al.*, 2015. Recommended implementation of arterial spin-labeled perfusion MRI for clinical applications: a consensus of the ISMRM perfusion study group and the European consortium for ASL in dementia. Magn. Reson. Med. 73 (1), 102–116.

Bajwa, A., Wesolowski, R., Patel, A., *et al.*, 2016. Blood oxygenation level-dependent CMR-derived measures in critical limb ischemia and changes with revascularization. J. Am. Coll. Cardiol. 67 (4), 420–431.

Baligand, C., Wary, C., Menard, J., Giacomini, E., Hogrel, J.Y., Carlier, P., 2011. Measuring perfusion and bioenergetics simultaneously in mouse skeletal muscle: a multiparametric functional-NMR approach. NMR Biomed. 24 (3), 281–290.

Baligand, C., Hirschler, L., Veeger, T.T., *et al.*, 2021. A split-label design for simultaneous measurements of perfusion in distant slices by pulsed arterial spin labeling. Magn. Reson. Med. 86 (5), 2441–2453.

Bayer, M.L., Hoegberget-Kalisz, M., Jensen, M.H., *et al.*, 2018. Role of tissue perfusion, muscle strength recovery, and pain in rehabilitation after acute muscle strain injury: a randomized controlled trial comparing early and delayed rehabilitation. Scand. J. Med. Sci. Sports 28 (12), 2579–2591.

Bertoldi, D., de Sousa, P.L., Fromes, Y., Wary, C., Carlier, P.G., 2008. Quantitative, dynamic and noninvasive determination of skeletal muscle perfusion in mouse leg by NMR arterial spin-labeled imaging. Magn. Reson. Imaging 26 (9), 1259–1265.

Boss, A., Heskamp, L., Breukels, V., Bains, L.J., van Uden, M.J., Heerschap, A., 2018. Oxidative capacity varies along the length of healthy human tibialis anterior. J. Physiol. 596 (8), 1467–1483.

Bradley, J.R., Anderson, J.R., Evans, D.B., Cowley, A.J., 1990. Impaired nutritive skeletal muscle blood flow in patients with chronic renal failure. Clin. Sci. (Lond.) 79 (3), 239–245.

Brandejsky, V., Kreis, R., Boesch, C., 2012. Restricted or severely hindered diffusion of intramyocellular lipids in human skeletal muscle shown by in vivo proton MR spectroscopy. Magn. Reson. Med. 67 (2), 310–316.

Brink, W.M., Versluis, M.J., Peeters, J.M., Bornert, P., Webb, A.G., 2016. Passive radiofrequency shimming in the thighs at 3 Tesla using high permittivity materials and body coil receive uniformity correction. Magn. Reson. Med. 76 (6), 1951–1956.

Buckley, D.L., Kershaw, L.E., Stanisz, G.J., 2008. Cellular-interstitial water exchange and its effect on the determination of contrast agent concentration in vivo: dynamic contrast-enhanced MRI of human internal obturator muscle. Magn. Reson. Med. 60 (5), 1011–1019.

Buxton, R.B., Frank, L.R., Wong, E.C., Siewert, B., Warach, S., Edelman, R.R., 1998. A general kinetic model for quantitative perfusion imaging with arterial spin labeling. Magn. Reson. Med. 40 (3), 383–396.

Callot, V., Bennett, E., Decking, U.K., Balaban, R.S., Wen, H., 2003. In vivo study of microcirculation in canine myocardium using the IVIM method. Magn. Reson. Med. 50 (3), 531–540.

Cameron, D., Bouhrara, M., Reiter, D.A., *et al.*, 2017. The effect of noise and lipid signals on determination of Gaussian and non-Gaussian diffusion parameters in skeletal muscle. NMR Biomed. 30 (7).

Caroca, S., Villagran, D., Chabert, S., 2021. Four functional magnetic resonance imaging techniques for skeletal muscle exploration, a systematic review. Eur. J. Radiol. 144, 109995.

Chappell, M.A., Groves, A.R., Whitcher, B., Woolrich, M.W., 2009. Variational Bayesian inference for a nonlinear forward model. IEEE Trans. Signal Process. 57 (1), 223–236.

Conlin, C.C., Layec, G., Hanrahan, C.J., *et al.*, 2019. Exercise-stimulated arterial transit time in calf muscles measured by dynamic contrast-enhanced magnetic resonance imaging. Phys. Rep. 7 (1), e13978.

Damon, B.M., Hornberger, J.L., Wadington, M.C., Lansdown, D.A., Kent-Braun, J.A., 2007. Dual gradient-echo MRI of post-contraction changes in skeletal muscle blood volume and oxygenation. Magn. Reson. Med. 57 (4), 670–679.

Duteil, S., Bourrilhon, C., Raynaud, J.S., *et al.*, 2004. Metabolic and vascular support for the role of myoglobin in humans: a multiparametric NMR study. Am. J. Phys. Regul. Integr. Comp. Phys. 287 (6), R1441–R1449.

Duteil, S., Wary, C., Raynaud, J.S., *et al.*, 2006. Influence of vascular filling and perfusion on BOLD contrast during reactive hyperemia in human skeletal muscle. Magn. Reson. Med. 55 (2), 450–454.

Dvoretskiy, S., Lieblein-Boff, J.C., Jonnalagadda, S., Atherton, P.J., Phillips, B.E., Pereira, S.L., 2020. Exploring the association between vascular dysfunction and skeletal muscle mass, strength and function in healthy adults: a systematic review. Nutrients 12 (3).

Edalati, M., Hastings, M.K., Muccigrosso, D., *et al.*, 2019. Intravenous contrast-free standardized exercise perfusion imaging in diabetic feet with ulcers. J. Magn. Reson. Imaging 50 (2), 474–480.

El Rafei, M., Teixeira, P., Norberciak, L., Badr, S., Cotten, A., Budzik, J.F., 2018. Dynamic contrast-enhanced MRI perfusion of normal muscle in adult hips: variation of permeability and semi-quantitative parameters. Eur. J. Radiol. 108, 92–98.

Englund, E.K., Langham, M.C., Ratcliffe, S.J., *et al.*, 2015. Multiparametric assessment of vascular function in peripheral artery disease: dynamic measurement of skeletal muscle perfusion, blood-oxygen-level dependent signal, and venous oxygen saturation. Circ Cardiovasc Imaging 8 (4), e002673.

Englund, E.K., Rodgers, Z.B., Langham, M.C., Mohler III, E.R., Floyd, T.F., Wehrli, F.W., 2016. Measurement of skeletal muscle perfusion dynamics with pseudo-continuous arterial spin labeling (pCASL): assessment of relative labeling efficiency at rest and during hyperemia, and comparison to pulsed arterial spin labeling (PASL). J. Magn. Reson. Imaging 44 (4), 929–939.

Englund, E.K., Reiter, D.A., Shahidi, B., Sigmund, E.E., 2022. Intravoxel incoherent motion magnetic resonance imaging in skeletal muscle: review and future directions. J. Magn. Reson. Imaging 55 (4), 988–1012.

Esposito, F., Wagner, P.D., Richardson, R.S., 2015. Incremental large and small muscle mass exercise in patients with heart failure: evidence of preserved peripheral haemodynamics and metabolism. Acta Physiol (Oxford) 213 (3), 688–699.

Federau, C., Kroismayr, D., Dyer, L., Farshad, M., Pfirrmann, C., 2020. Demonstration of asymmetric muscle perfusion of the back after exercise in patients with adolescent idiopathic scoliosis using intravoxel incoherent motion (IVIM) MRI. NMR Biomed. 33 (3), e4194.

Filli, L., Boss, A., Wurnig, M.C., Kenkel, D., Andreisek, G., Guggenberger, R., 2015. Dynamic intravoxel incoherent motion imaging of skeletal muscle at rest and after exercise. NMR Biomed. 28 (2), 240–246.

Frank, L.R., Wong, E.C., Haseler, L.J., Buxton, R.B., 1999. Dynamic imaging of perfusion in human skeletal muscle during exercise with arterial spin labeling. Magn. Reson. Med. 42 (2), 258–267.

Galanakis, N., Maris, T.G., Kontopodis, N., *et al.*, 2020. The role of dynamic contrast-enhanced MRI in evaluation of percutaneous transluminal angioplasty outcome in patients with critical limb ischemia. Eur. J. Radiol. 129, 109081.

Galbraith, S.M., Lodge, M.A., Taylor, N.J., *et al.*, 2002. Reproducibility of dynamic contrast-enhanced MRI in human muscle and tumours: comparison of quantitative and semi-quantitative analysis. NMR Biomed. 15 (2), 132–142.

Gao, Y., Goodnough, C.L., Erokwu, B.O., *et al.*, 2014. Arterial spin labeling-fast imaging with steady-state free precession (ASL-FISP): a rapid and quantitative perfusion technique for high-field MRI. NMR Biomed. 27 (8), 996–1004.

Goyault, G., Bierry, G., Holl, N., *et al.*, 2012. Diffusion-weighted MRI, dynamic susceptibility contrast MRI and ultrasound perfusion quantification of denervated muscle in rabbits. Skelet. Radiol. 41 (1), 33–40.

Groezinger, G., Pohmann, R., Schick, F., *et al.*, 2014. Perfusion measurements of the calf in patients with peripheral arterial occlusive disease before and after percutaneous transluminal angioplasty using MR arterial spin labeling. J. Magn. Reson. Imaging 40 (4), 980–987.

Gurney-Champion, O.J., Klaassen, R., Froeling, M., *et al.*, 2018. Comparison of six fit algorithms for the intra-voxel incoherent motion model of diffusion-weighted magnetic resonance imaging data of pancreatic cancer patients. PLoS One 13 (4), e0194590.

Haykowsky, M.J., Tomczak, C.R., Scott, J.M., Paterson, D.I., Kitzman, D.W., 2015. Determinants of exercise intolerance in patients with heart failure and reduced or preserved ejection fraction. J. Appl. Physiol. (1985) 119 (6), 739–744.

Heinonen, I., Koga, S., Kalliokoski, K.K., Musch, T.I., Poole, D.C., 2015. Heterogeneity of muscle blood flow and metabolism: influence of exercise, aging, and disease states. Exerc. Sport Sci. Rev. 43 (3), 117–124.

Heskamp, L., Lebbink, F., van Uden, M.J., *et al.*, 2021. Post-exercise intramuscular O2 supply is tightly coupled with a higher proximal-to-distal ATP synthesis rate in human tibialis anterior. J. Physiol. 599 (5), 1533–1550.

Hillestad, L.K., 1963. The peripheral blood flow in intermittent claudication. V. Plethysmographic studies. The significance of the calf blood flow at rest and in response to timed arest of the circulation. Acta Med Scand. 174, 23–41.

Houben, A., Martens, R.J.H., Stehouwer, C.D.A., 2017. Assessing microvascular function in humans from a chronic disease perspective. J. Am. Soc. Nephrol. 28 (12), 3461–3472.

Huang, W., Beckett, B.R., Tudorica, A., *et al.*, 2016. Evaluation of soft tissue sarcoma response to preoperative chemoradiotherapy using dynamic contrast-enhanced magnetic resonance imaging. Tomography 2 (4), 308–316.

Hudlicka, O., 2011. Microcirculation in skeletal muscle. Muscles Ligaments Tendons J. 1 (1), 3–11.

Huegli, R.W., Schulte, A.C., Aschwanden, M., *et al.*, 2009. Effects of percutaneous transluminal angioplasty on muscle BOLD-MRI in patients with peripheral arterial occlusive disease: preliminary results. Eur. Radiol. 19 (2), 509–515.

Hurley, D.M., Williams, E.R., Cross, J.M., *et al.*, 2019. Aerobic exercise improves microvascular function in older adults. Med. Sci. Sports Exerc. 51 (4), 773–781.

Jaspers, K., Aerts, H.J., Leiner, T., *et al.*, 2009. Reliability of pharmacokinetic parameters: small vs. medium-sized contrast agents. Magn. Reson. Med. 62 (3), 779–787.

Jaspers, K., Leiner, T., Dijkstra, P., *et al.*, 2010. Optimized pharmacokinetic modeling for the detection of perfusion differences in skeletal muscle with DCE-MRI: effect of contrast agent size. Med. Phys. 37 (11), 5746–5755.

Jerome, N.P., Orton, M.R., d'Arcy, J.A., Collins, D.J., Koh, D.M., Leach, M.O., 2014. Comparison of free-breathing with navigator-controlled acquisition regimes in abdominal diffusion-weighted magnetic resonance images: effect on ADC and IVIM statistics. J. Magn. Reson. Imaging 39 (1), 235–240.

Jezzard, P., Chappell, M.A., Okell, T.W., 2018. Arterial spin labeling for the measurement of cerebral perfusion and angiography. J. Cereb. Blood Flow Metab. 38 (4), 603–626.

Joyner, M.J., Casey, D.P., 2015. Regulation of increased blood flow (hyperemia) to muscles during exercise: a hierarchy of competing physiological needs. Physiol. Rev. 95 (2), 549–601.

Karampinos, D.C., King, K.F., Sutton, B.P., Georgiadis, J.G., 2010. Intravoxel partially coherent motion technique: characterization of the anisotropy of skeletal muscle microvasculature. J. Magn. Reson. Imaging 31 (4), 942–953.

Kim, S.-G., Tsekos, N.V., 1997. Perfusion imaging by a flow-sensitive alternating inversion recovery (Fair) technique: application to functional brain imaging. Magn. Reson. Med. 37 (3), 425–435.

Koopman, T., Martens, R., Gurney-Champion, O.J., *et al.*, 2021. Repeatability of IVIM biomarkers from diffusion-weighted MRI in head and neck: Bayesian probability versus neural network. Magn. Reson. Med. 85 (6), 3394–3402.

Kwong, K.K., Chesler, D.A., Weisskoff, R.M., *et al.*, 1995. MR perfusion studies with T1-weighted echo planar imaging. Magn. Reson. Med. 34 (6), 878–887.

Latroche, C., Matot, B., Martins-Bach, A., *et al.*, 2015. Structural and functional alterations of skeletal muscle microvasculature in dystrophin-deficient mdx mice. Am. J. Pathol. 185 (9), 2482–2494.

Le Bihan, D., 1988. Intravoxel incoherent motion imaging using steady-state free precession. Magn. Reson. Med. 7 (3), 346–351.

Lebon, V., Brillault-Salvat, C., Bloch, G., Leroy-Willig, A., Carlier, P.G., 1998. Evidence of muscle BOLD effect revealed by simultaneous interleaved gradient-echo NMRI and myoglobin NMRS during leg ischemia. Magn. Reson. Med. 40 (4), 551–558.

Ledermann, H.P., Schulte, A.C., Heidecker, H.G., *et al.*, 2006. Blood oxygenation level-dependent magnetic resonance imaging of the skeletal muscle in patients with peripheral arterial occlusive disease. Circulation 113 (25), 2929–2935.

Li, Z., Muller, M.D., Wang, J., *et al.*, 2017. Dynamic characteristics of T2*-weighted signal in calf muscles of peripheral artery disease during low-intensity exercise. J. Magn. Reson. Imaging 46 (1), 40–48.

Limberg, J.K., Casey, D.P., Trinity, J.D., *et al.*, 2020. Assessment of resistance vessel function in human skeletal muscle: guidelines for experimental design, Doppler ultrasound, and pharmacology. Am. J. Physiol. Heart Circ. Physiol. 318 (2), H301–H325.

Look, D.C., Locker, D.R., 1970. Time saving in measurement of NMR and EPR relaxation times. Rev. Sci. Instrum. 41 (2), 250–251.

Lopez, D., Pollak, A.W., Meyer, C.H., *et al.*, 2015. Arterial spin labeling perfusion cardiovascular magnetic resonance of the calf in peripheral arterial disease: cuff occlusion hyperemia vs exercise. J. Cardiovasc. Magn. Reson. 17 (1), 1–9.

Lu, H., Donahue, M.J., Van Zijl, P.C., 2006. Detrimental effects of BOLD signal in arterial spin labeling fMRI at high field strength. Magn. Reson. Med. 56 (3), 546–552.

Lutz, A.M., Weishaupt, D., Amann-Vesti, B.R., *et al.*, 2004. Assessment of skeletal muscle perfusion by contrast medium first-pass magnetic resonance imaging: technical feasibility and preliminary experience in healthy volunteers. J. Magn. Reson. Imaging 20 (1), 111–121.

McClatchey, P.M., Bauer, T.A., Regensteiner, J.G., Schauer, I.E., Huebschmann, A.G., Reusch, J.E.B., 2017. Dissociation of local and global skeletal muscle oxygen transport metrics in type 2 diabetes. J. Diabet. Its Complicat. 31, 1311–1317.

Mendelson, A.A., Milkovich, S., Hunter, T., *et al.*, 2021. The capillary fascicle in skeletal muscle: structural and functional physiology of RBC distribution in capillary networks. J. Physiol. 599 (8), 2149–2168.

Morifuji, T., Murakami, S., Fujita, N., Kondo, H., Fujino, H., 2012. Exercise training prevents decrease in luminal capillary diameter of skeletal muscles in rats with type 2 diabetes. ScientificWorldJournal 2012, 645891.

Murrant, C.L., Lamb, I.R., Novielli, N.M., 2017. Capillary endothelial cells as coordinators of skeletal muscle blood flow during active hyperemia. Microcirculation 24 (3).

Nguyen, A., Ledoux, J.B., Omoumi, P., Becce, F., Forget, J., Federau, C., 2016. Application of intravoxel incoherent motion perfusion imaging to shoulder muscles after a lift-off test of varying duration. NMR Biomed. 29 (1), 66–73.

Nguyen, A., Ledoux, J.B., Omoumi, P., Becce, F., Forget, J., Federau, C., 2017. Selective microvascular muscle perfusion imaging in the shoulder with intravoxel incoherent motion (IVIM). Magn. Reson. Imaging 35, 91–97.

Ogg, R.J., Kingsley, R., Taylor, J.S., 1994. WET, a T1-and B1-insensitive water-suppression method for in vivo localized 1H NMR spectroscopy. J. Magn. Reson. Ser. B 104 (1), 1–10.

Ohno, N., Miyati, T., Fujihara, S., Gabata, T., Kobayashi, S., 2020. Biexponential analysis of intravoxel incoherent motion in calf muscle before and after exercise: comparisons with arterial spin labeling perfusion and T2. Magn. Reson. Imaging 72, 42–48.

Padilla, D.J., McDonough, P., Behnke, B.J., et al., 2006. Effects of Type II diabetes on capillary hemodynamics in skeletal muscle. Am. J. Physiol. Heart Circ. Physiol. 291 (5), H2439–H2444.

Patlak, C.S., Blasberg, R.G., Fenstermacher, J.D., 1983. Graphical evaluation of blood-to-brain transfer constants from multiple-time uptake data. J. Cereb. Blood Flow Metab. 3 (1), 1–7.

Pekar, J., Moonen, C.T., van Zijl, P.C., 1992. On the precision of diffusion/perfusion imaging by gradient sensitization. Magn. Reson. Med. 23 (1), 122–129.

Pipe, J., 2009. Chapter 2—Pulse sequences for diffusion-weighted MRI. In: Johansen-Berg, H., Behrens, T.E.J. (Eds.), Diffusion MRI. Academic Press, San Diego, pp. 11–35.

Pollak, A.W., Meyer, C.H., Epstein, F.H., et al., 2012. Arterial spin labeling MR imaging reproducibly measures peak-exercise calf muscle perfusion: a study in patients with peripheral arterial disease and healthy volunteers. JACC Cardiovasc. Imaging 5 (12), 1224–1230.

Poole, D.C., Copp, S.W., Ferguson, S.K., Musch, T.I., 2013. Skeletal muscle capillary function: contemporary observations and novel hypotheses. Exp. Physiol. 98 (12), 1645–1658.

Prior, S.J., Ryan, A.S., Blumenthal, J.B., Watson, J.M., Katzel, L.I., Goldberg, A.P., 2016. Sarcopenia is associated with lower skeletal muscle capillarization and exercise capacity in older adults. J. Gerontol. A Biol. Sci. Med. Sci. 71 (8), 1096–1101.

Ran, J., Yin, C., Liu, C., et al., 2021. The diagnostic value of MR IVIM and T2 mapping in differentiating autoimmune myositis from muscular dystrophy. Acad. Radiol. 28 (6), e182–e188.

Raynaud, J., Duteil, S., Vaughan, J., et al., 2001. Determination of skeletal muscle perfusion using arterial spin labeling NMRI: validation by comparison with venous occlusion plethysmography. Magn. Reson. Med. 46 (2), 305–311.

Reiter, D.A., Adelnia, F., Cameron, D., Spencer, R.G., Ferrucci, L., 2021. Parsimonious modeling of skeletal muscle perfusion: connecting the stretched exponential and fractional Fickian diffusion. Magn. Reson. Med. 86 (2), 1045–1057.

Richardson, R.S., Haseler, L.J., Nygren, A.T., Bluml, S., Frank, L.R., 2001. Local perfusion and metabolic demand during exercise: a noninvasive MRI method of assessment. J. Appl. Physiol. (1985) 91 (4), 1845–1853.

Riexinger, A., Laun, F.B., Hoger, S.A., et al., 2021. Effect of compression garments on muscle perfusion in delayed-onset muscle soreness: A quantitative analysis using intravoxel incoherent motion MR perfusion imaging. NMR Biomed. 34 (6), e4487.

Sakkas, G.K., Ball, D., Mercer, T.H., Sargeant, A.J., Tolfrey, K., Naish, P.F., 2003. Atrophy of non-locomotor muscle in patients with end-stage renal failure. Nephrol. Dial. Transplant. 18 (10), 2074–2081.

Sanchez, O.A., Copenhaver, E.A., Elder, C.P., Damon, B.M., 2010. Absence of a significant extravascular contribution to the skeletal muscle BOLD effect at 3 T. Magn. Reson. Med. 64 (2), 527–535.

Schepers, J., van Osch, M.J., Bartels, L.W., Heukels, S.N., Viergever, M.A., Nicolay, K., 2005. The effect of B1 field inhomogeneity and the nonselective inversion profile on the kinetics of FAIR-based perfusion MRI. Magn. Reson. Med. 53 (6), 1355–1362.

Schewzow, K., Andreas, M., Moser, E., Wolzt, M., Schmid, A.I., 2013. Automatic model-based analysis of skeletal muscle BOLD-MRI in reactive hyperemia. J. Magn. Reson. Imaging 38 (4), 963–969.

Schewzow, K., Fiedler, G.B., Meyerspeer, M., et al., 2015. Dynamic ASL and T2-weighted MRI in exercising calf muscle at 7 T: a feasibility study. Magn. Reson. Med. 73 (3), 1190–1195.

Schmid, A.I., Schewzow, K., Fiedler, G.B., et al., 2014. Exercising calf muscle T(2)* changes correlate with pH, PCr recovery and maximum oxidative phosphorylation. NMR Biomed. 27 (5), 553–560.

Schulte, A.C., Aschwanden, M., Bilecen, D., 2008. Calf muscles at blood oxygen level-dependent MR imaging: aging effects at postocclusive reactive hyperemia. Radiology 247 (2), 482–489.

Sigmund, E.E., Vivier, P.H., Sui, D., et al., 2012. Intravoxel incoherent motion and diffusion-tensor imaging in renal tissue under hydration and furosemide flow challenges. Radiology 263 (3), 758–769.

Stejskal, E.O., Tanner, J.E., 1965. Spin diffusion measurements: spin echoes in the presence of time-dependent field gradient. J. Chem. Phys. 42, 288–292.

Sujlana, P., Skrok, J., Fayad, L.M., 2018. Review of dynamic contrast-enhanced MRI: Technical aspects and applications in the musculoskeletal system. J. Magn. Reson. Imaging 47 (4), 875–890.

Thompson, R.B., Aviles, R.J., Faranesh, A.Z., et al., 2005. Measurement of skeletal muscle perfusion during postischemic reactive hyperemia using contrast-enhanced MRI with a step-input function. Magn. Reson. Med. 54 (2), 289–298.

Tofts, P.S., 1997. Modeling tracer kinetics in dynamic Gd-DTPA MR imaging. J. Magn. Reson. Imaging 7 (1), 91–101.

Tofts, P.S., Brix, G., Buckley, D.L., et al., 1999. Estimating kinetic parameters from dynamic contrast-enhanced T(1)-weighted MRI of a diffusable tracer: standardized quantities and symbols. J. Magn. Reson. Imaging 10 (3), 223–232.

Tonson, A., Noble, K.E., Meyer, R.A., Rozman, M.R., Foley, K.T., Slade, J.M., 2017. Age reduces microvascular function in the leg independent of physical activity. Med. Sci. Sports Exerc. 49 (8), 1623–1630.

Towse, T.F., Elder, C.P., Bush, E.C., et al., 2016. Post-contractile BOLD contrast in skeletal muscle at 7 T reveals inter-individual heterogeneity in the physiological responses to muscle contraction. NMR Biomed. 29 (12), 1720–1728.

Veeger, T.T., Hirschler, L., Baligand, C., et al., 2022. Microvascular response to exercise varies along the length of the tibialis anterior muscle. NMR Biomed., e4796.

Versluis, B., Backes, W.H., van Eupen, M.G., et al., 2011. Magnetic resonance imaging in peripheral arterial disease: reproducibility of the assessment of morphological and functional vascular status. Investig. Radiol. 46 (1), 11–24.

Versluis, B., Dremmen, M.H., Nelemans, P.J., et al., 2012. Dynamic contrast-enhanced MRI assessment of hyperemic fractional microvascular blood plasma volume in peripheral arterial disease: initial findings. PLoS One 7 (5), e37756.

Wang, J., Li, Y.H., Li, M.H., Zhao, J.G., Bao, Y.Q., Zhou, J., 2011. Use of dynamic contrast-enhanced magnetic resonance imaging to evaluate the microcirculation of lower extremity muscles in patients with Type 2 diabetes. Diabet. Med. 28 (5), 618–621.

Wang, X.H., Mitch, W.E., Price, S.R., 2022. Pathophysiological mechanisms leading to muscle loss in chronic kidney disease. Nat. Rev. Nephrol. 18 (3), 138–152.

Wong, E.C., Buxton, R.B., Frank, L.R., 1998a. A theoretical and experimental comparison of continuous and pulsed arterial spin labeling techniques for quantitative perfusion imaging. Magn. Reson. Med. 40 (3), 348–355.

Wong, E.C., Buxton, R.B., Frank, L.R., 1998b. Quantitative imaging of perfusion using a single subtraction (QUIPSS and QUIPSS II). Magn. Reson. Med. 39 (5), 702–708.

Wray, D.W., Nishiyama, S.K., Monnet, A., et al., 2009a. Multiparametric NMR-based assessment of skeletal muscle perfusion and metabolism during exercise in elderly persons: preliminary findings. J. Gerontol. Ser. A Biomed. Sci. Med. Sci. 64 (9), 968–974.

Wray, D.W., Nishiyama, S.K., Monnet, A., et al., 2009b. Antioxidants and aging: NMR-based evidence of improved skeletal muscle perfusion and energetics. Am. J. Phys. Heart Circ. Phys. 297 (5), H1870–H1875.

Wu, W.-C., Wang, J., Detre, J.A., et al., 2008. Hyperemic flow heterogeneity within the calf, foot, and forearm measured with continuous arterial spin labeling MRI. Am. J. Phys. Heart Circ. Phys. 294 (5), H2129–H2136.

Wu, W.-C., Mohler, E., Ratcliffe, S.J., Wehrli, F.W., Detre, J.A., Floyd, T.F., 2009. Skeletal muscle microvascular flow in progressive peripheral artery disease: assessment with continuous arterial spin-labeling perfusion magnetic resonance imaging. J. Am. Coll. Cardiol. 53 (25), 2372–2377.

Wurnig, M.C., Donati OF, Ulbrich, E., et al., 2015. Systematic analysis of the intravoxel incoherent motion threshold separating perfusion and diffusion effects: proposal of a standardized algorithm. Magn. Reson. Med. 74 (5), 1414–1422.

Yao, L., Sinha, U., 2000. Imaging the microcirculatory proton fraction of muscle with diffusion-weighted echo-planar imaging. Acad. Radiol. 7 (1), 27–32.

Yao, J., Anjum, M.A.R., Swain, A., Reiter, D.A., 2021. Analytical and numerical connections between fractional fickian and intravoxel incoherent motion models of diffusion MRI. Mathematics 9 (16), 1963.

Yoon, M.A., Hong, S.J., Ku, M.C., Kang, C.H., Ahn, K.S., Kim, B.H., 2018. Multiparametric MR imaging of age-related changes in healthy thigh muscles. Radiology 287 (1), 235–246.

Zhang, J.L., Rusinek, H., Bokacheva, L., et al., 2008. Functional assessment of the kidney from magnetic resonance and computed tomography renography: impulse retention approach to a multicompartment model. Magn. Reson. Med. 59 (2), 278–288.

Zhang, J.L., Conlin, C.C., Carlston, K., et al., 2016. Optimization of saturation-recovery dynamic contrast-enhanced MRI acquisition protocol: Monte Carlo simulation approach demonstrated with gadolinium MR renography. NMR Biomed. 29 (7), 969–977.

Zhang, J.L., Layec, G., Hanrahan, C., et al., 2019. Exercise-induced calf muscle hyperemia: quantitative mapping with low-dose dynamic contrast enhanced magnetic resonance imaging. Am. J. Physiol. Heart Circ. Physiol. 316 (1), H201–H211.

Zheng, J., Hasting, M.K., Zhang, X., et al., 2014. A pilot study of regional perfusion and oxygenation in calf muscles of individuals with diabetes with a noninvasive measure. J. Vasc. Surg. 59 (2), 419–426.

Zheng, J., Hastings, M.K., Muccigross, D., et al., 2015. Non-contrast MRI perfusion angiosome in diabetic feet. Eur. Radiol. 25 (1), 99–105.

Index

Note: Page numbers followed by "*f*" indicate figures, "*t*" indicate tables, and "*b*" indicate boxes.

Printed in the United States
by Baker & Taylor Publisher Services